Biotechnological Advances for Microbiology, Molecular Biology, and Nanotechnology

An Interdisciplinary Approach to the Life Sciences

Biotechnological Advances for Microbiology, Molecular Biology, and Nanotechnology

An Interdisciplinary Approach to the Life Sciences

Edited by

Jyoti Ranjan Rout, PhD

Rout George Kerry

Abinash Dutta

AAP | APPLE ACADEMIC PRESS

First edition published 2022

Apple Academic Press Inc.
1265 Goldenrod Circle, NE,
Palm Bay, FL 32905 USA

4164 Lakeshore Road, Burlington,
ON, L7L 1A4 Canada

CRC Press
6000 Broken Sound Parkway NW,
Suite 300, Boca Raton, FL 33487-2742 USA

2 Park Square, Milton Park,
Abingdon, Oxon, OX14 4RN UK

Library and Archives Canada Cataloguing in Publication

Title: Biotechnological advances for microbiology, molecular biology, and nanotechnology : an interdisciplinary approach to the life sciences / edited by Jyoti Ranjan Rout, PhD, Rout George Kerry, Abinash Dutta.

Names: Rout, Jyoti Ranjan, editor. | Kerry, Rout George, editor. | Dutta, Abinash, editor.

Description: First edition. | Includes bibliographical references and index.

Identifiers: Canadiana (print) 20210335998 | Canadiana (ebook) 2021033603X | ISBN 9781771889995 (hardcover) | ISBN 9781774639474 (paperback) | ISBN 9781003161158 (ebook)

Subjects: LCSH: Biotechnology. | LCSH: Microbiology. | LCSH: Molecular biology. | LCSH: Nanotechnology. | LCSH: Life sciences.

Classification: LCC TP248.2 .B56 2022 | DDC 660.6—dc23

Library of Congress Cataloging-in-Publication Data

Names: Rout, Jyoti Ranjan, editor. | Kerry, Rout George, editor. | Dutta, Abinash, editor.

Title: Biotechnological advances for microbiology, molecular biology, and nanotechnology : an interdisciplinary approach to the life sciences / edited by Jyoti Ranjan Rout, Rout George Kerry, Abinash Dutta.

Description: First edition. | Palm Bay, FL : Apple Academic Press, 2022. | Includes bibliographical references and index. | Summary: "Biotechnological Advances for Microbiology, Molecular Biology, and Nanotechnology: An Interdisciplinary Approach to the Life Sciences presents cutting-edge research associated with the beneficial implications of biotechnology on human welfare. The volume mainly focuses on the highly demanding thrust areas of biotechnology that are microbiology, molecular biology, and nanotechnology. The book provides a detailed overview of the beneficial roles of microbes and nanotechnology-based engineered particles in biological developments. Also, it highlights the role of epigenetic machinery and redox modulators during the development of diseases. In addition, it provides research on nanotechnology-based applications in tissue engineering, stem cell, and regenerative medicines. Overall, the book provides an extended platform for acquiring the methodological knowledge needed for today's biotechnological applications, such as DNA methylation, redox homeostasis, CRISPR, nano-based drug delivery systems, proteomics, genomics, metagenomics, bioluminescence, bioreactors, bioremediation, biosensors, etc. Divided into three sections, the book first highlights some recent trends in applied microbiology used in different areas, such as crop improvement, wastewater treatment, drug delivery, healthcare management, and more. The volume goes on to cover some advances in cellular and molecular mechanisms, such as CRISPR technology in biological systems, induced stem cells in disease prevention, integrated omics technology, and others. The volume also explores the indispensable role of nanotechnology in the precisely modulating intricate functioning of an organism in diagnostic and therapy along its application in tissue engineering and regenerative medicine and in food science as well as its role in ecological sustainability. This multidisciplinary volume will be highly valuable for the researchers, scientists, biologists, and faculty and students striving to expand their horizon of knowledge in their respective fields"-- Provided by publisher.

Identifiers: LCCN 2021049600 (print) | LCCN 2021049601 (ebook) | ISBN 9781771889995 (hardcover) | ISBN 9781774639474 (paperback) | ISBN 9781003161158 (ebook)

Subjects: MESH: Microbiological Techniques | Biotechnology | Genetic Techniques | Nanotechnology

Classification: LCC QR41.2 (print) | LCC QR41.2 (ebook) | NLM QW 25 | DDC 579--dc23/eng/20211015

LC record available at https://lccn.loc.gov/2021049600

LC ebook record available at https://lccn.loc.gov/2021049601

ISBN: 978-1-77188-999-5 (hbk)
ISBN: 978-1-77463-947-4 (pbk)
ISBN: 978-1-00316-115-8 (ebk)

About the Editors

Jyoti Ranjan Rout, PhD, is Assistant Professor in the School of Biological Sciences, AIPH University, Bhubaneswar, Odisha, India. Before joining AIPH, he worked as Assistant Professor and Head of the Post Graduate Department of Biotechnology at the Academy of Management and Information Technology (AMIT), Bhubaneswar, India. Dr. Rout has more than 10 years of experience in both teaching and research in the field of biochemistry and molecular biology. He is the recipient of an NESA-Scientist of the Year Award and the Professor B.K. Nanda Memorial Award from the National Environmental Science Academy, India, and Orissa Botanical Society, India, respectively. Dr. Rout has published over 45 research articles in various national and international peer-reviewed journals and has contributed several book chapters to different edited book volumes with international publishers, including Springer and CRC Press. He is actively involved as an editorial board member and reviewer for many international journals. His area of research includes biochemical and molecular aspects of toxicology, protein and elemental profiling, gene expression of antioxidant enzymes, phytochemical screening and in vitro tissue and organ culture of medicinal plants. His recent area of interest is to understand the metal causing toxicity in cellular, biochemical, molecular level in plants which act as a potential bio-indicator of metal pollution to agriculture and public health. He earned his master's degree in Biotechnology from North Orissa University and his PhD (Biotechnology) from Utkal University in the area of stress biochemistry, molecular biology, and nutritional stress tolerance in plants.

Rout George Kerry, is a Research Scholar at Utkal University, Odisha, India. He has published about 27 articles in peer-reviewed international journals and edited books. He became acquainted with the importance of nanotechnology by one of his mentors, which triggered his interest to do scientific research in depth. Thereafter, Mr. Kerry was fully engaged in exploring the infinite potential that is possible from the crossover of biotechnology and nanotechnology. Basically, he emphasizes the formulation of nano-based therapeutics against infectious and noninfectious ailments, the impact of probiotics in health management, and the role of nanotechnology in health as well as for the sustainable development of agriculture. Presently, he is working on an array of organic and inorganic nanoparticle-based drug delivery systems for reversal of type-II diabetes mellitus. Mr. Kerry graduated from Utkal University and went on to pursue his master's degree in Biotechnology from Berhampur University, India.

Abinash Dutta, recently joined the Institute of Life Sciences, Odisha, India, and carried out his postdoctoral research work on zebrafish development. His research work is mainly focused on the improvement of larval as well as cocoon traits (commercial traits) of the Tasar silkworm, Antheraea mylitta, an important component of the Asian nonmulberry silk industry, through foliar supplementation of exogenous antioxidants. Moreover, he is also interested in finding out the molecular mechanism and epigenetic regulation of redox homeostasis in silkworms in response to exogenous antioxidants. The outcome of his research work has been highly appreciated among insect biologists, and he was awarded a best paper and poster presentation award at a national and international level symposium. Recently, he has also awarded a BBA Young Investigator Award. Some of his research findings are published in reputed peer-reviewed international journals. Mr. Dutta received his master's degree in Biotechnology from North Orissa University, Odisha, India. He has submitted his PhD thesis in Biotechnology at P.G. Department of Biotechnology, Utkal University, Odisha, India.

Contents

Contributors ... xi
Abbreviations ... xv
Preface ... xxi

PART I: Trends in Applied Microbiology ... 1

1. **Role of Endophytes in Crop Improvement ... 3**
Bicky Jerin Joseph, A. R. Nayana, and E. K. Radhakrishnan

2. **Omics Approach to Understanding Microbial Diversity ... 25**
Shilpee Pal, Arijit Jana, Keshab Chandra Mondal, and Suman Kumar Halder

3. **Role of Bioremediation in Wastewater Treatment ... 39**
Iqbal Ansari, Muniyan Sundararajan, Deblina Maiti, Anand Kumar, and Jyoti Ranjan Rout

4. **Usage of Engineered Virus-Like Particles in Drug Delivery ... 65**
Sushil Kumar Sahu, Ramakanta Rana, and Ashok Kumar Mallik

5. **Novel Microbial Compounds as a Boon in Health Management ... 77**
Shubha Rani Sharma, Rajani Sharma, and Debasish Kar

6. **Rise of the Microbial World: An Economic Point of View ... 119**
Binita Dev and R. Jayabalan

7. **Biosafety Principles for Microbial Culture Technologies ... 147**
Vidushi Abrol, Sundeep Jaglan, and Sharada Mallubhotla

PART II: Advances in Cellular and Molecular Mechanisms ... 177

8. **Intracellular Redox Status and Disease Development: An Overview of the Dynamics of Metabolic Orchestra ... 179**
Sharmi Mukherjee and Anindita Chakraborty

9. **Oxidative Stress as a Detrimental Factor in Various Clinical Pathology ... 227**
Aradhana Behura and Sanjaya Kumar Panda

10. **Implications of CRISPR Technology in Biological Systems ... 245**
Kikku Sharma and Souvik Sen Gupta

11. Revolutionary Approaches of Induced Stem Cells in Disease Prevention 283
Stanzin Ladol

12. Stem Cell Biology: An Overview 297
Sumit Siddharth

13. Recent Advances in Imaging and Analysis of Cellular Dynamics in Real Time 311
Chandra Bhan, Pankaj Dipankar, Shiba Prasad Dash, Papiya Chakraborty, Nibedita Dalpati, and Pranita P. Sarangi

14. Integrated Omics Technology for Basic and Clinical Research 351
Kuldeep Giri, Vinod Singh Bisht, Sudipa Maity, and Kiran Ambatipudi

15. Current State of Malaria Diagnosis: Conventional, Rapid, and Safety Diagnostic Methods 409
Barsa Baisalini Panda and Rupenangshu Kumar Hazra

PART III: Nanotechnological Intervention in Life Sciences 423

16. Current Perspective of Biofunctionalized Nanomaterials in Biology and Medicine 425
Namita Bhoi and Iswar Baitharu

17. Nano-System as Therapeutic Means 445
Ananya Ghosh and Aniruddha Mukherjee

18. Recent Developments in Nanoparticulate-Mediated Drug Delivery in Therapeutic Approaches 475
Janmejaya Bag, Swetapadma Sahu, and Monalisa Mishra

19. Beneficial Utility and Perspective of Nanomaterials Toward Biosensing 517
Ravindra Pratap Singh and Kshitij RB Singh

20. Benefits of Nanomaterials-Based Biosensors 551
Sourav Mishra, Rohit Kumar Singh, Uday Suryakanta, Bijayananda Panigrahi, and Dindyal Mandal

21. Role of Nanotechnology in Tissue Engineering and Regenerative Medicine 573
Bijayananda Panigrahi, Uday Suryakanta, Sourav Mishra, Rohit Kumar Singh, and Dindyal Mandal

22. Protein-Based Nanosystems as Emerging Bioavailability Enhancers for Nutraceuticals 605
Rohini Samadarsi and Debjani Dutta

23. Application of Nanomaterials in Environmental Pollution Abatement and Their Impact on Ecological Sustainability: Recent Status and Future Perspective 629

Syed Nikhat Ahmed, Subhashree Subhadarsini Mishra, Jayanta Kumar Sahu, Sabita Shroff, Prajna Paramita Naik, Iswar Baitharu, and Sanjat Kumar Sahu

Index...655

Contributors

Vidushi Abrol
Microbial Biotechnology Division, CSIR-Indian institute of integrative Medicine, Jammu 180001, India

Syed Nikhat Ahmed
P.G. Department of Environmental Sciences, Sambalpur University, Jyoti Vihar, Burla, Odisha, India

Kiran Ambatipudi
Department of Biotechnology, Indian institute of Technology Roorkee, Roorkee 247667, Uttarakhand, India

Kumar Anand
Department of Biotechnology, Vinoba Bhave University, Hazaribag, Jharkhand 825301, India

Iqbal Ansari
CSIR-Central institute of Mining and Fuel Research, Dhanbad, Jharkhand 826015, India

Janmejaya Bag
Neural Developmental Biology Lab, Department of Life Science, NIT Rourkela, Rourkela 769008, Odisha, India

Iswar Baitharu
P.G. Department of Environmental Sciences, Sambalpur University, Burla, Odisha

Chandra Bhan
Department of Biotechnology, Indian Institute of Technology, Roorkee, Roorkee 247667, Uttarakhand, India

Namita Bhoi
Nano Research Centre, School of Chemistry, Sambalpur University, Burla, Odisha

Vinod Singh Bisht
Department of Biotechnology, Indian institute of Technology Roorkee, Roorkee 247667, Uttarakhand, India

Anindita Chakraborty
UGC-DAE Consortium for Scientific Research, Kolkata Centre, Kolkata 700106, West Bengal, India

Papiya Chakraborty
Department of Biotechnology, Indian Institute of Technology, Roorkee, Roorkee 247667, Uttarakhand, India

Prakash Roy Choudhury
Department of Life Science & Bioinformatics, Assam University, Silchar 788011, Assam, India

Nibedita Dalpati
Department of Biotechnology, Indian Institute of Technology, Roorkee, Roorkee 247667, Uttarakhand, India

Shiba Prasad Dash
Department of Biotechnology, Indian Institute of Technology, Roorkee, Roorkee 247667, Uttarakhand, India

Binita Dev
Food Microbiology and Bioprocess Laboratory, Department of Life Science,
National institute of Technology, Rourkela 769008, Odisha

Pankaj Dipankar
Department of Biotechnology, Indian Institute of Technology, Roorkee, Roorkee 247667, Uttarakhand, India

Debjani Dutta
Department of Biotechnology, National institute of Technology Durgapur, Mahatma Gandhi Avenue, Durgapur 713209, West Bengal, India

Ananya Ghosh
P. G. Department of Biotechnology, Utkal University, Vani Vihar, Odisha, India

Kuldeep Giri
Department of Biotechnology, Indian institute of Technology Roorkee, Roorkee 247667, Uttarakhand, India

Souvik Sen Gupta
Division of Biological and Life Sciences, School of Arts and Sciences, Ahmedabad University, Central Campus, Navrangpura, Ahmedabad 380009, Gujarat

Suman Kumar Halder
Department of Microbiology, Vidyasagar University, Midnapore, West Bengal

Rupenangshu Kumar Hazra
ICMR-Regional Medical Research Centre, Chandrasekharpur, Bhubaneswar 751023, Odisha, India

Sundeep Jaglan
Microbial Biotechnology Division, CSIR-Indian institute of integrative Medicine, Jammu 180001, India
Academy of Scientific and innovative Research (AcSIR), Jammu Campus, Jammu 180001, India

Arijit Jana
Department of Chemical Engineering, Indian institute of Technology, Roorkee, Uttarakhand, India

R. Jayabalan
Food Microbiology and Bioprocess Laboratory, Department of Life Science, National institute of Technology, Rourkela 769008, Odisha, India

Bicky Jerin Joseph
School of Biosciences, Mahatma Gandhi University, Kottayam 686 560, Kerala, India

Debasish Kar
Department of Biotechnology, M.S. Ramaiah University of Applied Sciences, Bangalore 560054, India

Stanzin Ladol
Department of Zoology, Central University of Jammu, J & K, India

Deblina Maiti
CSIR-Central institute of Mining and Fuel Research, Dhanbad, Jharkhand 826015, India

Sudipa Maity
Department of Biotechnology, Indian institute of Technology Roorkee, Roorkee 247667, Uttarakhand, India

Ashok Kumar Mallik
Centre for Ecological Sciences, Indian Institute of Science, Bangalore, Karnataka 560012, India

Sharada Mallubhotla
School of Biotechnology, Shri Mata Vaishno Devi University, Katra 182320, India

Dindyal Mandal
School of Biotechnology, Kalinga Institute of Industrial Technology Deemed to be University, Campus 11, Patia, Bhubaneswar 751024 Odisha, India
School of Pharmacy, Chapman University, Irvine, CA, USA

Monalisa Mishra
Neural Developmental Biology Lab, Department of Life Science, NIT Rourkela, Rourkela 769008, Odisha, India

Sourav Mishra
School of Biotechnology, Kalinga Institute of Industrial Technology Deemed to be University, Campus 11, Patia, Bhubaneswar 751024 Odisha, India

Subhashree Subhadarsini Mishra
P.G. Department of Environmental Sciences, Sambalpur University, Jyoti Vihar, Burla, Odisha, India

Keshab Chandra Mondal
Department of Microbiology, Vidyasagar University, Midnapore, West Bengal

Aniruddha Mukherjee
Department of Microbiology, Gurudas College, Kolkata, West Bengal, India

Sharmi Mukherjee
UGC-DAE Consortium for Scientific Research, Kolkata Centre, Kolkata 700106, West Bengal, India

Prajna Paramita Naik
PG Department of Zoology, Vikram Deb Autonomous College, Jeypore, Odisha, India

Rajat Nath
Department of Life Science & Bioinformatics, Assam University, Silchar 788011, Assam, India

A. R. Nayana
School of Biosciences, Mahatma Gandhi University, Kottayam 686 560, Kerala, India

Shilpee Pal
Department of Microbiology, Vidyasagar University, Midnapore, West Bengal

Barsa Baisalini Panda
ICMR-Regional Medical Research Centre, Chandrasekharpur, Bhubaneswar 751023, Odisha, India

Bijayananda Panigrahi
School of Biotechnology, Kalinga Institute of Industrial Technology Deemed to be University, Campus 11, Patia, Bhubaneswar, Odisha 751024, India

E. K. Radhakrishnan
School of Biosciences, Mahatma Gandhi University, Kottayam 686 560, Kerala, India

Ramakanta Rana
Regional Medical Research Centre, Chandrasekharpur, Bhubaneswar, Odisha 751023, India

Jyoti Ranjan Rout
School of Biological Sciences, AIPH University, Bhubaneswar 752101, Odisha, India

Priyanka Saha
Department of Life Science & Bioinformatics, Assam University, Silchar 788011, Assam, India

Jayanta Kumar Sahu
School of Life Sciences, Sambalpur University, Jyoti Vihar, Burla, Odisha, India

Sanjat Kumar Sahu
P.G. Department of Environmental Sciences, Sambalpur University, Jyoti Vihar, Burla, Odisha, India

Sushil Kumar Sahu
Department of Pharmacology & Molecular Sciences, School of Medicine, Johns Hopkins University, Baltimore, Maryland 21205, USA

Swetapadma Sahu
Neural Developmental Biology Lab, Department of Life Science, NIT Rourkela, Rourkela 769008, Odisha, India

Rohini Samadarsi
Department of Biotechnology, National institute of Technology Durgapur, Mahatma Gandhi Avenue, Durgapur 713209, West Bengal, India

Pranita P. Sarangi
Department of Biotechnology, Indian Institute of Technology, Roorkee, Roorkee 247667, Uttarakhand, India

Kikku Sharma
Division of Biological and Life Sciences, School of Arts and Sciences, Ahmedabad University, Central Campus, Navrangpura, Ahmedabad 380009, Gujarat

Rajani Sharma
Department of Biotechnology, Amity University, Ranchi 834002, India

Shubha Rani Sharma
Department of Bio-Engineering, Birla Institute of Technology, Mesra, Ranchi 835215, India

Sabita Shroff
School of Chemistry, Sambalpur University, Jyoti Vihar, Burla, Odisha, India

Sumit Siddharth
Department of Oncology, Johns Hopkins University School of Medicine and the Sidney Kimmel Comprehensive Cancer Center at Johns Hopkins, Baltimore 21231 MD, USA

Ravindra Pratap Singh
Department of Biotechnology, Indira Gandhi National Tribal University, Anuppur, Amarkantak 484887, Madhya Pradesh, India

Kshitij RB Singh
Department of Biotechnology, Indira Gandhi National Tribal University, Anuppur, Amarkantak 484887, Madhya Pradesh, India

Rohit Kumar Singh
School of Biotechnology, Kalinga Institute of Industrial Technology Deemed to be University, Campus 11, Patia, Bhubaneswar 751024 Odisha, India

Muniyan Sundararajan
CSIR-Central institute of Mining and Fuel Research, Dhanbad, Jharkhand 826015, India

Uday Suryakanta
School of Biotechnology, Kalinga Institute of Industrial Technology Deemed to be University, Campus 11, Patia, Bhubaneswar 751024 Odisha, India

Anupam Das Talukdar
Department of Life Science & Bioinformatics, Assam University, Silchar 788011, Assam, India

Abbreviations

βLG	β-lactoglobulin
AAV	adeno-associated virus
ACC	aminocyclopropane-1-carboxylic acid
AD	Alzheimer's disease
AE	acridinium ester
AFM	atomic force microscopy
AGE	advanced glycosylation end
AIDS	acquired immunodeficiency syndrome
ALS	amyotrophic lateral sclerosis
AMD	age-related macular degeneration
AMPs	antimicrobial peptides
AO	acridine orange
AOA	antioxidant activity
AOPPs	advanced oxidation protein products
APT	3-aminopropyltrimethoxysilane
APV	avian polyomavirus
ASCs	adult stem cells
AuNPs	gold nanoparticles
Aβ	amyloid β
BAF	bioaccumulation factor
BBB	blood brain barrier
BCP	benzothio carboxypurine
BRCA1/2	breast cancer gene1/2
BSA	bovine serum albumin
BSC	biological safety cabinet
BSL	biosafety level
CAD	coronary artery disease
CAGR	compound annual growth rate
CCD	charge-coupled device
CDI	clostridium difficile infection
CI	cytoplasmic incapability
CKD	chronic kidney disease
CL	chemiluminescence
CM	cardiomyocytes
CN	chitin nanofibrils
CNS	central nervous system
CNT	carbon nanotube

CNV	copy number variants
COD	chemical oxygen demand
Con A	concanavalin A
COPD	chronic obstructive pulmonary disease
CP	conducting polymer
CPPs	cell-penetrating peptides
CRISPR	cluster regularly interspersed short palindromic repeat
CT	computed tomography
CTCs	circulating tumor cells
CuS	copper sulfide
CV	cardiovascular
CVD	cardiovascular diseases
CYP3A4	cytochrome p-450 3A4
Cys-AgNPs	cysteine modified silver nanoparticles
DBSs	dried blood spots
DDS	drug delivery system
Dex	dexamethasone
DIC	differential interference contrast
DLS	dynamic light scattering
DM	diabetes mellitus
DMD	Duchenne muscular dystrophy
DOX	doxorubicin
ds DNA	double-stranded DNA
DYS	dystrophin
ECM	extracellular matrix
ELISA	enzyme-linked immunosorbent assay
EMCCD	electron-multiplying charge-coupled device
EPF	endemic pemphigus foliaceus
EPR	enhanced permeability and retention
EPS	exopolysaccharides
ER	endoplasmic reticulum
ER	estrogen receptor
ESCs	embryonic stem cells
ESI	electrospray ionization
ETH	ethionamide
E-ZVI	emulsified zero-valent iron
FET	field-effect transistor
FLIM	fluorescence lifetime imaging
FMT	fecal microbiota transplant
Fn Cas9	Francisella novicida Cas9
FRAP	fluorescence recovery after photobleaching
FRET	fluorescence resonance energy transfer
g RNA	guide RNA

GA	glutaraldehyde
GCE	glassy carbon electrode
GFP	green fluorescent protein
GM	glioblastoma multiforme
GM	Goeppert–Mayer
GMOs	genetically modified organisms
GO	graphene oxide
Gox	glucose oxidases
GR	glutathione reductase
GRAS	generally considered as safe
GWAS	Genome-Wide Association Studies
HA	hyaluronic acid
HAC	human artificial chromosome
HapMap	haplotype map
HAS	human serum albumin
HBV	hepatitis B virus
hCas9	humanized Cas9
HCuS	hollow CuS
HD	Huntington's disease
HEPA	high-efficiency particulate air
hESCs	human embryonic stem cells
HGP	Human Genome Project
HIV	human immune deficiency virus
HPV	human papiloma virus
HRP	horseradish peroxidase
HSA	human serum albumin
HSCs	hematopoietic stem cells
HTS	high-throughput sequencing
IAA	indole-3-acetic acid
IBD	inflammatory bowel disease
INH	isoniazid
iNPG	injectable nanoparticle generator
iPSCs	induced pluripotent stem cells
IR	ischemia/reperfusion
iTRAQ	isobaric tags for relative and absolute quantification
IVM	intravital microscopy
IVMPM	multiphoton intravital microscopy
IWQI	integrated water quality index
JCV	John Cunningham virus
LAI	laboratory-acquired infections
LAPA	Lapatinib
LDL	low-density lipoprotein
LOD	limit of detection

LPS	lipopolysaccharides
LSPR	large surface plasmon resonance
LTR	long terminal repeats
LVH	left ventricular hypertrophy
MDA	malondialdehyde
MI	myocardial infarction
MMP	matrix material proteinases
MMP-7	matrix metalloproteinase matrilysin
MNP	magnetic nanoparticle
MPI	mean intensity projection
MPO	myeloperoxidase
MRI	magnetic resonance imaging
MRM	multiple reaction monitoring
MS	mass spectrometry
ncRNA	noncoding transcriptome
NGF	nerve growth factors
NGS	Next Generation Sequencing
NIH	National Institutes of Health
NMDA	*N*-methyl-D-aspartate
NMR	nuclear magnetic resonance
NO	nitric oxide
NOS	NO synthase
NOS	reactive nitrogen species
NP	nanoparticle
NSC	neuronal stem cells
NWs	nanowires
nZVI	nanoscale zero-valent iron
OET	organic electrochemical transistor
PAH	poly(allylamine)
PAH	polycyclic aromatic hydrocarbons
PAI	photoacoustic Imaging
PAMAM	poly(amidoamine)
PAMs	plant-associated microorganisms
PBS	phosphate buffer saline
PCB	polychlorinated biphenyls
PCBs	polychlorinated biphenyls
PCL	polycaprolactone
PCR	polymerase chain reaction
PCS	photon-correlation spectroscopy
PD	programmed cell death 1
PD	Parkinson's disease
PDDA	poly(diallyl dimethyl ammonium chloride)
PDI	polydispersity

PDI	protein disulfide isomerase
PEDOT	poly(3,4-ethyelenedioxythiophene)
PEG	polyethene glycol
PEM	polyelectrolyte multilayer
PET	positron emission tomography
Pf	plasmodium falciparum
PGP	plant growth promotion
PGPB	plant growth-promoting bacteria
PGPR	plant growth-promoting rhizobacteria
PGx	pharmacogenomics
PKC	protein kinase C
pLDH	parasite lactate dehydrogenase
PLGA	poly(D,L-lactide-*co*-glycolide)
PLL	poly-L-lysine
PLLA	poly(lactic acid)
PMTs	photomultiplier tubes
PNL	propranolol hydrochloride
PPE	poly(*p*-phenylene ethynylene)
PS-*b*-P4VP	poly(styrene-*b*-4-vinylpyridine)
PSS	poly(4-styrenesulfonic acid)
PSS	polystyrene sulfonate
PT	polythiophene
PTMs	post-translational modifications
PTX	Paclitaxel
PUFAs	polyunsaturated fatty acids
Pv	plasmodium vivax
PVA	poly(vinyl alcohol)
QD	quantum dot
RAGE	receptor for advanced glycation end
RDT	rapid diagnostic test
RGCs	retinal ganglion cells
RGN	RNA-guided nuclease
rGO	reduced graphene oxide
RNS	reactive nitrogen species
ROS	reactive oxygen species
RPE	retinal pigment epithelium
SAGE	serial analysis of gene expression
SARS	severe acute respiratory syndrome
SDG	sustainable development goals
SEM	scanning electron microscope
SILAC	stable isotope labeling with amino acid in cell culture
SJN	solid lipid nanoparticles
SN	substantia nigra

SNP	single nucleotide polymorphism
SOD	superoxide dismutase
SOP	standard operating procedure
SP	substance P
SPECT	single-photon emission computed tomography
SPION	superparamagnetic iron oxide nanoparticles
SRM	single reaction monitoring
ssDNA	single-stranded DNA
TAG	triacylglycerol
TB	tuberculosis
TEM	transmission electron microscope
TfR	transferrin receptor
TGF	transforming growth factor
TH	total hardness
TJs	tight junctions
TNF	tumor necrosis factor
TPP	tripolyphosphate
UC	ulcerative colitis
ULPA	ultra-low penetration air
US	ultrasonography/ultrasound
UV	ultraviolet
VCP	valosin-containing protein
VLP	virus-like particle
VMPM	intravital multiphoton microscopy
VOCs	volatile organic compounds
WES	whole-exome sequencing
WGS	whole-genome sequencing

Preface

During the last few decades, the global scientific community has witnessed tremendous advancement in the field of biological sciences. This has led to the emergence of several frontiers in the field, of which the areas of microbiological approaches for human welfare, cellular, and molecular mechanism underlying the biological processes, and nanotechnological intervention in biological sciences are adjudged to be the most prominent areas. The present book, *Biotechnological Advances for Microbiology, Molecular Biology, and Nanotechnology: An Interdisciplinary Approach to the Life Sciences,* is an attempt to address these areas through the compilation of various articles authored by experts who have made long-standing research contributions in such frontier areas. The chapters have been appropriately arranged in three different parts: Part I—*Trends in Applied Microbiology*, Part II—*Advances in Cellular and Molecular Mechanisms, and* Part IIII—*Nanotechnological Intervention in Life Sciences*.

Part I of the book focuses on microbiology, as it has proved to be among the other principal disciplines in biology, typically encompassing the study of various microorganisms. Investigations are being conducted to understand and mimic the physicochemical properties that compel them to survive under such hard and extreme environmental conditions. Therefore, this part describes the importance of novel microbial metabolites in health management, endophytes and their role in crop improvement, microbial techniques for wastewater treatment, biosafety principles for microbial culture, approaches for proper analysis of microbial diversity, usage of engineered virus-like particles in drug delivery, and economic importance of microbial world.

Additionally, understanding the biochemical properties and their utility for the amelioration of mankind can be harnessed through molecular biology, which is covered in Part II. It orchestrates an essential role in apprehending the developmental process, actions, and regulations of multiple cell organelles, for the efficient formulation of new drug targets, early disease diagnosis, and deeper understanding of the physiology of the cell. Moreover, this part also presents the significance of gene editing such as CRISPR, revolutionary approaches of induced stem cells in disease prevention, recent advances in imaging, and analysis of cellular dynamics in real time.

The final Part III of the book focuses on current nanotechnological advancements in life science research. Nanotechnology provides an option to custom-engineer a novel and improved approach to disease diagnosis and treatment. Nanoscale devices can interact with bulky biological macromolecules present on both extrinsic and intrinsic cell surfaces involved in a variety of diseases. Thus, this part of the book encompasses a perspective of biofunctionalized nanomaterials in medicine, beneficial utility in biosensing, role in tissue engineering and regenerative medicine, protein based nanosystems in nutraceuticals, application in environmental pollution abatement and its impact on ecological sustainability, etc. Thus, in a nutshell, this volume of the book highlights the scientific understanding of life science in every possible way based on cutting-edge scientific insight.

All the chapters in this book are presented with clarity in language along with the support of appropriate figures, tables, and illustrations that will help to generate keen interest and support for readers to gain broad and analytical knowledge on a particular subject matter. Overall, the book is expected to meet the requirement of readers or research workers from the divergent field of biological sciences. However, this book is primarily targeted to graduate and professional students of biological sciences and also provide concept and appropriate reference material for scientists, innovators, industries, and government agencies involved in microbial research, environmental biotechnology, cell and molecular biology, nanotechnology, and overall multidiscipline related to biotechnology.

Our gratefulness can be expressed to all of the contributing authors who aided us enormously with their contributions and recommendations in this volume. The editorial members profusely express their gratitude to all individual authors for their timely cooperation, critical thoughts, valuable suggestions, and contributing attributes to this book. We are very grateful to Apple Academic Press Inc., USA, for considering our wishes and for the affable cooperation. At last, we thank our family members for silently supporting us to do the compilation work for finishing this book in time.

—Jyoti R. Rout
Rout G. Kerry
Abinash Dutta

PART I

Trends in Applied Microbiology

CHAPTER 1

Role of Endophytes in Crop Improvement

BICKY JERIN JOSEPH, A. R. NAYANA, and E. K. RADHAKRISHNAN*

School of Biosciences, Mahatma Gandhi University, Kottayam 686560, Kerala, India

**Corresponding author. E-mail: radhakrishnanek@mgu.ac.in*

ABSTRACT

Among the various plant-associated microorganisms, endophytes occupy a relatively significant niche within the plant tissues that make them have a determining role in various functions of plants. For this, they have diverse evolutionarily adapted multimechanistic features to promote plant growth, improve plant defense, produce antiherbivory compounds, and to protect the plant from biotically and abiotically originated stress conditions. The plethora of secondary metabolites produced for the same has potential applications in the field of medicine, agriculture, and industry. Many recent studies have indicated the endophytic microbial distribution to be ubiquitous in almost all plants and most of the tissues. Since these microorganisms function as the second genome of the plant system, manipulating its natural association in crops will have immense opportunities to maximize crop productivity and to substitute the agrochemical usage in a sustainable way. These remarkable features of endophytes make the current chapter significant for crop improvement.

1.1 INTRODUCTION

The global population may reach to about 9 billion by the mid of this century itself (Godfray et al., 2010). With the increasing population, there will be a demand for 75% to 100% more food by 2050 (Oya, 2009). But, diverse factors such as climatic changes, nature of the soil, water, and nutrient availability as well as plant pathogens generate many challenges to crop productivity (Lopes et al., 2018). Plant pathologists all over the world are interested to implement novel methods to control the pathogens and thereby to accelerate agricultural production (Dillard, 2019). Agrochemical based method was one of the most exploited strategies to increase crop yield, but the trend is moving toward organic agricultural practices due to the lethal effects of chemicals used.

Microorganisms of plants form an intrinsic part of the plant system. Hence, plant-associated microorganisms (PAMs) are investigated in detail to develop newer methods to improve the overall functionality of plants. These PAMs constitute a complex system containing fungi, archaea, and bacteria, which collectively form the plant microbiome. These play a vital role in plant growth promotion (PGP) through their mutualistic, commensalistic, as well as saprophytic association with plants. Based on the localization, these microbes can be of epiphytic or endophytic nature. The endophytic microorganisms have the ability to inhabit within the plant system by promoting innate defense mechanisms, plant growth, yield, and also resistance to various stress conditions (Chen et al., 2019). Endophytes can be of obligate or facultative nature. Obligate endophytes strictly need a plant host for their survival (Singh et al., 2017) as in the case of mycorrhizal fungi and fungi of the family Clavicipitaceae (Strobel, 2018). Endophytes are considered to promote plant growth through mechanisms such as mineral solubilization (Liu et al., 2019), nitrogen fixation (Ke et al., 2019), siderophore production (Sahu et al., 2019), phytohormones synthesis (Asaf et al., 2019), and also those that provide tolerance to both abiotic and biotic stress conditions (Rajkumar et al., 2013). The various secondary metabolites produced for implementing these functions can also have significant applications in agriculture, medicine, and industry.

Bacteria belonging to various phyla such as Actinobacteria, Proteobacteria, Bacteroidetes, and Firmicutes are reported to mediate plant growth through endophytic association (Conn et al., 2008, Girsowicz et al., 2019). The Proteobacteria that occupies the majority of the total microbial endophytic community include Sphingomonadales, Xanthomonadales, Enterobacteriales, Rhizobiales, Burkholderiales, Actinomycetales, Flavobacteriales, and Pseudomonadales (Zhang et al., 2011). Endophytes of the genera *Bacillus* (Mehta et al., 2014), *Pseudomonas* (Weyens et al., 2010), *Serratia*, and *Paenibaccillus* are well reported to improve crop yield and provide stress tolerance to plants. Fungal endophytes have also been reported to have a significant role in crop improvement due to its biocontrol and stress control mechanisms. A wide variety of endophytic fungi have already been described for plant beneficial effects mediated through the nutrient and water uptake, stress tolerance, and pest resistance (Llorens et al., 2019). In a study, Bae et al. (2011) have demonstrated remarkable, features of *Trichoderma* species to be exploited to protect the pepper from biotic stress induced by *Phytophtora capsici*. Similarly, endophytic fungi such as *Cladosporium cladosporioides, Paraconiothyrium* sp., and *Penicillium resedanum* were also reported to promote plant growth with protective effect from pathogenic and environmental stress conditions (Halász et al., 2016). These indicate the remarkable role of diverse endophytic microorganisms in the normal functioning of the plant system. So the main focus of this chapter is to describe the importance of endophytes in crop improvement by alleviating stress response. This also describes various steps involved in exploring the endophytes for field application in the form of consortia and formulations.

1.2 ENDOPHYTIC COLONIZATION

Endophytic microorganisms can be localized at various parts of the plant occupying either at the intracellular or intercellular regions. These organisms are considered to complete most of their life cycle by mutualistic interaction with the plant system and through this the plant is benefited significantly (Jasim et al., 2016). The entry of microorganisms into the plant system to occupy as endophytes can be initiated through various routes. Rhizosphere can be one among the most common sites of entry for microorganisms. The soil niche around the root of the plant is commonly referred to as the rhizosphere and is a hot spot of plant beneficial microorganisms. From this region, plants are considered to recruit selected organisms as endophyte by permitting their entry through cracks and wounds in lateral root hairs and also through other tissue wounds (Santoyo et al., 2016). Signaling for the same is considered to be mediated by the chemical composition of the plant exudates. This is because, leakage of metabolites and other chemoattractants through the wound attracts microorganism toward them (Singh et al., 2017). Several chemicals are released by the host plant that facilitates the colonization of endophytes in its different tissues (Panichikkal et al., 2019). Among the different microbial candidates available at the rhizosphere, endophytic recruitment will be made as per the selective criteria defined by the plants. Apart from this, endophytes can also be transmitted through the seeds or can be recruited through leaf and stem as per the requirement of the plant (Santoyo et al., 2016). Further mutualistic interaction between the plant and the microbe can lead to the designation of entered microorganism as a specific type of endophyte.

Within plants, colonization of endophytes can occur at the apoplast, intercellular spaces, xylem vessels, or vegetative parts (Romero et al., 2019). On the basis of the location of colonization, endophytes are categorized as root endophyte, leaf endophyte, seed endophyte, and so on (Bamisile et al., 2018). In a previous report, *Fusarium* sp., *Metarhizium* sp., *Piriformospora indica*, and *Glomus* sp. were described to enter into roots through the rhizosphere as root endophytes (Wyrebek et al., 2011). In the case of foliar endophytes, they usually invade through the stem, leaves, or aerial part of the plant (Meyling et al., 2011). The various stages of endophytic colonization have been described through the reporter gene technique using *gfp* or *gus*, SEM, TEM, and fluorescence in situ hybridization methods (Liu et al., 2017).

1.3 ISOLATION AND SCREENING OF ENDOPHYTES

By residing within various plant parts, endophytes significantly influence the plant physiological functions. So they can be present in any tissue type and based on the specific objective of the study, any tissue can be selected for endophytic isolation. The various plant parts used for the isolation of endophytes includes fruits (Valencia et al., 2019), flowers (Bungtongdee et al., 2019), roots (Rigerte et al., 2019), leaves (Lateef et al., 2019), seeds (Martinez-Rodriguez et al., 2019), and shoot meristematic tissues (Koskimäki et al., 2015). The selected tissues for the same is usually processed by washing with sterile water followed by surface sterilization using ethanol, sodium hypochlorite (Rao et al., 2018), hydrogen peroxide (Balsanelli et al., 2019), Tween 20, sodium thiosulphate, or a mixture of these at different concentrations (Zhang et al., 2016). After the surface sterilization, the microbes can be isolated by plating the treated plant tissue on the surface of specific media. For the same, the samples can also be diluted serially following maceration and this can further be plated over media. Further screening can be done using both conventional and nonconventional methods. In the conventional method, culturing can be done on an appropriate medium. For example, Kings B medium can be used for the selective growth of *Pseudomonas* species (Andreolli et al., 2019). Rapid identification of the isolated endophytes can be done using PCR followed by the sequencing of 16s rDNA (Tuo et al., 2019). The basic criteria for selecting any organism for crop improvement are based on various growth promoting characteristics that are screened by in vitro and in vivo methods (Figure 1.1). Metagenomic and metatranscriptomic analysis are also popular in recent years for identifying unculturable endophytic microbial diversity and prediction of its plant growth promoting features.

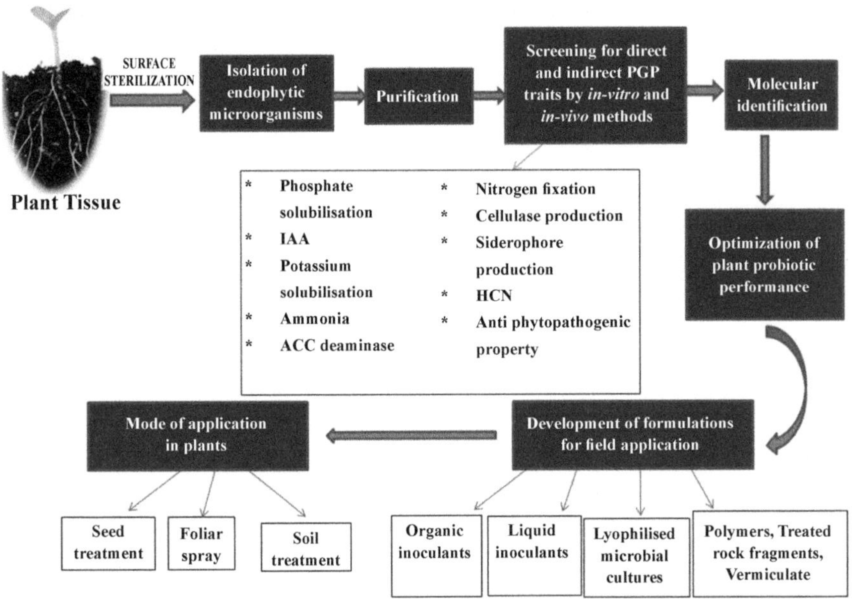

FIGURE 1.1 General steps involved in the isolation, screening and application of potential endophytic microorganisms

1.4 ENDOPHYTES-MECHANISM OF ACTION AND ANALYSIS OF PGP TRAITS

The mechanism of action of endophytes in plants can be direct or indirect. PGP is directly boosted up by the endophytes through mineral solubilization, nitrogen fixation, production of phytohormone, ACC deaminase, and others (Nelson, 2004). The indirect mechanisms mainly include phytopathogen interaction and antifungal mechanisms mediated through siderophore (Brígido et al., 2019), HCN, antimicrobial metabolites, and activation of induced systemic resistance (Figure 1.2). The indirect mechanisms contribute to the biocontrol properties of endophytes.

1.4.1 PHYTOHORMONE PRODUCTION AND REGULATION

1.4.1.1 INDOLE-3-ACETIC ACID

Indole-3-acetic acid (IAA) is the most popular candidate of the auxin family that plays an evident role in controlling the overall plant physiology including

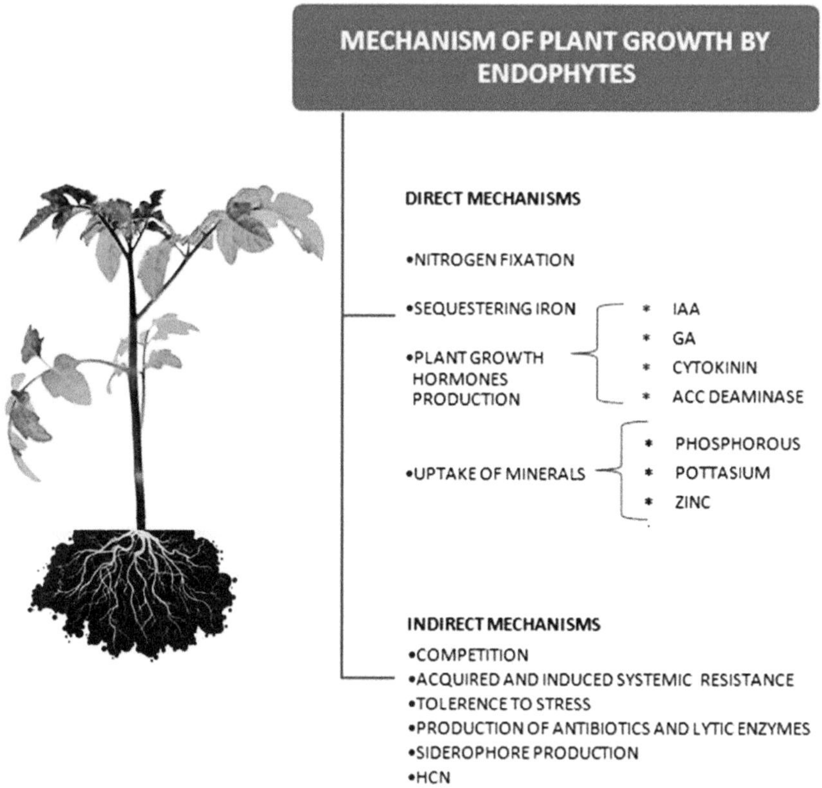

FIGURE 1.2 Mechanisms of plant growth promotion by endophytes

cell division, cell elongation, differentiation, and also tropism (gravity and light) (Vanneste and Friml, 2009). Numerous plant endophytes have the capability to synthesize IAA to mediate the growth promotion of respective host plant (Jasim et al., 2014b). In microbes, IAA can act as a regulating factor for the cell differentiation associated with spore germination and also in mycelial elongation (Matsukawa et al., 2007). The IAA produced by the endophytes triggers increased protection to plants from external stress by enhancing the cellular defense mechanisms (Egamberdieva et al., 2017). The biosynthetic pathways for IAA in plants and bacteria have been observed to be similar and the primary precursor for both is tryptophan (Li et al., 2018). Hence, tryptophan can have an important role in microbial IAA production. However, microorganisms are also considered to synthesize IAA in a tryptophan independent manner. IAA production has been reported for endophytes isolated from different plant materials and also from plants of different environmental conditions. These include endophytes from *Piper nigrum* (Jasim et al., 2014b), from saline dessert (Matsukawa et al., 2007) and also from tissue-like leaves (Hoffman et al., 2013). Endophytically associated *Paenibacillus* strains have previously been confirmed for IAA production by HPLC and are considered to have growth-promoting effects in *Curcuma longa*, from where these were isolated (Aswathy et al., 2012).

1.4.1.2 ACC DEAMINASE

1-Aminocyclopropane-1-carboxylic acid (ACC) deaminase produced by endophyte is mainly involved in the regulation of ethylene production by its utilization (Afridi et al., 2019). Such plant beneficial endophytes metabolize ACC and thus lowers the ethylene production in plants. This can favor plant growth and its lower susceptibility to stress conditions (Van de Poel and Van Der Straeten, 2014). Such mechanisms have also been explained to be useful during the continuous flooded conditions. Hence, ACC deaminase has been suggested to be involved in endophyte mediated root elongation in rice (Shen et al., 2019). *Burkholderia* sp. have been reported to have extensive ACC deaminase activity and hence are considered to enhance the plant growth through ethylene modulation (Onofre-Lemus et al., 2009). Endophytic bacteria from copper tolerant plants have also been reported for ACC deaminase activity and copper accumulation in *Brassica napus* (Zhang et al., 2011). Based on the available results, ACC deaminase producing organisms can be considered to have a pivotal role in plant growth under stressful environmental conditions.

1.4.2 NITROGEN FIXATION AND MINERAL SOLUBILIZATION BY ENDOPHYTES

After the green revolution, the chemical fertilizer usage has drastically increased as several major and minor minerals are essential for plant growth. The most prominently used fertilizers are the NPK fertilizers (Kandel et al., 2017). By considering the environmental impact of heavy fertilizer input into the agriculture field, methods to reduce fertilizer use is important because most of the applied fertilizers are getting immobilized in the soil. Microbes have an inevitable role to reduce the environmental damage caused by fertilizers, as some of them can function as biofertilizer by itself. Endophytes decrease the fertilizer load in agricultural soil due to their plant beneficial mechanisms.

From several studies related to N fertilizer usage, it is estimated that out of 100 Tg N only 17 Tg N is available for agriculture (Kandel et al., 2017). Thus the nitrogen-fixing bacteria have got a significant role in reducing the nitrogen fertilizer load of the soil. The diazotrophs associated with root nodules are known to help the plant to fix atmospheric nitrogen (Ke et al., 2019). Several genera of endophytic diazotrophs such as *Burkholderia* (Govindarajan et al., 2007), *Klebsiella* (Lin et al., 2019), and *Pseudomonas* (Ke et al., 2019) located at various plant parts are considered to promote the growth of crops under harsh conditions. This indicates the nitrogen-fixing organisms to have a remarkable influence on the plant system.

Phosphorus (P) is another important nutrient for plants and some microbes have the potential to mobilize P by converting the insoluble P to its soluble form. The important reservoir of insoluble phosphate is organic matter and only a basal portion of this is accessible to plants under normal conditions (Linu et al., 2019). Hence, phosphate solubilizing microorganisms can be used as an alternative for conventional chemical fertilizer. Microorganisms can release organic acids that are considered to convert the inorganic phosphate to its soluble forms and thereby make it available to plants. In a previous study, gluconic acid produced by endophytic bacteria has been described to solubilize the insoluble phosphate and thereby supported plant beneficial effects (Oteino et al., 2015). Endophytic fungi are also known to solubilize various insoluble forms of phosphates such as aluminum (AlP), iron (FeP), and tricalcium phosphate (TCP) at varying temperatures through the production of phytases, organic acids, and phosphatases. However, phosphate solubilization is considered to be higher in acidic condition (Adhikari and Pandey, 2019).

Potassium is an essential element for plant growth and development. Use of potassium solubilizing bacteria to increase the soluble potassium in the soil can have a promising effect to resolve the potassium deficiency of plants in a sustainable way. In a previous study, the diversity of potassium-solubilizing endophytic bacteria from Moso Bamboo (*Phyllostachys edulis*) has been described (Yuan et al., 2015). Many bacterial spp. such as *Paenibacillus* sp., *Acidithiobacillus ferrooxidans*, *Bacillus circulans*, *Bacillus mucilaginosus*, and *Bacillus edaphicus* have been described to have the potential to solubilize potassium present in the soil (Etesami et al., 2017).

As an essential micronutrient, zinc has an important role in regulating diverse plant mechanisms. Plant requires zinc for controlling key functions like phytohormone synthesis (e.g., auxin, ABA), protein synthesis, seedling vigor, sugar formation, membrane function, defense against diseases, and response to abiotic stress factors like drought (Cakmak and Kutman, 2018).

Due to its immobile nature in soil, zinc deficiency in crops has been reported even though only a small amount is required for plant growth. Various zinc solubilizing microorganisms have been reported to solubilize the immobile zinc present in the soil (Sharma et al., 2011). Utilization of these microbes will be a better option to meet the zinc deficiency in crops because other approaches such as chemical fertilizer application, conventional breeding, transgenic approaches, and genetic engineering are slower, expensive, and laborious (Kamran et al., 2017). Zinc solubilizing bacteria reported from the rice are considered to be functionally involved in the root development and growth improvement (Idayu Othman et al., 2017). Various legume endophytes are also reported to solubilize different sources of zinc such as ZnO and $Zn_3(PO_4)_2$ (Sharma et al., 2011).

1.5 ENDOPHYTES WITH BIOCONTROL PROPERTIES

Biological control reduces the chemical usage and thus are offered as an attractive and eco-friendly approach for effective pest management. Wide spectra of endophytes are shown to exhibit biocontrol properties. Both bacterial and fungal endophytes have an effective role against phytopathogens, which make them promising biocontrol agents. Production of inhibitory substances such as HCN, antibiotics, and siderophore and competition for space make the endophytic bacteria to control phytopathogens by direct antagonism and through the induction of systemic acquired resistance (Ramesh and Phadke, 2012). Some fungal endophytes produce bioactive metabolites and enzymes that directly act against pathogens through diverse mechanisms. Low molecular weight iron-chelating molecules called siderophores produced by microbes have got antagonistic effects against fungal pathogens (Pal et al., 2019). In a study, siderophore-producing *Burkholderia* has been described to have inhibitory effects against disease-causing fungi (Loaces et al., 2011).

Phytopathogens such as *Colletotrichum* sp., *Phytophthora* sp., *Fusarium* sp., *Rhizoctonia* sp., and *Sclerotium* sp. have been described to be inhibited by diverse endophytic organisms. This makes endophytes to have applications in sustainable agriculture practices (Sabu et al., 2017). *Rhizopycnis vagum*, an endophytic fungus, has shown to have remarkable antagonism against *Pythium myriotylum*, which makes it to have important application as a biocontrol agent (Anisha et al., 2018). *Pseudomonas aeruginosa*, an endophytic bacteria from ginger rhizome, has also been reported to produce antifungal compound phenazine 1-carboxylic acid with a protective effect on ginger rot disease (Jasim et al., 2014a).

1.6 ABIOTIC STRESS TOLERANCE AND CROP IMPROVEMENT

Endophytes promote the growth of plants under various stress conditions such as drought, heavy metal toxicity, flood, salinity, and heat (Yaish et al., 2015). Microbial endophytes especially the *Trichoderma* strains isolated from wheat plants have previously been demonstrated to possess crucial functions to enhance the plant growth by providing tolerance to salt, heavy metals, temperature (55 °C) and drought (Ripa et al., 2019). These organisms can also have the role to mediate phytoremediation that is important to reduce heavy metal toxicity and to reduce the environmental pollution (Syed and Prasad Tollamadugu, 2019). Many endophytes are capable of rendering habitat adaption to plants by providing tolerance to habitat-specific selective pressures such as soil, temperature, pH, and salinity (Rodriguez et al., 2009). Various secondary metabolites of endophytes have also been proved to improve habitat adaptation of plants. In the case of endophytic fungi from ginger, several bioactive metabolites like tyrosol were characterized. By occupying within the rhizome, these fungi and their respective metabolites can consider to have significant role in the metabolic, biosynthetic, and survival mechanisms of rhizome (Anisha and Radhakrishnan, 2017).

1.7 MOLECULAR ANALYSIS OF GENES IN PLANT GROWTH AND STRESS TOLERANCE

Molecular screening for plant growth promoting potential of endophytic bacteria can be done by analyzing the presence of respective genes in them (Table 1.1). The gene *acdS* encodes ACC deaminase and hence primers specific to them can be used for its molecular identification (Chandra et al., 2018). In a study, *acdS* gene mutant of *Burkholderia phytofirmans* has demonstrated to lose their ACC deaminase activity, which thereby showed a negative effect on the canola root elongation (Sun et al., 2009). In another study, whole-genome analysis of *Caulobacter flavus has resulted in the identification of* genes associated with PGP traits. These included genes related to nitrogen fixation, phosphate solubilization, IAA metabolism, volatile compounds, phenazine, cobalamin, and spermidine synthesis. Several genes exhibiting tolerance to heavy metals were also identified in the genome of *Caulobacter (Yang et al., 2019)*. In a recent study, several genes involved in the plant growth such as genes for biosynthesis of IAA, 2,3-butanediol and acetoin, and rhodanese, the catalase genes *bglX* and *katGE*, nitrogen fixation genes *nifJ-NifQ*, and cellulase *bglH* gene were reported to be present in endophytic *Klebsiella variicola* isolated from sugarcane stem (Reyna-Flores et al., 2018). With the advancements in DNA sequencing and whole-genome analysis, more and accurate prediction of genetic basis of plant growth enhancement by diverse endophytic bacteria is possible.

TABLE 1.1 Candidate Genes Related to Plant Growth Promotion and Metal Resistance Reported From Diverse Plant Associated Bacteria

Traits	Gene Abbreviation	Gene Annotation	References
Phosphate solubilization or transport genes	*ppk* *ppnk* *pit* *pst*B *phy* *gcd*	Polyphosphate kinase Inorganic polyphosphate kinase Phosphate inorganic transporter Phosphate ABC transporter ATP binding protein 3 phytase Glucose dehydrogenase	Yang et al. (2019) Chhabra et al. (2013) Rodríguez et al. (2006)
Nitrogen fixation genes	*fix*K *nif*U *nif*S1, *nif*S2 *gln*A *nifJ*,*nifQ*	Nitrogen fixation-regulating protein fix K nitrogen fixation protein Nitrogenase metallocluster biosynthesis protein Glutamine synthetase gene nitrogen fixation protein	Yang et al. (2019) Guttman et al. (2008) Reyna-Flores et al. (2018)
IAA related genes	*trp*A *trp*B *trp*D *trp*F *trpR* *ipdc* *trp*S *trpE* *trpG* *trpC* *PAI*	Tryptophan synthase subunit alpha Tryptophan synthase subunit beta Anthranilate phosphor-ribosyltransferase N-(5′-Phosphoribosyl) anthranilite isomerase Tryptophan- tRNA ligase TrpR-binding protein Indole-3-pyruvate decarboxylase Tryptophan–tRNA ligase Anthranilate synthase subunit I Anthranilate synthase Indole-3-glycerol phosphate synthase Phospho-ribosyl anthranilate Isomerase	Yang et al. (2019) Rodrigues et al. (2016) Jasim et al. (2014) Asaf et al. (2018)

TABLE 1.1 *(Continued)*

Traits	Gene Abbreviation	Gene Annotation	References
H2S Production	*cys*P, *cys*W, *cys*T, *cys*A	Cysteine	Asaf et al. (2018)
Volatile Signal Related Genes	*aco*R *aco* *ilv*H *ilv*B *ilv*X	Acetoin catabolism regulatory protein Acetoin utilization protein Acetolactate synthase small subunit Acetolactate synthase isozyme 3 large subunit Putative acetolactate synthase large subunit	Sharifi and Ryu, (2018) Schulz-Bohm et al. (2017) Yang et al. (2019)
Antibiotic Related Genes	*phz*F *lys*R *pot*D *pot*B, *pot*C *pot*A *cob*T *cob*D, *cob*P, *cob*W, *cob*S, *cbi*G	Phenazine biosynthesis protein PhzF family LysR transcriptional regulator Spermidine/putrescine ABC transporter substrate binding protein Spermidine/putrescine ABC transporter permease Putrescine/spermidine ABC transporter ATP binding protein Cobaltochelatase Cobalamin biosynthesis	Yang et al. (2019)
Metal resistance	*czcA* *czsB*/*cusB* *czcC* *pbrA* *pbrB* *pbrT* *czcD, czcR* *hmrR* *cueA* *catalase hpII* *copA* *copB* *copC* *arsB* *arsC*	Cobalt/zinc/cadmium resistance protein CzcA Cobalt/zinc/cadmium efflux RND transporter, membrane fusion protein, CzcB family Heavy metal RND efflux outer membrane protein, CzcC family Lead efflux transporter phosphatase Lead uptake protein Cobalt/zinc/cadmium resistance protein CzcD Transcriptional regulator, MerR family Copper-translocating P-type ATPase Catalase related to oxidative stress	Yang et al. (2019) Shin et al. (2012) Han et al. (2010)

TABLE 1.1 *(Continued)*

Traits	Gene Abbreviation	Gene Annotation	References
		Multi-copper oxidase	
		Copper resistance protein	
		Copper homeostasis	
		Arsenic efflux membrane protein	
		Arsenate reductase	

1.8 RICE ENDOPHYTES AND ITS ROLE IN CROP IMPROVEMENT

Rice is the largest cultivated crop in the world and hence the crop improvement studies related to it has got significant importance. The endophytic community of rice seed belonging to *Sphingomonas, Methylobacterium*, and *Enterobacteriaceae* family were reported recently to increase the plant growth (Eyre et al., 2019). Endophytic fungi of the genera *Alternaria, Hannaella*, and members of the *Pleosporales* order were also described as the valuable members of rice microbiome with the role to increase plant performance and yield. A study by Shen et al. (2019) showed rice root endophytes such as *Rhizobium larrymoorei, Bacillus aryabhattai, Pseudomonas granadensis*, and *Bacillus fortis* to have plant growth–promoting characteristics including the potential to perform IAA production, phosphorus-solubilizing, and nitrogen fixation. Among these, *B. aryabhattai* has been reported as a potent bacterial biofertilizer with pesticide tolerance and PGP promises (Shen et al., 2019).

1.9 EMERGING TRENDS OF ENDOPHYTIC APPLICATION—CONSORTIUM AND FORMULATION DEVELOPMENT

Having more information about the effect of endophytic microbiome on plant growth and development, the next phase is to make use of these microbes as potential consortia for field application. Under varying environmental conditions microbial consortia may perform more effectively than monocultures due to the several features of these organisms (Bell et al., 2005). Consortia preparation requires high compatibility between the microbes used and the environment to which it is applied (Kaminsky et al., 2019). Several studies indicate that the use of endophytes together with PGPR has an effective role in plant growth. In one of the previous investigations, the root endophytic fungus *P. indica* and the plant growth-promoting rhizobacteria (fluorescent Pseudomonads) formulated with inorganic carrier-based (vermiculite and talcum powder) formulations was shown to have a positive result on the growth of *Vigna mungo* (Kumar et al., 2011). In another study two fluorescent Pseudomonads together with endophytic *P. indica* were used for the development of vermiculite- and talcum-based bioinoculant formulations for yield increase and also for the effective control of Fusarium wilt of tomato plants. Here, talcum-based bioinoculant performed more effectively than vermiculite-based formulations (Sarma et al., 2011).

Recently endophytic *P. aeruginosa* along with the recommended dosage of *Bradyrhizobium* sp. were shown to have synergistic effects on the growth of plants and showed their promise as bioinoculants and bioenhancers. This can be due to the multifarious PGP characteristics such as IAA production, P and Zn solubilization, siderophore, and production of cell wall degrading enzymes (Kumawat et al., 2019). Nanoformulations using endophytes are less investigated and will be an interesting area likely to emerge in the coming years. An efficient nanobiocontrol delivery system formulated with silica and graphite nanoparticles along with endophytic *Lysinibacillus* and PGPR *Bacillus subtilis*, and *Pseudomonas fluorescens* antagonistic to *Ralstonia solanacearum* in potato plants was shown to retain the viability of bacteria (Djaya et al., 2019). Most recently, a novel idea of field-specific microbial consortia application by screening the performance of individual microbes at varying physicochemical conditions has been recommended as suitable for field application (Awasthi, 2019).

1.10 CONCLUSION

A biosphere is a complex functioning system in which all the living forms are interconnected in some way or the other. Diverse endophytic microbes with multiple plant growth promoting traits, tolerance to heavy metals and pathogenic resistance from diverse crop plants are demonstrated to have an important role in assisting sustainable agriculture practices by reducing the requirement for chemical fertilizers. Both biotic as well as abiotic stress factors limit the plant growth and hence endophytes have the promises to manage both these conditions. This review summarizes the importance of both fungal and bacterial endophytes for crop improvement. A broad range of research work in this area is essential to translate it to a promising tool for increasing the crop productivity.

KEYWORDS

- **endophytes**
- **plant growth promotion**
- **crop improvement**
- **beneficial microorganisms**

REFERENCES

Adhikari, P.; Pandey, A. Phosphate solubilization potential of endophytic fungi isolated from *Taxus wallichiana* Zucc. roots. *Rhizosphere* **2019,** *9*, 2–9.

Afridi, M.S.; Mahmood, T.; Salam, A.; Mukhtar, T.; Mehmood, S.; Ali, J.; Khatoon, Z.; Bibi, M.; Javed, M.T.; Sultan, T; Chaudhary, H.J. Induction of tolerance to salinity in wheat genotypes by plant growth promoting endophytes: Involvement of ACC deaminase and antioxidant enzymes. *Plant Physiol. Biochem.* **2019,** *139*, 569–577.

Andreolli, M.; Zapparoli, G.; Angelini, E.; Lucchetta, G.; Lampis, S.; Vallini, G. Pseudomonas protegens MP12: A plant growth-promoting endophytic bacterium with broad-spectrum antifungal activity against grapevine phytopathogens. *Microbiol. Res.* **2019,** *219*, 123–131.

Anisha, C.; Jishma, P.; Bilzamol, V.S.; Radhakrishnan, E.K. Effect of ginger endophyte *Rhizopycnis vagum* on rhizome bud formation and protection from phytopathogens. *Biocatal. Agric. Biotechnol.* **2018,** *14*, 116–119.

Anisha, C.; Radhakrishnan, E.K. Metabolite analysis of endophytic fungi from cultivars of *Zingiber officinale* Rosc. identifies myriad of bioactive compounds including tyrosol. *3 Biotech* **2017,** *7*,146.

Asaf, S.; Khan, A.L.; Khan, M.A.; Al-Harrasi, A.; Lee, I.J. Complete genome sequencing and analysis of endophytic *Sphingomonas* sp. LK11 and its potential in plant growth. *3 Biotech* **2018,** *8*,389.

Asaf, S.; Khan, A.L.; Waqas, M.; Kang, S.M.; Hamayun, M.; Lee, I.J.; Hussain, A. Growth-promoting bioactivities of Bipolaris sp. CSL-1 isolated from *Cannabis sativa* suggest a distinctive role in modifying host plant phenotypic plasticity and functions. *Acta Physiol. Plant* **2019,** *41*, 65.

Aswathy, A.J.; Jasim, B.; Jyothis, M.; Radhakrishnan, E.K. Identification of two strains of *Paenibacillus* sp. as indole 3 acetic acid-producing rhizome-associated endophytic bacteria from *Curcuma longa*. *3 Biotech* **2013,** *3*, 219–224.

Awasthi, A. Field-specific microbial consortia are feasible: a response to Kaminsky et al. *Trends Biotechnol.* **2019,** *37*, 569–572.

Bae, H.; Roberts, D.P.; Lim, H.S.; Strem, M.D.; Park, S.C.; Ryu, C.M.; Melnick, R.L.; Bailey, B.A. Endophytic Trichoderma isolates from tropical environments delay disease onset and induce resistance against *Phytophthora capsici* in hot pepper using multiple mechanisms. *Mol. Plant Microbe Interact.* **2011,** *24*,336–351.

Balsanelli, E.; Pankievicz, V.C.; Baura, V.A.; de Oliveira Pedrosa, F.; de Souza, E.M. A New strategy for the selection of epiphytic and endophytic bacteria for enhanced plant performance. In: *Plant Innate Immunity*, Humana, New York, NY, 2019; pp. 247–256.

Bamisile, B.S.; Dash, C.K.; Akutse, K.S.; Keppanan, R.; Wang, L. Fungal endophytes: beyond herbivore management. *Front. Microbiol.* **2018,** *9*, 544.

Bell, T.; Newman, J.A.; Silverman, B.W.; Turner, S.L.; Lilley, A.K. The contribution of species richness and composition to bacterial services. *Nature* **2005,** *436*,1157–1160.

Brígido, C.; Singh, S.; Menéndez, E.; Tavares, M.J.; Glick, B.R.; Félix, M.D.R.; Oliveira, S.; Carvalho, M. Diversity and functionality of culturable endophytic bacterial communities in chickpea plants. *Plants* **2019,** *8*, 42.

Bungtongdee, N.; Sopalun, K.; Laosripaiboon, W.; Iamtham, S. The chemical composition, antifungal, antioxidant and antimutagenicity properties of bioactive compounds from fungal endophytes associated with Thai orchids. *J. Phytopathol.* **2019,** *167*, 56–64.

Cakmak, I.; Kutman, U.B. Agronomic biofortification of cereals with zinc: a review. *Eur. J. Soil Sci.* **2018,** *69*, 172–180.

Chandra, D.; Srivastava, R.; Glick, B.R.; Sharma, A.K. Drought-tolerant *Pseudomonas* spp. improve the growth performance of finger millet (*Eleusine coracana* (L.) Gaertn.) under non-stressed and drought-stressed conditions. *Pedosphere* **2018,** *28*, 227–240.

Chhabra, S.; Brazil, D.; Morrissey, J.; Burke, J.I.; O'Gara, F.; N. Dowling, D. Characterization of mineral phosphate solubilization traits from a barley rhizosphere soil functional metagenome. *MicrobiologyOpen* **2013,** *2*, 717–724.

Chen, L.; Shi, H.; Heng, J.; Wang, D.; Bian, K. Antimicrobial, plant growth-promoting and genomic properties of the peanut endophyte *Bacillus velezensis* LDO2. *Microbiol. Res.* **2019,** *218*, 41–48.

Conn, V.M.; Walker, A.R.; Franco, C.M. Endophytic actinobacteria induce defense pathways in *Arabidopsis thaliana. Mol. Plant Microbe Interact.* **2008,** 21, 208–218.

Dillard, H.R. Global food and nutrition security: from challenges to solutions. *Food Security* **2019,** *11*, 249–252.

Djaya, L.; Istifadah, N.; Hartati, S.; Joni, I.M. In vitro study of plant growth promoting rhizobacteria (PGPR) and endophytic bacteria antagonistic to *Ralstonia solanacearum* formulated with graphite and silica nano particles as a biocontrol delivery system (BDS). *Biocatal. Agric. Biotechnol.* **2019,** *19*, 101153.

Egamberdieva, D.; Wirth, S.J.; Shurigin, V.V.; Hashem, A.; Abd Allah, E.F. Endophytic bacteria improve plant growth, symbiotic performance of chickpea (*Cicer arietinum* L.) and induce suppression of root rot caused by *Fusarium solani* under salt stress. *Front. Microbiol.* **2017,** *8*,1887.

Etesami, H.; Emami, S.; Alikhani, H.A. Potassium solubilizing bacteria (KSB):: Mechanisms, promotion of plant growth, and future prospects: a review. *J. Soil Sci. Plant Nutri.* **2017,** *17*, 897–911.

Eyre, A.W.; Wang, M.; Oh, Y.; Dean, R.A. Identification and characterization of the core rice seed microbiome. *Phytobiomes J.* **2019,** *3*, 148–157.

Girsowicz, R.; Moroenyane, I.; Steinberger, Y. Bacterial seed endophyte community of annual plants modulated by plant photosynthetic pathways. *Microbiol. Res.* **2019,** *223*, 58–62.

Godfray, H.C.J.; Beddington, J.R.; Crute, I.R.; Haddad, L.; Lawrence, D.; Muir, J.F.; Pretty, J.; Robinson, S.; Thomas, S.M.; Toulmin, C. Food security: the challenge of feeding 9 billion people. *Science* **2010,** *327*, 812–818.

Govindarajan, M.; Balandreau, J.; Kwon, S.W.; Weon, H.Y.; Lakshminarasimhan, C. Effects of the inoculation of *Burkholderia vietnamensis* and related endophytic diazotrophic bacteria on grain yield of rice. *Microb. Ecol.* **2008,** *55*, 1–37.

Fouts, D.E.; Tyler, H.L.; DeBoy, R.T.; Daugherty, S.; Ren, Q.; Badger, J.H.; Durkin, A.S.; Huot, H.; Shrivastava, S.; Kothari, S.; Dodson, R.J. Complete genome sequence of the N2-fixing broad host range endophyte *Klebsiella pneumoniae* 342 and virulence predictions verified in mice. *PLoS Genet.* **2008,** *4,* e1000141.

Halász, K.; Borbély, C.; Pós, V.; Gáspár, L.; Haddadderafshi, N.; Winter, Z.; Lukács, N. Effect of crop management and cultivar on colonization of *Capsicum annuum* L. by endophytic fungi. *Acta Univ. Sapientiae Agric. Environ.* **2016,** *8*, 5–15.

Han, J.I.; Choi, H.K.; Lee, S.W.; Orwin, P.M.; Kim, J.; LaRoe, S.L.; Kim, T.G.; O'Neil, J.; Leadbetter, J.R.; Lee, S.Y.; Hur, C.G. Complete genome sequence of the metabolically versatile plant growth-promoting endophyte *Variovorax paradoxus* S110. *J. Bacteriol.* **2011,** *193*, 1183–1190.

Hoffman, M.T.; Gunatilaka, M.K.; Wijeratne, K.; Gunatilaka, L.; Arnold, A.E. Endohyphal bacterium enhances production of indole-3-acetic acid by a foliar fungal endophyte. *PLoS One* **2013,** 8, e73132.

Othman, N.M.I.; Othman, R.; Saud, H.M.; Wahab, P.E.M. Effects of root colonization by zinc-solubilizing bacteria on rice plant (*Oryza sativa* MR219) growth. *Agric. Nat. Resour.* **2017,** *51*, 532–537.

Jasim, B.; Anisha, C.; Rohini, S.; Kurian, J.M.; Jyothis, M.; Radhakrishnan, E.K. Phenazine carboxylic acid production and rhizome protective effect of endophytic *Pseudomonas aeruginosa* isolated from Zingiber officinale. *World J. Microbiol. Biotechnol.* **2014a,** *30*, 1649–1654.

Jasim, B.; Jimtha John, C.; Shimil, V.; Jyothis, M.; Radhakrishnan, E.K. Studies on the factors modulating indole-3-acetic acid production in endophytic bacterial isolates from *Piper nigrum* and molecular analysis of ipdc gene. *J. Appl. Microbiol.* **2014b,** *117*, 786–799.

Jasim, B.; Jimtha, C.J.; Jyothis, M.; Radhakrishnan, E.K. Plant growth promoting potential of endophytic bacteria isolated from *Piper nigrum*. *Plant Growth Regul.* **2013,** *71*, 1–11.

Jasim, B.; Sreelakshmi, K.S.; Mathew, J.; Radhakrishnan, E.K. Surfactin, iturin, and fengycin biosynthesis by endophytic *Bacillus* sp. from *Bacopa monnieri*. *Microb. Ecol.* **2016,** *72*, 106–119.

Kaminsky, L.M.; Trexler, R.V.; Malik, R.J.; Hockett, K.L.; Bell, T.H. The inherent conflicts in developing soil microbial inoculants. *Trends Biotechnol.* **2019,** *37*, 140–151.

Kamran, S.; Shahid, I.; Baig, D.N.; Rizwan, M.; Malik, K.A.; Mehnaz, S. Contribution of zinc solubilizing bacteria in growth promotion and zinc content of wheat. *Front. Microbiol.* **2017,** *8*, 2593.

Kandel, S.L.; Joubert, P.M.; Doty, S.L. Bacterial endophyte colonization and distribution within plants. *Microorganisms* **2017,** *5*, 77.

Ke, X.; Feng, S.; Wang, J.; Lu, W.; Zhang, W.; Chen, M.; Lin, M. Effect of inoculation with nitrogen-fixing bacterium *Pseudomonas stutzeri* A1501 on maize plant growth and the microbiome indigenous to the rhizosphere. *Syst. Appl. Microbiol.* **2019,** *42*, 248–260.

Koskimäki, J.J.; Pirttilä, A.M.; Ihantola, E.L.; Halonen, O.; Frank, A.C. The intracellular scots pine shoot symbiont Methylobacterium extorquens DSM13060 aggregates around the host nucleus and encodes eukaryote-like proteins. *MBio* **2015,** *6*, e00039–15.

Kumar, V.; Sarma, M.V.R.K.; Saharan, K.; Srivastava, R.; Kumar, L.; Sahai, V.; Bisaria, V.S.; Sharma, A.K. Effect of formulated root endophytic fungus *Piriformospora indica* and plant growth promoting rhizobacteria fluorescent pseudomonads R62 and R81 on *Vigna mungo*. *World J. Microbiol. Biotechnol.* **2012,** *28*, 595–603.

Kumawat, K.C.; Sharma, P.; Sirari, A.; Singh, I.; Gill, B.S.; Singh, U.; Saharan, K. Synergism of *Pseudomonas aeruginosa* (LSE-2) nodule endophyte with Bradyrhizobium sp.(LSBR-3) for improving plant growth, nutrient acquisition and soil health in soybean. *World J. Microbiol. Biotechnol.* **2019,** *35*, 47.

Lateef, A.A.; Garuba, T.; Sa'ad, G.; Olesin, M.; Eperetun, G.G.; Tiamiyu, B.B. Isolation and molecular identification of dominant fungal endophytes from green leaves of physic nut (*Jatropha curcas*) from unilorin plantation, Ilorin, Nigeria. *Sri Lankan J. Biol.* **2019,** *4*.

Li, M.; Guo, R.; Yu, F.; Chen, X.; Zhao, H.; Li, H.; Wu, J. Indole-3-acetic acid biosynthesis pathways in the plant-beneficial bacterium *Arthrobacter pascens* ZZ21. *Int. J. Mol. Sci.* **2018,** *19*, 443.

Lin, B.; Song, Z.; Jia, Y.; Zhang, Y.; Wang, L.; Fan, J.; Lin, Z. Biological characteristics and genome-wide sequence analysis of endophytic nitrogen-fixing bacteria *Klebsiella variicola* GN02. *Biotechnol. Biotechnol. Equip.* **2019,** *33*, 108–117.

Linu, M.S.; Asok, A.K.; Thampi, M.; Sreekumar, J.; Jisha, M.S. Plant growth promoting traits of indigenous phosphate solubilizing *Pseudomonas aeruginosa* isolates from Chilli (Capsicum annuum L.) Rhizosphere. *Commun. Soil Sci. Plan.* **2019,** *50*, 444–457.

Liu, C.; Mou, L.; Yi, J.; Wang, J.; Liu, A.; Yu, J. The eno gene of *Burkholderia cenocepacia* strain 71–2 is involved in phosphate solubilization. *Curr. Microbiol.* **2019,** *76*, 495–502.

Liu, H.; Carvalhais, L.C.; Crawford, M.; Singh, E.; Dennis, P.G.; Pieterse, C.M.; Schenk, P.M. Inner plant values: diversity, colonization and benefits from endophytic bacteria. *Front. Microbiol.* **2017,** *8*, 2552.

Llorens, E.; Sharon, O.; Camañes, G.; García-Agustín, P.; Sharon, A. Endophytes from wild cereals protect wheat plants from drought by alteration of physiological responses of the plants to water stress. *Environ. Microbiol.* **2019,** *21*, 3299–3312.

Loaces, I.; Ferrando, L.; Scavino, A.F. Dynamics, diversity and function of endophytic siderophore-producing bacteria in rice. *Microb. Ecol.* **2011,** *61*, 606–618.

Lopes, R.; Tsui, S.; Gonçalves, P.J.; de Queiroz, M.V. A look into a multifunctional toolbox: endophytic Bacillus species provide broad and underexploited benefits for plants. *World J. Microbiol. Biotechnol.* **2018,** *34*, 94.

Martinez-Rodriguez, A.; Macedo-Raygoza, G.; Huerta-Robles, A.X.; Reyes-Sepulveda, I.; Lozano-Lopez, J.; García-Ochoa, E.Y.; Fierro-Kong, L.; Medeiros, M.H.; Di Mascio, P.; White, J.F.; Beltran-Garcia, M.J. Agave seed endophytes: ecology and impacts on root architecture, nutrient acquisition, and cold stress tolerance In: *Seed Endophytes*; Springer, Cham, 2019; pp. 139–170.

Matsukawa, E.; Nakagawa, Y.; Iimura, Y.; Hayakawa, M. Stimulatory effect of indole-3-acetic acid on aerial mycelium formation and antibiotic production in Streptomyces spp. *Actinomycetologica* **2007,** *21*, 32–39.

Mehta, P.; Walia, A.; Kakkar, N.; Shirkot, C.K. Tricalcium phosphate solubilisation by new endophyte *Bacillus methylotrophicus* CKAM isolated from apple root endosphere and its plant growth-promoting activities. *Acta Physiol. Plant.* **2014,** *36*, 2033–2045.

Meyling, N.V.; Thorup-Kristensen, K.; Eilenberg, J. Below-and aboveground abundance and distribution of fungal entomopathogens in experimental conventional and organic cropping systems. *Biol. Control* **2011,** *59*, 180–186.

Nelson, L.M. Plant growth promoting rhizobacteria (PGPR): prospects for new inoculants. *Crop Manage.* **2004,** *3*.

Onofre-Lemus, J.; Hernández-Lucas, I.; Girard, L.; Caballero-Mellado, J. ACC (1-aminocyclopropane-1-carboxylate) deaminase activity, a widespread trait in Burkholderia species, and its growth-promoting effect on tomato plants. *Appl. Environ. Microbiol.* **2009,** *75*, 6581–6590.

Otieno, N.; Lally, R.D.; Kiwanuka, S.; Lloyd, A.; Ryan, D.; Germaine, K.J.; Dowling, D.N. Plant growth promotion induced by phosphate solubilizing endophytic Pseudomonas isolates. *Front. Microbiol.* **2015,** *6*, 745.

Oya, C. The World Development Report 2008: inconsistencies, silences, and the myth of 'win-win'scenarios. *J. Peasant Stud.* **2009,** *36*, 593–601.

Pal, G.; Kumar, K.; Verma, A.; White, J.F.; Verma, S.K. Functional roles of seed-inhabiting endophytes of rice In: *Seed Endophytes*; Springer, Cham, 2019; 213–236.

Panichikkal, J.; Thomas, R.; John, J.C.; Radhakrishnan, E.K. Biogenic gold nanoparticle Supplementation to plant beneficial *Pseudomonas monteilii* was found to enhance its plant probiotic effect. *Curr. Microbiol.* **2019,** *76*, 503–509.

Rajkumar, M.; Prasad, M.N.V.; Swaminathan, S.; Freitas, H. Climate change driven plant–metal–microbe interactions. *Environ. Int.* **2013,** *53*, pp. 74–86.

Ramesh, R.; Phadke, G.S. Rhizosphere and endophytic bacteria for the suppression of eggplant wilt caused by *Ralstonia solanacearum*. *Crop Prot.* **2012,** *37*, 35–41.

Rao, A.; Ramakrishna, N.; Arunachalam, S.; Sathiavelu, M. Isolation, screening and optimization of laccase-producing endophytic fungi from *Euphorbia milii*. *Arab. J. Sci. Eng.* **2019,** *44*, 51–64.

Reyna-Flores, F.; Barrios-Camacho, H.; Dantán-González, E.; Ramírez-Trujillo, J.A.; Beltrán, L.F.L.A.; Rodríguez-Medina, N.; Garza-Ramos, U.; Suárez-Rodríguez, R. Draft genome sequences of endophytic isolates of *Klebsiella variicola* and *Klebsiella pneumoniae* obtained from the same sugarcane plant. *Genome Announc.* **2018,** *6*, e00147–18.

Rigerte, L.; Blumenstein, K.; Terhonen, E. New R-based methodology to optimize the identification of root endophytes against *Heterobasidion parviporum*. *Microorganisms* **2019,** *7*, 102.

Ripa, F.A.; Cao, W.D.; Tong, S.; Sun, J.G. Assessment of plant growth promoting and abiotic stress tolerance properties of wheat endophytic fungi. *BioMed Res. Int.* **2019,** 6105865.

Rodrigues, E.P.; Soares, C.D.P.; Galvão, P.G.; Imada, E.L.; Simões-Araújo, J.L.; Rouws, L.F.; de Oliveira, A.L.; Vidal, M.S.; Baldani, J.I. Identification of genes involved in indole-3-acetic acid biosynthesis by *Gluconacetobacter diazotrophicus* PAL5 strain using transposon mutagenesis. *Front. Microbial.* **2016,** *7*, 1572.

Rodríguez, H.; Fraga, R.; Gonzalez, T.; Bashan, Y. Genetics of phosphate solubilization and its potential applications for improving plant growth-promoting bacteria. *Plant Soil* **2006,** *287*, 15–21.

Rodriguez, R.J.; White Jr, J.F.; Arnold, A.E.; Redman, A.R.A. Fungal endophytes: diversity and functional roles. *New Phytol.* **2009,** *182*, 314–330.

Romero, F.M.; Rossi, F.R.; Gárriz, A.; Carrasco, P.; Ruíz, O.A. A bacterial endophyte from apoplast fluids protects canola plants from different phytopathogens via antibiosis and induction of host resistance. *Phytopathology* **2019,** *109*, 375–383.

Sabu, R.; Aswani, R.; Jishma, P.; Jasim, B.; Mathew, J.; Radhakrishnan, E.K. Plant growth promoting endophytic Serratia sp. ZoB14 protecting ginger from fungal pathogens. *Proc. Natl. Acad. Sci. India Sect. B Biol. Sci.* **2019,** *89*, 213–220.

Sahu, S.; Prakash, A.; Shende, K. Talaromyces trachyspermus, an endophyte from Withania somnifera with plant growth promoting attributes. *Environ. Sustain.* **2019,** *2*, 13–21.

Santoyo, G.; Moreno-Hagelsieb, G.; del Carmen Orozco-Mosqueda, M.; Glick, B.R. Plant growth-promoting bacterial endophytes. *Microbiol. Res.* **2016,** *183*, 92–99.

Sarma, M.V.R.K., Kumar, V., Saharan, K., Srivastava, R., Sharma, A.K., Prakash, A., Sahai, V. and Bisaria, V.S., 2011. Application of inorganic carrier-based formulations of fluorescent pseudomonads and *Piriformospora indica* on tomato plants and evaluation of their efficacy. *J. Appl. Microbiol.* **2011,** *111*, 456–466.

Schulz-Bohm, K.; Martín-Sánchez, L.; Garbeva, P. Microbial volatiles: small molecules with an important role in intra-and inter-kingdom interactions. *Front. Microbiol.* **2017,** *8*, 2484.

Sharifi, R.; Ryu, C.M. Revisiting bacterial volatile-mediated plant growth promotion: lessons from the past and objectives for the future. *Ann. Bot.* **2018,** *122*, 349–358.

Sharma, P.; Kumawat, K.C.; Kaur, S.; Kaur, N. Assessment of zinc solubilization by endophytic bacteria in legume rhizosphere. *Indian J. Appl. Res.* **2014,** *4*, 439–441.

Shen, F.T.; Yen, J.H.; Liao, C.S.; Chen, W.C.; Chao, Y.T. Screening of rice endophytic biofertilizers with fungicide tolerance and plant growth-promoting characteristics. *Sustainability* **2019,** *11*, 1133.

Shin, M.N.; Shim, J.; You, Y.; Myung, H.; Bang, K.S.; Cho, M.; Kamala-Kannan, S.; Oh, B.T. Characterization of lead resistant endophytic Bacillus sp. MN3–4 and its potential for promoting lead accumulation in metal hyperaccumulator *Alnus firma*. *J. Hazard. Mater.* **2012,** *199*, 314–320.

Singh, M.; Kumar, A.; Singh, R.; Pandey, K.D. Endophytic bacteria: a new source of bioactive compounds. *3 Biotech* **2017,** *7*, 315.

Strobel, G. The emergence of endophytic microbes and their biological promise. *J. Fungi.* **2018,** *4*, 57.

Sun, Y.; Cheng, Z.; Glick, B.R. The presence of a 1-aminocyclopropane-1-carboxylate (ACC) deaminase deletion mutation alters the physiology of the endophytic plant growth-promoting bacterium *Burkholderia phytofirmans* PsJN. *FEMS Microbiol. Lett.* **2009,** *296*, 131–136.

Syed, S.; Tollamadugu, N.P. Role of plant growth-promoting microorganisms as a tool for environmental sustainability In: *Recent Developments in Applied Microbiology and Biochemistry*; Academic Press, 2019; pp. 209–222.

Tuo, L.; Yan, X.R.; Xiao, J.H. Jiella endophytica sp. nov., a novel endophytic bacterium isolated from root of *Ficus microcarpa* Linn. f. *Antonie van Leeuwenhoek* **2019,** *112*, 1457–1463.

Valencia, A.L.; Gil, P.M.; Latorre, B.A.; Rosales, I.M. Characterization and pathogenicity of Botryosphaeriaceae species obtained from avocado trees with branch canker and dieback and from avocado fruit with stem end rot in chile. *Plant Dis.* **2019,** *103*, 996–1005.

Van de Poel, B.; Van Der Straeten, D. 1-aminocyclopropane-1-carboxylic acid (ACC) in plants: more than just the precursor of ethylene. *Front. Plant Sci.* **2014,** 5, 640.

Vanneste, S.; Friml, J. Auxin: a trigger for change in plant development. *Cell* **2009,** *136*, 1005–1016.

Weyens, N.; Truyens, S.; Dupae, J.; Newman, L.; Taghavi, S.; Van Der Lelie, D.; Carleer, R.; Vangronsveld, J. Potential of the TCE-degrading endophyte Pseudomonas putida W619-TCE to improve plant growth and reduce TCE phytotoxicity and evapotranspiration in poplar cuttings. *Environ. Pollut.* **2010,** *158*, 2915–2919.

Wyrebek, M.; Huber, C.; Sasan, R.K.; Bidochka, M.J. Three sympatrically occurring species of Metarhizium show plant rhizosphere specificity. *Microbiology* **2011,** *157*, 2904–2911.

Yaish, M.W.; Antony, I.; Glick, B.R. Isolation and characterization of endophytic plant growth-promoting bacteria from date palm tree (*Phoenix dactylifera* L.) and their potential role in salinity tolerance. *Antonie van Leeuwenhoek* **2015,** *107*, 1519–1532.

Yang, E.; Sun, L.; Ding, X.; Sun, D.; Liu, J.; Wang, W. Complete genome sequence of *Caulobacter flavus* RHGG3 T, a type species of the genus Caulobacter with plant growth-promoting traits and heavy metal resistance. *3 Biotech* **2019,** *9*, 42.

Yuan, Z.S.; Liu, F.; Zhang, G.F. Characteristics and biodiversity of endophytic phosphorus-and potassium-solubilizing bacteria in Moso Bamboo (Phyllostachys edulis). *Acta Biol. Hung.* **2015,** *66*, 449–459.

Zhang, X.; Gao, Z.; Zhang, M.; Jing, F.; Du, J.; Zhang, L. Analysis of endophytic actinobacteria species diversity in the stem of Gynura cusimbua by 16S rRNA gene clone library. *Microbiology* **2016,** *85*, 379–385.

Zhang, Y.F.; He, L.Y.; Chen, Z.J.; Wang, Q.Y.; Qian, M.; Sheng, X.F. Characterization of ACC deaminase-producing endophytic bacteria isolated from copper-tolerant plants and their potential in promoting the growth and copper accumulation of *Brassica napus*. *Chemosphere* **2011,** *83*, 57–62.

Zhang, Y.Z.; Wang, E.T.; Li, M.; Li, Q.Q.; Zhang, Y.M.; Zhao, S.J.; Jia, X.L.; Zhang, L.H.; Chen, W.F.; Chen, W.X. Effects of rhizobial inoculation, cropping systems and growth stages on endophytic bacterial community of soybean roots. *Plant Soil* **2011,** *347*, 147.

CHAPTER 2

Omics Approach to Understanding Microbial Diversity

SHILPEE PAL[1,3], ARIJIT JANA[2], KESHAB CHANDRA MONDAL[1], and SUMAN KUMAR HALDER[1*]

[1]*Department of Microbiology, Vidyasagar University, Midnapore, West Bengal, India*

[2]*Department of Chemical Engineering, Indian Institute of Technology, Roorkee, Uttarakhand, India*

[3]*CSIR-Institute of Microbial Technology (CSIR-IMTECH), Chandigarh, India*

[*]*Corresponding author. E-mail: sumanmic@mail.vidyasagar.ac.in*

ABSTRACT

The microbial biosphere is one of the most significant phylogenetically diverse living systems on our planet. Their evolutionary diversity, usefulness in human civilization, and importance in industrial organizations highlight the need of their genomic profiling. Enormous advancement of culture methods and sequencing technologies enable to identify a large number of microbes as well as their genome sequences. Depending upon this sequence data, *omics*-based study has been developed a new horizon in the biological research field, which not only explores the actual characteristics of microbial populations but also the fundamental basis of their diversities. *Omics* technology includes genomics that deals with structure and function of genomes; transcriptomics evaluates RNA transcripts produced by a genome; proteomics analyses a set of proteome. Furthermore, in-depth exploration of microbial communities within an environment, the *omics* technologies have been further extended to metagenomics, explicates taxonomic profiles of a microbial community; metatranscriptomics and metaproteomics estimate RNA activity and classification of expressed proteins of a bacterial population, respectively.

Metabolomics is another extensive *omics* technology that explores the metabolic network of different microbes in a complex ecosystem. All these platforms are culture-independent, depending upon statistical and computational procedures that have the potential to recognize the systems biology and functional dynamics of diverse microbial populations in a limited time. The present topic would explain various *omics* approaches and their importance in biological fields to understand microbial diversity within the natural environment.

2.1 INTRODUCTION

Microbial diversity reflects the heterogeneity of microorganisms present in the environment. Among living organisms, the highest diversity found among microbes, the oldest inhabitant of this planet. Depending upon the relative richness and niche adaptability, a widespread variability observed among genetically different groups such as bacteria, archaea, fungi, algae, and viruses. Generally, microbes are well known for their pathogenic behavior on the globe. Apart from playing the role of a villain against human and animal health, microbes also play indispensable roles in ecological processes. Moreover, not all microorganisms that interact with human or animal body are pathogenic. Host-associated nonpathogenic microorganisms can modulate the immune response of their host and thus can resist the establishment of pathogenic consequences. They also participate in the synthesis of essential amino acids and vitamins as well as help in toxin degradation within their hosts (Khoroshkin et al., 2016; LeBlanc et al., 2017). Not only living organisms, but the microbial community is also associated with environmental quality, that is, the physical and chemical properties of an environment. There is a dependent relationship between microbial diversity and the environment. Microbial growth ecology creates significant variations among metabolic networks of its inhabitants and allows them to generate their species variations.

The hidden features underlying the diversity observed within closely related but ecologically different species can be identified by analyzing their morphoanatomical features. It includes staining, biochemical tests, and microscopic appearance of cellular structure that explores the physiological characteristics of microbes. Furthermore, morphoanatomical identification typically requires verification by molecular identification. As a whole, the overall process is prolonged and challenging and requires refined observation of an experienced microbiologist, as well. Moreover, the uncultured microbial world is far from a culture based traditional technologies, thus cannot be identified by classical microbial techniques. Therefore, it is of great interest to the scientists to determine the microorganisms without collecting their morphoanatomical data. Now it has become a developing approach of molecular identification by analyzing and amplifying the genetic elements of microbes in a sample (He et al., 2016).

With the arrival of high-throughput sequencing (HTS) methods the vast amount of genomic data is now publicly available that has extensive use in exploring microbial diversity as well as their genetic variants associated with different environmental niches. In this context, *omics*-based studies have become an emerging field that comprises collective technologies to reveal the molecular compositions that make up microbial cells, as well as the modification of metabolic pathways in microorganisms of diverse locations. These techniques are beyond microbial cultivation technologies, thus free from plate count as well as colony differentiating difficulties. Among various omics experiments, researchers generally apply genomics, transcriptomics, and proteomics as well as their extended versions (Figure 2.1) for analysis of microbial diversity and their interaction with environments.

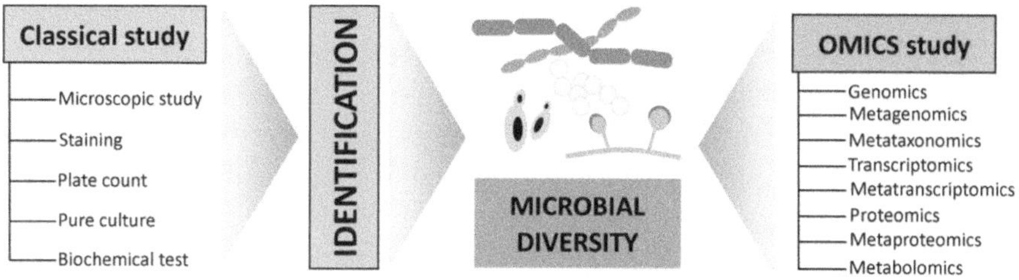

FIGURE 2.1 Overview of classical culture and omics based strategies for microbial diversity analysis.

2.2 GENOMICS AND METAGENOMICS

In an organism, genome includes the complete set of coding and noncoding regions, and the analysis of those parts by using various techniques is called genomics. In 1986, Tom Roderick, a geneticist (Bar Harbor, Maine), coined the term "genomics" for the mapping of the human genome in Maryland (Winkler 1920). In 1995, the publication of complete genome sequence of *Haemophilus influenzae* by J. Craig Venter and colleagues brought one of the most significant breakthroughs in genomics that benefitted most the field of microbiology (Fleishmann et al., 1995). Advancement of rapid, precise, and low-cost technologies to sequence, assemble, and analyze the microbial genomes has made it easier to study the coding and noncoding regions as well as microbial diversity (Hilton et al., 2016; Wei et al., 2002; Rajendhran and Gunasekaran, 2011). A very well-known approach for genome sequencing, assembly, and analysis is metagenomics.

Metagenomics provides the analysis of a large number of genomic data from similar but nonidentical members in a sample. In 1998, the word "metagenomics" was first used in a publication of Jo Handelsman and her colleagues from the University of Wisconsin, Madison and Cornell University, New York (Handelsman et al., 1998). Microbial world covers the largest proportion of organisms; including 10^6–10^8 genospecies (Sleator et al., 2008) present in different environments such as soil, air, and water. Relatively few of them have been characterized genetically or otherwise. Apart from the environmental sample, a vast range of microbes also exists as host-associated pathogens, nonpathogens, and opportunistic pathogens. Traditional genome-based approaches through cloning and culturing of individual microorganisms cannot be used to study whole microbial communities existing in a given environment. Moreover, all these organisms are not cultivable and demand a culture-independent approach for identification. Their genomes may encode novel enzymes and metabolic abilities. In this context, metagenomics is used for taxonomic classification of those microorganisms without applying classical methods, that is, isolation and cultivation of the microbes (Tringe et al., 2005). It can analyze the DNA sequences directly extracted from environmental samples, and thus can be used as a powerful tool to compare and explore the ecology and metabolic profiles of complex microbial communities (Steele et al., 2009; Simon and Daniel, 2010).

After extracting and sequencing the genomic fragments of microbes (Tyson et al., 2004), their taxonomic diversity is assessed by using two metagenomic approaches: untargeted (shotgun sequencing) and targeted (marker gene) metagenomics. Marker gene metagenomics is performed through sequencing of amplicons, that is, specific or targeted regions of marker genes such as bacterial 16S *rRNA*, *rpoA*, *atpB*, *dnaA*, *ftsZ*, *groEL*, *gltA*, *lux*, *pyrH*, and *rpoB* (Clarridge 2004). Phylogenomic studies generally use this approach, precisely termed as metataxonomics, a cost-effective means to identify a wide range of organisms, and suggested an essential tool for microbial ecology. Among these marker genes, 16S rRNA genes are widely used marker sequences for bacteria (Woese and Fox, 1977). Apart from their evolutionary stability, 16S rRNAs also contain nine hypervariable regions (V1) to (V9) (Klindworth et al., 2012). They vary in length, position, and taxonomic insights, which avow them as ideal markers for the characterization of microbial diversity.

On the other hand, shotgun sequencing includes the entire genomic content of individuals without selecting any particular gene (Scholz et al., 2016). Thus, it can be used to analyze the compositional and functional potential of microbial communities as well as taxonomic profiling. A typical shotgun metagenomics approach is comprised of five steps: (1) collection, processing, and sequencing of samples; (2) preprocessing of sequencing reads; (3) analysis of sequences for profiling the genomic, taxonomic, and functional features of the microbiome; (4) postprocessing analysis through statistics and biology; and (5) validation of the result. Several computational approaches are available to perform each step (Table 2.1). Shotgun metagenomics is becoming popular in biomedical and environmental applications, as well as to identify extremophiles and symbiotic interactions between microbes and hosts (Cowan et al., 2015; Belda-Ferre et al., 2012; Woyke et al., 2006). Shotgun sequencing can provide higher taxonomic information than targeted metagenomics. It can also identify many uncharacterized species, especially those present in low abundance.

TABLE 2.1 Computational Approaches for Metagenomics

Steps of Metagenomics	Tools and Databases
Sequencing technologies (NGS technologies)	454 pyrosequencing, Illumina, SOLiD, Pacific Biosciences (PacBio), Ion Torrent
Assembly of reads	
Referenced-based	Newbler (Roche), MIRA 4, MetaAMOS
De novo	EULER, Velvet, SOAP, Abyss
Combined binning	MetaVelvet, Meta-IDBA, IDBA-UD, PhymmBL, MetaCluster
Annotation of metagenomics sequences	
Coding DNA sequences (CDSs)	MetaGeneMark, Metagene, Prodigal, Orphelia, FragGeneScan
Clustered regularly interspaced short palindromic repeats (CRISPRs)	CRT, PILER-CR
Noncoding RNAs	IMG/MER, and MG-RAST
Function annotation	KEGG, SEED, eggNOG, COG/KOG, PFAM, TIGRFAM, BLAT
Taxonomic analysis	
16S sequence analysis	QIIME, Mothur, SILVAngs, AmpliconNoise, RDP classifier, RTAX,
Statistical analysis of metagenomic data	Primer-E package, vegan, phyloseq, Bioconductor
Statistical analysis and visualization	IMG/MER, CAMERA, MG-RAST, EBI, metagenomics, MEGAN 5

In both approaches, identification and removal of low-quality sequences, as well as host genome sequences, are essential steps, which can be performed by applying computational tools, viz., FastQC (Andrews, 2014), Cutadapt (Martin, 2011) programs. After quality control, reads are assembled into long contiguous sequences called contigs. In case of taxonomic classification, binning is performed, in which every read is grouped into bins as per their taxon ID.

In the case of uncultured microbes, the qualitative understanding of their physiology can be possible by analyzing binned draft genomes. Their taxonomy and complete genome can be evaluated by identifying conserved genes in the contig bins (Sangwan et al., 2016). Moreover, their metabolic pathways, functional potentials, and interactions with the environment can be gained by matching methanogenic (draft) assemblies against pathway or gene databases (Truong et al., 2017; Hahn et al., 2017). Sometimes, microbes isolated from unexplored ecological niches have no closely related species in the database. In that case, assembly and binning are of two essential steps in the metagenomics analysis.

After extracting DNA from the target environment, the structure of microbial communities, such as species diversity and their abundance, can be evaluated by using structural and functional metagenomics (Oulas et al., 2015). Structural metagenomics focuses on the study of uncultivated microbial population structure through a reconstruction of complex metabolic networks between members of the community (Andelsman, 2005). Whereas, functional metagenomic approaches are performed to identify genes of interest that provide novel insights into crucial metabolic processes of communities. The community structure can reveal the relationships between individual components of that community, which is essential to decipher the ecological or biological functions among its members (Vieites et al., 2009). Recently, in clinical microbiology, metagenomics sequencing technology is used to identify infectious pathogens (Dekker, 2018).

Thus, metagenomics has become an advantageous technique to analyze microbial diversity by exploring the structure, richness, and dynamics of the microbial communities as well as their surrounding environments. The methods are also employed to decipher the genetic potential of microbiome in different backgrounds, including soil, seawater, human gut, and food. Currently, the advent of cost-efficient, high-throughput microbial analysis by metagenomic technologies has explored extensive insights into the coexisting population by revealing the evolutionary forces shaping the genetic diversity of microbial communities in different ecosystems.

2.3 TRANSCRIPTOMICS AND METATRANSCRIPTOMICS

A transcriptome is the entire set of RNA transcribed from the genome of an organism under a particular physiological condition. Transcriptomics comprises high-throughput methods to comprehend the expression of genes that resides in the genome at the transcription level. Thus, it provides the information on genome dynamics of an organism, that is, gene structure, regulation of gene expression, and gene product function at specific physiological, developmental, and/or pathological processes. Transcriptomics can also reveal the regulation network of biological processes, which further helps in proteomics research. Moreover, transcriptomics has now been applied to explore the structure and function of nonprotein-coding RNA (ncRNA), which has essential roles in gene regulation (Eddy, 2001).

There are two primary techniques for transcriptome analysis: microarray and RNA-Seq. The microarray can profile only predefined transcripts or genes through the hybridization technique, whereas, RNA-Seq provides full sequencing of the whole transcriptome of an organism. Both approaches can provide the knowledge of individual's mRNAs, ncRNA, small RNAs, the transcriptional structure of genes, their start sites, 5′ and 3′ ends, splicing, and other posttranscriptional modifications. It can also quantify the expression of each transcript, thus plays significant roles in determining environmental tolerance of an organism resides in the challenging habitat as well as the microbial diversity in an environment. Among these two methods, RNA-Seq data analysis has become a vital transcriptomics approach. It is a combination of HTS technology with computational techniques to quantify transcripts (Ozsolak and Milos, 2011). In this technology, after sampling the short-read sequences, they are allowed to align with the reference genome and then are assembled to produce a transcriptome. Tools like RSEM, BitSeq, Sailfish are used to analyze and measure the expression levels of transcriptomes, and MISO is used to identify alternatively spliced transcripts (Katz et al., 2010) of a transcriptome. Thus RNA-Seq permits to identify active genes that influence microbial livelihood in a diverse environment.

In the case of microbial community study, metatranscriptomics is used to explore the key transcripts of significant functions in microbes that reside in a particular environment. It is an emerging approach that directly analyzes the mRNA from an environmental sample, thus allows to characterize the changes in mRNA abundance triggered by ecological variables like temperature, oxygen, and pH (Logan and Somero, 2010; Gracey et al., 2011; Evans et al.,

2013). Transcriptomics and metatranscriptomics are RNA-based profiling of microbial consortium that reflects the *active* functional profile of a microbial community through analyzing gene expression. These *omics*-based transcripts analyzing technologies have integrated the view of cellular life complications. Cost-effectiveness of the approaches has made them accessible to small laboratories for the experiments that compare transcriptomes from thousands of microorganisms in a particular environment.

2.4 PROTEOMICS AND METAPROTEOMICS

The proteome is a complete set of proteins of a specific biological system. It is highly dynamic and is continuously varying according to the environment. Proteomics is the study of proteome at a particular time in a specific environment. Microorganisms have to face expeditious and severe changing environmental factors such as temperature, moisture, nutrient availability, and predators. Modification of the protein expression profiles is essential to overcome those challenges. This niche adaptation policy can be estimated by proteomics that investigates the physiology of complex microbial associations at a molecular level. Overall proteomic approaches are used to analyze various proteomic aspects such as proteome profiling, comparative synthesis of two or more protein samples, identification and localization of posttranslational modifications, and for protein–protein interaction study. Thus proteomics includes structural and functional knowledge of proteins.

Structural proteomics reveals the high throughput characterization of the three-dimensional structure of proteins. Whereas, functional proteomics is dealing with the functional interactions between proteins (Karp et al., 2008). Among these two, functional approach is more appropriate and informative for microbial proteomics. For characterization of protein in a sample, the most useful technique is mass spectrometry (MS). It is an imaging method that allows mapping all the isoforms of proteins within a sample and has considered being an appropriate tool for functional proteomics. It also reveals the expression pattern of proteins in an organism grown in a particular environment. The activation and inhibition of proteins expressed by an organism grown in a specific condition (Keller and Zengler, 2004) are assessed by activity-based proteomics. It provides fast, sensitive, and selective solutions involving chemical, biological, and bioanalytical approaches for the identification and enrichment of "hidden" proteins from the complex mixtures of proteomes. This approach can explore the active forms of specific enzymes in vitro as well as in vivo by using small molecular probes, called activity-based probes. These probes with fluorescent dye target the active site of a particular protein and covalently label only the active enzyme. Another approach of functional proteomics is differential proteomics, employed to compare protein expressions of organisms incubated in two different conditions. This approach is used to identify the essential proteins in a key metabolic pathway of an organism that helps in their niche adaptation (Weiss et al., 2009; Li et al., 2016).

Functional proteomics, which is used to analyze the proteomes of a microbial community, is called metaproteomics or environmental proteomics. It provides information about microbial cooperation and competition that helps the microbe to survive in a complex natural ecosystem. It helps in mapping the protein complement at a specific time by evaluating protein expression, their localizations, and post-translational modifications which affect protein functionality, interactions as well as amino acid sequences. Thus, metaproteomics has potential applications in the identification of new enzymes or entire metabolic pathways of a microbial community, which allows recording the mechanisms for adaptive response of microorganisms to environmental stimuli or communications with other organisms or host cells (Wang et al., 2014). There are three critical steps in metaproteomics: separation of proteins by their charge to mass ratio (electrophoresis), their mass or density (centrifugation), their binding affinity with solid surface (liquid chromatography); identification of protein or peptide by tandem MS (MS/MS); and protein quantification by using PANDA (Cheng et al., 2019), ProtQuant (Bridges, 2007) to compare their expression patterns in a different environment. Metaproteomics data analysis provides an integrated aspect of biological objects, their roles, and relationships in a specific situation. Furthermore, such studies of microbial communities draw attention to investigate the roles of uncultivated microorganisms in an environment.

2.5 METABOLOMICS

Metabolomics comprises a group of methods that can identify, classify, and potentially quantify small molecules (<1.5 kD) produced by living cells or tissue of an organism that result from the interactions with its ecological niche. Metabolites are the end product of gene expression, thus regulate a system in an integrated manner. Microbial metabolomics explores the regulations of their metabolic compositions to survive in various environmental conditions. Therefore the characterization of their metabolome set is one of the best approaches to evaluate the molecular basis of microbial diversity. It is also useful for the phenotypic characterization of bioactive compounds that are useful to humans and animals. These low molecular weight (<1000 Da) compounds are generally the final products of essential biological pathways that serve in defense and competitiveness of an organism in its territory (Rochfort, 2005). Metabolomics can also explore environmental stress or seasonal variations that affect microbial diversity. As a phenotypic classifier, metabolomics is the ending tool in *omics* chain that can explore microbial systems through two aspects: exometabolomics and endometabolomics.

Exometabolomics, metabolic footprint analysis, reveals the excretion of cells into the extracellular surrounding. It can discriminate metabolic states of the cells of different microbial species, as well as different strains from the same species (Villas-Bôas et al., 2006). Thus exometabolomics is noteworthy in microbial species discrimination of a specific environment and has wide application in different areas like bioremediation, biomarker discovery, bioprocess monitoring as well as functional genomics. Whereas endometabolomics reveals the intracellular metabolites that are secreted within the mechanical barrier, that is, cell membrane or cell envelope of microbes. These metabolites are involved in cellular metabolism; thus enzymatic activity of a microbial cell can be measured by estimating endometabolites. It helps in understanding the upregulation and/or downregulation of genes or proteins corresponding to the changes in biological functions. Both aspects provide critical information to understand the microbial systems, thus have been considered essential for recognizing microbial diversity (Mashego, 2006).

Analysis of biological samples obtained from microbial systems involves the following steps: growing microorganism in an appropriate media; quenching of microbial cells to collect the sample at a desired stage of growth; and separation of cells and supernatant from growth media, where microbial cells are used for endometabolomics and supernatant is used for exometabolomics. There are numerous instrumental methods to analyze the extracted metabolite from microbial cells. Among them, NMR and liquid chromatography are commonly used techniques. The subsequent spectra generated from above techniques are further analyzed by using statistical methods like principal component analysis, orthogonal projection onto latent structures to know significant spectral features that define each metabolite group (Dai et al., 2010). By implementing these techniques, metabolomics approaches are divided into targeted and untargeted metabolomics. In untargeted metabolomics, the generated data are assigned to distinct metabolic pathways and metabolites by using spectral libraries of known metabolites (Schrimpe-Rutledge et al., 2016). Hence, it is a discovery-based approach to identify a novel target as well as reveals how a system tackles environmental or genetic stress. However, the major challenges of this method lie in the procedure and time necessary to process a broad spectrum of raw data, as well as problems to identify and characterize unknown small molecules, and biases to discover high-abundance molecules. Conversely, targeted metabolomics focuses on the measurement of a specific set of characterized and annotated metabolites in a system (Ribbenstedt et al., 2018). This approach offers a complete understanding of the metabolic enzymes, kinetics, end products and associated known biochemical pathways. Therefore targeted metabolomics is more sensitive and quantitative through the use of internal standards that provides higher reproducibility and lower false-positive rate than untargeted metabolomics. Thus, metabolomics can explore molecular networks between diverse microbial members as well as their biosynthetic potential to survive in a specific niche.

2.6 CONCLUSION

Overall *omics*-based approaches have allowed us to realize the structural and functional dynamics thoroughly of microbial diversity. It helps to understand the correlations between genotype and phenotype of the microbial communities. Now it is also feasible to analyze large-scale biological data within a considerate time with the application of *omics*-based techniques. Statistical analysis has provided more accurate and realistic predictions for previously unknown networks among morphological features of microbes. Microbial interactions with their environmental niches have also been well understood through *omics* approaches.

ACKNOWLEDGMENT

The authors are thankful to the Science and Engineering Research Board, Department of Science and Technology under the National Post-Doctoral Fellowship (File Number- PDF/2019/003065), Department of Biotechnology, Government of India, and Department of Science and Technology and Biotechnology, Government of West Bengal, India, for different mode of financial support.

KEYWORDS

- **genomics**
- **metagenomics**
- **transcriptomics**
- **proteomics**
- **metaproteomics**
- **metabolomics**
- **microbial diversity**
- **computational tools**

REFERENCES

Andrews, S. FastQC a quality-control tool for high-throughput sequence data http://www. *Bioinformaticsbabraham. ac. uk/projects/fastqc*. **2014.**

Belda-Ferre, P.; Alcaraz, L.D.; Cabrera-Rubio, R.; Romero, H.; Simon-Soro, A.; Pignatelli, M.; Mira, A. The oral metagenome in health and disease. *ISME J.* **2012**, *6*, 46–56.

Bridges, S.M.; Magee, G.B.; Wang, N.; Williams, W.P.; Burgess, S.C.; Nanduri, B. ProtQuant: a tool for the label-free quantification of MudPIT proteomics data. *BMC Bioinform.* **2007**, *8*, S24.

Chang, C.; Li, M.; Guo, C.; Ding, Y.; Xu, K.; Han, M.; He, F.; Zhu, Y. PANDA: A comprehensive and flexible tool for quantitative proteomics data analysis. *Bioinformatics.* **2019**, *35*, 898–900.

Clarridge, J.E. Impact of 16S rRNA gene sequence analysis for identification of bacteria on clinical microbiology and infectious diseases. *Clin. Microbiol. Rev.* **2004**, *17*, 840–862.

Cowan, D.A.; Ramond, J.B.; Makhalanyane, T.P.; De Maayer, P. Metagenomics of extreme environments. *Curr. Opin. Microbiol.* **2015**, *25*, 97–102.

Dai, H.; Xiao, C.; Liu, H.; Hao, F.; Tang, H. Combined NMR and LC– DAD-MS analysis reveals comprehensive metabonomic variations for three phenotypic cultivars of *Salvia miltiorrhiza* Bunge. *J. Proteome Res.* **2010**, *9*, 1565–1578.

Dekker, J.P. Metagenomics for clinical infectious disease diagnostics steps closer to reality. *J. Clin. Microbiol.* **2018**, *56*, e00850–18.

Eddy, S.R. Non–coding RNA genes and the modern RNA world. *Nat. Rev. Genet.* **2001**, *2*, 919–929.

Evans, T.G.; Chan, F.; Menge, B.A.; Hofmann, G.E. Transcriptomic responses to ocean acidification in larval sea urchins from a naturally variable pH environment. *Mol. Ecol.* **2013**, *22*, 1609–1625.

Fleischmann, R.D.; Adams, M.D.; White, O.; Clayton, R.A.; Kirkness, E.F.; Kerlavage, A.R.; Bult, C.J.; Tomb, J.F.; Dougherty, B.A.; Merrick, J.M. Whole-genome random sequencing and assembly of *Haemophilus influenzae* Rd. *Science* **1995**, *269*, 496–512.

Gracey, A.Y.; Lee, T.H.; Higashi, R.M.; Fan, T. Hypoxia-induced mobilization of stored triglycerides in the euryoxic goby *Gillichthys mirabilis*. *J. Exp. Biol.* **2011**, *214*, 3005–3012.

Hahn, A.S.; Altman, T.; Konwar, K.M.; Hanson, N.W.; Kim, D.; Relman, D.A.; Dill, D.L.; Hallam, S.J. A geographically-diverse collection of 418 human gut microbiome pathway genome databases. *Sci. Data* **2017**, *4*, 1–12.

Handelsman, J.; Rondon, M.R.; Brady, S.F.; Clardy, J.; Goodman, R.M. Molecular biological access to the chemistry of unknown soil microbes: a new frontier for natural products. *Chem. Biol.* **1998**, *5*, R245-R249.

Handelsman, J. Metagenomics: application of genomics to uncultured microorganisms. *Microbiol. Mol. Biol. Rev.* **2004**, *68*, 669–685. Modified

He, Z.; Wang, J.; Hu, J.; Zhang, H.; Cai, C.; Shen, J.; Xu, X.; Zheng, P.; Hu, B. Improved PCR primers to amplify 16S rRNA genes from NC10 bacteria. *Appl. Microbiol. Biotechnol.* **2016**, *100*, 5099–5108.

Hilton, S.K.; Castro-Nallar, E.; Pérez-Losada, M.; Toma, I.; McCaffrey, T.A.; Hoffman, E.P.; Siegel, M.O.; Simon, G.L.; Johnson, W.E.; Crandall, K.A. Metataxonomic and metagenomic approaches vs. culture-based techniques for clinical pathology. *Front. Microbiol.* **2016**, *7*, 484.

Karp, N.A.; Feret, R.; Rubtsov, D.V.; Lilley, K.S.; Comparison of DIGE and post-stained gel electrophoresis with both traditional and SameSpots analysis for quantitative proteomics. *Proteomics* **2008**, *8*, 948–960.

Katz, Y.; Wang, E.T.; Airoldi, E.M.; Burge, C.B. Analysis and design of RNA sequencing experiments for identifying isoform regulation. *Nat. Methods* **2010**, *7*, 1009.

Keller, M.; Zengler, K. Tapping into microbial diversity. *Nat. Rev. Microbiol.* **2004;** *2*, 141–150.

Khoroshkin, M.S.; Rodionov, D. Syntrophic metabolism of vitamins and amino acids in gut microbial community as revealed by in silico genomic analyses. *FASEB J.* **2016**, *30*, 819–5.

Klindworth, A.; Pruesse, E.; Schweer, T.; Peplies, J.; Quast, C.; Horn, M.; Glöckner, F.O. Evaluation of general 16S ribosomal RNA gene PCR primers for classical and next-generation sequencing-based diversity studies. *Nucleic Acids Res.* **2013**, *41*, e1-e1.

LeBlanc, J.G.; Chain, F.; Martín, R.; Bermúdez-Humarán, L.G.; Courau, S.; Langella, P. Beneficial effects on host energy metabolism of short-chain fatty acids and vitamins produced by commensal and probiotic bacteria. *Microb. Cell Fact.* **2017**, *16*, 79.

Li, J.; Ding, X.; Han, S.; He, T.; Zhang, H.; Yang, L.; Yang, S.; Gai, J. Differential proteomics analysis to identify proteins and pathways associated with male sterility of soybean using iTRAQ-based strategy. *J. Proteomics* **2016**, *138*, 72–82.

Logan, C.A.; Somero, G.N. Transcriptional responses to thermal acclimation in the eurythermal fish Gillichthys mirabilis (Cooper 1864). *Am. J. Physiol. Regul. Integr. Comp. Physiol.* **2010**, *299*, R843-R852.

Martin, M. Cutadapt removes adapter sequences from high-throughput sequencing reads. *EMBnet. J.* **2011**, *17*, 10–12.

Mashego, M.R.; Rumbold, K.; De Mey, M.; Vandamme, E.; Soetaert, W.; Heijnen, J.J. Microbial metabolomics: past, present and future methodologies. *Biotechnol. Lett.* **2007**, *29*, 1–16.

Oulas, A.; Pavloudi, C.; Polymenakou, P.; Pavlopoulos, G.A.; Papanikolaou, N.; Kotoulas, G.; Arvanitidis, C.; Iliopoulos, L. Metagenomics: tools and insights for analyzing next-generation sequencing data derived from biodiversity studies. *Bioinform. Biol. Insights* **2015**, *9*, BBI-S12462.

Ozsolak, F.; Milos, P.M. RNA sequencing: advances, challenges and opportunities. *Nat. Rev. Genet.* **2011**, *12*, 87–98.
Rajendhran, J.; Gunasekaran, P. Microbial phylogeny and diversity: small subunit ribosomal RNA sequence analysis and beyond. *Microbiol. Res.* **2011**, *166*, 99–110.
Ribbenstedt, A.; Ziarrusta, H.; Benskin, J.P. Development, characterization and comparisons of targeted and non-targeted metabolomics methods. *PLoS One* **2018**, *13*, e0207082.
Rochfort, S. Metabolomics reviewed: a new "omics" platform technology for systems biology and implications for natural products research. *J. Nat. Prod.* **2005**, *68*, 1813–1820.
Sangwan, N.; Xia, F.; Gilbert, J.A. Recovering complete and draft population genomes from metagenome datasets. *Microbiome* **2016**, *4*, 8.
Schrimpe-Rutledge, A.C.; Codreanu, S.G.; Sherrod, S.D.; McLean, J.A. Untargeted metabolomics strategies—challenges and emerging directions. *J. Am. Soc. Mass Spectrom.* **2016**, *27*, 1897–1905.
Scholz, M.; Ward, D.V.; Pasolli, E.; Tolio, T.; Zolfo, M.; Asnicar, F.; Truong, D.T.; Tett, A.; Morrow, A.L.; Segata, N. Strain-level microbial epidemiology and population genomics from shotgun metagenomics. *Nat. Methods* **2016**, *13*, 435–438.
Simon, C.; Daniel, R. Construction of small-insert and large-insert metagenomic libraries. *Methods Mol. Biol.* **2010**, *668*, 39–50.
Sleator, R.D.; Shortall, C.; Hill, C. Metagenomics. *Lett. Appl. Microbiol.* **2008**, *47*, 361–366.
Steele, H.L.; Jaeger, K.E.; Daniel, R.; Streit, W.R. Advances in recovery of novel biocatalysts from metagenomes. *J. Mol. Microbiol. Biotechnol.* **2009**, *16*, 25–37.
Tringe, S.G.; Von Mering, C.; Kobayashi, A.; Salamov, A.A.; Chen, K.; Chang, H.W.; Podar, M.; Short, J.M.; Mathur, E.J.; Detter, J.C.; Bork, P. Comparative metagenomics of microbial communities. *Science* **2005**, *308*, 554–557.
Truong, D.T.; Tett, A.; Pasolli, E.; Huttenhower, C.; Segata, N. Microbial strain-level population structure and genetic diversity from metagenomes. *Genome Res.* **2017**, *27*, 626–638.
Tyson, G.W.; Chapman, J.; Hugenholtz, P.; Allen, E.E.; Ram, R.J., Richardson, P.M.; Solovyev, V.V.; Rubin, E.M.; Rokhsar, D.S.; Banfield, J.F. Community structure and metabolism through reconstruction of microbial genomes from the environment. *Nature* **2004**, *428*, 37–43.
Vieites, J.M.; Guazzaroni, M.E.; Beloqui, A.; Golyshin, P.N.; Ferrer, M. Metagenomics approaches in systems microbiology. *FEMS Microbiol. Rev.* **2008**, *33*, 236–255.
Villas-Bôas, S.G.; Noel, S.; Lane, G.A.; Attwood, G.; Cookson, A. Extracellular metabolomics: a metabolic footprinting approach to assess fiber degradation in complex media. *Anal. Biochem.* **2006**, *349*, 297–305.
Wang, D.Z.; Xie, Z.X.; Zhang, S.F. Marine metaproteomics: current status and future directions. *J. Proteomics* **2014**, *97*, 27–35.
Weiss, S.; Carapito, C.; Cleiss, J.; Koechler, S.; Turlin, E.; Coppee, J.Y.; Heymann, M.; Kugler, V.; Stauffert, M.; Cruveiller, S.; Médigue, C. Enhanced structural and functional genome elucidation of the arsenite-oxidizing strain Herminiimonas arsenicoxydans by proteomics data. *Biochimie* **2009**, *91*, 192–203.
Winkler, H. Verbreitung und ursache der parthenogenesis im pflanzen-und tierreiche. *Jena: Verlag Fischer* **1920**.
Woese, C.R.; Fox, G.E. Phylogenetic structure of the prokaryotic domain: the primary kingdoms. *Proc. Natl. Acad. Sci.* **1977**, *74*, 5088–5090.
Woyke, T.; Teeling, H.; Ivanova, N.N.; Huntemann, M.; Richter, M.; Gloeckner, F.O.; Boffelli, D.; Anderson, I.J.; Barry, K.W.; Shapiro, H.J.; Szeto, E. Symbiosis insights through metagenomic analysis of a microbial consortium. *Nature* **2006**, *443*, 950–955.

CHAPTER 3

Role of Bioremediation in Wastewater Treatment

IQBAL ANSARI[1*], MUNIYAN SUNDARARAJAN[1], DEBLINA MAITI[1], ANAND KUMAR[2], and JYOTI RANJAN ROUT[3]

[1]*CSIR-Central Institute of Mining and Fuel Research, Dhanbad, Jharkhand 826015, India*

[2]*Department of Biotechnology, Vinoba Bhave University, Hazaribagh, Jharkhand 825301, India*

[3]*School of Biological Sciences, AIPH University, Bhubaneswar 752101, Odisha, India*

[*]*Corresponding author. E-mail: iqbal.cimfr@gmail.com*

ABSTRACT

Industrial activity has deteriorated the quality and diversity of aquatic life by discharging wastewater containing toxic heavy metals, pesticides, and hydrocarbons into the water bodies. The occurrences of heavy metals draw a primary concern due to their toxic nature and bioaccumulation affinity in the existing living organism. The elevated percentage of heavy metals enters in food chains results in high bioaccumulation levels in consumers, which pretenses a severe risk to their health. Thus bioremediation is an efficient process and is the only solution to mitigate pollutants from water through potential microbes such as bacteria, fungi, algae, and their consortium. Some higher plants also perform the process of bioremediation. Moreover, microbes like bacteria, fungi, algae, and genetically engineered microbes in immobilized form have drawn attention for the elimination of heavy metals. This chapter reviews various wastewater treatment studies done using wetland plants and algae. Experimental studies done at our institute have reported the efficiency of wetland plants and other different useful algal species in the treatment of mining water.

3.1 INTRODUCTION

Environmental pollution in developing country like India has frequently recognized adverse effect of technological developments, like speedy urbanization and industrialization, with deprived planning in waste disposal and management leading to heavy metal pollution, which is a big issue (Swarup and Dwivedi, 1998). Copious anthropogenic actions like tremendous use of fossil fuel, mining, and metallurgy activities are responsible for reallocating contaminated heavy metals into the environment that continue to stay for a longer period in different components of the ecosystem as they are not degradable (Kaplan et al., 2010). Some heavy metals such as arsenic and iron occur naturally in earth's crust could also contaminate the groundwater in vicinity. Some important anthropogenic spots of heavy metal origination are sewage sludges, municipal waste composts, chemicals from agricultural activities, and industrial waste (Kloke et al., 1984). Heavy metals are defined as a group of elements having atomic density >4 g cm^{-3} (Hawkes, 1997). Heavy metal toxicity not only damages the environment but also causes potential health problems due to bioaccumulation in living organisms (Aycicek et al., 2008; Lenntech, 2004). Metal toxicity is governed by the captivated dose, route of exposure, and period of exposure. Presence of high amount of heavy metals in soil, drinking water, food, and in animals has been reported throughout the world. Some other toxic metals that have been largely studied are cadmium, lead, and mercury, which affect the hormone system and growth of human beings. On long term exposure of heavy metals degradation occurs in physical, muscular, and neurological systems of the body. Some of these diseases are multiple sclerosis, Parkinson's disease, Alzheimer's disease, muscular dystrophy, and cancer (Jaishankar, et al., 2014). Cadmium and zinc are also freely absorbed by the plant roots. Growth reduction, lower biomass metabolism production, metal accumulation, and display altered in their metabolism have been noticed in plants grown in metal-polluted area (Nagajyoti et al., 2010). Excessive damage to the living tissues occurs when oxidative stress is being induced by free radical formation (Rajaganapathy et al., 2011). They are associated with catabolism activities of biological macromolecules due to oxidative process (Ercal et al., 2001). Redox inactive metal ions are critical components in many biological electron transfer reaction. ROS can engulf cell's intrinsic antioxidant defenses, results in "oxidative stress" displaying dysfunctions like lesions to proteins, lipids and DNA, and final cell death (Figure 3.1).

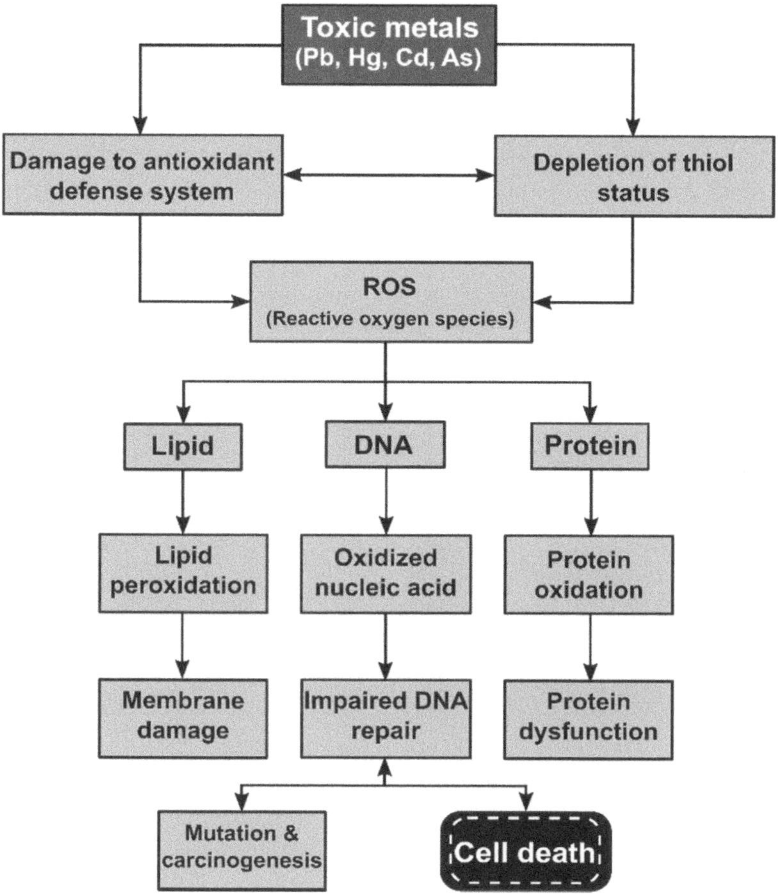

FIGURE 3.1 The heavy metal association in the production of ROS which further leads the cells under oxidative stress which parade several dysfunctions due to lesions caused by ROS to lipids, proteins and DNA and finally it leads to cell death (Adapted from Ercal et al., 2001).

3.2 HEAVY METAL TOXICITIES IN FISH AND HUMAN HEALTH

The element arsenic considered as the most hazardous heavy metals exposed to humans by means of water contamination. The toxicity of arsenic in potable water has been observed in more than in 30 countries in the world (Chowdhury et al., 2000). Most of the agricultural practices such as the use of pesticides, fertilizers, and four-footed animal feeding habits are also responsible for releasing arsenic to our ecosystem large quantities. The mineral forms of arsenic are arsenite and arsenate. Arsenic contamination causes abdominal pain, vomiting, nausea, and severe diarrhea, which occur through acute toxicity. It can also cause cancer of lung, bladder, liver, and skin (Ratnaike, 2003). Chronic arsenic toxicity causes pigmentation and keratosis on the skin (Martin and Griswold, 2009) along with liver diseases, polyneuropathy, peripheral vascular disease, hypertension, diabetes mellitus, non-pitting oedema, chronic lung disease, conjunctival congestion, weakness and anemia (Guha and Mazumder, 2008; Mazumder and Dasgupta, 2011). Arsenic facilitates its toxicity by generating oxidative stress, causing immune dysfunction, endorsing genotoxicity, hampering DNA repair, and disorderly signal transduction, which may observed in arsenicosis. The arsenic toxicity may also lead to severe skin lesions on human body that is shown in Figure 3.2.

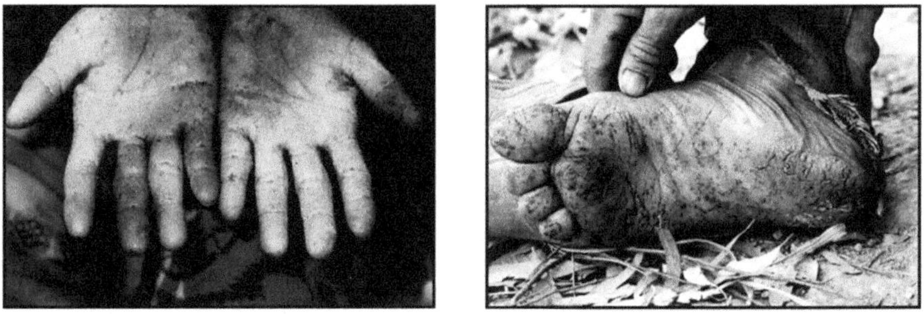

(a) (b)

FIGURE 3.2 Showing the effects of arsenic on human body. (a) Arsenic keratosis, so-called raindrops on a dusty road) and (b) skin lesions due to arsenicosis

Source: Reprinted from Jaishankar et al., 2014. https://creativecommons.org/licenses/by/4.0

Untreated industrial wastewater and domestic household water discharge into water bodies are degrading the water quality (Clover, 2000; Priyanka et al., 2017). The effects of toxicity of metals on human health and their probable consequences in fish consumption as diet has been reviewed (Castro-Gonzalez and Mendez-Armenta 2008). The fish *Tilapia nilotica* found in the lake ecosystem was badly pretentious with heavy metals. The chief levels of copper and zinc were found in all parts of fish body where liver was the most affected organ. The highest percentage of manganese was present in intestine and stomach. Their outer layers scales possess the maximum levels of cobalt, chromium; nickel and strontium, lowest level of concentrations were found in their gills and vertebral column (Rashed, 2001). A similar study by Hussien (2015) reports that cadmium, lead, iron, copper, and zinc were found in different tissues like muscles and gills of *Tilapia nilotica* in the south part of Manzala Lake (Egypt). Similarly, Abdel-Mohsien and

Mahmoud (2015) reported that the percentage of cadmium, lead, chromium and aluminum were found in the muscle of freshwater fish *Oreochromis niloticus* in Egypt. The concentrations of lead, cadmium, and chromium were numerous times greater in fish samples than their concentrations in water and the bioaccumulation factor (BAF) was ranged from 8 to 122. But for a long time of disposal of wastewater into such aquatic bodies where aquaculture is going on will have resulted in a higher concentration of heavy metals in the various parts like tissues and gills of the fish.

3.3 PHYTOREMEDIATION: AN INNOVATIVE TECHNIQUE FOR WASTEWATER TREATMENT

Phytoremediation is another thoughtful effort in which aquatic macrophytes remove the extreme nutrient burden from the water that otherwise it causes eutrophication in water. It fascinates nutrient, mineral ions from water column, and enhances metal holding indirectly by acting as traps for particulate matter, by slowing the water current and favoring sedimentation of suspended particles (Jatin et al., 2008). According to Ilya et al. (1994), phytoremediation is a promising technology that uses plants to clean up soil and water contaminated with heavy metals through phytostabilization, phytoextraction, rhizofiltration, phytotransformation, and rhizosphere bioremediation. Metal accumulation inside the vascular system of the plants occurs through the process of chelation, precipitation, compartmentalization, and translocation. The technology practices for cleaning contaminated locations with a very low contamination of organic or metal pollutants. Additionally, it is cost operative and has long term applicability along with esthetic advantages (Chhotu and Fuleka, 2009). The process rhizoflltration eliminates lethal metals from polluted waters by the usage of plant roots. Throughout the phytoremediation process, the roots of plant absorb metals from soil and allocate them to the top ground shoots. Some plants that are capable to accumulate high levels of metals in the shoot without incurring any toxic effect to the photosynthetic system are known as metal tolerant plants (Nandakumar et al., 1995). They can be subsequently be harvested after appropriate plant growth and locked in various end uses based on the usefulness of the plants; thus resulting in the permanent eradication of metals from the site. Phytoremediation is ideal methods as compared to other biological agents due to easy availability; handling and frequent abundance of the plants in terrestrial; as well as aquatic ecosystems (Rai, 2008). This technique has been widely used during Ramsar convention and proposed at Ramsar, Iran in 1975. About 400 plants species have been potentially identified in the remediation of soil and water. The plant species like, *Azolla, Eichhornia, Typha, Phragmites, Lemna, Thlaspi, Sedum Alfredii, Arabidopsis,* and another aquatic macrophytes show potent wetland plants whose potential can be harnessed for removal of heavy metal toxicity. The widespread rhizosphere of wetland plants provides a nutritive medium culture zone for microbial activities and later degradation of pollutants. This wetland sediment zone provides reducing conditions which is favorable to metal removal. For example, constructed wetlands has shown efficiency in attenuation of heavy metals from acid mine drainage, thermal power plants, landfill leachates and municipal, agricultural or refinery effluents. Recent advances in biotechnology such as transfer of metal hyperaccumulating genes from low biomass producing wild species to the higher biomass producing cultivated species can make the process more fruitful (David et al., 1995; Lone et al., 2008).

The natural remediation and uptake process of metals by plant roots is depicted in Figure 3.3. The whole process is divided into four phases. Metal partitioning among solid and liquid phase is guided by processes such as adsorption, precipitation, complexation, and redox reactions. This completes phase A that encompasses the base for bioavailability of metals. Phase (B, B′) involves the transport of metal to the organism in soluble or colloidal form. Colloidal form of metal transportation involves organic matter and is highly reactive. It contains higher metal concentration in comparison to those in the solution. Phase C involves the passing of the metals through a biological membrane or root membranes. Roots also serve as a biofilter for contaminants. The last phase D implicates circulation and assimilation of metals in their metabolic pathway of the organisms for a biological response such as growth and biomass of the organisms. Transfer of metals from soil to plant is a multi-faceted process and is governed by natural as well as anthropogenic factors. Moreover, uptake, translocation and bioaccumulation mechanisms differed for various heavy metals and for the plant species (Maiti and Prasad, 2016).

3.4 BIOREMEDIATION BY AQUATIC PLANTS

In an experimental study three aquatic macrophytes such as

Lemna minor, Spirodela polyrrhiza, and Eichhornia crassipe have removed a considerable amount of arsenic and mercury in 21 days. Their results favored some selected species as promising accumulator of metals (Virendra et al., 2007). Rai (2008) showed phytoremediation capability of *Azolla pinnata*, by growing it in 24–40 L aquariums containing mercury and

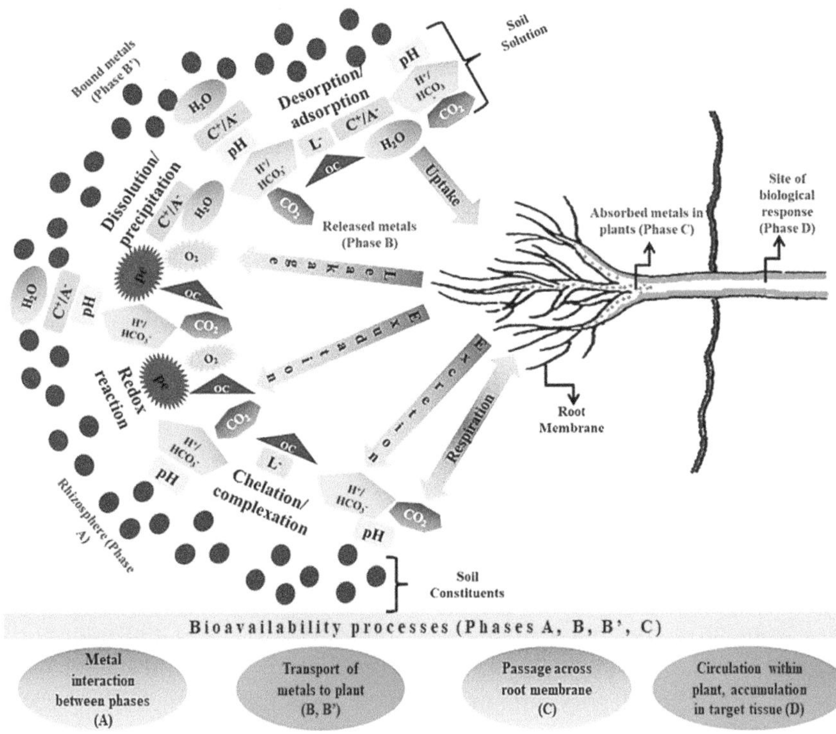

FIGURE 3.3 Bioavailability processes for metals to the plants. Legends: OC = organic carbon; C+ = cation; A+ = anion; L- = ligand; pe = redox potential.

Source: Reprinted from Maiti and Prasad, 2016.

cadmium that lies in between 0.5, 1.0, and 3.0 mg L^{-1} concentrations. The fern accumulated 310–740 mg kg^{-1} metal on dry mass basis and decreased metal concentration 70%–94% after 13 days. Putra et al. (2016) found that *E. crassipes* could hyper accumulate copper in its root and remove up to 95% zinc as well as 84% chromium after 11days from solutions; further, the metals can also be mined out from plant bodies (Mishra and Tripathi, 2009). It is also suitable for elimination of organic pollutants from wastewater (Shahabaldin et al., 2015). Priyanka et al. (2017) reported that *E. crassipes* could remove 99% chromium in 15 days from mine water along with removal of total dissolved solids, biological oxygen demand, and chemical oxygen demand (COD). Thus aquatic plants can be potentially utilized for mine water treatment as shown in a study done by Prasad and Maiti (2016). The study investigates with metal (copper, manganese, lead, cadmium) concentrations in sediment, water, and analogous leaf samples of *E. crassipes* taken from ponds situated in nonmining and coal mining regions. Manganese, copper, and cadmium deposited in the maximum range of allowable concentration for plants whereas lead deposited above the limit. BAF in plants corresponding to their sediments values for manganese was highest followed by copper and lead whereas BAF in plants with respect to water was highest for copper (428–3205) followed by manganese (285–1100), lead (242–506) and cadmium (7–130). This study reveals that *E. crassipes* plays a significant role in removal of metals from the pond ecosystem. Further, it was observed that the leaves of *E. crassipes* can be used efficiently for biomonitoring, phytoremediation of polluted wetlands through proper management strategies. The study has been detailed in the following paragraphs.

3.4.1 CASE STUDY 1: BIOREMEDIATION POTENTIAL OF *E. crassipes*

3.4.1.1 STUDY SITE, SAMPLING, AND ANALYSIS

Leaf, water, and sediment samples had been collected during April–May, 2013 from different ponds situated in mining (P2, P3, P4) and nonmining areas (P1) of Dhanbad, Jharkhand, India. All these ponds mitigated water requirement by the nearby human settlement. Rainfall is the chief source of water for these ponds. However, railways discharge and drain water released into these systems. Sediment samples were taken at 7 random points from each pond and leaves of *E. crassipes* was collected from rooted as well as free-floating plants from inside 1 × 1 m quadrats. The samples were collected in clean, labeled plastic bags and immediately transported for analysis in laboratory. Water samples were collected in cleaned plastic bottles.

All the analysis was done as per standard methods reported in previous studies. Sediments were analyzed for pH in water in the ratio (w/v) 1:2 (Mclean, 1982), loss on ignition (Heiri et al, 2001), metals (copper, manganese, lead, cadmium) through Atomic Absorption Spectrometer after digestion in acid mixture (Anilava and Das, 1999). All water samples had been analyzed for pH, total hardness (TH), calcium hardness, chloride hardness, total dissolved solids, sulfate, and heavy metals through Atomic Absorption Spectrometer (Clesceri, 1998). Water quality was analyzed on the basis standard drinking water specifications (BIS 10500 2012). Leaves were washed, cut, and dried at 80 °C, grinded, burnt to ashes, and digested before metal analyses through Atomic Absorption Spectrometer.

BAF of metals in leaf was determined as a ratio of metal concentration in the leaf to than of sediment (BAF_{sl}) and water (BAF_{wl}). When BAF values are greater than 1 indicate the potent of the plant to be used for the phytoremediation process (Kumari et al., 2013). The ecological risk for the presence of such heavy metals in the sediments was calculated by sediment quality guideline index (SQG-I) (Fairey et al., 2001). This index provides a measure of the risk to living organisms from metal pollution. The magnitude of values signifies the degree of risk as follows (Das and Chakrapani, 2011):

- SQG-I < 0.5 = low toxicity level.
- $0.5 <$ SQG-I < 1 = moderate toxicity level.
- SQG-I > 1 = site is fatal to metal toxicity for living organism.
- SQG-I > 1.5 = high probability of toxicity to living organisms.

Sediments are a significant sink for metals and contain higher concentrations than water. Metal accumulation in sediments occurs through slow exchange rates (Rai, 2008) and progressively gets associated with organic matter to settle down at the bottom. These metals get mobilized later in specific conditions to lead to increased concentration in the water. This phenomenon may have adverse impacts on water quality but also ensures self purification of the aquatic environment at times due to the reverse mechanism promoted by low levels of dissolved oxygen and pH (Rai, 2009). Water consumption also increases secondary water pollution through a factor of 1.5–3 during the summer season (Linnik and Zubenko, 2000). Lead, iron, chromium, and nickel are strongly linked to the sediments, whereas cadmium, zinc, and copper are mobile metals (Morillo et al. 2002). Parameters analyzed for sediments and their corresponding values are shown in Tables 3.1 and 3.3. All the sites had a pH near neutral. Loss on ignition in control or pond from nonmining area was 3.3% compared to other ponds (10%–17%), which could be due to deposition of mine waste discharge in the latter ponds. Metals in sediments are usually present in a reduced state and are thus are nonbioavailable. Moreover, bioavailability depends on water solubility, exchangeability, complex forming, and adsorption tendency of metals (Weis and Weis, 2004). The metal concentration of the sediments has been compared to maximum allowable concentration given in Kabata-Pendias and Sadurski (2004) and has been found within the ranges. However, pond sediments from mining areas had higher concentration than nonmining areas. The high concentration of manganese and copper in the former could be due to the discharge of water from the residential complexes into it. The cadmium was not detected in sediment samples, but in contrast, found in leaf and water samples. This might be due to the interference of other metals present in high concentration in the sediment samples. The SQG-I values as shown in Figure 3.4a show the low level of toxicity to living organisms. Moreover, the concentrations of the studied metals in sediments, as well as water, did not follow any significant linear relationships and correlation as shown in Figure 3.4b.

TABLE 3.1 Analysis of Sediment From Ponds in Mining and Nonmining Regions

Parameters (n = 7)	Nonmining		Mining	
	P1	P2	P3	P4
pH_{H2O} (1:2; w/v)	7.52	7.32	6.62	7.12
Moisture air dried (%)	0.86	1.60	2.74	2.35
Loss of ignition (%)	3.29	10.82	17.13	15.96

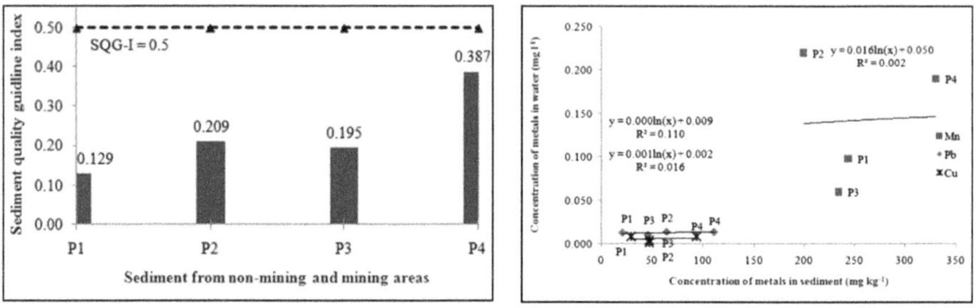

FIGURE 3.4 (a) Sediment quality indices in pond sediments and (b) variations in lead, manganese, copper concentrations (mg kg−1) in sediment with respect to the concentration in pond water.

All the values in water samples were (Tables 3.2 and 3.3) under desirable limits. Ponds had significantly similar pH near neutral. Smolyakov (2012) studied the removal of zinc, copper, lead, and cadmium by water hyacinth and found that metal concentration decreased in the order lead ~ copper > cadmium ~ zinc, which proved that removal is more efficient at higher pH (7–8). TH, calcium and magnesium, sulfate, and chloride were not alarming. Total dissolved solids of the pond water were >500 mg L^{-1} indicating them unfit for drinking. Water hyacinth can efficiently decrease water pH, total dissolved solids, and hardness in 25%–50% dilutions of wastewater (Shah et al., 2010). Similar to sediments, manganese in some pond water from mining regions were higher than desirable limit (0.1 mg L^{-1}) might be due to discharge from municipal residential complexes; yet this water can be used in absence of any alternate source as the values are below the maximum allowable limit (0.3 mg L^{-1}). As per all the metal concentrations, water from the ponds could be used for drinking uses after an advance treatment for removal of total dissolved solids and manganese.

TABLE 3.2 Analysis of Pond Water From Mining and Nonmining Regions

Parameters	Nonmining		Mining		
	P1	P2	P3	P4	Desirable[a]
pH	7.29	7.14	6.64	7.11	7.5
Total hardness (mg L^{-1})	82	81.5	76.0	117.5	300
Calcium (mg L^{-1})	49.5	52.5	48.5	62.5	75
Magnesium (mg L^{-1})	7.89	7.04	6.68	13.36	30
Total dissolved solid (mg L^{-1})	587.5	717	520	675	500
Sulfate (mg L^{-1})	30.9	9.88	–	–	200
Chloride (mg L^{-1})	–	–	26	18	250

"–": not detected.

; [a]*Drinking water-specification BIS 10500 2012.*

TABLE 3.3 Metal Concentration in Surface Water, Sediment, Leaf Samples From Different Ponds. Data is Compared to Maximum Allowable Concentration for Soil, Leaves, and Standards for Drinking Water

Metals (*n* = 7)	Sampling Stations	Pond Water (mg L^{-1})	Sediment (mg kg^{-1})	Leaf (mg kg^{-1})
Mn	P1	0.098	243.85	67.9
	P2	0.22	199.34	62.9
	P3	0.06	234.15	66.04
	P4	0.19	330.65	65.97
Standard limit		0.1[a]	1500–3000[b]	20–500[c]
Cu	P1	0.009	29.46	3.86
	P2	0.002	47.12	6.41
	P3	0.004	47.19	9.68
	P4	0.009	93.76	13.54
Standard limit		0.05[a]	60–150[b]	15–20[c]
Pb	P1	0.013	20.87	4.38
	P2	0.014	64.3	4.06
	P3	0.01	46.03	5.06
	P4	0.014	111.18	3.4
Standard limit		0.05[a]	20–300[b]	2.5[c]
Cd	P1	0.001	<	0.042
	P2	0.001		0.13
	P3	0.005		0.037
	P4	0.001		0.049
Standard limit		0.01[a]	1–5[b]	1.5[c]

[a]*Drinking water-specification BIS-10500-2012.*

[b]*Kabata-Pendias and Sadurski (2004).*

[c]*Alloway (2013).*

Metal concentrations in leaves (Table 3.3) showed that manganese, copper, cadmium were below maximum allowable concentration for plants except lead (Kabata-Pendias and Sadurski, 2004). Manganese accumulated in the range 63–68 mg kg^{-1}. Heavy metal removal via plants depends on treatment time and metal concentration.

Uptake is at a faster rate at lower concentrations because high level of metals induces toxic effect. For example, in the initial few hours of plant–metal interaction, biosorption helps in accumulating 40% Cd from substrate followed by its migration to stems and gradual deposition and transformation in leaves (Shuvaeva et al., 2013). High concentrations of Mn in water hyacinth leaves corresponding to the other metals might be due to higher concentrations in the sediments and greater affinity of Mn toward bioaccumulation. BAF_{sl} (0.20–0.27) and BAF_{wl} (285–1100) of manganese (Figure 3.5a and b) depicted decrease in the accumulation with increasing concentration in substrate. Copper in leaves from mining areas was significantly higher than nonmining area; the values accounted for 13%–21% of the total Cu in sediments. Lokeshwari and Chandrappa, (2006) reported a Cu concentration of 4–16 μg g^{-1} in Hyacinth plant leaves growing in a wetland in receipt of urban run-off. Although copper was in least concentrations in sediment water yet it had highest BAF_{wl} followed by manganese (285–1100), lead (242–506), and cadmium (7–130). Lead in leaves had been in a narrow range irrespective of the wide range in the sediments which was 3%–21% of the total concentration. The concentration of lead in leaves decreased sharply with growth in metal concentration in the substrates (Figure 3.5a and c). BAFwl for lead was six times higher than that of cadmium that showed higher affinity (Mahamadi and Nharingo, 2010). Cadmium in leaves from mining area had been 42% higher concentration than nonmining regions.

Sediments had significantly higher metal concentrations than water, but the accumulation in leaves varied. Roots of the plants are most efficient in accumulating metals at higher levels. Moreover, the leaves constitute 75% of the plant aboveground biomass. The whole plant can accumulate a huge amount of metal in bulk cultivation that can process out through proper management strategies.

3.5 BIOREMEDIATION BY ALGAE

Biosorption is a process of fastening and concentration of adsorbate from aqueous solutions by certain types of stationary and dead microbial biomass. This is a capable technology used as bioremediation for heavy metal

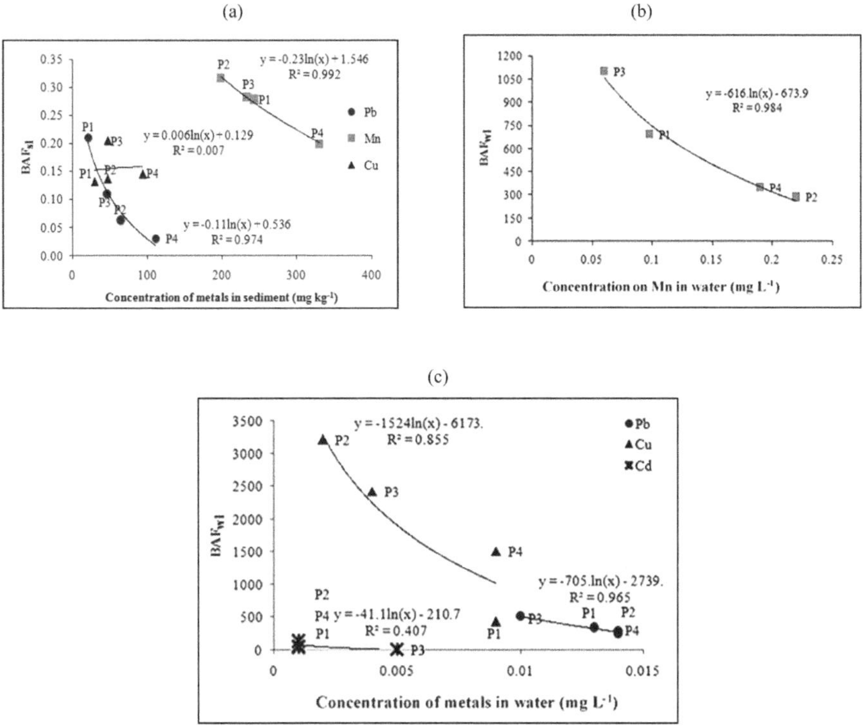

FIGURE 3.5 (a) Scatter plots for bioaccumulation factor of metals in leaves (BAF_{sl}) with respect to their concentration in the sediments (mg kg^{-1}) from the ponds, (b) bioaccumulation factor of Mn in leaves (BAF_{wl}) with respect to Mn in pond water, (c) bioaccumulation factor of Pb, Cu, Cd in leaves (BAF_{wl}) with respect to their concentration in pond water (mg L^{-1}).

pollutants from tainted natural waters and wastewaters (Ackmez et al., 2012). Microalgae have great potential to resolve energy and ecological challenges around the world. Wastewater treatment by microalgae is a more eco-friendly approach that reduces nitrogen and phosphorous and also removes heavy metals from wastewater. Microalgae are capable in absorbing a substantial amount of nutrients since they need of large amounts of nitrogen and phosphorous for proteins (45%–60% microalgae dry weight) and metals as micronutrients for their growth. Oswald (1957) firstly developed the idea of wastewater treatment by using microalgae and performed photosynthesis in sewage treatment. Microalgae have been considered as an environmentally and prudently sustainable approach in bioremoval of heavy metal ions using toxic metals from wastewaters (Mata et al., 2009; Oswald and Gotaas, 1957). When *Chlorella vulgaris* was grown in swine wastewater it removed

96% and 95.3% of phosphorus and nitrogen, respectively (Kim et al., 1998). Similarly, Travieso et al. (2008) reported that in laboratory-scale microalgae pond, the removal of organic nitrogen, ammonia, and total phosphorus was observed 90.2%, 84.1%, and 85.5%, respectively from distillery wastewater with an anaerobic fixed-bed reactor. Renuka et al. (2013) have investigated *Calthrix* sp. for phosphates and nitrates removal from sewage wastewater. *Calthrix* sp. has removed 44%–91% PO_4–P and 57%–58% NO_3–N, and a maximum dry cell weight of 0.97 mg L^{-1} can be attained in sewage wastewater using this species. Hala and Laila (2014) found that culturing *C. vulgaris* has removed color as well as COD in waste effluent from the textile industry. Bioremediation of wastewater using various microalgae species is also capable of removing nitrogen and phosphorus as shown in Table 3.4.

TABLE 3.4 Potential of Microalgae in Removal of Nitrogen and Phosphorus (Saikumar, 2014)

Type of Wastewater	Microalgae Species	Removal (%)		
		Ammonia Nitrogen	Nitrate	Phosphate
Domestic	Botryococcus braunii	–	79	79
Primary treated	Haematococcus pluvialis	–	100	100
Secondary treated	*Botryococcus braunii*	–	80	–
Synthetic	*Chlorella kessleri*	–	19	–
	Chlorella vulgaris	97	–	96
	Neochloris oleoabundas		99	100
Soyabean processing	*Chlorella pyrenoidosa*	89	–	70
Agro-industrial	*Scenedesmus dimorphus*	95	–	–
Pig	*Spirulina*	84–96		72–87

Deviller (2004) reported the uptake efficiency of *Ulva* sp. to treat the effluents of a recirculation system can reach 0.5 g N m^{-2} day^{-1} and 0.03 g P m^{-2} day^{-1} for nitrate and phosphate, respectively during optimal climatic conditions for algal growth. A similar study reported by (Neori, 2003) with the same algae, the nitrogen removal rate may reach 2.9 g m^{-2} day^{-1} in a flow-through system effluent containing mainly ammonia–nitrogen, with the protein contents of algal biomass up to 44% dry weight. A study investigated the elimination capability of *C. vulgaris* microalga as feasible for removal of phosphorous along with organic matter from wastewater. Phosphorus concentration has been removed by 99% and the COD was decreased by 71%. The biomass concentration has also found to be increased from 0.05 to 0.57 g L^{-1} after cultivation of 9 days (Salgueiro et al., 2016). An experimental study has shown that *C. vulgaris* and *Chlorella salina* have potentially reduced pH, total dissolved solids, biological oxygen demand, COD, sulfate, calcium, ammonia, nitrate, sodium, magnesium, potassium, heavy metals (Zn, Cu, Mn, Ni, Co, Fe and Cr), and the total coliform bacteria after 10 days of treatment from untreated water samples. The heavy metals removal was observed 13%–100% (Mostafa et al., 2016). A study reviewed 14 different species of algae that are capable in eliminating heavy metal from wastewater such as gold, copper, nickel, lead, zinc, cadmium, mercury, arsenic, and iron (Iqbal et al., 2018).

Several authors have also reported that numerous species of algae such as *C. vulgaris, Scenedesmus dimorphous, Neochloris oleoabundans, Nannochloroposis, Spirulina, Botrycoccus braunii,* and *Dunaliella salina* can grow healthy in wastewater. Moreover, they are capable to accrue cadmium and chromium for treating wastewater and removed heavy metals (Ting et al., 1995; Zhou et al., 1998; Chong et al., 2000). Similarly, Kidgell et al. (2014) investigated biomass of freshwater macroalga *Oedogonium* sp. (Chlorophyta) as a valuable substrate for the manufacture of biosorbents, which remediates metals and metalloids from a multifaceted industrial effluent. Chaisuksant (2003) has investigated the adsorption properties of copper (n) and cadmium(II) in chemically pretreated biomass of *Gracilaria fisheri*. For batch equilibrium exposed the maximum adsorption capacity values for cadmium and copper was 0.63 and 0.72 mmol g^{-1}, respectively, of the pretreated biomass. In conditions of species differences, in copper sorption capacity the microalgal species follow as *Fucus spiralis*>*Ascophyllum nodosum*>*Chondrus crispus*>*Asparagopsis armata*>*Spirogyra insignis*>*Codium vermilara* has been found (Romera et al., 2007). Pera-Castro et al. (2004) have grown *Scenedesmus incrassatulus* in a laboratory scale that removed chromium(VI) metals up to 25%–78% in continuous cultures as compared to batch culture owing to the uptake of chromate that favored by viable growing algae. Moreover, algae cultures deal with an operative solution to wastewater treatments owing to its capability by up taking inorganic nitrogen and the phosphorous for their growth and also in removal of other noxious compounds nearby in the wastewater.

3.5.1 CASE STUDY 2: BIOREMEDIATION POTENTIAL OF ALGAE

Ansari and Sundararajan (2015) have attempted an experimental study that evaluates the efficiency of different algae species in treating both metallic and nonmetallic toxicities in mine wastewater using a mathematical approach.

3.5.1.1 STUDY AREA

The samples of mine water were collected from open pits of five different opencast projects of Bharat Coking Coal Limited namely Dahibari-Basantimata, New Laikdih, Chaptoria, Borira, and Damagoria and brought them for physicochemical analysis in the laboratory. The sampling locations coordinate and location map of the study area have been shown in Table 3.5 and Figure 3.6. The area experiences a tropical climate and is characterized by very hot summer and cold winters. The months of May and June are very hot. The cold weather experiences in the month of November–February and temperature varies from 8.3 to 34.4 °C.

TABLE 3.5 Different Sampling Locations Used in Case Study 2

Name of the Opencast Mine	Sampling Code
Dahibari Basantimata	S1
New Laikdih	S2
Chaptoria	S3
Borira	S4
Damagoria OCP	S5

During summer month, which runs from March to June, experiences highest maximum temperature of 47.0 °C. During the remaining months, from July to October, the rainy season, the temperature varies from the lowest minimum of 15–36 °C. The relative humidity is high in the rainy days being about 94% in June and 36% low in May. Thunderstorms usually occur in the month of May and June accompanied by a temporary fall in temperature by a little degree. The area receives an annual rainfall of about 1100–1200 mm, out of which 75%–80% observed during the three months of June to September with smaller amounts during winter months.

3.5.1.2 MICROSCOPIC STUDY OF ALGAE SPECIES

The algal species were isolated from the mine water and analyzed microscopically for physical and structural characterization. They are speculated to be *spirogyra*, *Oscillataria*, *Chara*, and *Diatoms*. These algae were freshly isolated and collected for their efficiency to be used as bioremediation agents for the treatment of mine wastewater. A total of 10 gm (moist weight) of uniform suspension of full algal species as initial inoculums of 10 days old culture was taken in each flask containing 1000 mL of mine wastewater sample. Standard methodology was adopted for the mine water sampling and chemical analysis. Standard microscopic methodology was adopted for identifying algae based on its shape, size, and type as shown in Table 3.6.

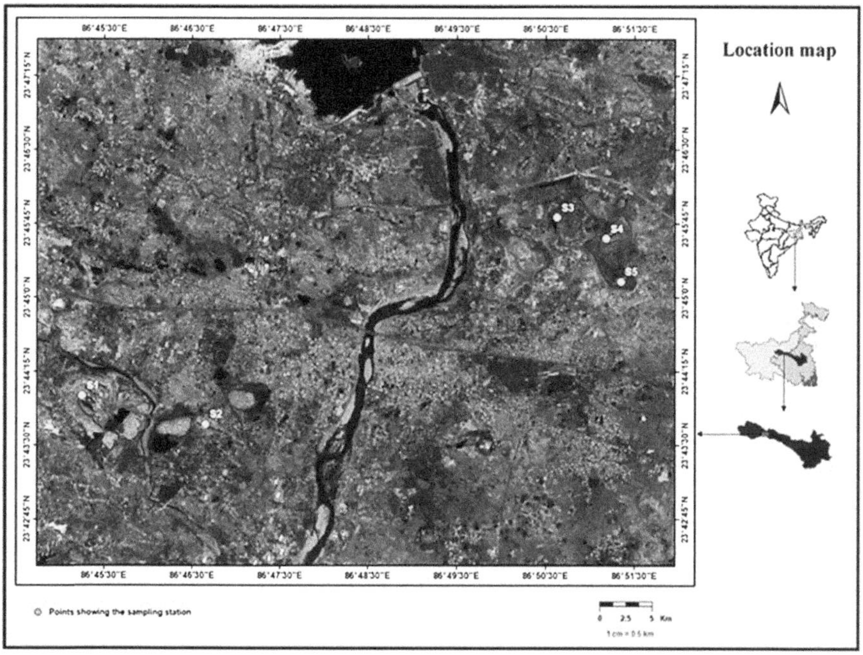

FIGURE 3.6 Location map of the study area in case study 2.

TABLE 3.6 Naturally Dominant Species of Algae Found in Open-Pit Water

Algae Species	Spirogyra	Diatoms	Ossilatoria	Chara
Shape	Filamentous	Round (Semicircular)	Filamentous (Cyanobacteria)	Rod shape
Size(μm)	20–40	20–60	2.54–38	25–55
Type	BGA (Blue green algae)	BGA	Red algae	BGA

3.5.1.3 EXPERIMENTAL SETUP

The experiment was performed at laboratory scale by taking mine wastewater from opencast coal mining located at five different sampling stations in Dhanbad district of Jharkhand state, India. The mine wastewater sample was collected in sterile bottle for chemical analysis such as pH, temperature, TH, nitrate, iron, sulfate, fluoride calcium, and manganese. The algae samples found nearby these open cast project were also analyzed for shape, size, and types. For that, about 10 g of algae was obtained in the beaker with 1000 mL of mine wastewater. The major objective of this experiment was to check the changes occurring in the chemical parameters of mine water after culturing algae in the same mine wastewater. The bioremediation process was continued for 10 days to observe the significant changes occurred in the physciochemical parameters of mine wastewater. All the nine (09) physciochemical parameters were studied at the initial stage (before putting algal biomass in mine wastewater) and algal growth after 10 days. Both the data were compared which is shown in Table 3.7.

3.5.1.4 PHYSICOCHEMICAL CHARACTERISTICS OF MINE WATER

The experiment was conducted for analysis of the physical and chemical variations in mine wastewater through culturing of algal species biomass from five different sampling stations. The detail findings are given in Table 3.7.

3.5.1.5 MODELING AND COMPUTER SIMULATION

The rate of changes in mine water quality after algal treatment was estimated and further the water quality of the mine water in which the algae is being cultivated was predicted using a linear model for up to 100 days. The integrated water quality indices were computed by using a multicriteria decision-making approach (Sundararajan and Loveson, 2002). The formula for estimating the integrated water quality index (IWQI) with respect to n parameters is given as follows:

$$IWQI = \frac{1}{2}\left[\frac{\sum_{i=1}^{n} q_i}{n} + \left(\prod_{i=1}^{n} q_i\right)^{1/n}\right] XM$$

where q_i = the quality index of the parameter I, which is estimated using the following sensitivity function $f_p(x)$ of a parameter "P" when the concentration is x:

TABLE 3.7 Physicochemical Analysis of Mine Water Before and After Algal Treatment

Sampling Code		pH	Hardness	Chloride	Fluoride	Calcium	Nitrate	Sulfate	Iron	Manganese
S1	Before	7.7	804	16	1.15	321.6	7.6	375	0.01	0.035
	After	7.4	632	10	0.61	252.8	0.8	300	BDL	0.008
S2	Before	7.5	668	31	1.65	267.2	2.2	245	BDL	0.038
	After	7.1	584	20	–	233.6	0.30	190	BDL	BDL
S3	Before	7.6	208	41	1.15	83.2	1.4	175	BDL	0.231
	After	7.3	152	32	0.81	60.8	0.8	150	BDL	0.12
S4	Before	7.9	416	50	1.0	166.4	0.02	170	0.03	0.903
	After	7.7	304	27	0.68	121.60	–	140	BDL	0.508
S5	Before	7.5	936	90	1.25	374.4	5.33	325	0.045	0.203
	After	7.2	892	70	0.35	356.8	1.0	300	BDL	0.172

$$f_P(x) = \begin{cases} -1 & \text{if } x < l \\ \dfrac{x - r_1}{r_1 - l} & \text{if } 1 \le x < r_1 \\ 0 & \text{if } r_1 \le x \le r_2 \\ \dfrac{x - r_2}{u - r_2} & \text{if } r_2 < x \le u \\ +1 & \text{if } x > u \end{cases}$$

where

l and u are the minimum and maximum permissible limits, respectively.

r_1 and r_2 are the lower and upper limits of recommended or desirable range.

M is the microscopic number to expand the interval by taking the values 100, 1000, 10,000, and so on.

The impact index (p) is calculated as an absolute value of sensitivity number and the quality index (q) = $1 - p$. Therefore, it can be noted that $p+q=1$.

3.5.1.6 WATER QUALITY PREDICTION

The mine water quality for each day starting from the initial day of algal treatment up to 100 days was assessed with the Indian Standard IS:10500 (2012) applying the above-described model using the predicted water quality parameters. The quality changes corresponding to the individual parameters in each mine water samples have been presented in Figure 3.7. The variation in integrated water quality indices while treating the mine wastewater by means of algae is visualized in Figure 3.8.

In water, there are certain parameters that favor the suitability of water for drinking purpose whereas a few were nonfavorable. However, the integrated quality can be estimated using multicriteria decision-making analysis using the above-discussed formula. It has been pragmatic that the integrated water quality indices increase initially after the algal treatment and then it starts decreasing gradually. It happens because the algae species uptake the nonfavorable parameters, such as trace elements, rapidly than the favorable parameters. This experiment was conducted in the laboratory. But the scenario of water quality changes in mine pits will be different as the mine wastewater flows frequently. As a result, the integrated water quality may increase and may be more suitable after algal treatment. More experimental data has to be analyzed through various case studies in different mines in order to establish the fact.

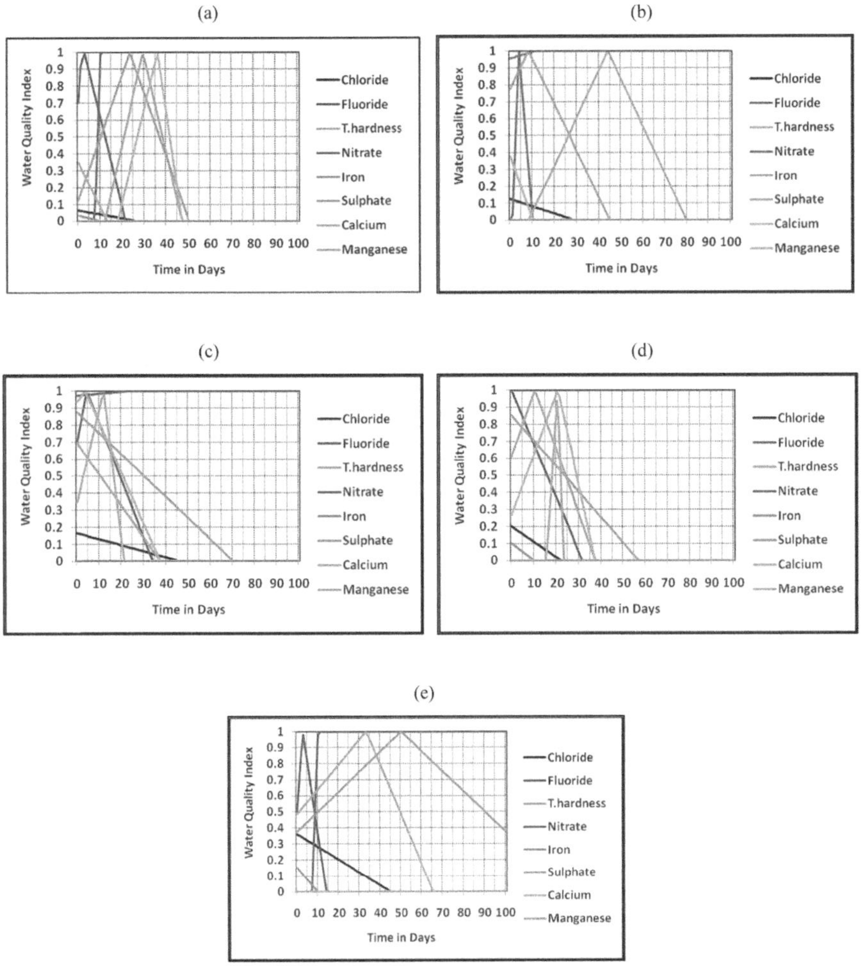

FIGURE 3.7 Variations in parameter wise water quality taken from various sampling station: (a) Dahibari-Basantimata, (b) New Laikdih, (c) Chaptoria, (d), Borira, and (e) Damagoria.

3.6 CONCLUSION

Algae have become a potent organism for biological remediation of wastewater as they are capable to accumulate plant nutrients, pesticides, heavy metals, organic and inorganic noxious substances, and even radioactive matters in their cells, which is due to their bioaccumulation ability. The studies conclude that wetland plants and algae reduce the percentage of contaminants from the wastewater. Properly designed coupled wetland-algal wastewater treatment systems can be more effective than other conventional treatment systems, due to cost-effectiveness and environmental sustainability.

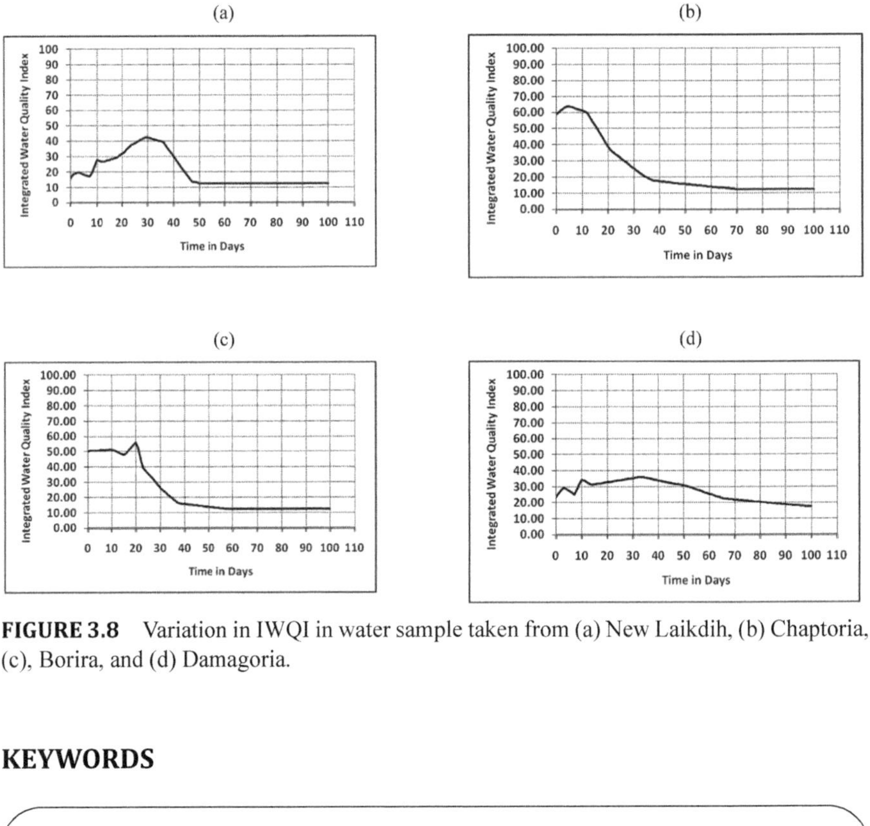

FIGURE 3.8 Variation in IWQI in water sample taken from (a) New Laikdih, (b) Chaptoria, (c), Borira, and (d) Damagoria.

KEYWORDS

- **water pollution**
- **organic pollutants**
- **heavy metal toxicity**
- **microorganism**
- **algae**
- **wetland plants**
- **treatment efficiency**

REFERENCES

Abdel-Mohsien, H.S.; Mahmoud, M.A. Accumulation of some heavy metals in Oreochromis niloticus from the Nile in Egypt: potential hazards to fish and consumers. *J. Environ. Prot.* **2015,** *6*, 1003.

Alloway, B.J. Heavy metals in soils. Trace metals and metalloids in soils and their bioavailability, The Netherlands, **2013**.

Anilava, K.; Das, S. Effects of fertilization on the deposition, partitioning and bioavailability of copper, zinc and cadmium in four perennial ponds of an industrial town. *Indian J. Environ. Health* **1999,** *41*, 6–15.

Ansari, M.I.; Sundararajan, M. Waste to wealth. In: *Proceedings of the India International Science Festival-Young Scientists Meet*, Department of Science and Technology, Government of India, Paper Code: Waste-12. December 4–8, **2015**.

Aycicek, M.; Kaplan, O.; Yaman, M. Effect of cadmium on germination, seedling growth and metal contents of sunflower (*Helianthus annus* L.). *Asian J. Chem.* **2008,** *20*, 2663.

Bunting, S.W. Appropriation of environmental goods and services by aquaculture: a reassessment employing the ecological footprint methodology and implications for horizontal integration. *Aquac. Res.* **2001,** *32*, 605–609.

Castro-González, M.I.; Méndez-Armenta, M. Heavy metals: implications associated to fish consumption. *Environ. Toxicol. Pharmacol.* **2008,** *26*, 263–271.

Chaisuksant, Y. Biosorption of cadmium(II) and copper(II) by pretreated biomass of marine alga *Gracilaria fisheri*. *Environ. Technol.* **2003,** *24*, 1501–1508.

Chalivendra, S. *Bioremediation of Wastewater Using Microalgae*, Doctoral dissertation, University of Dayton, **2014**.

Clover,C. Pollution from fish farms as bad as sewage. *The Telegraph* https://www.telegraph.co.uk/news/uknews/1355936/Pollution-from-fish-farms-as-bad-as-sewage.html 1/2 (accessed on 8/10/2018), 2000.

Chong, A.M.Y.; Wong, Y.S.; Tam, N.F.Y. Performance of different microalgal species in removing nickel and zinc from industrial wastewater. *Chemosphere* **2000,** *41*, 251–257.

Chowdhury, U.K.; Biswas, B.K.; Chowdhury, T.R.; Samanta, G.; Mandal, B.K.; Basu, G.C.; Chanda, C.R.; Lodh, D.; Saha, K.C.; Mukherjee, S.K.; Roy, S. Groundwater arsenic contamination in Bangladesh and West Bengal, India. *Environ. Health Perspect.* **2000,** *108*, 393–397.

Clesceri, L.S. Standard methods for the examination of water and waste water. In: eds. Arnold, E., Greenbergy and Eaton, AD, APHA, AWWA, WEF. *Collection and Preservation of Samples and Metals* **1998,** pp.1–27.

Das, S.K.; Chakrapani, G.J. Assessment of trace metal toxicity in soils of Raniganj Coalfield, India. *Environ. Monit. Assess.* **2011,** *177*, 63–71.

Deviller, G.; Aliaume, C.; Nava, M.A.F.; Casellas, C.; Blancheton, J.P. High-rate algal pond treatment for water reuse in an integrated marine fish recirculating system: effect on water quality and sea bass growth. *Aquaculture* **2004,** *235*, 331–344.

El-Kassas, H.Y.; Mohamed, L.A. Bioremediation of the textile waste effluent by *Chlorella vulgaris*. *Egypt. J. Aquat. Res.* **2014,** *40*, 301–308.

El-Sheekh, M.M.; Farghl, A.A.; Galal, H.R.; Bayoumi, H.S. Bioremediation of different types of polluted water using microalgae. *Rend. Fis. Acc. Lincei.* **2016,** *27*, 401–410.

Ercal, N.; Gurer-Orhan, H.; Aykin-Burns, N. Toxic metals and oxidative stress part I: mechanisms involved in metal-induced oxidative damage. *Curr. Top. Med. Chem.* **2001,** *1*, 529–539.

Fairey, R.; Long, E.R.; Roberts, C.A.; Anderson, B.S.; Phillips, B.M.; Hunt, J.W.; Puckett, H.R.; Wilson, C.J. An evaluation of methods for calculating mean sediment quality guideline quotients as indicators of contamination and acute toxicity to amphipods by chemical mixtures. *Environ. Toxicol. Chem.* **2001,** *20,* 2276–2286.

Hawks, J.S. Heavy metals. *J. Chem. Educ.* **1997,** *74,* 1374.

Heiri, O.; Lotter, A.F.; Lemcke, G. Loss on ignition as a method for estimating organic and carbonate content in sediments: reproducibility and comparability of results. *J. Paleol.* **2001,** *25,* 101–110.

Hunt, K.M.; Srivastava, R.K.; Elmets, C.A.; Athar, M. The mechanistic basis of arsenicosis: pathogenesis of skin cancer. *Cancer Lett.* **2014,** *354,* 211–219.

Iqbal, A.; Maiti, D.; Sundararajan, M.; Singh, S.K. Utilization of algae for remediation of metals from wastewater. In: *Sustainable Technologies for Better Environment.* Bharti Publications: New Delhi, India, **2018**, pp 42–52.

Jadia, C.D.; Fulekar, M.H. Phytoremediation of heavy metals: recent techniques. *Afr. J. Biotechnol.* **2009,** *8,* 921–928.

Jaishankar, M.; Tseten, T.; Anbalagan, N.; Mathew, B.B.; Beeregowda, K.N. Toxicity, mechanism and health effects of some heavy metals. *Interdiscip. Toxicol.* **2014,** *7,* 60–72.

Kabata-Pendias, A.; Sadurski, W. *Elements and Their Compounds in the Environment.* Wiley-VCH: Weinheim, Germany, **2004**.

Kaplan, O.; Yildirim, N.C.; Yildirim, N.; Cimen, M. Toxic elements in animal products and environmental health. *Asian J. Anim. Vet. Adv.* **2011**, *6,* 228–232.

Kidgell, J.T.; de Nys, R.; Hu, Y.; Paul, N.A.; Roberts, D.A. Bioremediation of a complex industrial effluent by biosorbents derived from freshwater macroalgae. *PLoS One* **2014,** *9,* e94706.

Kim, S.B.; Lee, S.J.; Kim, C.K.; Kwon, G.S.; Yoon, B.D.; Hee-Mock, O.H. Selection of microalgae for advanced treatment of swine wastewater and optimization of treatment condition. *Korean J. Appl. Microbiol. Biotechnol.* **1998,** *26,* 76–82.

Kloke, A.; Sauerbeck, D.R.; Vetter, H. The contamination of plants and soils with heavy metals and the transport of metals in terrestrial food chains. In: *Changing Metal Cycles and Human Health,* Springer: Berlin, Heidelberg, **1984**, pp. 113–141.

Kumar, P.N.; Dushenkov, V.; Motto, H.; Raskin, I. Phytoextraction: the use of plants to remove heavy metals from soils. *Environ Sci.Technol.* **1995,** *29,* 1232–1238.

Kumari, A.; Pandey, V.C.; Rai, U.N. Feasibility of fern *Thelypteris dentata* for revegetation of coal fly ash landfills. *J. Geochem. Explor.* **2013,** *128,* 147–152.

Lokeshwari, H.; Chandrappa, G.T. Heavy metals content in water, water hyacinth and sediments of Lalbagh Tank, Bangalore (India). *J. Environ. Sci. Eng.* **2006,** *48,* 183–188.

Lone, M.I.; He, Z.L.; Stoffella, P.J.; Yang, X.E. Phytoremediation of heavy metal polluted soils and water: progresses and perspectives. *J. Zhejiang Univ. Sci. A* **2008,** *9,* 210–220.

Mahamadi, C.; Nharingo, T. Utilization of water hyacinth weed (*Eichhornia crassipes*) for the removal of Pb (II), Cd (II) and Zn (II) from aquatic environments: an adsorption isotherm study. *Environ. Technol.* **2010,** *31,* 1221–1228.

Maiti, D.; Prasad, B. Revegetation of fly ash–a review with emphasis on grass-legume plantation and bioaccumulation of metals. *Appl. Ecol. Environ. Res.* **2016,** *14,* 185–212.

Martin, S.; Griswold, W. Human health effects of heavy metals. *Environ. Sci. Technol. Briefs Citizens* **2009,** *15,* 1–6.

Mata, Y.N.; Blázquez, M.L.; Ballester, A.; González, F.; Munoz, J.A. Biosorption of cadmium, lead and copper with calcium alginate xerogels and immobilized *Fucus vesiculosus. J. Hazard. Mater.* **2009,** *163,* 555–562.

Mazumder, D.G. Chronic arsenic toxicity & human health. *Indian J. Med. Res.* **2008,** *128*, 436–447.
Mazumder, D.G.; Dasgupta, U.B. Chronic arsenic toxicity: studies in West Bengal, India. *Kaohsiung J. Med. Sci.* **2011,** *27*, 360–370.
McLean, E.O. Soil pH and lime requirement. *Methods of Soil Analysis: Part 2 Chemical and Microbiological Properties*, **1983,** *9*, 199–224.
Mishra, V.K.; Tripathi, B.D. Accumulation of chromium and zinc from aqueous solutions using water hyacinth (*Eichhornia crassipes*). *J. Hazard. Mater.* **2009,** *164*,1059–1063.
Mishra, V.K.; Upadhyay, A.R.; Pathak, V.; Tripathi, B.D. Phytoremediation of mercury and arsenic from tropical opencast coalmine effluent through naturally occurring aquatic macrophytes. *Water Air Soil Pollut.* **2008,** *192*, 303–314.
Morillo, J.; Usero, J.; Gracia, I. Partitioning of metals in sediments from the Odiel River (Spain). *Environ. Int.* **2002,** *28*, 263–271.
Mudhoo, A.; Garg, V.K.; Wang, S. Removal of heavy metals by biosorption. *Environ. Chem. Lett.* **2012,** *10*, 109–117.
Nagajyoti, P.C.; Lee, K.D.; Sreekanth, T.V.M. Heavy metals, occurrence and toxicity for plants: a review. *Environ. Chem. Lett.* **2010,** *8*, 199–216.
Neori, A.; Msuya, F.E.; Shauli, L.; Schuenhoff, A.; Kopel, F.; Shpigel, M. A novel three-stage seaweed (Ulva lactuca) biofilter design for integrated mariculture. *J. Appl. Phycol.* **2003,** *15*, 543–553.
Oswald, W.; Gotass, H. Photosynthesis in sewage treatment. *Trans. Am. Soc. Civil Eng.* **1957,** *122*, 73–105.
Pena-Castro, J.M.; Martınez-Jerónimo, F.; Esparza-Garcıa, F.; Canizares-Villanueva, R.O. Heavy metals removal by the microalga *Scenedesmus incrassatulus* in continuous cultures. *Bioresour. Technol.* **2004,** *94*, 219–222.
Prasad, B.; Maiti, D. Comparative study of metal uptake by *Eichhornia crassipes* growing in ponds from mining and nonmining areas—a field study. *Bioremediat. J.* **2016**, *20*, 144–152.
Putra, R.S.; Novarita, D.; Cahyana, F. Remediation of lead (Pb) and copper (Cu) using water hyacinth [*Eichornia crassipes* (Mart.) Solms] with electro-assisted phytoremediation (EAPR). In: *AIP Conference Proceedings*, AIP Publishing LLC, **2016**, 1744, pp. 020052.
Rai, P.K. Heavy metal pollution in aquatic ecosystems and its phytoremediation using wetland plants: an ecosustainable approach. *Int. J. Phytoremediat.***2008,** *10*,133–160.
Rai, P.K. Heavy metals in water, sediments and wetland plants in an aquatic ecosystem of tropical industrial region, India. *Environ. Monit. Assess.* **2009,** *158*, 433–457.
Rai, P.K. Mercury pollution from a chloralkali source in a tropical Lake and its biomagnification in aquatic biota: link between chemical pollution, biomarkers, and human health concern. *Hum. Ecol. Risk Assess.* **2008,** *14*, 1318–1329.
Rai, P.K. Phytoremediation of Hg and Cd from industrial effluents using an aquatic free floating macrophyte *Azolla pinnata*. *Int. J. Phytoremediat.* **2008,** *10*, 430–439.
Rajaganapathy, V.; Xavier, F.; Sreekumar, D.; Mandal, P.K. Heavy metal contamination in soil, water and fodder and their presence in livestock and products: a review. *J. Environ. Sci. Technol.* **2001,** *4*, 234–249.
Rashed, M.N. Monitoring of environmental heavy metals in fish from Nasser Lake. *Environ. Int.***2001,** *27*, 27–33.
Raskin, I.; Kumar, P.N.; Dushenkov, S.; Salt, D.E. Bioconcentration of heavy metals by plants. *Curr. Opin. Biotechnol.* **1994,** *5*, 285–290.
Ratnaike, R.N. Acute and chronic arsenic toxicity. *Postgrad. Med. J.* **2003,** *79*, 391–396.

Renuka, N.; Sood, A.; Ratha, S.K.; Prasanna, R.; Ahluwalia, A.S. Nutrient sequestration, biomass production by microalgae and phytoremediation of sewage water. *Int. J. Phytoremediat.* **2013,** *15*, 789–800.

Rezania, S.; Ponraj, M.; Talaiekhozani, A.; Mohamad, S.E.; Din, M.F.M.; Taib, S.M.; Sabbagh, F.; Sairan, F.M. Perspectives of phytoremediation using water hyacinth for removal of heavy metals, organic and inorganic pollutants in wastewater. *J. Environ. Manage.* **2015,** *163*, 125–133.

Romera, E.; González, F.; Ballester, A.; Blázquez, M.L.; Munoz, J.A. Comparative study of biosorption of heavy metals using different types of algae. *Bioresour. Technol.* **2007,** *98*, 3344–3353.

Saha, P.; Shinde, O.; Sarkar, S. Phytoremediation of industrial mines wastewater using water hyacinth. *Int. J. Phytoremediat.* **2017,** *19*, 87–96.

Salgueiro, J.L.; Perez, L.; Maceiras, R.; Sanchez, A.; Cancela, A. Bioremediation of wastewater using *Chlorella vulgaris* microalgae: phosphorus and organic matter. *Int. J. Environ. Res.* **2016,** *10*, 465–470.

Salt, D.E.; Blaylock, M.; Kumar, N.P.; Dushenkov, V.; Ensley, B.D.; Chet, I.; Raskin, I. Phytoremediation: a novel strategy for the removal of toxic metals from the environment using plants. *Nat. Biotechnol.* **1995,** *13*, 468–474.

Shah, R.A.; Kumawat, D.M.; Singh, N.; Wani, K.A. Water hyacinth (*Eichhornia crassipes*) as a remediation tool for dye effluent pollution. *Int. J. Sci. Nat.* **2010,** *1*, 172–178.

Shuvaeva, O.V.; Belchenko, L.A.; Romanova, T.E. Studies on cadmium accumulation by some selected floating macrophytes. *Int. J. Phytoremediat.* **2013,** *15*, 979–990.

Smolyakov, B.S. Uptake of Zn, Cu, Pb, and Cd by water hyacinth in the initial stage of water system remediation. *Appl.Geochem.* **2012,** *27*, 1214–1219.

Srivastava, J.; Gupta, A.; Chandra, H. Managing water quality with aquatic macrophytes. *Rev. Environ. Sci. Biotechnol.* **2008,** *7*, 255–266.

Standard, I. *Bureau of Indian Standards Drinking Water Specifications IS 10500: 1991*. New Delhi, India, **1991**.

Sundararajan, M.; Loveson, V.J. An approach towards the development of environmental quality index for evaluation and categorization of environmental impacts. *J. Sci. Ind. Res.* **2002,** 61,219–223.

Swarup, D.; Dwivedi, S.K. Research on effects of pollution in livestock. *Indian J. Animal Sci.* **1998,** 68, 814–824.

Ting, Y.P.; Teo, W.K.; Soh, C.Y. Gold uptake by *Chlorella vulgaris*. *J. Appl. Phycol.* **1995,** *7*, 97–100.

Travieso, L.; Benítez, F.; Sánchez, E.; Borja, R.; León, M.; Raposo, F.; Rincón, B. Assessment of a microalgae pond for post-treatment of the effluent from an anaerobic fixed bed reactor treating distillery wastewater. *Environ. Technol.* **2008,** *29*, 985–992.

Treatment, L. *Water Treatment*, Lenntech Water Treatment and Air Purification Rotterdam Sewage, Netherlands, **2004**.

Weis, J.S.; Weis, P. Metal uptake, transport and release by wetland plants: implications for phytoremediation and restoration. *Environ. Int.* **2004,** *30*, 685–700.

Zhou, J.L.; Huang, P.L.; Lin, R.G. Sorption and desorption of Cu and Cd by macroalgae and microalgae. *Environ. Pollut.* **1998,** *101*, 67–75.

CHAPTER 4

Usage of Engineered Virus-Like Particles in Drug Delivery

SUSHIL KUMAR SAHU[1*], RAMAKANTA RANA[2], and ASHOK KUMAR MALLIK[3]

[1]*Department of Pharmacology & Molecular Sciences, School of Medicine, Johns Hopkins University, Baltimore, MD 21205, USA*

[2]*Regional Medical Research Centre, Chandrasekharpur, Bhubaneswar, Odisha 751023, India*

[3]*Centre for Ecological Sciences, Indian Institute of Science, Bangalore, Karnataka 560012, India*

[*]*Corresponding author. E-mail: sahu.sushil@gmail.com*

ABSTRACT

Virus-like particles (VLPs) are assembled from viral proteins. These particles resemble the structure of native virion but are devoid of genetic materials. VLPs are generated by expressing the viral proteins in the laboratory strain of prokaryotic cells, eukaryotic cells, or in cell-free conditions followed by their assembly. Also, they can be constructed from proteins of different sources and are called chimeric VLPs. VLPs maintain their tropism and immunogenicity, unlike native viruses which attract investigators in the field of drug delivery, vaccine trial, and gene-targeted therapeutic approach. However, this chapter focuses mostly on the use of VLPs as a device in the perspective of drug delivery.

4.1 INTRODUCTION

A virus consists of either DNA or RNA as genetic material, which is packaged inside a protein structure known as a capsid. Some viruses possess an envelope that is made up of lipid membranes and multiple viral proteins. The envelope encloses the nucleocapsid (nucleic acid plus capsid). The capsid is made up of several identical subunits of a peptide called capsomere. Capsomeres have the capacity to combine together without any viral genetic materials in an in vitro condition, driven by intra- and inter-molecular interactions. This forms a shell that is empty in the center and is designated as a virus-like particle (VLP). VLPs are highly stable, nonreplicating, and are very small in size (20–100 nM in diameter). Synthesis and purification of VLPs are performed in the biosafety level 3 (BSL 3) or BSL 4 facility like native/wild strain of viruses. This is usually done in the following three different ways: (1) express all structural proteins of VLPs encoded in a single plasmid by transfection into host cells (e.g., 293T cell line) or (2) coexpress all structural proteins of VLPs by cotransfection of the required number of plasmid by transfection into host cells, or (3) infect recombinant baculoviruses that encode all structural proteins of VLPs into insect cells (e.g., Sf9 cell line). Then, the cell culture media containing VLPs are harvested, cell debris is removed by passing through a filter (e.g., 0.45-micron filter). Finally, VLPs are pellet down by ultracentrifugation on a 20% sucrose cushion. VLPs generated in this way maintain their tropism and immunogenicity, unlike native viruses. These properties have made them attractive to be used as tools for drug delivery (Jalaguier et al., 2011; Zeltins, 2013). VLPs have been successfully produced in vitro from proteins of numerous viruses including but not limited to adeno-associated virus (AAV), Hepatitis B virus (HBV), and human papillomavirus (HPV) (Zeltins, 2013). AAV-VLPs were constructed in which ovalbumin derived b-cell epitope was expressed as a fusion protein with the VP3 protein (a capsid protein out of three capsid protein VP1, VP2, and VP3) of AAV2. These were successfully delivered into a mouse model (Manzano-Szalai et al., 2014). HBV-VLPs were generated by altering the gene sequence that encodes for the viral capsid. This change made it successful to skip the attack of host immunity (Lu et al., 2015). HPV-VLPs were generated by using the L1 protein (major capsid protein) of HPV-16. These VLPs can package and deliver foreign plasmid DNA to cells of eukaryotes (Touze and Coursaget, 1998).

4.2 PRODUCTION OF VLPS

For the production of VLPs, the following properties are taken into account:

1. Suitable vector for cloning of desired viral genes for production of constituent proteins of VLPs.

2. Appropriate expression systems like prokaryote (e.g., bacteria strain), eukaryotic cells (e.g., HEK 293T human cell lines, sf9 insect cell lines), or yeast for production of desired viral genes.
3. Bond between capsomere which are bound to one another by covalent (e.g., disulfide bond) or noncovalent (e.g., hydrogen bond, van der Waals forces) interactions. Hence, for the production of VLPs, it is needed to express all viral proteins required for capsomere formation followed by their assembly.

VLPs can be targeted into a particular site depending on the type of VLPs (Zeltins, 2013; Jalaguier et al., 2011; Zdanowicz and Chroboczek, 2016; Kushnir et al., 2012; Lua et al., 2014).

4.2.1 EXPRESSION SYSTEM FOR VIRAL PROTEINS

There are various cell types used for viral protein expression among which laboratory strains of bacteria, insect cell lines, and yeast are preferred. The structural protein component of John Cunningham virus (JCV), avian polyomavirus (APV), murine polyomavirus, and monkey B-Lymphatic papovavirus has been generated in insect cells by using a baculovirus expression vector (Gillock et al., 1998; Chang et al., 1997; Sandalon and Oppenheim, 1997; Pawlita et al., 1996). For example, VP1 (major capsid protein) of JCV or APV was cloned into an appropriate plasmid and expressed in yeast cells, and then purified from yeast cell lysates (Chen et al., 2001; Palkova et al., 2000). Also, VP1 of JCV was expressed in insect cells using a recombinant baculovirus system (Goldmann et al., 1999).

4.2.2 PURIFICATION OF VIRAL PROTEINS

It is needed to check the expression of viral proteins for which VLPs are produce. For purification of viral proteins, lyse the host cells expressing them and follow the appropriate purification methods as below. Cell lysate is usually filtered and purified by using CsCl-density gradient centrifugation followed by affinity column chromatography. Desired protein can be express as a glutathione S-transferases fusion protein and is purified using an immobilized glutathione column (Smith and Johnson, 1988). Similarly, histidine-fusion proteins are purified in a Nickel column. The fusion proteins cleaved from the carrier by digesting with site-specific proteases (e.g., thrombin). Further purification is achieved by passing the proteins through the phospho cellulose column (Leavitt et al., 1985) or cationic or anionic exchange column and dialysis. Purity of the proteins is checked by SDS-PAGE. Proteins are stored at −80 °C in presence of salt (NaCl or KCl) and about 10% glycerol until further use (Gillock et al., 1998).

4.2.3 ASSEMBLY OF VIRAL PROTEINS

Viral proteins assemble together by covalent and noncovalent interactions. For example, VP1 monomers are assembled together to form pentamer, and pentamers are assembled together to give rise to the capsid structure of the virus (Salunke et al., 1986; Braun et al., 1999). Disulfide bonds stabilize the capsid of VLP and are needed for virus assembly (Schmidt et al., 2000). Deletion of around 50 amino acids at the C-terminus of the VP1 cannot block the formation of capsomeres, but these pentamers are unable to associate and form a capsid (Garcea et al., 1987).

4.2.4 LOADING OF VLPS

VLPs are considered as useful biological tools to transfer drugs and desired genetic materials into target cells. Polyoma-VLP is composed of empty capsids and modified polyomavirus DNA (Barr et al., 1979). The DNA uptake happened by an osmotic shock. First, a complex was formed between the empty capsids and the DNA. Then, incubated in the water bath at 37 °C to load the capsids (Ou et al., 2001; Braun et al., 1999). The integrated DNA was protected from DNase degradation (Bertling et al., 1991). This VLP is composed of VP1 and has the potential to incorporate DNA fragments up to 3 kbp (Soeda et al., 1998). Linear, circular, supercoiled, single-stranded DNA, double-stranded DNA, rRNA, and synthetic polymers can be encapsulated into VLPs (Henke et al., 2000). The DNA is complexed with polylysine and is encapsulated in VLPs to protect the DNA against degradation (Soeda et al., 1998). The major structural protein VP1 of the Aleutian disease virus bind to DNA in contrast to minor structural proteins VP2 and VP3 (Willwand and Kaaden, 1988). It has been reported that VP1 of Simian virus 40 (SV40) binds to the DNA tightly, but the affinity of binding is stronger for single-stranded DNA than for double-stranded DNA (Soussi, 1986). VLPs consisting of VP1 (VP1-VLP) are more efficient in transporting heterologous DNA into host cells or tissues than DNA on its own (Slilaty and Aposhian, 1983). Using VLPs loaded with desired DNA, a long-term and higher level expression was achieved (Krauzewicz et al., 2000). Murine polyomavirus VP1-VLPs showed an enhanced immune response in the mice model. After VP1-VLPs infection, antibodies were accumulated and cells were started to proliferate which may be helpful for vaccination (Clark et al., 2001). VP1-VLPs are shown to be useful tools to incorporate not only DNA but also small biological molecules and to improve the specific delivery of desired molecules to their target cells. Delivery of DNA was experimentally proven by staining DNA with propidium iodide, which intercalates into DNA and was detected under microscopy (Goldmann et al., 2000).

4.2.5 TARGETING OF VLPS

Like native viruses, VLPs are classified as nonenveloped or enveloped (Melnick et al., 1974; Ou et al., 2001). Nonenveloped VLPs are composed of one or many capsid proteins. For nonenveloped VLPs production, a bacterial system is preferred. The presence of bacterial endotoxins and the absence of post-translational modifications (PTMs) are disadvantages of bacterial systems. For avoiding these problems, eukaryotic systems were used (Goldmann et al., 1999). VLPs' proteins that need PTMs, usually produced in eukaryotic hosts like yeast and insects. Especially, insects are suitable systems for large scale production of VLPs. Moreover, insect systems perform PTMs of target proteins similar to mammalian cells, but this similarity is less in yeast systems (Goldmann et al., 1999). Mammalian cells are used to produce VLPs with specific properties. Enveloped VLPs obtained envelope from their host cell where they are generated. They are more complex in comparison to nonenveloped VLPs. Eukaryotic cells (i.e., virus packaging cells) are considered as a host for the production enveloped VLPs (Ou et al., 2001). VLPs were successfully used for transfection into mammalian cells (Smith and Johnson, 1988; Braun et al., 1999; Henke et al., 2000). During the transfection procedure, VLPs have a tendency to bind sialic acid residues present on most of the higher eukaryotes. For cell-specific targeting of this vector system, novel domains need to be discovered and established on the VP1-VLP surface. One approach is the insertion of Fv fragments (antigen-binding antibody fragments). But, this is limited to the viral coat protein modification and development of single-chain Fv fragments. Another approach is the insertion of an immunoglobulin binding domain, isolated from *Staphylococcus aureus*, the protein Z (Gleiter and Lilie, 2001; May et al., 2002). This allows working with a modified viral coat protein in multiple applications.

4.3 MODIFICATIONS OF VLPS

Modification in VLPs to gain extra functionalities is possible by applying a genetic or chemical approach. Peptide, epitopes, and drugs can be introduced on VLPs surface by genetic engineering to use VLPs as carriers. Disassembly and reassembly feature VLPs provide an opportunity for encapsulation of desired components like genomic material, peptide fragments, and drugs. These are advantageous characteristics of VLPs over the synthetic particles developed earlier (Yildiz et al., 2011). Modification on the outer and inner surface of VLP by chemical approach brings changes in physiological properties and provide a novel function (Smith and Johnson, 1988). For example, VLPs with modification in the inner surface were done by six histidines tag (e.g., HBc144-VLPs) which developed an affinity to bind nickel-nitrilotriacetic acid (Ni-NTA). In terms of cellular uptake of VLPs (i.e., Fe_3O_4–NTA–Ni2+ core having HBc-144-His VLPs) are more efficient than Fe_3O_4 nanoparticles (Shen et al., 2015).

4.4 VLPS AS DELIVERY SYSTEMS

Delivery of macromolecules (e.g., nucleic acids, proteins, drugs, siRNA, miRNA, and synthetic liposomes) into the living cells is possible in many ways. However, there is no method that is universal and completely perfect. Each method has its own limitations such as inefficient uptake, endosomal entrapment, restricted delivery to certain cell types, undesirable immunological properties, and damage to cells in the delivery process (Zeltins, 2013; Kaczmarczyk et al., 2011). Investigators keep on working to find new/updated methods for efficient delivery. Delivery of macromolecules by using VLPs as carriers is one promising way among them. The properties of viral capsid allowed to generate engineered VLPs which encapsulate macromolecules for delivery into target cells, tissue, and organ (Lua et al., 2014).

VLPs possess the following characteristics which make them an efficient tool for delivery systems: (1) VLPs contain wild type or engineered functional proteins that help in the attachment and penetration into host cells; (2) VLPs are safe because they do not contain any genetic material, which is preset in the native virion. Therefore, no genetic material integrate into the host genome for maintenance or replication (Zdanowicz and Chroboczek, 2016; Kaczmarczyk et al., 2011); (3) macromolecule of interest for delivery into host cells can be packed inside VLPs or attached to the inner or outer surface of VLPs by chemical conjugation (Zdanowicz and Chroboczek, 2016).

If target molecules for VLPs are DNA and RNA, it can be examined by polymerase chain reaction (PCR), real time-PCR, and EMSA. If target molecules are protein, effect of target protein on the target cell will verify encapsulation. If target molecules are ligand, VLPs can deliver it on their surfaces to a target cell. Antigens and ligands are fuses to VLPs through either covalent (e.g., bond between cysteine and lysine) or noncovalent bonds (e.g., creating an attachment between streptavidin as linkers and biotinylated antigens) (Zdanowicz and Chroboczek, 2016; Kaczmarczyk et al., 2011). To confirm that ligand binds to VLP, electron microscopy and dynamic light scattering are used. The size difference between VLPs before binding and after binding will approve binding of ligands to VLPs (Thrane et al., 2015). To ensure the transmission of the VLPs into the target cell, GFP gene cassette is used as controls. GFP fluorescence is detected by fluorescence of confocal microscopy. JCV genome contains the early region and late region. The late region encodes capsid proteins VP1, VP2, and VP3. For safe delivery of siRNA, JCV-VLPs reduced the expression of receptor activator for NF-kB ligand in osteoblast cells of rats (Hoffmann et al., 2016). Cell-penetrating peptides (CPPs) are able to transport different molecules such as peptides and oligonucleotides into specific cells, but these peptides have restrictions. It is sensitive to proteases and also cannot make a high effective crosslink with target molecules. For the elimination of these limitations, a study used bacteriophage PP7-VLPs as a delivery system to protect CPPs and mRNA from degradation. They insert low molecular weight protamine DNA as the model CPP into the cDNA of PP7 coat protein and also encapsulate GFP mRNA as a foreign mRNA into chimeric PP7 VLPs. Examinations showed CPPs were successfully presented on the surface of PP7-VLPs and also maintain the cell-penetrating function. As a result, it was demonstrated that chimeric PP7-VLPs are a safe and efficient for peptide and RNA delivery (Sun et al., 2016). Bacteriophage-MS2 VLPs possess many properties to be used as a delivery system. They can be obtained by the method of recombinant-protein technology. Also, the MS2 capsid binds and encapsulates the target, that is, pac site. This inhibits the degradation of the target DNA or RNA by nucleases. Epitopes can be delivered by MS2 VLPs. MS2 CP gene has a unique site for insertion of DNA and the peptide generated from here (i.e., epitope) is presented to the immune system. In addition, VLPs display viral antigens at the surface capsid as epitope which has a higher immunologic response. Major advantages of MS2-VLPs are due to their appropriate size and stability for presenting viral epitopes. Moreover, MS2 VLPs can be used as carriers for targeted drug delivery. Drugs can be packaged into the MS2 CP during the step of self-assembly (Fu and Li, 2016). The efficiency of this strategy depends on the structure, type of drug that is loaded, and the environment of the targeted nanoparticle (Molino and Wang, 2014). Targeted drug delivery is an attractive strategy for the treatment of solid tumors. Multiple VLPs were developed depending on the properties of the drugs. For example, MS2 bacteriophage-VLP for 5-fluorouracil and doxorubicin, Murine polyomavirus-VLP for methotrexate, and JC polyomavirus-VLP for paclitaxel delivery were reported by investigators (Abbing et al., 2004; Ashley et al., 2011; Niikura et al., 2013). In the previous study, it was reported that modified adenovirus can be strongly targeted to cancer tissue and has a less toxic effect on normal cells (Shan et al., 2012). In a further study, paclitaxel (an anticancer drug) was conjugated to adenovirus nanoparticles to form prodrugs. The targeting of paclitaxel in cancer site and the residence time was improved by using a paclitaxel-conjugated vector. In vitro and in vivo studies reported that coxsackie adenovirus receptor-mediated uptake of paclitaxel showed high anticancer action. Chemically modified adenovirus loaded with a drug can be used as a delivery system for targeting cancer site (Shan et al., 2012). During chemotherapy, prodrugs are used in combination with enzymes. This results in the conversion of the inactive prodrugs into their active form. In a study, 5-Fluorocytosine was used as a prodrug. Fcy protein converts 5-fluorocytosine into 5-fluorouracil. Then, 5-FU is converted to 5-F UMP by Fur protein and blocks DNA synthesis. This method depends upon the expression and act of the enzymes required to convert the prodrug into its active form. This issue was solved by VLPs that used Gag–Fcy–Fur fusion gene, which could convert prodrug into its active form inside the host cells (Kaczmarczyk et al., 2011). Bleomycin is used for the treatment of many types of cancer, such as lymphoma, testicular cancer, penile cancer, etc. Adenovirus VLP was used for its enhanced delivery. Doxorubicin is another anticancerous drug for the treatment of lymphoma, leukemia, and breast cancer can be encapsulated into different VLPs. For instance, doxorubicin encapsulated cucumber mosaic VLP was targeted to folate-expressing cancer cells. The results showed an increase in antitumor responses and a decrease in toxicity to normal cells as compared to the free drug (Zeng et al., 2013). Another strategy for cancer is to silence the target gene expression by using RNAi. But, delivering RNAi sequences in vivo is not easy. To solve this issue, JC virus VLPs was tested to deliver RNAi against IL-10 which resulted in the reduction of IL-10 expression by >80%, when compared with only VLPs (Chou et al., 2010). In an in vitro study, it was reported that VLPs were assembled with the RNA bacteriophage MS2 coat protein, an RNA conjugate having a siRNA sequence and a capsid assembly signal. This can be targeted into HeLa cell lines by protecting from nuclease (Galaway and Stockley, 2013).

4.5 CONCLUSIONS

A large number of studies explored VLPs as an excellent tool for vaccine research, drug delivery, and multiple biological applications because of their intrinsic properties, such as particulate structure, presence of multimeric antigens, and tropism to the specific target. Here, we discussed the basic research and application area of VLPs mostly in the area of drug delivery. The VLPs are very efficient in cell entry, avoid endosomal sequestering, possess multivalency, and are biocompatible. A variety of molecules can be attached to the membrane of VLPs or encapsulated inside it for delivery into the host cells. VLPs constructed without genetic material is safe for delivery and cause single round infection which aid risk-free condition both for laboratory and clinical studies. VLPs are useful for in vivo applications with the least immune response. The features like self-assembly and disassembly of VLPs play a key point to design the particle with desired modifications and encapsulation for extra immunogenicity and function in comparison to native virions. Especially, when VLPs are delivered to a therapeutic site, the drugs inside VLPs are better effective than the free drug. Recombinant VLPs can be formed to gain extra functionalities to present synthetic small molecules, disease-associated biomolecules on the target site. VLPs mediated drug delivery opens a new avenue to overcome the toxicity effect of the chemotherapeutic drug on normal cells and treatment of drug-resistant solid tumor.

KEYWORDS

- **virus**
- **VLPs**
- **capsid**
- **tropism**
- **immunogenicity**
- **vaccine**
- **gene therapy**
- **drug delivery**

REFERENCES

Abbing, A.; Blaschke, U. K.; Grein, S.; Kretschmar, M.; Stark, C. M.; Thies, M. J.; Walter, J.; Weigand, M.; Woith, D. C.; Hess, J.; Reiser, C. O. Efficient intracellular delivery of a protein and a low molecular weight substance via recombinant polyomavirus-like particles. *J. Biol. Chem.* **2004,** 279, 27410–27421.

Ashley, C. E.; Carnes, E. C.; Phillips, G. K.; Durfee, P. N.; Buley, M. D.; Lino, C. A.; Padilla, D. P.; Phillips, B.; Carter, M. B.; Willman, C. L.; Brinker, C. J.; Caldeira Jdo, C.; Chackerian, B.; Wharton, W.; Peabody, D. S. Cell-specific delivery of diverse cargos by bacteriophage MS2 virus-like particles. *ACS Nano.* **2011,** 5, 5729–5745.

Barr, S. M.; Keck, K.; Aposhian, H. V. Cell-free assembly of a polyoma-like particle from empty capsids and DNA. *Virology* **1979,** 96, 656–659.

Bertling, W. M.; Gareis, M.; Paspaleeva, V.; Zimmer, A.; Kreuter, J.; Nurnberg, E.; Harrer, P. Use of liposomes, viral capsids, and nanoparticles as DNA carriers. *Biotechnol. Appl. Biochem.* **1991,** 13, 390–405.

Braun, H.; Boller, K.; Lower, J.; Bertling, W. M.; Zimmer, A. Oligonucleotide and plasmid DNA packaging into polyoma VP1 virus-like particles expressed in *Escherichia coli*. *Biotechnol. Appl. Biochem.* **1999,** 29, 31–43.

Chang, D.; Fung, C. Y.; Ou, W. C.; Chao, P. C.; Li, S. Y.; Wang, M.; Huang, Y. L.; Tzeng, T. Y.; Tsai, R. T. Self-assembly of the JC virus major capsid protein, VP1, expressed in insect cells. *J. Gen. Virol.* **1997,** 78, 1435–1439.

Chen, P. L.; Wang, M.; Ou, W. C.; Lii, C. K.; Chen, L. S.; Chang, D. Disulfide bonds stabilize JC virus capsid-like structure by protecting calcium ions from chelation. *FEBS Lett.* **2001,** 500, 109–113.

Chou, M. I.; Hsieh, Y. F.; Wang, M.; Chang, J. T.; Chang, D.; Zouali, M.; Tsay, G. J. In vitro and in vivo targeted delivery of IL-10 interfering RNA by JC virus-like particles. *J. Biomed. Sci.* **2010,** 17, 51.

Clark, B.; Caparros-Wanderley, W.; Musselwhite, G.; Kotecha, M.; Griffin, B. E. Immunity against both polyomavirus VP1 and a transgene product induced following intranasal delivery of VP1 pseudocapsid-DNA complexes. *J. Gen. Virol.* **2001,** 82, 2791–2797.

Fu, Y.; Li, J. A novel delivery platform based on Bacteriophage MS2 virus-like particles. *Virus Res.* **2016,** 211, 9–16.

Galaway, F. A.; Stockley, P. G. MS2 viruslike particles: a robust, semisynthetic targeted drug delivery platform. *Mol. Pharm.* **2013,** 10, 59–68.

Garcea, R. L.; Salunke, D. M.; Caspar, D. L. Site-directed mutation affecting polyomavirus capsid self-assembly in vitro. *Nature* **1987,** 329, 86–87.

Gillock, E. T.; An, K.; Consigli, R. A. Truncation of the nuclear localization signal of polyomavirus VP1 results in a loss of DNA packaging when expressed in the baculovirus system. *Virus Res.* **1998,** 58, 149–160.

Gleiter, S.; Lilie, H. Coupling of antibodies via protein Z on modified polyoma virus-like particles. *Protein Sci.* **2001,** 10, 434–444.

Goldmann, C.; Petry, H.; Frye, S.; Ast, O.; Ebitsch, S.; Jentsch, K. D.; Kaup, F. J.; Weber, F.; Trebst, C.; Nisslein, T.; Hunsmann, G.; Weber, T.; Luke, W. Molecular cloning and expression of major structural protein VP1 of the human polyomavirus JC virus: formation of virus-like particles useful for immunological and therapeutic studies. *J. Virol.* **1999,** 73, 4465–4469.

Goldmann, C.; Stolte, N.; Nisslein, T.; Hunsmann, G.; Luke, W.; Petry, H. Packaging of small molecules into VP1-virus-like particles of the human polyomavirus JC virus. *J. Virol. Methods* **2000,** 90, 85–90.

Henke, S.; Rohmann, A.; Bertling, W. M.; Dingermann, T.; Zimmer, A. Enhanced in vitro oligonucleotide and plasmid DNA transport by VP1 virus-like particles. *Pharm. Res.* **2000,** 17, 1062–1070.

Hoffmann, D. B.; Boker, K. O.; Schneider, S.; Eckermann-Felkl, E.; Schuder, A.; Komrakova, M.; Sehmisch, S.; Gruber, J. In vivo siRNA delivery using JC virus-like particles decreases the expression of RANKL in Rats. *Mol. Ther. Nucleic Acids* **2016,** 5, e298.

Jalaguier, P.; Turcotte, K.; Danylo, A.; Cantin, R.; Tremblay, M. J. Efficient production of HIV-1 virus-like particles from a mammalian expression vector requires the N-terminal capsid domain. *PloS One* **2011,** 6, e28314.

Kaczmarczyk, S. J.; Sitaraman, K.; Young, H. A.; Hughes, S. H.; Chatterjee, D. K. Protein delivery using engineered virus-like particles. *Proc. Natl. Acad. Sci. USA* **2011,** 108, 16998–17003.

Krauzewicz, N.; Cox, C.; Soeda, E.; Clark, B.; Rayner, S.; Griffin, B. E. Sustained ex vivo and in vivo transfer of a reporter gene using polyoma virus pseudocapsids. *Gene Ther.* **2000,** 7, 1094–1102.

Kushnir, N.; Streatfield, S. J.; Yusibov, V. Virus-like particles as a highly efficient vaccine platform: diversity of targets and production systems and advances in clinical development. *Vaccine* **2012,** 31, 58–83.

Leavitt, A. D.; Roberts, T. M.; Garcea, R. L. Polyoma virus major capsid protein, VP1. Purification after high level expression in *Escherichia coli*. *J. Biol. Chem.* **1985,** 260, 12803–12809.

Lu, Y.; Chan, W.; Ko, B. Y.; Vanlang, C. C.; Swartz, J. R. Assessing sequence plasticity of a virus-like nanoparticle by evolution toward a versatile scaffold for vaccines and drug delivery. *Proc. Natl. Acad. Sci. USA* **2015,** 112, 12360–12365.

Lua, L. H.; Connors, N. K.; Sainsbury, F.; Chuan, Y. P.; Wibowo, N.; Middelberg, A. P. Bioengineering virus-like particles as vaccines. *Biotechnol. Bioeng.* **2014,** 111, 425–440.

Manzano-Szalai, K.; Thell, K.; Willensdorfer, A.; Weghofer, M.; Pfanzagl, B.; Singer, J.; Ritter, M.; Stremnitzer, C.; Flaschberger, I.; Michaelis, U.; Jensen-Jarolim, E. Adeno-associated virus-like particles as new carriers for B-cell vaccines: testing immunogenicity and safety in BALB/c mice. *Viral Immunol.* **2014,** 27, 438–448.

May, T.; Gleiter, S.; Lilie, H. Assessment of cell type specific gene transfer of polyoma virus like particles presenting a tumor specific antibody Fv fragment. *J. Virol. Methods* **2002,** 105, 147–157.

Melnick, J. L.; Allison, A. C.; Butel, J. S.; Eckhart, W.; Eddy, B. E.; Kit, S.; Levine, A. J.; Miles, J. A.; Pagano, J. S.; Sachs, L.; Vonka, V. Papovaviridae. *Intervirology* **1974,** 3, 106–120.

Molino, N. M.; Wang, S. W. Caged protein nanoparticles for drug delivery. *Curr. Opin. Biotechnol.* **2014,** 28, 75–82.

Niikura, K.; Sugimura, N.; Musashi, Y.; Mikuni, S.; Matsuo, Y.; Kobayashi, S.; Nagakawa, K.; Takahara, S.; Takeuchi, C.; Sawa, H.; Kinjo, M.; Ijiro, K. Virus-like particles with removable cyclodextrins enable glutathione-triggered drug release in cells. *Mol. Biosyst.* **2013,** 9, 501–507.

Ou, W. C.; Hseu, T. H.; Wang, M.; Chang, H.; Chang, D. Identification of a DNA encapsidation sequence for human polyomavirus pseudovirion formation. *J. Med. Virol.* **2001,** 64, 366–373.

Palkova, Z.; Adamec, T.; Liebl, D.; Stokrova, J.; Forstova, J. Production of polyomavirus structural protein VP1 in yeast cells and its interaction with cell structures. *FEBS Lett.* **2000,** 478, 281–289.

Pawlita, M.; Muller, M.; Oppenlander, M.; Zentgraf, H.; Herrmann, M. DNA encapsidation by viruslike particles assembled in insect cells from the major capsid protein VP1 of B-lymphotropic papovavirus. *J. Virol.* **1996,** 70, 7517–7526.

Salunke, D. M.; Caspar, D. L.; Garcea, R. L. Self-assembly of purified polyomavirus capsid protein VP1. *Cell* **1986,** 46, 895–904.

Sandalon, Z.; Oppenheim, A. Self-assembly and protein-protein interactions between the SV40 capsid proteins produced in insect cells. *Virology* **1997**, 237, 414–421.

Schmidt, U.; Rudolph, R.; Bohm, G. Mechanism of assembly of recombinant murine polyomavirus-like particles. *J. Virol.* **2000,** 74, 1658–1662.

Shan, L.; Cui, S.; Du, C.; Wan, S.; Qian, Z.; Achilefu, S.; Gu, Y. A paclitaxel-conjugated adenovirus vector for targeted drug delivery for tumor therapy. *Biomaterials* **2012,** 33, 146–162.

Shen, L.; Zhou, J.; Wang, Y.; Kang, N.; Ke, X.; Bi, S.; Ren, L. Efficient encapsulation of Fe(3) O(4) nanoparticles into genetically engineered hepatitis B core virus-like particles through a specific interaction for potential bioapplications. *Small* **2015,** 11, 1190–1196.

Slilaty, S. N.; Aposhian, H. V. Gene transfer by polyoma-like particles assembled in a cell-free system. *Science* **1983,** 220, 725–727.

Smith, D. B.; Johnson, K. S. Single-step purification of polypeptides expressed in *Escherichia coli* as fusions with glutathione S-transferase. *Gene* **1988,** 67, 31–40.

Soeda, E.; Krauzewicz, N.; Cox, C.; Stokrova, J.; Forstova, J.; Griffin, B. E. Enhancement by polylysine of transient, but not stable, expression of genes carried into cells by polyoma VP1 pseudocapsids. *Gene Ther.* **1998** 5, 1410–1419.

Soussi, T. DNA-binding properties of the major structural protein of simian virus 40. *J. Virol.* **1986,** 59, 740–742.

Sun, Y.; Zhao, R.; Gao, K. Intracellular delivery of messenger RNA by recombinant PP7 virus-like particles carrying low molecular weight protamine. *BMC Biotechnol.* **2016,** 16, 46.

Thrane, S.; Janitzek, C. M.; Agerbaek, M. O.; Ditlev, S. B.; Resende, M.; Nielsen, M. A.; Theander, T. G.; Salanti, A.; Sander, A. F. A novel virus-like particle based vaccine platform displaying the placental malaria antigen VAR2CSA. *PloS One* **2015,** 10, e0143071.

Touze, A.; Coursaget, P. In vitro gene transfer using human papillomavirus-like particles. *Nucleic Acids Res.* **1998,** 26, 1317–1323.

Willwand, K.; Kaaden, O. R. Capsid protein VP1 (p85) of Aleutian disease virus is a major DNA-binding protein. *Virology* **1988,** 166, 52–57.

Yildiz, I.; Shukla, S.; Steinmetz, N. F. Applications of viral nanoparticles in medicine. *Curr. Opin. Biotechnol.* **2011,** 22, 901–908.

Zdanowicz, M.; Chroboczek, J. Virus-like particles as drug delivery vectors. *Acta Biochim. Pol.* **2016,** 63, 469–473.

Zeltins, A. Construction and characterization of virus-like particles: a review. *Mol. Biotechnol.* **2013,** 53, 92–107.

Zeng, Q.; Wen, H.; Wen, Q.; Chen, X.; Wang, Y.; Xuan, W.; Liang, J.; Wan, S. Cucumber mosaic virus as drug delivery vehicle for doxorubicin. *Biomaterials* **2013,** 34, 4632–4642.

CHAPTER 5

Novel Microbial Compounds as a Boon in Health Management

SHUBHA RANI SHARMA[1*], RAJANI SHARMA[2], and DEBASISH KAR[3]

[1]*Department of Bio-Engineering, Birla Institute of Technology, Mesra, Ranchi 835215, India*

[2]*Department of Biotechnology, Amity University, Ranchi 834002, India*

[3]*Department of Biotechnology, M.S. Ramaiah University of Applied Sciences, Bangalore 560054, India*

[*]*Corresponding author. E-mail: srsharma@bitmesra.ac.in*

ABSTRACT

The variety of chemicals present in the various species of plants, animals, marine organisms, and microorganisms in nature serve as an attractive reservoir of new medically beneficial therapeutic candidates. Microorganisms can serve as potential natural sources of drugs for the management and prevention of diseases such as diabetes, anemia, cancer, etc. Apart from being used as antibacterial, antifungal, and antiviral, the secondary metabolites obtained from the microbes are now being exploited for medical applications also. Microbial natural products may possess the property of bioactivity which can be used for protection against diseases. The secondary metabolites produced by the microorganisms possess anticancerous, anti-tumor, anti-inflammatory, and antiviral activity. Not only proteins, but also the exopolysaccharides of pseudoplastic nature produced by the microorganisms also find exceptional application in the industry as a thickening, coagulating, adhesion, stabilizing, and gelling agent. They possess the unique property of the resistance against temperature, pH, and salinity. For the purpose of dental impression mold for tablets, etc., polysaccharides such alginates, are used which the products of microbial fermentation. With the advancement in the research in the scientific field, we have discovered a plethora of applications of the microbial product in the field of medical sciences. The microbial products have been used directly as drugs like antibiotics; immunomodulators, as biopolymers, etc., and still a bunch of unraveled applications is in the stage of innovation. In this chapter, we aim to throw some light on all the possible microbial biochemicals and their relevance in health management and application of microbial products.

5.1 INTRODUCTION

The presence of incredible biochemical diversity in various living entities like plants, animals, marine organisms, and microorganisms has proposed nature as an attractive resource of novel curative answers for different diseases. The natural legacies prove to be an excellent source of novel drugs, novel drug leads, as well as innovative biochemicals. The microorganisms have been exploited since time immemorial for the manufacture of an array of biomolecules including alcohol, antibiotics, and also in the disposition of food products. Microorganisms have proved to have immense potential to produce natural drugs to use for treatment and curing diseases such as atopic dermatitis, anemia, obesity, cancer, diabetes, Crohn's disease, diarrhea, etc. They are also the plausible origin for the production of naturally occurring antioxidants, enzyme inhibitors, immunosuppressants, hypocholesterolemic agents, etc. The secondary metabolites from the microbes have found a variety of applications including antibacterial, antifungal, and antiviral activity. Microalgae, bacteria, and yeast have been exploited for their bioactivity in the production of valuable food products, nutraceuticals, proteins, vitamins, organic acids, antibiotics, and enzymes. These microbes produce secondary metabolites for self-defense or against some adverse conditions so that the microbes are alive and healthy in spite of the harsh conditions. These secondary metabolites act as anticancerous, antitumor, anti-inflammatory, and antiviral agents. Apart from their above-mentioned anticancerous, antitumor activity some of the metabolites also exhibit immunosuppressing activity, for example, cyclosporin is of immense help in organ transplantation which otherwise would suffer from rejection in the body. With the discovery of the first antibiotic, penicillin soil microorganisms in about 1920, a plethora of microorganisms have been exploited for the production of certain products which have proved to be very beneficial for mankind. The marine microbes have been utilized extensively as a source of anti-HIV protein as well as cosmeceuticals with ultraviolet (UV) protectant activity as they sequester the free radicals produced by the UV radiations. The microorganisms are also a great source of EPS which find their application in the different industries where they can be used as agents for coagulation, gelling adhesion, etc. These EPS act as artificial plastics that are endowed with the property of resistance against temperature pH, etc., so are very versatile. This chapter aims to recapitulate the substances produced by the microbes and their application in health management.

The microbial products are most commonly exploited for the fields of medical science and agricultural sciences or as an initial substrate for microbial derivatization or as novel constituents in the formulation of drugs. Alginates produced by the microbes are being utilized as a material medium for the formation of tablets as well as for taking the dental impression for the creation of dentures. Hyaluronic acid, an immunoneutral polysaccharide, and its derivatives are very essential for the functions of various cells and tissues. When hyaluronic acid is modified chemically, it can be altered into different types of materials like viscoelastic solutions, soft or stiff hydrogels, nanoparticulate fluids, etc., which can be used in diverse biomedical applications like in surgery, treatment of arthritis, as well as healing of the wound. The cellulose obtained from microbes has been used in regular dressings of the wound as well as in the form of scaffolds to be utilized in tissue engineering. The biopolymers are being extensively used in various medical applications like the emergence of drug controlled-release systems which prove to be biodegradable and nontoxic for human beings. The various systems for drug release have been fabricated in such a way so that they are soluble and hence their transport easily in any part of the human body. The symbiotic alliance of the microbes in the different areas of our body exhibits a chief role in the enhancement of our health. The utilization of the natural fermentation yields from microbes like *Saccharomyces cerevisiae* is being used as supplements that can replace the antibiotics one day. One more example of the use of microbial products is the use of short-chain fatty acids in the gut of the poultry so as to improvise gut health. A novel method of collection of fecal matter from a healthy donor and transplantation of the same in a patient after mixing with saline solution is termed as fecal microbiota transplant (FMT) or fecal transplantation. This chapter is a recapitulation of all the various types of applications of the microbes and microbial products in enhancing the health of an organism including animals, human beings, etc. Let us deal with each, one by one, and ponder over the utilities of the microbial products which otherwise was previously considered as a source of ill health only. The microbial products have to be used in various fields, health science being the most prominent as shown in Figure 5.1. The word antibiotic has given a new platform as well as dimension to the world of medical science where it has produced a very great impact on the curing of various diseases. It all started with the great discovery by Alexander Fleming in 1928 when he discovered the first antibiotic penicillin which is commonly known as the wonder drug. Other microbial metabolites like cyclosporin A which acts as an immunosuppressant have gained importance in the field of medical science where they are used in organ transplantation procedures. There are innumerable applications of microbial metabolites in the health management area, for example, antidiabetic products, anticancer drugs, etc. The secondary metabolites produced by the microbes not only influence the microbial growth but also produce bioactive compounds which are very essential for cell metabolism. The antihelminthic activity of the avermectins from microbes is being commercialized (Jansson and Dybas, 1998). This was the result of a collaboration between the Campbell group of Merk Research Laboratories as well as the Kitasato Institute.

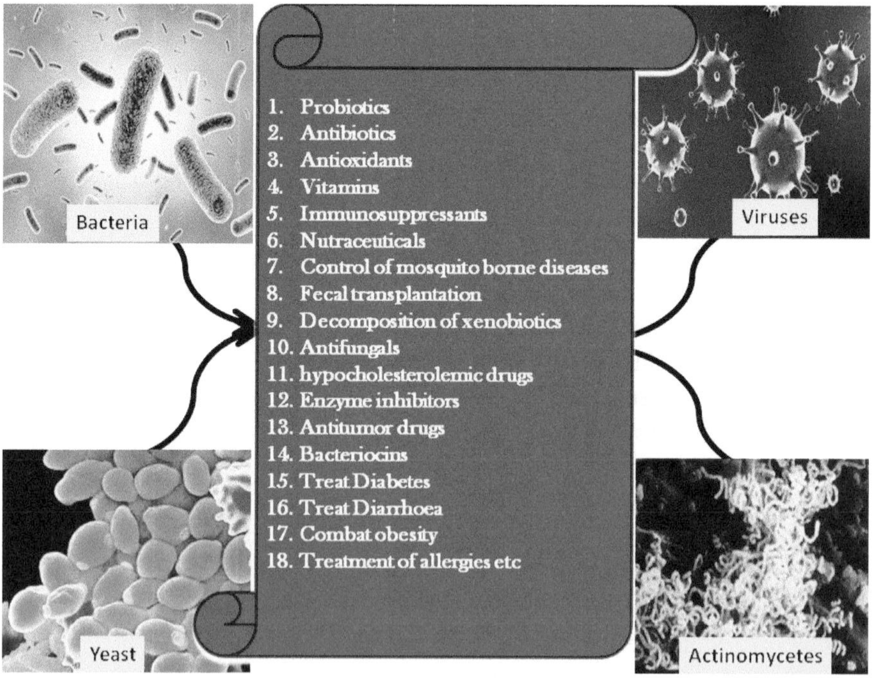

FIGURE 5.1 Significant microbial products in healthcare.

5.2 MICROBIAL PRODUCTS AS PROSPECTIVE SOURCES OF NATURAL ANTIOXIDANTS

Microbes are one of the most diverse and profuse species on earth and their exploitation to obtain different beneficial products have been found since time immemorial. Until the 1980s, the antioxidants from the microbes did not attract the attention of scientists in the medical area. The significance of antioxidants in the health improvement of human has gained a lot of importance in recent times. *Actinobacteria* can produce a wide variety of secondary metabolites which form a rich source of antibiotics and other pharmaceuticals. *Streptomyces* sp. also produces certain bioactive compounds, which can be exploited as potential pharmacological drugs. To date, a large number of microbial antioxidants have been characterized, as shown in Table 5.1. Surface-associated marine bacteria have proven to be an extensive resource of valuable secondary metabolites. It has been established that an epiphytic bacteria *Bifurcaria bifurcate* produces novel natural antioxidant compounds along with antimicrobials that act on the Gram-positive bacteria. *Streptococcus thermophilus*, a well-known probiotic bacteria, has been found to have a potent antioxidant activity (AOA), which protects the body from hazardous free radicals that are produced in the body as a result of stress, age factors, etc. *Staphylococcus aureus* produces a series of carotenoids that are well-known for their AOA as they sequester free radicals produced in the body.

An increased intracellular level of free oxygen radicals is said to have oxidative stress that leads to damage proteins, lipids, and DNA (Schieber and Chandel, 2014). Potentially active oxygen free radicals include hydroxyl radicals, superoxide anion radicals, and hydrogen peroxide. During evolution, the living organisms have been able to produce a degree of enzymatic defenses with products like glutathione peroxidase, glutathione reductase (GR), etc., and nonenzymatic antioxidant defenses by the virtue of Vitamin C, Vitamin E, glutathione, etc., to protect them from oxidative stress (Mishra et al., 2015). The use of butylated hydroxyanisole and butylated hydroxytoluene as synthetic antioxidants has been extensively used in treating lipid oxidation, but the side-effects produced by them like liver damage, as well as carcinogenicity have restricted their use. Therefore, in recent times, there is a search for safer and natural antioxidant from biological sources, to find an alternative of synthetic antioxidants, has acknowledged a great degree of attention. Recently, a tremendous amount of interest is grown in probiotics, which are believed to have health benefits. Probiotics denote

TABLE 5.1 Various Microbes Producing Antioxidants and Their Applications

S. No.	Name of Microorganisms	Application	Reference
1	*Lactobacillus rhamnosus*	To increase antioxidant levels and neutralize the effects of reactive oxygen species	Wang et al. (2017)
2.	Probiotic *Bifidobacterium*	Promote antitumor immunity relieve irritable bowel syndrome in women	Sivan et al. (2015)
3.	*Lactobacillus plantarum* P-8	Curtailing the accumulation of liver lipids and protecting healthy liver function	Bao et al. (2012)
4.	*Lactobacillus rhamnosus*	Strong antioxidant activity in situations of elevated physical stress	Martarelli et al. (2011)
5.	*Streptococcus thermophiles Lactobacillus casei* KCTC 3260	Chelating ability for both Fe^{2+}& Cu^{2+}	Ashraf and Shah (2011)
6.	*Pseudomonas koreensis* (JX915782) a *Sargassum* associated yellowish brown pigmented bacterium	High antioxidant activity against DPPH scavenging	Pawar et al. (2015)
7.	*Aspergillus candidus* CCRC 31543	Inhibition of peroxidation [IP]	Yen and Chang (1999)
8.	Filamentous fungi like *Mucor circinelloides*	β-Carotene/linoleic acid bleaching, radical scavenging, reduction of metal ions, and chelating abilities against ferrous ions	Hameed et al. (2017)
9.	Bacteria associated with sponge *Stylotella sp.*	Scavenging of DPPH and superoxide free radicals	Yoghiapiscessa et al. (2016)
10.	Fungi like *Circinella muscae* and bacteria like *Bacillus brevis*	Free-radical scavengers or hydrogen donors	Hameed et al. (2015)
11.	Aspergillus species	Prevented oxidation of methyl linoleate	Esaki et al. (1999)
12.	Epiphytic bacteria from the marine brown alga, *Bifurcaria bifurcata*	Antimicrobial compounds against Gram-positive bacteria	Horta et al. (2014)

live nonpathogenic microorganisms, which, when given in enough doses, deliberate microbial balance, mainly in the gastrointestinal tract (Rauch and Lynch, 2012). Probiotic bacteria showed antioxidant capabilities both in vivo and in vitro (Lin and Yen, 1999; Wang et al., 2017).

5.3. PROBIOTIC BACTERIA AS A SOURCE OF VITAMINS

Vitamins are vital substances that are classified into two categories—fat-soluble vitamins (A, D, E, and K) and water-soluble (B and C). Humans are not able to synthesize most of the vitamins, so these vitamins can be considered to be essential vitamins that need to be supplemented from outside. Vitamins are micronutrients and are needed in minute quantities in the body but their role in metabolic activities is immense. Their deficiency can create havoc that can lead to unusual symptoms. Fat-soluble vitamins form a very important component of the cell membranes while the water-soluble vitamins prove to be essential for the activity of enzymes and work as cofactors. Instead of using the chemically synthesized vitamins which are also referred pseudo-vitamins, it is always a better option to use naturally synthesized vitamins by the microbes. The talent of certain microorganisms to manufacture vitamins can be exploited to replace the chemical production process of vitamins which prove to be expensive. These vitamins are being used to enrich the foods as well as in situ enrichment of fermented foods. Most of the probiotic bacteria from *Lactobacillus* and *Bifidobacterium*, bestow several health benefits to human beings one of them being vitamin production. They are capable of synthesizing vitamin K needed for blood clotting as well as B vitamins, like biotin, cobalamin, folates, etc. The application of a plethora of microbes for the biosynthesis of microbes is innumerable. Table 5.2 gives a vivid account of the various microorganisms and the vitamins synthesized by them.

A novel method for the production of riboflavin has been devised by employing *A gossypii* with oil waste as a substrate (Park et al., 2007). When *L. lactis* and *Propionicbacterium freudenreichii* ssp *shermanii* when used for fermentation of the cow's milk, it was witnessed that there was a high increase in riboflavin production (Singh and Deodhar, 1994).

5.4 POTENTIAL OF MICROBES IN IMMUNOSUPPRESSION

Immunosuppression can be defined as the suppression of the immune response by employing techniques like the administration of medicines or by therapy due to radiation to avoid the rejection of the graft and also to control autoimmune diseases. There has been the discovery of numerous microbial compounds that are capable of immunosuppression. The most important immunosuppressants evolved in the form of Cyclosporin A. These immunosuppressants were produced by a mold *Tolypocladium nivenum* which was supposed to be a narrow spectrum antifungal by the process of aerobic fermentation (Isaac et al., 1990). Cyclosporins which were approved to be used as immunosuppressive drugs during transplants way back in 1983 are a family of microbial products. The mechanism of action of cyclosporin is seen to be its binding to immunophilin of T lymphocytes which is a cytosolic protein cyclophilin. This aids in the inactivation of transcription of interleukin-2, as well as inhibition of lymphokine production and, thus reduced the function of effector T cells.

TABLE 5.2 List of Vitamins and the Microbes Used for Their Biosynthesis

S. No.	Vitamin Produced	Microorganism	References
1.	Vitamin B1 or Thiamine pyrophosphate	Lactic Acid Bacteria Saccharomyces cerevisiae (yeast) Methanococcus jannaschii (Archaebacteria)	Begley et al. (2012), Eser et al. (2016)
2.	Riboflavin or vitamin B2	Bacillus subtilis and Escherichia coli, A gossapii, Candida famata	Schwechheimer et al. (2016), (Wang et al. 2018)
3.	Vitamin B3	S. cerevisiae	Belenky et al. (2011)
4.	Vitamin B5	*C. glutamicum, B. subtilis*	Hüser et al. (2005)
5.	Vitamin B6	*E.coli, S. meliloti, B.subtilis*	Rosenberg et al. (2017)
6.	Vitamin B7	*S. marcescens, Agrobacterium, Rhizobium, B. subtilis, E.coli*	Van Arsdell et al. (2005)
7.	Vitamin B9	gossypii	Serrano-Amatriain et al. (2016)
8.	Folic acid or vitamin B11	Human gut commensals like Lactobacilli	Wegkamp et al. (2004)
9.	Vitamin B12 or cobalamine	*Pseudomonas dentrificans, Bacillus megaterium, Propionibacterium freudenreichii, Lactobacillus reuteri*	Warren (2006), Escalante-Semerena (2007)
10.	Vitamin K	Lactic acid bacteria producing menaquinones	Suttie (1995)

The drugs like calcineurin inhibitors, for example, tacrolimus as well as cyclosporine decrease the proliferation and activation of lymphocytes. The major causative agents include *Coxiella burnetti, Bartonella* sp, *Brucella melitensis, Nocardia farcinica, Mycobacterium tuberculosis*, and *Ehrlichia chaffeensis*. However, the causative agents of pneumonia are very often noticed in medicinal practice in immunocompromised patients. In AIDS patients, *Mycobacterium avium* or Mycobacterium tuberculosis infections are profoundly noticed (Douek et al., 2002). Besides, an immunocompromised patient having diabetes and extreme renal failure has been reported with a lethal infection of *N. farcinica* (Sonesson et al., 2004). Therefore, infection with unscrupulous pathogens might be more alarming than to the host than infection caused by pathogens that cause immunosuppression. The immunosuppression caused due to bacteria varies with the type of host–parasite interactions as well as elicited host immune mediators that make a swing in the steadiness between Th1 and Th2 responses. For example, the continued making of transforming growth factor (TGF)-β has been connected to immunosuppression in patients infected with chronic brucellosis (Elfaki and Al-Hokail, 2009) or tuberculosis (Toossi et al., 1995). However, the most commonly administered calcineurin inhibitors (cyclosporine and tacrolimus) repress the immune system by inhibiting interleukin-2 production in T cells. The human gut is the residence of more than trillions of microbes that can stimulate several aspects of host physiology. Precisely, the bacteria present in the intestinal aid in the hydrolysis of orally administered drugs thus affecting the efficacy and/or toxicity of drugs, for example, a gut bacterium *Eggerthella lenta* helps in the transformation of digoxin into an inactive metabolite, dihydrodigoxin, (Haiser et al., 2014). Although the gut microbiome is a current promising field in medicine, several reports are found on the alteration of this complex system in patients with renal transplants (Lee et al., 2014). Recently, the recipients of renal transplant established post-transplant diarrhea, a recurrent impediment with a significant medicinal impact on graft survival was noticed to hold lowered saccharolytic bacteria (e.g., *Ruminococcus, Dorea, Bacteroides, Coprococcus*) commonly related to intestinal homeostasis. Tacrolimus is routinely used as an immunosuppressant to patients with a kidney transplant, glomerular diseases like membranous nephropathy, and glomerulosclerosis (Krzyżowska et al., 2018). A list of the immunosuppressants obtained from the microbes has been enumerated in Table 5.3.

5.5 MICROBIALPRODUCTIONOFVALUE-ADDEDNUTRACEUTICALS

The bioactive compounds produced by the microbes naturally and assists in attaining good health are categorized as nutraceuticals. The need to fulfill the high market demands for nutraceuticals, the microbes are being employed

TABLE 5.3 List of Immunosuppressants and Their Microbial Sources

No.	Immunosuppresants	Microbial Source	References
1.	Metacycloprodigiosin	*Streptomyces & Serratia*	Magae et al. (1996)
2.	L-asparaginase	*Streptomyces variabilis ASU319*	Abd-Alla et al. (2013)
3.	Tacrolimus	*Streptomyces tsukubaensis*	Haddad et al. (2006)
4.	Sirolimus or rapamycin	*Streptomyces hygroscopicus*	Seto (2012)
5.	Cyclosporin A or Ciclosporin	Fungus *Tolypocladium inflatum*	Bushley et al. (2013)
6.	4-Phenyl-3-butenoic acid	Streptomyces koyangensis	Lee et al. (2005)
7.	Mizoribine (INN, trade name Bredinin)	fungus *Penicillium brefeldianum*	Ishikawa (1999)
8.	Mycophenolic acid, or mycophenolate	*Penicillium stoloniferum*	Maripuri and Kasiske (2014)
9.	Myriocin or FTY720	*Mycelia sterilia*	Abd-Alla1 et al. (2016)
10.	Gliotoxin	*Gliocladium fimbriatum, Aspergillus fumigatus, Trichoderma,* and *Penicillium.*	Scharf et al. (2012)
11.	Subglutinols	*Fusarium subglutinans*	Kim et al. (2010)

to produce such value-added nutraceuticals to help in combating various diseases. Recently the microbial production of these compounds via the process of metabolic engineering is being treated as an ecologically sound approach. The microorganisms such as *Escherichia coli* and *S. cerevisiae* have been used for the production of various nutraceuticals like prebiotics, phytochemicals, etc. The property of fastidious growth and the easy manipulation of the *E. coli* and *S. cerevisiae* make them versatile for the production of nutraceuticals. The mushrooms have a lot of nutritional value and are very important for therapeutics. The mushroom known as *Hygrophorus burners* produces fatty acids that act as antibiotics as well as fungicides. *Tricholoma magnivelare* commonly known as the pine mushroom produces certain nutraceuticals that are instrumental in increasing fertility and activating the immune system. They are also supposed to be anticancerous as well as an aphrodisiac. *Amanita phalloides*, commonly known as "death cup," a local mushroom is used in treating tumors. Cancer and coronary heart disease are treated by peroxidases and superoxide dismutase present in fungus-like *Flammulina velutipes.* Some phenolic compounds like melanin, lanostane-type terpenoids, etc., present in *I. obliquus* commonly known as Chaga mushroom has been exploited for curing diseases like cancer, tuberculosis of bones, diabetes, etc. Cultivated mushroom fruit bodies or mycelia aid in the manufacture of certain substances that could be exploited as nutraceuticals. One of the most novel strategies utilized by fungi is exploited is their selenium accumulating potential so that these fungi could be consumed as Se-fortified food. Some of the fungi like *Agaricus bisporus*, *Pleurotus ostreatus,* etc., are known to be selenium accumulators. The Basidiomycetes and some of the ascomycetes produce certain fungal bioactive polysaccharides that are utilized as a part of the local customary diet and as medicines. Lactic acid bacteria like *Lactococcus lactis* has been exploited for the production of certain nutraceuticals (Hugenholtz and Smid, 2002). Probiotics are produced by the microorganisms like *Bacillus bulgaricus* which when consumed prove to be essential for detoxifying the intestinal flora and creating an environment-friendly atmosphere in the intestine (Holzapfel et al., 2001) The marine microalgae are reservoirs of food constituents like β-carotene, various vitamins, polysaccharides, and polyunsaturated fatty acids (Yap and Chen, 2001). These types of bioactive molecules are being used as food supplements to be used in different types of food commodities as well as the milk formulations for infants. Macroalgae is also very extensively used for its products hydrocolloids like agarose, carrageenan, and agar. Red and brown seaweeds are the major sources of various bioactive compounds like vitamins, minerals, proteins, and essential fatty acids (Giavasis, 2014; Plaza et al., 2008). Antioxidants like β-carotene and astaxanthin have been commercially produced by using *Dunaliella* species and *Haematococcus*.

5.6 ROLE OF BACTERIA IN THE CONTROL OF MOSQUITO-BORNE DISEASES

Mosquito belongs to the class insects. Mosquito is sanguinivorous. During the blood-sucking process, they release saliva which causes itching and rashes. They prove to be the vectors for the transmission of diseases such as dengue, malaria, yellow fever, chikungunya, etc. So, it is quite preferable to sleep under the net. Another challenge is that such mosquitoes bite during day time. It is not possible to keep an eye on mosquitoes all the time. Several measures have been taken to control the mosquito-borne disease. The first step to prevent such diseases is to prevent the mosquito from breeding. The best way for this is to use insecticides. On the other hand, the use of insecticides is not cost-effective. It also causes environmental hazards. Even the regular application of insecticides may create a resistant breed of mosquitoes (World Health Organization, 2008; Zaim and Guillet, 2002).

Instead of targeting mosquitoes, researchers have targeted the microflora in their gut, salivary glands, and reproductive organs. Both bacteria and fungi show a symbiotic relationship with the vector. Bacteria may be present in the reproductive organ and can be transmitted to the next-generation (Damiani et al., 2010). A symbiotic control mechanism was used to control mosquito-borne diseases. Symbiotic control can also lead to a change in the mating behavior of insects (Miller et al., 2010). This method is environmentally friendly and highly efficient. Table 5.4 shows the list of mosquito-borne diseases and their vectors. Targeting the symbiotic nature of bacteria researcher has developed the following technique to control the mosquito as a vector for the transmission of disease:

1. *By creating paratransgenic insect*: In this approach, the researcher genetically modify the microbes so that pathogen transformation can be inhibited. The modification can bring about a reduction in vector's competence and can interfere with oogenesis and embryogenesis (Wilke and Marrelli, 2015).
2. *Cytoplasmic incapability (CI)*: This method is quite instrumental in controlling the breeding habitat of mosquitoes. Matting with sterilized male controls the spread of the vector very effectively. The sterile insect technique is highly preferred even in the agricultural field (Alphey et al., 2010). A mosquito can be sterilized using ionizing radiation (Ahmadi et al., 2018). Another effective alternative method is cytoplasmic incompatibility. *Wolbachia* strain is quite desirable for the technique. It can manipulate and distort reproduction leading to CI (Beckmann and Fallon, 2013).

Many fungal pathogens such as *Lagenidium*, *Coelomomyces,* and *Culicinomyces* in their zoospore stage kill the mosquitoes (Scholte et al., 2004). The fungus can kill mosquitoes in their larval as well as in the adult stage. The fungal spores, toxins enzymes, and other proteins at its zoospore stage target the developmental stages of mosquitoes (Singh and Prakash, 2014).

5.7 FMT OR FECAL TRANSPLANTATION

Our gut has multiple of microflora. They have a symbiotic relationship in our gut. They also prevent the infection from any foreign microflora (Pickard et al., 2017). Many infectious diseases like clostridium difficile infection (CDI), inflammatory bowel disease (IBD), and Crohn's disease develops when the microflora of the gut is destroyed. *C. difficile* spores are quite resistant to

TABLE 5.4 List of Mosquito-borne Diseases and Their Vectors: The Endosymbiotic Relation is Used to Cure the Mentioned Diseases Using Genetic Editing Techniques

Diseases	**Vector**	**Endosymbiont**	**Parasite**	**Technique**	**References**
Chagas disease	Triatominae or "kissing bugs"	*Rhodnius prolixus*	*Trypanosoma cruzi.*	Transformed to express cecropin A, a peptide lethal to the parasite cecropin A expression results in elimination or reduction in number of *T. cruzi*	Durvasula et al. (1997)
Malaria	Anopheles mosquitoes	*Pantoea agglomerans*	*Plasmodium* spp.	Engineered to express and secrete anti-*Plasmodium* effector proteins (SM1, anti-Pbs21, and PLA2)	Bisi and Lampe (2011)
Chikungunya	Aedes aegypti mosquito	*Wolbachia pipientis*		Hinder replication of viruses	Moreira et al. (2009)
Dengue	Aedes aegypti mosquito	*Wolbachia* strain *w*Mel		Feeded to A. aegypti through blood meal	Audsley et al. (2017)

adverse conditions. Their infection can be symptomatic or asymptomatic. Symptoms include abdominal pain, ileus, diarrhea, or even death can occur in severe conditions (McDonald et al., 2007). *C. difficile* also encodes two toxins TcdA and TcdB binds to host cell receptors and act as monoglucosyltransferases to inactivate Rho family GTPases (Just et al., 1995). Many risk factors can cause CDI. The use of broad-spectrum antibiotics like clindamycin and cephalosporins which can target even guts bacteria is one of them. The killing of gut microflora leads to CDI (McDonald et al., 2005; McFarland, 2008). Alteration of gut microflora also inhibits the proton pump which increases the risk of CDI (Imhann et al., 2016). CDI can be treated by a prescribed dose of metronidazole, vancomycin, or the combination of two. Fidaxomicin has a similar effect to that of vancomycin but it is not cost-effective. Fidaxomicin is used for the treatment of reoccurring cases of CDI. Reoccurrence is very common in all types of mentioned treatment. It has been estimated that almost 45% of the patient has to suffer from CDI (Huebner and Surawicz, 2006). While seeking the solution, FMT has been proved to be the best solution for CDI as this is quite economic. In this technique, the healthy stool is implanted to restore the gut microflora. This technique has been in use since the 1950s. Firstly, this was used to treat pseudomembranous colitis, inflammation due to CDI. Initially, this was administered through an enema (Eiseman et al., 1958). Recently it can be administered through a nasogastric or the nasojejunal tube, colonoscopy, or via gastroscopy (Kassam et al., 2013).

In 2017, at European Consensus Conference, 28 experts from 10 different countries collaborated to discuss the general issues related to FMT (Cammarota et al., 2017). Selection of donor, preparation of fecal material, its management, and delivery was the main matter of concern. The donor is selected based on both blood and stool inspection. The testing is done at least 4 weeks. Fresh as well as the frozen sample can be used for FMT. A minimum of 30 g fresh samples of stool should be used within 6 h after defecation. During this time, it should be stored between 20 and 30 °C. Special care should be taken to protect anaerobic bacteria during their processing. The collected fecal is quite prone to coagulate. To prevent this, it should be ground and sieved. This process avoids the clogging of infusion syringes and tubes. The processed fecal is mixed with 10% glycerol and stored at –80 °C to be used as a frozen sample. The frozen sample should be thawed before the application. FMT has also been proved effective against ulcerative colitis (UC), a severe IBD. Like CDI, UC has also decreased the prevalence of *Bacteroideted* and *Firmicutes* while there is increased prevalences of pathogenic bacteria like *Actinobacteria* and *Proteobacteria* (Sartor and Mazmanian, 2012). To treat this, FMT helps to restore the lost microflora from the patient's gut. Crohn's disease is also an intestinal inflammatory disease that can be treated by simply re-establishing the healthy microbiomes in the gut, which can be achieved by FMT (Bak et al., 2017).

5.8. CHEMICAL TRANSFORMATION OF XENOBIOTICS BY THE HUMAN GUT MICROBIOTA

In our day-to-day life, our gut interacts with so many molecules that act as a foreign particle for them. Such compounds or particles are known as xenobiotics. Such particles are not always harmful. These include even food and nutritious particles and medicines. The microflora presents over there directly alters the chemical structure of such compounds with the help of their digestive enzyme. Few of the enzymes present in microflora are listed in Table 5.5. Such modification facilitates bioavailability and modifies the lifetimes. The microbial action also alters the pharmacodynamics and pharmacokinetics of the drug. Many anti-inflammatory drugs like sulfasalazine which is a very efficient drug against *C. difficile* depend on gut microbiota to be converted into sulfapyridine (active form) (Peppercorn and Goldman, 1972). In our body, there is a variability of cell physiology. This also affects the microbial distribution and their interaction with xenobiotics (Sender et al., 2016). Xenobiotics can be administered through different routes. The one taken orally needs to travel a large distance through the digestive tract. On their way, they interact with digestive enzymes and metabolized while on the route. Digested parts are absorbed and transported to the target site. The undigested parts of xenobiotics pass to the large intestine where they are transformed by gut microbes using hydrolytic and reductive reactions (Sousa et al., 2008). The transformed substances get absorbed by the epithelial cells and distributed into tissues through systemic circulation. The xenobiotics which are administered intravenously also interact with the microbes in the gut through biliary excretion. The microbial interaction is ultimately excreted out through feces or as kidney filtrate in the form of urine (Koppel et al., 2017). Sometimes such chemical alteration can be hazardous. Microbial hydrolysis of artificial sweeteners produces carcinogenic compound cyclohexylamine (Bopp et al., 1986). Microbial enzymatic action successfully affects the bioavailability of drugs and some nutrients (Table 5.6).

TABLE 5.5 Enzymes Involved by Microbes and Their Mode of Action

S. No.	Enzymes	Mode of Action
1.	Hydrolytic enzyme	Catalyses the addition of a water molecule to a substrate followed by bond cleavage
a.	Protease	Cleave the peptide bond
b.	Glycosidases	Targets glycosidic bonds
c.	Sulfatases	Hydrolyse sulfate esters
2.	Lyases	Cleaves C–C, C–O, C–N bonds by other means than by hydrolysis or oxidation resulting in the formation of double bond or a new ring
a.	Microbial polysaccharide lyases (PLs)	Modifies glysosidic bond at the a glycosidic bond
b.	Microbial C–S b-lyases	Cleave C–S bonds found in both dietary compounds and cysteine-S-conjugates of xenobiotics
3.	Reductive transformation	Reduces functional groups
4.	Transferase	Transfers methyl and acyl groups to or from xenobiotic scaffolds

5.9 USE OF PROBIOTICS IN THE TREATMENT OF SEVERE ACUTE PANCREATITIS

The pancreas is the organ situated behind the stomach. Its inflammation is quite painful and the condition is known as acute pancreatitis. It can be mild, moderate, severe to critically acute. In the case of severe and critical acute pancreatitis, there is persistence organ failure (Kylänpää et al., 2012). If this persists, it can lead to death. The main challenging issue with the disease is its diagnosis. It is quite difficult to access the organ without surgery. It has been observed that the sufferer has an increased level of amylase or lipase. But this is not always the case. In such a situation, abdominal imaging which includes computed tomography, transabdominal ultrasound, endoscopic ultrasound, and magnetic resonance cholangiopancreatography is preferred. Gallstones, idiopathic, chronic alcohol consumption, drug-induced hypertriglyceridemia, and idiopathic are the major risk factors for acute pancreatitis (Pang et al., 2018). Of this, gallstone is the main cause followed by alcohol consumption. Inappropriate trypsin activation is also one of the major causes of acute pancreatitis (Whitcomb, 2006). Its treatment also needs a specification for the risk factor. Acute pancreatitis due to gallstone needs to be operated on first. In the case of necrosis, antibiotics are to be given to avoid morbidity (Büchler et al., 2000). Another effective way of treatment is the introduction of probiotics. In most cases, it was not proved much effective (Gou et al., 2014) while some have shown a positive effect of probiotics on acute pancreatitis treatment (Hooijmans et al., 2012).

TABLE 5.6 Gut Microflora Alters the Nature of Drugs and Certain Compounds that can Contribute to Certain Diseases Either Positively or Negatively. (Some of Such Diseases are Listed Below)

Sr. No.	Disease	Effect of Chemical Transformation	References
1.	Inflammatory bowel disease (IBD)	Gutmicrobes reduce Sulfasalazine (Prodrug) into sulfapyridine and the active antiinflammatory agent 5-ASA	van Hogezand et al. (1992)
2.	Parkinson's disease	L-dopa is the drug against Parkinson's disease. This is decarboxylated by host as well as microbial enzymes to restore dopamine levels by effecting its bioavailability	Bergmark et al. (1972)
3.	Cancer	Drugs like gemcitabine, fludarabine, and CB1954 are used as chemotherapy. Microbial action on this drugs may increase or decrease the effectiveness which effect the interindividual variation in cancer therapy	Bopp, et al. (1986), Lehouritis et al. (2015)
4.	Celiac disease	This is an autoimmune disorder responds due to gluten found in wheat-based foods. During digestion, it releases high molecular weight immunogenic peptides. Gut bacteria can alter gluten proteolysis through FMT	Caminero et al. (2015, 2016)
5.	Decreasing serum cholesterol	Gut microbes reduces the cholesterol to coprostanol. As this product is not absorbed can be excreted out. Hence help to reduce it, while ingested cholesterol is absorbed in the cholesterol from the serum	Gérard et al. (2004), Ren et al. (1996)
6.	Reduction of heavy metals	Enzymes like demethylating, organomercuric lyase (MerB) and mercuric reductase (MerA) present in gut microbes can reduce the heavy metals in human isolates	Liebert et al. (1997)

5.10 MARINE MICROBIAL BIOACTIVE COMPOUNDS

A major area of the earth is covered with water. Hydrosphere also has a diverse niche. Many of the organisms dwell even under adverse conditions. Such organisms differ in their structure and bioactive compounds as secondary metabolites. These bioactive compounds are synthesized only by aquatic organisms. Such compounds can be exploited for the well being and are enumerated in Table 5.7. Most of the compounds are demanding in the field of pharmacy and cosmetics. The bioactive

TABLE 5.7 Marine Bioactive Compounds and Their Function

Microbes	Bioactive Compound	Function	References
Antimicrobial activity			
Brevibacillus laterosporus PNG276	Tauramamide contains 2D amino acids and is acylated at the N-terminus	*Enterococcus* sp., *Staphylococcus aureu*	Desjardine et al. (2007)
Marinispora (strain NPS008920	2-Alkylidene-5-alkyl-4-oxazolidinones, lipoxazolidinone A	*Haemophilus influenzae*	Macherla et al. (2007)
Actinomycete, NPS12745	Chromopyrrolic acid, Lynamicins A–E	*Staphylococcus epidermidis* and *Enterococcus faecalis*,	McArthur et al. (2008)
Actinomycete strain CNQ-418	Marinopyrroles A (23) and B	*S. aureus*	Hughes et al. (2008)
Pseudomonas stutzeri (CMG 1030)	Zafrin (4β-methyl-5,6,7,8–tetrahydro-1 (4β-H)–phenanthrenone)	*Bacillus subtilis* (more effective than ampicillin, vancomycin or tetracycline)	Uzair et al. (2008)
Nocardia sp. ALAA 2000	Ayamycin [1,1-dichloro-4-ethyl-5-(4-nitrophenyl)- hexan-2-one]	*Candida albicans, Aspergillus niger* and *Botrytis fabae*	El-Gendy et al. (2008)
Nigrospora sp.	Nigrospoxydons A–C	*S. aureus* ATCC 25923 (SA) and methicillin-resistant *S. aureus* (MRSA)	Trisuwan et al. (2009)
Anticancerous			
Mycelial extract of the bacterium *Micromonospora marina*	Thiocoraline	Inhibits DNA polymerase-a	Newman and Cragg (2004)
Curvularia sp. (strain no. 768)	Macrolide apralactone A	Tumor treatment	Greve et al. (2008)
Spicellum roseum	Spicellamide A and spicellamide B	Against rat neuroblastoma B104 cell line. 49.83 mM	Kralj et al. (2007)
Petriella sp.	Three new infectopyrone derivatives together with the cyclic tetrapeptide WF-3161	Strongly cytotoxic against the L5178Y mouse lymphoma cell line	Proksch et al. (2008)
Marine brown algae	Phlorotannins	HIV treatment	Pal Singh and Bharate (2006)

TABLE 5.7 *(Continued)*

Microbes	Bioactive Compound	Function	References
Brown algae *Ecklonia cava*	8,8′-Bieckol and 8,4‴-dieckol	Effect on HIV-1 reverse transcriptase and protease	Mi-Jeong Ahn et al. (2004)
Ishige okamurae	Diphlorethohydroxycarmalol	Effect on HIV-1 reverse transcriptase and protease	Ahn et al. (2006)
Red algae *Grateloupia filicina*	Sulfated galactans	Antiretroviral activity	Wang et al. (2007)
Marine fungi Phaeotheca triangularis	Mycosporine	UV protecting	Kogej et al. (2006)
Marine fungus, Epicoccum sp.	4,5,6-trihydroxy-7-methylphthalide	Radical scavenging activity	Abdel-Lateff et al. (2003)

compounds of bacteria and fungi are used by other organisms also who themselves do not produce such compounds. Earlier very few of the compounds were known. With the realization of their importance, many were discovered. In a year between 2006 and 2007, a 24% increment in bioactive compounds from aquatic microorganisms was seen (Blunt et al., 2009).

5.11 MICROBIAL ANTIBIOTICS IN THE TREATMENT OF DISEASES

The era from 1945 to 1955 can also be considered as the "Antibiotic age." During this period, many of the known antibiotics including penicillin, streptomycin, chloramphenicol, streptomycin, and tetracycline were discovered. Although the name antibiotic was coined in 1941 itself by Selman Waksman. He considered antibiotic as a molecule which has an antagonistic effect in the growth of other life. During his further study, in 1961, he found that antibiotics are produced as a survival molecule during stress conditions. Antibiotics are low molecular weight, bioactive secondary metabolites produce by bacteria and fungi. Such molecules have antibacterial, antiviral, and antifungal activity. Such products can be utilized both as natural and semisynthetic forms (Chater et al., 2010; Clardy et al., 2009). In nature, bioactive compounds are derived from filamentous and nonfilamentous bacteria and fungi in the form of mycotoxins or bacteriocins (Falkinham et al., 2009). Different strains of the same species produce different types of antibiotics (Table 5.8).

TABLE 5.8 List of Antibiotics Produced by Bacteria

Bacteria	Antibiotics
Bacillus subtilis	Polymyxin and bacitracin
Streptomyces griseus	Streptomycin
Streptomyces venezuelae	Chloromycetin
Streptomyces coelicolor	Actinorhodin and undecylprodiginines
Streptomyces niveus	Novobiocin
Penicillium notatum	Penicillins
Penicillium chrysogenum	Cephalosporins
Amycolatopsis mediterranei	Rifampicin
Paenibacillus polymyxa	Polymyxins

5.12 ANTIMICROBIAL ADJUVANTS TO COMBAT ANTIMICROBIAL RESISTANCE

The vaccine is the inactivated or attenuated and killed form of micro-organisms, which has the tendency to activate the immune system but is not pathogenic. Bacterial derived products are used as an adjuvant. They increase the release of antibodies from plasma cells and also increase the immunogenic response. Adjuvant also minimized the amount of antigen to generate an immunogenic response. Bacteria-based adjuvant compromised of the empty bacterial outer structure. They include membrane proteins, lipopolysaccharides (LPS), adhesions, and peptidoglycan. Such compounds, when injected as a part of live bacteria, can cause toxicity. The creation of bacterial envelop persist the immunogenic response but are nonpathogenic (Hajam et al., 2017).

5.13 MICROBES AS A SOURCE OF ANTIFUNGALS

The fungal infection affects not only the animal kingdom rather the plant kingdom is equally affected. The fungal infection may affect only a part of the body or may have life-threatening a result. Lots of chemicals have been formulated having antifungal activity. But with the demand for eco-friendly action, microbes were exploited to extract antifungal compounds. Microbes produce such compounds as secondary metabolites. Some of them use it as their defense mechanism and also inhibit the growth of other microbes nearby. This provides them an optimal amount of space and nutrition. A different mechanism is applied to inhibit or kill the fungus. Based on their action mechanism, compounds are classified as cell wall synthesis inhibitors, sphingolipid synthesis inhibitors, and protein inhibitors.

Fungal cell wall mainly contains glucan, chitin, and mannoproteins. Glucan is the polymer of glucose. The monomer units of glucose are linked by (1–3)-β or (1,6)-β bonds. Echinocandins inhibit noncompetitively β-1,3-glucan synthase and disrupt the cell wall formation in fungus (Emri et al., 2013). Pneumocandins that belong to the echinocandins family also show a similar mechanism of inhibitory action on the growth of fungus (Chen et al., 2015). Chitin is also a polysaccharide made of β-(1,4)-linked *N*-acetylglucosamine monomers. Secondary metabolites belonging to a peptide-nucleoside family, which are similar to UDP-*N*-acetylglucosamine, inhibit competitively chitin synthesis. Few of the antifungal compounds have been noted in Table 5.9.

TABLE 5.9 Antifungal Compounds and Their Mode of Action

Antifungal Compound	Source	Mode of Action	References
Echinocandins	Ascomycota fungi.	Noncompetitive inhibitors of β-1,3-glucan synthase	Emri et al. (2013)
Pneumocandins	Glarea lozoyensis	Noncompetitive inhibitors of β-1,3-glucan synthase	
Nikkomycins	*Streptomyces ansochromogenes.*	Competitive inhibition of chitin synthase results in inhibition of cuticle synthesis	Liao et al. (2010), Mothes and Seitz (1982)
Polyoxins	*Streptomyces cacaoi*	Due to their structural similarity to UDP-*N*-acetylglucosamine, polyoxins act as competitive inhibitors of chitin synthetases	Li et al. (2012)
Phellinsin A	Phellinus sp.	Inhibited chitin synthase I and II of Saccharomyces cerevisiae	Hwang et al. (2001)
Arundifungin	*Arthrinium arundinis*	Glucan synthase inhibitor	Cabello et al. (2001)

5.14 MICROBES FOR LOWERING THE CHOLESTEROL

Atherosclerosis is a dreaded disease that is caused due to accumulation of atheromatous plaque within the arterial wall. A variety of antiatherosclerotic therapies have been introduced in the last two decades. The level of cholesterol is lowered by statins by inhibition of 3-hydroxy-3-methylglutaryl-coenzyme A reductase (HMG-CoA reductase) (Nicholls et al., 2007). Currently, there are several statins prescribed as antihypocholesterolemic drugs. Certain statins like compactin and mevastatin were isolated from *Penicillium brevicompactum* and *Penicillium citrinum* and used as promising antibiotics. Also, the derivates of compactin witnessed overwhelming medical success. Lovastatin was isolated from the broths of the *Monascus rubber* and *Aspergillus terreus* as a secondary metabolite (Alberts et al., 1980). Lovastatin was the first commercially available statin, approved by the FDA in 1987. Simvastatin, a major cholesterol-lowering drug that is a semisynthetic derivative of lovastatin is also a microbial product. Another promising molecule, pravastatin (US$ 3.6 billion per year), was synthesized from compactin by *Streptomyces acidophilus* (Serizawa and Matsuoka, 1991) and *Actinomadura* sp. (Peng and Demain, 1998) through several biotransformation processes. A synthetic drug, obtained from natural statins, atorvastatin, inhibits HMG-CoA reductase, has been the principal antihypocholesterolemic drug of the entire pharmaceutical industry for many years in terms of market share. Additionally, there were several genera that were very commonly involved in the production of statins are Hypomyces, Paecilomyces Doratomyces, Gymnoascus, Pleurotus, Eupenicillium, Phoma, and Trichoderma (Alarcón et al., 2003).

5.15 MICROBES AS ENZYME INHIBITORS

There was always a search for naturally occurring enzyme inhibitors from plants and animal sources for many decades but now, there are brilliant reports on proteolytic enzyme inhibitors (Fossum, 1970). The existence of the microbial origin of enzyme inhibitors, however, was suggested very recently by the ability of microorganisms to produce antibiotics, which sequentially prevent enzyme actions required for multiplication (Göbel, 1976). Microorganisms producing proteases got the ability to defend themselves from the effect of the enzymes by producing protease inhibitors. In 1962, a research report suggested the existence of trypsin inhibitors containing culture supernatants of Clostridium botulinum, types A, B, and E. An inhibitor for alkaline protease from *Aspergillus sojae* was reported but the acid and neutral proteases produced by the same organism were not so effective. A bunch of inhibitor-producing microbial strains were discovered, namely, Penicillium, Rhizopus, and Aspergillus. A heat-stable and nondialyzable, an inhibitor of trypsin, and chymotrypsin enzyme were reported in the 90,000 g supernatant *E. coli* cells that were undergone sonication (Brecher and Pugatch, 1969). An enzyme inhibition report demonstrated against chymotrypsin, trypsin, and some extracellular proteinases of a group of microorganisms by some fragmented cells of gram-negative bacteria. Also, such inhibitors are thought to be produced widely by various species of the same genus, as shown in Table 5.10, obtained from a study, especially on leupeptin which inhibits papain and trypsin, because leupeptin is being produced by more than 17 microbial species of actinomycetes (Wingender, 1974).

5.16 MICROBES AS A SOURCE OF ANTITUMOR DRUGS

Despite the excellent progress of medical science to detect and treat cancer, it has remained a major menace and one of the foremost causes of deaths across the globe. The use of traditional chemotherapeutic agents to treat cancer often shows nonspecific toxicity to normal body cells. Also, the uncovered body cells during chemotherapy become resistant to drugs due to increased ability to repair defective DNA in cellular machinery which subsequently interferes with apoptosis (Raguz and Yagüe, 2008). There are several incidences where bacteria are considered carcinogens and tumor promoters (Lax, 2005). Bacteria release toxins that interrupt the cellular signal, and subsequently disturb the growth regulation of the cell. Apart from their toxicity, they are excellent promoters of the tumor through the induction of inflammation. Bacteria have demonstrated tremendous potential for cancer therapy. There are several species of bacteria that display their astonishing ability to attack and colonize solid tumors, which frequently consequences in neoplasm growth obstruction, and in few cases, results in promising tumor clearance (Leschner and Weiss, 2010). Certain strains of Salmonella, Clostridia, and Bifidobacteria can colonize in the hypoxic area of the tumor and naturally extinguish the tumor cells. Therefore, they are high potential strains to be selected for target-based tumor therapy (Wei et al., 2007). The microbes used in anticancer therapy are listed in Table 5.11.

TABLE 5.10 Inhibitors of Proteolytic Enzymes and Their Origin

Proteolytic Enzyme Inhibitors and their Origin	
Enzyme Inhibitors	**Microbial Species**
Leupeptin	*S. roseus, S. roseochromogenes, S. lavandulae, S. chartreusis S. thioluteus, S. albireticuli,* other than 11 species of *Actinomycetes*
Antipain	*S. michigaenis, S. yokosukaensis, S. violascens,* etc.
Chymostatin	*S. hygroscopicus, S. lavendulae,* etc.
Pepstatin	*S. testaceus, S. S. argenteolus,* etc.

TABLE 5.11 A Representative List of Microorganisms Used/Planned to be Used in Anticancer Therapy

Microorganism	Strain/antigen	Cancer
Mycobacterium bovis	Atenuated strain Calmette-Guerin	Superficial bladder cancer
Streptococcous pyogenes	OK-432	Lymphangioma
Clostridium novyi	Strain NT	Solid tumors
Salmonella enterica Serovar Typhymurium	Strain VNP20009	Melanoma
Magnetococcus marinus	MC1	Solid tumors and some metabolic tumors
Toxoplasma gondii	CPS/TLA	Pancreas, lung and ovarian cancer, and melanoma
Plasmodium falciparum	rVAR2-DT	Melanoma expressing CS

5.17 BACTERIAL BACTERIOCINS TO BE EXPLOITED AS PRESERVATIVES

There are several antibacterial components isolated from plants, animals, insects, and bacteria, namely, fatty acids, antibiotics, and bacteriocins, etc. Antimicrobial peptides (AMPs) or proteins isolated from bacteria are termed as bacteriocins. A scanty environment causes the production of a group of bacteriocins due to a lack of resources and space for growth. Microcins are a low molecular mass that is <10 kDa, hydrophobic AMPs. Microcins are classified based on their molecular masses, disulfide bonds, and the nature of post-translational modifications. The bacteriocins obtained from gram-positive bacteria exhibit similar features to microcins. Their molecular mass generally, lesser than 60 amino acids. They are categorized into different classes like class I consisting of modified peptides, the lantibiotics, class II consists of unmodified peptides the nonlanthionine, and class III which have large proteins and are heat sensitive.

Almost all groups of bacteria produce bacteriocins, which are biodegradable as well as nonimmunogenic and possess toxicities against cancer cells. Therefore, the bacteriocins have shown a potential effect to act as synergistic agents to traditional cancer drugs (Kaur and Kaur, 2015). Colicins, the bacteriocin released from Enterobacteriaceae, namely, *E. coli*. These molecules have great anticancer activities in human tumor cell lines in-vitro which include bone cancer, colon cancer, breast cancer, and uteri cell line HeLa (Chumchalová and Šmarda, 2003). Pediocin, a bacteriocin extracted from Pediococcus acidilactici K2a2-3 showed cytotoxic activities against HeLa and HT29 cell lines (Villarante et al., 2011). Also, Nisin, a bacteriocin isolated from Lactobacillus lactis holds the cytotoxic effect on MCF-7 (Villarante et al., 2011), HNSCC (Joo et al., 2012), and HepG2 (Paiva et al., 2012), both in-vivo and in-vitro. Phenazine 1,6-dicarboxylic acid extracted from *Streptomyces species* controls the biofilm formation and even metabolism in Candida albicans (Morales et al., 2013). Biofilm is a syntrophic consortium of microorganisms that help bacteria to develop resistance against antibiotics (Kinnari, 2015). Biofilms are very common to unscrupulous bacterial pathogens, namely, *Salmonella typhimurium*. They prove to maintain the pathogenesis of chronic infectious diseases (Komor et al., 2012). For example, polysaccharides liberated by *Streptococcus agalactiae* prevent the adhesion of cancer cells to endothelial cells, a crucial step in cancer metastasis. *Mariprofundus ferroxydans* produce waste material from which iron oxide nanowire was formed that can act as a multifunctional drug carrier in therapy for cancer and cancer hyperthermia (Kumeria et al., 2016).

5.18 MICROBES INVOLVED IN CURING ANAEMIA

The conventional methods of folic acid production prove to be expensive by chemical processes and thus are being replaced by microbial synthesis. The raw materials that are used are very expensive as compared to the product, the folic acid. Thus several microbes that are capable of producing folic acid are being exploited to treat anaemia. The examples of bacteria that assist the uptake of folic acid are *Lactococcus lactis* sub sp. *cremoris*, etc.; yeasts like *Candida famata, Yarrowia lipolytica, S. cerevisiae*, etc. *Bacteria like Pseudomonas denitrificans* and *Propionibacterium shermanii* produce vitamin B12 and can be exploited to prevent diseases like megaloblastic anaemia. They are involved in the production of lactic acid fermented foods that are instrumental in increasing the absorption of iron by altering the pH of the digestive tract which in turn activates the enzyme-like phytases and enhances the production of organic acids. Probi, a Swedish firm that fabricated a bacterium, *Lactobacillus plantarum* 299v was witnessed to double the iron absorption from food in women. This bacteria also helps in digestion, minimizes the ill effects of an antibiotic drug on colonic fermentation that improves the immune system and bowel moments. Microbial synthesis of erythropoietin is being used to treat anaemia but the negative aspect of this type of treatment is that very frequent injection is needed. Probiotics produce various effective metabolites with antioxidant capabilities, such as butyrate, glutathione (GSH), and folate (Pompei et al., 2007). From various research works, it was found that Bifidobacteria was a good source of folate that was able to increase the concentration of folate in both rats and human (Strozzi and Mogna, 2008). Also, folate-producing probiotic *Lactobacillus helveticus* CD6 exhibited antioxidant potentials (Ahire et al., 2013). In another report, it was found that E-3 and E-18 were the two antioxidative *Lactobacillus fermentum* strains, which were rich in Glutathione (Kullisaar et al., 2002). The MIYAIRI 588 strain of *Clostridium butyricum* was able to reduce oxidative stress in the liver in rats suffering from nonalcoholic fatty liver disease (Endo et al., 2013). The levels of antioxidants of the host can also be controlled by probiotics action. Vitamin B12 and folate deficiency endorsed oxidative stress in type 2 diabetes of adults (Al-Maskari et al., 2012). Regular consumption of the *Lactobacillus acidophilus* La1 yogurt expressively improved the levels of vitamin B12 and plasma folate in children, thus signifying an enhanced oxidative status (Mohammad et al., 2006).

5.19 MICROBES AS A SOURCE OF ANTIOXIDANTS

The rapid upsurge in immune-related disorders like allergic disease is sturdily linked to abridged early exposure to microbes (Williams et al., 2008). The intestine is the largest immune organ of the body; most of the cells producing antibodies exist in the intestine (Brandtzaeg, 2002). The intestinal microbiota may often mark the jeopardy of emerging allergic diseases (Penders et al., 2007). The properties of beneficial bacteria which selectively proliferate certain beneficial bacteria are found to be effective in the treatment of allergies (Tang et al., 2010). There are several pieces of evidence for probiotics and prebiotics to deal with allergic disease. The probiotics like lactic acid bacteria include *Lactobacillus bulgaricus, Lactobacillus casei, L. acidophilus*, etc. The Lactobacillus species exert many crucial properties such as effective adherence to epithelial cells of the intestine to decrease or inhibit growth, colonization of pathogens, and synthesis of metabolites to prevent or destroy nonpathogens and pathogens. Besides, other bacterial species, namely, Bifidobacterium spp., Bacillus, and Propionibacterium spp. are also reported as probiotic strains in many commercial products. The use of L. casei strain Shirota is reported as a potential probiotic element for stimulating immune responses and subsequently inhibiting enterobacterial infections, similarly, Lactobacillus GG is used against rotaviruses as a potential oral vaccine.

5.20 MICROBES FOR TREATMENT OF ALLERGIES

The microbial population is extremely vibrant wherein each species attempts to surpass the intraspecies rivalry. Microbes, by releasing chemicals, fight with each other to overcome competition. Humans had harnessed these biochemicals to treat different types of infectious diseases, for example, the discovery of penicillin by Alexander Fleming paved the way for the treatment of various infectious diseases. Various types of antibiotics are discovered but the excessive use of antibiotics has helped bacteria to emerge resistance toward antibiotics also known as "superbugs," which has become a global threat (Dong et al., 2007). There are few life-threatening examples of microbial drug resistance that include fluoroquinolone-resistant *S. aureus* and methicillin-resistant *S. aureus*, vancomycin-resistant enterococci, and erythromycin-resistant Streptococcus pyogenes and *S. pneumonia*. Incidentally, there are some surprising reports of killing factors that are released by bacterial cells to attack sibling cells during the phase of starvation. This incident received a lot of interest in the scientific community as it can be described as cannibalism in the higher organism. A set of genes for "cannibalism" is found in *B. subtilis* that accelerates the hydrolysis of the surrounding cells during starvation (Claverys and Håvarstein, 2007). The released nutrients from the broken and lysed cells were taken up by the killer cells for their own survival and spore development. The *B. subtilis* predation includes *A. Lwoff, P. aeruginosa, E. coli, X. oryzae*, and *X. campestris* (Nandy et al., 2007). The concept of cannibalism in Bacillus *subtilis* is accredited to sporulation killing factor which consists of two peptides and sporulation suspending protein. Spo0A, a crucial transcriptional regulator, believed to regulate biofilm formation and sporulation thus regulates the production of these peptides (Claverys and Håvarstein, 2007). Scientific reports are claiming that killer B. subtilis cells preferentially attack and destroy non-*B. subtilis* cells thereby establish the potential of killer peptides as an effective antibiotic agent.

5.21 ANTIVIRULENCE FACTORS FROM MICROBES FOR TREATMENT OF INFECTIOUS DISEASES

The existence of more than 250,000 microbial species is reported in the ocean, and it is the largest reservoir for the diversity of microbes (Mora et al., 2011). Currently, the scientific investigation of the oceans has witnessed the finding of well-diversified unknown microbial habitats featured by extreme conditions (Danovaro et al., 2014). These environments allow a group of organisms adapted to these conditions and help to produce an extensive choice of active biomolecules (Panno et al., 2013). Numerous marine microorganisms have emerged a set of mechanisms to defend themselves from the lethal effects of UV radiation, making UV-absorbing elements, namely, cyclosporine, mycosporine-like amino acids, scytonemins (mainly in cyanobacteria), melanin, and carotenoids (Carreto and Carignan, 2011). These elements promise a potential development for a novel UV filter to be comprised of sunscreen products. Almost all of the common sunscreen products have organic and/or inorganic filters. UV filters which were produced by certain components are derived from microbes. Certain photo-protective compounds derived from bacterial and fungal sources and their applications have been studied (Corinaldesi et al., 2017).

5.22 MICROBE-BASED APPROACHES FOR THE TREATMENT OF DIABETES

Diabetes mellitus is considered the most dreadful metabolic disorder which distressing the life of millions of people globally. Hyperglycemia is the characteristic feature of all types of diabetes, be it Type 1, Type 2, or gestational diabetes. Hyperglycemia due to diabetes results in various types of disorders like cardiovascular diseases, problems related to the kidney, as well as retina, and various other complications. Various therapeutic techniques have been employed to treat diabetes disease through the use of products synthesized chemically or biologically from different sources like microbes. Microbes are now gut microbiota that has been observed to play a chief role in the control of sugar levels in diabetes. The concept of personal bioremediation, in which the microbes of a healthy individual are transplanted into the gut of the diabetic patient, has been applied to treat diabetics. This is done to heal the imbalance of the microbes that lead to the cause of the disease. It has been found that acarbose (pseudotetrasaccharide) a product from *Actinoplanes* sp. SE has valienamine which is an aminocyclitol moiety. The product interferes with the activity of α-glucosidase and sucrase produced in the intestine and inhibits them. Now, this activity decreases the rate of starch breakdown thus controlling diabetes (Youmans et al., 2015). Paim, the α-amylase inhibitors from *Streptomyces corchorushii* and some oligosaccharide compounds like TAI-A, TAI-B produced by *Streptomyces calvus* TM-521 are found to be very instrumental in controlling diabetes (Hirayama et al., 1987). An inhibitor of pancreatic lipase known as Lipstatin produced by *Streptomyces toxytricini* intervenes with the fat absorption in the gastrointestinal tract. This strategy has been exploited to treat obesity and diabetes (Weibel et al., 1987).

5.23 USING GUT BACTERIA TO FIGHT DIARRHEA

A combination of microbes found in the gut may prove to be a superb method to treat diarrhea. In one of the research works, it was found that a cocktail of certain microbes proved to be a wonderful weapon to treat a diarrheal infection in mice caused by a stubborn bacterium. This research has paved the way for the treatment of certain gastrointestinal diseases where this can replace the treatment procedure where we use the transplantation of fecal matter to the similar bacterial mix will replace the treatment involving the intake of fecal matter to restore the correct blend of microorganisms in the gut (Youmans et al., 2015).

Antibiotics that destroy the harmful microbes in our body also kill some of the beneficial bacteria which disturb the balance of the microbe present in the gut which leads to diarrhea. There has been ample research on this issue of diarrhea showing that if the patients who have taken antibiotics are given probiotics in the form of good microbes, the chances of occurrence of diarrhea can be minimized. *Saccharomyces boulardii* as well as certain strains of lactobacillus are found to be instrumental in controlling Diarrhea (Gaón et al., 2003).

5.24 GUT BACTERIA COULD ONE-DAY COMBAT OBESITY

A vast number of research works have proved that a high-fat diet can destroy the bacteria present in the gut thus leading to profuse weight gain. Several functions of the microbes in the gut have been recognized like digestion, altering hunger, etc. Thus, several scientists are involved in the investigation of various microbiomes present in obese people, as well as underweight people, and what is the role of these microbes in maintaining body weight? They have established that the food that we eat directly affects the presence of microbes in our gut. Some of the food items promote the growth of the bacteria that increase the bodyweight of the person while some support the growth of those bacteria which decrease the body weight. Now it is a matter of ponderance as to which of the fats serve to increase the microbial flora instrumental in promoting obesity. The common bad fats which increase the bodyweight are refined omega 6 vegetable oils like soybean oil. The polyunsaturated fats obtained from soybean, canola, and other seed oils cause inflammation and need to be avoided. Though we are actually misguided by the personals involved in marketing in the food industry that these types of vegetable fats are good for our health and do not cause any harm to our body, but the fact is that if we want to stay healthy and fit we need to totally boycott the use of these harmful oils. Omega 3 fats and monounsaturated fats like extra-virgin olive oil are involved in improving the growth of good microbes which keep our body healthy and do not cause obesity. The harmful microbes or the bad bacteria cause the production of toxins like the LPS which result in the inflammation of the whole body, also cause diabetes, and thus cause weight gain. Turnbaugh et al. (2009) furnished the first evidence that the transfer of the microbiota of obese mice into mice that is germfree or devoid of any microbes, led to the increase in obesity of the mice. This fact has not only been established in mice but in rodents and even in human beings. The research was performed with the genetically obese mice that were deficient in leptin receptor and came to a conclusion that the microbiota present in the caecum were having about 50% fewer Bacteroidetes that were present in the lean mice also found that a large number of Firmicutes were found in the obese as compared to the unhealthy mice (Ley et al., 2006).

5.25 CONCLUSION

Nature has provided us with a plethora of natural products that can be exploited to treat a lot of dreadful diseases, as well as can be used to lead a healthy life. We know that our health is the most important aspect of our life and we need to take care of our health properly. To date, we have used and we are still using dreaded chemicals to cure several ailments. Although we are getting cured, still we are facing a lot of challenges due to the after-effects of those chemicals which produce permanent ill effects on us. Thus, it is a very wise decision to use natural products obtained from the microbiome present on this earth as they are eco-friendly as well as can be treated as naturopathy. Here in this chapter, we have tried to compile a maximum of the microbial products that are being used in healthcare to treat different kinds of diseases. Though it seemed to be a vast saga of microbial products yet we have made a concise review of most of the microbial products and their uses in the healthcare sector.

KEYWORDS

- **antibiotics**
- **bioactive compounds**
- **cosmeceuticals**
- **immunomodulators**
- **microbialnutraceuticals**
- **probiotics**

REFERENCES

Abd-Alla, M.H.; El-Sayed, E.S.A.; Rasmey, A.H.M. Biosynthesis of L-glutaminase by Streptomyces variabilis ASU319 isolated from rhizosphere of triticum vulgaris. *Univ. J. Microbiol. Res.* **2013,** *1*, 27–35.

Abd-Alla, M.H.; Rasmey, A.H.M.; El-Sayed, E.S.A.; El-Kady, I.A.; Yassin, I.M. Biosynthesis of anti-inflammatory immunosuppressive metabolite by Streptomyces variabilis ASU319. *Eur. J. Biol. Res.* **2016,** *6*, 152–69.

Abdel-Lateff, A.; Fisch, K.M.; Wright, A.D.; König, G.M. A new antioxidant isobenzofuranone derivative from the algicolous marine fungus *Epicoccum sp. Planta Med.* **2003,** *69*, 831–4.

Ahire, J.J.; Mokashe, N.U.; Patil, H.J.; Chaudhari, B.L. Antioxidative potential of folate producing probiotic Lactobacillus helveticus CD6. *J. Food Sci. Technol.* **2013,** *50*, 26–34.

Ahmadi, M.; Salehi, B.; Abd-Alla, A.M.M.; Babaie, M. Feasibility of using the radiation-based sterile insect technique (SIT) to control the olive fruit fly, Bactrocera oleae Gmelin (Diptera: Tephritidae) in Iran. *Appl. Radiat. Isot.* **2018,** *139*, 279–84.

Ahn, M.J.; Yoon, K.D.; Kim, C.Y.; Kim, J.H.; Shin, C.G.; Kim, J. Inhibitory activity on HIV-1 reverse transcriptase and integrase of a carmalol derivative from a brown Alga, Ishige okamurae. *Phytother. Res.* **2006,** *20*, 711–3.

Ahn, M.J.; Yoon, K.D.; Min, S.Y.; Lee, J.S.; Kim, J.H.; Kim, T.G.; Kim, S.H.; Kim, N.G.; Huh, H.; Kim, J. Inhibition of HIV-1 reverse transcriptase and protease by phlorotannins from the brown alga Ecklonia cava. *Biol. Pharm. Bull.* **2004,** *27*, 544–7.

Al-Maskari, M.Y.; Waly, M.I.; Ali, A.; Al-Shuaibi, Y.S.; Ouhtit, A. Folate and vitamin B12 deficiency and hyperhomocysteinemia promote oxidative stress in adult type 2 diabetes. *Nutrition* **2012,** *28*, e23–6.

Alarcón, J.; Aguila, S.; Arancibia-Avila, P.; Fuentes, O.; Zamorano-Ponce, E.; Hernández, M. Production and purification of statins from *Pleurotus ostreatus* (Basidiomycetes) strains. *Z. Naturforsch. C.* **2003,** *58*, 62–4.

Alberts, A.W.; Chen, J.; Kuron, G.; Hunt, V.; Huff, J.; Hoffman, C.; Rothrock, J.; Lopez, M.; Joshua, H.; Harris, E.; Patchett, A.; Monaghan, R.; Currie, S.; Stapley, E.; Albers-Schonberg, G.; Hensens, O.; Hirshfield, J.; Hoogsteen, K.; Liesch, J.; Springer, J. Mevinolin: a highly potent competitive inhibitor of hydroxymethylglutaryl-coenzyme A reductase and a cholesterol-lowering agent. *Proc. Natl. Acad. Sci. USA.* **1980,** *77*, 3957–61.

Alphey, L.; Benedict, M.; Bellini, R.; Clark, G.G.; Dame, D.A.; Service, M.W.; Dobson, S.L. Sterile-insect methods for control of mosquito-borne diseases: an analysis. *Vector-Borne Zoonotic Dis.* **2010,** *10*, 295–311.

Ashraf, R.; Shah, N.P. Selective and differential enumerations of Lactobacillus delbrueckii subsp. bulgaricus, *Streptococcus thermophilus*, *Lactobacillus acidophilus*, *Lactobacillus casei* and Bifidobacterium spp. in yoghurt-a review. *Int. J. Food Microbiol.* **2011,** *149*, 194–208.

Audsley, M.D.; Ye, Y.H.; McGraw, E.A. The microbiome composition of Aedes aegypti is not critical for Wolbachia-mediated inhibition of dengue virus. *PLoS Negl. Trop. Dis.* **2017,** *11*, e0005426.

Bak, S.H.; Choi, H.H.; Lee, J.; Kim, M.H.; Lee, Y.H.; Kim, J.S.; Cho, Y.S. Fecal microbiota transplantation for refractory Crohn's disease. *Intestinal Res.* **2017,** *15*, 244.

Bao, Y.; Wang, Z.; Zhang, Y.; Zhang, J.; Wang, L.; Dong, X.; Su, F.; Yao, G.; Wang, S.; Zhang, H. Effect of Lactobacillus plantarum P-8 on lipid metabolism in hyperlipidemic rat model. *Eur. J. Lipid Sci. Technol.* **2012,** *114*, 1230–36.

Beckmann, J.F.; Fallon, A.M. Detection of the Wolbachia protein WPIP0282 in mosquito spermathecae: implications for cytoplasmic incompatibility. *Insect Biochem. Mol. Biol.* **2013,** *43*, 867–78.

Begley, T.P.; Ealick, S.E.; McLafferty, F.W. Thiamin biosynthesis: still yielding fascinating biological chemistry. *Biochem. Soc. Trans.* **2012,** *40*, 555–60.

Belenky, P.; Stebbins, R.; Bogan, K.L.; Evans, C.R.; Brenner, C. Nrt1 and Tna1-independent export of NAD+ precursor vitamins promotes NAD+ homeostasis and allows engineering of vitamin production. *PLoS One* **2011,** *6*, e19710.

Bergmark, J.; Carlsson, A.; Granerus, A.K.; Jagenburg, R.; Magnusson, T.; Svanborg, A. Decarboxylation of orally administered L-dopa in the human digestive tract. *Naunyn. Schmiedebergs. Arch. Pharmacol.* **1972,** *272*, 437–40.

Bisi, D.C.; Lampe, D.J. Secretion of anti-plasmodium effector proteins from a natural pantoea agglomerans isolate by using PelB and HlyA secretion signals. *Appl. Environ. Microbiol.* **2011,** *77*, 4669–75.

Blunt, J.W.; Copp, B.R.; Hu, W.P.; Munro, M.H.G.; Northcote, P.T.; Prinsep, M.R. Marine natural products. *Nat. Prod. Rep.* **2009,** *26*, 170.

Bopp, B.A.; Sonders, R.C.; Kesterson, J.W. Toxicological aspects of cyclamate and cyclohexylamine. *Crit. Rev. Toxicol.* **1986,** *16*, 213–306.

Brandtzaeg, P.E.R. Current understanding of gastrointestinal immunoregulation and its relation to food allergy. *Ann. N. Y. Acad. Sci.* **2002,** *964*, 13–45.

Brecher, A.S.; Pugatch, R.D. A non-dialyzable inhibitor of proteolytic activity in soluble extracts of *Escherichia coli*. *Experientia* **1969,** *25*, 251–2.

Büchler, M.W.; Gloor, B.; Müller, C.A.; Friess, H.; Seiler, C.A.; Uhl, W. Acute necrotizing pancreatitis: treatment strategy according to the status of infection. *Ann. Surg.* **2000,** *232*, 619–26.

Bushley, K.E.; Raja, R.; Jaiswal, P.; Cumbie, J.S.; Nonogaki, M.; Boyd, A.E.; Owensby, C.A.; Knaus, B.J.; Elser, J.; Miller, D.; Di, Y.; McPhail, K.L.; Spatafora, J.W. The genome of tolypocladium inflatum: evolution, organization, and expression of the cyclosporin biosynthetic gene cluster. *PLoS Genet.* **2013,** *9*, e1003496.

Cabello, M.A.; Platas, G.; Collado, J.; Díez, M.T.; Martín, I.; Vicente, F.; Meinz, M.; Onishi, J.C.; Douglas, C.; Thompson, J.; Kurtz, M.B.; Schwartz, R.E.; Bills, G.F.; Giacobbe, R.A.; Abruzzo, G.K.; Flattery, A.M.; Kong, L.; Peláez, F. Arundifungin, a novel antifungal compound produced by fungi: biological activity and taxonomy of the producing organisms. *Int. Microbiol.* **2001,** *4,* 93–102.

Caminero, A.; Galipeau, H.J.; McCarville, J.L.; Johnston, C.W.; Bernier, S.P.; Russell, A.K.; Jury, J.; Herran, A.R.; Casqueiro, J.; Tye-Din, J.A.; Surette, M.G.; Magarvey, N.A.; Schuppan, D.; Verdu, E.F. Duodenal bacteria from patients with celiac disease and healthy subjects distinctly affect gluten breakdown and immunogenicity. *Gastroenterology* **2016,** *151,* 670–83.

Caminero, A.; Nistal, E.; Herrán, A.R.; Pérez-Andrés, J.; Ferrero, M.A.; Vaquero Ayala, L.; Vivas, S.; Ruiz de Morales, J.M.G.; Albillos, S.M.; Casqueiro, F.J. Differences in gluten metabolism among healthy volunteers, coeliac disease patients and first-degree relatives. *Br. J. Nutr.* **2015,** *114*, 1157–67.

Cammarota, G.; Ianiro, G.; Tilg, H.; Rajilić-Stojanović, M.; Kump, P.; Satokari, R.; Sokol, H.; Arkkila, P.; Pintus, C.; Hart, A.; Segal, J.; Aloi, M.; Masucci, L.; Molinaro, A.; Scaldaferri, F.; Gasbarrini, G.; Lopez-Sanroman, A.; Link, A.; de Groot, P.; de Vos, W.M.; Högenauer, C.; Malfertheiner, P.; Mattila, E.; Milosavljević, T.; Nieuwdorp, M.; Sanguinetti, M.; Simren, M.; Gasbarrini, A. European consensus conference on faecal microbiota transplantation in clinical practice. *Gut* **2017,** *66*, 569–80.

Carreto, J.I.; Carignan, M.O. Mycosporine-like amino acids: relevant secondary metabolites. Chemical and ecological aspects. *Mar. Drugs* **2011,** *9,* 387–446.

Chater, K.F.; Biró, S.; Lee, K.J.; Palmer, T.; Schrempf, H. The complex extracellular biology of Streptomyces. *FEMS Microbiol. Rev.* **2010,** *34*, 171–98.

Chen, L.; Yue, Q.; Li, Y.; Niu, X.; Xiang, M.; Wang, W.; Bills, G.F.; Liu, X.; An, Z. Engineering of Glarea lozoyensis for exclusive production of the pneumocandin B0 precursor of the antifungal drug caspofungin acetate. *Appl. Environ. Microbiol.* **2015,** *81*, 1550–58.

Chumchalová, J.; Šmarda, J. Human tumor cells are selectively inhibited by colicins. *Folia Microbiol. (Praha)*. **2003,** *48*, 111–15.

Clardy, J.; Fischbach, M.A.; Currie, C.R. The natural history of antibiotics. *Curr. Biol.* **2009,** *19,* R437–41.

Claverys, J.P.; Håvarstein, L.S. Cannibalism and fratricide: mechanisms and raisons d'être. *Nat. Rev. Microbiol.* **2007,** *5*, 219–29.

Corinaldesi, C.; Barone, G.; Marcellini, F.; Dell'Anno, A.; Danovaro, R. Marine microbial-derived molecules and their potential use in cosmeceutical and cosmetic products. *Mar. Drugs* **2017,** *15,* 118.

Damiani, C.; Ricci, I.; Crotti, E.; Rossi, P.; Rizzi, A.; Scuppa, P.; Capone, A.; Ulissi, U.; Epis, S.; Genchi, M.; Sagnon, N.; Faye, I.; Kang, A.; Chouaia, B.; Whitehorn, C.; Moussa, G.W.; Mandrioli, M.; Esposito, F.; Sacchi, L.; Bandi, C.; Daffonchio, D.; Favia, G. Mosquito-bacteria symbiosis: the case of anopheles gambiae and Asaia. *Microb. Ecol.* **2010,** *60*, 644–54.

Danovaro, R.; Snelgrove, P.V.R.; Tyler, P. Challenging the paradigms of deep-sea ecology. *Trends Ecol. Evol.* **2014,** *29*, 465–75.

Desjardine, K.; Pereira, A.; Wright, H.; Matainaho, T.; Kelly, M.; Andersen, R.J. Tauramamide, a lipopeptide antibiotic produced in culture by Brevibacillus laterosporus Isolated from a marine habitat: structure elucidation and synthesis. *J. Nat. Prod.* **2007,** *70*, 1850–53.

Dong, Y.; Wang, L.; Zhang, L.H. Quorum-quenching microbial infections: mechanisms and implications. Philos. *Trans. R. Soc. Lond. B. Biol. Sci.* **2007,** *362*, 1201–11.

Douek, D.C.; Brenchley, J.M.; Betts, M.R.; Ambrozak, D.R.; Hill, B.J.; Okamoto, Y.; Casazza, J.P.; Kuruppu, J.; Kunstman, K.; Wolinsky, S.; Grossman, Z.; Dybul, M.; Oxenius, A.; Price, D.A.; Connors, M.; Koup, R.A. HIV preferentially infects HIV-specific CD4+ T cells. *Nature* **2002,** *417*, 95–8.

Durvasula, R.V; Gumbs, A.; Panackal, A.; Kruglov, O.; Aksoy, S.; Merrifield, R.B.; Richards, F.F.; Beard, C.B. Prevention of insect-borne disease: an approach using transgenic symbiotic bacteria. *Proc. Natl. Acad. Sci. USA.* **1997,** *94*, 3274–8.

Eiseman, B.; Silen, W.; Bascom, G.S.; Kauvar, A.J. Fecal enema as an adjunct in the treatment of pseudomembranous enterocolitis. *Surgery* **1958,** *44,* 854–9.

El-Gendy, M.M.A.; Hawas, U.W.; Jaspars, M. Novel bioactive metabolites from a marine derived bacterium Nocardia sp. ALAA 2000. *J. Antibiot. (Tokyo)* **2008,** *61*, 379–86.

Elfaki, M.G.; Al-Hokail, A.A. Transforming growth factor beta production correlates with depressed lymphocytes function in humans with chronic brucellosis. *Microbes Infect.* **2009,** *11*, 1089–96.

Emri, T.; Majoros, L.; Tóth, V.; Pócsi, I. Echinocandins: production and applications. *Appl. Microbiol. Biotechnol.* **2013,** *97*, 3267–84.

Endo, H.; Niioka, M.; Kobayashi, N.; Tanaka, M.; Watanabe, T. Butyrate-producing probiotics reduce nonalcoholic fatty liver disease progression in rats: new insight into the probiotics for the gut-liver axis. *PLoS One* **2013,** *8*, e63388.

Esaki, H.; Watanabe, R.; Onozaki, H.; Kawakishi, S.; Osawa, T. Formation mechanism for potent antioxidative o-dihydroxyisoflavones in soybeans fermented with *Aspergillus saitoi*. *Biosci. Biotechnol. Biochem.* **1999,** *63*, 851–8.

Escalante-Semerena, J.C. Conversion of cobinamide into adenosylcobamide in bacteria and archaea. *J. Bacteriol.* **2007,** *189*, 4555–60.

Eser, B.E.; Zhang, X.; Chanani, P.K.; Begley, T.P.; Ealick, S.E. From suicide enzyme to catalyst: the iron-dependent sulfide transfer in *Methanococcus jannaschii* thiamin thiazole biosynthesis. *J. Am. Chem. Soc.* **2016,** *138*, 3639–42.

Falkinham, J.O.; Wall, T.E.; Tanner, J.R.; Tawaha, K.; Alali, F.Q.; Li, C.; Oberlies, N.H. Proliferation of antibiotic-producing bacteria and concomitant antibiotic production as the basis for the antibiotic activity of Jordan's red soils. *Appl. Environ. Microbiol.* **2009,** *75*, 2735–41.

Fossum, K. Proteolytic enzymes and biological inhibitors. 3. Naturally occurring inhibitors in some animal and plant materials and their effect upon proteolytic enzymes of various origin. *Acta Pathol. Microbiol. Scand. B. Microbiol. Immunol.* **1970,** *78*, 741–54.

Gaón, D.; García, H.; Winter, L.; Rodríguez, N.; Quintás, R.; González, S.N.; Oliver, G. Effect of *Lactobacillus strains* and *Saccharomyces boulardii* on persistent diarrhea in children. *Medicina (B. Aires).* **2003,** *63*, 293–8.

Gérard, P.; Béguet, F.; Lepercq, P.; Rigottier-Gois, L.; Rochet, V.; Andrieux, C.; Juste, C. Gnotobiotic rats harboring human intestinal microbiota as a model for studying cholesterol-to-coprostanol conversion. *FEMS Microbiol. Ecol.* **2004,** *47*, 337–43.

Giavasis, I. Bioactive fungal polysaccharides as potential functional ingredients in food and nutraceuticals. *Curr. Opin. Biotechnol.* **2014,** *26,* 162–73.

Göbel, H.H. Umezawa: enzyme inhibitors of microbial origin. University Park Press, Baltimore, 1972. *Food/Nahrung* **1976,** *20*, 87.

Gou, S.; Yang, Z.; Liu, T.; Wu, H.; Wang, C. Use of probiotics in the treatment of severe acute pancreatitis: a systematic review and meta-analysis of randomized controlled trials. *Crit. Care* **2014,** *18*, R57.

Greve, H.; Schupp, P.J.; Eguereva, E.; Kehraus, S.; Kelter, G.; Maier, A.; Fiebig, H.H.; König, G.M. Apralactone A and a new stereochemical class of curvularins from the marine-derived fungus *Curvularia* sp. *European J. Org. Chem.* **2008,** *2008*, 5085–92.

Haddad, E.M.; McAlister, V.C.; Renouf, E.; Malthaner, R.; Kjaer, M.S.; Gluud, L.L. Cyclosporin versus tacrolimus for liver transplanted patients. *Cochrane database Syst. Rev.* **2006,** CD005161.

Haiser, H.J.; Seim, K.L.; Balskus, E.P.; Turnbaugh, P.J. Mechanistic insight into digoxin inactivation by Eggerthella lenta augments our understanding of its pharmacokinetics. *Gut Microbes.* **2014,** *5*, 233–8.

Hajam, I.A.; Dar, P.A.; Won, G.; Lee, J.H. Bacterial ghosts as adjuvants: mechanisms and potential. *Vet. Res.* **2017,** *48,* 37.

Hameed, A.; Hussain, S.A.; Yang, J.; Ijaz, M.U.; Liu, Q.; Suleria, H.A.R.; Song, Y. Antioxidants potential of the Filamentous Fungi (*Mucor circinelloides*). *Nutrients* **2017,** *9*, 1101.

Hameed, S.R.; Ss, M.; Al-Wasify, R.S.; MS, S. Production of secondary metabolites as antioxidants from marine-derived fungi and bacteria. *Int. J. ChemTech. Res.* **2015**, *8,* 92–99.

Hirayama, K.; Takahashi, R.; Akashi, S.; Fukuhara, K.; Oouchi, N.; Murai, A.; Arai, M.; Murao, S.; Tanaka, K.; Nojima, I. Primary structure of Paim I, an alpha-amylase inhibitor from *Streptomyces corchorushii*, determined by the combination of Edman degradation and fast atom bombardment mass spectrometry. *Biochemistry* **1987,** *26*, 6483–8.

Holzapfel, W.H.; Haberer, P.; Geisen, R.; Björkroth, J.; Schillinger, U. Taxonomy and important features of probiotic microorganisms in food and nutrition. *Am. J. Clin. Nutr.* **2001,** *73*, 365S–373S.

Hooijmans, C.R.; de Vries, R.B.M.; Rovers, M.M.; Gooszen, H.G.; Ritskes-Hoitinga, M. The effects of probiotic supplementation on experimental acute pancreatitis: a systematic review and meta-analysis. *PLoS One* **2012,** *7*, e48811.

Horta, A.; Pinteus, S.; Alves, C.; Fino, N.; Silva, J.; Fernandez, S.; Rodrigues, A.; Pedrosa, R. Antioxidant and antimicrobial potential of the Bifurcaria bifurcata epiphytic bacteria. *Mar. Drugs* **2014,** *12*, 1676–89.

Huebner, E.S.; Surawicz, C.M. Treatment of recurrent clostridium difficile diarrhea. *Gastroenterol. Hepatol. (N. Y).* **2006,** 2, 203–08.

Hugenholtz, J.; Smid, E.J. Nutraceutical production with food-grade microorganisms. *Curr. Opin. Biotechnol.* **2002,** *13*, 497–507.

Hughes, C.C.; Prieto-Davo, A.; Jensen, P.R.; Fenical, W. The marinopyrroles, antibiotics of an unprecedented structure class from a marine Streptomyces sp. *Org. Lett.* **2008,** *10*, 629–31.

Hüser, A.T.; Chassagnole, C.; Lindley, N.D.; Merkamm, M.; Guyonvarch, A.; Elisáková, V.; Pátek, M.; Kalinowski, J.; Brune, I.; Pühler, A.; Tauch, A. Rational design of a *Corynebacterium glutamicum* pantothenate production strain and its characterization by metabolic flux analysis and genome-wide transcriptional profiling. *Appl. Environ. Microbiol.* **2005,** *71*, 3255–68.

Hwang, E.I.; Yun, B.S.; Kim, Y.K.; Kwon, B.M.; Kim, H.G.; Lee, H.B.; Jeong, W.J.; Kim, S.U. Phellinsin A, a novel chitin synthases inhibitor produced by Phellinus sp. PL3. *J. Antibiot.* **2000,** *53*, 903–11.

Imhann, F.; Bonder, M.J.; Vich Vila, A.; Fu, J.; Mujagic, Z.; Vork, L.; Tigchelaar, E.F.; Jankipersadsing, S.A.; Cenit, M.C.; Harmsen, H.J.M.; Dijkstra, G.; Franke, L.; Xavier, R.J.; Jonkers, D.; Wijmenga, C.; Weersma, R.K.; Zhernakova, A. Proton pump inhibitors affect the gut microbiome. *Gut* **2016,** *65*, 740–48.

Isaac, C.E.; Jones, A.; Pickard, M.A. Production of cyclosporins by *Tolypocladium niveum* strains. *Antimicrob. Agents Chemother.* **1990,** *34*, 121–27.

Ishikawa, H. Mizoribine and mycophenolate mofetil. *Curr. Med. Chem.* **1999,** *6*, 575–97.

Jansson, R.K.; Dybas, R.A. Avermectins: biochemical mode of action, biological activity and agricultural importance. In: *Insecticides with Novel Modes of Action*. Springer: Berlin. **1998**, 152–70.

Joo, N.E.; Ritchie, K.; Kamarajan, P.; Miao, D.; Kapila, Y.L. Nisin, an apoptogenic bacteriocin and food preservative, attenuates HNSCC tumorigenesis via CHAC1. *Cancer Med.* **2012,** *1*, 295–305.

Just, I.; Wilm, M.; Selzer, J.; Rex, G.; von Eichel-Streiber, C.; Mann, M.; Aktories, K. The enterotoxin from *Clostridium difficile* (ToxA) monoglucosylates the Rho proteins. *J. Biol. Chem.* **1995,** *270*, 13932–936.

Kassam, Z.; Lee, C.H.; Yuan, Y.; Hunt, R.H. Fecal microbiota transplantation for *Clostridium difficile* infection: systematic review and meta-analysis. *Am. J. Gastroenterol.* **2013,** *108*, 500–8.

Kim, H.; Baker, J.B.; Park, Y.; Park, H.B.; DeArmond, P.D.; Kim, S.H.; Fitzgerald, M.C.; Lee, D.S.; Hong, J. Total synthesis, assignment of the absolute stereochemistry, and structure-activity relationship studies of subglutinols A and B. *Chem. Asian J.* **2010,** *5*, 1902–10.

Kinnari, T.J. The role of biofilm in chronic laryngitis and in head and neck cancer. *Curr. Opin. Otolaryngol. Head Neck Surg.* **2015,** *23*, 448–53.

Kogej, T.; Gostinčar, C.; Volkmann, M.; Gorbushina, A.A.; Gunde-Cimerman, N. Mycosporines in extremophilic fungi—novel complementary osmolytes? *Environ. Chem.* **2006,** *3*, 105.

Komor, U.; Bielecki, P.; Loessner, H.; Rohde, M.; Wolf, K.; Westphal, K.; Weiss, S.; Häussler, S. Biofilm formation by *Pseudomonas aeruginosa* in solid murine tumors—a novel model system. *Microbes Infect.* **2012,** *14*, 951–8.

Koppel, N.; Maini Rekdal, V.; Balskus, E.P. Chemical transformation of xenobiotics by the human gut microbiota. *Science* **2017,** *356*, eaag2770.

Kralj, A.; Kehraus, S.; Krick, A.; van Echten-Deckert, G.; König, G.M. Two new depsipeptides from the marine fungus *Spicellum roseum*. *Planta Med.* **2007,** *73*, 366–71.

Krzyżowska, K.; Kolonko, A.; Giza, P.; Chudek, J.; Więcek, A. Which kidney transplant recipients can benefit from the initial tacrolimus dose reduction? *Biomed. Res. Int.* **2018,** *4573452*, 1–9.

Kullisaar, T.; Zilmer, M.; Mikelsaar, M.; Vihalemm, T.; Annuk, H.; Kairane, C.; Kilk, A. Two antioxidative lactobacilli strains as promising probiotics. *Int. J. Food Microbiol.* **2002,** *72*, 215–24.

Kumeria, T.; Maher, S.; Wang, Y.; Kaur, G.; Wang, L.; Erkelens, M.; Forward, P.; Lambert, M.F.; Evdokiou, A.; Losic, D. Naturally derived iron oxide nanowires from bacteria for magnetically triggered drug release and cancer hyperthermia in 2D and 3D culture environments: bacteria biofilm to potent cancer therapeutic. *Biomacromolecules* **2016,** *17*, 2726–36.

Kylänpää, L.; Rakonczay, Z.; O'Reilly, D.A. The clinical course of acute pancreatitis and the inflammatory mediators that drive it. *Int. J. Inflam.* **2012,** 360685, 1–10.

Lax, A.J. Opinion: bacterial toxins and cancer—a case to answer? *Nat. Rev. Microbiol.* **2005,** *3*, 343–9.

Lee, J.M.; Wagner, M.; Xiao, R.; Kim, K.H.; Feng, D.; Lazar, M.A.; Moore, D.D. Nutrient-sensing nuclear receptors coordinate autophagy. *Nature* **2014,** *516*, 112–5.

Lee, J.Y.; Lee, J.Y.; Jung, H.W.; Hwang, B.K. Streptomyces koyangensis sp. nov., a novel actinomycete that produces 4-phenyl-3-butenoic acid. *Int. J. Syst. Evol. Microbiol.***2005,** *55*, 257–62.

Lehouritis, P.; Cummins, J.; Stanton, M.; Murphy, C.T.; McCarthy, F.O.; Reid, G.; Urbaniak, C.; Byrne, W.L.; Tangney, M. Local bacteria affect the efficacy of chemotherapeutic drugs. *Sci. Rep.* **2015,** *5*, 14554.

Leschner, S.; Weiss, S. Salmonella-allies in the fight against cancer. *J. Mol. Med. (Berl).* **2010,** *88*, 763–73.

Ley, R.E.; Turnbaugh, P.J.; Klein, S.; Gordon, J.I. Microbial ecology: human gut microbes associated with obesity. *Nature* **2006,** *444*, 1022–3.

Li, J.; Li, L.; Feng, C.; Chen, Y.; Tan, H. Novel polyoxins generated by heterologously expressing polyoxin biosynthetic gene cluster in the sanN inactivated mutant of Streptomyces ansochromogenes. Microb. *Cell Fact.* **2012,** *11*, 135.

Liao, G.; Li, J.; Li, L.; Yang, H.; Tian, Y.; Tan, H. Cloning, reassembling and integration of the entire nikkomycin biosynthetic gene cluster into Streptomyces ansochromogenes lead to an improved nikkomycin production. *Microb. Cell Fact.* **2010,** *9*, 6.

Liebert, C.A.; Wireman, J.; Smith, T.; Summers, A.O. Phylogeny of mercury resistance (mer) operons of gram-negative bacteria isolated from the fecal flora of primates. *Appl. Environ. Microbiol.* **1997**, *63*, 1066–76.

Lin, M.Y.; Yen, C.L. Antioxidative ability of lactic acid bacteria. *J. Agric. Food Chem.* **1999,** *47*, 1460–6.

Macherla, V.R.; Liu, J.; Sunga, M.; White, D.J.; Grodberg, J.; Teisan, S.; Lam, K.S.; Potts, B.C.M. Lipoxazolidinones A, B, and C: antibacterial 4-Oxazolidinones from a marine actinomycete isolated from a guam marine sediment. *J. Nat. Prod.* **2007,** *70*, 1454–57.

Magae, J.; Miller, M.W.; Nagai, K.; Shearer, G.M. Effect of metacycloprodigiosin, an inhibitor of killer T cells on murine skin and heart transplants. *J. Antibiot. (Tokyo).* **1996,** *49*, 86–90.

Maripuri, S.; Kasiske, B.L. The role of mycophenolate mofetil in kidney transplantation revisited. *Transplant. Rev. (Orlando).* **2014,** *28*, 26–31.

Martarelli, D.; Verdenelli, M.C.; Scuri, S.; Cocchioni, M.; Silvi, S.; Cecchini, C.; Pompei, P. Effect of a probiotic intake on oxidant and antioxidant parameters in plasma of athletes during intense exercise training. *Curr. Microbiol.* **2011,** *62*, 1689–96.

McArthur, K.A.; Mitchell, S.S.; Tsueng, G.; Rheingold, A.; White, D.J.; Grodberg, J.; Lam, K.S.; Potts, B.C.M. Lynamicins A−E, chlorinated bisindole pyrrole antibiotics from a novel marine actinomycete†. *J. Nat. Prod.* **2008,** *71*, 1732–37.

McDonald, L.C.; Coignard, B.; Dubberke, E.; Song, X.; Horan, T.; Kutty, P.K.; Ad Hoc Clostridium difficile surveillance working group. Recommendations for surveillance of Clostridium difficile-associated disease. *Infect. Control Hosp. Epidemiol.* **2007,** *28*, 140–5.

McDonald, L.C.; Killgore, G.E.; Thompson, A.; Owens, R.C.; Kazakova, S. V.; Sambol, S.P.; Johnson, S.; Gerding, D.N. An epidemic, toxin gene-variant strain of *Clostridium difficile. N. Engl. J. Med.* **2005,** *353*, 2433–41.

McFarland, L. V. Update on the changing epidemiology of *Clostridium difficile*-associated disease. *Nat. Clin. Pract. Gastroenterol. Hepatol.* **2008,** *5*, 40–48.

Miller, W.J.; Ehrman, L.; Schneider, D. Infectious speciation revisited: impact of symbiont-depletion on female fitness and mating behavior of *Drosophila paulistorum. PLoS Pathog.* **2010,** *6*, e1001214.

Mishra, V.; Shah, C.; Mokashe, N.; Chavan, R.; Yadav, H.; Prajapati, J. Probiotics as potential antioxidants: a systematic review. *J. Agric. Food Chem.* **2015,** *63*, 3615–26.

Mohammad, M.A.; Molloy, A.; Scott, J.; Hussein, L. Plasma cobalamin and folate and their metabolic markers methylmalonic acid and total homocysteine among Egyptian children before and after nutritional supplementation with the probiotic bacteria *Lactobacillus acidophilus* in yoghurt matrix. *Int. J. Food Sci. Nutr.* **2006,** *57*, 470–80.

Mora, C.; Tittensor, D.P.; Adl, S.; Simpson, A.G.B.; Worm, B. How many species are there on earth and in the ocean? *PLoS Biol.* **2011,** *9*, e1001127.

Morales, D.K., Grahl, N., Okegbe, C., Dietrich, L.E.P., Jacobs, N.J., Hogan, D.A., 2013. Control of *Candida albicans* metabolism and biofilm formation by Pseudomonas aeruginosa phenazines. *Am. Soc. Microbiol.* **2013**, *4*, e00526–12.

Moreira, L.A.; Iturbe-Ormaetxe, I.; Jeffery, J.A.; Lu, G.; Pyke, A.T.; Hedges, L.M.; Rocha, B.C.; Hall-Mendelin, S.; Day, A.; Riegler, M.; Hugo, L.E.; Johnson, K.N.; Kay, B.H.; McGraw, E.A.; van den Hurk, A.F.; Ryan, P.A.; O'Neill, S.L. A Wolbachia symbiont in Aedes aegypti limits infection with dengue, Chikungunya, and Plasmodium. *Cell* **2009,** *139*, 1268–78.

Mothes, U.; Seitz, K.A. Action of the microbial metabolite and chitin synthesis inhibitor nikkomycin on the mite *Tetranychus urticae*; an electron microscope study. *Pestic. Sci.* **1982,** *13*, 426–41.

Nandy, S.K.; Bapat, P.M.; Venkatesh, K.V. Sporulating bacteria prefers predation to cannibalism in mixed cultures. *FEBS Lett.* **2007,** *581*, 151–6.

Newman, D.J.; Cragg, G.M. Marine natural products and related compounds in clinical and advanced preclinical trials. *J. Nat. Prod.* **2004,** *67*, 1216–38.

Nicholls, S.J.; Tuzcu, E.M.; Sipahi, I.; Grasso, A.W.; Schoenhagen, P.; Hu, T.; Wolski, K.; Crowe, T.; Desai, M.Y.; Hazen, S.L.; Kapadia, S.R.; Nissen, S.E. Statins, high-density lipoprotein cholesterol, and regression of coronary atherosclerosis. *JAMA* **2007,** *297*, 499–508.

Paiva, A.D.; de Oliveira, M.D.; de Paula, S.O.; Baracat-Pereira, M.C.; Breukink, E.; Mantovani, H.C. Toxicity of bovicin HC5 against mammalian cell lines and the role of cholesterol in bacteriocin activity. *Microbiology* **2012,** *158*, 2851–8.

Pal Singh, I.; Bharate, S.B. Phloroglucinol compounds of natural origin. *Nat. Prod. Rep.* **2006,** *23*, 558–91.

Pang, Y.; Kartsonaki, C.; Turnbull, I.; Guo, Y.; Yang, L.; Bian, Z.; Chen, Y.; Millwood, I.Y.; Bragg, F.; Gong, W.; Xu, Q.; Kang, Q.; Chen, J.; Li, L.; Holmes, M. V.; Chen, Z. Metabolic and lifestyle risk factors for acute pancreatitis in Chinese adults: a prospective cohort study of 0.5 million people. *PLOS Med.* **2018,** *15*, e1002618.

Panno, L.; Bruno, M.; Voyron, S.; Anastasi, A.; Gnavi, G.; Miserere, L.; Varese, G.C. Diversity, ecological role and potential biotechnological applications of marine fungi associated to the seagrass Posidonia oceanica. *New Biotechnol.* **2013,** *30*, 685–94.

Park, E.Y.; Zhang, J.H.; Tajima, S.; Dwiarti, L. Isolation of Ashbya gossypii mutant for an improved riboflavin production targeting for biorefinery technology. *J. Appl. Microbiol.* **2007,** *103*, 468–76.

Pawar, R.; Mohandass, C.; Sivaperumal, E.; Sabu, E.; Rajasabapathy, R.; Jagtap, T. Epiphytic marine pigmented bacteria: a prospective source of natural antioxidants. *Braz. J. Microbiol.* **2015,** *46*, 29–39.

Penders, J.; Stobberingh, E.E.; van den Brandt, P.A.; Thijs, C. The role of the intestinal microbiota in the development of atopic disorders. *Allergy* **2007,** *62*, 1223–36.

Peng, Y.; Demain, A.L. A new hydroxylase system in Actinomadura sp cells converting compactin to pravastatin. *J. Ind. Microbiol. Biotechnol.* **1998,** *20*, 373–75.

Peppercorn, M.A.; Goldman, P. The role of intestinal bacteria in the metabolism of salicylazosulfapyridine. *J. Pharmacol. Exp. Ther.* **1972,** *181*, 555–62.

Pickard, J.M.; Zeng, M.Y.; Caruso, R.; Núñez, G. Gut microbiota: role in pathogen colonization, immune responses, and inflammatory disease. *Immunol. Rev.* **2017,** *279*, 70–89.

Plaza, M.; Cifuentes, A.; Ibanez, E. In the search of new functional food ingredients from algae. *Trends Food Sci. Technol.* **2008,** *19*, 31–39.

Pompei, A.; Cordisco, L.; Amaretti, A.; Zanoni, S.; Matteuzzi, D.; Rossi, M. Folate production by bifidobacteria as a potential probiotic property. *Appl. Environ. Microbiol.* **2007,** *73*, 179–85.

Proksch, P.; Ebel, R.; Edrada, R.; Riebe, F.; Liu, H.; Diesel, A.; Bayer, M.; Li, X.; Han Lin, W.; Grebenyuk, V.; Müller, W.E.G.; Draeger, S.; Zuccaro, A.; Schulz, B. Sponge-associated fungi and their bioactive compounds: the *Suberites* case. *Bot. Mar.* **2008,** *51*, 209–218.

Raguz, S.; Yagüe, E. Resistance to chemotherapy: new treatments and novel insights into an old problem. *Br. J. Cancer* **2008,** *99*, 387–91.

Rauch, M.; Lynch, S.V. The potential for probiotic manipulation of the gastrointestinal microbiome. *Curr. Opin. Biotechnol.* **2012,** *23*, 192–201.

Ren, D.; Li, L.; Schwabacher, A.W.; Young, J.W.; Beitz, D.C. Mechanism of cholesterol reduction to coprostanol by Eubacterium coprostanoligenes ATCC 51222. *Steroids* **1996,** *61*, 33–40.

Rosenberg, J.; Ischebeck, T.; Commichau, F.M. Vitamin B6 metabolism in microbes and approaches for fermentative production. *Biotechnol. Adv.* **2017,** *35*, 31–40.

Sartor, R.B.; Mazmanian, S.K. Intestinal microbes in inflammatory bowel diseases. *Am. J. Gastroenterol. Suppl.* **2012,** 1, 15–21.

Scharf, D.H.; Heinekamp, T.; Remme, N.; Hortschansky, P.; Brakhage, A.A.; Hertweck, C. Biosynthesis and function of gliotoxin in *Aspergillus fumigatus*. *Appl. Microbiol. Biotechnol.* **2012,** *93*, 467–72.

Schieber, M.; Chandel, N.S. ROS function in redox signaling and oxidative stress. *Curr. Biol.* **2014,** *24*, R453–62.

Scholte, E.J.; Knols, B.G.J.; Samson, R.A.; Takken, W. Entomopathogenic fungi for mosquito control: a review. *J. Insect Sci.* **2004,** *4*, 19.

Schwechheimer, S.K.; Park, E.Y.; Revuelta, J.L.; Becker, J.; Wittmann, C. Biotechnology of riboflavin. *Appl. Microbiol. Biotechnol.* **2016,** *100*, 2107–19.

Sender, R.; Fuchs, S.; Milo, R. Revised estimates for the number of human and bacteria cells in the body. *PLoS Biol.* **2016,** *14*, e1002533.

Serizawa, N.; Matsuoka, T. A two component-type cytochrome P-450 monooxygenase system in a prokaryote that catalyzes hydroxylation of ML-236B to pravastatin, a tissue-selective inhibitor of 3-hydroxy-3-methylglutaryl coenzyme A reductase. *Biochim. Biophys. Acta* **1991,** *1084*, 35–40.

Serrano-Amatriain, C.; Ledesma-Amaro, R.; López-Nicolás, R.; Ros, G.; Jiménez, A.; Revuelta, J.L. Folic acid production by engineered *Ashbya gossypii. Metab. Eng.* **2016,** *38*, 473–82.

Seto, B. Rapamycin and mTOR: a serendipitous discovery and implications for breast cancer. *Clin. Transl. Med.* **2012,** *1*, 29.

Singh, G.; Prakash, S. New prospective on fungal pathogens for mosquitoes and vectors control technology. *J. Mosq. Res.* **2014,** *4*, 36–52.

Singh, R.; Deodhar, A.D. Relative bioavailability of riboflavin in cows' milk and fermented milk using rat bioassay. *Int. Dairy J.* **1994,** *4*, 59–71.

Sivan, A.; Corrales, L.; Hubert, N.; Williams, J.B.; Aquino-Michaels, K.; Earley, Z.M.; Benyamin, F.W.; Lei, Y.M.; Jabri, B.; Alegre, M.L.; Chang, E.B.; Gajewski, T.F. Commensal bifidobacterium promotes antitumor immunity and facilitates anti-PD-L1 efficacy. Science **2015,** 350, 1084–9.

Sonesson, A.; Oqvist, B.; Hagstam, P.; Björkman-Burtscher, I.M.; Miörner, H.; Petersson, A.C. An immunosuppressed patient with systemic vasculitis suffering from cerebral abscesses due to *Nocardia farcinica* identified by 16S rRNA gene universal PCR. *Nephrol. Dial. Transplant.* **2004,** *19*, 2896–900.

Sousa, T.; Paterson, R.; Moore, V.; Carlsson, A.; Abrahamsson, B.; Basit, A.W. The gastrointestinal microbiota as a site for the biotransformation of drugs. *Int. J. Pharm.* **2008,** *363*, 1–25.

Strozzi, G.P.; Mogna, L. Quantification of folic acid in human feces after administration of Bifidobacterium probiotic strains. *J. Clin. Gastroenterol.* **2008,** *42 Suppl 3*, S179–84.

Suttie, J.W. The importance of menaquinones in human nutrition. *Annu. Rev. Nutr.* **1995,** *15*, 399–417.

Tang, M.L.K.; Lahtinen, S.J.; Boyle, R.J. Probiotics and prebiotics: clinical effects in allergic disease. *Curr. Opin. Pediatr.* **2010,** *22*, 626–34.

Toossi, Z.; Gogate, P.; Shiratsuchi, H.; Young, T.; Ellner, J.J. Enhanced production of TGF-beta by blood monocytes from patients with active tuberculosis and presence of TGF-beta in tuberculous granulomatous lung lesions. *J. Immunol.* **1995,** *154*, 465–73.

Trisuwan, K.; Rukachaisirikul, V.; Sukpondma, Y.; Phongpaichit, S.; Preedanon, S.; Sakayaroj, J. Lactone derivatives from the marine-derived fungus *Penicillium* sp. PSU-F44. *Chem. Pharm. Bull. (Tokyo).* **2009,** *57*, 1100–2.

Turnbaugh, P.J.; Hamady, M.; Yatsunenko, T.; Cantarel, B.L.; Duncan, A.; Ley, R.E.; Sogin, M.L.; Jones, W.J.; Roe, B.A.; Affourtit, J.P.; Egholm, M.; Henrissat, B.; Heath, A.C.; Knight, R.; Gordon, J.I. A core gut microbiome in obese and lean twins. *Nature* **2009,** *457*, 480–4.

Uzair, B.; Ahmed, N.; Ahmad, V.U.; Mohammad, F.V.; Edwards, D.H. The isolation, purification and biological activity of a novel antibacterial compound produced by *Pseudomonas stutzeri. FEMS Microbiol. Lett.* **2008,** *279*, 243–50.

Van Arsdell, S.W.; Perkins, J.B.; Yocum, R.R.; Luan, L.; Howitt, C.L.; Chatterjee, N.P.; Pero, J.G. Removing a bottleneck in the *Bacillus subtilis* biotin pathway: bioA utilizes lysine rather than S-adenosylmethionine as the amino donor in the KAPA-to-DAPA reaction. *Biotechnol. Bioeng.* **2005,** *91*, 75–83.

van Hogezand, R.A.; Kennis, H.M.; van Schaik, A.; Koopman, J.P.; van Hees, P.A.; van Tongeren, J.H. Bacterial acetylation of 5-aminosalicylic acid in faecal suspensions cultured under aerobic and anaerobic conditions. *Eur. J. Clin. Pharmacol.* **1992,** *43*, 189–92.

Villarante, K.I.; Elegado, F.B.; Iwatani, S.; Zendo, T.; Sonomoto, K.; de Guzman, E.E. Purification, characterization and in vitro cytotoxicity of the bacteriocin from Pediococcus acidilactici K2a2-3 against human colon adenocarcinoma (HT29) and human cervical carcinoma (HeLa) cells. *World J. Microbiol. Biotechnol.* **2011,** *27*, 975–80.

Wang, G.; Shi, T.; Chen, T.; Wang, X.; Wang, Y.; Liu, D.; Guo, J.; Fu, J.; Feng, L.; Wang, Z.; Zhao, X. Integrated whole-genome and transcriptome sequence analysis reveals the genetic characteristics of a riboflavin-overproducing Bacillus subtilis. *Metab. Eng.* **2018,** *48*, 138–49.

Wang, S.C.; Bligh, S.W.A.; Shi, S.S.; Wang, Z.T.; Hu, Z.B.; Crowder, J.; Branford-White, C.; Vella, C. Structural features and anti-HIV-1 activity of novel polysaccharides from red algae *Grateloupia longifolia* and *Grateloupia filicina. Int. J. Biol. Macromol.* **2007,** *41*, 369–75.

Wang, Y.; Wu, Y.; Wang, Y.; Xu, H.; Mei, X.; Yu, D.; Wang, Y.; Li, W. Antioxidant properties of probiotic bacteria. *Nutrients* **2017,** *9*, 521.

Warren, M.J. Finding the final pieces of the vitamin B12 biosynthetic jigsaw. *Proc. Natl. Acad. Sci.* **2006,** *103*, 4799–800.

Wegkamp, A.; Starrenburg, M.; de Vos, W.M.; Hugenholtz, J.; Sybesma, W. Transformation of folate-consuming *Lactobacillus gasseri* into a folate producer. *Appl. Environ. Microbiol.* **2004,** *70*, 3146–8.

Wei, M.Q.; Ellem, K.A.O.; Dunn, P.; West, M.J.; Bai, C.X.; Vogelstein, B. Facultative or obligate anaerobic bacteria have the potential for multimodality therapy of solid tumours. *Eur. J. Cancer* **2007,** *43*, 490–6.

Weibel, E.K; Hadvary, P.; Hochuli, E.; Kupfer, E.; Lengsfeld, H. Lipstatin, an inhibitor of pancreatic lipase, produced by *Streptomyces toxytricini*. I. Producing organism, fermentation, isolation and biological activity. *J. Antibiot. (Tokyo).* **1987,** *40*, 1081–5.

Whitcomb, D.C. Acute pancreatitis. *New Engl. J. Med.* **2006,** *354*, 2142–50.

World Health Organization. World malaria report 2008. World Health Organization, Switzerland, **2008**, 99–101.

Wilke, A.B.B.; Marrelli, M.T. Paratransgenesis: a promising new strategy for mosquito vector control. *Parasit. Vectors* **2015,** *8*, 342.

Williams, H.; Stewart, A.; von Mutius, E.; Cookson, W.; Anderson, H.R.; International study of asthma and allergies in childhood (ISAAC) phase one and three study groups. Is eczema really on the increase worldwide? *J. Allergy Clin. Immunol.* **2008,** *121*, 947–54.e15.

Wingender, W. Proteinase inhibitors of microbial origin a review. In: *Proteinase Inhibitors*. Springer: Berlin. **1974**, 548–59.

Yap, C.Y.; Chen, F. Polyunsaturated fatty acids: biological significance, biosynthesis, and production by microalgae and microalgae-like organisms. In *Algae and Their Biotechnological Potential,* Springer: Dordrecht. **2001**, 1–32.

Yen, G.C.; Chang, Y.C. Medium Optimization for the production of antioxidants from *Aspergillus candidus*. *J. Food Prot.* **1999,** *62*, 657–61.

Yoghiapiscessa, D.; Batubara, I.; Wahyudi, A.T. Antimicrobial and antioxidant activities of bacterial extracts from marine bacteria associated with Sponge Stylotella sp. *Am. J. Biochem. Biotechnol.* **2016,** *12*, 36–46.

Youmans, B.P.; Ajami, N.J.; Jiang, Z.D.; Campbell, F.; Wadsworth, W.D.; Petrosino, J.F.; DuPont, H.L.; Highlander, S.K. Characterization of the human gut microbiome during travelers' diarrhea. *Gut Microbes.* **2015,** *6*, 110–9.

Zaim, M.; Guillet, P. Alternative insecticides: an urgent need. *Trends Parasitol.* **2002,** *18*, 161–3.

CHAPTER 6

Rise of the Microbial World: An Economic Point of View

BINITA DEV and R. JAYABALAN*

Food Microbiology and Bioprocess Laboratory, Department of Life Science, National Institute of Technology, Rourkela 769008, Odisha, India

Corresponding author. E-mail: jayabalanr@nitrkl.ac.in

ABSTRACT

With the demographic rise of population and depletion of nonrenewable resources, the sustainability of the environment is at risk and demands urgent empowerment of sustainable growth of the economy and national infrastructure. The sustainable development goals (SDG) set for 2030 by United Nations warrant the achievement of socioeconomic and environmental growth through greener processes and cleaner technologies intending to accomplish people's needs. This would also create enormous employment opportunities through sustainable techniques without encouraging the overexploitation of natural resources. The use of microorganisms offers several advantages in terms of sustainability, although it remains mostly underestimated. Therefore, the scope of the utilization of diversified microbes in the fulfillment of the 2030 agenda of SDG can contribute immensely to sustainable growth. According to BCC research reports, the market of microbial products in 2018 is globally estimated to be approximately \$186.3 billion and is expected to reach as high as \$302.4 billion by 2023. The prevalence of diseases, as well as the elevating demands of healthcare products and rising lifestyle patterns, are the driving factors for the growth in the market of microbial products. This chapter focuses on the advent of the microbial world in food, healthcare, energy production (bioethanol, biodiesel, and biogas), medicine, industry, agriculture, bioremediation, wastewater treatment, and others.

6.1 INTRODUCTION

The indiscriminate exploitation of fossil fuel resources, energy crisis, overpopulation, and global climate changes have jeopardized environmental sustainability which can adversely affect our future generations. The adoption of sustainable goals, waste minimization, pollution prevention, and eco-efficiency is currently the need of the hour for the conservation of earth's natural resources. In order to protect the world from the current menace, 17 sustainable development goals (SDG) were adopted by the United Nations in January 2016 to which 173 countries have agreed. These goals encapsulate 5 P's—planet, people, peace, prosperity, and partnerships with special emphasis on the preservation of the environment (Akinsemolu, 2018; Timmis et al., 2017) and judicial use of microbes. Although microbes are often associated with causing diseases, they also play several essential roles in maintaining the liveliness on earth, and hence in the current scenario, the need to discover the wonders of the microbial world has become a prime focus (Akinsemolu, 2018).

6.2 APPLICATION OF MICROBES

Microbes such as yeast, bacteria, fungus are minute, microscopic living organisms and have been used since centuries for the production of bread, vinegar, wine, beer, and other common products. The microbial world is still an unexplored and undiscovered reservoir on earth. They constitute a large fraction of earth's biomass and are well-known for their diverse functional roles ranging in agriculture, environment sustainability, wastewater treatment as well as the development of biosensors, probiotics, and so on (Bhattacharyya et al., 2016; Ahmad et al., 2011). In addition to the industrial production of a vast array of products, microbes play a crucial role in modulating various biogeochemical cycles and green bioconversions. Despite the endless commercial prospects of microbes, their use in the empowerment of SDG goals has been given little emphasis (Akinsemolu, 2018). Summarized below are the list of varied uses of microbes and their contribution to the economic progress of the world (Figure 6.1).

6.2.1 MICROBES IN AGRICULTURAL SECTOR

The rise in the world's population and global climate change has posed a challenge to agriculture production. Microbiomes associated with plants are known to promote growth, provide nutrients, and modulate phytohormone levels to combat biotic and abiotic stress (Timmusk et al., 2017). Microbes play an immense role in increasing crop productivity, improving soil fertility, and antagonizing plant-pathogen interaction and therefore it can be utilized as a green technology (Johansson et al., 2004). Microbes mostly colonize the plant parts above ground and the rhizosphere region. The colonization includes bacteria in the majority (95%) and also consists of actinomycetes, fungi, algae, and protozoa (Glick 2012). These colonizing microbes are often termed as plant growth-promoting bacteria (PGPB) or rhizobacteria (PGPR) and in association with plants, they exhibit multifunctional potential in the form of biopesticides or biocontrol agents and biofertilizer (Pereg and McMillan, 2015).

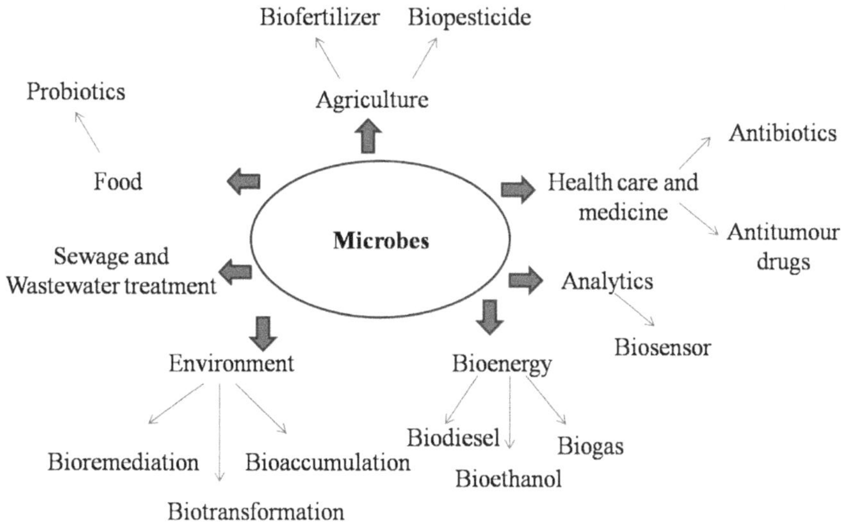

FIGURE 6.1 Potential role of microbes.

Biocontrol agents or biopesticides suppress the occurrence and development of various diseases. In order to circumvent these diseases, plants implement several strategies that include the production of siderophores, antibiotics, and cell wall lysis enzymes, such as chitinase, β-1,3-glucanase (Ahmad et al., 2008; Glick, 2012).

Biofertilizers such as *Azotobacter, Rhizobium,* and *Azospirillum* help in plant growth by providing or increasing the availability of primary nutrients to the plants (Timmusk et al., 2017). The mechanism includes phosphate solubilization, nitrogen fixation, sequestration of iron, and production of plant growth regulators (indole acetic acid, gibberellins, cytokinin, and ethylene) (Ahmad et al., 2008; Glick 2012).

Some of the commercialized strains used as adjuncts in agriculture include *Agrobacterium radiobacter, Azotobacter chroococcum, Bacillus licheniformis, Pseudomonas fluorescens, Serratia entomophilia*, and several *Rhizobia* spp. (Lucy et al., 2004; Rai, 2006). Unlike chemical fertilizers, biopesticides and biofertilizers are eco-friendly renewable resources that maintain the integrity of soil biology and lessen the effect of abiotic stress (Sharma et al., 2016; Bharti et al., 2016) Therefore, the use of biofertilizers is envisioned to provide a viable alternative to the use of chemicals in agriculture in the near future.

Commercially, the PGPB and PGPR are addressed as biofertilizers and biopesticides. According to the report of Markets and markets, the biopesticide market valuation is projected to reach US$ 6.4 billion in 2023 with a compound annual growth rate (CAGR) of 15.99% between 2017 and 2023 (Anonymous, 2019a).

The biofertilizer market valuation is also projected to increase up to $2.27 billion by 2023 with a CAGR of 13.6% as per Market Data forecast reports (Anonymous, 2018a).

6.2.2 HEALTHCARE AND MEDICINE

Though few microbes are pathogenic, most of them are beneficial for human health. The population of microbes is 10 times more compared to that of the human cells inside our body and a vast majority of them reside in the intestinal tract and others in the skin, airway, and genital tract (Chen and Tsao, 2013; Lamont et al., 2011; Grice and Segre, 2012). About 1000 different known bacterial species reside within the human body and carry approximately 150 genes more than that of humans (Ursell et al., 2014). With the Human Genome Project, a total of 178 genomes from these microbes residing both on the surface and inside our body were analyzed and collectively termed as microbiome (Turnbaugh et al., 2007). The bacteria growing symbiotically metabolize indigestible dietary fibers such as xyloglucans, synthesizing, and delivering essential vitamins to the host and foster humoral and cellular mucosal immune system (Round et al., 2009).

The serendipitous discovery of the world's first antibiotic penicillin in 1929 has driven the interest toward the remedial use of microbes and microbial products in treating life-threatening diseases (Singh et al., 2017). Subsequent research had resulted in the discovery of several antibiotics from fungi and actinomycetes such as cephalosporin, tetracycline, glycopeptides, aminoglycosides, and macrolides, which have a wide application in food preservation, chemotherapy, and their use as lifesaving drugs (Hasan et al., 2012).

However, antimicrobial drug resistance is a serious menace of the inappropriate use of antibiotics in microbes such as *E. coli*, *Klebsiella*, *Streptococcus*, and *Enterococcus*. Also, the emergence of superbugs can be checked and controlled by immunization, restricted intake of nonprescribed antimicrobials, and promoting awareness toward the legal use of antimicrobials (Planta, 2007).

Due to the rise of infectious diseases, the consumption of antibiotics has escalated which has led to the increased growth of the global antibiotic market at a CAGR of 2.1% (2018–2025) and is expected to reach $50,374 million by 2025 as per Allied market research reports (Sumant et al., 2018)

Microbes are extensively studied for vaccine production. Vaccines are introduced in the body as suspensions of an avirulent form of a pathogen that can be live-attenuated or used as toxoids for inducing a specific and adaptive immune response in the host against a number of diseases. When the vaccine carrying an epitope similar to that of the antigen present on the surface of the disease-causing microbe enters the body, it triggers the host immune response by the generation of antigen-specific antibodies. Subsequently, upon exposure to the disease-causing microbe, the immune system identifies the antigen and responds rapidly to it by producing the same antibodies made against the vaccine and thereby preventing the development of the disease (Lee et al., 2012). The first vaccine discovery dates back to 1796 by British physician Edward Jenner who used cowpox virus, vaccinia like lesions on a milkmaid's hand to prevent infection against the smallpox virus (Riedel, 2005). Further research resulted in the discovery of a series of vaccines against mumps, measles, plague, rabies, polio, typhoid fever, cholera, tetanus, tuberculosis, and others.

The importance of immunization through vaccines has resulted in the growth of the global vaccine market with an expectation to reach $103.57 billion by 2028 at a CAGR of 11.02% from 2018 to 2028 (Anonymous, 2018b).

The microbes have gained significant interest in the development of antitumor drugs due to the nonspecific toxicity of the conventional methods to eradicate metastatic cancers (Demain and Vaishnav, 2011). The first microbial-based anticancer agent, Actinomycin D isolated from *Streptomyces antibioticus* was used against Wilms tumor in children which resulted in a 90% survival rate (Chung, 2009). Antitumor drugs derived from microbes are obtained from different forms of bacteria live, attenuated, nonpathogenic, or genetically modified variants. Reports suggest that bacteria perform the targeted antitumor effect in various ways such as by completely removing the tumor and antineoplastic growth retardation by *Clostridia, Bifidobacteria*, and *Salmonella* (Leschner and Weiss, 2010). Further, it causes depletion of nutrients required for cancer metabolism (Danino et al., 2013) which enhances the immunity of the host (Song et al., 2018) and also inhibits tumor growth by releasing substances such as enzymes, bacteriocolicins from *Escherichia coli* (Kaur and Kaur, 2015). Biofilm mediated metastasis arrest by *Salmonella typhimurium* (Weitao 2009) and synthesis of immunosuppressant pigments such as Prodigiosin from *Serratia marcescens, Vibrio psychroerythrus, Hahella chejuensis* has also been reported (Chang et al., 2011).

The demand for antitumor drugs is on the rise with leading pharmaceutical companies such as Novartis, Roche, Celgene, and others. The global antitumor drug market is expected to reach USD 110 billion by 2022 at a CAGR of >6% (Anonymous 2018c).

6.2.3 MICROBES IN BIOGAS PRODUCTION

The major crisis in the energy resources, environmental consequences of the use of fossil fuels, and soaring crude oil prices have accentuated the need for renewable fuels for sustainable development. One such prominent renewable energy resource is biogas that can be harnessed for power production and used as vehicle fuel after its conversion into biomethane (Rasi et al., 2011). Biogas is an eco-friendly fuel produced by anaerobic bacteria consortium from the decomposition of organic matter particularly biological wastes such as municipal solid waste, cattle dung, vegetable waste in an oxygen-free environment to produce a mixture of gases such as methane (50%–70%), carbon dioxide (20%–40%), nitrogen, hydrogen, ammonia, and smaller amounts of hydrogen sulfide (Christy et al., 2014; Hidayati et al., 2018; Gopinath et al., 2014). Biogas production is a four-step process mediated by the microbial population which in the first step acts synergistically to hydrolyze the complex organic compound into a simpler form (hydrolysis). This is followed by its decomposition into organic acids (acidogenesis), then into acetate (acetogenesis), and ultimately into biogas (methanogenesis) (Angelidaki et al., 1993; Christy et al., 2014).

The hydrolysis step is facilitated by hydrolytic enzymes secreting bacteria such as *Escherichia, Enterobacter, Pseudomonas, Acetobacter, Bifidobacter* that secrete cellulase, cellobiase, xylanase, protease, amylase, and lipase to hydrolyze the organic waste into simpler biomolecules (Cirne et al., 2007).

In the next step of biogas production, the long-chain fatty acids and amino acids obtained from the hydrolysis is used by fermentative microorganisms or anaerobic oxidizers such as *Streptococcus*, *Lactobacillus*, *Bacillus*, and *Salmonella* as the substrate which is converted into short-chain fatty acids, organic acids namely acetic acid, propionic acid, lactic acid, butyric acid along with hydrogen and carbon dioxide (Kalyuzhnyi et al., 2000; Gujer and Zehnder, 1983). Acidogenesis often leads to the production of hydrogen that results in the accumulation of electron sinks with more than two carbon-containing fatty acids and with more than one carbon-containing compound namely lactate, ethanol, propionate, and butyrate. These molecules cannot be consumed directly by methanogens thus hindering the process of methanogenesis and is degraded by acetogenic bacteria to produce acetate, carbon dioxide, hydrogen in a process known as acetogenesis (Björnsson et al., 2000). This step is carried out by hydrogen-producing acetogenic bacteria namely *Syntrophomonas wolfeii*, *Syntrophobacter wolinii* (Mah, 1982). The fourth and the last step comprises methanogenesis, the metabolic pathway in which methanogens such as *Methanobacterium*, *Methanobacillus*, *Methanoscarcina*, *Methanococcus* uses acetate, carbon dioxide as carbon sources and their conversion into methane (Hidayati et al., 2018).

According to a research report of Technavio, the global biogas market value is predicted to be close to 6% during 2019–2023 (Anonymous, 2019b) which emphasizes the role of microbes in global biogas production.

6.2.4 MICROBES IN BIOETHANOL PRODUCTION

The greenhouse gas emissions from conventional fuel and their finite availability have boosted the demand for sustainable, eco-friendly, and cost-effective renewable fuel that can satisfy global future energy needs. Among the promising biofuels namely bioethanol, biodiesel, biobutanol, bioethanol has gained prominence in research, industry, and government (Thangavelu et al., 2016). The use of gasoline-ethanol blend has reduced carbon dioxide emissions from the transportation sector by about 43.5 million metric tons in 2016 which is approximately equal to the removal of 9.3 million cars for 1 year (Robak and Balcerek, 2018). Microbes possess a unique metabolic diversity by virtue of which it can utilize various substrates and therefore have been widely exploited for renewable biofuels production from biomass and biological wastes (Liao et al., 2016). Lignocellulosic biomass such as agricultural wastes and forestry residues, herbaceous crops, and other wastes are the most preferred substrate for bioethanol production because their use is not related to food security and are abundantly available. So bioethanol production from the lignocellulosic substrate can be the best option of their disposal (Spyridon et al., 2016). The production of bioethanol from lignocellulosic waste biomass comprises of three steps—pretreatment, saccharification, and fermentation with the usage of different microbes in each step.

The pretreatment unwinds the cellulose from the hemicellulose and lignin accompanied by an increase in biomass digestibility and slight loss of hemicellulose (Wyman et al., 2005). The various physical, chemical, and physicochemical methods of pretreatment have competent conversion potential but involves the use of harmful chemicals, consume energy, and generate inhibitors which makes it infeasible (Li et al., 2015). The biological pretreatment exploits the natural resources, the potential microbes that are enzymatically capable of delignifying recalcitrant lignocellulosic biomass in an eco-friendly way. It employs microbes such as white-rot, brown-rot, and soft-rot fungi to degrade lignin, hemicellulose, and little cellulose with low energy input, no waste generation, and less inhibitor production (Keller et al., 2003; Kumar et al., 2009; Shi et al., 2009). Unlike brown-rot fungi that degrade cellulose, white-rot fungi possess lignin degradation enzymes such as polyphenol oxidase, lignin peroxidase, and laccase that selectively degrade lignin and are preferred for microbial pretreatments (Perie and Gold, 1991). Several white-rot fungi such as *Ceriporiopsis subvermispora, Phanerochaete chrysosporium, Phlebia subserialis,* and *Pleurotus ostreaus* have shown high delignification efficiency when used on different lignocellulosic biomass (Keller et al., 2003). Certain bacteria such as *Cellulomonas cartae, Bacillus macerans, Cellulomonas uda,* and *Zymomonas mobilis* have also delignification potential up to 50% (Singh et al., 2008).

The next stage is bioethanol production that involves hydrolysis of polymeric sugars-cellulose and hemicellulose into fermentable sugars-hexoses such as glucose, mannose, galactose, and pentoses such as ribose, arabinose, and xylose using hydrolytic enzyme obtained from microbes (das Neves et al., 2007). These enzymes possess the cellulolytic ability and are derived mostly from filamentous fungi such as *Trichoderma reesei, Aspergillus niger, Fusarium oxysporum*, and basidiomycytes such as *Fomitopsis*, *P. chrysosporium* (Sajith et al., 2016; Chandel et al., 2012) and cellulase producing eubacteria belonging to genera such as *Cellulomonas, Clostridium, Bacillus, Thermomonospora, Ruminococcus, Bacteriodes, Erwinia, Microbispora, Acetovibrio*, and *Streptomyces* (Bisaria1998).

The final step involves the fermentation of monomeric sugars into ethanol by the metabolic activity of yeast, bacteria, and fungi (Katahira et al., 2006).

Theoretically, 1 g of glucose produces 0.51 g of ethanol and 0.48 g of carbon dioxide. As the microbe uses some of the glucose for cellular growth, the actual yield is always <100% (Demirbas 2005).

Saccharomyces cerevisiae possesses high ethanol producing ability with theoretical specific productivity of around 2 g ethanol/g cells/h and is most frequently used in microbiological fermentation of hexose sugars into ethanol (Skoog and Hahn-Hägerdal, 1988). The constraint in using *Saccharomyces* in ethanol production is its inefficiency in the fermentation of polysaccharides such as starch, cellulose, and hemicellulose, disaccharides such as cellobiose, xylobiose, and even pentose sugars such as xylose and arabinose that renders the ethanol production to be exorbitant (Boulton et al., 1999; Zhang and Lynd, 2005; Lynd et al., 2002). Therefore, its inability to ferment lignocellulosic hydrolysate containing a wide range of mono-, di-, and tri-saccharides makes it uneconomical. Certain strains like *Pichia stiptis*, *Candida shehatae,* and *Pachysolen tannophilus* are capable of fermentation of xylose into ethanol (Hahn-Hägerdal et al., 2007). The use of concentrated acids and high temperatures often generate fermentable sugars and inhibitors that ultimately decrease the yield of ethanol. These inhibitors mostly belong to furan derivatives such as furfural and 5-hydroxy methyl furfural, weak organic acids, such as formic acid, acetic acid, and levulinic acid, and phenolic compounds (Palmqvist et al., 2000; Chandel et al., 2010). The detoxification of inhibitors is time-consuming and incurs additional costs, thereby increasing the cost of second-generation bioethanol production (Robak and Balcerek, 2018). Various isolated nonconventional yeasts which include osmotolerant *Zygosaccharomyces rouxii*, furan tolerant *Pichia kudriavzevii*, acetic acid-tolerant *Zygosaccharomyces bailii*, ethanol tolerant *Dekkera bruxellensis,* and thermotolerant *Hansenula polymorpha* (Radecka et al., 2015) have been reported to tolerate stress and resist inhibitors. Recently, certain bacteria have gained much interest because of their higher growth rate than *Saccharomyces*. *Z. mobilis* is one such familiar organism with a high growth rate and specific ethanol production but the fermentation ability is restricted to glucose, fructose, and sucrose. Another bacteria *Zymobacter palmae* can ferment broad carbohydrate substrate from hexose, α-linked di- and tri-saccharides, and sugar alcohols (Busic et al., 2018). Some filamentous fungi such as *F. oxysporum*, *Neurospora*, *Monilia, Paecilomyces* also have the ability to ferment cellulose directly into ethanol (Panagiotou et al., 2005; Sanchez et al., 2008).

A research report "India Ethanol Market By Source, By Application, By Purity, Competition, Forecast & Opportunities, 2024" has predicted the rise of the Indian ethanol market from 2019 to 2024 at a CAGR of 14.50%

(Anonymous, 2019c). Recent news by reports and data have stated the global fuel ethanol market to reach USD 122.35 billion in 2026 (Anonymous 2019d) which reiterates the significance of the microbial world from a global perspective.

6.2.5 MICROBES IN BIODIESEL PRODUCTION

Burgeoning crude oil prices and reduced oil supplies have drawn attention toward biodiesel as a prominent source of renewable fuel (Kemp, 2006). To replace fossil fuel, a biofuel should not only be beneficial to the environment but also must be cost-effective than other fossil fuel substitutes. Biodiesels are monoalkyl esters of long-chain fatty acids namely fatty acid methyl ester and fatty acid ethyl esters which are derived from transesterification of renewable feedstock such as organic waste (municipal swedge sludge), animal oil and fats, vegetable oils (palm, soybean, oilseed), and oleaginous microorganisms (yeasts, microalgae, and bacteria) (Antolin et al., 2002; Vicente et al., 2004). Biodiesel is not a carbon dioxide emission fuel as the carbon dioxide released on burning is captured by the biomass during its growth and is therefore regarded as carbon-neutral fuel (Meng et al., 2009). Biodiesel has a huge potential in replacing petroleum-based diesel in internal combustion engines with little or no modification. Also, it lacks any unwanted sulfur emissions and has large prospects toward reducing the greenhouse gases in Earth's atmosphere. Its low flash point makes it a suitable alternative to petroleum-based diesel (Meng et al., 2009; Bajpai and Tyagi, 2006).

Biodiesel production is mainly facilitated by microbes like microalgae, yeasts, and bacteria that accumulate microbial lipid content in excess of 20% by metabolizing organic carbon sources such as sugars and organic acids (Meng et al., 2009; Cho and Park, 2018). These microbes are called oleaginous microorganisms and the lipids are termed as single-cell oil having 16–18 carbon atoms. The oleaginous organisms offer advantages in terms of reduced land requirement, increased lipid accumulation, small generation time, and low-cost cultivation substrate (Cho and Park, 2018). Oleaginous microalgae such as *Scenedesmus quadricauda*, *Chlorella vulguris*, *Chlorella protothecoides*, *Schizochytrium mangrovei* uses organic wastes in the form of municipal sewage, food waste, sugarcane bagasse hydrolysate, domestic wastewater as the substrate for lipid production (Meng et al., 2009).

Yeasts are known as favorable oleaginous organisms as they produce lipids consisting of C_{16}–C_{18} fatty acids and have a high C/N ratio which is favorable for lipid synthesis and lipid accumulation respectively (Kolochova et al., 2016; Economou et al., 2011). Yeasts such as *Cryptococcus curvatus*, *Rhodotorula glutinis*, and *Yarrowia lipolytica* uses food waste leachates as various carbon sources for lipid production (Meng et al., 2009).

Oleaginous bacteria also synthesize a large amount of fatty acids and accumulate them in the form of triacylglycerol from simpler carbon sources like glucose under growth-limiting conditions. They possess the ability to tolerate toxic compounds and have a shorter generation time which makes them a better candidate for biodiesel production compared to microalgae. These bacteria belong to the genera *Bacillus* sp., *Arthrobacter* sp., *Rhodococcus* sp., and *Acinetobacter* sp. strains that can use dairy wastewater as well as lignocellulosic waste to convert it into lignin (Wei et al., 2015; Sriwongchai et al., 2012; Kosa and Ragauskas 2012). They are also capable of growing in organic wastes containing toxic chemical constituents such as aromatic compounds, phenols, and heavy metals (Meng et al., 2009).

The global biodiesel market size is estimated to a value of US$ 33748.2 in 2016 and is projected to expand at a CAGR of over 4% during 2017–2026 (Anonymous 2019e).

6.2.6 MICROBES AS PROBIOTICS

Probiotics refer to live microbes that provide potential health benefits when administered in sufficient amounts to the host. These may be similar or different from the naturally occurring beneficial microbes present in our gut and are supplied in the form of manufactured products such as food additives, dietary supplements, drugs, etc. (Song et al., 2012). Although humans have been aware of numerous beneficial effects of consuming fermented foods like milk since Vedic times even before the discovery of microbes, the scientific evidence came when Metchnikoff in 1908 published the book entitled- "The prolongation of life" where the consumption of fermented milk containing healthy bacteria, *Lactobacilli* was emphasized for a prolonged life span (Raghuwanshi et al., 2015). He stated that the bacteria dwelling in the human bowel degrades protein (putrefaction) and releases some toxic substances such as amine, indoles, ammonia into vascular and nervous systems which accelerate the aging process and these were identified for the cause of autointoxication (Mackowiak, 2013; Tannock, 2004). Gram-positive bacteria *Lactobacillus* and *Bifidobacterium* are some of the notable probiotic groups recognized to date (Behnsen et al., 2013). The most popular probiotics are *Lactobacillus acidophilus*, *L. casei*, *L. helviticus*, *L. plantarum*, *L. fermentum*, *L. bulgaricus*, *Bifidobacterium breve*, *B. bifidum*, *B. longum*, *Enterococcus faecalis*. The probiotic strains are predominantly of bacterial origin; however, it could also belong to yeasts (Raghuwanshi et al., 2015). For instance, certain yeasts like *Saccharomyces boulardii* have been reported to exhibit probiotic properties (Stamatova and Meurman, 2009). These probiotics have not been reported to cause any harm or disease and are categorized under "generally regarded as safe" organisms. The potential health impact of microbes ranges from immunomodulation, respiratory, gastrointestinal functions, and releasing antimicrobial substances such as bacteriocins or metabolites such as acetic acid and lactic acid and has shown intrinsic properties of antigenotoxicity, antimutagenicity, and anticarcinogenicity (Song et al., 2012; Behnsen et al., 2013). It prevents and/or reduces the prevalence of gastrointestinal and extraintestinal disorders, infectious diarrhea, irritable bowel syndrome, inflammatory bowel disease, vaginal infections, and lactose intolerance (de Vrese et al., 2001; Marteau et al., 2002; Reid et al., 2001; Isolauri et al., 2001). Probiotics also inhibit oral pathogens in childhood (Twetman et al., 2008). Probiotics containing foods are mostly available in the markets as yogurt, fermented milk, miso, tempeh, cheese and vegetable, fruit, meat-based products (Song et al., 2012). Consumption of fermented dairy products is associated with reduced occurrence of colon cancer due to inhibition of bacterial growth responsible for the conversion of procarcinogen into a carcinogen (Vasiljevic and Shah, 2008).

The awareness of the nutritional health benefits of probiotics has raised market growth and is projected to reach USD 66.03 billion by 2024. According to Statistics MRC, the global probiotics dietary supplements market is projected to achieve $6.95 billion in 2022 with a CAGR of 9.1% from $4.11 billion in 2016 (Anonymous, 2017).

6.2.7 MICROBES AS BIOSENSORS

Microbial biosensors employ microorganisms as a biological sensing agent immobilized with a transducer for recognition of a target analyte (Dai and Choi, 2013). The bio-element is usually a biomolecule such as microbial-derived enzymes, antibodies, receptors, or organelles that generate a specific and selective response with analytes into a measurable signal which is directly proportional to the analyte concentration. The signal can be a result of the metabolism of the target analyte by the bio-element and is generated in the form of gaseous exchange (uptake and release), change in proton concentration, photoemission, absorption, and so on (Lei et al., 2006). The transducer changes the signal into a quantifiable response in the form of an electrical signal, electrochemical, or optical response that is amplified and can be recorded for further analysis. Among various biological sensing agents, microbes have gained prominence as ideal biological sensing material due to their various advantages like the ability to detect a large range of chemicals, amenable to genetic manipulation, and functional at a broader range of temperature and pH (Nakamura and Karube, 2003; Park et al., 2013). To construct biosensor devices, immobilization plays a pivotal role where microorganisms have been immobilized onto a variety of transducers with amperometric, potentiometric, calorimetric, conductimetric, fluorescent, or luminescent properties (Dsouza, 2001a). The conventional method of immobilization such as entrapment, adsorption, encapsulation, cross-linkage via covalent bonds lacks long term stability (Dsouza 2001b). Nanomaterials such as nanoparticles, nanotubes, and fiber optics are some of the advanced immobilization techniques that offer high reliability and stability (Tuncagil et al., 2011).

Biosensors find their application in different fields including environmental and bioprocess control, food and fermentation industry, agriculture, clinical diagnostics, and military purposes (Dai and Choi, 2013). Pollutants in the atmosphere pose a serious threat to human beings. The exposure to nonbiodegradable heavy metal accumulated in water resources can be detrimental for us and cause severe diseases. The microbial biosensor monitors the presence of pollutants by immobilizing microbes such as *Chlorella*, *Pseudomonas aeruginosa*, *Pseudomonas putidaX4*, *P. fluorescens* for the detection of toxic heavy metal-containing pollutants like mercury, lead Pb(II), zinc, phenol, and nitrophenols (Gammoudi et al., 2010; Dai and Choi, 2013).

Biosensors with amperometric properties are widely used for the determination of biochemical oxygen demand which is essential for measuring the biodegradable organic contaminants in water bodies. The microbial strains mostly used for sensing oxygen consumption/respiration are *Torulopsis candida, P. putida*, *Bacillus subtilis*, *Trichosporon cutaneum* and yeast (Liu and Mattiason, 2002; Lei et al., 2006).

The first biosensor was developed for ethanol determination using bacterial cells of *Acetobacter xylinum* by Divies in 1974 (Divies, 1975). Other biosensors include *Trichosporon brassicae*, *Clostridium butyricum,* and *P. fluorescens* which were used for the identification of ethanol in the range of 2–22.5 mg/L, formic acid in the range of 10–1000 mg/L, and glucose, respectively (Hikuma et al., 1981). Food quality and freshness are one of the major concerns and interests of both the consumer and the food industry. In order to monitor the nutritional requirement and food adulterants, rapid and affordable microbial-based biosensors are necessary. For instance, to ensure the quality of milk, genetically modified *S. cerevisiae* is used as bio-element for detecting a common contaminant, mycotoxin (Välimaa et al., 2010). Further, the determination of caffeine in coffee is conducted using an amperometric biosensor by attaching *Pseudomonas alcaligenes* MTCC 5264 to a cellophane membrane (Dai and Choi, 2013).

Microbial biosensors have also been shown to be highly efficient for a rapid, inexpensive, and accurate diagnosis of various diseases. A white-rot fungus, *P. chrysosporium* ME446 immobilized on the gelatin matrix was shown to be useful for detecting epinephrine levels. Detection of pathogens can be done by using a microbial biosensor fabricated with rat basophilic leukemia cells that produce an exocytotic response upon the addition of antigen (Curtis et al., 2008). The global biosensor market is projected to excel at US$ 30 billion by 2024 at a CAGR of 8.5% (Anonymous, 2019f).

6.2.8 MICROBES AS BIODEGRADER

Since the last few decades, direct or indirect exposure to highly toxic organic compounds has adversely affected the environment. Eradication of a wide range of contaminants and pollutants from the environment has intensified the demand for sustainable ways of clean up with diminished environmental impact. The use of microbes provides a viable solution in the present scenario as the microbes are considered to be the ultimate cleaners and recyclers in the world (Diez, 2010; Abatenh et al., 2017). Biodegradation involves the application of biological methodologies-bioremediation and biotransformation in the eradication or conversion of complex organic compounds into simpler smaller compounds. The process of biodegradation requires the xenobiotic ability of living microbes to transform the substance enzymatically or metabolically into a less toxic form. These substances usually include common environmental pollutants-hydrocarbons (oil), polyaromatic hydrocarbon, heterocyclic compounds (pyridine, quinoline), heavy metal, pesticide, dye. Bioremediation and biotransformation exploit the catabolic versatility of bacteria, yeast, fungi for degradation, transformation, or accumulation of organic pollutants into a form that can be reused by other microbes (Ogilvie, 2012; Joutey et al., 2013).

Structurally hydrocarbon consists of carbon and hydrogen that can be linear, branched, or cyclic. Bacteria capable of degrading hydrocarbon are referred to as hydrocarbon-degrading microorganisms. Bacterial strains *Pseudomonas* sp., *Brevibacillus* sp., *Enterobacter.*, *Mycobacterium* sp., *Streptococcus* sp.,

Marinobacter sp., and *Streptobacillus* sp. are potent petroleum hydrocarbon degraders (Joutey et al., 2013; Yakimov et al., 2007).

Industrially produced polycyclic aromatic hydrocarbons (PAH) can flow through soils, sediments to fish, and eventually into humans through seafood intake. Bacterial strains, namely, *Pseudomonas* sp., *Mycobacterium* sp., *Corynebacterium* sp., *Rhodococcus* sp., and *Bacillus* sp., have been reported to be capable of metabolizing PAH (Mrozik et al., 2003). Polychlorinated biphenyls (PCB) are synthetically produced organic chemicals that impart a toxic effect on endocrine receptors and are carcinogenic in nature. Bacterial strains *Pseudomonas* sp., *Burkholderia* sp., *Achromobacter sp.*, *Rhodococcus* sp., *Microbacterium* sp. are capable of biotransforming PCB (Seeger et al., 2010; Petric et al., 2007). Further, pesticides used in agricultural lands for destroying pests are degraded by soil microbes, such as bacteria and fungi. Soil isolates of *Bacillus*, *Staphylococcus*, and *Stenotrophomonas* have been outlined to degrade Dichlorodiphenyltrichloroethane (Vargas, 1975; Kanade et al., 2012).

Heavy metals cannot be degraded and, therefore, the bioremediation by bacteria involves its conversion of one oxidation state into another via different mechanisms such as bioleaching, biosorption, biomineralization, intra-accumulation, and transformation. This has evolved as a protective mechanism in the bacterial system (Lloyd and Lovely, 2001; Joutey et al., 2013). In the process of biosorption, the heavy metal is physicochemically bound to the cell surface which results in the intracellular entrapment of heavy metals and their subsequent adsorption in the cellular structures (Joutey et al., 2013).

Methylation of the heavy metal by bacteria makes the methylated heavy metal volatile. For example, biomethylation of Hg(II) is facilitated by different bacteria such as *Bacillus pumilus*, *P. aeruginosa*, *Brevibacterium iodinium*, *Acidithiobacillus ferrooxidans* by converting it into volatile methyl mercury (De et al., 2008)

Biomineralization is another mechanism in which the heavy metal gets immobilized either by forming polymeric complexes or by precipitation of heavy metals into insoluble sulfides by sulfate-reducing bacteria (Lloyd and Lovely, 2001).

Azo dyes with aromatic bonds are hard to degrade biologically because of their structures. Bacterial discoloration of dyes is an enzyme-mediated transformation in which reduction of azo bond forms colorless amine which is further oxidized into simple and nontoxic forms. *B. subtilis*, *Aeromonas hydrophila*, *S. marcescens*, and *Bacillus cereus* are some of the strains reported to have dye decolorization property (Verma and Madamwar, 2003; Mahmood et al., 2015).

Radionuclides are another set of pollutants with an unstable nucleus released as a waste product from different worldwide atomic energy and nuclear power plants (Joutey et al., 2013). These are of major concern to the environment and human health as dumping of radioactive waste into the environment and its exposure brings life-threatening disorders in human beings causing nausea, headache, vomiting in minor doses. With increased exposure, it could lead to radiation poisoning that causes fatigue, fever, hair loss, weakness, blood in stool, decline in blood pressure, and ultimately death. Long-time exposure to radionuclides also causes transmissible lethal health problems such as kidney damage, leukemia, leucopenia, and an enhanced risk of cancers of the colon, liver, lung, esophagus, and stomach (Prakash et al., 2013; Möhner et al., 2006). Most commonly encountered radionuclides include cobalt-60 (60Co), plutonium-239 (239Pu), radium-226 (226Ra), radon-222 (222Rn), technetium-99 (99Tc), thorium-232 (232Th), and uranium-238 (238U). Typical radionuclides released from the nuclear reactors via splitting of atoms eventually have a long half-life which includes thallium-201 (201Tl), iridium-192 (192Ir), caesium-137 (137Cs), and strontium-90 (90Sr) (Prakash et al., 2013). Excavation of radioactively contaminated soil and transporting to distant disposal area is the most common way of eliminating radioactively contaminated soil but the cleanup process involves a huge investment in the long run (Lloyd and Renshaw, 2005). Eradication of radionuclides from ecological niches such as sediments, soil, and water by exploiting the versatility of bioremedial properties of microbes is an effective approach that makes them an ideal candidate for bioremediation of environmental pollutants including radionuclides. Microbes employ fascinating strategies for the eradication of radionuclides by regulating the mobility, solubility which has made microbial-based remediation of radionuclides an ecologically responsible alternative over physical-based remediation. Once the radionuclides come in contact with the site-specific microbe, the mobility of the radionuclide is altered by electron addition or removal that results in a changed oxidization state thereby accelerating the mobility of the contaminant and its subsequent removal from that area. Also, the radionuclide can be immobilized in a place by extracting electrons from organic compounds and transferring it to a radioactive electron acceptor that immobilizes it and prevents its leakage from the sites (Kumar et al., 2007; Amachi et al., 2010; Prakash et al., 2013). The various mechanisms employed for the eradication of radionuclide include solubilization directly or indirectly by enzymatic reduction, alteration of pH and electron state, biosorption, bioaccumulation, and biomineralization (Holker et al., 2002; Hegazy et al., 2010). An array of microbes namely

Desulfovibrio desulfuricans, *Geothrix fermentans*, *Delta proteobacteria*, and *Clostridium* have been outlined to detoxify the soluble radionuclide uranium(VI), technetium(VII), and chromium(VI) through direct enzymatic reduction that changes it to insoluble precipitate (Kumar et al., 2007). In the indirect enzymatic reduction of radionuclide, metal-reducing bacteria capture electrons via the synthesis of organic compounds such as lactate and acetate. Such acids upon oxidation reduce iron and other metals along with the radionuclides that form insoluble precipitate as oxide and hydroxide minerals. Sulfate-reducing bacteria produce hydrogen peroxide that solubilizes the radionuclide and eventually results in bioleaching. *Microbacterium flavescens* grows in the presence of radioisotopes such as uranium, thorium, plutonium, americium releases organic acids, extracellular metabolites, and aid in the solubilization and mobilization of radionuclides in soil (John et al., 2001, Prakash et al., 2013). Further, *P. aeruginosa* produces chelating agents that interact with uranium and thorium (Prakash et al., 2013). Biosorption works via sequestration of positively charged radioisotopes by using negatively charged lipopolysaccharides (LPS) on the cell membrane. This process is accompanied by active uptake of radioactive isotopes into the cell where it is retained in the form of complexes in association with negatively charge intracellular components, resulting in its precipitation or granule formation. *Citrobacter freudii* and *Firmicutes* have significant biosorption properties that are evident from the interaction of the phosphates of LPS in *Citrobacter* and uranium (Langley and Beveridge, 1999).

In the method of biomineralization, microbes produce ligands that precipitate radionuclides and form biogenic minerals that aid in retaining radioactive contaminants. Microbes can accumulate precipitated radionuclides several times larger than their weight. *Serratia* accumulates chernikovite, a uranium-rich compound by the generation of biofilms (Prakash et al., 2013).

6.2.9 MICROBES IN SEWAGE AND WASTEWATER TREATMENT

Our daily activities generate a huge volume of sewage and wastewater from industries and municipality that requires treatment prior to its disposal into natural water bodies. The generated water contains a pathogen, toxic compounds, and chemical wastes that may lead to aquatic water pollution (Reemtsma et al., 2006). The wastewater often contains excess nutrients such as nitrogen, phosphorous that promote and support the growth of algal blooms resulting in eutrophication of water bodies. Untreated sewage and wastewater contain organic pollutants that upon decomposition by microorganisms, consume the dissolved oxygen with concomitant depletion of the same (Seow et al., 2016). The direct discharge of untreated or primary treated domestic and industrial wastewater into water bodies and its impact on humans as well as the environment has made the proper disposal as one of the greatest challenges of the 21st century (Nascimento et al., 2018).

Wastewater treatment can be broadly classified into chemical, physical, and biological. Chemical treatment of wastewater employs chemicals that aid in settling and removal of contaminants present in wastewater. It includes common techniques namely coagulation/flocculation, chlorination, chloramination, ozonation, and treatment with ultraviolet light. But the chemical wastewater treatment imparts negative effects to the user as well as the environment due to the use of large volumes of hazardous chemicals which also renders a part of the pollutant material to be unaffected thereby making this method deficient (Akpor et al., 2014; Samer 2015). On the other hand, the physical method of separation of pollutants from wastewater works on the application of natural forces such as gravity, electrostatic attraction, and vanderwaal forces. Different methods include floatation, sedimentation, and adsorption, physical barriers such as bar racks, screens, sieves, etc. (Rahel and Bhatnagar, 2017). However, in spite of the physical treatment, the wastewater contains a huge amount of dissolved and colloidal material in small amounts which is difficult to sediment into larger particles. A possible alternative is a biological treatment that utilizes the use of microbes such as bacteria, yeast, and fungi (Samer, 2015). Biological treatment offers several advantages in contrast to chemical and physical treatment since it is relatively cheaper, with no production of secondary pollutants, less damage to the environment, and less maintenance cost and capital investment (Dadrasnia et al., 2017). Most microorganisms are present in the form of activated sludge in biological wastewater treatment systems which, contains numerous microbes embedded within an extracellular polymeric substance composed of protein, polysaccharides, humic acids, and lipids (Ni and Yu, 2012). The activated sludge efficiently removes 75%–90% of BOD from sewage and is considered to be one of the most widely used processes of wastewater treatment (Schmidt et al., 2002). The microbes have significant sedimentation properties with bacteria present in the highest majority in biological floc. Floc forming bacteria are mostly categorized into (1) heterotrophs such as traditional aerobic bacteria—*Zoogloea*, *Flavobacterium*, *Alcaligenes*, *Aeromonas*, *Pseudomonas*; denitrifiers—*Pseudomonas*, *Bacillus*, *Xanthomonas*; Glycogen accumulating bacteria-*Canditatus*; (2) autotrophs such as ammonia oxidizers—*Nitrosomonas*, *Nitrosococcus*, *Nitrosospira*; Nitrite oxidizers such as *Nitrobacter*, *Nitrococcus*, *Nitrospina*;

(3) filamentous bacteria such as *Thiothrix, Nostoc oidalimicola, Sphaerotilusnatans*. The high suspension of active organisms is continuously stirred to convert the carbonaceous organic substrates from wastewater by oxidation into carbon dioxide, sulfates, nitrates, and phosphates (Akpor et al., 2014).

6.3 CONCLUSION

The emergence of diseases, fossil fuel depletion, climate deterioration, pollution, increased carbon footprints, food security, and water stress has become a major issue in the present times. In this context, the natural source of microbes can be harnessed for their significant contribution in the eradication of diseases, enhancing greener technologies, climate change alleviation with reduced greenhouse gas emissions, and improving agricultural productivity. Only 3% of the microbes available on our planet are identified and utilized. However, the remaining 97% of microbes are unexplored due to our inability to grow them in laboratory conditions. The metagenomics approach has been found to be a suitable solution to get benefits from even uncultured microbes. Based on the facts and findings presented in this chapter, it is evident that the microbial world can be used ubiquitously in a wide range of environmental applications and it could be the key factor in propelling global economic and industrial progress in the future.

ACKNOWLEDGEMENT

B.D. acknowledges the CSIR-UGC grant for Senior Research fellowship.

KEYWORDS

- **agriculture**
- **bioethanol**
- **biosensor**
- **healthcare and medicine**
- **microbial products**
- **probiotics**
- **wastewater treatment**

REFERENCES

Abatenh, E.; Gizaw, B.; Tsegaye, Z.; Wassie, M. The role of microorganisms in bioremediation-A review. *Open J. Environ. Biol.* **2017**, *2*, 038–046.

Ahmad, F.; Ahmad, I; Khan, M.S. Screening of free-living rhizospheric bacteria for their multiple plant growth promoting activities. *Microbiol. Res.* **2008**, *163*, 173–181.

Ahmad, I.; Khan, M.S.A.; Aqil, F.; Singh, M. Microbial applications in agriculture and the environment: a broad perspective. In: *Microbes and Microbial Technology*, Springer: New York, NY, **2011**, 1–27.

Akinsemolu, A.A. The role of microorganisms in achieving the sustainable development goals. *J. Clean. Prod.* **2018**, *182*, 139–155.

Akpor, O.B.; Ogundeji, M.D.; Olaolu, D.T.; Aderiye, B.I. Microbial roles and dynamics in wastewater treatment systems: an overview. *Int. J. Pure Appl. Biosci.* **2014**, *2*, 156–168.

Amachi, S.; Minami, K.; Miyasaka, I.; Fukunaga, S. Ability of anaerobic microorganisms to associate with iodine: 125I tracer experiments using laboratory strains and enriched microbial communities from subsurface formation water. *Chemosphere* **2010**, *79*, 349–354.

Angelidaki, I.; Ellegaard, L; Ahring, B.K. A mathematical model for dynamic simulation of anaerobic digestion of complex substrates: focusing on ammonia inhibition. *Biotechnol. Bioeng.* **1993**, *42*, 159–166.

Anonymous. Reuters. **2017**. https://www.reuters.com/brandfeatures/venture-capital/article?id=5780.

Anonymous. Biofertilizers market. market data forecast. **2018a**. https://www.marketdata-forecast.com/market-reports/global-biofertilizers-market.

Anonymous. Global vaccine market analysis & forecast (2018–2028): a $103.57 billion opportunity, growing at a CAGR of 11.02 %. Research and markets. **2018b**. https://www.globenewswire.com/news-release/2018/12/10/1664583/0/en/Global-Vaccine-Market-Analysis-Forecast-2018–2028-A-103–57-Billion-Opportunity-Growing-at-a-CAGR-of-11-02.html.

Anonymous. Head and neck cancer drugs market size worth $1.41 billion by 2022: Grand View Research, Inc. MedIndia. **2018c**. https://www.medindia.net/health-press-release/head-and-neck-cancer-drugs-market-size-worth-141-billion-by-2022-grand-view-research-inc-365748-1.htm.

Anonymous. Biopesticides market worth $6.4 billion by 2023. Marketsandmarkets. **2019a**. https://www.marketsandmarkets.com/PressReleases/biopesticide.asp.

Anonymous. Global biogas market 2019–2023. Businesswire. **2019b**. https://www. business-wire.com /news /home /20190130005536 /en/Global-Biogas-Market-2019–2023-Adoption-Integrated-Waste.

Anonymous. India ethanol market by source, by application, by purity, competition, forecast & opportunities, 2024. Research and Markets. **2019c**. https://www.researchandmarkets.com/reports/4758449/india-ethanol-market-by-source-by-application.

Anonymous. Fuel ethanol market to reach USD 122.35 billion by 2026. Reports and Data. **2019d**. https://www.globenewswire.com/news-release/2019/04/10/1802223/0/en/Fuel-Ethanol-Market-To-Reach-USD-122–35-Billion-By-2026-Reports-And-Data.html.

Anonymous. Marketwatch. **2019e**. https://www.marketwatch.com/press-release/cagr-of-4-biodiesel-market-boosting-with-cagr-of-4-by-2026-2019-05-21.

Anonymous. Marketwatch. **2019f**. https://www.marketwatch.com/press-release/biosensor-market-is-determined-to-reach-us-30-billion-by-2024-2019-03-04.

Antolın, G.; Tinaut, F.V.; Briceno, Y.; Castano, V.; Perez, C.; Ramırez, A.I. Optimisation of biodiesel production by sunflower oil transesterification. *Bioresour. Technol.* **2002**, *83*, 111–114.

Bajpai, D.; Tyagi, V.K. Biodiesel: source, production, composition, properties and its benefits. *J. Oleo Sci.* **2006**, *55*, 487–502.

Behnsen, J.; Deriu, E.; Sassone-Corsi, M.; Raffatellu, M. Probiotics: properties, examples, and specific applications. *Cold Spring Harb. Perspect. Med.* **2013**, *3*, a010074.

Bharti, N.; Pandey, S.S.; Barnawal, D.; Patel, V.K.; Kalra, A. Plant growth promoting rhizobacteria Dietzia natronolimnaea modulates the expression of stress responsive genes providing protection of wheat from salinity stress. *Sci. Rep.* **2016**, *6*, 34768.

Bhattacharyya, P.N.; Goswami, M.P.; Bhattacharyya, L.H. Perspective of beneficial microbes in agriculture under changing climatic scenario: a review. *J. Phytol.* **2016**, *8*, 26–41.

Bisaria, V.S. Bioprocessing of agro-residues to value added products. In: *Bioconversion of Waste Materials to Industrial Products*, Boston, MA: Springer, **1998**, 197–246.

Björnsson, L.; Murto, M.; Mattiasson, B. Evaluation of parameters for monitoring an anaerobic co-digestion process. *Appl. Microbiol. Biotechnol.* **2000**, *54*, 844–849.

Boulton, R.B.; Singleton, V.L.; Bisson, L.F.; Kunkee, R.E. Yeast and biochemistry of ethanol fermentation. In: *Principles and Practices of Winemaking,* Boston, MA: Springer, **1999**; 102–192.

Bušić, A.; Marđetko, N.; Kundas, S.; Morzak, G.; Belskaya, H.; Šantek, M.I.; Komes, D.; Novak, S.; Šantek, B. Bioethanol production from renewable raw materials and its separation and purification: a review. *Food Technol. Biotechnol.* **2018**, *56*, 289.

Chandel, A.K.; Chandrasekhar, G.; Silva, M.B.; Silvério da Silva, S. The realm of cellulases in biorefinery development. *Crit. Rev. Biotechnol.* **2012**, *32*, 187–202.

Chandel, A.K.; Singh, O.V.; Rao, L.V. Biotechnological applications of hemicellulosic derived sugars: state-of-the-art. In *Sustainable Biotechnology*, Dordrecht: Springer, **2010**, 63–81.

Chang, C.C.; Chen, W.C.; Ho, T.F.; Wu, H.S.; Wei, Y.H. Development of natural anti-tumor drugs by microorganisms. *J. Biosci. Bioeng.* **2011**, *111*, 501–511.

Chen, Y.E.; Tsao, H. The skin microbiome: current perspectives and future challenges. *J. Am. Acad. Dermatol.* **2013**, *69*, 143–155.

Cho, H.U.; Park, J.M. Biodiesel production by various oleaginous microorganisms from organic wastes. *Bioresour. Technol.* **2018**, *256*, 502–508.

Christy, P.M.; Gopinath, L.R.; Divya, D. A review on anaerobic decomposition and enhancement of biogas production through enzymes and microorganisms. *Renew. Sustainable Energy Rev.* **2014**, *34*, 67–173.

Chung, K.T. Harold Boyd Woodruff (b. 1917): antibiotics hunter and distinguished soil microbiologist. *SIM News* **2009**, *59*, 178–185.

Cirne, D.G.; Lehtomäki, A.; Björnsson, L.; Blackall, L.L. Hydrolysis and microbial community analyses in two-stage anaerobic digestion of energy crops. *J. Appl. Microbiol.* **2007**, *103*, 516–527.

Curtis, T.; Naal, R.M.Z.; Batt, C.; Tabb, J.; Holowka, D. Development of a mast cell-based biosensor. *Biosens. Bioelectron.* **2008**, *23*, 1024–1031.

D'souza, S.F. Immobilization and stabilization of biomaterials for biosensor applications. *Appl. Biochem. Biotechnol.* **2001**, *96*, 225–238.

Dadrasnia, A.; Usman, M.M.; Lim, K.T.; Velappan, R.D.; Shahsavari, N.; Vejan, P.; Ismail, S. Microbial aspects in wastewater treatment—a technical review. *Environ. Pollut. Prot.* **2017**, *2*, 75–84.

Dai, C.; Choi, S. Technology and applications of microbial biosensor. *Open J. Appl. Biosen.* **2013**, *2*, 83.

Danino, T.; Prindle, A.; Hasty, J.; Bhatia, S. Measuring growth and gene expression dynamics of tumor-targeted S. Typhimurium bacteria. *J. Vis. Exp.* **2013**, 77, e50540.

das Neves, M.A.; Kimura, T.; Shimizu, N.; Nakajima, M. State of the art and future trends of bioethanol production. *Dyn. Biochem. Process Biotechnol. Mol. Biol.* **2007**, *1*, 1–14.

De, J.; Ramaiah, N.; Vardanyan, L. Detoxification of toxic heavy metals by marine bacteria highly resistant to mercury. *Mar. Biotechnol.* **2008**, *10*, 471–477.

Demain, A.L.; Vaishnav, P. Natural products for cancer chemotherapy. *Microb. Biotechnol.* **2011,** *4*, 687–699.

DEMİRBAŞ, A. Bioethanol from cellulosic materials: a renewable motor fuel from biomass. *Energy Sources* **2005,** *27*, 327–337.

deVrese, M.; Stegelmann, A.; Richter, B.; Fenselau, S.; Laue, C.; Schrezenmeir, J. Probiotics—compensation for lactase insufficiency. *Am. J. Clin. Nutr.* **2001**, *73*, 421s–429s.

Diez, M.C. Biological aspects involved in the degradation of organic pollutants. *J. Soil Sci. Plant Nutr.* **2010**, *10*, 244–267.

Divies, C. Remarks on ethanol oxidation by an "Acetobacterxylinum" microbial electrode (author's transl). *Ann. de Microbiologie*, **1975**, 126(2), 175–186.

D'souza, S.F. Microbial biosensors. *Biosens. Bioelectron.* **2001**, *16*, 337–353.

Economou, C.; Aggelis, G.; Pavlou, S.; Vayenas, D.V. Modeling of single-cell oil production under nitrogen-limited and substrate inhibition conditions. *Biotechnol. Bioeng.* **2011**, *108*, 1049–1055.

Gammoudi, I.; Tarbague, H.; Othmane, A.; Moynet, D.; Rebière, D.; Kalfat, R.; Dejous, C. Love-wave bacteria-based sensor for the detection of heavy metal toxicity in liquid medium. *Biosens. Bioelectron.* **2010**, *26*, 1723–1726.

Glick, B.R. Plant growth-promoting bacteria: mechanisms and applications. *Scientifica* **2012**, *2012*, 963401.

Gopinath, L.R.; Christy, P.M.; Mahesh, K.; Bhuvaneswari, R.; Divya, D. Identification and evaluation of effective bacterial consortia for efficient biogas production. *IOSR J. Environ. Sci. Toxicol. Food Technol.* **2014**, *8*, 80–86.

Grice, E.A.; Segre, J.A. The human microbiome: our second genome. *Annu. Rev. Genom. Hum. G.* **2012**, *13*, 151–170.

Gujer, W.; Zehnder, A.J. Conversion processes in anaerobic digestion. *Water Sci. Technol.* **1983**, *15*, 127–167.

Hahn-Hägerdal, B.; Karhumaa, K.; Fonseca, C.; Spencer-Martins, I.; Gorwa-Grauslund, M.F. Towards industrial pentose-fermenting yeast strains. *Appl. Microbiol. Biotechnol.* **2007**, *74*, 937–953.

Hassan, M.; Kjos, M.; Nes, I.F.; Diep, D.B.; Lotfipour, F. Natural antimicrobial peptides from bacteria: characteristics and potential applications to fight against antibiotic resistance. *J. Appl. Microbiol.* **2012**, *113*, 723–736.

Hegazy, A.K.; Emam, M.H. Accumulation and soil-to-plant transfer of radionuclides in the Nile Delta coastal black sand habitats. *Int. J. Phytorem.* **2010**, *13*, 140–155.

Hidayati, Y.A.; Kurnani, T.B.A.; Marlina, E.T.; Rahmah, K.N.; Harlia, E.; Joni, I.M. The production of anaerobic bacteria and biogas from dairy cattle waste in various growth mediums. *AIP Conference Proceedings*, AIP Publishing, **2018**, *1927*(1): 030021.

Hikuma, M.; Yasuda, T.; Karube, I.; Suzuki, S. Application of microbial sensors to the fermentation process. *Ann. NY Acad. Sci.* **1981**, *369*, 307–320.

Hölker, U.; Schmiers, H.; Grosse, S.; Winkelhöfer, M.; Polsakiewicz, M.; Ludwig, S.; Dohse, J.; Höfer, M. Solubilization of low-rank coal by Trichoderma atroviride: evidence for the involvement of hydrolytic and oxidative enzymes by using 14C-labelled lignite. *J. Ind. Microbiol. Biotechnol.* **2002**, *28*, 207–212.

Isolauri, E.; Sütas, Y.; Kankaanpää, P.; Arvilommi, H.; Salminen, S. Probiotics: effects on immunity. *Am. J. Clin. Nutr.* **2001**, *73*, 444s–450s.

Johansson, J.F.; Paul, L.R.; Finlay, R.D. Microbial interactions in the mycorrhizosphere and their significance for sustainable agriculture. *FEMS Microbiol. Ecol.* **2004**, *48*, 1–13.

John, S.G.; Ruggiero, C.E.; Hersman, L.E.; Tung, C.S.; Neu, M.P. Siderophore mediated plutonium accumulation by Microbacteriumflavescens (JG-9). *Environ. Sci. Technol.* **2001**, *35*, 2942–2948.

Joutey, N.T.; Bahafid, W.; Sayel, H.; El Ghachtouli, N. Biodegradation: involved microorganisms and genetically engineered microorganisms. *Biodegrad.-Life Sci.* **2013**, 289–320.

Kalyuzhnyi, S.; Veeken, A.; Hamelers, B. Two-particle model of anaerobic solid state fermentation. *Water Sci. Technol.* **2000**, *41*, 43–50.

Kanade, S.N.; Ade, A.B.; Khilare, V.C. Malathion degradation by Azospirillum lipoferum Beijerinck. *Sci. Res. Reporter* **2012**, *2*, 94–103.

Katahira, S.; Mizuike, A.; Fukuda, H.; Kondo, A. Ethanol fermentation from lignocellulosic hydrolysate by a recombinant xylose-and cellooligosaccharide-assimilating yeast strain. *Appl. Microbiol. Biotechnol.* **2006**, *72*, 1136–1143.

Kaur, S.; Kaur, S. Bacteriocins as potential anticancer agents. *Front. Pharmacol.* **2015**, *6*, 272.

Keller, F.A.; Hamilton, J.E.; Nguyen, Q.A. Microbial pretreatment of biomass. In: *Biotechnology for Fuels and Chemicals*, Totowa, NJ: Humana Press, **2003**, 27–41

Kemp, W.H. *Biodiesel: Basics and Beyond: A Comprehensive Guide to Production and Use for the Home and Farm*, Tamworth: Aztext Press, **2006**.

Kolouchová, I.; Maťátková, O.; Sigler, K.; Masák, J.; Řezanka, T. Lipid accumulation by oleaginous and non-oleaginous yeast strains in nitrogen and phosphate limitation. *Folia Microbiol.* **2016**, *61*, 431–438.

Kosa, M.; Ragauskas, A.J. Bioconversion of lignin model compounds with oleaginous Rhodococci. *Appl. Microbiol. Biotechnol.* **2012**, *93*, 891–900.

Kuhad, R.C. Microbes and their role in sustainable development. *Indian J. Microbiol.* **2012**, *52*, 309–313.

Kumar, P.; Barrett, D.M.; Delwiche, M.J.; Stroeve, P. Methods for pretreatment of lignocellulosic biomass for efficient hydrolysis and biofuel production. *Ind. Eng. Chem. Res.* **2009**, *48*, 3713–3729.

Kumar, R.; Singh, S.; Singh, O.V. Bioremediation of radionuclides: emerging technologies. *Omics: J. Integr. Biol.* **2007**, *11*, 295–304.

Lamont, R.F.; Sobel, J.D.; Akins, R.A.; Hassan, S.S.; Chaiworapongsa, T.; Kusanovic, J.P.; Romero, R. The vaginal microbiome: new information about genital tract flora using molecular based techniques. *BJOG-Int. J. Obstet. Gynaecol.* **2011**, *118*, 533–549.

Langley, S.; Beveridge, T.J. Effect of O-side-chain-lipopolysaccharide chemistry on metal binding. *Appl. Environ. Microbiol.* **1999**, *65*, 489–498.

Lee, N.H.; Lee, J.A.; Park, S.Y.; Song, C.S.; Choi, I.S.; Lee, J.B. A review of vaccine development and research for industry animals in Korea. *Clin. Exp. Vaccine* **2012**, *1*, 18–34.

Lei, Y.; Chen, W.; Mulchandani, A. Microbial biosensors. *Anal. Chim. Acta* **2006**, *568*, 200–210.

Leschner, S.; Weiss, S. Salmonella—allies in the fight against cancer. *J. Mol. Med.* **2010**, *88*, 763–773.

Li, H.; Wu, M.; Xu, L.; Hou, J.; Guo, T.; Bao, X.; Shen, Y. Evaluation of industrial *Saccharomyces cerevisiae* strains as the chassis cell for second-generation bioethanol production. *Microb. Biotechnol.* **2015**, *8*, 266–274.
Liao, J.C.; Mi, L.; Pontrelli, S.; Luo, S. Fuelling the future: microbial engineering for the production of sustainable biofuels. *Nat. Rev. Microbiol.* **2016**, *14*, 288.
Liu, J.; Mattiasson, B. Microbial BOD sensors for wastewater analysis. *Water Res.* **2002,** *36*, 3786–3802.
Lloyd, J.R.; Lovley, D.R. Microbial detoxification of metals and radionuclides. *Curr. Opin. Biotechnol.* **2001,** *12*, 248–253.
Lloyd, J.R.; Renshaw, J.C. Microbial transformations of radionuclides: fundamental mechanisms and biogeochemical implications. *Met. Ions Biol. Syst.* **2005,** *44*, 205.
Lucy, M.; Reed, E.; Glick, B.R. Applications of free living plant growth-promoting rhizobacteria. *Anton Leeuw.* **2004**, *86*, 1–25.
Lynd, L.R.; Weimer, P.J.; Van Zyl, W.H.; Pretorius, I.S. Microbial cellulose utilization: fundamentals and biotechnology. *Microbiol. Mol. Biol. Rev.* **2002**, *66*, 506–577.
Mackowiak, P.A. Recycling Metchnikoff: probiotics, the intestinal microbiome and the quest for long life. *Front. Public Health* **2013**, *1*, 52.
Mah, R.A. Methanogenesis and methanogenic partnerships. *Philos. T. R. Soc. B.* **1982**, *297*, 599–616.
Mahmood, R.; Sharif, F.; Ali, S.; Hayyat, M.U. Enhancing the decolorizing and degradation ability of bacterial consortium isolated from textile effluent affected area and its application on seed germination. *Sci. World J.* **2015**, *2015*.
Marteau, P.; Seksik, P.; Jian, R. Probiotics and intestinal health effects: a clinical perspective. *Br. J. Nutr.* **2002**, *88*, s51–s57.
Meng, X.; Yang, J.; Xu, X.; Zhang, L.; Nie, Q; Xian, M. Biodiesel production from oleaginous microorganisms. *Renew. Energy* **2009**, *34*, 1–5.
Möhner, M.; Lindtner, M.; Otten, H.; Gille, H.G. Leukemia and exposure to ionizing radiation among German uranium miners. *Am. J. Ind. Med.* **2006**, *49*, 238–248.
Mrozik, A.; Piotrowska-Seget, Z.; Labuzek, S. Bacterial degradation and bioremediation of polycyclic aromatic hydrocarbons. *Pol. J. Environ. Stud.* **2003**, *12*.
Nakamura, H.; Karube, I. Current research activity in biosensors. *Anal. Bioanal. Chem.* **2003**, *377*, 446–468.
Nascimento, A.L.; Souza, A.J.D.; Andrade, P.A.M.D.; Dini-Andreote, F.; Gomes, A.R.C.; Oliveira, F.C.; Regitano, J.B. Sewage sludge microbial structures and relations to their sources, treatments, and chemical attributes. *Front. Microbiol.* **2018**, *9*, 1462.
Ni, B.J.; Yu, H.Q. Microbial products of activated sludge in biological wastewater treatment systems: a critical review. *Crit. Rev. Env. Sci. Technol.* **2012**, *42*, 187–223.
Ogilvie, L.A.; Overall, A.D.; Jones, B.V. The human-microbe coevolutionary continuum. *Microbial Ecol. Theory* **2012**, *25*, 42.
Sumant, O.; Kunsel, T.; Pandey, S. Antibiotics market by class (beta lactam & beta lactamase, quinolones, macrolides, and others), drug origin (natural, semisynthetic, and synthetic), spectrum of activity (broad-spectrum antibiotic and narrow-spectrum antibiotic), and route of administration (oral, intravenous, and others): global opportunity analysis and industry forecast, 2018–2025. *Allied Market Res.* **2018**.
Palmqvist, E.; Hahn-Hägerdal, B. Fermentation of lignocellulosic hydrolysates. II: inhibitors and mechanisms of inhibition. *Bioresour. Technol.* **2000**, *74*, 25–33.

Panagiotou, G.; Villas-Bôas, S.G.; Christakopoulos, P.; Nielsen, J.; Olsson, L. Intracellular metabolite profiling of *Fusarium oxysporum* converting glucose to ethanol. *J. Biotechnol.* **2005**, *115*, 425–434.

Park, M.; Tsai, S.L.; Chen, W. Microbial biosensors: engineered microorganisms as the sensing machinery. *Sensors* **2013**, *13*, 5777–5795.

Pereg, L.; McMillan, M. Scoping the potential uses of beneficial microorganisms for increasing productivity in cotton cropping systems. *Soil Biol. Biochem.* **2015**, *80*, 349–358.

Périé, F.H.; Gold, M.H. Manganese regulation of manganese peroxidase expression and lignin degradation by the white rot fungus Dichomitussqualens. *Appl. Environ. Microbiol.* **1991**, *57*, 2240–2245.

Petric, I.; Hrsak, D.; Fingler, S.; Voncina, E.; Cetkovic, H.; Kolar, A.B.; Kolic, N.U. Enrichment and characterization of PCB-degrading bacteria as potential seed cultures for bioremediation of contaminated soil. *Food Tech. Biotechnol.* **2007**, *45*, 11–20.

Planta, M.B. The role of poverty in antimicrobial resistance. *J. Am. Board Family Med.* **2007**, *20*, 533–539.

Prakash, D.; Gabani, P.; Chandel, A.K.; Ronen, Z.; Singh, O.V. Bioremediation: a genuine technology to remediate radionuclides from the environment. *Microb. Biotechnol.* **2013**, *6*, 349–360.

Radecka, D.; Mukherjee, V.; Mateo, R.Q.; Stojiljkovic, M.; Foulquié-Moreno, M.R.; Thevelein, J.M. Looking beyond Saccharomyces: the potential of non-conventional yeast species for desirable traits in bioethanol fermentation. *FEMS Yeast Res.* **2015**, *15*.

Raghuwanshi, S.; Misra, S.; Bisen, P.S. Indian perspective for probiotics: a review. *Indian J. Dairy Sci.* **2015**, *68*, 3.

Rahel, C.; Bhatnagar, M. Study on the removal characteristics of heavy metals from aqueous solution by fly ash collected from Suratgarh and Kota thermal power stations. *Int. J. Innov. Res. Sci. Eng. Technol.* **2017**, *6*(2), 2455–2478.

Rai, M. Plant-growth-promoting rhizobacteria as biofertilizers and biopesticides. In: *Handbook of Microbial Biofertilizers*, Boca Raton, FL: CRC Press, **2006**, 165–210.

Rasi, S.; Läntelä, J.; Rintala, J. Trace compounds affecting biogas energy utilisation—a review. *Energy Convers. Manage.* **2011**, *52*, 3369–3375.

Reemtsma, T.; Weiss, S.; Mueller, J.; Petrovic, M.; González, S.; Barcelo, D.; Ventura, F.; Knepper, T.P. Polar pollutants entry into the water cycle by municipal wastewater: a European perspective. *Environ. Sci. Technol.* **2006**, *40*, 5451–5458.

Reid, G.; Howard, J.; Gan, B.S. Can bacterial interference prevent infection? *Trends Microbiol.* **2001,** *9*, 424–428.

Riedel, S. Edward Jenner and the history of smallpox and vaccination. In: *Baylor University Medical Center Proceedings*, Abingdon: Taylor & Francis, **2005**, 21–25.

Robak, K.; Balcerek, M. Review of second-generation bioethanol production from residual biomass. *Food Technol. Biotechnol.* **2018;** *56*, 174.

Round, J.L.; Mazmanian, S.K. The gut microbiota shapes intestinal immune responses during health and disease. *Nat. Rev. Immunol.* **2009**, *9*, 313.

Sajith, S.; Priji, P.; Sreedevi, S.; Benjamin, S. An overview on fungal cellulases with an industrial perspective. *J. Nutr. Food. Sci.* **2016**, *6*, 461.

Samer, M. Biological and chemical wastewater treatment processes. In: *Wastewater Treatment Engineering*, Hamilton, NJ: IntechOpen, **2015**.

Sanchez, O.J.; Cardona, C.A. Trends in biotechnological production of fuel ethanol from different feedstocks. *Bioresour. Technol.* **2008**, *99*, 5270–5295.

Schmidt, I.; Sliekers, O.; Schmid, M.; Cirpus, I.; Strous, M.; Bock, E.; Kuenen, J.G.; Jetten, M.S. Aerobic and anaerobic ammonia oxidizing bacteria–competitors or natural partners? *FEMS Microbiol. Ecol.* **2002**, *39*, 175–181.

Seeger, M.; Hernández, M.; Méndez, V.; Ponce, B.; Córdova, M.; González, M. Bacterial degradation and bioremediation of chlorinated herbicides and biphenyls. *J. Soil Sci. Plant Nutr.* **2010**, *10*, 320–332.

Seow, T.W.; Lim, C.K.; Nor, M.H.M.; Mubarak, M.F.M.; Lam, C.Y.; Yahya, A.; Ibrahim, Z. Review on wastewater treatment technologies. *Int. J. Appl. Environ. Sci.* **2016**, *11*, 111–126.

Sharma, S.; Kulkarni, J.; Jha, B. Halotolerant rhizobacteria promote growth and enhance salinity tolerance in peanut. *Front. Microbiol.* **2016**, *7*, 1600.

Shi, J.; Sharma-Shivappa, R.R.; Chinn, M.; Howell, N. Effect of microbial pretreatment on enzymatic hydrolysis and fermentation of cotton stalks for ethanol production. *Biomass Bioenerg.* **2009**, *33*, 88–96.

Singh, P.; Suman, A.; Tiwari, P.; Arya, N.; Gaur, A.; Shrivastava, A.K. Biological pretreatment of sugarcane trash for its conversion to fermentable sugars. *World J. Microb. Biotechnol.* **2008,** *24*, 667–673.

Singh, R.; Kumar, M.; Mittal, A.; Mehta, P.K. Microbial metabolites in nutrition, healthcare and agriculture. *3 Biotech* **2017,** *7*, 15.

Skoog, K.; Hahn-Hägerdal, B. Xylose fermentation. *Enzyme Microb. Technol.* **1988**, *10*, 66–80.

Song, D.; Ibrahim, S.; Hayek, S. Recent application of probiotics in food and agricultural science. In: *Probiotics*, Hamilton, NJ: IntechOpen, **2012**.

Song, S.; Vuai, M.S.; Zhong, M. The role of bacteria in cancer therapy–enemies in the past, but allies at present. *Infect. Agent. Cancer* **2018**, *13*, 9.

Spyridon, A.; Euverink, W.; Jan, G. Consolidated briefing of biochemical ethanol production from lignocellulosic biomass. *Electron. J. Biotechnol.* **2016**, *19*, 44–53.

Sriwongchai, S.; Pokethitiyook, P.; Pugkaew, W.; Kruatrachue, M.; Lee, H. Optimization of lipid production in the oleaginous bacterium *Rhodococcus erythropolis* growing on glycerol as the sole carbon source. *Afr. J. Biotechnol.* **2012**, *11*, 14440–14447.

Stamatova, I.; Meurman, J.H. Probiotics: health benefits in the mouth. *Am. J. Dentistry* **2009**, *22*, 329.

Tannock, G.W. A special fondness for lactobacilli. *Appl. Environ. Microbiol.* **2004**, *70*, 3189–3194.

Thangavelu, S.K.; Ahmed, A.S.; Ani, F.N. Review on bioethanol as alternative fuel for spark ignition engines. *Renew. Sust. Energy Rev.* **2016**, *56*, 820–835.

Timmis, K.; De Lorenzo, V.; Verstraete, W.; Ramos, J.L.; Danchin, A.; Brüssow, H.; Singh, B.K.; Timmis, J.K. The contribution of microbial biotechnology to economic growth and employment creation. *Microb. Biotechnol.* **2017**, *10*, 1137–1144.

Timmusk, S.; Behers, L.; Muthoni, J.; Muraya, A.; Aronsson, A.C. Perspectives and challenges of microbial application for crop improvement. *Front. Plant Sci.* **2017**, *8*, 49.

Tuncagil, S.; Ozdemir, C.; Demirkol, D.O.; Timur, S.; Toppare, L. Gold nanoparticle modified conducting polymer of 4-(2, 5-di (thiophen-2-yl)-1H-pyrrole-1-l) benzenamine for potential use as a biosensing material. *Food Chem.* **2011**, *127*, 1317–1322.

Turnbaugh, P.J.; Ley, R.E.; Hamady, M.; Fraser-Liggett, C.M.; Knight, R.; Gordon, J.I. The human microbiome project. *Nature* **2007**, *449*, 804.

Twetman, S.; STECKSÉN-BLICKS, C.H.R.I.S.T.I.N.A. Probiotics and oral health effects in children. *Int. J. Paediatr. Dentistry* **2008**, *18*, 3–10.

Ursell, L.K.; Haiser, H.J.; Van Treuren, W.; Garg, N.; Reddivari, L.; Vanamala, J.; Dorrestein, P.C.; Turnbaugh, P.J.; Knight, R. The intestinal metabolome: an intersection between microbiota and host. *Gastroenterology* **2014**, *146*, 1470–1476.

Välimaa, A.L.; Kivistö, A.T.; Leskinen, P.I.; Karp, M.T. A novel biosensor for the detection of zearalenone family mycotoxins in milk. *J. Microbiol. Methods* **2010**, *80*, 44–48.

Vargas, J.M. Pesticide degradation. *J. Arboric.* **1975**, *1*, 232–233.

Vasiljevic, T.; Shah, N.P. Probiotics—from Metchnikoff to bioactives. *Int. Dairy J.* **2008**, *18*, 714–728.

Verma, P.; Madamwar, D. Decolourization of synthetic dyes by a newly isolated strain of Serratiamarcescens. *World J. Microb. Biotechnol.* **2003**, *19*, 615–618.

Vicente, G.; Martınez, M.; Aracil, J. Integrated biodiesel production: a comparison of different homogeneous catalysts systems. *Bioresour. Technol.* **2004**, *92*, 297–305.

Wei, Z.; Zeng, G.; Kosa, M.; Huang, D.; Ragauskas, A.J. Pyrolysis oil-based lipid production as biodiesel feedstock by *Rhodococcus opacus*. *Appl. Biochem. Biotechnol.* **2015**, *175*, 1234–1246.

Weitao, T. Bacteria form biofilms against cancer metastasis. *Med. Hypotheses* **2009**, *72*, 477–478.

Wyman, C.E.; Dale, B.E.; Elander, R.T.; Holtzapple, M.; Ladisch; M.R.; Lee, Y.Y. Coordinated development of leading biomass pretreatment technologies. *Bioresour. Technol.* **2005**, *96*, 1959–1966.

Yakimov, M.M.; Timmis, K.N.; Golyshin, P.N.; Obligate oil-degrading marine bacteria. *Curr. Opin. Biotechnol.* **2007**, *18*, 257–266.

Zhang, Y.H.P.; Lynd, L.R. Cellulose utilization by Clostridium thermocellum: bioenergetics and hydrolysis product assimilation. *Proc. Nat. Acad. Sci. USA* **2005**, *102*, 7321–7325.

CHAPTER 7

Biosafety Principles for Microbial Culture Technologies

VIDUSHI ABROL[1,3], SUNDEEP JAGLAN[1,2], and SHARADA MALLUBHOTLA[3*]

[1]*Microbial Biotechnology Division, CSIR-Indian Institute of Integrative Medicine, Jammu 180001, India*

[2]*Academy of Scientific and Innovative Research (AcSIR), Jammu Campus, Jammu 180001, India*

[3]*School of Biotechnology, Shri Mata Vaishno Devi University, Katra 182320, Himachal Pradesh, India*

[*]*Corresponding author. E-mail: sharda.p@smvdu.ac.in*

ABSTRACT

Microbial cultures, especially bacteria, fungi, and yeasts, have been earlier recognized as laboratory hazards dealing with various infectious agents. Microbiological laboratories are often unique, exceptional, and their working environment poses special contagious disease risks to individual workers who are in constant contact with or near them. So far, various reports are available regarding associated laboratory cases of typhoid, glanders, brucellosis, cholera, polio, tetanus, and other such diseases. A published survey of voluntary responses from 88 facilities that were associated with laboratory infection concluded that *Shigella* was the most common infecting agent, followed by *Brucella*, *Salmonella*, and *Staphylococcus aureus*. *Neisseria meningitidis* was also perceived as a commonly acquired infection, and therefore the handling of microbial cultures or specimens or the inhalation of dust in the microbial system is strikingly dangerous to laboratory workers. Realizing the need for tackling these hazards, specific biosafety guidelines had been developed nationally and internationally to protect laboratory researchers in microbiological laboratories through a medley of biosecurity safeguards. They also include procedures for risk assessment and management techniques besides adopting appropriate laboratory experimental practices. The details of such systems and guidelines shall be elucidated herewith.

7.1 INTRODUCTION

The modern development in the fields of biotechnology and research often carries legal implications and ethical controversies. Generally, biotechnological research requires expensive investments for building up sophisticated laboratories for performing experiments and also requires time, labor, and dedication. It is a well-known fact that biotechnological processes and products offer diverse applications in the fields of medicine, agriculture, and industrial applications. From the past few decades, biotechnological research has been played a significant role in the welfare of humanity. Apart from this, there are few biological disadvantages that are associated with the genetic modification of plants and animals that result in modern revolutionizing applications. Moreover, there is a potential risk of biological agents used during scientific research and possess probability for being discharged into the surroundings and posing an unknown risk on the ecosystem. Industries and organizations are still skeptical of openly accept genetically modified products for personal, philosophical, cultural, religious, or bioethical reasons. Henceforth, biosafety measures have been introduced as methods and techniques that can be applied to biological unity for hindering maximum drawbacks due to biotechnological processes in the surroundings. The main aim of these steps is to minimize or even reduce any kind of accidental exposure to infectious agents and their inadvertent release into the environment.

The backbone of the implementation of biosafety principles is the biological risk and biohazard evaluation. Microbial laboratories are a storehouse of infections due to a wide variety of bacteria, fungi, viruses, and parasites available for experimental purposes. Approximately 500,000 scientists, researchers, workers are working in such workspaces in the USA alone (Sewell, 1995). While many precautionary tools are available in their standardized experimental protocols; however, still professional expert audits are mandatory. Workers in the laboratories are mostly exposed to a variety of hazardous toxic microorganisms that may bring them at a considerable risk of infections. Risk factors should be taken care of by the individuals who are going to handle these pathogens and other hazardous material with the specific peculiarities of the microorganisms being examined for multiple purposes, various handling equipment, and protocols to be implemented. Likewise, for in vivo studies, mice are used as a model, and the other containment equipment available could pose a risk to the researcher (Brooks et al., 2004). The head of the laboratory is answerable for securing that sufficient and timely risk analysis is performed, and for coordinating meticulously with the institution's safety team and biosafety cadre to ensure that proper equipment and resources are available to promote the work being considered and other relevant new information from the literature. However, the actual risk posed to individual laboratory researchers after direct/indirect contact is difficult to resolve due to the lack of systematic reporting.

Researchers frequently come in contact accidentally through formerly unexpected modes of transmission. These were discussed in detail in 2004, initially by the laboratory-acquired case study of the severe acute respiratory syndrome (SARS) (Lim et al., 2004). It was a case of a researcher whose age was 27 years, who was working with a nonattenuated strain of West Nile virus, which was identified for flu-like symptoms in a microbiology laboratory in Singapore. The patient was unaware of any exposure to SARS and disagreed and also had no travel record. The patient was gradually discharged from the emergency, but reoccurrence was observed after a few days with relentless fever. So based on the previous records of the Singapore lab for SARS alerts, he was prescribed for a polymerase chain reaction assay with a sputum specimen that ended up with a SARS coronavirus positive result. Additionally, epidemiologic investigations also betray that the laboratory where the patient performed his research work was also associated with research on SARS coronavirus and that one of the cell cultures of West Nile virus was adulterated with the same infecting strain of SARS coronavirus. However, these types of cases are very exceptional, but it points out the inherent risk present to laboratory researchers, workers, employees, students with respect to their occupation.

Therefore, one of the most beneficial techniques for conducting a risk assessment of microbiology is the enlisting of risk groups as per their microbiological infections (Figure 7.1). Moreover, for microbial laboratory agents, the primary purpose of risk identification is to distinguish the microorganisms or the microbial toxins of concern with food. Preferably, the analysis should be done based on qualitative processes. Risk can be analyzed or prevented from relevant data sources. Detailed information of the chemicals, solutions, microorganisms, laboratory apparatus, various tools, techniques, strategies, and other relevant information regarding risk groups, risk assessment, and management can be acquired from literature studies, available databases like in food and chemical manufacturing, government representatives, and international authorities and through experts advice. Other information comprises clinical and epidemiological analysis and vigilance, several aspects of microbes, microbe–microbe synergy, microorganisms–environment interaction, and various studies on analogous microorganisms and situations (Commission, 1999; Weinstein and Singh, 2009).

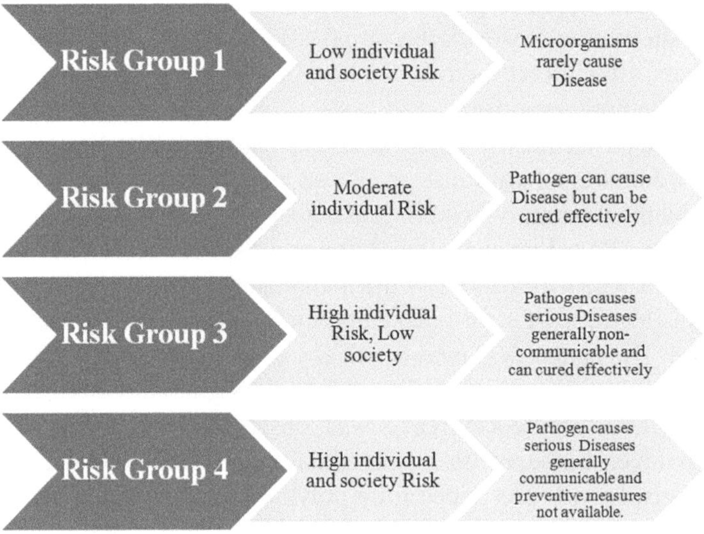

FIGURE 7.1 Classification of risk groups based on parthenogenesis.

However, some risk grouping for a particular agent is sometimes inadequate in the conduct of a risk assessment. Multiple or other risk factors that should also be considered include pathogenic nature of the agent and infectious dose, natural and different route of infections, adherence of the agent in the environmental conditions, concentration absorption of the agent and volume of concentrated material to be used and handled, suitable monitoring, reports of laboratory-acquired infections (LAI), laboratory standard operating procedure (SOP) (sonication, centrifugation, etc.) (Voysey and Brown, 2000). Therefore, basic biosafety principles are essential for recognizing hazardous substances and the safe handling of toxic and/or genetically modified organisms (GMOs). There are different safety parameters laid down by various monitoring agencies worldwide, which can be distinguished based on protection to the workers, experiments, human beings, and the associated environmental conditions.

7.2 BIOSAFETY

There are some standard microbiological activities that are not restricted in laboratories like eating, drinking, smoking, handling contact lenses, pipetting by mouth, etc. (Chosewood, 2007). The biosafety of laboratory work should be the central focal point during laboratory practices. Biosafety generally describes a safe mode of handling and managing the intrinsic infectious agents of living organisms in the experimental environment and the incorporation of biosafety measures to minimize the risk factors within the laboratory premises (Fleming and Hunt, 2006). Some researchers have to deal with genetic material as such ("naked" DNA), which can be dangerous to mankind as well. Even in a few cases, the use of vaccines may also provide an increased level of personal protection. But, before initiating any kind of work with pathogens or GMOs in a laboratory every individual should be made aware so that they can plan their work accordingly taking into consideration about all possible hazards of these organisms, their outcomes, and also need to take appropriate biosafety measures to decrease any risks for mankind and the surroundings (Karlsson, 2003). It was proved by Adel Hussein Elduma, who evaluated the biosafety measures that were practiced in Khartoum state diagnostic laboratories in 2009, which showed that the standards of biosafety precautions approved by the laboratory were shallow. Also, the awareness of the laboratory cadre toward biosafety principles and exercises was next to negligible (Elduma, 2012).

7.2.1 LABORATORY TECHNIQUES AND PRACTICES

Primarily, the most critical component of safety is strict adherence to standard microbial methods and practices. Individuals handling the hazardous agents or potentially infected substances should be familiar with all the toxic elements and must have hands-on experience under expert guidance with all the laboratory environment, tools, and techniques for handling such material with proper prevention (Washington, 2012). All the staff and the head of the division/person in charge of the laboratory should take responsibility and make appropriate arrangements for the training of cadre. Moreover, every laboratory should promote or accept a biosafety manual that catalogs the potential hazards which they may come across, and that particular techniques and protocols should be shaped in such a way so as to diminish exposures to toxic components. Organizations should be advised of unique hazards, and it should be made mandatory for all concerned to read and follow the required practices and procedures. All the knowledgeable staff, scientists, trained employees, researchers working with suitable laboratory techniques with safety procedures, and hazards affiliated with handling toxic agents must be answerable for the plan of work with any poisonous agents. Additionally, other measures may also be required when standard laboratory practices are not ample to handle the hazards related to specific agents or with any protocol. Moreover, relevant laboratory and facility design and features, safety tools and equipment, and handling practices must be provided to the laboratory cadre along with safety proceedings and approaches.

7.2.2 MICROBIOLOGICAL RISK ASSESSMENT

Risk assessment is the backbone of the biosafety practices in any laboratory (Songer, 1995). However, the most essential and significant method is professional judgment (Chosewood, 2007). Nowadays, many modern tools and techniques are available to analyze risks from a given protocol or experiment, and as a matter of routine, risk assessments should be executed by the laboratory worker/project supervisor before initiating and performing the tests. All individuals must be well aware of the organism's specification that they may be going to handle, the equipment and procedures to be implemented, animal models that may be managed, and the containment apparatuses and all other experimental facilities (Brown and Stringer, 2002). The head of the laboratory or director/principal investigator is solely responsible for ensuring that adequate and timely risk assessments must be executed. Additionally, for working on the institution premises, a duly constituted institutional safety committee and biosafety team must ensure the excellent condition of equipment and modern facilities to help and promote the experiment being conducted (Fleming and Hunt, 2006). Moreover, once the investigation is initiated and conducted, risk assessments must be analyzed and revealed routinely and periodically and must be updated with relevant revised information from the reported literature when required.

Therefore, a microbiological risk assessment must be executed based upon specific parameters and renewed literature data like:

1. There should be a functional partitioning between risk assessment and risk management.
2. This microbiological risk assessment must be performed in a systematic pattern, which includes hazard identification, characterization, exposure, and risk factor characterization.

3. Risk assessment should clearly mention the basic principle of the experiment to be conducted, which also includes outcome benefits.
4. Microbiological risk assessment should be transparently performed.
5. The microbiological risk assessment must analyze the motion of microbial growth, survival, and death in media culture, available nutrients, and hormones.
6. Assessment parameters must find the complexity of the interaction between individual and agent if consumed or splashed accidentally.
7. Risk analysis must be reassessed within short intervals of time in comparison with independent human illness data, if possible.
8. Reappraisement of microbiological risk assessment must be required whenever updated, relevant information, or data becomes available in the scientific literature.

7.2.2.1 CHARACTERISTICS OF MICROORGANISMS BASED ON THEIR RISK GROUP

Risk assessment is a well-suited strategy for identifying the fundamental biosafety level (BSL) and safety practices for any type of laboratory proceedings. Although various tools and techniques are possible to help in the estimation of risk for available methods, still the most decisive one is expert assessment. The judgment of risk should be done by the person who is well-known with the specifications of the microbes. The apparatus and protocol to be employed, mice like animal models may be handled; the containment apparatus and facilities available should be audited regularly. The laboratory head or principal investigator is responsible for providing proper and appropriate risk judgment reports. Once performed, risk assessments should be verified after regular intervals and amended when needed.

1. Risk Group 1 (no or low individual and community risk)—In this category, the microorganism that did not cause human or animal disease and included.
2. Risk Group 2 (moderate individual risk, low community risk)—Microbes can stimulate human or animal diseases, but they are not dangerous to the workers, surrounding, and the environment and can be medicated by accurate diagnosis and treatment.
3. Risk Group 3 (high individual risk, low community risk)—This group of microbes usually causes hazardous diseases but generally

noncommunicable. Adequate medical care and safety measures are available for cure.

4. Risk Group 4 (high individual and community risk)—This group of microbes is highly risky and cause hazardous diseases, which are not curable. These diseases are communicable diseases that can transfer from one individual to individual by any means.

Microbiological risk assessment and risk groups have also been assigned to biohazardous agents. They indicate how dangerous a particular microorganism or other biohazard is. Firstly, it determines the risk level of the agent with which researchers are working. Other factors include pathogenicity/virulence, route of infection, mode of transmission, survival in the environment, infectious dose, availability of effective preventative and therapeutic treatments, host range, natural distribution, impact of discharge into the surroundings, aerosol generation, concentration of the pathogen, illness severity and duration, antitoxins and vaccines availability, chemicals biosafety measures for technical use like solvents, acids, etc. (Kimman et al., 2008; Pauwels et al., 2009).

7.2.3 BIOSAFETY MATERIALS

Biosafety materials include a biological safety cabinet (BSC), which was used as primary barrier equipment against exposure to infectious biological agents (Goel and Parashar, 2013). They are framed to protect the individual/researcher, the environment of the laboratory, and other working materials available in the laboratory from contamination and exposure to hazardous airborne particulates and splashes that may be induced while handling experimental materials like microbial cultures, stocks, and other diagnostic species (Chosewood, 2009). Airborne fragments are created by any laboratory activity that can transform liquid or semiliquid material during procedures like pouring, stirring, shaking onto a working bench or surface, or transfer from one into another liquid. Moreover, other laboratory movements like streaking, inoculating cell culture, pipetting single, or using a multichannel pipette, homogenizing and vortexing and centrifugation of infectious fluids, heating, etc., may produce toxic aerosols. Generally, secondary aerosol particles cannot be visualized through naked eyes, and the individuals working in the laboratory are most of the time unaware of the aerosol particles present around that may be accidentally inhaled or may cross-contaminate other laboratory materials (Noble, 2011). Therefore,

BSCs were proposed and, when used meticulously, have been observed to be highly effective against LAIs and cross-contaminations of cultures due to these airborne particle exposures and play a crucial role in environment protection (Arroyo Sanchez, 2001).

This BSC has high-efficiency particulate air (HEPA) filters onto the contagious exhaust system. These HEPA filter pore size for particle entrapment is 0.3 μm in diameter size, which allows the HEPA filter to create a microbe-free environment and trap all unwanted infectious agents. The airflow used in a BSC was laminar, that is, the air travels with uniform velocity in one direction along parallel flow lines, and the direction of airflow depends on the design of the cabinet. The BSCs are not to be confused with chemical fume hoods, which have a very different mechanism of operation (Stuart et al., 2006).

Usually, there are three types of BSCs, that is, BSC class I, BSC class II, BSC class III used in microbiological laboratories for handling microbial materials (Figure 7.2) (World Health Organization, 2004). Open benched BSCs class I and class II are primary barriers that offer symbolic levels of prevention to laboratory cadre and to the ambiance with useful microbiological handling tools and techniques (Table 7.1). The BSC class II also gives security against external contamination of the materials like microbial cultures, stocks, etc., which are mostly being handled inside the cabinet. The BSC class III is gas-tight and provides the maximum obtainable level of safety to both workers and the environment in the microbiological laboratory.

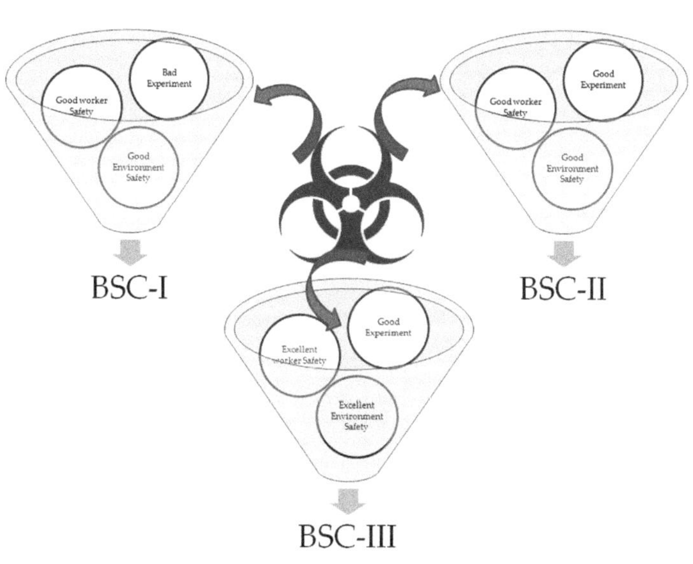

FIGURE 7.2 Good lab practices in various biosafety cabinets.

TABLE 7.1 Selection of Biosafety Cabinet Based on Specifications and Protection Type

Specifications	Type of Protection	BSC Selection
Microbiological safety cabinet	Personnel protection, microorganisms in Risk Groups 1–3	Class I, Class II, Class III
Autoclave	Personnel protection, microorganisms in Risk Group 4, glove-box laboratory	Class III
Double-ended autoclave	Personnel protection, microorganisms in Risk Group 4, suit laboratory	Class I, Class II
Centrifuge in the containment zone	Product protection	Class II, Class III
Vacuum generator fitted with a HEPA filter	Volatile radionuclide/chemical protection	Class IIB1, Class IIA2 vented to the outside

7.2.3.1 BSCS CLASS I

The BSC Class I has the most fundamental and straightforward outlay amongst all well-known to date. It has a ventilated cabinet with an inward air movement having splashes produced during microbes handling. Further, air moves via a filtration unit and traps all aerosol particles and unwanted contaminants. The outflow air through HEPA filters is clean and decontaminated from the cabinet.

7.2.3.2 BSCS CLASS II (TYPES A AND B)

The basic design of Class II safety cabinets is similar to Class I cabinets having an incoming air stream moves into the cabinet, which prevents the movement of airborne particles that are produced during microbiological manipulations. But, unlike basic Class, I safety cabinets, the inward flow in BSCs class II moves via a front inlet grille nearer to the worker working in the chamber. Therefore, with this, none of the unfiltered air enters/inflows into the working chamber of the cabinet, hence the operator can work in an uncontaminated sterilized condition to get better results.

The laminar airflow of the BSC protects and maintains the cadre, product, and material and environment biological safety. The exclusive character of this Class II chamber is a unidirectional HEPA-filtered air stream into the laminar that collapses downward from inside of the chamber. This protects the samples and prevents the aerosol particle inside the chamber (downflow), for the safe handling of microbes inside the cabinet from exposure to biohazards.

This BSC Class II type is subdivided into two varieties (A and B) based on cabinet design, construction, airflow velocities movement and arrangement, and exhaust units.

Actually, Types A1 and A2 cabinets were generally satisfactory for experimental microbiological purposes in the vacancy or minimum availability of volatile or toxic chemicals and radionuclides because cabinet air rotates the inner side of the chamber itself.

All BSC cabinet Class II type A chambers may be run down inside the laboratory premises or to the outdoors through a "thimble" linkage to the building exhaust unit.

BSC cabinet Class II type B chambers are also again categorized into types B1 and B2. These type B cabinets systems are hard fitted to the construction exhaust system and possess negative pressure plena. Due to this characteristic feature with a face velocity of 100 fpm, the facility permits workers to work with toxic chemicals or radionuclides.

Moreover, it is important that Classes I and II BSCs be verified and certified in situ at the time of initiation and installation of the chambers within the laboratory. Both Classes I and II, cabinets must be placed away from crowded areas in separate chambers and should be located away from doors. There should be strict adherence to approved practices for the handling of BSCs and their appropriate induction in the laboratory for experimental utilization as a mandate in securing the maximum containment capability of the equipment for better performance and life of the equipment itself.

7.2.3.3 BSCS CLASS III

The BSC Class III maintains a complete level of biosafety and protection, which is much better and trustworthy than the earlier two classes of cabinets. This Class III chamber is a completely embedded, open chamber with gas-tight construction, and gives the highest degree of care and environmental safety from infectious aerosols, along with the protection of innovative products and materials from microbiological contaminants. All the work/manipulations are performed through glove ports in the chamber (Stuart et al., 2004). While performing the experiment by the operator, negative pressure analogous to the ambient environment is sustained within the cabinet, which gives an additional fail-safe system in case any physical containment is avoided.

Comparatively, in Class III cabinets, a supply of HEPA filtered air gives product safety and protection against cross-contamination of materials. Alternatively, the source of HEPA filtered air and the chamber emitted air is filtered through double HEPA filters consecutively taken for the incineration, before releasing it outside in the open surroundings so as to disinfect all materials exiting the cabinet. They must be linked to a double door autoclave system to allow sterile supplies to enter the cabinet. All the contents are relocated into the cabinet using a pass-through unit installed at the side of the work area. Moreover, all equipment necessary for the laboratory activity, for example, incubators, refrigerators, and centrifuges, should be an integral part of the cabinet unit. Many Class III cabinets were generally set up as an interconnected system. When a committed ductwork unit is employed, they are also suitable for work applying hazardous chemicals that may be a supplement to microbiological processes.

Additionally, BSC Class III BSCs were advisable for handling microbiological agents authorized to BSLs 1, 2, 3, and 4. They are commonly particularized for experimental manipulations involving the most necrotic biological hazards.

The HEPA filter is the major component of the BSC called the heart of the BSC. They are disposable dry-type filters, designed with borosilicate microfibers adjusted into a thin sheet, likewise in a piece of paper. The basic dimensions of the filters used in BSC to restrict dust, smoke, bacterial range size 0.3–500 microns, soot, pollen, radioactive material, etc. are as follows:

- HEPA with 0.3 microns diameter of Modern "American-convention" HEPA: 99.99%
- Ultra-low penetration air (ULPA) with 0.12 microns diameter of Modern "American-convention" ULPA: 99.999%.

7.2.4 HANDLING OF THE BIOSAFETY CABINET

Initially, before starting work in the BSC wearing gloves for hand protection, then surface decontamination with cavicide, 70% ethanol onto the work surface, sidewalls, and inner back walls are recommended. Further, it is essential to permit the operator zone air to be evacuated for a few minutes before the start-up of work. The work area should never be overcrowded and measures to restrict the drain valve before the initialization of action should be ensured (Multanen, 2014). All the required materials should be loaded in the hood before initiating the work and surface decontamination.

Measures for efficient experimental performance handling are as follows:

1. Never arrest the front or back air grilles.
2. Always work away from the cabinet as much as possible.
3. Be careful with the elbow and arm movements during the operation.
4. Always perform slow and cautious handling of the equipment and never disturb the laminar airflow chamber.
5. After every interval or finishing a task, sterilize the surface and then remove the arms out very slowly.
6. Make sure that there should be minimum external airflow disruption.
7. Always work unidirectionally from clean to dirty.
8. Keep biohazard collection bin inside the cabinet for better performance rather than outside.
9. Take care when the glass apparatus is being handled like pasteur pipettes and ampoules. If glass pieces are found at the bottom, remove them with metal tongs and never with gloved hands.
10. Cleaning of the materials should be done far away from airborne generating objects to diminish the chances of cross-contamination.
11. Always try to place centrifuge, blender, or sonicator in the back one-fourth of the cabinet.
12. Appropriate handling and maintenance of the BSC will ensure better functioning, as well as the accuracy of the outcome

Hence, after shutting down the experiment, always seal biohazard bags if used during the operation of the experimental manipulations. Then, decontaminate the surface cabinet inner sidewalls, back wall, work zone, drain area, and the other hinged windows. Further, permit work area air to purge and activate UV lamp (if desired) after installation of the front sash. Maintenance/replacement of the prefilter is required once every 3 months, UV lamp substitution once a year.

7.3 BSLS FOR HANDLING BIOLOGICAL RISKS

There are four significant BSLs for microbiological laboratories (Li, 2007): basic level-1, containment level-2, level-3, and highly containment level-4 (Table 7.2).

TABLE 7.2 Brief Outlay of Biosafety Levels Based on Microbiological Laboratory Handling

BSL	Risk Group	Laboratory Type	Laboratory Practices	Safety Equipment	Laboratory Requirements	Agents	Examples	References
1	I	Primitive teaching and research	Standard management practices, including appropriate medical care programs	None	Basic laboratory requirements with hand wash basin, environmental, and working separation from general crowd	Not mentioned to frequently cause infection in healthy adults	*Bacillus subtilis, E.coli, Naegleria gruberi*, infectious canine, hepatitis virus	Levine (1987)
2	II	Principal health benefits; diagnostic advantage, research performances	BSL-1 plus protective clothing, biohazards sign, limited access to laboratory, sharps precaution, Biosafety updated manual	BSC class I/ II used for experimental manipulations, personal protective equipment like lab coats, gloves, face masks, goggles as needed	Desirable inward flow ventilation with controlled ventilation unit, BSC and autoclave on site	Agents linked with human infections; toxics like percutaneous injury, swallow, mucous membrane exposure	Toxoplasma spp, measles virus, Hepatitis B virus, salmonellae	Burnett et al. (2009)
3	III	Specialized diagnostic benefits, advance research performances	BSL-2 plus limited access, discard, and decontamination of waste material, sterilization of lab clothing, measures for serum samples of lab cadre	BSCs I/ II BSCs for all open manipulations of agents. Personal protective equipment like lab coats, gloves, face masks, goggles, respiratory protection as needed	Isolated laboratory with sealed rooms, double door entry, anteroom with shower, BSC, and autoclave on site, desirable inside the laboratory and double ended, controlled ventilation system along with inward flow, individual safety monitoring capabilities like communication systems, windows, etc.	Indigenous or exotic component may cause serious lethal infection via inhalation exposure	*M. tuberculosis, Bacillus anthracis, Coxiella burnetii*, St. Louis *Encephalitis* virus	Lietman and Blower (2000)

TABLE 7.2 *(Continued)*

BSL	Risk Group	Laboratory Type	Laboratory Practices	Safety Equipment	Laboratory Requirements	Agents	Examples	References
4	**IV**	Life threatening hazardous units	BSL-3 plus clothing change and shower before entry into lab and after exit from laboratory, airlock entry, special waste disposal	BSC III or I/II BSCs in combination with full-body, double ended autoclave, air-supplied, positive- pressure personnel suit	Isolated laboratory with sealed rooms, Double door entry, anteroom, airlock with shower, BSC and autoclave on site, inside the laboratory and double ended, controlled ventilation system along with inward flow, individual safety monitoring capabilities like communication systems, windows, etc.	Threatening agents with high risk of splashes of airborne particulates frequently deadly and incurable, also components with an identical antigenic relationship requiring BSL-4 until complete information data are available and other agents with unknown risk	Sin Nombre Virus, Rift Valley fever, *Ebola zaire*	Günther et al. (2011)

7.3.1 BIOSAFETY LEVEL 1

BSL-1 laboratory practices, safety equipment, and suitable facilities for experiments with well mentioned and characterized culture strains of viable microbe that is not known to infect or cause disease in healthy mankind. Laboratory experimental work is usually implemented on open bench slabs using basic microbiological laboratory practices. For this type of laboratory practice, mostly BSC is not involved. For handling the materials and equipment, the laboratory cadre must have specialized training in the protocols organized in the laboratory for performing experiments and is mentored by supervisors with general training in microbiology study. Microbes like *Bacillus subtilis*, dangerous canine hepatitis virus, *Nigeria gruberi*, and excluded microbes mentioned under the NIH Guidelines are the best representative organism categories for this level (World Health Organization, 2004). BSL-1 symbolizes a rudimentary level of containment that is based on simple standard microbiological techniques with no such specific barriers approved, except hand washing after every performance.

7.3.1.1 SPECIFICATION FOR HANDLING MICROBIOLOGICAL OPERATIONS IN A BSL-1 FACILITY

1. An individual should wash their hands before and after performing laboratory practices with toxic components.
2. Any type of eating material, drinking, smoking, managing eye contact lenses, using or applying cosmetics, and a different kind of storage material for personal human consumption should be restricted in laboratory areas.
3. Mechanical pipetting devices must be used in the laboratory while pipetting from the mouth is totally prohibited.
4. The supervisor of the laboratory should maintain the decorum of the institutional policies with the restricted access to the laboratory.
5. All the experiments should be performed in the laboratory very cautiously with the limited generation of airborne particulates and splashes.
6. Always decontaminate the working area at the end of the experiment and clean all types of spill or splash generated during the performance with appropriate disinfectant.
7. An effective pest control system should be implemented in the workspace to maintain the aseptic condition of the laboratory.

8. All the available strategies for the safe handling of sharps tools, like needles, pipettes, scalpels, and broken glassware, must be generated and employed.
9. Always ensure to use safety equipment while handling and performing the experimental manipulations, that is:
   - Usage of protective laboratory coats, gowns, or uniforms is mandatory for the prevention of direct contact of contaminated material with the individual clothing.
   - Always wear protective goggles when performing experiments that have the potential to create airborne particulates of micro-organisms or other toxic components, especially those who wear contact lenses.
   - Gloves must be used while performing protocols to avoid direct contact and exposure to toxic products. The selection of gloves must be made as per the level of the risk. Do not wash or reuse disposable gloves.

7.3.2 BIOSAFETY LEVEL 2

BSL-2 laboratory practices, biosafety equipment, and other facilities designed and building structures are appropriate to various clinical, diagnostic, teaching, and other working laboratories with vast native cautious risk causing agents that are available in the group and analogous with human infection of different extremity.

Comparatively, it differs from BSL-I based on the fact that laboratory cadre has specialized courses and training in handling and managing hazardous agents that are needed and were directed by competent supervisors. Additionally, the general admittance to the laboratory is restricted when experimental work is being operated. Also, strict precautions are needed to be taken with hazardous sharp equipment available in the work zone. However, various protocols where infectious airborne particles or splashes may be generated are handled explicitly in the BSCs. For this type of work Class I/II, BSC is highly approved. Handling of pathogenic materials like Hepatitis B virus, *Salmonella*, HIV, and *Toxoplasma* are assigned to this BSL-2 containment. This level is also recommended for appropriate handling with any human-derived blood samples, body fluids like saliva, tissues, or primary human cell lines where infectious agents may be unknown in the sample. For handling these products, personal protective equipment should be employed perfectly, like splash shields, face protection, lab coats/gowns, safety masks, and hand gloves. Also, washbasins for hands and incinerators like waste material discard/decontamination amenity must be accessible to reduce potential environmental contamination.

7.3.2.1 SPECIFICATION FOR HANDLING MICROBIOLOGICAL OPERATIONS

1. BSL-1 specifications; plus.
2. Every laboratory individual should be provided medical supervision, must be offered adequate medical care for hazardous agents safely supervised in the laboratory area.
3. The microbiology laboratory-specific biosafety manual should be well maintained and adopted as mandatory and easily accessible.
4. All toxic materials should be kept in good quality, leak-proof containers during collection, processing, storage, handling, or transport within a laboratory amenity.
5. All the protocols required for modifications and manipulation of hazardous materials that may produce airborne particulates must be operated inside the BSC.

7.3.3 BIOSAFETY LEVEL 3

BSL-3 laboratory working and practices, safety equipment, and appropriate facility structure and building for work done are appropriate for diagnostic, teaching, clinical trials and research, or production facilities with which native or foreign agents with a prospective for worker respiratory transmission, and may also cause severe and potentially mortal infection. Pathogenic diseases like the Saint Louis Encephalitis virus, *Coxiella burnetii*, and *Mycobacterium tuberculosis* depicts to this level. Initial toxicity to cadre performing with these agents compares with ingestion, autoinoculation, and exposure to infectious airborne particles. At this level, more significance is given on primary and secondary barriers to protect cadre in the infection, causing zones, the area, and the environment from manifestation to potentially infectious airborne particles. Likewise, various laboratory modifications and manipulations must be implemented in a BSC, for example, a gas-tight airborne fragment generation cabinet. Additionally, barriers for this particular level include limited entry to the laboratory area, and open ventilation is a must which releases dangerous airborne particulates from the experimental zone. BSC class I/II was necessary for performing innovative work involving these hazardous agents.

7.3.3.1 SPECIFICATION FOR HANDLING MICROBIOLOGICAL OPERATIONS

1. BSL-2 specifications; plus.
2. Every individual entering the laboratory area must be advised and be aware of the potential hazards.
3. One should meet personal protective entry/exit requirements inside the working zone.
4. The head of the laboratory must ensure that every worker should be aware of the standard and specific microbiological practices before handling BSL-3 agents.
5. Any animals and plants or another microorganism that is not associated with the experiment that is being executed should not be permitted in the working area of the laboratory.
6. No experiment with open vessels should be performed on the public bench.
7. When an experiment cannot be executed within a BSC, proper use of personal protective equipment and other devices, such as a centrifuge safety cup or sealed rotor must be applied to avoid contamination.
8. Always ensure to use safety equipment which handling and performing the experimental manipulations, that is,
   - Usage of protective laboratory coats, gowns, or uniforms is mandatory for the prevention of direct contact of contaminated material with the proper clothing.
   - Always wear protective goggles when performing experiments that have the potential to create airborne particulates of microorganisms or other toxic components, especially those who wear contact lenses.
   - Gloves should be used while performing experiments to cover hands from direct contact and exposure to toxic products. The selection of gloves must be adequate as per the level of risk assessment (Crooks, 2004). Do not rinse or reuse disposable gloves.
   - Every protocol involving the modifications and manipulation of hazardous materials must be conducted inside a BSC class II/III.

7.3.4 BIOSAFETY LEVEL 4

BSL-4 laboratory working and handling, safety equipment, and amenity design and building construction is appropriate for handling hazardous and foreign agents that carry an acute individual risk of life-threatening disease, that may be transferred from individual to individual via airborne particle transmission and also for this no cure/treatment is available like vaccine or therapy, etc. (Nisii et al., 2009). Components with similar or alike antigenic relationship to BSL-4 agents also should be managed and handled at this particular level only. Based on the data obtained from reported literature, performing experiments, and manipulations with these agents may continue at this very level or maybe lower than this. Marburg or Congo–Crimean Hemorrhagic fever-like viruses are represented and modified at BSL-4 (Linthicum et al., 2016). The BSL-4 amenity itself is designed as an independent building or completely isolated zone with a sophisticated, specialized opening for air ventilation requirements, and waste management systems to protect the discharge of feasible agents to the environment. The particular facility operations (SOPs) manual must be prepared or adopted by every individual working in the laboratory. BSC-III is required for work involving these types of hazardous agents. Basically, there are two well-known models for setting up BSL-4 laboratories, that is, Cabinet Laboratory in which laboratory manipulation of harmful agents should be taken place inside a BSC class-III and another one is the worn of a laboratory suit with a positive pressure supplied protective air suit or a Hazmat suit.

7.3.4.1 SPECIFICATION FOR HANDLING MICROBIOLOGICAL OPERATIONS

1. BSL-3 specifications; plus.
2. Every person who is entering the BSL-4 laboratory must be instructed for the severity of hazards and guided to meet significant entry needs according to the organizational policies.
3. Each entry into the amenity must be limited in the sense of secure, locked doors. Every record files, logbook, and other means of documenting with date and time of all persons entering and leaving the workspace must be well maintained.
4. At this level, during the laboratory experiments are in process, cadre should change the clothing and must pass-through shower rooms

with entry and exit from the laboratory. All these clothing and other essential items must be treated and sectioned as contaminated materials and decontaminated before laundering passage.

5. The head supervisor or the person who is immediately mentoring the laboratory is solely culpable for protecting that laboratory cadre. BSL-4 containment needs detailed demonstration with high proficiency in basic and specialized microbiological practices, tools, and techniques for handling and managing with agents and equipment.
6. Circumstances and situations that may be induced in contact with dangerous substances must be immediately assessed and treated according to techniques mentioned in the updated laboratory biosafety manual. All reasonable medical judgments, surveillance, and treatments must be implemented, and proper records should be maintained.
7. Any animals and plants or another microorganism that is not associated with the experiment that is being executed should not be permitted in the working area of the laboratory.
8. No experiment with open vessels should be performed on the public bench. When the research cannot be executed within a BSC, proper use of personal protective equipment and other containment devices, such as a centrifuge safety cup or sealed rotor must be applied to avoid contamination.
9. Inside the laboratory, every worker should wear separate laboratory clothing, like scrub suits, prior to arriving at the area used for clothing positive pressure suits.
10. All laboratories should be withdrawn into one side of the changing room prior to arriving in the personal shower area.
11. On the inner side, disposable gloves must be used to avoid break, tears, or damage in the outer suit gloves.
12. Disposable gloves must not be worn or taken outside the changing zone.
13. Never wash or reuse disposable gloves. Every time discard the used gloves with other contaminated waste materials.
14. Always decontamination must be performed of outer suit gloves during/after experimental laboratory handling to eliminate cross-contamination and minimize the risk of future contamination also.
15. A suitable communication unit must be available in between the workspace and the out space like voice, fax, and internet. Must have emergency communication provisions, and emergency access should be monitored and materialized.

16. The BSL-4 amenity structure parameters and working procedures must be appropriately documented. Available facilities must be verified while designing and also be reverified annually. The entire verification pattern should be improved as necessary by working experience.
17. The laboratory head/director should have the prime responsibility for the safe experimental operations to be conducted in the laboratory. The approved BSL depicts those conditions under which the hazardous and risk causing materials can be cautiously handled and manipulated in the available laboratory zone.

7.4 MICROBIAL AGENTS AND TOXINS

While handling any microbial agents in the laboratory, the researcher comes in direct contact with various toxic materials that are highly harmful and hazardous to mankind and the environment. These toxic substances are available in different forms in the working area of the laboratory, like cultures of blood or any type of fluids, specimens, tissues, and mice as a sample model and also other workers.

As per Occupational Safety and Health Act (OSHA) (1970) biosafety guidelines, there are various regulated microbial agents like viruses, bacteria, fungi, prions, and pathogens that have the capability to cause hazardous effects to mankind and safety, also to the environment and their products are also referred to as high-consequence livestock pathogens and toxins, nonoverlap agents and toxins, and cataloged plant pathogens.

The OSHA Safety and Health topics page entitled "Biological Agents" can be accessed at www.osha.gov/SLTC/biologicalagents/index.html, which requires the management to provide a workspace, which is free of recognized hazard (Biosafety Blue Ribbon Panel, 2012).

Anthrax: Anthrax, a disease induced by sporulating bacteria *Bacillus anthracis*, acquired by anthrax-infected animals and its infected products. *B. anthracis* is an HHS and USDA select agent.

Hantavirus: Hantaviruses are acquired from the dried droppings, urine, or saliva of mice and rats. Animal laboratory workers and individuals, that are in contact, are at increased risk.

Molds and Fungi: Millions of spores are released from microbes like molds and fungi, which may have adverse effects on human health, including allergic reactions, asthma, and other respiratory problems.

Plague: The pneumonic disease can be introduced by the release of a plague that could lead to dangerous consequences. The inventive agent of the epidemic is *Yersinia pestis*, an HHS/CDC agent. The World Health Organization mentioned more than 3000 incidences of the disease annually.

7.4.1 FOOD BORNE DISEASES

Food-borne pathogens such as viruses, bacteria, parasites, toxins, metals, and prions are root causes of nonirritating gastroenteritis to dangerous neurologic, hepatic, and renal syndromes.

7.5 BIOSAFETY OFFICER AND BIOSAFETY COMMITTEE FOR LABORATORY BIOSAFETY

This was made mandatory that each working area and laboratory in the organization and institution must have extensive safety rules and policies, a complete biosafety manual, and other various supporting methods for their employment. The accountability for this generally lies with the director or project investigator or laboratory supervisor, who may depute specific roles and responsibilities as a biosafety officer (Biosafety Blue Ribbon Panel, 2012). Complete laboratory safety and security of the workers are also the sole liability of all mentors and laboratory employees. Every single employee is solely liable for their self-safety and security, along with their neighboring colleagues. Therefore, those who are working in the laboratory are expected to perform their experiment safely and must report on time if found any unsafe acts, situations, or accidents to their supervisor. After short intervals, internal or external cadre are required and desired to conduct safety audits for the secure monitoring and safety of all personnel.

7.5.1 OFFICERS RESPONSIBLE FOR BIOSAFETY

It is also mandatory in all laboratories that well-trained/experienced biosafety officers should be recruited who shall ensure that biosafety rules and regulations, policies, and techniques regularly take place throughout the laboratory. On behalf of the head of the department or the institution, the biosafety officers implement and look after these responsibilities and duties. It depends on the laboratory system and units. The obligation must be allotted accordingly, like it may be a microbiologist working in the laboratory or the technical staff/faculty, who can look after these duties on a defined/ part-time basis. It not the matter of degrees of education, qualifications, etc., slightly, it is generally based on the extent of entanglement in biosafety, the individual who is designated for a particular job must be competent enough to acquire necessary expertise in the area to suggest, review, and approve specific activities such as biological containment and biosafety handling and protocols (Van Houten, 2000). The biosafety officer allotted for a particular position must apply required national and international norms, regulations, and guidelines as well as create SOPs in the laboratory. Moreover, the person enrolled or working in the laboratory should be trained or possess a technical background in microbiology, biochemistry, biotechnology, basic instrumentation, or biological sciences, etc. Knowledge of laboratory and apparatus handling, techniques practices, and safety, including principles relevant to the design of the instrument, working, and maintenance of availabilities, is essential, highly advisable, and desirable for providing these safety issues. The communication skills, demonstration, and administrative capabilities of the biosafety officer should also be highly effective.

7.5.2 BIOSAFETY TEAM

A management team of biosafety committees should be established to promote organizational biosafety rules, regulations, policies, and codes of practices in the laboratory (Chamberlain et al., 2009). This biosafety committee has the sole responsibility to review research protocols and all the techniques and instruments used in the laboratory for performing experiments involving infectious agents, animal use, recombinant DNA, genetically modified materials, etc. (NIH office, 2002). Moreover, many other activities involving risk assessments, creating physical and biological biosafety containment strategies, formulation of new safety policies, and judgments are in conflict over safety matters by the biosafety committee. The team members of the biosafety committee should belong to diverse occupational areas of the organization as well as be able to demonstrate their scientific expertise and must initiate measures to prevent accidents/hazards.

The biosafety committee members must include biosafety officer, scientific staff, medical care team, veterinarian (if animal handling is required), technical staff representatives, and other laboratory management members. The biosafety committee should be cooperative and update with the present research scenario and must seek advice from different divisions and experienced safety officers and sometimes require assistance from various independent authorities and national/international regulatory bodies (Huber, 2007). Every community member, staff, faculty, student, researcher, the employee may also be helpful when required for particularly sensitive issues.

7.6 BIOSAFETY, BIOETHICS, AND BIOSECURITY

Laboratory biosafety and biosecurity have different risks, but they serve a common goal. Biosafety is directly related to contamination-free unique environmental principles, tools, and techniques. Microbiologists working in their environment require individual practices that are considered to reduce unintentional accidental exposures to toxic materials and various hazardous pathogens.

Bioethics is majorly concerned with controversial ethical issues emerging from new laboratory practices and possible working situations. Typically, it relates to various medical techniques, practices, and research, which understands and follows health care laws, moral structures of our practices both ethically and legally perceive the need to confront partiality and unfairness, also understand our own moral values and ethics which also help in appreciating the variations in meaningful reasoning among individuals and groups of working individuals (Macer, 2001). Similarly, biosafety is a prevention strategy for large-scale harmful biological synergy, targeting both ecology, as well as human health.

Likewise, laboratory biosecurity usually administrates and coordinates with various regulatory bodies. Numerous laboratory experiments, tools, and techniques used in the experimental laboratory area are considered as secured for the implementation of different laboratory practices (World Health Organization, 2006). Therefore, biosafety and biosecurity complement and adjust each other's activities, including the execution of various laboratory tasks and responsibilities.

Additionally, particular laboratory biosecurity methods are also involved in handling the biological risks such as:

1. Appropriate designing and preparing the facilities.
2. Selective conduction of laboratory work.
3. Work must fulfill the mandate of the institution.
4. Proper recommendation of concerned authorities and heads of the institution.
5. Inventories should be updated with correct information.

6. Documentation of internal and external transfers within and between facilities should be maintained.
7. Proper discards/disposals of material should be done.

7.7 CONCLUSION

Amongst the recent studies, we examined the influence of measures aimed at preventing and protecting humans and the environment against the biological risks of handling and manipulating the microorganisms based on our interest and society's needs using genetically and nongenetically modified microorganisms, as well as hazardous pathogens/microorganisms (Kimman et al., 2008). It has been observed that very few principles and techniques handling the present biosafety practices are available like risk assessment, microbial containment, application and building, minimization exposure, physical/biological containment, and toxicity minimization. Moreover, no such latest practices are reported related to understanding and expert assessment on particular biosafety management. Therefore, the potency of biosafety practices has been well evaluated in this chapter. Individual containment apparatuses and procedures, levels of biosafety of the research area altogether, or at the microbiological level, that is, BSL-1 does not cause infection in healthy adults, microbes includes *B. subtilis, Escherichia coli, Naegleria gruberi*, infectious canine, hepatitis virus; BSL-2 agents linked with human diseases; toxics like percutaneous injury, swallowing, mucous membrane exposure, microbes includes: *Toxoplasma* spp, measles virus, Hepatitis B virus, *Salmonella*; BSL-3 are indigenous or exotic component may cause serious lethal infections via inhalation exposure, microbes involved: *M. tuberculosis*, *B. anthracis*, *C. burnetii*, Saint Louis Encephalitis Virus; BSL-4 means threatening agents with a high risk of splashes of airborne particulates frequently deadly and incurable, also components with an identical antigenic relationship requiring BSL-4 until complete information, examples are Sin Nombre Virus, rift valley fever, *Zaire ebola* virus. Data are also elucidated about the containment apparatuses and laboratory procedures, which in literature are scanty and disintegrated. Additionally, for analyzing LAIs (Weinstein and Singh, 2009), it is very much decisive for assessing the influence and suggesting the failures of biosafety, which were not yet noticed and reported. Due to the mishandling and careless attitude of the biosafety committee members and laboratory supervisors and staff, it is revealed that the number of reported laboratory issues related to genetically modified microorganisms is considerably lower than that of those pertaining to nongenetically modified microbes. Individuals are unaware to a great extent of specific measures that should contribute to the overall level of biosafety that should be maintained. So from this above statement, we strongly advocate that the information based on biosafety handling and practice needs to be further strengthened. Also, the intent of this chapter is to call for action with the development of sound efficient, and economical methods to reduce the accountability of biological risks and institute suitable biosafety committees and trained personnel for intermittent monitoring.

ACKNOWLEDGMENTS

The authors acknowledge Hon'ble Vice-Chancellor, SMVDU, Katra and Director, CSIR-IIIM, Jammu for their kind support and facilitation. This work was supported by grant HCP0007 from CSIR, and VA is also thankful to CSIR for Project Assistant Level-II Fellowship.

CONFLICT OF INTEREST

Author disclosures: V.A., S.M., and S.J. declare no conflicts of interest

KEYWORDS

- **biohazard**
- **risk**
- **biosafety level**
- **universal precautions**
- **disinfection**
- **sterilization**

REFERENCES

Arroyo Sanchez, M. C. *Biological Safety: Principles and Practices*, 43, 3rd. ed. Washington, DC: ASM Press, **2001**, 784.

Brooks, G.; Butel, J.; Morse, S. The Neisserieae. In: *Jawetz, Melnick, and Adelberg's Medical Microbiology.* 23rd ed. New York, NY: Lange/Mc Graw-Hill, **2004**, 295–304.
Brown, M.; Stringer, M. *Microbiological Risk Assessment in Food Processing*, Brown, M. and Stringer, M.; Eds, 1st ed. Sawston: Woodhead Publishing, **2002**, 320.
Burnett, L. C.; Lunn, G.; Coico, R. Biosafety: guidelines for working with pathogenic and infectious microorganisms. *Curr. Protoc. Microbiol.* **2009,** 13, 1A-1.
Chamberlain, A. T.; Burnett, L. C.; King, J. P.; Whitney, E. S.; Kaufman, S. G.; Berkelman, R. L. Biosafety training and incident-reporting practices in the United States: a 2008 survey of biosafety professionals. *Appl. Biosaf.* **2009,** 14, 135–143.
Chosewood, L. C. *Biosafety in Microbiological and Biomedical Laboratories*, Deborah, E.; Wilson, D. E. and Chose wood, L. C. Eds, 5th ed. Darby, PA: Diane Publishing, **2007**, 438.
Commission, C. A. Principles and guidelines for the conduct of microbiological risk assessment (CAC/GL 30–1999). In: *Codex Alimentarius, Food Hygiene Basic Texts, 4th ed.* Joint FAO/WHO Food Standards Programme. Rome: FAO/WHO, **2009**, 43–50.
Crooks, S. American glovebox society. *Chem. Health Saf.* **2004,** 11, 27–28.
Elduma, A. H. Assessment of biosafety precautions in Khartoum state diagnostic laboratories, Sudan. *Pan Afr. Med. J.* **2012,** 11, 19.
Fleming, D. O.; Hunt, D. L. *Biological Safety: Principles and Practices*, Fleming, D. O. and Hunt, D. L. Eds, 4th ed. Washington, DC: ASM Press, **2006**, 642.
Goel, D.; Parashar, S. *IPR, Biosafety and Bioethics*, Parashar, S. and Goel, D. Eds. Pearson Education India, Delhi, India, **2013**.
Gunther, S.; Feldmann, H.; Geisbert, T. W.; Hensley, L. E.; Rollin, P. E.; Nichol, S. T.; Ströher, U.; Artsob, H.; Peters, C. J.; Ksiazek, T. G. Management of accidental exposure to Ebola virus in the biosafety level 4 laboratory, Hamburg, Germany. *J. Infect. Dis.* **2011,** 204, S785–S790.
NIH office. NIH guidelines for research involving recombinant DNA molecules (NIH guidelines). NIH office, Maryland, USA, **2002**, **1**–86.
Huber, L. *Validation and Qualification in Analytical Laboratories*, Huber, L. Ed, 2nd ed. Boca Rato, FL: CRC Press, **2007**, 288.
Karlsson, M. Biosafety principles for GMOs in the context of sustainable development. *Int. J. Sustainable Dev. World Ecol.* **2003,** 10, 15–26.
Kimman, T. G.; Smit, E.; Klein, M. R. Evidence-based biosafety: a review of the principles and effectiveness of microbiological containment measures. *Clin. Microbiol. Rev.* **2008,** 21, 403–425.
Levine, M. M. *Escherichia coli* that cause diarrhea: enterotoxigenic, enteropathogenic, enteroinvasive, enterohemorrhagic, and enteroadherent. *J. Infect. Dis.* ***1987,*** 155, 377–389.
Li, J. Laboratory biosafety of pathogenic microorganisms in China. In: *Beijing on Biohazards: Chinese Experts on Bioweapons Nonproliferation Issues,* Smithson, A. E.; Monterey, C. A. Eds. Monterey, Cl: Monterey Institute of International Studies, **2007**, 1–137.
Lietman, T.; Blower, S. Potential impact of tuberculosis vaccines as epidemic control agents. *Clin. Infect. Dis.* **2000,** 30, S316–S322.
Lim, P. L.; Kurup, A.; Gopalakrishna, G.; Chan, K. P.; Wong, C. W.; Ng, L. C.; Se-Thoe, S. Y.; Oon, L.; Bai, X.; Stanton, L. W. Laboratory-acquired severe acute respiratory syndrome. *New Engl. J. Med.* **2004,** 350, 1740–1745.
Linthicum, K. J.; Britch, S. C.; Anyamba, A. Rift valley fever: an emerging mosquito-borne disease. *Annu. Rev. Entomol.* **2016,** 61, 395–415.
Macer, D. Bioethics: perceptions of biotechnology and policy implications. *Int. J. Biotechnol.* **2001,** 3, 116–133.

Multanen, J. P. Safety cabinets in cleanroom environment. *45th R 3* Nordic Symposium. **2014**, 49.

Nisii, C.; Castilletti, C.; Di Caro, A.; Capobianchi, M.; Brown, D.; Lloyd, G.; Gunther, S.; Lundkvist, A.; Pletschette, M.; Ippolito, G. The European network of Biosafety-Level-4 laboratories: enhancing European preparedness for new health threats. *Clin. Microbiol. Infect.* **2009,** 15, 720–726.

Noble, M. A. Prevention and control of laboratory-acquired infections. In: *Manual of Clinical Microbiology,* 10th ed. Washington, DC: American Society of Microbiology, **2011**.

World Health Organization. *Laboratory Biosafety Manual*, 3rd ed. Geneva: World Health Organization, **2004**, 1–170.

World Health Organization. *Biorisk Management: Laboratory Biosecurity Guidance*. Geneva: World Health Organization, **2006**, 1–33.

Biosafety Blue Ribbon Panel. Guidelines for safe work practices in human and animal medical diagnostic laboratories. *Morb. Mortal. Wkly. Rep.* **2012,** 61, 1–101.

Pauwels, K.; Gijsbers, R.; Toelen, J.; Schambach, A.; Willard-Gallo, K.; Verheust, C.; Debyser, Z.; Herman, P. State-of-the-art lentiviral vectors for research use: risk assessment and biosafety recommendations. *Curr. Gene Ther.* **2009,** 9, 459–474.

Sewell, D. L. Laboratory-associated infections and biosafety. *Clin. Microbiol. Rev.* **1995,** 8, 389–405.

Songer, J. Laboratory safety management and the assessment of risk. In: *Laboratory Safety: Principles and Practices*, 2nd ed. Washington, DC: American Society for Microbiology, **1995**, 257–268.

Stuart, D.; Kiley, M.; Ghidoni, D.; Zarembo, M. The class III biological safety cabinet. In J. Y. Richmond (Ed.), Anthology of biosafety VII: Biosafety level 3. Mundelein, IL: American Biological Safety Association.

Stuart, D. G.; Eagleson, D. C.; Quint, Jr, C. W. Primary barriers: biological safety cabinets, fume hoods, and glove boxes. In: *Biological Safety.* Washington, DC: American Society of Microbiology, **2006**, 303–323.

Van Houten, I. Leadership and management. In: *Biological Safety: Principles and Practices*. Fleming, D.O. and Hunt, D. L. (Eds), ASM Press, **2000**.

Voysey, P.; Brown, M. Microbiological risk assessment: a new approach to food safety control. *Int. J. Food Microbiol.* **2000,** 58, 173–179.

Washington, J. A. *Laboratory Procedures in Clinical Microbiology*, Tang, W. E. and Stratton, C. W. Eds, 2nd ed. Berlin: Springer Science & Business Media, **2012**.

Weinstein, R. A.; Singh, K. Laboratory-acquired infections. *Clin. Infect. Dis.* **2009,** 49, 142–147.

PART II

Advances in Cellular and Molecular Mechanisms

CHAPTER 8

Intracellular Redox Status and Disease Development: An Overview of the Dynamics of Metabolic Orchestra

SHARMI MUKHERJEE and ANINDITA CHAKRABORTY*

UGC-DAE Consortium for Scientific Research, Kolkata Centre, Kolkata 700106, West Bengal, India

Corresponding author. E-mail: anindita.iuc@gmail.com

ABSTRACT

Normal aerobic cellular metabolic processes promote free radical generation wherein the activation of the antioxidant defense mechanism maintains redox homeostasis. A state of imbalance generates when cells fail to detoxify the excess of free radicals formed, resulting in the development of oxidative stress. Oxidation of various cellular components stimulates structural, biological alterations, and homeostatic imbalances in cellular metabolism. The products of complex biomolecular interactions between free radicals and the molecular targets represent the oxidative markers of lipid, protein, and DNA damages and serve as the basis for the onset of various disease complications. In recent years, there has been a significant advancement in research regarding the contributory role of oxidative stress in disease pathophysiology. Oxidative stress promotes the development and progression of acute pathologies, chronic illness, cognitive deteriorations and neurodegenerative disorders. Also, the impact of oxidative stress is established for alcohol and drug abuse cases, multiple seizures, and complications in transplant recipients. The clinical manifestations highlight that oxidative stress rarely occurs in isolation but involves complex interactions with other forms of stress. Oxidative stress during pregnancy has been linked to miscarriage, pre-eclampsia, foetal morbidity, congenital disabilities with prolonged consequences for the mature organism. The contribution of oxidative stress in the pathophysiology of different disease warrants further exploration to establish a firm correlation as there are many diseases which present an apparent indirect association with oxidative stress. Prognosis and diagnosis of various disease complications depend on the successful identification of biomarkers of oxidative modifications and their realization as crucial therapeutic targets.

8.1 INTRODUCTION

The maintenance of redox equilibrium accomplishes normal physiological homeostasis in aerobic organisms. Any shift in the balance triggers feedback reactions that regulate the level of the two crucial constituents of cell redox homeostasis, pro-oxidants, and antioxidants (Brigelius-Flohe and Flohé, 2011; Ursini et al., 2016). Pro-oxidants like reactive oxygen species (ROS) have a short life span and are generated because of normal physiological, metabolic processes, including cytochrome P-450 and hypoxanthine-xanthine oxidase systems as well as cell proliferation, differentiation, inflammatory responses, circulation, immune system regulation, phagocytosis, prostaglandin synthesis, and vascular remodeling, mediated by various enzymes like lipoxygenase, cyclooxygenase, xanthine oxidase (Hussain et al., 2003; Li et al., 2016). Similarly, nitric oxide (NO) is a significant regulator of vascular homeostasis, and NO synthase (NOS) along with NADPH oxidase catalyzes reactive nitrogen species (RNS) production. Both ROS and RNS pro-oxidants can be classified into two groups: radicals and nonradicals. Radicals can exist independently and are highly reactive due to the presence of unpaired electrons in shells (Engwa, 2018). Superoxide ($O•^{-2}$), oxygen radicals (O_2), hydroxyl (OH•), alkoxy-radical (RO•), peroxyl radical (ROO•), nitric oxide (NO•), and nitrogen dioxide ($NO_2•$) are examples of free radicals. Singlet oxygen ($^1O_2$), ozone (O_3), hydrogen peroxide (H_2O_2), hypobromous acid (HOBr), hypochlorous acid (HOCl), nitrous acid (HNO_2), nitroxyl anion (NO^-), nitrosyl cation (NO^+), nitronium (nitryl) cation (NO_2^+), dinitrogen trioxide (N_2O_3) and tetra-oxide (N_2O_4), peroxy-nitrite (ONOOH), organic peroxides (ROOH), and aldehydes (HCOR) (i.e., all the nonradical species) can easily elicit free radical reactions in living organisms (Halliwell, 2005; Kohen and Nyska, 2002; Phaniendra et al., 2015). Different chemical, physical, and biological challenges promote the production of highly reactive pro-oxidants from cellular organelles with high oxygen consumption, such as the endoplasmic reticulum (ER), peroxisomes, and most importantly, mitochondria (Medzhitov, 2008; Panieri and Santoro, 2016). As such, microbial infections stimulate NADPH oxidase-mediated ROS generation in phagocytic cells (Leto et al., 2009). The enormous production of cellular oxidants at the rate of 50 hydroxyl radicals/sec/cell or 1.7 kg superoxide radicals/year in a 70 kg weighing adult brings a shift in the redox homeostasis toward excess oxidation. However, the involvement of a feedback system under normal conditions restore balance (Lane, 2002; Maltepe and Saugstad, 2009). However, a highly efficient system of antioxidants aids in counteracting the damaging effects of free radicals on cells (Rahman, 2007). Transcription factors like nuclear factor erythroid 2-related factor 2 (Nrf2), stimulate gene expression of enzymatic, and nonenzymatic antioxidant scavengers, namely, catalases (CAT), superoxide dismutases (SOD), glutathione (GSH), glutathione peroxidases (GPx) (Smith et al., 2016; Ursini et al., 2016). As the name suggests, antioxidants minimize pro-oxidant reactivity by donating electrons and thus maintain redox homeostasis. A permanent shift in homeostasis due to inadequate antioxidant feedback or the presence of a harmful stimulus establishes an altered redox state. Cigarette smoking, exposure to particulate air pollution, radiation, and industrial chemicals in urban areas trigger free radical generation while depleting *in vivo* antioxidant levels (Lobo et al., 2010). Severe infections, circadian rhythm dysregulation, as well as chronic psychological stress, promote sustained activation of the hypothalamic-pituitary-adrenal axis (Spiers et al., 2015; Spiga et al., 2014). Thus, disturbances of redox circuitries regulating ROS turnover and related events lead to abnormal signaling, toxic byproduct accumulation, cytotoxicity, and oxidative damage induction having implications in many pathological conditions (Holmström and Finkel, 2014; Panieri and Santoro, 2016). Moreover, oxidative stress is associated with damage induction in a wide range of crucial biomolecules, including membranes, lipids, proteins, and nucleic acids. Figure 8.1 depicts a scheme of generation and the impact of oxidative imbalance on biomolecules and disease aetiology. These interactions impart detrimental alterations in membrane fluidity, ion transport, protein cross-linking, enzyme activity loss, protein synthesis inhibition, ultimately bringing about cell death (Sharma et al., 2012). The clinical manifestations are indicative of the complex interactions of oxidative stress with other forms of stress, like ER stress (Bhandary et al., 2012). During the unfolded protein response and protein misfolding in the ER, electron transfer from thiol group of proteins to oxygen by ER oxido-reduction (ERO-1) and protein disulfide isomerase (PDI) induces ROS generation due to GSH depletion (Higa and Chevet, 2012; Malhotra and Kaufman, 2007; Masui et al., 2011; Zhang, 2010). Consequently, the series of chain reactions stimulate oxidative stress development. ER unfolded proteins promote the cytosolic Ca^{2+} release, which also stimulates mitochondrial ROS generation (Masuda et al., 2017; Moserova and Kralova, 2012). Oxidative stress has a profound effect on mitochondria causing the defective synthesis of mitochondrial electron transport chain subunits and decreased transmembrane potential, which accelerates ROS generation and contributes to the redox imbalance (Sureshbabu and Bhandari, 2013).

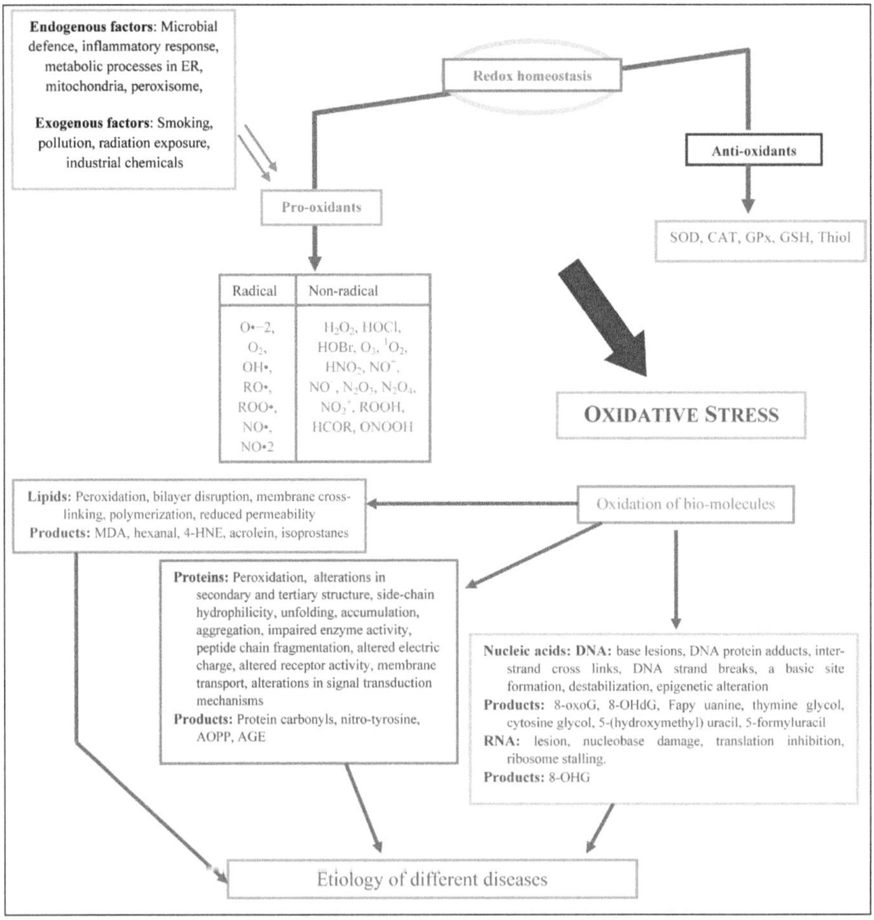

FIGURE 8.1 Generation of oxidative imbalance and oxidation of bio-molecules form the basis for the onset of different diseases.

8.2 MOLECULAR BASIS FOR THE ONSET OF DISEASE COMPLICATIONS

8.2.1 OXIDATIVE DAMAGE TO LIPIDS

Elevated ROS levels are associated with enhanced lipid peroxidation of cellular and subcellular membranes, resulting in lipid bilayer disruption affecting membrane-bound receptors and membrane permeability (Birben et al., 2012; Girotti, 1985). Highly reactive lipid-derived radicals are generated because of the peroxidation of polyunsaturated fatty acids (PUFAs) residues present in membrane triglycerides and phospholipids (Pizzimenti et al., 2013). These radicals damage proteins and DNA and aggravate oxidative stress with concomitant membrane polymerization and cross-linking. The presence of a double bond between two carbon atoms and ester linkage between fatty acid and glycerol makes PUFAs more vulnerable to ROS attack (Sharma et al., 2012). A single radical (R•) by initiating cyclic chain reactions can induce peroxidation of several PUFAs. The radical causes transfer of hydrogen from the methyl residue of PUFA, forming pentadienyl lipid radical with carbon center (L•) (Shichiri et al., 2014). Lipoperoxyl radical (LOO•) is produced by subsequent oxidation, which, on reaction with lipid yield unstable products like lipid hydroperoxides (LOOH) (Abuja and Albertini, 2001; Porter, 1984).

$$R\bullet + LH = RH + L\bullet,\ O_2 + L\bullet = LOO\bullet$$

$$LOO\bullet + LH = LOOH + L\bullet$$

Acyl chain lipid peroxidation due to oxidation of these lipid radicals induce the bond rearrangement of membrane phospholipids promoting further chain propagation by the capture of radicals (Hauck and Bernlohr, 2016). Depending upon the lipid species oxidized and level of unsaturation, final products like mutagenic MDA, hexanal, highly toxic 4-hydroxynonenal (4-HNE), or acrolein are produced which serve as the second messengers of oxidative stress due to their high reactivity (Barrera, 2012; Weber et al., 2013).

$$LOO\bullet + LOO\bullet = \text{Nonradical products (such as MDA, 4-HNE)}$$

4-HNE, a significant biomarker of lipid peroxidation regulates various stress-sensitive transcription factors like nuclear factor-kappa B (NF-κB), Nrf2, activating protein-1 (AP-1), and receptor activity (peroxisome proliferator-activated receptors, PPAR), promoting cell survival, proliferation, differentiation, autophagy, or senescence, apoptosis, and necrosis (Braithwaite et al., 2010; Camandola et al., 2000; Huang e al., 2012; Lee, 2017; Wang et al., 2012). Reactive carbonyl compound MDA, a crucial marker of oxidative stress in the clinical scenario can lead to lifestyle and dietary factor related cancers by forming DNA adducts and protein cross-links. Their reaction with nucleosides like cytidine and deoxyguanosine result in the formation of heterocyclic pyrimido-purinone called pyrimido [1, 2-a]purin-10(3H-)one (M1G) (Ayala et al., 2014; Marnett, 1999). Peroxidation of arachidonic acid generates prostaglandin like isoprostanes, whose increasing level in biological fluids has clinical implications. The products such as 4-HNE and MDA are involved in the aetiology of various metabolic disorders, including obesity, cardiovascular disease, neurodegenerative disease, diabetes, respiratory diseases, and even cancer (Hauck and Bernlohr, 2016; Shichiri, 2014).

8.2.2 OXIDATIVE DAMAGE TO PROTEINS

In cells manifesting oxidative imbalance, almost 70% of affected biomolecules constitute proteins, whose multiple side-chain and backbone sites are the main targets of oxidants (Davies, 2016; Schöneich, 2016; Woods et al., 2003). Sulfur-containing side chains of methionine and cysteine residues, histidine, and arginine are highly susceptible to oxidation (Dahl et al., 2015; Gray et al., 2013; Pattison and Davies, 2006; Schöneich, 2005). Oxidative stress can modify protein structure, alter the electric charge, increase side-chain hydrophilicity, fragmentation, cross-linking, induce protein unfolding, proteolysis, enhance accumulation, and aggregation affecting protein stability with impaired enzyme activity and altered interactions with biological partners (Pizzino et al., 2017). Destabilizing protein oxidations speed up secondary damage on the proteostasis complex of cells (Dahl et al., 2015; Roth and Balch, 2011; Trougakos et al., 2013). Protein oxidation-damaged products also affect receptors, membrane transport, with alterations in the mechanism of signal transduction (Lobo et al., 2010). ROS modifies transcription factors like hypoxia-inducible factor 1 (HIF-1), NF-κB, AP-1, nuclear factor of activated T cells (NFAT), various receptors of growth factor, tyrosine kinase receptors, protein tyrosine phosphatases, and serine/threonine kinases. Members of mitogen-activated protein kinase family-like ERK, JNK, and p38 involved in critical cellular processes of proliferation, differentiation, and apoptosis, are also regulated by oxidants (Birben et al., 2012; Filomeni et al., 2002; Johnson and Lapadat, 2002; Poli et al., 2004; Schieber and Chandel, 2014; Sun and Oberley, 1996). Alterations of these crucial intracellular pathways and the accumulation of damaged proteins contribute to multiple human pathologies. The interactions of free radicals with proteins involve the generation of high levels of peroxyl radicals with protein peroxidation constituting up to 70% of initial oxidation effects (Davies, 2016; Luxford et al., 1999). Peroxyl radicals take part in chain termination reactions yielding alcohols, aldehydes, or ketones involving ROO• with HOO•/O_2−• or two ROO• (Davies, 2016). Peroxyl (ROO•) radicals react with aliphatic side-chain while O_2−• and $^1O_2$ interact with aromatic side chains of amino acids (Davies, 2004, 2016). Tertiary alkoxyl radicals can abstract hydrogen-atom from C–H (or S–H with Cys) bonds in protein and other targets producing alcohol or via β-scission produce ketone and another radical (Dalle-Donne et al., 2005; Davies, 2005). Radicals like HO•, alkoxyl RO•, and peroxyl ROO• form adducts at electron-rich sulfur of cysteine and methionine. On the other hand, phenyl and carbon-centered radicals, mainly attack electron-deficient sites at the side-chain of amino acids. The functional group and distance from the amine group within free amino acids regulate the site of radical attack (Watts and Easton, 2009). These reactions trigger the production of protein carbonyls and nitro-tyrosine, relevant biomarkers of protein oxidation in diseases, and aging. ROS-mediated tyrosine nitration generates nitrotyrosine involved in inflammation and NO production (Sabuncuoğlu et al., 2012). Oxidation-induced protein carbonyl group production on protein side chains has been observed in rheumatoid arthritis, Alzheimer's disease (AD), diabetes, lung damages, chronic renal failure, and sepsis (Dalle-Donne et al., 2003; Ghosh et al., 2018). Cysteine residue oxidations form various protein adducts like s-nitrosylation, s-glutathionylation, s-sulfonation, and disulfide formation (Cai and Yan, 2013; Jacob et al., 2012; Janssen-Heininger et al., 2008; Poole, 2004; Salsbury et al., 2008).

8.2.3 OXIDATIVE DAMAGE TO NUCLEIC ACIDS

8.2.3.1 DNA

Interactions of free radicals with DNA are associated with oxidative modifications of purines and pyrimidines, purine 5′,8-cyclonucleoside formation, base lesions, DNA protein adducts, interstrand cross-links along with the development of DNA strand breaks and abasic sites (Cadet and Wagner, 2014). Nearly 100 different oxidatively modified bases and sugar molecules, like diastereomeric nucleosides and hydroperoxides of thymidine, have been observed under the influence of oxidative stress (Cadet and Wagner, 2013). Of all the natural bases, guanine is highly susceptible to oxidation, producing hydantoin lesion via generation of oxidation sensitive products

8-oxo-7,8-dihydroguanine (8-oxoG), 8-oxo-2′-deoxyguanosine (8-OHdG), and 2,6-diamino-4-hydroxy-5-formamidopyrimidine (Fapy guanine) (Barnes et al., 2019; Fleming and Burrows, 2017; Zhou et al., 2015). The interaction of $^1O_2$ or OH• radical with C8 position of the guanine ring and the redox environment favors the formation of these products (Bergeron et al., 2010; Cadet and Wagner, 2013, 2016; Douki et al., 2002; Kropachev et al., 2014). OH• radical interaction with positions 5 or 6 of pyrimidine (thymine and cytosine) ring or methyl group of thymine generate different products like 5,6-dihydroxy-5,6-dihydrothymine (thymine glycol, Tg), 5, 6-dihydroxy-5, 6-dihydrocytosine (cytosine glycol), 5-(hydroxymethyl) uracil and 5-formyluracil (Kryston et al., 2011; Radak and Boldogh, 2010). The generation of modified bases can perturb local stacking forces, H_2 bonds, leading to destabilization of DNA. Hydrolysis of *N*-glycosidic bonds in DNA generates AP sites that block DNA polymerases and are highly mutagenic. Again, oxidative stress-induced hypomethylation of the CpG site because of inhibition of methyl-CpG binding protein (MBP-2) binding results in the conversion of guanine to 8-oxodG (Donkena et al., 2010). Base oxidation products, including 5-methyl-cytosine, hydroxymethyl cytosine also lower MBP binding affinity inducing progressive demethylations in DNA. These malignant DNA lesions are biomarkers of tissue oxidative stress-inducing mutagenesis and epigenetic alterations involved in various pathophysiological conditions (Pizzino et al., 2017; Yasui et al., 2014). Carbon-based radicals formed due to the abstraction of hydrogen from 2-deoxyribose generate peroxyl radicals (ROO•), which then abstract hydrogen from sugar moieties leading to DNA strand breaks. The ratio between total oxidative lesions and oxybase clusters ranges between 5:1 and 10:1, producing an expected frequency of 4 oxidative lesions/Mbp and 400–800 oxidative clusters/Gbp in an average ~6 Gbp sized genome (Chan et al., 2010; Chastain et al., 2006; Dedon, 2008; Kryston et al., 2011; Sutherland et al., 2000; Yu et al., 2003). Hydrogen abstraction from amino acid and oxidation of thymine-amino acid adduct produce DNA-histone cross-links promoting cellular disorders by replication and transcription inhibition (Jena, 2012). Additionally, oxidative stress can also modify the pool of dNTPs present in a cell. Incorporation of damaged dNTPs in nascent DNA during repair or replication has a severe impact on genomic integrity. In some cases, as in guanine, dGTP has higher oxidation sensitivity than G (Markkanen, 2017; Rudd et al., 2016). The amount of G in a sequence may determine its susceptibility to oxidative modifications. The highest reactivity of OH• radical is observed with the consensus sequence, TTAGGG, found in telomeric DNA. At TTAGGG repeat sites, OH• radicals induce cleavage of 5′ GGG producing SSBs. The DNA damaging effects due to oxidation are more prominent in telomeres than microsatellite repeats and bulk genomic DNA due to G rich strand (Barnes et al., 2019; Coluzzi et al., 2014). As mtDNA manifests more sensitivity toward oxidative stress in comparison to nuclear DNA, rapid DNA loss, and mitochondrial function decline are evident in the case of mtDNA. As such, mitochondrial DNA damage constitutes an essential biomarker of oxidant injury (Malik and Czajka, 2013; Salazar and Van Houten, 1997; Yakes and Van Houten, 1997, Van Houten et al., 2018).

8.2.3.2 RNA

The high cellular abundance of RNA and lesser association with proteins makes them more frequent targets for oxidative modifications. Oxidative stress induces up to 10-fold higher accumulation of lesions in RNA compared to unstressed conditions modifying RNA functions which contribute to different neurodegenerative conditions such as Parkinson's disease (PD), AD, and amyotrophic lateral sclerosis (ALS) (Hofer et al., 2005; Ramanan and Saykin, 2013; Simms and Zaher, 2016). rRNA and mRNA are the primary targets for oxidative modification of nucleobases and are involved in the pathogenesis of several diseases. Oxidized 23S rRNA induces ribosomal damages and disturbs the process of translation, causing cellular dysfunctions or even death (Willi et al., 2018). Oxidations in mRNA are selective, with some mRNAs being more vulnerable to oxidative damages (Bazin et al., 2011). Oxidized mRNA affects fidelity and speed of protein biosynthesis and is associated with the stalling of the ribosome (Honda et al., 2005; Kong and Lin, 2010; Shan et al., 2007; Wurtmann and Wolin, 2009). Oxidative stress induces tRNA cleavage, forming tRNA fragments that induce stress granule formation, loss of function, and modification of nucleotide bases (Nawrot et al., 2011). 8-hydroxyguanosine (8-OHG) makes up the most common oxidatively modified base in RNA (Fimognari, 2015).

8.3 OXIDATIVE STRESS-INDUCED COMPLICATIONS

8.3.1 OXIDATIVE STRESS IN DISEASES

Oxidative stress has implications in the pathophysiology of both acute and chronic diseases (Figure 8.2).

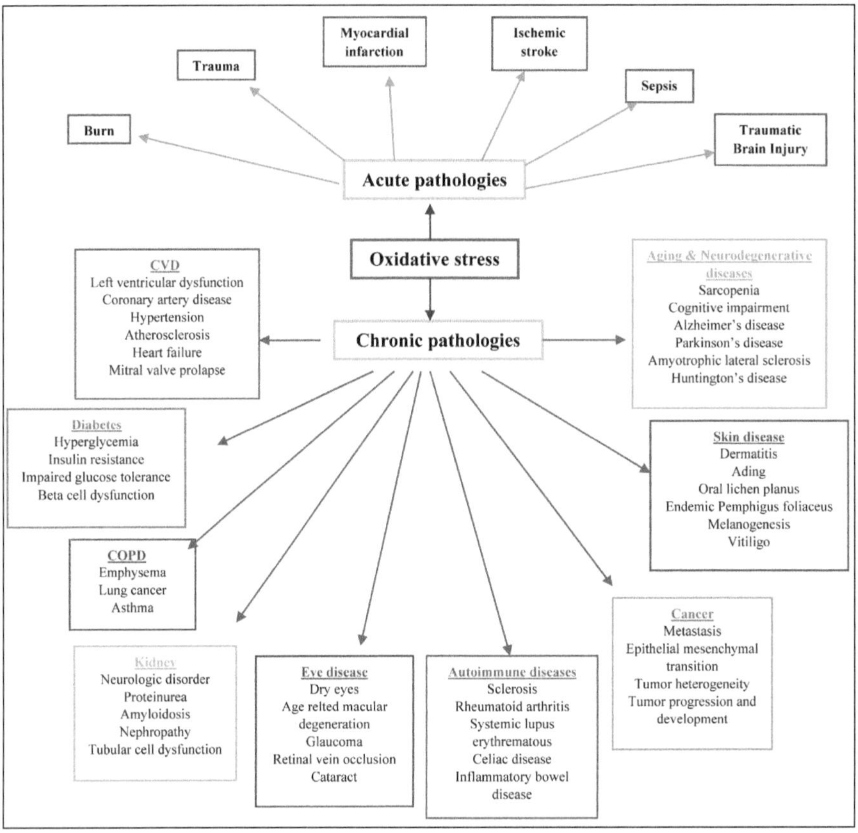

FIGURE 8.2 Impact of oxidative stress in the pathophysiology of different acute and chronic diseases.

8.3.1.1 OXIDATIVE STRESS IN ACUTE PATHOLOGIES

In the prognosis and progression of acute diseases, oxidative stress constitutes an essential regulatory factor. In extreme cases of cardiovascular diseases (CVD) where lack of blood supply in heart muscles initiates myocardial infarction (MI), the generation of ROS and RNS is triggered, which is again responsible for some MI associated extracellular injuries. The increased ROS generation under hypoxic conditions is related to hypoxia-inducible factor 1 (HIF 1) transcription factor stabilization at low pressures of oxygen that elevate superoxide and hydrogen peroxide release (Chandel et al., 1998; Guzy and Schumacker, 2006; Sena and Chandel, 2012). ROS induced increased Nox-2 (nicotinamide adenine dinucleotide phosphate oxidase) expression in cardiomyocytes is responsible for post-MI cardiac remodeling by myocardial apoptosis, necrosis, or hypertrophy (Krijnen et al., 2003). Reperfusion of damaged myocardium also triggers oxidative stress by stimulating the peroxidation of lipids (Palace et al., 1999; Dhalla et al., 2000). This situation is more prominent in patients with diabetes mellitus (DM), where MI induces severe atherosclerosis, frequent infarcts, and worse outcomes than nondiabetic patients (Di Filippo et al., 2006). During cardiopulmonary bypass, oxidative stress activates proapoptotic and proinflammatory signaling pathways (Zakkar et al., 2015). The inflammatory cycle generated due to sepsis induces oxidative stress with elevated lipid and protein oxidized products (Goode et al., 1995). ROS-induced mitochondrial respiratory impairment, reduced vascular tone, increased vascular permeability, vasodilation, and decreased cardiac output are responsible for multiorgan failure and mortality in septic patients (Mantzarlis et al., 2017; Prauchner, 2017). Studies have shown that postischemic stroke, increased ROS levels induce brain cell apoptosis, disruption of the blood–brain barrier, vasoconstriction, microvascular constriction which enhance the ischemic insult (Atochin et al., 2007; Crack and Taylor, 2005; Neumann-Haefelin et al., 2000; Shirley et al., 2014; Sandoval and Witt, 2008; Szczepańska-Szerej et al., 2011; Tsai et al., 2014). Traumatic brain injury is characterized by an immediate increase in ROS generation resulting in elevated lipid peroxidation, peroxy-nitrite production, reduction in antioxidant reserve, and development of secondary brain injury (Abou-El-Hassan et al., 2017; Bayir et al., 2002; Hall et al., 1993; Smith et al., 1994; Wang et al., 2016). In severe brain injury scenarios, elevated levels of nitro-tyrosine characterize the cerebrospinal fluid (Bar-Or et al., 2015; Darwish et al., 2007). In other cases of trauma like burns, ROS, and NO generation trigger systemic inflammatory reactions and the development of lipid peroxidation (Parihar et al., 2008; Wagener et al., 2013). The accumulation of polymorphonuclear leukocytes at trauma sites affects the recovery of different organs and regulates clinical outcomes (Henrich et al., 2011; Pepper et al., 1994; Zakkar et al., 2015).

8.3.1.2 OXIDATIVE STRESS IN CHRONIC PATHOLOGIES

8.3.1.2.1 CVD and Atherosclerosis

Oxidative stress has associations with various types of chronic cardiovascular disorders like left ventricular dysfunction, arrhythmia, coronary artery disease (CAD), mitral valve prolapse, hypertension, ischemia/reperfusion (IR) injury, heart failure, and peripheral vascular diseases like arterial remodeling, restenosis, and thrombosis (Gracia et al., 2017; Goyal and Sharma, 2013; Münzel et al., 2017; Pignatelli et al., 2018). Most cardiovascular ischemic events are associated with the complex, multifactorial disorder of arteries, atherosclerosis which is characterized by atherogenesis, that is the formation of a fibro-fatty plaque with a fibrous cap surrounding central lipid-rich atheroma, or atheromatous necrotic core (Schwartz et al., 2000; Yang et al., 2017). Oxidative imbalance due to acute and chronic ROS generation along with oxidative stress-induced autophagy in the different associated cell types and inflammation regulate disease pathogenesis and are considered markers of the disease (Martinet and de Meyer, 2009). Disruption of homeostasis in endothelial cells, vascular inflammation, and dysfunctions due to ROS generation by NOX enzyme system and lipooxygenase present in vascular wall marks, the initiation of atherogenesis (Cross and Segal, 2004). Damage initiation results in inflammatory tissue cascade characterized by increased production of inflammatory mediators, cytokines, and chemokines that infiltrate monocytes at the inflammation sites in vessel wall exacerbating ROS generation from endothelial (ECs) and smooth muscle cells (SMCs) (Harja et al., 2008; Nowak et al., 2017). Inflammation is mainly responsible for atherosclerotic plaque formation, plaque rupture, relapsed thrombosis, dyslipidemia, metabolic disorders, and atherothrombosis leading to disease manifestations (Iyer et al., 2010; Kassi et al., 2013; Xu et al., 2010). At the initial stages of atherosclerosis, angiogenesis is accelerated by inflammation and oxidative stress that increases plaque vulnerability by invading the intima and enlarging the core (Lozhkin et al., 2017). Wnt-β-catenin-WISP-1 (WNT1-inducible-signaling pathway protein 1) promotes survival of SMCs, and oxidative stress enhances migration of SMCs, leading to intimal thickening and formation of fibrous cap (Mill et al., 2014; Williams et al., 2016). Vascular injury prolongs ROS1 receptor tyrosine kinase activation, which also enhances SMC migration and proliferation (Ali et al., 2014). Myeloperoxidases (MPOs) induce modification of low-density lipoprotein (LDL) into cytotoxic oxidized forms (oxLDL), which are ingested by scavenger receptors of macrophages resulting in the formation of foam cells laden with lipids (Aviram, 1993; Ding et al., 2013). Increased autophagy and apoptosis in SMCs, foam cells, and ECs because of excessive ROS generation reduce collagen synthesis, decrease fibrous cap, induce thrombosis causing plaque destabilization and rupture that result in acute clinical events (Badimon and Vilahur, 2014; Jin and Choi, 2012). Enhanced oxidative imbalance in macrophages is associated with cholesterol oxidation, suppressed autophagy, and thiol oxidative stress (Brown and Jessup, 2009; Li et al., 2010). The lack of apoptotic cell clearance by phagocytosis accelerates plaque necrosis leading to atherothrombotic cardiovascular pathogenesis (Tabas, 2010). The accumulated necrotic debris forms a sclerotic plaque affecting vessel conformity, and via rupturing of the plaque elevates embolus or thrombus development possibilities increasing multiple comorbidities risks (Nowak et al., 2017).

8.3.1.2.2 Chronic Obstructive Pulmonary Disorder and Asthma

Oxidative stress has an association with the pathophysiology of inflammatory lung diseases like chronic obstructive pulmonary disease (COPD) and asthma. Elevated supply of blood, high oxygen levels, and exposure to environmental toxins make lungs highly susceptible to oxidative attack (McGuinness and Sapey, 2017). Exogenous ROS from environmental stressors like cigarette smoking or endogenous ROS from the inflammatory process and mitochondrial dysfunction triggers oxidative stress and associated with the progression of disease complications (Salama et al., 2014). Oxidative stress increase bronchial inflammation, arachidonic acid release, vascular permeability with airway oedema, hypersecretion and hyperplasia of the mucous gland, corticosteroid resistance by inhibiting glucocorticoid receptor expression, senescence, and the release of neurokinins and tachykinins with an elevation of neurogenic inflammation, impaired bronchodilator responses, and small airway fibrosis (Babusikova et al., 2012; Domej et al., 2014; Kirkham and Barnes, 2013). Oxidative stress in the lungs involves activation of alveolar macrophages and lung resident epithelial cells, generating chemotactic molecules and cytokines that recruit neutrophils, monocytes, and lymphocytes (Van Eeden, 2013). These lead to persistent inflammation, chronic oxidative stress, leading to lung cell apoptosis, autophagy, and repair mechanism defects (Rahman and Adcock, 2006).

COPD is associated with ROS-induced pulmonary neutrophilic inflammation via the release of critical mediators such as matrix metalloproteinase, neutrophil elastase, and altered migration of neutrophils (Barnes, 2016; Gupta et al., 2016; Hobbins et al., 2017). Oxidative stress damages the protective antiprotease screen in the lungs, facilitating the degradation of elastic fibers by neutrophil elastase (Domej et al., 2014). Up-regulation of the iNOS and ROS induced lipid peroxidation, RNA oxidations damage alveolar cell wall leading to emphysema (Boukhenouna et al., 2018) and the risk of acquiring lung cancer is high in these patients (Durham and Adcock, 2015). Oxidative stress-modified protein epitopes are responsible for autoimmune responses in COPD (Kirkham et al., 2011). The underlying molecular mechanisms also involve NF-κB activation, gene polymorphisms, with altered oxidative homeostasis in lungs and circulation (Barnes, 2000; Domej et al., 2014).

A common chronic respiratory disease in children is bronchial asthma, in which oxidative stress constitutes a vital component. Exposure to allergens (i.e., pollutants, infection, physical endurance, mites in house dust) accelerates ROS/RNS production leading to airway remodeling and inflammation (Jesenak et al., 2017). Metals on the surface of particulate matter participate in the redox cycling process producing free radicals, and PM2.5 induced ROS generation can lead to the development of asthma (Valko et al., 2007). Increased activation of cytokines like TNF-α, IL-1, IL-5, IL-33 in immune cells stimulate ROS production and inflammation. The allergic reactions are enhanced by increased activation of NADPH oxidase, metalloproteinase-9 (MMP-9), vascular cell adhesion molecule-1 (VCAM-1), intracellular adhesion molecule-1 (ICAM-1), and EGF-R (Hristova et al., 2016; Polonikov et al., 2015). Oxidative stress injuries affect airway epithelial cell barrier functions and promote inflammation via proteinase-activated receptor 2 (PAR-2) activated DUOX-2/ROS pathway (Qu et al., 2017). ROS also promotes contraction in smooth muscle cells (HASMC) causing hyper-responsiveness, increased dendritic cell activation induced T-helper cell activation, enhanced the release of serotonin and histamine from mast cells, and development of peripheral cutaneous anaphylaxis (Kim et al., 2013; Tuo et al., 2013).

8.3.1.2.3 Diabetes

Oxidative stress is a crucial factor involved in the pathogenesis of macrovascular and microvascular diabetic complexities inducing both apoptotic and autophagic cell injury (Giacco et al., 2010; Mahajan et al., 2013; Maiese, 2015). Besides, oxidative stress has an association with the onset of several secondary diabetes complications like retinopathy, neuropathy, nephropathy, cardiomyopathy, atherosclerosis, stroke, CAD (Newsholme et al., 2016; Ullah et al., 2016). Hyperglycemia, hyperinsulinemia, lipid peroxidation, glucose oxidations, nonenzymatic glycation of proteins, dyslipidemia, macroangiopathies stimulate free radical generation via NADPH-oxidase and diacylglycerol, protein kinase C (PKC) pathways that hinder enzyme action, augment insulin resistance, promote hyperglycemia, impair glucose tolerance, generate insoluble oxidative aggregates of LDL, induce cell death, and tissue damage adding to the complications and progression of the disease in the body (Matough et al., 2012; Ullah et al., 2016; Volpe et al., 2018; Wright et al., 2006). Oxidative stress-induced development of vascular complications in type 2 diabetes has an association with diabetic mortalities. Hyperglycemia induced oxidative stress that affects various enzymes like poly[ADP-ribose] polymerase-1, glyceraldehyde 3-phosphate dehydrogenase decreasing glucose metabolism efficiency via impaired glycolysis, ATP synthesis, and electron transport (Ceriello, 2003; Newsholme et al., 2016; Gero, 2017; Yan, 2014). Diabetic metabolic abnormalities induce endothelial dysfunction in blood vessels via alterations of p65 subunit of NF-κB, increased superoxide production that stimulates crucial pathways including polyol pathway flux, enhanced AGEs-products formation, increased AGE receptor expression, PKC isoform activation, hexosamine pathway hyperactivity, glyceraldehyde auto-oxidation, inactivation of eNOS, prostacyclin synthase, and antiatherosclerotic enzymes associated with diabetes pathophysiology. These complications result in vascular inflammation, retinal hyperpermeability as well as chronic diseases like cancer and **CVD** (Giacco et al., 2010; Ighodaro, 2018; Newsholme et al., 2016; Paneni et al., 2013; Treps et al., 2016; Yan, 2014). Via activation of these pathways, ROS induce angiogenic defects under the influence of ischemia, stimulate proinflammatory pathway activation, and induce persistent epigenetic alterations leading to long-term expression of inflammatory genes (Giacco et al., 2010). Increased ROS levels trigger the thickening of the glomerular basement membrane as well as mesangial expansions (Bohlender et al., 2005; Park et al., 2019). Oxidative stress, in combination with inflammatory responses, is also associated with β-cell dysfunctions leading to type 1 DM via ROS induced autoantibody attack against ox-LDL, leading to compromised insulin production and sensitivity (Matough et al., 2012). Impaired mitochondrial functions and increased production of mtROS enhance the vulnerability of β-cells to autoimmunity (Chen et al., 2018; Kitada et al., 2019). NOS induces proapoptotic effects in β-cells via JNK pathway activation (Newsholme et al., 2016).

8.3.1.2.4 Kidney Diseases

High metabolic activity and increased mitochondrial oxidation reactions in the kidney make them highly vulnerable to oxidative stress-induced complications (Daenen et al., 2019). Chronic inflammation associated with chronic kidney disease (CKD) promotes oxidative stress development resulting in prolonged complications like anemia (shortened lifespan of erythrocytes), neurologic disorders (myelin oxidation), accelerated aging with high malignancy rates in end-stage renal disease patients (Krata et al., 2018; Shang et al.,

2016; Sung et al., 2013). Increased ROS levels are also associated with renal blood flow alterations, sodium/fluid retention, fibrotic changes, proteinuria onset, increasing the disease burden (Nistala et al., 2008). Additionally, oxidative stress deteriorates renal functions via protein carbonylation and thiol oxidation leading to hypertension, inflammation (NF-κB activation), and damage of the glomerular filtration barrier (Kim and Vaziri, 2010). ROS induced structural changes in β2-microglobulin, along with inflammatory processes, are also associated with the development of amyloidosis. Enhanced oxidative stress alters biological properties of circulating lipoproteins and lipids in the uremic milieu producing different end products like oxidized thiol compounds, ox-LDL, fatty acid peroxidation products, MPO, oxysterols, nitrotyrosine, advanced glycosylation end (AGE) product, F2-isoprostanes (Duni et al., 2017; Florens et al., 2016; Krata et al., 2018; Ling and Kuo, 2018). Oxidative stress is the factor that links CKD with CVD and inflammation via mitochondrial dysfunctioning, eNOS uncoupling, and increased activity of NADPH oxidase (Sibal et al., 2010; Yu et al., 2011). Altered NO levels increase cardiovascular disease through endothelial dysfunctions by decreased vasorelaxation, resulting in the initiation of atherosclerosis in CKD (Popolo et al., 2013). Lipid peroxidation product MDA triggers HDL dysfunctions leading to cardiovascular morbidity and left ventricular hypertrophy (LVH) (Shroff et al., 2014). Oxidative stress also promotes hypertrophy of cardiomyocytes and related signaling molecules like Ras, PKC, Src, MAPK, Jun-nuclear kinase, transcription factors (AP-1), apoptosis signaling kinase-1, and NF-κB that enhances the expression of MMPs (Paoletti et al., 2005; Sundaram et al., 2014). CKD progression is also associated with oxidative stress-induced diabetic nephropathy, proximal tubular cell dysfunction, mitochondrial defects with impaired complex IV activity, and polycystic kidney disease (Gailly et al., 2008; Granata et al., 2009; Kashihara et al., 2010). The use of bioincompatible dialyzer membranes or exposure to dialysate endotoxins contributes to oxidative stress development producing complications like platelet aggregation and thrombogenesis (Esper et al., 2006; Ling and Kuo, 2018). Via the oxidation of parathyroid hormones, oxidative stress can impair Ca^{2+} homeostasis, leading to the onset of multiple diseases.

8.3.1.2.5 Cancer

Oxidative stress plays a crucial role in inducing malignant alterations at cellular and molecular levels, promoting tumor initiation, progression, and development (Calaf et al., 2018). The complex interactions of the cellular immune system and cytokines stimulate oxidative stress-induced tumor development at sites of chronic inflammation (Andrisic et al., 2018). Oxidative stress induces genotoxic effects like genomic instability, proliferation, epithelial-mesenchymal transition, antiprotease inhibition, local tissue injury, leading to the development of carcinogenesis, metastasis, and tumor heterogeneity (Andrisic et al., 2018; Obrador et al., 2019; Palmirotta et al., 2016). Redox signaling can regulate metastasis by modifying the different steps of the cascade, namely, invasion, extravasation, and intravasation. Oxidative stress stimulates cell migration rates by decreasing tumor cell attachment to the basal lamina or increasing expression of cell motility associated with proteins by modulating Rac1, intercellular adhesion protein-1 (ICAM-1), and NF-κB levels (Reuter et al., 2010). Increased ROS levels in cancer-associated fibroblasts activate MMP-3 involved in ECM breakdown, collagen, fibronectin, and laminin degradation leading to metastasis (Nourazarian et al., 2014; Sosa et al., 2013). Activation of growth factors, mitogenesis, and increased transcription due to oxidative imbalance regulate cancer onset and development, cell transition from quiescent to proliferative status, activation of cell death or growth arrest associated with the onset of the tumor, and progression toward a malignant phenotype (Ameziane El Hassani et al., 2019; Calaf et al., 2018; Farhood et al., 2018; Leone et al., 2017; Obrador et al., 2019). The initiation of cancer is associated with oxidative stress-induced gene mutations and DNA structural alterations, causing genetic lesions resulting in tumorigenicity (Ahmed, 2016). Oxidative stress-induced anomalous gene expression, cell–cell communication obstruction, and second messenger system alterations stimulate cell proliferation (Reuter et al., 2010). Further oxidative alterations in DNA lead to the progression stage of carcinogenesis, causing benign/malignant neoplasm formation (Ahmed, 2016). Increased ROS levels lead to increased metabolism in cancer cells to facilitate angiogenesis, survival, and proliferation via modulation of signaling pathways like EGFR, Nrf2, Raf, HIF1α, MAPK, PI3K, PKC, COX-2, PAK-1, VEGF, Smad2, ERK1/2 along with alterations in p53 gene expression (Leone et al., 2017; Matschke et al., 2019; Nourazarian et al., 2014; Sosa et al., 2013). Lifestyle-related factors like dietary fat consumption can contribute to the development of oxidative stress associated with an increased rate of mortality in different cancer types (Pizzino et al., 2017).

8.3.1.2.6 Autoimmune Diseases and IBD

Oxidative stress-induced structural alterations of proteins may result in the exposure of some cryptic determinants as neo-antigens that are internalized by antigen-presenting cells and presented to the immune system causing initiation, amplification, and stimulation of autoimmune reactions (Hoffmann and Griffiths, 2018; Wargo et al., 2015). ROS induce cleavage of thyroglobulin to expose novel epitopes that direct the activation of autoantibodies and impart tissue damage (Dalmazi et al., 2016). In the case of multiple sclerosis, ROS dysregulates blood–brain barrier, promotes increased infiltration of monocytes, lymphocytes, and inflammation, activation of microglia modifying myelin sheaths that promote autoimmune attack. Increased ROS levels in epidermis damage melanocyte, modify autoantigen structures like melan A and tyrosinase and stimulate autoimmune responses against them, which form the basis of the generation of vitiligo (Dalmazi et al., 2016). In rheumatoid arthritis, increased ROS in synovial fluid is associated with hyaluronic acid oxidative damages and depolymerization, lipid peroxidation, synovial fluid degradation, low-density-lipid proteins (LDL) oxidation, protein carbonylation, damages to type II collagen and cells of the cartilage.

Moreover, increased ECM degradation, increased T cell activation and differentiation, p53 mutations in the synovium, proteoglycan synthesis, glycosaminoglycan sulfation, accumulation of advanced oxidation protein products (AOPPs) lead to the progression of the disease and viscosity loss in the joint (Smallwood et al., 2018; Yamazaki et al., 2003; Yoo et al., 2016). RNS inhibits interactions between chondrocytes and ECM and promotes apoptosis; ROS affect the response of chondrocyte to growth factors and migration to injury sites. Chronic oxidative stress in the synovial environment amplifies signaling mediators and prolongs synovitis (Filippin et al., 2008; Hitchon and El-Gabalawy, 2004; Moodley et al., 2008). In systemic lupus erythematosus, dysregulation of the immune system, macrophage apoptosis, serum protein chemical modifications, and comorbidities are associated with damages of vital organs due to the action of autoantibody against oxidized proteins like 3-nitrotyrosine. Apoptotic cells are autoantigen reservoir comprising oxidation modified proteins. Secondary necrosis, membrane integrity loss causes an extracellular release of cellular proteins that activate T and B cell-induced inflammations (Hoffmann and Griffiths, 2018; Shah et al., 2014). Oxidative stress induces redox-dependent mTOR activation in T cells, elicit T cell dysfunctions leading to abnormal immune responses, and production of autoantigens stimulating inflammatory reactions (Manda et al., 2015; Perl, 2013). α-gliadin toxicity in intestinal cells induces oxidative imbalances altering cell morphology, proliferation, and apoptosis by regulating the expression of HSP-70, HIF-1α, and BAX molecules that influence the intestinal barrier integrity leading to celiac disease (Ferretti et al., 2012; Piatek-Guziewicz et al., 2017).

Oxidative stress regulates the progression and pathogenesis of IBD (Piechota-Polanczyk and Fichna, 2014). ROS generation in the GI tract stimulates mucosal layer damages, epithelial barrier disruptions, cytoskeletal protein, and tight junction damages, intestinal dysbiosis, accelerated cell damages, causing increased intestinal permeability and disruption in the intestinal barrier (Liu et al., 2018; Tian et al., 2017). This causes alterations in GI tract commensal microbiota composition and promotes pathogenic bacteria invasion that initiates an immune response and IBD in the intestinal epithelium (Liu et al., 2018). *Helicobacter pylori* bacteria invasion induces neutrophilic ROS generation, elevated platelet-activating factor, and leukotriene B4 (LTB4) levels, activation of NOX, the release of LC8 peptides activating NF-κB causing inflammation and increasing IBD risk (Liu et al., 2018; Tian et al., 2017). Oxidation of cholesterol produces oxysterols that induce apoptosis, activate various fibrogenic cytokines like, IL-6, IL-8, IL-23, IL-1β, TNF-α, monocyte chemotactic protein-1 (MCP-1), TLR2, and TLR9 causing prolonged inflammation of the intestines stimulating mucosal barrier damage, and abnormal intestinal motor function (Balmus et al., 2016; Dessing et al., 2007; Guan and Lan, 2018; Mete et al., 2013).

8.3.1.2.7 Skin Diseases

Skin, the largest organ of the body aid in protection against different biological, physical, and chemical damaging agents, as well as regulate temperature and metabolic waste product excretion (Kolarsick et al., 2011; Kruk and Duchnik, 2014). The high PUFA content in the skin makes them one of the main targets of ROS attack. ROS-induced elevated MMP-1, and MMP-3 levels initiate collagen type I fragmentation and disorganization in the dermis altering integrin expression, signal transduction, and other mechanical properties of the skin (Fisher et al., 2009). By modulating nuclear transcription factors like NF-κB, AP-1, MAPK, JAK-STAT, p38; ROS mediates the generation of proinflammatory cytokines and COX-2 molecules leading to chronic dermal inflammation, weakened skin barrier functions, oedema, and altered vascular permeability ultimately leading to various skin diseases including cancer (Bickers and Athar, 2006). The skin aging process is also associated with oxidative stress, leading to ECM degradation along with an increased accumulation of oxidation damaged biomolecules (Kruk and Duchnik, 2014). ROS is also involved in skin allergic reactions causing dermatitis. Oxidative imbalance in epidermal keratinocytes is associated with epidermal spongiosis, increased pruritus, Th2 polarization with disrupted stratum corneum, and altered keratinization process resulting in skin barrier dysfunctions and aggravated atopic dermatitis (Ahn, 2014; Ji and Li, 2016). Diseases like oral lichen planus and chronic inflammatory mucosal disease involve oxidative lipoperoxidation (Aly and Shahin, 2010; Mansourian et al., 2017). Hydrogen peroxide generation via degradation of scavenger receptor B1 modulates cholesterol trafficking and thereby contributes to the generation of permeability barrier (Sticozzi et al., 2012). Oxidative stress has essential contributions to the pathophysiology of various autoimmune skin disorders. Oxidative stress-induced T cell activation, proliferation, alongside the production of proinflammatory cytokines, most importantly, TNF-α drive the immune-pathology of psoriasis (Small et al., 2018). Overexpressed cytokine production promotes hyperplasia with altered differentiation of epidermal keratinocytes as well as cytokine, chemokine, and AMP production. A self-amplifying loop is generated when these inflammatory molecules act back on T cells, dendritic cells, and neutrophils, prolong the inflammatory process and result in the formation of psoriatic lesions (Lowes et al., 2014). Besides, oxidative stress is associated with increased angiogenic processes and a lack of Th2 ability to counter-regulate Th1 responses in psoriasis (Lin and Huang, 2016). Increased ROS generation and lipid peroxidation bring about oxidative alterations in the structures of desmosomal glycoproteins. Desmogleins 1 and 3 (Dsg1 and Dsg3) modifying them into targets of circulating autoantibodies mainly IgG4 inducing acantholysis. This results in autoimmune skin disorder endemic pemphigus foliaceus (EPF) or Fogo Selvagem characterized by superficial blisters in the subcorneal region (Gutierrez et al., 2018). ROS/RNS generation can also affect skin pigmentation and depigmentation via interaction with melanocyte. By the upregulation of tyrosine genes, ROS and NO can stimulate melanogenesis.

Conversely, ROS induced melanocyte degeneration generates depigmented macules and patches on the skin known as vitiligo. This is due to the disruption of the Nrf2-p62 signaling pathway that causes autophagy dysregulation and enhances the sensitivity of melanocytes to oxidative stress (He et al., 2017). The tyrosine present in melanocyte gets oxidized into reactive o-quinones generating autoreactive T cells that selectively attack the melanocyte (Singh et al., 2016). Oxidation of sebum increases the level of oxidized lipids like squalene and creates ideal sites for the growth of propionibacterium, which via enhanced ROS production promotes inflammation and add to the severities of acne (Garem et al., 2014). Sebaceous gland oxidations are also associated with seborrheic dermatitis characterized by erythematous patches and scaling on the scalp (Toruan et al., 2017; Trueb et al., 2018). Alopecia areata characterized by nonscarring hair loss are also associated with increased levels of lipid peroxidation products in the scalp, plasma, and erythrocytes (Pektaş et al., 2018; Prie et al., 2015). Increased serum lipid hydroperoxide levels with a shift in thiol/disulfide balance toward disulfides indicate rosacea, an inflammatory dermatosis associated with remissions and flare-ups (Pektas, 2018; Sener et al., 2019).

8.3.1.2.8 Eye Diseases

The exposition of the eye to different environmental factors, for example, light, ionizing radiation, pollutants, ultraviolet rays, high oxygen pressure, and microbes, can alter redox balance (Kruk et al., 2015). This imbalance can bring about alterations in the morphology and functions of retinal ganglion cells (RGCs), retinal pigment epithelium (RPE), endothelial cells, corneal epithelial cells, optic nerve as well as ocular surface leading to a plethora of eye disorders (Masuda et al., 2017). Dry eyes and dry age-related macular degeneration (AMD) are the two most common age-related eye disorders associated with ROS induced ocular surface inflammation, lipid peroxidation, mitochondrial DNA damage in the retina, glycation, hyperosmolarity of the tear film with decreased tear production (Deng et al., 2015; Gordois et al., 2012; Taskintuna et al., 2016). In the case of dry eyes, oxidative stress induce acini atrophy, fibrosis, immune cell infiltration in the lachrymal gland, decrease in total lacrimal derived tear protein, increased dysfunctions of the meibomian gland affecting tear stability leading to visual disturbance and discomfort. Oxidative stress-induced ocular diseases also include corneal inflammation and aging, bullous keratopathy, keratoconus, and Fuchs' endothelial dystrophy. Lipid peroxidation or peroxynitrite production causes progressive thinning of the cornea, causing vision impairment (Cejka and Cejkova, 2015). In dry-AMD patients, lipid peroxidation in RPE is associated with lipofuscin accumulation resulting in drusen formation. Drusen induce degeneration of RPE and atrophy in the eye fundus, ultimately leading to a severe loss of vision (Sparrow and Boulton, 2005). Wet type AMD is present in 10%–15% of cases characterized by the neo-vascularization of the choroid. The high oxygen consumption, elevated PUFA proportion, and continuous light exposure account for the high susceptibility of RPE to ROS. Optic neuropathy, glaucoma is associated with a chronic oxidative stress-induced increase in excitatory amino acids (e.g., glutamate and glycine), elevated intraocular pressure due to impaired trabecular meshwork cells causing an outflow of aqueous humor as well as RGC death resulting in a cupping of optic nerve head and impaired vision (Saccà and Izzotti, 2008; Vieira-Potter et al., 2016). Hydroxyl radical (•OH) generated in the retina also promotes RGC death during conditions of retinal ischemia and remains elevated during the reperfusion period (Becatti et al., 2016). A major complication associated with DM is diabetic retinopathy, which is a leading cause of blindness and is regulated by oxidative stress (Brownlee, 2005). The oxidative imbalance and complications associated with hyperglycemia induce senescence of endothelial cells and by upregulation of angiogenic genes (like VEGF) in retinal cells promote angiogenesis leading to proliferative diabetic retinopathy (Izuta et al., 2010). This can result in retinal ischemia, retinal detachment, and macular oedema due to vascular leakage or hemorrhage. Synaptic transmitter damages and neurotrophic factor degradation due to ROS induce visual impairment by apoptosis of neural cells (Ozawa et al., 2011). Retinal vein occlusion is a complication of diabetic retinopathy associated with fundus hemorrhage and blood hyperviscosity (Masuda et al., 2017). Oxidative imbalance in lens epithelium cells has a vital role in another age-related visual problem, cataract characterized by oxidations of glutathione, and methionine residues of nuclear proteins (Chang et al., 2013; Wang et al., 2003). The resulting aggregation and denaturation processes increase lens opacity by influencing the solubility of lens proteins in cataractous lenses (Chang et al., 2013; Kaur et al., 2012; Vinson, 2006). Oxidative stress-induced mitochondrial dysfunctions lead to the degeneration of photoreceptors resulting in the development of retinitis pigmentosa (Cejka and Cejkova, 2015).

8.3.1.2.9 Geriatric Diseases

The aging process has a direct association with oxidative stress-induced telomere shortening, mitochondrial damages, metabolic alterations (Cui et al., 2012). Oxidative damage accumulations with age lead to functional loss, senescence, and generation of the senescence-associated secretory phenotype, which forms the basis for the onset of various acute and chronic pathological processes (Ogrunc and di Fagagna, 2011). The loss of muscle strength, sarcopenia, is associated with oxidative stress-induced posttranscriptional modifications, increased proteolysis, causing muscle quantity decrease (Liguori et al., 2018; Powers et al., 2011). Inhibition of action potential in the sarcolemma, altered neuromuscular junction morphology, reduced release of calcium, modifications of actin-myosin structures is some of the processes by which oxidative stress reduces muscle quality and strength with age. The oxidative imbalance is also associated with reduced memory performances and age-related cognitive decline (Hajjar et al., 2018). Oxidative stress-induced stress granules form stable aggregates that hinder neuronal functioning, leading to cognitive impairment (Chen and Liu, 2017; Liguori et al., 2018).

8.3.1.2.10 Neurodegenerative Diseases

Higher PUFA levels in the neuronal cell membrane increased oxygen consumption, and weakened antioxidant defenses make the neurons sensitive to oxidative stress. The oxidative imbalance is associated with the pathogenesis of different neurodegenerative diseases like AD, PD, Huntington's disease (HD), and ALS (Liu et al., 2018; Manoharan et al., 2016). Additionally, oxidative stress-induced protein misfolding has associations with Bovine Spongiform Encephalopathy, Kuru, Creutzfeldt-Jakob disease, Fatal Familial Insomnia, as well as Gerstmann–Straussler–Scheinker syndrome (Islam, 2017; Moore et al., 2009).

In AD, increased activation of *N*-methyl-D-aspartate (NMDA)-type glutamate receptors trigger ROS generation that mediates activation of stress-activated protein kinase pathways like JNK, p38MAPK leading to tau protein hyperphosphorylation, polymerization, and accumulation causing intracellular neurofibrillary tangle formation and generation of the fibrillar pathology (Liu et al., 2018). In combination with neuroinflammation, oxidative stress promotes amyloid-β (Aβ) generation, deposition and accumulation, activating microglia and stimulating a cascade of proinflammatory reactions inducing neuronal cell apoptosis, mitochondrial dysfunction, and energy failure (Islam, 2017; Patten et al., 2010; Singh et al., 2019). Increased ROS levels are also associated with astrocytes modifications, synaptic loss, increased transcription of proinflammatory genes, IL-1,6, TNF-α cytokine, and chemokine release that stimulates neuroinflammatory processes which if persistent trigger neuronal loss and damage immune sensitivity contributing to degeneration in nerve functioning (Chen et al., 2012).

In HD, oxidative imbalance induced misfolded protein triggers the formation of inclusion bodies, which inhibit neurotransmitter transmission by forming aggregates at axons and dendrites (Islam, 2017; Lim and Yue, 2015). ROS production is associated with increasing CAG repeat length, leading to mutated huntingtin protein formation and amplified mtDNA damage (Radi et al., 2014). Oxidative imbalance also decreases glucose transporter-3 expression, inhibiting glucose uptake and lactate accumulation, inhibits glyceraldehyde-3-phosphate dehydrogenase catalytic activity causing mitochondrial damage and its accumulation in mutant HTT-expressed cells, leading to cell death. This generates positive feedback loops increasing oxidative stress and loss of neurons from the striatum and cortex (Liot et al., 2017).

In the case of PD, increased ROS levels promote auto-oxidation of dopamine, causing neuromelanin formation, proteasomal impairment, modification of α-synuclein, oligomerization, and fibrillization. Consequently, this causes dopaminergic neuron degeneration in the substantia nigra of the concerned patients (Chen et al., 2012; Herrera et al. 2005). Increased oxidative stress is also associated with restricting enzyme activities, oxidative deamination of primary monoamine oxidase, peroxidation of the mitochondrial cardiolipin, cytochrome c release from mitochondria causing neurotoxicity and cell death (Singh et al., 2019).

In ALS, oxidative stress induces mtDNA mutations, gliosis, and aggregation of mutant and wild-type SOD1. In 5% of cases, mutations in the SOD1 gene are associated with ALS development. Mutant SOD1 causes dysregulation of signaling molecules in motor neurons and regulates the activity of glial cells. Increased ROS levels from damaged motor neurons cause a hindrance to glutamate uptake into astrocytes, elevating extracellular glutamate levels, and excitotoxicity (Shichiri et al., 2014). ROS also activates glial cells leading to the release of further ROS/RNS and proinflammatory cytokines and inhibition of IGF-I/AKT neuroprotective pathway resulting in neurodegeneration (Liu et al., 2017; Radi et al., 2014). Redoxosomes are also known to regulate proinflammatory signals via NF-κB modulation (Li et al., 2011; Liu et al., 2017). Oxidative stress-induced mitochondrial alterations in the Purkinje cells characterize spinocerebellar ataxia (Liu et al., 2017; Stucki et al., 2016).

8.3.1.3 PREGNANCY AND DEVELOPMENTAL ABNORMALITY

Oxidative stress has implications in the pathophysiology of various pregnancy-related complications like miscarriage, pre-eclampsia, infertility, foetal growth restrictions, as well as preterm labor. Pathological and physiological alterations in the female reproductive tract and chronic complications in the uterus such as increased production of glycoproteins from granulosa cells, lipid peroxidation in oocytes, and consequent damages in embryonic development are associated with oxidative stress (Silveira et al., 2018). Intrauterine oxidative stress influences the mortality and morbidity of premature newborns (Bak and Roszkowski, 2013). High levels of ROS have an association with reduced sperm motility in men and oocyte maturation in the female. ROS induced lipid peroxidation causes placental membrane disruption, and women with recurrent miscarriages have elevated ROS levels in blood granulocytes (Duhig et al., 2016). Oxidative stress-induced placental disruptions also contribute to the preterm prelabor rupture of membranes caused by infections (Tchirikov et al., 2017). Bacterial infections stimulate ROS generation from immune cells, which increase MMP activity and degrade collagen within chorioamnion, rupturing the foetal membrane, and resulting in amniotic fluid leakage (Gupta et al., 2009; Duhig et al., 2016). Oxidative stress generated in the ischemic placenta can induce cytotoxic factor release into circulation, giving rise to systemic inflammatory response activating maternal endothelial cells that give rise to hypertensive disorder. Vasoconstriction, proinflammatory, and prothrombotic tendencies characterize microvascular endothelial dysfunctions. Under severe conditions, these complications can induce organ damage, seizures, and even maternal death (Anonymous, 2018). Abundant mitochondrial mass, increased availability of iron as an essential cofactor, and oxygen richness of the placenta make it ideal for ROS generation. Again, iron deficiency can induce defects in mitochondrial function and mitochondrial DNA damage, causing ROS leakage from mitochondria (Mannaerts et al., 2018). ROS generation is also associated with reperfusion injury following ischemia along with increased xanthine oxidase activity in cytotrophoblasts. Oxidative stress increases the formation of misfolded proteins in the placenta that may contribute to fetal growth restrictions during the intrauterine period (Duhig et al., 2016).

Maternal oxidative stress can have direct or indirect associations with foetal and obstetric complications like developmental and programming defects (Dennery, 2007; Mannaerts et al., 2018). Elevated ROS levels during organ formation affect vital signaling pathways inducing functional loss, structural aberration, or spontaneous abortion of the foetus (Laforgia et al., 2018). Preterm infants are more susceptible to risks due to insufficient antioxidants, hyperoxia, enhanced free radical generation due to abundance of free iron, poor nutritional status, increased glucocorticoid level, insufficiency of the placenta resulting in an increased risk of CVD, malformed heart, high blood pressure, hypertension, arrhythmias, LVH, nephrogenesis alterations, neural tube defects, morbidity, vascular and endothelial dysfunction leading to hypoxia and hypoperfusion (Laforgia et al., 2018; Rodríguez-Rodríguez et al., 2018). Intraventricular hemorrhage, broncho-pulmonary dysplasia, respiratory distress syndrome, hypoxic-ischemic encephalopathy, chronic lung disease, periventricular leukomalacia necrotizing enterocolitis, and retinopathy of prematurity causing vision loss are some organ-specific manifestations of oxidative stress in infants (Marseglia et al., 2014; Ozsurekci and Aykac, 2016). Increased oxidative stress has associations with disruption of NF-κB functioning, affecting the outgrowth of limb bud and limb reduction defects in the foetus. In severe cases, like alcohol abuse, oxidative stress can induce retardation of growth, facial abnormalities, and impaired central nervous system in the offspring (Hansen et al., 2018). Increased ROS generation is also associated with the pathophysiology of Down's syndrome, triggering an accumulation of amyloid beta-peptide (Aβ), resulting in nervous system abnormalities (Laforgia et al., 2018). Figure 8.3 depicts the oxidative stress-induced complications in pregnancy and development.

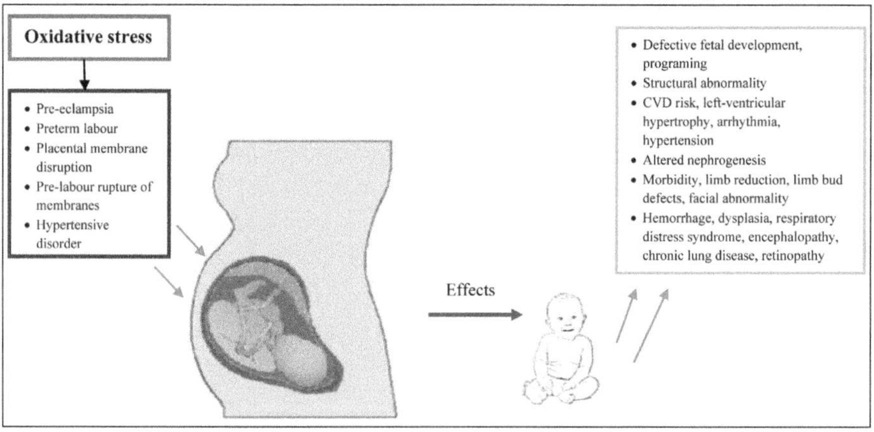

FIGURE 8.3 Effect of oxidative stress on pregnancy and development.

8.3.1.4 SOME OTHER COMPLICATIONS

Oxidative stress is also associated with complications related to alcohol, drug abuse, seizures, and transplantation. The oxidative metabolites of drugs like cocaine which include nor cocaine and derivatives, drugs like amphetamine and their derivatives, morphine, heroin contribute to oxidative stress-induced toxicity in the liver, nervous system, psychic excitement, convulsions in the brain, respiratory arrest, cardiac dysfunctions, increased ROS levels in the left ventricle and WBC, myocardial hypertrophy and heart failure, liver and kidney lipid peroxidation, myofilament tyrosine nitration, peroxidation of foetal membrane, increased H_2O_2 in prefrontal cortex and striatum (Kovacic, 2005). Chronic administration of cocaine can induce severe myocardial oxidative stress through MAPK and NOX-2 activation (Fan et al., 2009). Drug exposure can overstimulate NMDA glutamate receptors, increasing extracellular glutamate levels (Cunha-Oliveira et al., 2013). Alcohol stimulates ROS generation particularly in the detoxifying organ liver by cytochrome P450 enzyme activation, alterations of certain metal levels in the body, and antioxidant level reduction (Darbandi et al., 2018; Wu and Cederbaum, 2003). This activates inflammatory processes leading to cell death, bone destruction, modified expression of proinflammatory signals like ILs, MMPs, Rho, ERK, GSK-β pathways in nervous and connective tissues (Barcia et al., 2017). Oxidative stress induces severe brain damage, neuronal hyperexcitability, neuronal cell death in epilepsy by promoting the increased expression of heme oxygenase-I and metallothionein-I mRNA, IL-1β, c-fos, and heat shock protein-70 mRNA causing an inflammatory response and oxidative modifications of membrane receptors, ion channels, and neurotransmitter (Badawy et al., 2009; Cardenas-Rodriguez et al., 2013; Pearson-Smith and Patel, 2017). Oxidative stress is linked to the pathophysiology of delayed graft function, primary nonfunction, and acute allograft rejection in marginal donors (Chen et al., 2015). During organ transplantation, IR injury generates ROS leading to various post-transplant complications as well as long-term outcomes by increasing vascular permeability, parenchymal and endothelial cell injury, and inflammatory response (Shi and Xue, 2016). In addition to graft rejection, inflammation along with oxidative stress can cause cancer, metabolic syndrome, cardiovascular disease, and various disorders in transplant recipients. Oxidative stress also contributes to chronic allograft nephropathy, causing the allograft kidney functions to deteriorate slowly over the years after transplantation and associated formation of arteriosclerotic lesions (Nafar et al., 2011; Yılmaz and Berdeli, 2009).

8.4 CONCLUSION

Oxidative stress affects various pathways that initiate disorders or perturbations in healthy cellular functioning. The affected functional components may serve as biomarkers, which could aid in designing novel therapeutic strategies against oxidative stress. Identification of precise biomarkers of oxidative stress is crucial for a proper assessment of the nature and extent of severity of such complications. ROS and RNS induced redox alterations in the complex network of proteins, lipids, and nucleic acids directly influence molecular functions (like enzyme inhibition) and highlight oxidative stress intensity in the microenvironment, finally reflected in various diseases. The ability to impart oxidative modifications in organs and system functions determine the validity of the markers (Ho et al., 2013). For application as a clinical diagnostic marker, the oxidatively modified products must be stable and abundant and also indicate molecular mechanisms associated with several pathologies to help classify the patient population based on such mechanisms rather than just disease severity. Products like carbonyls, AGE, AOPP indicate protein oxidation, oxidized LDL is associated with atherosclerosis, MDA, and HNE indicate lipid peroxidation. DNA oxidation products excreted via urine like 8oxodG indicate lung or breast cancer, diabetes, atherosclerosis, while RNA oxidation products like 8oxoGuo are associated with neurodegeneration, diabetes, breast cancer development (Marrocco et al., 2017). The method used for clinical detection of modifications is restricted to antibody-based methods/protocols which have problems of specificity and sample amount in case of immune-histochemistry. The oxidative burst in conditions like autoimmune neutropenia and asymptomatic HIV+ individuals are determined using flow cytometry techniques. However, rigorous testing of the antibody-based markers is the need of the day for achieving technical specificity and sensitivity (Frijhoff et al., 2015; Torlakovic et al., 2015). Advanced molecular profiling techniques can help to probe into the molecular insights of various diseases and identify the potential targets of therapy (Marra et al., 2011). Some of the mammalian forkhead transcription factors and cytokine erythropoietin associated with various cell functions like survival, proliferation, differentiation, destruction, immune functions, and even metastasis have shown prospects as critical therapeutic targets as revealed in the case of AD (Maiese et al., 2010; Maiese, 2016).

ACKNOWLEDGMENT

The authors acknowledge UGC-DAE Consortium for Scientific Research, Kolkata Centre for providing the necessary facilities and Indian Council of Medical Research (ICMR) for fellowships and contingency grant.

KEYWORDS

- **redox homeostasis**
- **oxidative stress**
- **free radical**
- **disease complications**
- **developmental abnormality**

REFERENCES

Abuja, P. M.; Albertini, R. Methods for monitoring oxidative stress, lipid peroxidation and oxidation resistance of lipoproteins. *Clin. Chim. Acta* **2001,** *306*, 1–17.

Abou-El-Hassan, H.; Dia, B.; Choucair, K.; Eid, S. A.; Najdi, F.; Baki, L.; Talih, F.; Eid, A. A.; Kobeissy, F. Traumatic brain injury, diabetic neuropathy and altered-psychiatric health: the fateful triangle. *Med. Hypotheses* **2017,** *108*, 69–80.

Ahmed, O. M. Relationships between oxidative stress, cancer development and therapeutic interventions. *Can. Med. Anti Cancer Drug* **2016,** *1*, 2.

Ahn, K. The role of air pollutants in atopic dermatitis. *J. Allergy Clin. Immunol.* **2014,** *134*, 993–999.

Ali, Z. A.; de Jesus Perez, V.; Yuan, K.; Orcholski, M.; Pan, S.; Qi, W.; Chopra, G.; Adams, C.; Kojima, Y.; Leeper, N. J.; Qu, X.; Zaleta-Rivera, K.; Kato, K.; Yamada, Y.; Oguri, M.; Kuchinsky, A.; Hazen, S. L.; Jukema, J. W.; Ganesh, S. K.; Nabel, E. G.; Channon, K.; Leon, M. B.; Charest, A.; Quertermous, T.; Ashley, E. A. Oxido-reductive regulation of vascular remodelling by receptor tyrosine kinase ROS1. *J. Clin. Invest.* **2014,** *124*, 5159–5174.

Aly, D. G.; Shahin, R. S. Oxidative stress in lichen planus. *Acta Dermatoven APA* **2010,** *19*, 3–11.

Ameziane El Hassani, R.; Buffet, C.; Leboulleux, S.; Dupuy, C. Oxidative stress in thyroid carcinomas: biological and clinical significance. *Endocr.-Related Cancer* **2019,** *26*, R131–R143.

Andrisic, L.; Dudzik, D.; Barbas, C.; Milkovic, L.; Grune, T.; Zarkovic, N. Short overview on metabolomics approach to study pathophysiology of oxidative stress in cancer. *Redox Biol.* **2018,** *14*, 47–58.

Atochin, D. N.; Wang, A.; Liu, V. W.; Critchlow, J. D.; Dantas, A. P.; Looft-Wilson, R.; Murata, T.; Salomone, S.; Shin, H. K.; Ayata, C.; Moskowitz, M. A.; Michel, T.; Sessa, W. C.; Huang, P. L. The phosphorylation state of eNOS modulates vascular reactivity and outcome of cerebral ischemia in vivo. *J. Clin. Invest.* **2007,** *117*, 1961–1967.

Aviram, M. Modified forms of low density lipoprotein and atherosclerosis. *Atherosclerosis* **1993,** *98*, 1–9.

Ayala, A.; Muñoz, M. F.; Argüelles, S. Lipid peroxidation: production, metabolism, and signalling mechanisms of malondialdehyde and 4-hydroxy-2-nonenal. *Oxid. Med. Cell Longevity* **2014,** *2014*, 360438.

Babusikova, E.; Jurecekova, J.; Evinova, A.; Jesenak, M.; Dobrota, D. Oxidative damage and bronchial asthma. In: Respiratory Diseases. Rijeka: InTech, **2012**, 151–176.

Badawy, R. A. B.; Harvey, A. S.; Macdonell, R. A. L. Cortical hyper excitability and epileptogenesis: understanding the mechanisms of epilepsy—part 1. *J. Clin. Neurosci.* **2009,** *16*, 355–365.

Badimon, L.; Vilahur, G. Thrombosis formation on atherosclerotic lesions and plaque rupture. *J. Internal Med.* **2014,** *276*, 618–632.

Bak, A.; Roszkowski, K. Oxidative stress in pregnant women. *Arch. Perinat. Med.* **2013,** *19*, 150–155.

Balmus, I. M.; Ciobica, A.; Trifan, A.; Stanciu, C. The implications of oxidative stress and antioxidant therapies in inflammatory bowel disease: clinical aspects and animal models. *Saudi. J. Gastroenterol.* **2016,** *22*, 3–17.

Barcia, J. M.; Portolés, S.; Portolés, L.; Urdaneta, A. C.; Ausina, V.; Pérez-Pastor, G. M.; Romero, F. J.; Villar, V. M. Does oxidative stress induced by alcohol consumption affect orthodontic treatment outcome? *Front. Physiol.* **2017,** *8*, 22.

Barnes, P. J. Chronic obstructive pulmonary disease. *N. Engl. J. Med.* **2000,** *343*, 269–280.

Barnes, P. J. Inflammatory mechanisms in patients with chronic obstructive pulmonary disease. *J. Allergy Clin. Immunol.* **2016,** *138*, 16–27.

Barnes, R. P.; Fouquerel, E.; Opresko, P. L. The impact of oxidative DNA damage and stress on telomere homeostasis. *Mech. Ageing Dev.* **2019,** *177*, 37–45.

Bar-Or, D.; Bar-Or, R.; Rael, L. T.; Brody, E. N. Oxidative stress in severe acute illness. *Redox Biol.* **2015,** *4*, 340–345.

Barrera, G. Oxidative stress and lipid peroxidation products in cancer progression and therapy. *ISRN Oncol.* **2012,** *2012*, 137289.

Bayir, H.; Kagan, V. E.; Tyurina, Y. Y.; Tyurin, V.; Ruppel, R. A.; Adelson, P. D.; Graham, S. H.; Janesko, K.; Clark, R. S. B.; Kochanek, P. M. Assessment of antioxidant reserves and oxidative stress in cerebrospinal fluid after severe traumatic brain injury in infants and children. *Pediatr. Res.* **2002,** *51*, 571–578.

Bazin, J.; Langlade, N.; Vincourt, P.; Arribat, S.; Balzergue, S.; El-Maarouf-Bouteau, H.; Bailly C. Targeted mRNA oxidation regulates sunflower seed dormancy alleviation during dry after-ripening. *Plant Cell* **2011,** *23*, 2196–208.

Becatti, M.; Marcucci, R.; Gori, A. M.; Mannini, L.; Grifoni, E.; Alessandrello Liotta, A.; Sodi, A.; Tartaro, R.; Taddei, N.; Rizzo, S.; Prisco, D.; Abbate, R.; Fiorillo, C. Erythrocyte oxidative stress is associated with cell deformability in patients with retinal vein occlusion. *J. Thromb. Haemost.* **2016,** *14*, 2287–2297.

Bergeron, F.; Auvré, F.; Radicella, J. P.; Ravanat J. L. HO* radicals induce an unexpected high proportion of tandem base lesions refractory to repair by DNA glycosylases. *Proc. Nat. Acad. Sci. USA* **2010,** *107*, 5528–5533.

Bickers, D. R.; Athar, M. Oxidative stress in the pathogenesis of skin disease. *J. Invest. Dermatol.* **2006,** *126*, 2565–2575.

Birben, E.; Sahiner, U. M.; Sackesen, C.; Erzurum, S.; Kalayci, O. Oxidative stress and antioxidant defense. *World Allergy Organ. J.* **2012,** *5*, 9–19.

Bhandary, B.; Marahatta, A.; Kim, H. R.; Chae, H. J. An involvement of oxidative stress in endoplasmic reticulum stress and its associated diseases. *Int. J. Mol. Sci.* **2012,** *14*, 434–456.

Bohlender, J. M.; Franke, S.; Stein, G.; Wolf, G. Advanced glycation end products and the kidney. *Am. J. Physiol.-Renal.* **2005,** *289*, F645–F659.

Boukhenouna, S.; Wilson, M. A.; Bahmed, K.; Kosmider, B. Reactive oxygen species in chronic obstructive pulmonary disease. *Oxid. Med. Cell Longevity* **2018,** *2018*, 5730395.

Braithwaite, E. K.; Mattie, M. D.; Freedman, J. H. Activation of metallothionein transcription by 4-hydroxynonenal. *J. Biochem. Mol. Toxic.* **2010,** *24*, 330–334.

Brigelius-Flohe, R.; Flohe, L. Basic principles and emerging concepts in the redox control of transcription factors. *Antioxid. Redox Signalling* **2011,** *15*, 2335–2381.

Brown, A. J.; Jessup, W. Oxysterols: sources, cellular storage and metabolism, and new insights into their roles in cholesterol homeostasis. *Mol. Aspects Med.* **2009,** *30*, 111–122.

Brownlee, M. The pathobiology of diabetic complications: a unifying mechanism. *Diabetes* **2005,** *54*, 1615–1625.

Cadet, J.; Wagner, J. R. DNA base damage by reactive oxygen species, oxidizing agents, and UV radiation. *CSH Perspect. Biol.* **2013,** *5*, a012559.

Cadet, J.; Wagner, J. R. Oxidatively generated base damage to cellular DNA by hydroxyl radical and one-electron oxidants: similarities and differences. *Arch. Biochem. Biophys.* **2014,** *557*, 47–54.

Cadet, J.; Wagner, J. R. Radiation-induced damage to cellular DNA: chemical nature and mechanisms of lesion formation. *Radiat. Phys. Chem.* **2016,** *128*, 54–59.

Cai, Z.; Yan, L. J. Protein oxidative modifications: beneficial roles in disease and health. *J. Biochem. Pharmacol. Res.* **2013,** *1*, 15–26.

Calaf, G. M.; Urzua, U.; Termini, L.; Aguayo, F. Oxidative stress in female cancers. *Oncotarget* **2018,** *9*, 23824–23842.

Camandola, S.; Poli, G.; Mattson, M. P. The lipid peroxidation product 4-hydroxy-2, 3-nonenal increases AP-1- binding activity through caspase activation in neurons. *J. Neurochem.* **2000,** *74*, 159–168.

Cardenas-Rodriguez, N.; Huerta-Gertrudis, B.; Rivera-Espinosa, L.; Montesinos-Correa, H.; Bandala, C.; Carmona-Aparicio, L.; Coballase-Urrutia, E. Role of oxidative stress in refractory epilepsy: evidence in patients and experimental models. *Int. J. Mol. Sci.* **2013,** *14*, 1455–1476.

Cejka, C.; Cejkova, J. Oxidative stress to the cornea, changes in corneal optical properties, and advances in treatment of corneal oxidative injuries. *Oxid. Med. Cell Longevity* **2015,** *2015*, 591530.

Ceriello, A. New insights on oxidative stress and diabetic complications may lead to a "causal" antioxidant therapy. *Diabetes Care* **2003,** *26*, 1589–1596.

Chan, W.; Chen, B.; Wang, L.; Taghizadeh, K.; Demott, M. S.; Dedon, P. C. Quantification of the 2-deoxyribonolactone and nucleoside 5′-aldehyde products of 2-deoxyribose oxidation in DNA and cells by isotope-dilution gas chromatography mass spectrometry: differential effects of γ-radiation and Fe^{2+}–EDTA. *J. Am. Chem. Soc.* **2010,** *132*, 6145–6153.

Chandel, N. S.; Maltepe, E.; Goldwasser, E.; Mathieu, C. E.; Simon, M. C.; Schumacker, P. T. Mitochondrial reactive oxygen species trigger hypoxia-induced transcription. *Proc. Nat. Acad. Sci USA* **1998,** *95*, 11715–11720.

Chang, D.; Zhang, X.; Rong, S.; Sha, Q.; Liu, P.; Han, T.; Pan, H. Serum antioxidative enzymes levels and oxidative stress products in age-related cataract patients. *Oxid. Med. Cell. Longevity* **2013,** *2013*, 587826.

Chastain, P. D.; Nakamura, J.; Swenberg, J.; Kaufman, D. Nonrandom AP site distribution in highly proliferative cells. *FASEB J.* **2006,** *20*, 2612–2614.

Chen, C. C.; Chapman, W. C.; Hanto, D. W. Ischemia-reperfusion injury in kidney transplantation. *Front. Biosci. (Elite Edn)* **2015,** *7*, 134–154.

Chen, X.; Guo, C.; Kong, J. Oxidative stress in neurodegenerative diseases. *Neural Regen. Res.* **2012,** *7*, 376–385.

Chen, L.; Liu, B. Relationships between stress granules, oxidative stress, and neurodegenerative diseases. *Oxid. Med. Cell. Longevity* **2017,** *2017*, 1809592.

Chen, J.; Stimpson, S. E.; Fernandez-Bueno, G. A.; Mathews, C. E. Mitochondrial reactive oxygen species and type 1 diabetes. *Antioxidants Redox Signal.* **2018,** *29*, 1361–1372.

Coluzzi, E.; Colamartino, M.; Cozzi, R.; Leone, S.; Meneghini, C.; O'Callaghan, N.; Sgura, A. Oxidative stress induces persistent telomeric DNA damage responsible for nuclear morphology change in mammalian cells. *PLoS One* **2014,** *9*, e110963.

Crack, P. J.; Taylor, J. M. Reactive oxygen species and the modulation of stroke. *Free Radical Biol. Med.* **2005,** *38*, 1433–1444.

Cross, A. R.; Segal, A. W. The NADPH oxidase of professional phagocytes—prototype of the NOX electron transport chain systems. *Biochim. Biophys. Acta Bioenergy* **2004,** *1657*, 1–22.

Cui, H.; Kong, Y.; Zhang, H. Oxidative stress, mitochondrial dysfunction, and aging. *J. Signal Transduction* **2012,** *2012*, 646354.

Cunha-Oliveira, T.; Rego, A. C.; Oliveira, C. R. Oxidative stress and drugs of abuse: an update. *Mini Rev. Org. Chem.* **2013,** *10*, 321–334.

Daenen, K.; Andries, A.; Mekahli, D.; Van Schepdael, A.; Jouret, F.; Bammens, B. Oxidative stress in chronic kidney disease. *Pediatr. Nephrol.* **2019,** *34*, 975–991.

Dahl, J. U.; Gray, M. J.; Jakob, U. Protein quality control under oxidative stress conditions. *J. Mol. Biol.* **2015,** *427*, 1549–1563.

Dalle-Donne, I.; Giustarini, D.; Colombo, R.; Rossi, R.; Milzani, A. Protein carbonylation in human diseases. *Trends Mol. Med.* **2003,** *9*, 169–176.

Dalle-Donne, I.; Scaloni, A.; Giustarini, D.; Cavarra, E.; Tell, G.; Lungarella, G.; Colombo, R.; Rossi, R.; Milzani, A. Proteins as biomarkers of oxidative/nitrosative stress in diseases: the contribution of redox proteomics. *Mass Spectrom. Rev.* **2005,** *24*, 55–99.

Dalmazi, G. D.; Hirshberg, J.; Lyle, D.; Freij, J. B.; Caturegli, P. Reactive oxygen species in organ-specific autoimmunity. *AutoImmun. Highlights* **2016,** *7*, 11.

Darbandi, M.; Darbandi, S.; Agarwal, A.; Sengupta, P.; Durairajanayagam, D.; Henkel, R.; Sadeghi, M. R. Reactive oxygen species and male reproductive hormones. *Reprod. Biol. Endocrinol.* **2018,** *16*, 1–14.

Darwish, R.; Amiridze, N.; Aarabi, B. Nitro tyrosine as an oxidative stress marker: evidence for involvement in neurologic outcome in human traumatic brain injury. *J. Trauma.* **2007,** *63*, 439–442.

Davies, M. J. Reactive species formed on proteins exposed to singlet oxygen. *Photochem. Photobiol. Sci.* **2004,** *3*, 17–25.

Davies, M. J. The oxidative environment and protein damage. *Biochim. Biophys. Acta* **2005,** *1703*, 93–109.

Davies, M. J. Protein oxidation and peroxidation. *Biochem. J.* **2016,** *473*, 805–825.

Dedon, P. C. The chemical toxicology of 2-deoxyribose oxidation in DNA. *Chem. Res. Toxicol.* **2008,** *21*, 206–219.

Deng, R.; Hua, X.; Li, J.; Chi, W.; Zhang, Z.; Lu, F.; Zhang, L.; Pflugfelder, S. C.; Li, D. Q. Oxidative stress markers induced by hyperosmolarity in primary human corneal epithelial cells. *PLoS One* **2015,** *10*, e0126561.

Dennery, P. A. Effects of oxidative stress on embryonic development. *Birth Defects Res. C Embryo Today* **2007,** *81*, 155–162.

Dessing, M. C.; van der Sluijs, K. F.; Florquin, S.; van der Poll, T. Monocyte chemo-attractant protein 1 contributes to an adequate immune response in influenza pneumonia. *Clin. Immunol.* **2007,** *125*, 328–336.

Dhalla, N. S.; Elmoselhi, A. B.; Hata, T.; Makino, N. Status of myocardial antioxidants in ischemia–reperfusion injury. *Cardiovasc. Res.* **2000,** *47*, 446–456.

Di Filippo, C.; Cuzzocrea, S.; Rossi, F.; Marfella, R.; D'Amico, M. Oxidative stress as the leading cause of acute myocardial infarction in diabetics. *Cardiovasc. Drug Rev.* **2006,** *24*, 77–87.

Ding, Z.; Wang, X.; Schnackenberg, L.; Khaidakov, M.; Liu, S.; Singla, S.; Dai, Y.; Mehta, J. L. Regulation of autophagy and apoptosis in response to ox-LDL in vascular smooth muscle cells, and the modulatory effects of the microRNA hsa-let-7g. *Int. J. Cardiol.* **2013,** *168*, 1378–1385.

Domej, W.; Oettl, K.; Renner, W. Oxidative stress and free radicals in COPD—implications and relevance for treatment. *Int. J. Chron. Obstruct. Pulmon. Dis.* **2014,** *9*, 1207–1224.

Donkena, K. V.; Young, C. Y. F.; Tindall, D. J. Oxidative stress and DNA methylation in prostate cancer. *Obstet. Gynecol. Int.* **2010,** *2010*, 302051.

Douki, T.; Rivière, J.; Cadet, J. DNA tandem lesions containing 8-oxo-7, 8-dihydroguanine and formamido residues arise from intramolecular addition of thymine peroxyl radical to guanine. *Chem. Res. Toxicol.* **2002,** *15*, 445–454.

Duhig, K.; Chappell, L. C.; Shennan, A. H. Oxidative stress in pregnancy and reproduction. *Obstet. Med.* **2016,** *9*, 113–116.

Duni, A.; Liakopoulos, V.; Rapsomanikis, K. P.; Dounousi, E. Chronic kidney disease and disproportionally increased cardiovascular damage: does oxidative stress explain the burden? *Oxid. Med. Cell. Longevity* **2017,** *2017*, 9036450.

Durham, A. L.; Adcock, I. M. The relationship between COPD and lung cancer. *Lung Cancer* **2015,** *90*, 121–127.

Esper, R. J.; Nordaby, R. A.; Vilariño, J. O.; Paragano, A.; Cacharrón, J. L.; Machado, R. A. Endothelial dysfunction: a comprehensive appraisal. *Cardiovasc. Diabetol.* **2006,** *5*, 4.

Engwa, G. A. Free radicals and the role of plant phytochemicals as antioxidants against oxidative stress-related diseases. In: Phytochemicals—Source of Antioxidants and Role in Disease Prevention, London: IntechOpen, **2018**.

Fan, L.; Sawbridge, D.; George, V.; Teng, L.; Bailey, A.; Kitchen, I.; Li, J. M. Chronic cocaine-induced cardiac oxidative stress and mitogen-activated protein kinase activation: the role of Nox2 oxidase. *J. Pharmacol. Exp. Ther.* **2008,** *328*, 99–106.

Farhood, B.; Goradel, N. H.; Mortezaee, K.; Khanlarkhani, N.; Salehi, E.; Nashtaei, M. S.; Shabeeb, D.; Musa, AE.; Fallah, H.; Najafi, M. Intercellular communications-redox interactions in radiation toxicity; potential targets for radiation mitigation. *J. Cell Commun. Signal.* **2019,** *13*, 3–16.

Ferretti, G.; Bacchetti, T.; Masciangelo, S.; Saturni, L. Celiac disease, inflammation and oxidative damage: a nutrigenetic approach. *Nutrients* **2012,** *4*, 243–257.

Filomeni, G.; Rotilio, G.; Ciriolo, M. R. Cell signalling and the glutathione redox system. *Biochem. Pharmacol.* **2002,** *64*, 1057–1064.

Filippin, L. I.; Vercelino, R.; Marroni, N. P.; Xavier, R. M. Redox signalling and the inflammatory response in rheumatoid arthritis. *Clin. Exp. Immunol.* **2008,** *152*, 415–422.

Fimognari, C. Role of oxidative RNA damage in chronic-degenerative diseases. *Oxid. Med. Cell. Longevity* **2015,** *2015*, 358713.

Fisher, G. J.; Quan, T.; Purohit, T.; Shao, Y.; Cho, M. K.; He, T.; Varani, J.; Kang, S.; Voorhees, J. J. Collagen fragmentation promotes oxidative stress and elevates matrix metalloproteinase-1 in fibroblasts in aged human skin. *Am. J. Pathol.* **2009,** *174*, 101–114.
Fleming, A. M.; Burrows, C. J. Formation and processing of DNA damage substrates for the hNEIL enzymes. *Free Radical Biol. Med.* **2017,** *107*, 35–52.
Florens, N.; Calzada, C.; Lyasko, E.; Juillard, L.; Soulage, C. O. Modified lipids and lipoproteins in chronic kidney disease: a new class of uremic toxins. *Toxins (Basel)* **2016,** *8*, 376.
Frijhoff, J.; Winyard, P. G.; Zarkovic, N.; Davies, S. S.; Stocker, R.; Cheng, D.; Knight, A. R.; Taylor, E. L.; Oettrich, J.; Ruskovska, T.; Gasparovic, A. C.; Cuadrado, A.; Weber, D.; Poulsen, H. E.; Grune, T.; Schmidt, H. H.; Ghezzi, P. Clinical relevance of biomarkers of oxidative stress. *Antioxid. Redox Signal.* **2015,** *23*, 1144–1170.
Gailly, P.; Jouret, F.; Martin, D.; Debaix, H.; Parreira, K. S.; Nishita, T.; Blanchard, A.; Antignac, C.; Willnow, T. E.; Courtoy, P. J.; Scheinman, S. J.; Christensen, E. I.; Devuyst, O. A novel renal carbonic anhydrase type III plays a role in proximal tubule dysfunction. *Kidney Int.* **2008,** *74*, 52–61.
Garem, Y. F. E.; Ahmed, R. A. M.; Ragab, M. A.; AbouZeid, A. A. Study of oxidative stress in different clinical severities of acne vulgaris. *Egypt. J. Dermatol. Venereol.* **2014,** *34*, 53–57.
Gero, D. Hyperglycemia-induced endothelial dysfunction. In: Endothelial Dysfunction-Old Concepts and New Challenges. London: IntechOpen, **2017**.
Ghosh, N.; Das, A.; Chaffee, S.; Roy, S.; Sen, C. K. Reactive oxygen species, oxidative damage and cell death. In: Immunity and Inflammation in Health and Disease, Cambridge, MA: Academic Press, **2018**, 45–55.
Giacco, F.; Brownlee, M.; Schmidt, A. M. Oxidative stress and diabetic complications. *Circ. Res.* **2010,** *107*, 1058–1070.
Girotti, A. W. Mechanisms of lipid peroxidation. *J. Free Radical Biol. Med.* **1985,** *1*, 87–95.
Goode, H. F.; Cowley, H. C.; Walker, B. E.; Howdle, P. D.; Webster, N. R. Decreased antioxidant status and increased lipid peroxidation in patients with septic shock and secondary organ dysfunction. *Crit. Care Med.* **1995,** *23*, 646–651.
Gordois, A.; Cutler, H.; Pezzullo, L.; Gordon, K.; Cruess, A.; Winyard, S.; Hamilton, W.; Chua, K. An estimation of the worldwide economic and health burden of visual impairment. *Glob. Public Health* **2012**, 7, 465–481.
Goyal, S.K.; Sharma, A. Atrial fibrillation in obstructive sleep apnea. *World J. Cardiol.* **2013,** *5*, 157–163.
Gracia, K. C.; Llanas-Cornejo, D.; Husi, H. CVD and oxidative stress. *J. Clin. Med.* **2017,** *6*, 22.
Granata, S.; Zaza, G.; Simone, S.; Villani, G.; Latorre, D.; Carella, M.; Pontrelli, P.; Schena, F. P.; Grandaliano, G.; Pertosa, G. Mitochondrial dysregulation and oxidative stress in patients with chronic kidney disease. *BMC Genomics* **2009,** *10*, 388.
Gray, M. J.; Wholey, W. Y.; Jakob, U. Bacterial responses to reactive chlorine species. *Annu. Rev. Microbiol.* **2013,** *67*, 141–160.
Guan, G.; Lan, S. Implications of antioxidant systems in inflammatory bowel disease. *BioMed Res. Int.* **2018,** *2018*, 1290179.
Gupta, V.; Khan, A.; Higham, A.; Lemon, J.; Sriskantharajah, S.; Amour, A.; Hessel, E.M.; Southworth, T.; Singh, D. The effect of phosphatidylinositol-3 kinase inhibition on matrix metalloproteinase-9 and reactive oxygen species release from chronic obstructive pulmonary disease neutrophils. *Int. Immunopharmacol.* **2016,** *35*, 155–162.
Gupta, S.; Surti, N.; Metterle, L.; Chandra, A.; Agarwal, A. Antioxidants and female reproductive pathologies. *Arch. Med. Sci.* **2009,** *5*, S151–S173.

Gutierrez, E. L.; Ramo, W.; Seminario-Vida, L.; Tello, M.; Roncero, G.; Ortega-Loayz, A. G. Oxidative stress in patients with endemic pemphigus foliaceus and healthy subjects with anti-desmoglein 1 antibodies. *Anais Brasileiros de Dermatologia* **2018,** *93*, 212–215.

Guzy, R. D.; Schumacker, P. T. Oxygen sensing by mitochondria at complex III: the paradox of increased reactive oxygen species during hypoxia. *Exp. Physiol.* **2006,** *91*, 807–819.

Hajjar, I.; Hayek, S. S.; Goldstein, F. C.; Martin, G.; Jones, D. P.; Quyyumi, A. Oxidative stress predicts cognitive decline with aging in healthy adults: an observational study. *J. Neuroinflammation* **2018,** *15*, 17.

Hall, E. D.; Andrus, P. K.; Yonkers P. A. Brain hydroxyl radical generation in acute experimental head injury. *J. Neurochem.* **1993,** *60*, 588–594.

Halliwell, B. Free radicals and other reactive species in disease. In Encyclopedia of Life Sciences, 2005 (pp. 1– 7). Chichester, UK: John Wiley & Sons, Ltd. http://doi.wiley.com/10.1038/npg.els.0003913

Hansen, J. M.; Jacob, B. R.; Piorczynski, T. B. Oxidative stress during development: chemical-induced teratogenesis. *Curr. Opin. Toxicol.* **2018,** *7*, 110–115.

Harja, E.; Bu, D. X.; Hudson, B. I.; Chang, J. S.; Shen, X.; Hallam, K.; Kalea, A. Z.; Lu, Y.; Rosario, R. H.; Oruganti, S.; Nikolla, Z.; Belov, D.; Lalla, E.; Ramasamy, R.; Yan, S. F.; Schmidt, A. M. Vascular and inflammatory stresses mediate atherosclerosis via RAGE and its ligands in apoE-/- mice. *J. Clin. Invest.* **2008,** *118*, 183–194.

Hauck, A. K.; Bernlohr, D. A. Oxidative stress and lipotoxicity. *J. Lipid Res.* **2016,** *57*, 1976–1986.

He, Y.; Li, S.; Zhang, W.; Dai, W.; Cui, T.; Wang, G.; Gao, T.; Li, C. Dysregulated autophagy increased melanocyte sensitivity to H_2O_2-induced oxidative stress in vitiligo. *Sci. Rep.* **2017,** *7*, 42394.

Henrich, D.; Zimmer, S.; Seebach, C.; Frank, J.; Barker, J.; Marzi, I. Trauma-activated polymorphonucleated leukocytes damage endothelial progenitor cells: probable role of CD11b/CD18-CD54 interaction and release of reactive oxygen species. *Shock* **2011,** *36*, 216–222.

Herrera, A. J.; Tomás-Camardiel, M.; Venero, J. L.; Cano, J.; Machado, A. Inflammatory process as a determinant factor for the degeneration of substantia nigra dopaminergic neurons. *J. Neural Transm. (Vienna)* **2004,** *112*, 111–119.

Higa, A.; Chevet, E. Redox signalling loops in the unfolded protein response. *Cell. Signal.* **2012,** *24*, 1548–1555.

Hitchon, C. A.; El-Gabalawy, H. S. Oxidation in rheumatoid arthritis. *Arthritis Res. Ther.* **2004,** *6*, 265–278.

Ho, E.; Karimi Galougahi, K.; Liu, C. C.; Bhindi, R.; Figtree, G. A. Biological markers of oxidative stress: applications to cardiovascular research and practice. *Redox Biol.* **2013,** *1*, 483–491.

Hobbins, S.; Chapple, I. L.; Sapey, E.; Stockley, R. A. Is periodontitis a comorbidity of COPD or can associations be explained by shared risk factors/behaviors? *Int. J. Chron. Obstruct. Pulmon. Dis.* **2017,** *12*, 1339–1349.

Hofer, T.; Badouard, C.; Bajak, E.; Ravanat, J. L.; Mattsson, A.; Cotgreave, I. A. Hydrogen peroxide causes greater oxidation in cellular RNA than in DNA. *Biol. Chem.* **2005,** *386*, 333–337.

Hoffmann, M. H.; Griffiths, H. R. The dual role of reactive oxygen species in autoimmune and inflammatory diseases: evidence from preclinical models. *Free Radical Biol. Med.* **2018,** *125*, 62–71.

Holmström, K. M.; Finkel, T. Cellular mechanisms and physiological consequences of redox-dependent signalling. *Nat. Rev. Mol. Cell Biol.* **2014,** *15*, 411–421.

Honda, K.; Smith, M. A.; Zhu, X.; Baus, D.; Merrick, W. C.; Tartakoff, A. M.; Hattier, T.; Harris, P. L.; Siedlak, S. L.; Fujioka, H.; Liu, Q.; Moreira, P. I.; Miller, F. P.; Nunomura, A.; Shimohama, S.; Perry, G. Ribosomal RNA in Alzheimer disease is oxidized by bound redox-active iron. *J. Biol. Chem.* **2005,** *280*, 20978–20986.

Hristova, M.; Habibovic, A.; Veith, C.; Janssen-Heininger, Y. M.; Dixon, A. E.; Geiszt, M.; van der Vliet, A. Airway epithelial dual oxidase 1 mediates allergen-induced IL-33 secretion and activation of type 2 immune responses. *J. Allergy Clin. Immunol.* **2016,** *137*, 1545–1556.

Huang, Y.; Li, W.; Kong, A. T. Anti-oxidative stress regulator NF-E2-related factor 2 mediates the adaptive induction of antioxidant and detoxifying enzymes by lipid peroxidation metabolite 4-hydroxynonenal. *Cell Biosci.* **2012,** *2*, 40.

Hussain, S. P.; Hofseth, L. J.; Harris, C. C. Radical causes of cancer. *Nat. Rev. Cancer* **2003,** *3*, 276–285.

Ighodaro, O. M. Molecular pathways associated with oxidative stress in diabetes mellitus. *Biomed. Pharmacother.* **2018,** *108*, 656–662.

Islam, M. T. Oxidative stress and mitochondrial dysfunction-linked neurodegenerative disorders. *Neurol. Res.* **2017,** *39*, 73–82.

Iyer, A.; Fairlie, D. P.; Prins, J. B.; Hammock, B. D.; Brown, L. Inflammatory lipid mediators in adipocyte function and obesity. *Nat. Rev. Endocrinol.* **2010,** *6*, 71–82.

Izuta, H.; Matsunaga, N.; Shimazawa, M.; Sugiyama, T.; Ikeda, T.; Hara, H. Proliferative diabetic retinopathy and relations among antioxidant activity, oxidative stress, and VEGF in the vitreous body. *Mol. Vis.* **2010,** *16*, 130–136.

Jacob, C.; Battaglia, E.; Burkholz, T.; Peng, D.; Bagrel, D.; Montenarh, M. Control of oxidative posttranslational cysteine modifications: from intricate chemistry to widespread biological and medical applications. *Chem. Res. Toxicol.* **2012,** *25*, 588–604.

Janssen-Heininger, Y. M.; Mossman, B. T.; Heintz, N. H.; Forman, H. J.; Kalyanaraman, B.; Finkel, T.; Stamler, J. S.; Rhee, S. G.; van der Vliet, A. Redox-based regulation of signal transduction: principles, pitfalls, and promises. *Free Radical Biol. Med.* **2008,** *45*, 1–17.

Jena, N. R. DNA damage by reactive species: mechanisms, mutation and repair. *J. Biosci.* **2012,** *37*, 503–517.

Jesenak, M.; Zelieskova, M.; Babusikova, E. Oxidative stress and bronchial asthma in children—causes or consequences? *Front. Pediatr.* **2017,** *5*, 162.

Ji, H.; Li, X. K. Oxidative stress in atopic dermatitis. *Oxid. Med. Cell. Longevity* **2016,** *2016*, 2721469.

Jin, Y.; Choi, A. M. Cross talk between autophagy and apoptosis in pulmonary hypertension. *Pulm. Circ.* **2012,** *2*, 407–414.

Johnson, G. L.; Lapadat, R. Mitogen-activated protein kinase pathways mediated by ERK, JNK, and p38 protein kinases. *Science* **2002,** *298*, 1911–1912.

Kassi, E.; Adamopoulos, C.; Basdra, E. K.; Papavassiliou, A. G. Role of Vitamin D in Atherosclerosis. *Circulation* **2013,** *128*, 2517–2531.

Kashihara, N.; Haruna, Y.; Kondeti, V. K.; Kanwar, Y. S. Oxidative stress in diabetic nephropathy. *Curr. Med. Chem.* **2010,** *17*, 4256–4269.

Kaur, J.; Kukreja, S.; Kaur, A.; Malhotra, N.; Kaur, R. The oxidative stress in cataract patients. *J. Clin. Diagn. Res.* **2012,** *6*, 1629–1632.

Kim, D. K.; Kim, H. S.; Kim, A. R.; Kim, J. H.; Kim, B.; Noh, G.; Kim, H. S.; Beaven, M. A.; Kim, Y. M.; Choi, W. S. DJ-1 regulates mast cell activation and IgE-mediated allergic responses. *J. Allergy Clin. Immunol.* **2013,** *131*, 1653–1662.

Kim, H. J.; Vaziri, N. D. Contribution of impaired Nrf2-Keap1 pathway to oxidative stress and inflammation in chronic renal failure. *Am. J. Physiol. Renal Physiol.* **2010,** *298,* F662–F671.

Kirkham, P. A.; Barnes, P. J. Oxidative stress in COPD. *Chest* **2013,** *144,* 266–273.

Kirkham, P. A.; Caramori, G.; Casolari, P.; Papi, A. A.; Edwards, M.; Shamji, B.; Triantaphyllopoulos, K.; Hussain, F.; Pinart, M.; Khan, Y.; Heinemann, L.; Stevens, L.; Yeadon, M.; Barnes, P. J.; Chung K. F.; Adcock, I. M. Oxidative stress induced antibodies to carbonyl-modified protein correlate with severity of chronic obstructive pulmonary disease. *Am. J. Resp. Crit. Care Med.* **2011,** *184,* 796–802.

Kitada, M.; Ogura, Y.; Monno, I.; Koya, D. Sirtuins and type 2 diabetes: role in inflammation, oxidative stress, and mitochondrial function. *Front. Endocrinol.* **2019,** *10,* 187.

Kohen, R.; Nyska, A. Oxidation of biological systems: oxidative stress phenomenon: antioxidants, redox reactions, and methods for their quantification. *Toxicol. Pathol.* **2002,** *30,* 620–650.

Kolarsick, P. A. J.; Kolarsick, M. A.; Goodwin, C. Anatomy and physiology of the skin. *J. Dermatol. Nurses' Assoc.* **2011,** *3,* 203–213.

Kong, Q.; Lin, C. L. Oxidative damage to RNA: mechanisms, consequences, and diseases. *Cell. Mol. Life Sci.* **2010,** *67,* 1817–1829.

Kovacic, P. Role of oxidative metabolites of cocaine in toxicity and addiction: oxidative stress and electron transfer. *Med. Hypotheses* **2005,** *64,* 350–356.

Krata, N.; Zagożdżon, R.; Foroncewicz, B.; Mucha, K. Oxidative stress in kidney diseases: the cause or the consequence? *Arch. Immunol. Ther. Exp. (Warsz)* **2018,** *66,* 211–220.

Krijnen, P. A.; Meischl, C.; Hack, C.; Meijer, C. J.; Visser, C. A.; Roos, D.; Niessen, H. W. Increased Nox2 expression in human cardiomyocytes after acute myocardial infarction. *J. Clin. Pathol.* **2003,** *56,* 194–199.

Kropachev, K.; Ding, S.; Terzidis, M. A.; Masi, A.; Liu, Z.; Cai, Y.; Kolbanovskiy, M.; Chatgilialoglu, C.; Broyde, S.; Geacintov, N. E.; Shafirovich, V. Structural basis for the recognition of diastereomeric 5′, 8-cyclo-2′-deoxypurine lesions by the human nucleotide excision repair system. *Nucleic Acids Res.* **2014,** *42,* 5020–5032.

Kruk, J.; Duchnik, E. Oxidative stress and skin diseases: possible role of physical activity. *Asian Pac. J. Cancer Prev.* **2014,** *15,* 561–568.

Kruk, J.; Kubasik-Kladna, K.; Aboul-Enein, H. Y. The role oxidative stress in the pathogenesis of eye diseases: current status and a dual role of physical activity. *Mini Rev. Med. Chem.* **2015,** *16,* 241–257.

Kryston, T. B.; Georgiev, A. B.; Pissis, P.; Georgakilas, A. G. Role of oxidative stress and DNA damage in human carcinogenesis. *Mutat. Res.* **2011,** *711,* 193–201.

Laforgia, N.; Mauro, A. D.; Guarnieri, G. F.; Varvara, D.; Cosmo, L. D.; Panza, R.; Capozza, M.; Baldassarre, M. E.; Resta, N. The role of oxidative stress in the pathomechanism of congenital malformations. *Oxid. Med. Cell. Longevity* **2018,** *2018,* 7404082.

Lane, N. Oxygen: The Molecule That Made the World, Oxford: Oxford University Press, 2002.

Lee, C. Collabourative power of Nrf2 and PPARγ activators against metabolic and drug-induced oxidative injury. *Oxid. Med. Cell. Longevity* **2017,** *2017,* 1378175.

Leone, A.; Roca, M. S.; Ciardiello, C.; Costantini, S.; Budillon, A. Oxidative stress gene expression profile correlates with cancer patient poor prognosis: identification of crucial pathways might select novel therapeutic approaches. *Oxid. Med. Cell. Longevity* **2017,** *2017,* 2597581.

Leto, T. L.; Morand, S.; Hurt, D.; Ueyama, T. Targeting and regulation of reactive oxygen species generation by Nox family NADPH oxidases. *Antioxid. Redox Signal.* **2009,** *11*, 2607–2619.

Li, G.; Scull, C.; Ozcan, L.; Tabas, I. NADPH oxidase links endoplasmic reticulum stress, oxidative stress, and PKR activation to induce apoptosis. *J. Cell Biol.* **2010**, *191*, 1113–1125.

Li, Q.; Spencer, N. Y.; Pantazis, N. J.; Engelhardt, J. F. Alsin and SOD1 (G93A) proteins regulate endosomal reactive oxygen species production by glial cells and proinflammatory pathways responsible for neurotoxicity. *J. Biol. Chem.* **2011,** *286*, 40151–40162.

Li, J. K.; Liu, X. D.; Shen, L.; Zeng, W. M.; Qiu, G. Z. Natural plant polyphenols for alleviating oxidative damage in man: current status and future perspectives. *Trop. J. Pharm. Res.* **2016,** *15*, 1089–1098.

Liguori, I.; Russo, G.; Curcio, F.; Bulli, G.; Aran, L.; Della-Morte, D.; Gargiulo, G.; Testa, G.; Cacciatore, F.; Bonaduce, D.; Abete, P. Oxidative stress, aging, and diseases. *Clin. Intervention Aging* **2018,** *13*, 757–772.

Lim, J.; Yue, Z. Neuronal aggregates: formation, clearance, and spreading. *Dev. Cell* **2015,** *32*, 491–501.

Lin, X.; Huang, T. Oxidative stress in psoriasis and potential therapeutic use of antioxidants. *Free Radical Res.* **2016,** *50*, 585–595.

Ling, X. C.; Kuo, K. L. Oxidative stress in chronic kidney disease. *Renal Replacement Ther.* **2018,** *4*, 53.

Liot, G.; Valette, J.; Pépin, J.; Flament, J.; Brouillet, E. Energy defects in Huntington's disease: Why "in vivo" evidence matters. *Biochem. Biophys. Res. Commun.* **2017,** *483*, 1084–1095.

Liu, Z.; Ren, Z.; Zhang, J.; Chuang, C. C.; Kandaswamy, E.; Zhoum, T.; Zuo, L. Role of ROS and nutritional antioxidants in human diseases. *Front. Physiol.* **2018,** *9*, 477.

Liu, Z.; Zhou, T.; Ziegler, A. C.; Dimitrion, P.; Zuo, L. Oxidative stress in neurodegenerative diseases: from molecular mechanisms to clinical applications. *Oxid. Med. Cell. Longevity* **2017,** *2017*, 2525967.

Lobo, V.; Patil, A.; Phatak, A.; Chandra, N. Free radicals, antioxidants and functional foods: impact on human health. *Pharmacogn. Rev.* **2010,** *4,* 118–126.

Lowes, M. A.; Suárez-Fariñas, M.; Krueger, J. G. Immunology of psoriasis. *Annu. Rev. Immunol.* **2014,** *32*, 227–255.

Lozhkin, A.; Vendrova, A. E.; Pan, H.; Wickline, S. A.; Madamanchi, N. R.; Runge, M. S. NADPH oxidase 4 regulates vascular inflammation in aging and atherosclerosis. *J. Mol. Cell. Cardiol.* **2017,** *102*, 10–21.

Luxford, C.; Morin, B.; Dean, R. T.; Davies, M. J. Histone H1- and other protein- and amino acid-hydroperoxides can give rise to free radicals which oxidize DNA. *Biochem. J.* **1999,** *344*, 125–134.

Maiese, K.; Chong, Z. Z.; Hou, J.; Shang, Y. C. Oxidative stress: biomarkers and novel therapeutic pathways. *Exp. Gerontol.* **2010,** *45*, 217–234.

Maiese, K. New insights for oxidative stress and diabetes mellitus. *Oxid. Med. Cell. Longevity* **2015,** *2015*, 875961.

Maiese, K. Forkhead transcription factors: new considerations for Alzheimer's disease and dementia. *J. Transl. Sci.* **2016,** *2*, 241–247.

Malhotra, J. D.; Kaufman, R. J. Endoplasmic reticulum stress and oxidative stress: a vicious cycle or a double-edged sword? *Antioxid. Redox Signal.* **2007,** *9*, 2277–2293.

Malik, A. N.; Czajka, A. Is mitochondrial DNA content a potential biomarker of mitochondrial dysfunction? *Mitochondrion* **2013,** *13*, 481–492.

Maltepe, E.; Saugstad, O. D. Oxygen in health and disease: regulation of oxygen homeostasis–clinical implications. *Pediatr. Res.* **2009,** *65*, 261–268.

Manda, G.; Isvoranu, G.; Comanescu, M. V.; Manea, A.; Butuner, B. D.; Korkmaz, K. S. The redox biology network in cancer pathophysiology and therapeutics. *Redox Biol.* **2015,** *5*, 347–357.

Mannaerts, D.; Faes, E.; Cos, P.; Briedé, J. J.; Gyselaers, W.; Cornette, J.; Gorbanev, Y.; Bogaerts, A.; Spaanderman, M.; Van Craenenbroeck, E.; Jacquemyn, Y. Oxidative stress in healthy pregnancy and pre-eclampsia is linked to chronic inflammation, iron status and vascular function. *PLoS One* **2018,** *13*, e0202919.

Manoharan, S.; Guillemin, G. J.; Abiramasundari, R. S.; Essa, M. M.; Akbar, M.; Akbar, M. D. The role of reactive oxygen species in the pathogenesis of Alzheimer 's disease, Parkinson's Disease, and Huntington's Disease: a mini review. *Oxid. Med. Cell. Longevity* **2016,** *2016*, 8590578.

Mansourian, A.; Agha-Hosseini, F.; Kazemi, H. H.; Mortazavi, N.; Moosavi, M. S.; Beytollahi, J.; Mirzaii-Dizgah, I. Salivary oxidative stress in oral lichen planus treated with triamcinolone mouth rinse. *Dental Res. J.* **2017,** *14*, 104–110.

Mantzarlis, K.; Tsolaki, V.; Zakynthinos, E. Role of oxidative stress and mitochondrial dysfunction in sepsis and potential therapies. *Oxid. Med. Cell. Longevity* **2017,** *2017*, 5985209.

Markkanen, E. Not breathing is not an option: How to deal with oxidative DNA damage. *DNA Repair* **2017,** *59,* 82–105.

Marnett, L. J. Lipid peroxidation-DNA damage by malondialdehyde. *Mutat. Res.* **1999,** *424*, 83–95.

Marra, M.; Sordelli, I. M.; Lombardi, A.; Lamberti, M.; Tarantino, L.; Giudice, A.; Stiuso, P.; Abbruzzese, A.; Sperlongano, R.; Accardo, M.; Agresti, M.; Caraglia, M.; Sperlongano, P. Molecular targets and oxidative stress biomarkers in hepatocellular carcinoma: an overview. *J. Transl. Med.* **2011,** *9*, 171.

Marrocco, I.; Altieri, F.; Peluso, I. Measurement and clinical significance of biomarkers of oxidative stress in humans. *Oxid. Med. Cell. Longevity* **2017,** *2017*, 6501046.

Marseglia, L.; D'Angelo, G.; Manti, S.; Arrigo, T.; Barberi, I.; Reiter, R. J.; Gitto, E. Oxidative stress-mediated aging during the fetal and perinatal periods. *Oxid. Med. Cell. Longevity* **2014,** *2014*, 358375.

Martinet, W.; de Meyer, G. R. Y. Autophagy in atherosclerosis: a cell survival and death phenomenon with therapeutic potential. *Circul. Res.* **2009,** *104*, 304–317.

Masuda, T.; Shimazawa, M.; Hara, H. Retinal diseases associated with oxidative stress and the effects of a free radical scavenger (Edaravone). *Oxid. Med. Cell. Longevity* **2017,** *2017*, 9208489.

Masui, S.; Vavassori, S.; Fagioli, C.; Sitia, R.; Inaba, K. Molecular bases of cyclic and specific disulfide interchange between human ERO1alpha protein and protein-disulfide isomerase (PDI). *J. Biol. Chem.* **2011,** *286*, 16261–16271.

Matough, F. A.; Budin, S. B.; Hamid, Z. A.; Alwahaibi, N.; Mohamed, J. The role of oxidative stress and antioxidants in diabetic complications. *Sultan Qaboos Uni. Med. J.* **2012,** *12*, 5–18.

Matschke, V.; Theiss, C.; Matschke, J. Oxidative stress: the lowest common denominator of multiple diseases. *Neural Regen. Res.* **2019,** *14*, 238–241.

McGuinness, A. J. A.; Sapey, E. Oxidative stress in COPD: sources, markers, and potential mechanisms. *J. Clin. Med.* **2017,** *6*, 21.

Medzhitov, R. Origin and physiological roles of inflammation. *Nature* **2008,** *454*, 428–435.

Mete, R.; Tulubas, F.; Oran, M.; Yılmaz, A.; Avci, B. A.; Yildiz, K.; Turan, C. B.; Gurel, A. The role of oxidants and reactive nitrogen species in irritable bowel syndrome: a potential etiological explanation. *Med. Sci. Monit.* **2013,** *19*, 762–766.

Mill, C.; Monk, B. A.; Williams, H.; Simmonds, S. J.; Jeremy, J. Y.; Johnson, J. L.; George, S. J. Wnt5a-induced Wnt1-inducible secreted protein-1 suppresses vascular smooth muscle cell apoptosis induced by oxidative stress. *Arterioscler. Thromb. Vasc. Biol.* **2014,** *34*, 2449–2456.

Moodley, D.; Mody, G.; Patel, N.; Chuturgoon, N. A. Mitochondrial depolarization and oxidative stress in rheumatoid arthritis patients. *Clin. Biochem.* **2008,** *41*, 1396–1401.

Moore, R. A.; Taubner, L. M.; Priola, S. A. Prion protein misfolding and disease. *Curr. Opin. Struct. Biol.* **2009,** *19*, 14–22.

Moserova, I.; Kralova, J. Role of ER stress response in photodynamic therapy: ROS generated in different sub cellular compartments trigger diverse cell death pathways. *PLoS One* **2012,** *7*, e32972.

Münzel, T.; Camici, G. G.; Maack, C.; Bonetti, N. R.; Fuster, V.; Kovacic, J. C. Impact of oxidative stress on the heart and vasculature part 2 of a 3-part series. *J. Am. Coll. Cardiol.* **2017,** *70*, 212–229.

Nafar, M.; Sahraei, Z.; Salamzadeh, J.; Samavat, S.; Vaziri, N. D. Oxidative stress in kidney transplantation: causes, consequences, and potential treatment. *Iran. J. Kidney Dis.* **2011,** *5*, 357–372.

Nawrot, B.; Sochacka, E.; Düchler, M. tRNA structural and functional changes induced by oxidative stress. *Cell. Mol. Life Sci.* **2011,** *68*, 4023–4032.

Neumann-Haefelin, T.; Kastrup, A.; de Crespigny, A.; Yenari, M. A.; Ringer, T.; Sun, G. H.; Moseley, M. E. Serial MRI after transient focal cerebral ischemia in rats: dynamics of tissue injury, blood-brain barrier damage, and oedema formation. *Stroke* **2000,** *31*, 1965–1973.

Newsholme, P.; Cruzat, V. F.; Keane, K. N.; Carlessi, R.; de Bittencourt, P. I. Jr. Molecular mechanisms of ROS production and oxidative stress in diabetes. *Biochem. J.* **2016,** *473*, 4527–4550.

Nistala, R.; Whaley-Connell, A.; Sowers, J. R. Redox control of renal function and hypertension. *Antioxid. Redox Signal.* **2008,** *10*, 2047–2089.

Nourazarian, A. R.; Kangari, P.; Salmaninejad, A. Roles of oxidative stress in the development and progression of breast cancer. *Asian Pac. J. Cancer Prev.* **2014,** *15*, 4745–4751.

Nowak, W. N.; Deng, J.; Ruan, X. Z.; Xu, Q. Reactive oxygen species generation and atherosclerosis. *Arterioscler. Thromb. Vasc. Biol.* **2017,** *37*, 41–52.

Obrador, E.; Liu-Smith, F.; Dellinger, R. W.; Salvador, R.; Meyskens, F. L.; Estrela, J. M. Oxidative stress and antioxidants in the pathophysiology of malignant melanoma. *Biol. Chem.* **2019,** *400*, 589–612.

Ogrunc, M.; di Fagagna, F. D. Never-ageing cellular senescence. *Eur. J. Cancer* **2011,** *47*, 1616–1622.

Ozawa, Y.; Kurihara, T.; Sasaki, M.; Ban, N.; Yuki, K.; Kubota, S.; Tsubota, K. Neural degeneration in the retina of the streptozotocin-induced type 1 diabetes model. *Exp. Diabetes Res.* **2011,** *2011*, 108328.

Ozsurekci, Y.; Aykac, K. Oxidative stress related diseases in newborns. *Oxid. Med. Cell. Longevity* **2016,** *2016*, 2768365.

Palace, V. P.; Hill, M. F.; Farahmand, F.; Singal, P. K. Mobilization of antioxidant vitamin pools and hemodynamic function after myocardial infarction. *Circulation* **1999,** *99*, 121–126.

Palmirotta, R.; Cives, M.; Della-Morte, D.; Capuani, B.; Lauro, D.; Guadagni, F.; Silvestris, F. Sirtuins and cancer: role in the epithelial-mesenchymal transition. *Oxid. Med. Cell. Longevity* **2016,** *2016*, 3031459.

Paneni, F.; Beckman, J. A.; Creager, M. A.; Cosentino, F. Diabetes and vascular disease: pathophysiology, clinical consequences, and medical therapy: part I. *Eur. Heart J.* **2013,** *34*, 2436–2443.

Panieri, E.; Santoro, M. M. ROS homeostasis and metabolism: a dangerous liaison in cancer cells. *Cell Death Dis.* **2016,** *7*, e2253.

Paoletti, E.; Bellino, D.; Cassottana, P.; Rolla, D.; Cannella, G. Left ventricular hypertrophy in nondiabetic predialysis CKD. *Am. J. Kidney Dis.* **2005,** *46*, 320–327.

Parihar, A.; Parihar, M. S.; Milner, S.; Bhat, S. Oxidative stress and anti-oxidative mobilization in burn injury. *Burns* **2008,** *34*, 6–17.

Park, S.; Kang, H. J.; Jeon, J. H.; Kim, M. J.; Lee, I. K. Recent advances in the pathogenesis of micro vascular complications in diabetes. *Arch. Pharm. Res.* **2019,** *42*, 252–262.

Patten, D. A.; Germain, M.; Kelly, M. A.; Slack, R. S. Reactive oxygen species: stuck in the middle of neurodegeneration. *J. Alzheimers Dis.* **2010,** *20*, S357–367.

Pattison, D. I.; Davies, M. J. Reactions of myeloperoxidase-derived oxidants with biological substrates: gaining chemical insight into human inflammatory diseases. *Curr. Med. Chem.* **2006,** *13*, 3271–3290.

Pearson-Smith, J. N.; Patel, M. Metabolic dysfunction and oxidative stress in epilepsy. *Int. J. Mol. Sci.* **2017,** *18*, 2365.

Pektas, S. D. Rosacea and oxidative stress: mini review. *COJ Rev. Res.* **2018,** COJRR.000512. 2018. https://doi.org/10.31031/COJRR.2018.01.000512

Pepper, J. R.; Mumby, S.; Gutteridge, J. M. Sequential oxidative damage, and changes in iron-binding and iron-oxidizing plasma antioxidants during cardiopulmonary bypass surgery. *Free Radical Res.* **1994,** *21*, 377–385.

Perl, A. Oxidative stress in the pathology and treatment of systemic lupus erythematosus. *Nat. Rev. Rheumatol.* **2013,** *9*, 674–686.

Phaniendra, A.; Jestadi, D. B.; Periyasamy, L. Free radicals: properties, sources, targets, and their implication in various diseases. *Indian J. Clin. Biochem.* **2015,** *30*, 11–26.

Piatek-Guziewicz, A.; Ptak-Belowska, A.; Przybylska-Felus, M.; Pasko, P.; Zagrodzki, P.; Brzozowski, T.; Mach, T.; Zwolinska-Wcislo, M. Intestinal parameters of oxidative imbalance in celiac adults with extra intestinal manifestations. *World J. Gastroenterol.* **2017,** *23*, 7849–7862.

Piechota-Polanczyk, A.; Fichna, J. Review article: the role of oxidative stress in pathogenesis and treatment of inflammatory bowel diseases. *Naunyn-Schmiedeberg's Arch. Pharmacol.* **2014**, *387*, 605–620.

Pignatelli, P.; Menichelli, D.; Pastori, D.; Violi, F. Oxidative stress and cardiovascular disease: new insights. *Kardiol. Pol.* **2018,** 76, 713–722.

Pizzimenti, S.; Ciamporcero, E.; Daga, M.; Pettazzoni, P.; Arcaro, A.; Cetrangolo, G.; Minelli, R.; Dianzani, C.; Lepore, A.; Gentile, F.; Barrera, G. Interaction of aldehydes derived from lipid peroxidation and membrane proteins. *Front. Physiol.* **2013,** *4*, 242.

Pizzino, G.; Irrera, N.; Cucinotta, M.; Pallio, G.; Mannino, F.; Arcoraci, V.; Squadrito, F.; Altavilla, D.; Bitto, A. Oxidative stress: harms and benefits for human health. *Oxid. Med. Cell. Longevity* **2017,** *2017*, 8416763.

Poli, G.; Leonarduzzi, G.; Biasi, F.; Chiarpotto, E. Oxidative stress and cell signalling. *Curr. Med. Chem.* **2004,** *11*, 1163–1182.

Polonikov, A. V.; Ivanov, V. P.; Bogomazov, A. D.; Solodilova, M. A. Genetic and biochemical mechanisms of involvement of antioxidant defense enzymes in the development of bronchial asthma. *Biomed. Khim.* **2015,** *61*, 427–439.

Poole, L. B. Formation and functions of protein sulfenic acids. *Curr. Protoc. Toxicol.* **2004,** *18*, 17.1.1–17.1.15.

Popolo, A.; Autore, G.; Pinto, A.; Marzocco, S. Oxidative stress in patients with cardiovascular disease and chronic renal failure. *Free Radical Res.* **2013,** *47*, 346–356.

Porter, N. A. Chemistry of lipid peroxidation. *Methods Enzymol.* **1984,** *105*, 273–282.

Powers, S. K.; Ji, L. L.; Kavazis, A. N.; Jackson, M. J. Reactive oxygen species: impact on skeletal muscle. *Compr. Physiol.* **2011,** *1*, 941–969.

Prauchner, C. A. Oxidative stress in sepsis: pathophysiological implications justifying antioxidant co-therapy. *Burns* **2017,** *43*, 471–485.

Prie, B. E.; Voiculescu, V. M.; Ionescu-Bozdog, O. B.; Petrutescu, B.; Liviu, I.; Gaman, L.; Clatici, V. G.; Stoian, I.; Giurcaneanu, C. Oxidative stress and alopecia areata. *J. Med. Life* **2015,** *8*, 43–46.

Qu, J.; Li, Y.; Zhong, W.; Gao, P.; Hu, C. Recent developments in the role of reactive oxygen species in allergic asthma. *J. Thorac. Dis.* **2017,** *9*, E32–E43.

Radak, Z.; Boldogh, I. 8-Oxo-7, 8-dihydroguanine: links to gene expression, aging, and defense against oxidative stress. *Free Radical Biol. Med.* **2010,** *49*, 587–596.

Radi, E.; Formichi, P.; Battisti, C.; Federico, A. Apoptosis and oxidative stress in neurodegenerative diseases. *J. Alzheimers Dis.* **2014,** *42*, S125–S152.

Rahman, I.; Adcock, I. M. Oxidative stress and redox regulation of lung inflammation in COPD. *Eur. Respir. J.* **2006,** *28*, 219–242.

Rahman, K. Studies on free radicals, antioxidants, and co-factors. *Clin. Interv. Aging* **2007,** *2*, 219–236.

Ramanan, V. K.; Saykin, A. J. Pathways to neurodegeneration: mechanistic insights from GWAS in Alzheimer's disease, Parkinson's disease, and related disorders. *Am. J. Neurodegener. Dis.* **2013,** *2*, 145–175.

Reuter, S.; Gupta, S. C.; Chaturvedi, M. M.; Aggarwal, B. B. Oxidative stress, inflammation, and cancer: how are they linked? *Free Radical Biol. Med.* **2010,** *49*, 1603–1616.

Roth, D. M.; Balch, W. E. Modeling general proteostasis: proteome balance in health and disease. *Curr. Opin. Cell Biol.* **2011,** *23*, 126–134.

Rudd, S. G.; Valerie, N. C. K.; Helleday, T. Pathways controlling dNTP pools to maintain genome stability. *DNA Repair (Amst)* **2016,** *44*, 193–204.

Rodríguez-Rodríguez, P.; Ramiro-Cortijo, D.; Reyes-Hernández, C. G.; de Pablo, A. L. L.; González, M. C.; Arribas, S. M. Implication of oxidative stress in fetal programming of cardiovascular disease. *Front. Physiol.* **2018,** *9*, 602.

Sabuncuoğlu, S.; Öztaş, Y.; Çetinkaya, D. U.; Özgüneş, N.; Özgüneş, H. Oxidative protein damage with carbonyl levels and nitro tyrosine expression after chemotherapy in bone marrow transplantation patients. *Pharmacology* **2012,** *89*, 283–286.

Saccà, S. G.; Izzotti, A. Oxidative stress and glaucoma: injury in the anterior segment of the eye. *Prog. Brain Res.* **2008,** *173*, 385–407.

Salama, S. A.; Arab, H. H.; Omar, H. A.; Maghrabi, I. A.; Snapka, R. M. Nicotine mediates hypochlorous acid-induced nuclear protein damage in mammalian cells. *Inflammation* **2014,** *37*, 785–792.

Salazar, J. J.; Van Houten, B. Preferential mitochondrial DNA injury caused by glucose oxidase as a steady generator of hydrogen peroxide in human fibroblasts. *Mutat. Res.* **1997,** *385*, 139–149.

Salsbury, F. R. Jr.; Knutson, S. T.; Poole, L. B.; Fetrow, J. S. Functional site profiling and electrostatic analysis of cysteines modifiable to cysteine sulfenic acid. *Protein Sci.* **2008,** *17*, 299–312.

Sandoval, K. E.; Witt, K. A. Blood-brain barrier tight junction permeability and ischemic stroke. *Neurobiol. Dis.* **2008,** *32*, 200–219.
Schöneich, C. Methionine oxidation by reactive oxygen species: reaction mechanisms and relevance to Alzheimer's disease. *Biochim. Biophys. Acta* **2005,** *1703*, 111–119.
Schöneich, C. Thiyl radicals and induction of protein degradation. *Free Radical Res.* **2016,** *50*, 143–149.
Schieber, M.; Chandel, N. S. ROS function in redox signalling and oxidative stress. *Curr. Biol.* **2014,** *24*, R453–462.
Schwartz, S. M.; Virmani, R.; Rosenfeld, M. E. The good smooth muscle cells in atherosclerosis. *Curr. Atheroscler. Rep.* **2000,** *2*, 422–429.
Sena, L. A.; Chandel, N. S. Pulmonary circulation. *Mol. Cell* **2012,** *48*, 158–167.
Sener, S.; Akbas, A.; Kilinc, F.; Baran, P.; Erel, O.; Aktas, A. Thiol/disulfide homeostasis as a marker of oxidative stress in rosacea: a controlled spectrophotometric study. *Cutan. Ocul. Toxicol.* **2019,** *38*, 55–58.
Shah, D.; Mahajan, N.; Sah, S.; Nath, S. K.; Paudyal, B. Oxidative stress and its biomarkers in systemic lupus erythematosus. *J. Biomed. Sci.* **2014,** *21*, 23.
Shan, X.; Chang, Y.; Lin, C. L. Messenger RNA oxidation is an early event preceding cell death and causes reduced protein expression. *FASEB J.* **2007,** *21*, 2753–2764.
Shang, W.; Huang, L.; Li, L.; Li, X.; Zeng, R.; Ge, S.; Xu, G. Cancer risk in patients receiving renal replacement therapy: a meta-analysis of cohort studies. *Mol. Clin. Oncol.* **2016,** *5*, 315–325.
Sharma, P.; Jha, A. B.; Dubey, R. S.; Pessarakli, M. Reactive oxygen species, oxidative damage, and antioxidative defense mechanism in plants under stressful conditions. *J. Botany* 2012, *2012*, 217037.
Shi, S.; Xue, F. Current antioxidant treatments in organ transplantation. *Oxid. Med. Cell. Longevity* **2016,** *2016*, 8678510.
Shichiri, M. The role of lipid peroxidation in neurological disorders. *J. Clin. Biochem. Nutr.* **2014,** *54*, 151–160.
Shichiri, M.; Yoshida, Y.; Niki, E. Unregulated lipid peroxidation in neurological dysfunction, In: Omega-3 Fatty Acids in Brain and Neurological Health, Cambridge, MA: Academic Press, **2014**, 31–55.
Shirley, R.; Ord, E. N.; Work, L. M. Oxidative stress and the use of antioxidants in stroke. *Antioxidants* **2014,** *3*, 472–501.
Shroff, R.; Speer, T.; Colin, S.; Charakida, M.; Zewinger, S.; Staels, B.; Chinetti-Gbaguidi, G.; Hettrich, I.; Rohrer, L.; O'Neill, F.; McLoughlin, E.; Long, D.; Shanahan, C. M.; Landmesser, U.; Fliser, D.; Deanfield, J. E. HDL in children with CKD promotes endothelial dysfunction and an abnormal vascular phenotype. *J. Am. Soc. Nephrol.* **2014,** *25*, 2658–2668.
Sibal, L.; Agarwal, S. C.; Home, P. D.; Boger, R. H. The role of asymmetric dimethylarginine (ADMA) in endothelial dysfunction and cardiovascular disease. *Curr. Cardiol. Rev.* **2010,** *6*, 82–90.
Silveira, A. S.; Aydos, R. D.; Ramalho, R. T.; Silva, I. S.; Caldas, R. A.; Santos Neto, A. T. D.; Rodrigues, C. T. Oxidative stress effects in the uterus, placenta and fetus of pregnant rats submitted to acute and chronic stress. *Acta Cir. Bras.* **2018,** *33*, 806–815.
Simms, C. L.; Zaher, H. S. Quality control of chemically damaged RNA. *Cell. Mol. Life Sci.* **2016,** *73*, 3639–3653.
Singh, A.; Kukreti, R.; Saso, L.; Kukreti, S. Oxidative stress: a key modulator in neurodegenerative diseases. *Molecules* **2019,** *24*, 1583.

Singh, D.; Malhotra, S. K.; Gujral, U. Role of oxidative stress in autoimmune pathogenesis of vitiligo. *Pigm. Int.* **2016,** *3*, 90–95.

Small, H. Y.; Migliarino, S.; Czesnikiewicz-Guzik, M.; Guzik, T. J. Hypertension: focus on autoimmunity and oxidative stress. *Free Radical Biol. Med.* **2018,** *125*, 104–115.

Smallwood, M. J.; Nissim, A.; Knight, A. R.; Whiteman, M.; Haigh, R.; Winyard, P. G. Oxidative stress in autoimmune rheumatic diseases. *Free Radical Biol. Med.* **2018,** *125*, 3–14.

Smith, S. L.; Andrus, P. K.; Zhang, J. R.; Hall, E. D. Direct measurement of hydroxyl radicals, lipid peroxidation, and blood-brain barrier disruption following unilateral cortical impact head injury in the rat. *J. Neurotrauma.* **1994,** *11*, 393–404.

Smith, R. E.; Tran, K.; Smith, C. C.; McDonald, M.; Shejwalkar, P.; Hara, K. The role of the Nrf2/ARE antioxidant system in preventing cardiovascular diseases. *Diseases* **2016,** *4*, 34.

Sosa, V.; Moliné, T.; Somoza, R.; Paciucci, R.; Kondoh, H.; LLeonart, M. E. Oxidative stress and cancer: an overview. *Ageing Res. Rev.* **2013,** *12*, 376–390.

Sparrow, J. R.; Boulton, M. RPE lipofuscin and its role in retinal pathobiology. *Exp. Eye Res.* **2005,** *80*, 595–606.

Spiers, J. G.; Chen, H. J.; Sernia, C.; Lavidis, N. A. Activation of the hypothalamic-pituitary-adrenal stress axis induces cellular oxidative stress. *Front. Neurosci.* 2015, *8*, 456.

Spiga, F.; Walker, J. J.; Terry, J. R.; Lightman, S. L. HPA axis-rhythms. *Compr. Physiol.* 2014, *4*, 1273–1298.

Sticozzi, C.; Belmonte, G.; Pecorelli, A.; Arezzini, B.; Gardi, C.; Maioli, E.; Miracco, C.; Toscano, M.; Forman, H. J.; Valacchi, G. Cigarette smoke affects keratinocytes SRB1 expression and localization via H_2O_2 production and HNE protein adducts formation. *PLoS One* **2012,** *7*, e33592.

Stucki, D. M.; Ruegsegger, C.; Steiner, S.; Radecke, J.; Murphy, M. P.; Zuber, B.; Saxena, S. Mitochondrial impairments contribute to Spinocerebellar ataxia type 1 progression and can be ameliorated by the mitochondria-targeted antioxidant MitoQ. *Free Radical Biol. Med.* **2016,** *97*, 427–440.

Sun, Y.; Oberley, L. W. Redox regulation of transcriptional activators. *Free Radical Biol. Med.* **1996,** *21*, 335–348.

Sung, C. C.; Hsu, Y. C.; Chen, C. C.; Lin, Y. F.; Wu, C. C. Oxidative stress and nucleic acid oxidation in patients with chronic kidney disease. *Oxid. Med. Cell. Longevity* **2013,** *2013*, 301982.

Sundaram, S. B. P.; Nagarajan, S.; Devi, A. J. M. Chronic kidney disease—effect of oxidative stress. *Chin. J. Biol.* **2014,** *2014*, 216210.

Sureshbabu, A.; Bhandari, V. Targeting mitochondrial dysfunction in lung diseases: emphasis on mitophagy. *Front. Physiol.* **2013,** *4*, 384.

Sutherland, B. M.; Bennett, P. V.; Sidorkina, O.; Laval, J. Clustered damages and total lesions induced in DNA by ionizing radiation: oxidized bases and strand breaks. *Biochemistry* **2000,** *39*, 8026–8031.

Szczepańska-Szerej, A.; Kurzepa, J.; Wojczal, J.; Stelmasiak, Z. Simvastatin displays an antioxidative effect by inhibiting an increase in the serum 8-isoprostane level in patients with acute ischemic stroke: brief report. *Clin. Neuropharmacol.* **2011,** *34*, 191–194.

Tabas, I. Macrophage death and defective inflammation resolution in atherosclerosis. *Nat. Rev. Immunol.* **2010,** *10*, 36–46.

Taskintuna, I.; Elsayed, M. E. A. A.; Schatz, P. Update on clinical trials in dry age-related macular degeneration. *Middle East Afr. J. Ophthalmol.* **2016,** *23*, 13–26.

Tchirikov, M.; Schlabritz-Loutsevitch, N.; Maher, J.; Buchmann, J.; Naberezhnev, Y.; Winarno, A. S.; Seliger, G. Mid-trimester preterm premature rupture of membranes (PPROM): aetiology, diagnosis, classification, international recommendations of treatment options and outcome. *J. Perinat. Med.* **2018,** *46*, 465–488.

Tian, T.; Wang, Z.; Zhang, J. Pathomechanisms of oxidative stress in inflammatory bowel disease and potential antioxidant therapies. *Oxid. Med. Cell. Longevity* **2017,** *2017*, 4535194.

Torlakovic, E. E.; Nielsen, S.; Vyberg, M.; Taylor, C. R. Getting controls under control: the time is now for immunohistochemistry. *J. Clin. Pathol.* **2015,** *68*, 879–882.

Toruan, T. L.; Nopriyati, T.; Sari, Y. M. The relationship between serum lipid profile and sebum secretion in seborrheic dermatitis patients. *Int. J. Health Sci. Res.* **2017,** *7*, 138–143.

Treps, L.; Conradi, L. C.; Harjes, U.; Carmeliet, P. Manipulating angiogenesis by targeting endothelial metabolism: hitting the engine rather than the drivers—a new perspective? *Pharmacol. Rev.* **2016,** *68*, 872–887.

Trougakos, I. P.; Sesti, F.; Tsakiri, E.; Gorgoulis, V. G. Non-enzymatic post-translational protein modifications and proteostasis network deregulation in carcinogenesis. *J. Proteomics* **2013,** *92*, 274–298.

Trueb, R. M.; Henry, J. P.; Davis, M. G.; Schwartz, J. R. Scalp condition impacts hair growth and retention via oxidative stress. *Int. J. Trichol.* **2018,** *10*, 262–270.

Tsai, N. W.; Chang, Y. T.; Huang, C. R.; Lin, Y. J.; Lin, W. C.; Cheng, B. C.; Su, C. M.; Chiang, Y. F.; Chen, S. F.; Huang, C. C.; Chang, W. N.; Lu, C. H. Association between oxidative stress and outcome in different subtypes of acute ischemic stroke. *BioMed Res. Int.* **2014,** *2014*, 256879.

Tuo, Q. R.; Ma, Y. F.; Chen, W.; Luo, X. J.; Shen, J.; Guo, D.; Zheng, Y. M.; Wang, Y. X.; Ji, G.; Liu, Q. H. Reactive oxygen species induce a Ca (2+)-spark increase in sensitized murine airway smooth muscle cells. *Biochem. Biophys. Res. Commun.* **2013,** *434*, 498–502.

Ullah, A.; Khan, A.; Khan, I. Diabetes mellitus and oxidative stress—a concise review. *Saudi Pharm. J.* **2016,** *24*, 547–553.

Ursini, F.; Maiorino, M.; Forman, H. J. Redox homeostasis: The golden mean of healthy living. *Redox Biol.* **2016,** *8*, 205–215.

Valko, M.; Leibfritz, D.; Moncol, J.; Cronin, M. T. D.; Mazur, M.; Telser, J. Free radicals and antioxidants in normal physiological function and human disease. *Int. J. Biochem. Cell Biol.* **2007,** *39*, 44–84.

Van Eeden, S. F.; Sin, D. D. Oxidative stress in chronic obstructive pulmonary disease: a lung and systemic process. *Can. Respir. J.* **2013,** *20*, 27–29.

Van Houten, B.; Santa-Gonzalez, G. A.; Camargo, M. DNA repair after oxidative stress: current challenges. *Curr. Opin. Toxicol.* **2018,** *7*, 9–16.

Vieira-Potter, V. J.; Karamichos, D.; Lee, D. J. Ocular complications of diabetes and therapeutic approaches. *BioMed Res. Int.* **2016,** *2016*, 3801570.

Vinson, J. A. Oxidative stress in cataracts. *Pathophysiology* **2006,** *13*, 151–162.

Volpe, C. M. O.; Villar-Delfino, P. H.; dos Anjos, P. M. F.; Nogueira-Machado, J. A. Cellular death, reactive oxygen species (ROS) and diabetic complications. *Cell Death Dis.* **2018,** *9*, 119.

Wagener, F. A. D. T. G.; Carels, C. E.; Lundvig, D. M. S. Targeting the redox balance in inflammatory skin conditions. *Int. J. Mol. Sci.* **2013,** *14*, 9126–9167.

Wang, Z.; Dou, X.; Gu, D.; Shen, C.; Yao, T.; Nguyen, V.; Braunschweig, C.; Song, Z. 4-Hydroxynonenal differentially regulates adiponectin gene expression and secretion via activating PPARγ and accelerating ubiquitin-proteasome degradation. *Mol. Cell. Endocrinol.* **2012,** *349*, 222–231.

Wang, H. C.; Lin, Y. J.; Shih, F. Y.; Chang, H. W.; Su, Y. J.; Cheng, B. C.; Su, C. M.; Tsai, N. W.; Chang, Y. T.; Kwan, A. L.; Lu, C. H. The role of serial oxidative stress levels in acute traumatic brain injury and as predictors of outcome. *World Neurosurg.* **2016,** *87*, 463–470.

Wang, X.; Simpkins, J. W.; Dykens, J. A.; Cammarata, P. R. Oxidative damage to human lens epithelial cells in culture: estrogen protection of mitochondrial potential, ATP, and cell viability. *Invest. Ophthalmol. Vis. Sci.* **2003,** *44*, 2067.

Wargo, J. A.; Reuben, A.; Cooper, Z. A.; Oh, K. S.; Sullivan, R. J. Immune effects of chemotherapy, radiation, and targeted therapy and opportunities for combination with immunotherapy. *Semin. Oncol.* **2015,** *42*, 601–616.

Watts, Z. I.; Easton, C. J. Peculiar stability of amino acids and peptides from a radical perspective. *J. Am. Chem. Soc.* **2009,** *131*, 11323–11325.

Weber, D.; Milkovic, L.; Bennett, S. J.; Griffiths, H. R.; Zarkovic, N.; Grune, T. Measurement of HNE-protein adducts in human plasma and serum by ELISA—Comparison of two primary antibodies. *Redox Biol.* **2013,** *1*, 226–233.

Anonymous. What are the risks of pre-eclampsia & eclampsia to the mother? Eunice Kennedy Shriver National Institute of Child Health and Human development. **2018**. (viewed May 22, 2019). https://www.nichd.nih.gov/health/topics/pre-eclampsia/conditioninfo/risk-mother.

Willi, J.; Küpfer, P.; Evéquoz, D.; Fernandez, G.; Katz, A.; Leumann, C.; Polacek, N. Oxidative stress damages rRNA inside the ribosome and differentially affects the catalytic center. *Nucleic Acids Res.* **2018,** *46*, 1945–1957.

Williams, H.; Mill, C. A.; Monk, B. A.; Hulin-Curtis, S.; Johnson, J. L.; George, S. J. Wnt2 and WISP-1/CCN4 induce intimal thickening via promotion of smooth muscle cell migration. *Arterioscler Thromb. Vasc. Biol.* **2016,** *36*, 1417–1424.

Woods, A. A.; Linton, S. M.; Davies, M. J. Detection of HOCl-mediated protein oxidation products in the extracellular matrix of human atherosclerotic plaques. *Biochem. J.* **2003,** *370*, 729–735.

Wright Jr, E.; Scism-Bacon, J. L.; Glass, L. C. Oxidative stress in type 2 diabetes: the role of fasting and postprandial glycaemia. *Int. J. Clin. Pract.* **2006,** *60*, 308–314.

Wu, D.; Cederbaum, A. I. Alcohol, oxidative stress, and free radical damage. *Alcohol Res. Health* **2003,** *27*, 277–284.

Wurtmann, E. J.; Wolin, S. L. RNA under attack: cellular handling of RNA damage. *Crit. Rev. Biochem. Mol. Biol.* **2009,** *44*, 34–49.

Xu, J.; Lupu, F.; Esmon, C. T. Inflammation, innate immunity and blood coagulation. *Hamostaseologie* **2010,** *30*, 5–6, 8–9.

Yakes, F. M.; Van Houten, B. Mitochondrial DNA damage is more extensive and persists longer than nuclear DNA damage in human cells following oxidative stress. *Proc. Nat. Acad. Sci. USA* **1997,** *94*, 514–519.

Yamazaki, K.; Fukuda, K.; Matsukawa, M.; Hara, F.; Matsushita, T.; Yamamoto, N.; Yoshida, K.; Munakata, H.; Hamanishi, C. Cyclic tensile stretch loaded on bovine chondrocytes causes depolymerization of hyaluronan involvement of reactive oxygen species. *Arthritis. Rheum.* **2003,** *48*, 3151–3158.

Yan, L. J. Pathogenesis of chronic hyperglycaemia: from reductive stress to oxidative stress. *J. Diabetes Res.* **2014,** *2014*, 137919.

Yang, X.; Li, Y.; Li, Y.; Ren, X.; Zhang, X.; Hu, D.; Gao, Y.; Xing, Y.; Shang, H. Oxidative stress-mediated atherosclerosis: mechanisms and therapies. *Front. Physiol.* **2017,** *8*, 600.

Yasui, M.; Kanemaru, Y.; Kamoshita, N.; Suzuki, T.; Arakawa, T.; Honma, M. Tracing the fates of site-specifically introduced DNA adducts in the human genome. *DNA Repair (Amst)* **2014,** *15*, 11–20.

Yılmaz, E.; Berdeli, S. M. A. Endothelial nitric oxide synthase (eNOS) gene polymorphism in early term chronic allograft nephropathy. *Transpl. Proc.* **2009,** *41*, 4361–4365.
Yoo, S. J.; Go, E.; Kim, Y. E.; Lee, S.; Kwon, J. Roles of reactive oxygen species in rheumatoid arthritis pathogenesis. *Int. J. Rheum. Dis.* **2016,** *23*, 340–347.
Yu, S. L.; Lee, S. K.; Johnson, R. E.; Prakash, L.; Prakash, S. The stalling of transcription at abasic sites is highly mutagenic. *Mol. Cell. Biol.* **2003,** *23*, 382–388.
Yu, M.; Kim, Y. J.; Kang, D. H. Indoxyl sulfate-induced endothelial dysfunction in patients with chronic kidney disease via an induction of oxidative stress. *Clin. J. Am. Soc. Nephrol.* **2011,** *6*, 30–39.
Zakkar, M.; Guida, G.; Suleiman, M. S.; Angelini, G. D. Cardiopulmonary bypass and oxidative stress. *Oxid. Med. Cell Longevity* **2015,** *2015*, 189863.
Zhang, K. Integration of ER stress, oxidative stress and the inflammatory response in health and disease. *Int. J. Clin. Exp. Med.* **2010,** *3,* 33–40.
Zhou, J.; Fleming, A. M.; Averill, A. M.; Burrows, C. J.; Wallace, S. S. The NEIL glycosylases remove oxidized guanine lesions from telomeric and promoter quadruplex DNA structures. *Nucleic Acids Res.* **2015,** *43*, 4039–4054.

CHAPTER 9

Oxidative Stress as a Detrimental Factor in Various Clinical Pathology

PRIYANKA SAHA, ANUPAM DAS TALUKDAR*, and RAJAT NATH

Department of Life Science & Bioinformatics, Assam University, Silchar 788011, Assam, India

Corresponding author. E-mail: anupam@bioinfoaus.ac.in

ABSTRACT

Oxidative stress, chiefly contributed by reactive oxygen species (ROS), has been the role model in pathogenesis of numerous diseases. For the precedent, 40 years oxidative stress has been progressively more recognized as a causative factor in almost all forms of clinical manifestations. Presently, we tried to highlight oxidative stress. In the preceding years approach to oxidative stress has broadened, as a result now there are addressed as the condition occurring in the gene level and the ways the gene regulation are carried out. On the footprint of this thought, novel focus is on various transcription factors. It is designated as—master regulator of the antioxidant response that regulates the prime expression of responsible genes in the oxidation process, antioxidant enzymes, genes that manage various processes. In cancer, ROS reports for its resistance toward apoptosis, genomic instability, and uncontrolled proliferation. The ancestral connecting link of the genes under the regular direct of Nrf2 suggests that the inflammatory responses may represent the highest demand for amplified antioxidant defence, apart from continuous oxidative stress. This altogether hits for a well-to-do research field straddling from and biochemistry to cell biology landing into redox knowledge, environmental medicine, and molecular-based redox medicine. Use of bioenzymes, antioxidant supplements, and ROS-generating NADPH oxidases inhibitors is a rational restorative intercession for many diseases. We thus here try to review such attempts, progress, and challenges. Lastly, we focus on main phytochemicals with natural antioxidant potentialities, which have a promising therapeutic alternative, considering additional element to combat oxidative stress.

9.1 INTRODUCTION

Reactive oxygen species (ROS) have become the most significant element, which is the reason for almost all the major diseases these days. In simple terms, they are basically molecular oxygen evolved from various free radicals and reactive molecules. Free radical creates a toxic environment that subsequently hinders the normal physiological status of an individual. Such reactive molecules are generally produced from various biochemical and crucial metabolic pathways such as aerobic respiration that ultimately leads to deleterious events leading to various clinical conditions. Initially, the macrophages were solely considered to be the reason for producing various kinds of ROS. Recent researches have brought into light many other aspects for the production of the ROS. ROS has diverse role in gene expression; apoptosis, cell signaling, and also cell signaling cascades activation (Hancock et al., 2001). ROS interestingly exhibit both as intracellular and intercellular messengers.

9.1.1 TYPES OF ROS

Majority of the ROS are generally the product of the oxidation process taking place during aerobic mitochondrial electron transport system. Atomic oxygen contains two main unpaired electrons in their separate orbits in its outermost electron shell. This electronic structural configuration makes the oxygen prone to formation of various excited radicals. The chronological decrease of oxygen molecule is mediated by addition of subsequent electrons to the formation of various ROS members mainly: hydrogen peroxide; hydroxyl ion; superoxide, and hydroxyl radical (Figure 9.1).

9.1.2 CELLULAR DEFENSE AGAINST ROS

Clearance of ROS has become an integral issue for the survival of the aerobic species. Such clearance is essential for the normal sustainance of the living organism and to maintain a healthy condition. As a result of various endogenous mechanism, build-up toxic radicals are flushed out of the system either by neutralization, chelating effects, or by hydrolyzing effects. This arising imbalance occurring in prooxidant state and free radicals give rise to a condition called as oxidative stress.

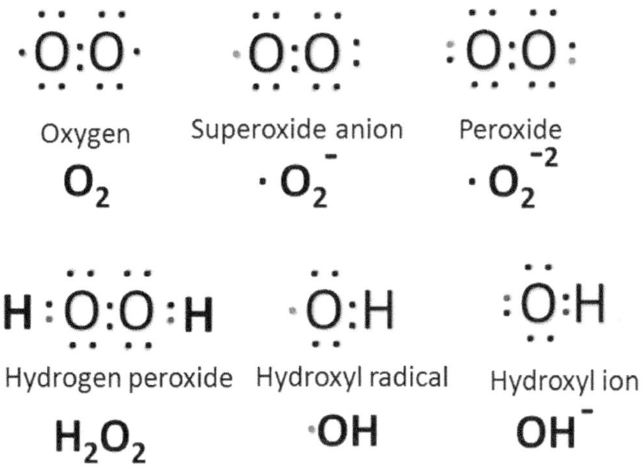

FIGURE 9.1 Various forms of free radicals leading to oxidative stress.

Cells have the immense potentialities by virtue of which it tends to eliminate harmful build-up endogenous free radicals in our system. Superoxide dismutase (SOD), an important character catalyzes the superoxide anions converted into a single hydrogen peroxide (H_2O_2) molecule and also a single molecule of oxygen (O_2) (McCord and Fridovich, 1968). In most eukaryotic peroxisome cells, the catalases enzyme converts molecules of H_2O_2 to water and O_2, and in this way generally detoxification process is initiated by SOD. Glutathione peroxidase is another well-known group of enzymes that contains selenium, which catalyzes the breakdown of the hydrogen peroxide molecule, also to the organic peroxide resulting in production of alcohol. A number of small molecules that are nonenzymatic in nature, plays a pivotal role in the antioxidant activity. Glutathione, another antioxidant is the most important intracellular defence alongside the toxic effects of the produced ROS. This glutamyl-cysteinyl-glycine sequenced tripeptide provides a rare sulphydryl group, that behaves as an huge volumptuous target for interaction. ROS molecules react and thus oxidize the endogenous glutathione, but the reduced form is regenerated in a redox by an NADPH-dependent reductase. Ascorbic acid better known as vitamin C is generally a water-soluble molecule that is competent of reducing the build-up concentrated ROS in the system, while tocopherols commonly called as vitamin E that is a lipid-soluble molecule has been suggested as a same, playing almost comparable role in cellular membranes. Ratio of the oxidized form of glutathione and the reduced form (GSH) is a dynamic marker of the extent of the oxidative stress of an organism (Jones, 2002).

9.1.3 OXIDATIVE STRESS AND DISEASES

Cardiovascular diseases (CVDs) have become the cause of mortality and morbidity in the elderly age group people with 50–60 ages and specifically atherosclerosis that turns out to have a crucial role (Testa et al., 2010). Many scientific reports have suggested that older age is susceptible to oxidative stress-mediated diseases. It is due to their inability to fight back to the built-up oxidant molecule. When the reason for this action was studied it was seen that a decrease in heart tolerance to oxidative stress is possibly the reason and due to reduction in the antioxidant enzymes concentration, this cardiovascular (CV) alterations takes place.

Diabetes mellitus is another metabolic disease directly related to free radicals where reduced antioxidant potential results in macrovascular and microvascular complications (Bashan et al., 2009). The actual mode of action by which it renders oxidative stress is possibly by the development of diabetic complications is only known to some far extent. Oxidant stress in type 2 diabetes (T2D) further accelerates prothrombotic reactions, leading to most fatal CV complications (De Cristofaro et al., 2003). Diabetes damages can thus be considered as tissue-oxidative-damaging effects of chronic hyperglycemia (Figure 9.2).

Chronic obstructive pulmonary disease is one important disease that is responsible for maximum number of deaths worldwide (Murray and Lopez, 1997) and in each passing days (Chapman et al., 2006). Many signaling pathways and mechanistic studies revealed several clinical conditions like shortening of the telomere length, immune senescence, oxidative stress, and cellular inflammation (Venkataraman et al., 2013).

Oxidative stress also imparts its implication in diseases like chronic kidney diseases (CKD), which is carried by conditions like renal ischemia and glomerular damage and indirectly in situations like hypertension, inflammation, and endothelial dysfunction (Balasubramanian, 2013). CKD patients generally are associated with chronic inflammation and are characterized by the activation of certain classes of monocytes and natural killer cells. Such active inflammatory cells will thereby increase the secretion of NADPH

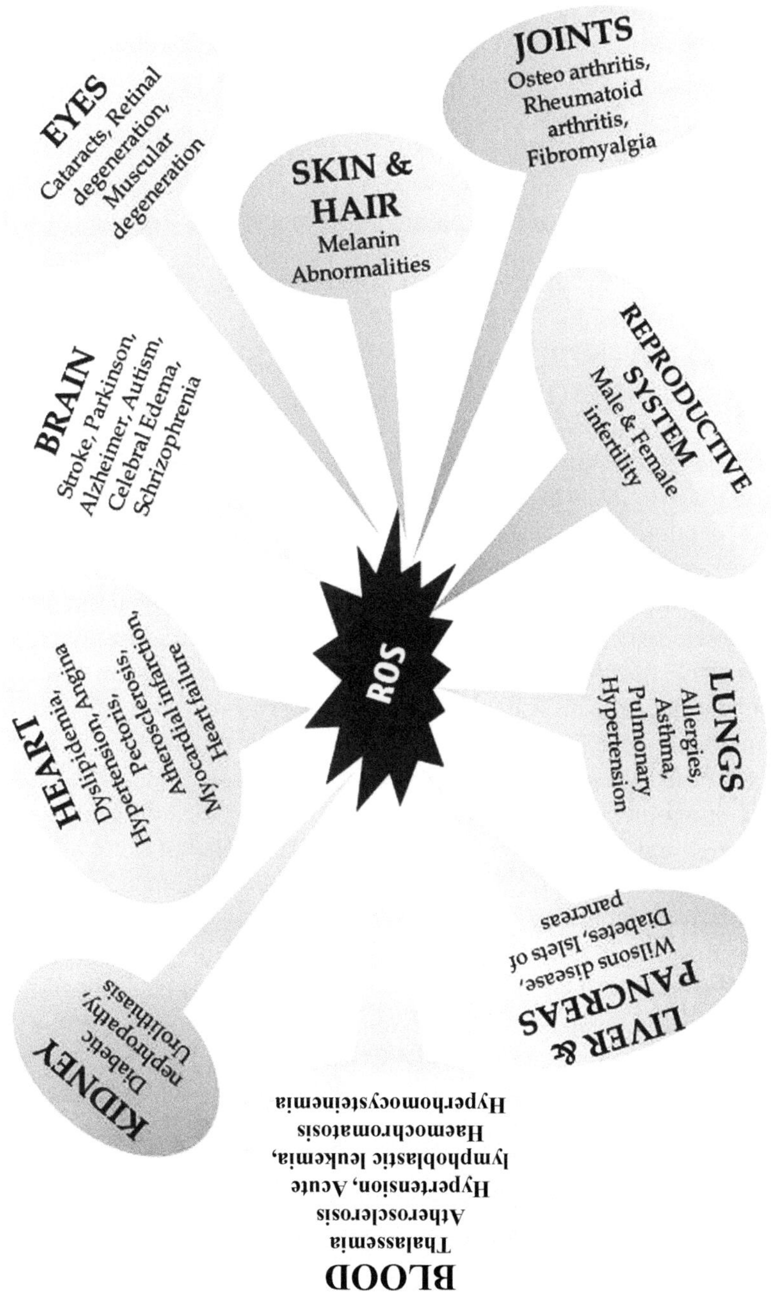

FIGURE 9.2 Implications of oxidative stress.

oxidase and MPO that enhances the formation of ROS (Putri and Thaha, 2016). Leukocytes of such victim produce radicals like superoxide anions, which inactivate NO, dropping the ability of blood vessels dilatation that contributes to hypertension.

Many research reports have established the direct association flanked by carcinogenesis and chronic inflammation (Khansari et al., 2009). The main chemical effectors of the inflammatory response are free-radical species resulting from oxidative stress.

9.2 OXIDATIVE STRESS AND CENTRAL NERVOUS SYSTEM (CNS) (NEURODEGENERATIVE DISEASES)

Neurodegenerative diseases affect millions of people worldwide. Neurodegenerative diseases are due to neurodegeneration that is the enlightened loss of function or structure of nerve cells, including neuronal death. Some common neurological disorders occurred due to the oxidative stress that are Alzheimer's disease, Parkinsons disease, amyolotrophic lateral sclerosis, and multiple sclerosis. Oxidative stress due to excessive ROS, reactive nitrogen species (NOS) formation or dysfunction of antioxidant system is a leading root of neurodegenerative diseases amongst the other various pathophysiologies involved. ROS and NOS are also formed in the mitochondria during ATP synthesis by oxidative phosphorylation (Thanan et al., 2015). Though in CNS or in other organs, a little amount of these ROS and NOS are considered as essential for normal development and function, but excessive amount have the ill effects. On CNS excessive amount of these radicals having the harmful effects including anxiety disorders and depression, is beginning to be recognized. The brain is vastly inclined to oxidative stress as it having rich lipid content and high oxygen consumption capacity (Fischer and Maier, 2015). Therefore, damages due to oxidative stress to the brain have a sturdy potential to negatively affect inusual CNS functions. High ROS concentrations apparently reduce long-term potentiation and synaptic signaling and brain plasticity mechanisms. Oxidative stress induced the excessive release of ROS and in result it fastens up the cellular damage (Chen et al., 2016; Parimisetty et al., 2016). That is why it is also often called as a self-propagating phenomenon; damaged macromolecules themselves may behave as and/or become ROS. Consequently, the brain, with its rich lipid content, high energy demand, and weak antioxidant capacity becomes an easy target of excessive oxidative insult (Huang et al., 2016). In the brain, phospholipids are the predominantly susceptible entities for ROS-mediated peroxidation, but DNA and proteins too are targeted by ROS, which becomes mostly problematic during aging, as aged brains have been conveyed to shows high levels of oxidative stress-induced mutations in the mitochondrial DNA. Therefore, ROS gathering is a cellular threat that, if it exceeds or bypasses neutralizing mechanisms, can results in substantial neuronal impairment (Leszek et al., 2016). Biochemically, it is obvious that different neurons have diverse levels of susceptibility to oxidative stress. As an example, amygdala, hippocampus, and cerebellar granule cells have been described as the most vulnerable to oxidative stress in some research findings. Hippocampus looks to be at the hot seat, and it seems that this brain province undergoes major biochemical alterations that eventually regulate neuronal function and connections (Islam, 2017; Salim, 2017). Within the hippocampus, it is well known that the dentate gyrus–cornu ammonis system exhibits structural plasticity with regenerative/remodeling capacity (Figure 9.3).

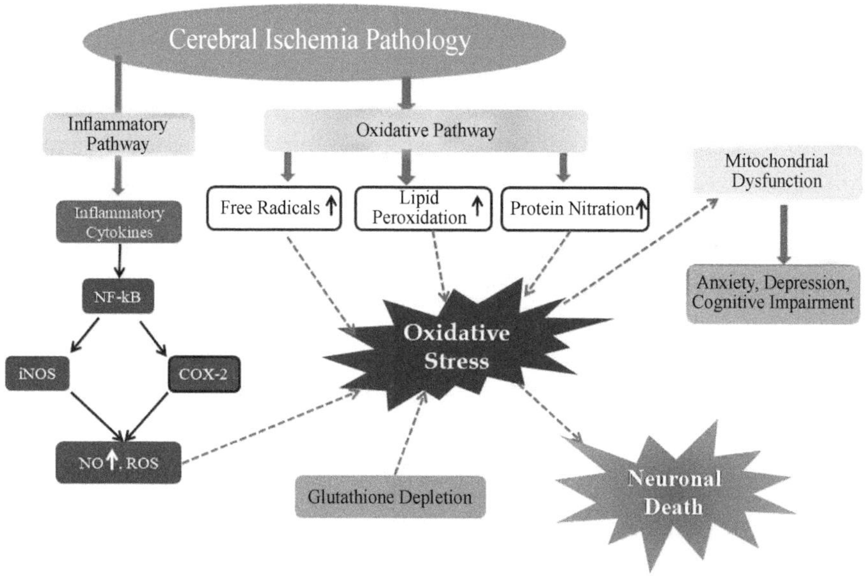

FIGURE 9.3 Oxidative stress-mediated neuronal death and other neuropsychiatric behavioral depletion: there are two pathways involved for the formation of oxidative stress in brain. In inflammatory pathway, inflammatory cytokines activate NF-kB that mediates to produce inducible nitric oxide synthase and icyclooxygenase-2, which produces NOS and ROS, the main elements responsible for oxidative stress. In oxidative pathway, free radicals, lipid peroxidation, and protein nitration occurs, which results in oxidative stress, results to neuronal death. Apart from these, glutathione depletion also results oxidative stress in brain and due to that mitochondrial dysfunction occurs, which results psychomotor abnormalities along with anxiety, depression, and cognitive impairments.

9.3 OXIDATIVE STRESS AND CVDS

CVDs are the diseases associated with blood vessels and heart. Myocardial infarction and angina caused in result of coronary artery disease are also includes under the CVD. Other CVDs include heart failure, stroke, hypertensive heart disease, heart arrhythmia, rheumatic heart disease, peripheral artery disease, cardiomyopathy, valvular heart disease, congenital heart disease, carditis, thromboembolic disease, aortic aneurysms, and venous thrombosis (Frijhoff et al., 2015; Skibska and Goraca, 2015). High blood pressure, obesity, aging, and insulin resistance are concomitant with the development of CVDs, and all these aspects are associated with metabolic syndrome. CVDs are the foremost cause of death world widely. CVD give rise to 17.9 million deaths (32.1%) in 2015, up from 12.3 million (25.8%) in 1990 (He and Zuo, 2015;Niemann et al., 2017). Deaths due to CVD are more frequent and have been growing in much of the under developing countries. Stroke and coronary artery disease excuses 75% of CVD deaths in females and 80% of CVD deaths in males. It has been widely verified that oxidative stress and inflammation play crucial roles in the pathogenesis of CVDs (Santilli et al., 2015). Various studies have revealed that progression of CVDs occurs mainly due to ROS and inflammation. NADPH oxidase, XOX, NOS, ROS, and NOS-mediated oxidative stress are directly related to Dyslipidaemia, T2D, Hypertension, Sympathetic activation, and so many other pathophysiological disorders (Santilli et al., 2015; Niemann et al., 2017). These diseases are indirectly or directly related to inflammation in various cells, which are related to blood vessels and heart. Inflammation to those cells leads to several kinds of CVDs. This is reinforced and supported by the fact that several threats such as diabetes mellitus, aging, hypercholesterolemia, and smoking are responsible for CVDs, and unavoidably lead to augmented generation of ROS, causing in oxidative stress. Moreover, tenacious oxidative stress can encourage disturb blood flow, inflammation, and remodel arterial wall structures, leading to CVDs (Espinosa-Diez et al., 2015; Münzel et al., 2015). In complex syndrome of heart failure as well as cardiac muscle dysfunctions, proinflammatory cytokines are also involved. Inflammation is the main factor in all phases of coronary disease including the initiation and progression of thrombosis (Robson et al., 2018), plaque rupture (Siti et al., 2015), atherosclerotic plaque especially in recurrent thrombosis; oxidative stress is the main culprit who plays a major roles (Figure 9.4).

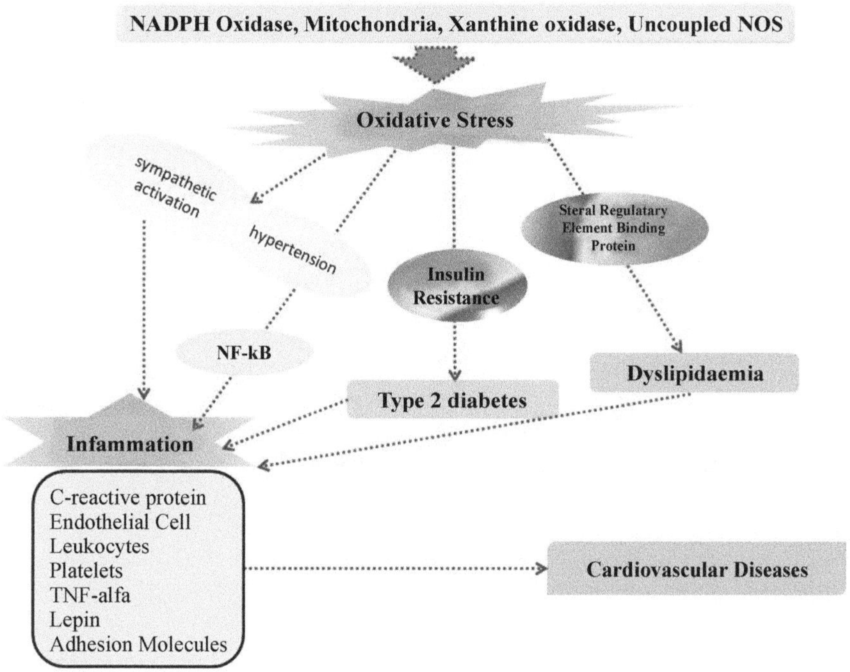

FIGURE 9.4 Oxidative stress-mediated cardiovascular diseases (CVDs): NADPH oxidase, XOX, NOS, ROS, and NOS-mediated oxidative stress are directly related to Dyslipidaemia, Type 2 Diabetes, Hypertension, Sympathetic activation, and so many other pathophysiological disorders. These diseases directly or indirectly related to inflammation in various cells, which are related to heart and blood vessels. Inflammation to those cells result in various kinds of CVDs.

9.4 OXIDATIVE STRESS AND AGING

ROS are the product of various biochemical reactions taking place in the body. Accumulation of those species create an environment that thereby promotes a series of expressive modulators pathways for abnormal cells proliferation (Reinisalo et al., 2015; Zhang et al., 2017). If such continues the situation, there occur sequential cellular damages. Such damages give rise to the process of aging implicated by localized cell death (Bonomini et al., 2015; Rinnerthaler et al., 2015; Zhang et al., 2015). So, basically, the aging and ultimately degenerative diseases come into play with prolonged severe oxidative stress. A detailed relation of the oxidative stress with that of the diseases has been clearly depicted in Figure 9.5.

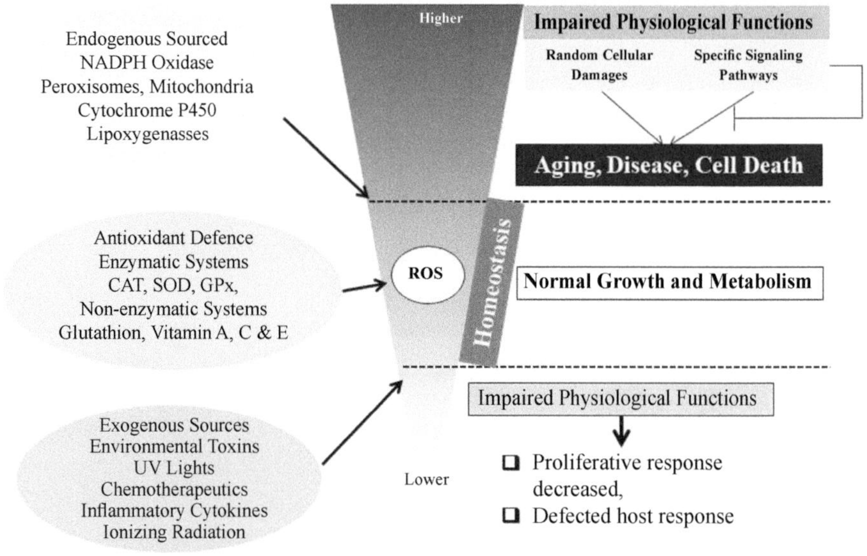

FIGURE 9.5 Relation between oxidative stress and aging.

9.5 OXIDATIVE STRESS AND CANCER

Very well-known Denham Harman expressed the view of "free-radical theory" narrating the upraise of free radicals in the process of various endogenous biochemical interaction in the living system and cumulative accumulation of the deleterious products that are responsible for various lethal diseases (Chen et al., 2015; Chikara et al., 2018). Initially, the concept of endogenous free radicals was not accepted until the discovery of a pronounced free radical, SOD, an enzyme came into the system. It's role in removal of superoxide anions provided a mechanistic approach for Harman's laid hypothesis. As presently it is very obvious that mitochondria being the power of cell consumes maximum number of intercellular oxygen for its necessary biochemical reactions and so there comes into play the free-radical theory of aging (Rani et al., 2016). The rate-of-living hypothesis is now regarded as the direct proportional elements with the free-radical theory of aging. So, it has been implicated that the more amount is the metabolic process, bigger is the production of ROS and simultaneously, higher is the disease rate that lands to cancer with shortening of the life span (Poprac et al., 2017). The free-radical theory of aging originally signified that the ROS targets were cumulative resulting in proliferation of the cells leading to transformation and ultimately cancer (Mileo and Miccadei, 2016). Though oxidants may in general function in various uncontrolled and basic physiological pathways yet several existing evidences considers ROS as a particular significant signaling element under both pathophysiological and physiological conditions. Increase in endogenous levels of oxidants has basically two important effects: cell components detioriation and the signaling pathway activation, all leading to a state of transformation subsequently, which is cancerous (Figure 9.6).

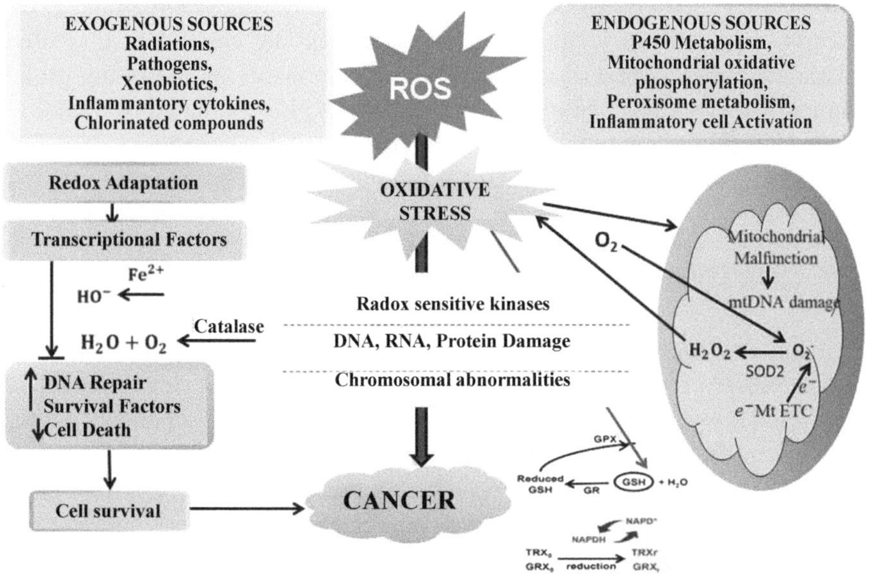

FIGURE 9.6 Relation of oxidative stress and cancer. Various endogenous and exogenous agent produces range of free radicals ultimately resulting in chromosomal aberrations, protein damage, DNA mutation, and formation of adducts. All of which give rise to transformed cells and subsequently cancer.

9.6 ROLE OF NATURAL PRODUCT IN OXIDATIVE STRESS

Polyphenols are among the finest source of phyto-based natural products, which serve as an significant groups of natural antioxidants having immense medicinal properties that serve as potent chemopreventive agents. They are generally prime Mediterranean diet, which comprises of fruits, nuts, grains, leafy vegetables, tea, and their derived products and also various other beverages. Various important clinical, nutritional, and epidemiological studies support the indication that dietary phenols boost up tremendously the human system possibly by lowering risk and minimizing the occurrence of various degenerative diseases like CVDs, metabolic disorders, cancer, and many more. Polyphenols, with their unique potent ring structure, have their immense bioactivity. Due to their potent aromatic ring and conjugated structure with various hydroxyl groups make such candidate hydrogen atom or good electron donors. Such property aids in neutralizing excited free radicals and several other ROS and also active as antiinflammatory and antioxidant molecules. Dietary intake of phenolic compound is generally recommended to about 1 g per day for a normal healthy build up. Recent in vitro and in vivo research reports have focused on having appropriate amount of phenolics enhancing antioxidant enzyme activities most probably by inhibiting proinflammatory cytokines and directly attenuating NFkB-mediated or oxidative stress-induced inflammatory signaling pathways. A cleared list of a few phytochemicals with its antioxidant activity is depicted in Table 9.1.

9.7 CONCLUSION

Growing evidence from investigations and cumulative data from clinical studies clearly pictorizes the role oxidative stress in the development of diseases. However, a better understanding of the ROS-dependent signal mechanism pathway, their integration, and localization of ROS-dependent signaling pathways in pathophysiology is a precondition for effectual pharmacological interventions. The development of a new lead of antioxidants that are targeted to specific cellular organelle may help in combating diseases. Another significant step toward effectual treatment will be the development of specific biomarkers that can be used to assess the oxidative stress that underlie a range of pathologies.

KEYWORDS

- **oxidative stress**
- **antioxidant**
- **NF-κB/IκB, Nrf2/Keap1**
- **free radical**

TABLE 9.1 Phytochemicals Showing Antioxidant Activity

Sl No.	Free Radicals	Disease	Phytochemicals	Plant Name	References
	Hydrogen peroxide	Alzheimer's disease (AD)	Allithiamin	*Allium sativum*	(Frijhoff et al., 2015; He and Zuo, 2015)
			Tocopherol	*Anacardium occidentale, Capsicum annuum, Ficus carica*	(He and Zuo, 2015; Huang et al., 2016)
			Thiamine	*Abelmoschus esculentus, Achilleamille folium, Acacia farnesiana*	(Münzel et al., 2015; Niemann et al., 2017)
		Parkinson's disease (PD)	Harmaline	*Passiflora incarnate, Peganum harmala, Banisteriopsis caapi*	(Parimisetty et al., 2016;Niemann et al., 2017)
			Hyoscyamine	*Lactuca sativa, Datura innoxia, Hyoscyamus niger*	(Reinisalo et al., 2015; Santilli et al., 2015; Robson et al., 2018)
		Arteriosclerosis	Ruscoside-a Silicon	*Actaearacemosa, Allium sativum*	(Salim, 2017; Robson et al., 2018)
		Thrombosis	Ferulic acid	*Achilleamillefolium, Actaeadahurica*	(Siti et al., 2015; Sies et al., 2017)
			Resveratrol	*Eucalyptus wandoo, Vaccinium corymbosum*	(Skibska and Goraca, 2015; Thanan et al., 2015)
	Superoxide anion	liver cancer Haemolytic anaemia	Fumaric acid	*Averrhoa carambola, Brassica oleracea*	(Thanan et al., 2015) (Mileo and Miccadei, 2016)
		cataracts	Ascorbic acid	*Acacia farnesiana, Achilleamillefolium, Allium sativum*	(Espinosa-Diez et al., 2015; He and Zuo, 2015; Zhang et al., 2017)
			Cysteine	*Allium cepa, Aloe vera, Citrulluslanatus*	(Chen et al., 2015)
	Nitric oxide (NO)	Alzheimer's disease (AD)	Allithiamin	*Allium sativum*	(Huang et al., 2016)

TABLE 9.1 *(Continued)*

Sl No.	Free Radicals	Disease	Phytochemicals	Plant Name	References
			Tocopherol	*Anacardiumoccidentale, Capsicum annuum, Ficuscarica*	(He and Zuo, 2015)
			Thiamin	*Abelmoschusesculentus, Achilleamillefolium, Acacia farnesiana*	(Huang et al., 20160; Leszek et al., 2016)
		Parkinson's disease (PD)	Harmaline	*Passiflora incarnate, Peganum harmala, Banisteriopsiscaapi*	(Fischer and Maier, 2015)
			Hyoscyamine	*Lactuca sativa, Datura innoxia, Hyoscyamus niger*	(Islam, 2017; Chikara et al., 2018)

- **redox balance**
- **physiology**
- **carcinogenesis**
- **fibrosis**
- **redox signaling**

REFERENCES

Balasubramanian, S. Progression of chronic kidney disease: mechanisms and interventions in retardation. *Apollo Med.* 2013, 10, 19–28.

Bashan, N., Kovsan, J., Kachko, I., Ovadia, H. and Rudich, A. Positive and negative regulation of insulin signaling by reactive oxygen and nitrogen species. *Physiol. Rev.* 2009, 89, 27–71.

Bonomini, F., Rodella, L.F., and Rezzani, R. Metabolic syndrome, aging, and involvement of oxidative stress. *Aging Dis.* 2015, 6, 109.

Chapman, K., Mannino, D., Soriano, J., Vermeire, P., Buist, A.S., Thun, M., Connell, C., Jemal, A., Lee, T. and Miravitlles, M. Epidemiology and costs of chronic obstructive pulmonary disease. *Eur. Respir. J.* 2006, 27, 188–207.

Chen, W.T., Ebelt, N.D., Stracker, T.H., Xhemalce, B., Van Den Berg, C.L. and Miller, K.M. ATM regulation of IL-8 links oxidative stress to cancer cell migration and invasion. *Elife* 2015, 4, e07270.

Chen, W.W., Zhang, X. and Huang, W.J. Role of neuroinflammation in neurodegenerative diseases. *Mol. Med. Rep.* 2016, 13, 3391–3396.

Chikara, S., Nagaprashantha, L.D., Singhal, J., Horne, D., Awasthi, S. and Singhal, S.S. Oxidative stress and dietary phytochemicals: Role in cancer chemoprevention and treatment. *Cancer Lett.* 2018, 413, 122–134.

De Cristofaro, R., Rocca, B., Vitacolonna, E., Falco, A., Marchesani, P., Ciabattoni, G., Landolfi, R., Patrono, C. and Davi, G. Lipid and protein oxidation contribute to a prothrombotic state in patients with type 2 diabetes mellitus. *J. Thromb. Haemost.* 2003, 1, 250–256.

Espinosa-Diez, C., Miguel, V., Mennerich, D., Kietzmann, T., Sánchez-Pérez, P., Cadenas, S., and Lamas, S. Antioxidant responses and cellular adjustments to oxidative stress. *Redox Biol.* 2015, 6, 183–197.

Fischer, R. and Maier, O. Interrelation of oxidative stress and inflammation in neurodegenerative disease: role of TNF. *Oxid. Med. Cell. Longev.* 2015, 2015, 545–589.

Frijhoff, J., Winyard, P.G., Zarkovic, N., Davies, S.S., Stocker, R., Cheng, D., Knight, A.R., Taylor, E.L., Oettrich, J. and Ruskovska, T. Clinical relevance of biomarkers of oxidative stress. *Antioxid. Redox Sign.* 2015, 23, 1144–1170.

Hancock, J., Desikan, R., and Neill, S. *Role of Reactive Oxygen Species in Cell Signalling Pathways*. Portland Press Limited, London. 2001. pp. 345–350.

He, F., and Zuo, L. Redox roles of reactive oxygen species in cardiovascular diseases. *Int. J. Mol. Sci.* 2015, 16, 27770–27780.

Huang, W.J., Zhang, X., and Chen, W.W. Role of oxidative stress in Alzheimer's disease. *Biomed. Rep.* 2016, 4, 519–522.

Islam, M.T. Oxidative stress and mitochondrial dysfunction-linked neurodegenerative disorders. *Neurol. Res.* 2017, 39, 73–82.

Jones, D.P. Redox potential of GSH/GSSG couple: assay and biological significance. In: *Methods in Enzymology*. Elsevier, the Netherlands. 2002. vol. 348. pp. 93–112.

Khansari, N., Shakiba, Y. and Mahmoudi, M. Chronic inflammation and oxidative stress as a major cause of age-related diseases and cancer. *Recent Pat. Inflamm. Allergy Drug Discov.* 2009, 3, 73–80.

Leszek, J., E Barreto, G., Gasiorowski, K., Koutsouraki, E. and Aliev, G. Inflammatory mechanisms and oxidative stress as key factors responsible for progression of neurodegeneration: role of brain innate immune system. *CNS Neurol. Disord-Dr.* 2016, 15, 329–336.

McCord, J.M. and Fridovich, I. The reduction of cytochrome c by milk xanthine oxidase. *J. Biol. Chem.* 1968, 243, 5753–5760.

Mileo, A.M. and Miccadei, S. Polyphenols as modulator of oxidative stress in cancer disease: new therapeutic strategies. *Oxid. Med. Cell Longev.* 2016, 2016, 455–462.

Münzel, T., Gori, T., Keaney Jr, J.F., Maack, C. and Daiber, A. Pathophysiological role of oxidative stress in systolic and diastolic heart failure and its therapeutic implications. *Eur. Heart J.* 2015, 36, 2555–2564.

Murray, C.J. and Lopez, A.D. Alternative projections of mortality and disability by cause 1990–2020: global burden of disease study. *The Lancet.* 1997, 349, 1498–1504.

Niemann, B., Rohrbach, S., Miller, M.R., Newby, D.E., Fuster, V. and Kovacic, J.C. Oxidative stress and cardiovascular risk: obesity, diabetes, smoking, and pollution: part 3 of a 3-part series. *J. Am. Coll. Cardiol.* 2017, 70, 230–251.

Parimisetty, A., Dorsemans, A.C., Awada, R., Ravanan, P., Diotel, N. and D'hellencourt, C.L. Secret talk between adipose tissue and central nervous system via secreted factors—an emerging frontier in the neurodegenerative research. *J. Neuroinfl.* 2016, 13, 67.

Poprac, P., Jomova, K., Simunkova, M., Kollar, V., Rhodes, C.J. and Valko, M. Targeting free radicals in oxidative stress-related human diseases. *Trends Pharmacol. Sci.* 2017, 38, 592–607.

Putri, A.Y. and Thaha, M. Role of oxidative stress on chronic kidney disease progression. *Acta Med. Indones.* 2016, 46.

Rani, V., Deep, G., Singh, R.K., Palle, K. and Yadav, U.C. Oxidative stress and metabolic disorders: Pathogenesis and therapeutic strategies. *Life Sci.* 2016, 148, 183–193.

Reinisalo, M., Kårlund, A., Koskela, A., Kaarniranta, K. and Karjalainen, R.O. Polyphenol stilbenes: molecular mechanisms of defence against oxidative stress and aging-related diseases. *Oxid Med. Cell Longev.* 2015, 2015, 73–82.

Rinnerthaler, M., Bischof, J., Streubel, M., Trost, A., and Richter, K. Oxidative stress in aging human skin. *Biomolecules* 2015, 5, 545–589.

Robson, R., Kundur, A.R., and Singh, I. Oxidative stress biomarkers in type 2 diabetes mellitus for assessment of cardiovascular disease risk. *Diabetes Metab. Syndrome: Clin. Res. Rev.* 2018, 12, 455–462.

Salim, S. Oxidative stress and the central nervous system. *J. Pharmacol. Exp. Ther.* 2017, 360, 201–205.

Santilli, F., Guagnano, M., Vazzana, N., La Barba, S., and Davi, G. Oxidative stress drivers and modulators in obesity and cardiovascular disease: from biomarkers to therapeutic approach. *Curr. Med. Chem.* 2015, 22, 582–595.

Siti, H.N., Kamisah, Y. and Kamsiah, J. The role of oxidative stress, antioxidants and vascular inflammation in cardiovascular disease (a review). *Vasc. Pharmacol.* 2015, 71, 40–56.

Skibska, B. and Goraca, A. The protective effect of lipoic acid on selected cardiovascular diseases caused by age-related oxidative stress. *Oxidative Med. Cell. Longev.* 2015, 2015, 404–406.

Testa, G., Cacciatore, F., Galizia, G., Della-Morte, D., Mazzella, F., Langellotto, A., Russo, S., Gargiulo, G., De Santis, D., and Ferrara, N. Waist circumference but not body mass index predicts long-term mortality in elderly subjects with chronic heart failure. *J. Am. Geriatr. Soc.* 2010, 58, 1433–1440.

Thanan, R., Oikawa, S., Hiraku, Y., Ohnishi, S., Ma, N., Pinlaor, S., Yongvanit, P., Kawanishi, S. and Murata, M. Oxidative stress and its significant roles in neurodegenerative diseases and cancer. *Int. J. Mol. Sci.* 2015, 16, 193–217.

Venkataraman, K., Khurana, S. and Tai, T. Oxidative stress in aging-matters of the heart and mind. *Int. J. Mol. Sci.* 2013, 14, 17897–17925.

Zhang, H., Davies, K.J. and Forman, H.J. Oxidative stress response and Nrf2 signaling in aging. *Free Radic. Biol. Med.* 2015, 88, 314–336.

Zhang, Y., Unnikrishnan, A., Deepa, S.S., Liu, Y., Li, Y., Ikeno, Y., Sosnowska, D., Van Remmen, H. and Richardson, A. A new role for oxidative stress in aging: the accelerated aging phenotype in $Sod1^{-/-}$ mice is correlated to increased cellular senescence. *Redox Biol.* 2017, 11, 30–37.

CHAPTER 10

Implications of CRISPR Technology in Biological Systems

KIKKU SHARMA and SOUVIK SEN GUPTA*

Division of Biological and Life Sciences, School of Arts and Sciences, Ahmedabad University, Central Campus, Navrangpura, Ahmedabad, Gujarat 380009, India

**Corresponding author. E-mail: souvik.sengupta@ahduni.edu.in*

ABSTRACT

The Clustered Regularly Interspaced Short Palindromic Repeats (CRISPR)-CRISPR-associated protein (Cas) system is a highly efficient specific genome-editing tool working on RNA-guided platform. The CRISPR system is an adaptive immune mechanism present in many bacteria and in a majority of Archaea, which memorizes prior infection by integrating short remnants of the invading genome (termed as spacers) into its CRISPR locus. Subsequently, the invading DNA or RNA is cleaved selectively by the CRISPR-Cas system. This system makes genome engineering easier, and in doing so it opens up too many paths of biology that give a new vision to the modern scientist and stakeholders. Due to the high accuracy in cutting and pasting DNA, the high degree of flexibility and relatively low cost, the CRISPR-Cas9 system has evolved as the most preferred genetic engineering tool. The application of this technique has been explored for a wide range of fields including agriculture and human health. The applications of this genome-editing technology are manifold, which consequently leads to the safety and ethical aspect of the same. In this article, we will discuss the advantages of this genomic tool and its implications in various biological fields. We will discuss its applications in microbes, plants (with special reference to agriculture), and human health. We will also emphasize the promising and versatile application of the CRISPR-Cas9 system in modern gene therapy and how this system can help to cure many serious diseases.

10.1 INTRODUCTION

After the discovery of DNA double helix, technology for manipulating DNA that enables advances in biology have taken the center stage. Biological research is now transformed by the invention of RNA programmable genome engineering that is the clustered regularly interspaced short palindromic repeats (CRISPR) CAS9 genome-editing system (Doudna and Charpentier, 2014). CRISPR/CAS bacterial immune system is a RNA-guided genome-editing tool that is efficient and can be applied in a diverse group of organisms (Wang et al., 2016a). This adaptive immune system is a characteristic feature of many bacteria as well as a few archaea. The foreign nucleic acid fragment from invading bacteriophages or plasmids is acquired by organisms containing the CRISPR system, which is subsequently transcribed into the CRISPR RNA thus guiding the cleavage of the invading RNA or DNA (Wang et al., 2016b). In this mechanism, the invading foreign DNA is recognized by the CAS protein followed by the insertion of a short sequence of invader DNA into the CRISPR array of the bacterial genome (Charpentier et. al., 2015). These short sequences lead to the formation of strong genetic memory that prevents reinfection (Charpentier et. al., 2015). The CRISPR CAS system is an example of how the bacterial immune system may help to solve the modern biotechnological problems. In the CRISPR/CAS9 system, the CAS9 enzyme is very important as it has a unique pathogen specificity. Many other genome-editing tools have been discovered earlier like site-directed zinc finger nucleases and TAL effector nucleases, which use the principles of DNA protein recognition. The main barrier to these techniques is the design and validation of the specific protein. In comparison, CRISPR/CAS9 is the simplest genome-editing tool as it has a very specific DNA cleaving mechanism and recognizes multiple targets. This system also has a natural variant of type II CAS9 and all the properties of this system supports the development of cost effective and easy to use technology, which is very precise and effective for targeting, editing, modification, and marking of a particular genomic locus of various organisms (Doudna and Charpentier, 2014). Genome editing by CRISPR-Cas9 introduces DNA double-strand break and repairing the break with nonhomologous end joining (Irion et al., 2014). Thus, gene knockout by CRISPR Cas9 may result in DNA double strand break and subsequent repair by nonhomologous end joining, which may result in alteration of reading frames. However, homology directed repair mediated by exogenously supplied nucleotide leads to precise alteration of the reading frame sequence (Aguirre et al., 2016). The CRISPR-CAS9 system has a wide range of applications including alteration metabolic pathways, improvement of crops, and new-age drug development namely gene mutation and silencing. CRISPR technique is also applicable in biomedical engineering and molecular biology (Liu et al., 2017a). Double-strand breaks in the genomic sequence are required for genome editing, which leads to activation of multiple repair pathways (Hartenian and Doench, 2015). There is considerable evidence that the CRISPR in bacteria and archaea are horizontally transferred between the two domains (Manica et al., 2013). By definition, the CRISPR-CAS9 system is the association of crRNA and Cas protein into a Cr RNA protein complex, which helps to identify DNA target and destroy matching sequences of foreign nucleic acid (Jiang and Doudna, 2017). The CRISPR CAS9-mediated genome engineering has shown promise to treat and cure many different genetic disorders, cancer, neuronal disorders, and immunological disorders. Although there are many advantages and effectivity in CRISPR technology, still there are many problems and obstacles in the actual implementation of this technology for clinical or therapeutic applications (Jiang and Doudna, 2017). The origin of the CRISPR-CAS9 system was the type II CRISPR-CAS9 system, which was first demonstrated experimentally in *Streptococcus thermophiles* (Barrangou et al., 2007). Consequently, the revelation of the role of tracer RNA in human pathogen *Streptococcus pyrogenes* has led to a more clear understanding of the mechanism of the type II CRISPR system (Charpentier et al., 2015). Recently, path-breaking research in this field has revealed a different type of CRISPR protein (CAS13) that can recognize and target RNA and is capable to edit them, although its application has many ethical issues (Brokowski and Adli, 2019). Now it is expected that this highly effective tool will further develop into the automated and rapid stain development process in industries (Mougiakos et al., 2018). Unlike other bacterial defense mechanisms that impart a generalized defense to the foreign invaders, the CRISPR system can create a memory that protects the organism from reinfection, which is analogous to the immune system of vertebrates (Wright et. al., 2016). Although CRISPR CAS9 is a naturally occurring system, in the laboratory CRISPR-CAS9 system is applied to the cells in conjunction to specifically designed guide RNA (g RNA) to facilitated specific alteration.

10.2 HISTORICAL ASPECTS OF CRISPR/CAS9

In the year 1987, a Japanese investigator first described short interspaced genome sequences in *Escherichia coli* and the basic idea of CRISPR originated from this study (Doudna and Charpentier, 2014). The first comprehensive proof of the CRISPR system was found when prokaryotic Gene Bank was accumulated. In 2002, isolation and sequencing of small noncoding RNA from *Archaeoglobus fulgidus* proved that CRISPR loci were transcribed to small RNA. In the same year, the Cas gene was also discovered, which belongs to the CRISPR loci. In 2005, spacer genomic sequences were discovered in plasmids and phages (Marraffini, 2015). In 2011, the type II CRISPR was transferred from *Streptococcus thermophiles* to *E. coli* and *E. coli* efficiently reconstituted this CRISPR system in another bacterial cell. Genetically, engineered CRISPR from *Streptococcus thermophiles* and *Streptococcus pyrogens* was used for genome editing in humans in the year 2013 (Hsu et al., 2014). In 2012, a team led by Jennifer Doudna at the University of California and Emmanuelle Charpentier from the University of Umea revealed how natural CRISPR/CAS9 may be harnessed as a tool to cut any DNA strand in a test tube. The first patent by the US patent office for editing the eukaryotic genome using CRISPR was awarded in April 2014. The CRISPR/CAS9 system was first exploited by the company Danisco to improve the immunity of bacterial culture in the year of 2008. Between 2014 and 2015, the CRISPR/CAS9 technique was applied in mice successfully (Table 10.1).

10.3 MECHANISM OF ACTION OF CRISPR/CAS9 SYSTEM

The CRISPR CAS9 has a defined molecular organization that enables its effective functioning. Specific adaptation and effector molecules constitute the principal components in the organization of the CRISPR-cas9 system. The system consists of a Cas gene whose product is necessary for obtaining the spacer DNA fragment and the processing of the pre RNA to mature Cr-RNA. CAS1 and CAS2 protein complex are found in most of the CRISPR system, and it acts as an endonuclease that performs cleavage in both the source of spacer containing DNA and the CRISPR array (Koonin, 2019).

The CRISPR system works through the involvement of CAS9 protein, which is encoded by the CRISPR loci. In general, the CRISPR system works in three different stages to generate a full immune response against foreign DNA particles (Wang et al., 2016b). The first stage is known as the acquisition stage in which a distinct CAS protein complex binds to the target DNA and then the protein migrates along the length of the DNA until it finds a distinct 2–4 bp motif known as PAM (Protospacer Adjacent Motif). Consequently, a portion of the DNA is cleaved that is further incorporated into the CRISPR array in between two repeats and thus a spacer is formed. Some

TABLE 10.1 Types CRISPR CAS9 System

Type name	Signature Gene	Encoded Product	Comments	Distribution
Type I CRISPR CAS System	This CRISPR loci contains *Cas3* gene	Encodes a large protein with separate helicase and DNase activity. This protein forms a cascade like complex with other protein and most often it contains RAMP superfamily protein.	This type I CRISPR domain targets DNA and the cleavage is catalyzed by the nuclease domain. It helps in the processing of long spacer repeat to convert it to mature Cr RNA.	This system is generally found in bacteria and archaea (Makarova and Koonin, 2015)
Type II CRISPR CAS system	Type II is unique because it contains HNH system and *Cas9* is the signature gene. In addition there is ubiquitous Cas1 and *Cas2* gene	Encodes restriction endonuclease type protein	The HNH and RuvC system cut the DNA and creates double strand break. A 20 nucleotide sequence is formed which is associated with the transcription of the mature Cr RNA. *Cas9* helps in gene silencing.	Most widely spread in bacteria (e.g.,: *Neisseria meningitides*) (Yingjun Li et al., 2016)
Type III CRISPR CAS system (can be further divided into type III A and type III B)	Type III A encodes *csm2* gene product and type III B produces *cmr5* gene product.	Type III contain polymerase and RAMP cascade complexes	It assembles into a cascade-like interference complex for target search and destruction. The RAMP complex is required for the processing of the spacer transcripts	Bacteria, also in archaea (Yingjun Li et al., 2016)
Type IV CRISPR CAS system	It usually lacks any Cas protein	Csf5 (a Cas6 variant protein) has been identified that helps in Cr RNA processing. This Cr RNA is associated with Cr RNP complexes.	Uncharacterized protein is proposed to form part of a Cascade-like complex. Presence of isolated *Cas* genes without an associated CRISPR array	Not known (Özcan et al., 2019)

TABLE 10.1 *(Continued)*

Type name	Signature Gene	Encoded Product	Comments	Distribution
Type V CRISPR CAS system. Further subdivided into type V A and type V B	The gene is *cpf1* in type V A which encodes a product similar to Cas9. The gene *c2c1* is encoded in type V B	Cpf1 is a RNA-guided endonuclease	Cpf1 contains RuvC domain as well as another nuclease domain that can target both specific and non specific strand. Break is generated which helps in mature Cr RNA formation. This Cr RNA forms complex with Cas protein and helps in self versus non-self discrimination. *c2c1* cleaves the DNA with the help of Cr RNA and tracer DNA. It also contains RuvC domain along with a separate endo nuclease domain. However, the *c2c1* mechanism is still elusive.	*c2c1* endonuclease activity is found in human cell lysate. (Özcan et al., 2019)

CRISPR also has an alternative mechanism for acquisition; they adapt RNA via reverse transcription with the help of reverse transcriptase that is encoded by the CRISPR locus. The protospacer or PAM that is incorporated in the CRISPR loci are not the component of the host; rather it is the conserved sequence of invader virus or plasmids. CAS9 protein can recognize the target only when conserved dinucleotide PAM sequence present upstream to the Cr-RNA binding site. PAM help to recognize bacterial self and nonself DNA and thus protect their DNA from cleavage (Demirci et al., 2017). The acquisition process is of different types-type I acquisition was first discovered in *E. coli*, and it is done by two mechanisms—naive and prime mechanism. In Naive acquisition, cas1 and cas2 integrase recognize and acquire spacer DNA fragments from a previously encountered foreign invader. After that, the CrRNA and CAS protein complex are capable of targeting the PAM DNA sequence to match with the Cr RNA. Then CAS3 is recruited to the target site progressively disrupting the foreign DNA. The recruitment of CAS3 and its translocation is so perfect that it does not degrade host DNA. The primed acquisition was experimentally observed in *Pectobacterium atrosepticum* where the memory of previously invading bacteria is present in the CRISPR loci of the host bacteria. Here CAS2 and CAS3 form a large polypeptide in association with CAS1 although CAS1 and CAS2 levels are minimal in this acquisition step. Type II acquisition also has some key aspects on the spacer acquisition; its acquisition is dependent on infection by a defective phage (Wright et al., 2016). The key factor for spacer integration is CAS1 and CAS2, which are in high concentration when integration of spacer is been done (Hsu et al., 2014). The next stage is the expression or biogenesis stage where CRISPR array is transcribed into a single long pre-cr (CRISPR) RNA that is processed to a mature cr-RNA with spacer and repeats. In type I and type II, the cas6 gene family process the pre-Cr RNA into an intermediated form (Hille and Charpentier, 2016). In type II, Tracer RNA is used as a complementary sequence so that RNA duplex can easily form with each repeat. This structure is stabilized by CAS9 and these duplexes are recognized by RNase III of the host, which yields an intermediate of mature Cr RNA, although the exact process of maturation is still not clear (Hille and Charpentier, 2016). In class I CRISPR, a protein complex helps in target degradation whereas in class II system a single effector protein is involved in target degradation (Hille and Charpentier, 2016). In the last and final stage, cr-RNA is closely associated with the processing complex, which subsequently identifies the protospacer or nearly similar sequence of invading genome or plasmid or other foreign DNA and cleaves them causing inactivation (Figure 10.1). As the CRISPR system modifies the host bacterial genome and stores memory of such invader genome so it works very efficiently and protects the host from unwanted intruders (Koonin, 2019).

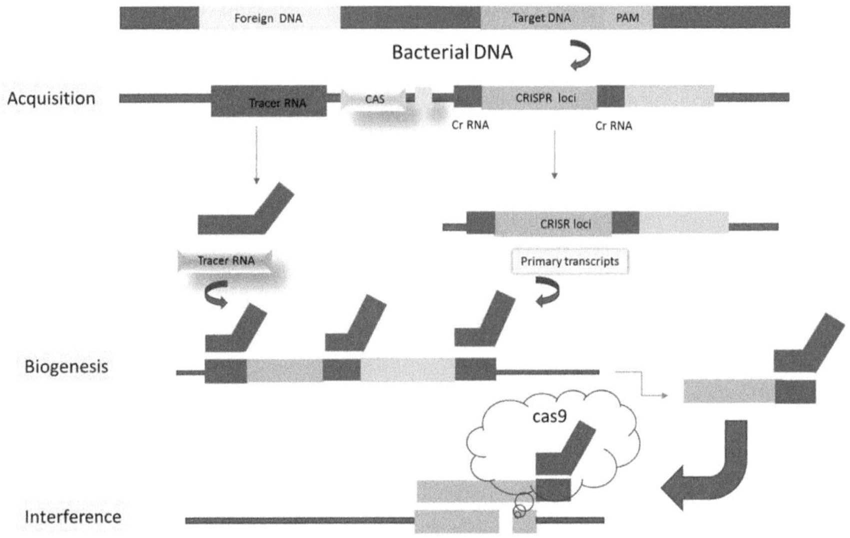

FIGURE 10.1 Mechanism of CRISPR Cas9. Step 1 - Acquisition, Step 2 - biogenesis of Cr RNA, Step 3 - interference and cleavage of foreign DNA mediated by Cas9 nuclease activity.

10.4 ANTI-CRISPR MECHANISM

As bacteria have developed the mechanism to memorize the invaders and destroying the foreign DNA, viruses have also developed strategies against the CRISPR-CAS9 system. Many measures of defense against the CRISPR-CAS system have been reported. The most frequent mechanism is random mutagenesis that alters the interaction of key bases with Cr RNA or PAM region. CRISPR-CAS system has also been reported to promote infection by viruses in certain cases. In *Pseudomonas aeruginosa*, the viruses encode several proteins that inhibit the function of type I-E and type I-F system which is the most well-known anti-CRISPR strategy. *Vibrio cholerae* carries a type I-F CRISPR system that targets the antiphage host locus. In this anti-CRISPR mechanism, the viral genome enters into the host cell and viral Cr RNA targets the host locus (Rath et al., 2015). Initially, scientists discovered that anti-CRISPR mechanism may be established by *P. aeruginosa* and recently the structure and the functional route of this anti-CRISPR protein have been tracked by researchers. This protein named AcrF1 is a potent inhibitor of the type 1-F CRISPR-CAS9 system as it inhibits the DNA binding activity. By using NMR spectroscopy, the structure of AcrF1 has been solved and the functional surface of this protein has been identified by site-directed mutagenesis (Maxwell et al., 2016).

10.5 APPLICATION AND EFFECTIVITY OF CRISPR/CAS9 SYSTEM IN MODERN BIOLOGY

In today's world, genome editing is relatively easy to handle due to the advent of different genome engineering techniques. The recent advances in genome engineering techniques are based on CRISPR and its associated CAS9 protein has led to its application in genome editing so easy that many of the genetic challenges can be overcome by its application. This technology that is derived from the microbial immune system is driving innovative applications in the field of basic biology, biotechnology, and medicine.

10.5.1 APPLICATION OF CAS9 IN GENOME EDITING

Cas9 genome editing has successfully enabled the generation of many transgenic animals like mice, zebrafish, bacteria, yeast, fruit flies, *Caenorhabditis elegans*, frogs, and even humans. Many studies have shown successful gene editing in an adult cell under in vivo conditions using the CRISPR-Cas9 system (Figure 10.2). A recent study has shown efficient restoration of the expression of dystrophin protein and consequent enhancement of muscle strength in DMD mice using the CRISPR-Cas9 system (Pandey et al., 2017).

10.5.2 CRISPR/CAS9 TECHNOLOGY USE IN HUMAN STEM CELL TO OVERCOME GENETIC DISORDERS

The CRISPR-CAS9 system has been used to cure many serious genetic disorders. For this purpose, the production and characterization of the iPS cells from patients with a specific genetic disorder has to be done first. Recently, the combination of CRISPR and iPS cells has made some serious breakthroughs in disease analysis and modeling. CRISPR mechanism perfectly helps in modeling disorders by complex manipulation of a gene in an easy way as compared to conventional processes which are very expensive and also often do not lead to the desired results. Human iPS cells have been mostly used for gene editing as they have the same gene expression and methylation pattern as in ESC cells and the reprogramming of Oct3/4, Sox2, Klf4, etc. factors are easy via retroviral infection. One of the potential utility of iPS cells is that it may lead to the creation of personalized treatment for specific individuals as the iPS cells have the same genetic makeup as the person from whom it is derived. Stem cell genome modifications with the help of CRISPR have made the gene-editing technology more developed and efficient. Toward the way of promising drug discovery, nowadays CRISPR Cas9-mediated insertion and deletion is done to know the gene responsible for a particular disease. By this, both the selective mutation of the candidate gene and the organoid phenotype of the disease can be studied (Freiermuth et al., 2018). There are several other strategies of somatic cell reprogramming and iPS cells are used as an in vitro model for several diseases like Duchenne and Becker muscular dystrophy, Parkinson's disease, Huntington's disease, and Down syndrome/trisomy. The advantage of the iPS cells is that it can be easily be adapted for drug delivery (Table 10.2).

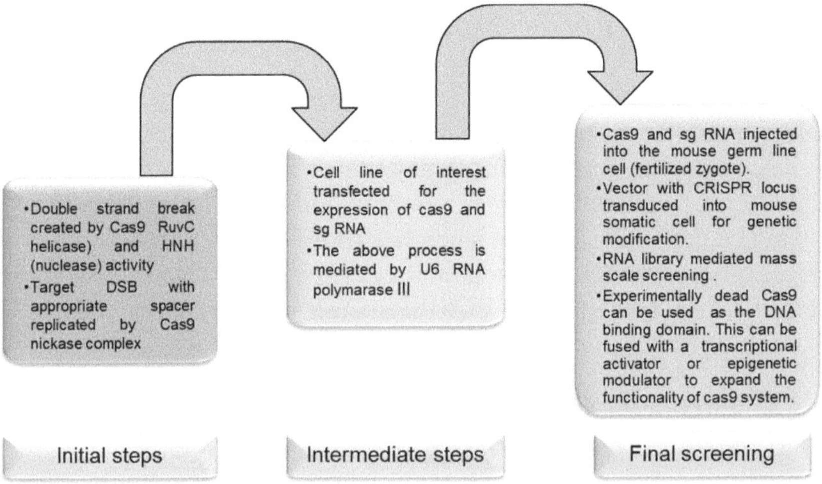

FIGURE 10.2 Process flow for genome editing of DMD gene in mouse and its screening.

10.5.3 ROLE OF CRISPR IN CANCER RESEARCH AND THERAPY

Cancer leads to mortality worldwide. After the invention of the CRISPR-Cas9 technique and its consequent clinical application, a new avenue for

TABLE 10.2 Use of CRISPR in Genetic Disorders

Disease Name	Gene	Role of CRISPR	Description	Animal Mode	Reference
Duchenne muscular dystrophy (DMD)	Dystrophin gene	CRISPR mediated deletion of 725 Kb in hiPsc of DMD. Reframed hiPsc in cardiac and skeletal muscles. This restores the function of dystrophin	Engineered hiPsc was used as therapeutic strategy for 60% Duchenne muscular dystrophy patients	Human (DMD Paiteints)	Young et al., 2016
Sickle cell anemia	β-globin	CRISPR *Cas9* gene editing used to edit the HBB gene by homologous recombination. In HSC, AAV was used for HR donor delivery and it is associated with Cas9 ribonucleoprotein.	Hematopoietic Stem Cell (HSC) containing HBB Gene Can Be targeted By CRISPR Associated with AAV. This results in the change of nucleotide at The position of mutation that causes Sickle cell anemia	Human	Dever et al., 2016
β-Thalassemia	β-globin	CRISPR technology combined with piggy Bac transposons to edit iPS cells derived from patients of β-thalassemia. No off target effect reported.	Human induced pluripotent stem (iPS) cells from β-thalassemia patients were edited by CRISPR tool	Human	Xie et al., 2019
Cystic fibrosis	CFTR	CRISPR *Cas9* was used to edit the F508 deletion mutation of CFTR by isolating adult intestinal stem cell from CF patients.	Human iPS cells from intestine was used for gene editing by CRISPR	Human	Colemeadow et al., 2016
α1-antitrypsin deficiency (A1ATD)	SERPINA1	CRISPR Cas9 effectively used to edit the mutated SERPINA1 in α1-antitrypsin deficiency affected mouse to restore the normal levels of α1-antitrypsin.	hSERPINA1 gene was selected for efficient editing by the use of CRISPR Cas9. Specific g RNA was designed for this purpose.	Mouse	Bjursell et al., 2018

TABLE 10.2 *(Continued)*

Disease Name	Gene	Role of CRISPR	Description	Animal Mode	Reference
Osteosarcoma	CDK11	CRISPR Cas9 used to target endogenous CDK11 gene of osteosarcoma cell line and total silencing of the CDK11 was reported	CRISPR Cas9 expression vector guided by sg RNA was used to silence the CDK11 gene. Reduction of CDK11 expression resulted in reduced cell proliferation and viability of osteosarcoma cell line	Human osteosarcoma cell line	Feng et al., 2015
Barth syndrome	Tafazzin (TAZ)	TAZ gene was targeted by CRISPR Cas9 in hiPsc cell using piggy Bac transposons and transfection near the mutation of barth syndrome patients.	Due to Cas9 activity, deletion was observed at the target site.	Human	Yang et al., 2014
Cataracts	crygc	A dominant mutation in crygc gene can be reduced by transfecting the mice zygote with Cas9 m RNA and a sg RNA.	Correction occurs via homology directed repair based on exogenously supplied oligonucleotide	Mouse	Yuxuan Wu et al., 2013
Hereditary tyrosinemia type I (HTI)	Fumaryl acetoacetate hydrolase (Fah)	For correction of mutated Fah gene in mouse model of human disease, CRISPR Cas9 system was injected into the liver cells	After the editing, normal expression of wild type Fah was observed	Mouse	Yin et al., 2014
Epstein-Barr virus (EBV)	EBV	CRISPR Cas9 mediated editing of EBV genome in human cell line leads 580 bp deletion near the promoter region which contain BART (BamHI A rightward transcript)	Due to deletion of BART region which contains micro RNA of viral particle, no recovery of virus reported	Human	Yuen et al., 2015

TABLE 10.2 *(Continued)*

Disease Name	Gene	Role of CRISPR	Description	Animal Mode	Reference
Hepatitis B virus (HBV)	P53 and PTEN	CRISPR Cas9 creates indel mutation in p53 and pten gene in adult HBV infected mouse cell as a result of which loss of function type mutation occurs	As a result of this mutation, accelerated hepatocarcinogenesis was observed in mouse cells	Mouse	Liu et al., 2017
Huntington's disease	huntingtin (*HTT*) gene	CRISPR Cas9 PAM dependent g RNA was designed by SNP allele in mutant chromosome and selective deletion of 44 kb region of transcription start site and CAG mutation in the mutant HTT gene	As a result of deletion of mutant genome, complete prevention of formation of m RNA from HTT gene was observed and protein synthesis was inhibited. There was no impact of this deletion on the normal allele	Human	Shin et al., 2016
Parkinson's disease	DJ-1/parkin/ PINK1 LRRK2	CRISPR Cas9 targeting perkin, PINK1 LRRK2, DJ-1 in the genome of pigs	By injecting cas9 m RNA and multiplex sg RNA targeting PARKIN, DJ-1, And Pink1 Genes, pigs remain healthy. CRISPR Cas9 resulted in genome modification with high medical value	Pigs	Wang et al., 2016
Down Syndrome	GATA1	CRISPR Cas9 used to design truncated GATA1 similar to myeloid leukemia In Down syndrome	CRISPR Cas9 was used to cleave GATA1. The broken DNA reanneals by non-homologous end joining resulting in production of truncated variant of GATA1 that ultimately inhibit normal GATA1 expression	Human	Bloh et al., 2017

TABLE 10.2 *(Continued)*

Disease Name	Gene	Role of CRISPR	Description	Animal Mode	Reference
Retinitis pigmentosa	RGPR	iPS cells derived from patients were transduced by CRISPR Cas9, g RNA and a donor homology template	CRISPR could precisely edit the pathogenic mutation and produce gene-corrected iPS cells for eventual use in autologous transplantation	Human	Bassuk et al., 2016

the treatment of cancer have opened up. CRISPR Cas9 is a powerful genetic tool for the discovery of novel targets for cancer therapy. In this process, CRISPR Cas9 screening is carried out by the formation of a population of a cell line with a series of gene knockouts. To achieve this, Sg RNA libraries that are predicted to be efficient for every gene are synthesized in a pool of oligonucleotides and replicated in lentiviral plasmids. Next, Cas9-expressing cells are infected by the lentiviral particles at a low multiplicity of infection so that every cell has a specific Sg RNA and specific gene knockout. This knockout population of cells is then kept in culture for a particular period in a constant environment after that environmental condition is changed and DNA is extracted from the cells. By amplification and sequencing of the Sg RNA, the population of cells carrying the knockouts is determined. This CRISPR Cas9 screening technology is used to detect essential genes in a systematic way (Figure 10.3). In this context, large scale and different modified applications of CRISPR Cas9 have been performed like a gain of function screening and investigation of drug–gene interaction, etc. (Zhan et al., 2019).

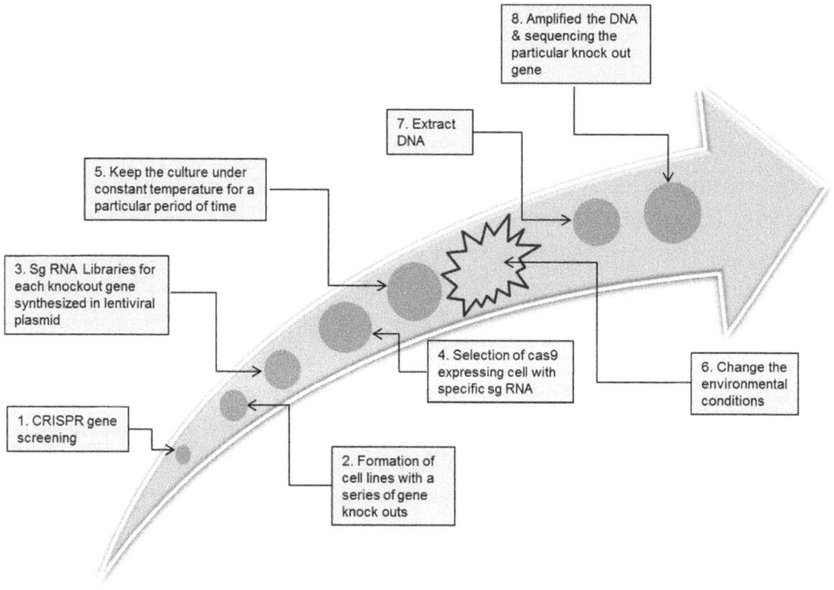

FIGURE 10.3 A generalized scheme of steps for the use of CRISPR CAS9 technology in generating gene knockouts and subsequent screening.

The human genome majorly consists of noncoding regions that contain regulatory elements like enhancer or noncoding RNA, etc. The transcription of an oncogene is regulated by near or distant enhancer elements and thus understanding the noncoding RNA may give new information about cancer. Nowadays, CRISPR/CAS9 technology provides useful data of noncoding RNA. By using this technology, scientists have revealed the enhancer regions of CUL3, NF1, NF2, which are cancer-causing genes. Scientists have successfully altered the up and downstream sequence of the CUL3 gene so that the expression of the gene is less. The g RNA can be recruited to a specific target DNA site with the help of catalytically inactive dCas9. The targeted DNA can be fused with either a transcriptional activator or inhibitor domain to either activate or repress a specific target gene. In another case, dCas9 has been used as a modifier of histone and other specific proteins that are related to DNA methylation, which is also known as epigenetic editing. By epigenetic regulatory mechanisms, many different types of cancer like leukemia, Erwin sarcoma, etc. can be dysregulated. Though it is difficult to effectively deliver CRISPR component into the cell, still cancer immunotherapy is emerging as a promising approach in the area of cancer treatment. Cancer immunotherapy is more effective compared to chemotherapy or radiotherapy as it causes low life risk and has durable activity. The CRISPR/CAS9 technology can be applied toward therapeutic exploration in human cells due to its promising efficacy. This technique has been used to generate immune cells and oncolytic viruses for cancer therapy (Martinez-Lage, 2018). The oncolytic viruses are the antitumor virus that selectively kills cancer forming cells. Recently, CRISPR Cas9 was used to design a modified oncolytic virus that can help to disrupt cancer forming tumor cells. In this case, the Cas9 protein targeted a green fluorescence protein (GFP) gene of the recombinant adenoviral vector and it was edited by three different RNA. The accuracy of mutation in the target EGFP gene is nearly about 47% (Yuan et al., 2016). In the hospital of Sichuan University (West China), the first clinical trial of CRISPR toward a cancer patient was done in the year of 2016. In this experiment, T cells from the patient were collected and engineered in which a programmed cell death protein 1 (PD1) was knocked out, and its efficacy was evaluated in the treatment of metastatic lung cancer, which progressed after all standard treatments. PD1 here acted as an inhibitor of T-cell activation thereby regulating immune tolerance and decreasing autoimmune reactions. It also allowed the immune escape of cancer. In other patients, the antibody that neutralized PD1 and its ligand PD L1 was used for the treatment. The patients who enrolled for genome-editing treatment were provided with peripheral lymphocytes in which the PD1 protein was knocked out using the CRISPR CAS9 ex vivo technique. This type of PD1 knockout editing was also used in four other types of cancers like prostate, bladder, etc. Since propagating and modifying T cells are laborious, there is a considerable limitation regarding the use of PD1 knocked out T cell over PD1 and PD L1 antibodies. However, this can be considered as the first proof of concept that CRISPR CAS9 may be used as a genetic tool for cancer treatment clinically. CRISPR technology provides a simple but versatile way of genome editing in a variety of organisms. However, there is still more scope for improvement of the CRISPR/CAS9 technique for discovering new ways toward clinical application in cancer research. In the future, essential genes of cancer cell lines may be comprehensively screened by the use of CRISPR technology. However, the data obtained by the CRISPR mechanism for gene interaction require proper expertise to understand. CRISPR Cas9 has been used to manipulate the noncoding genome that surely provides a much-needed impetus to the field of cancer research. The genetic sequence of many cancer cells has been copied by using the combination of CRISPR and organoid culture. The use of CRISPR Cas9 as a translational drug depends on the development of the variety of cas9 variants in the future. There is also room for improvisations in the viral and nonviral methods to improve the in vivo applications of CRISPR. Furthermore, the development of CRISPR research will give us a new path for tackling the problem of cancer in many different ways (Zhan et al., 2019).

The simplicity and vast majority of studies in the field of CRISPR point toward its amenability in cancer research but it also come with many challenges. The formation of cancer due to the over or under expression of many genes and the subsequent study of those genes namely CRISPR CAS9 requires the maintenance of the sg RNA libraries of these genes, which complicates the situation. Often, it is found that different individuals affected by the same type of cancer have different types of mutations in genes that create a huge problem in clinical therapy of cancer. Oncolytic viruses that are considered to be biotherapeutic agents can selectively target and destroy cancer cells but they have also been reported to produce a humoral immune response to the distant tumor cells. This type of biotherapy requires an alternative approach for an invasion, which is not applicable for a particular cancer cell type (Chira et al., 2017).

10.5.4 IMPLICATIONS OF CRISPR IN HIV

It has been suggested that CRISPR CAS9 can be used to cleave proviral HIV genetic material in a sequence-specific manner using g RNA. Many studies revealed that CRISPR Cas9 and its associated g RNA can be used to inhibit many pathogenic viruses like Epstein–Barr virus, hepatitis virus, etc. The CAS9 and the g RNA can also be used to inhibit the replication of retroviruses like HIV1, simply by targeting its reverse transcriptase (Wang et al., 2016c). It has also been reported that *Streptococcus pyrogens* Cas9 and corresponding sg RNA can successfully eliminate HIV1 and its proviral DNA, whereas the pretreatment with the same leads to resistance to new HIV1 infection in vitro.

Concerning the application of the CRISPR/CAS9 technology to HIV, several experimental setups may be created to eradicate HIV1 and its proviral DNA. By disrupting proviral DNA receptors like CCR5 and CXCR4, the entry of HIV1 or proviral DNA may be prohibited. Targeting the 5′ and 3′ terminal repeats of the proviral DNA may also inhibit the entry of HIV in the host cell. In this system, the designed g RNA should have a sequence identical to the 5′ or 3′ end of the desired excision site. The g RNA is not homologous to the host sequence as a result of which the off-target effect is reduced. However, as the proviral DNA has a high rate of mutation and also has intra and inter-patient variability, this creates a huge problem in the application of the CRISPR genome-editing technique for curing HIV. So, it has been suggested that targeting the long terminal repeats (LTRs) is a good option as they have the transcription factor binding site that is similar to the ones present in the human promoter. Previously, it was thought that the g RNA requires exactly 20 nucleotide repeats with NGG PAM motif but recent research shows that a very complex relationship exists between the g RNA and the targeted site of HIV1. There is no perfect matching between the g RNA and the target site of the HIV1 genome and thus the effect of subsequent binding in these two may lead to the off-target effect. This suggests that for reducing the off-target effect, a more efficient search is required than the normal BLAST technique (Dampier et al., 2017). Cas9 and g RNA have also been delivered successfully by the lentiviral-mediated delivery to the CD4+ T cells, which has led to the excision of HIV (Yin et al., 2014).

CRISPR CAS9 can directly target the lentiviral RNA during its life cycle into the host cell and disrupt its reverse transcriptase enzyme. For this, multiple potential g RNA site in the HIV genome has been identified as the optimized target which would provide long-term efficient protection against HIV1 in T cells and pluripotent stem cells. In the normal life cycle of the lentivirus, the RNA is reverse transcribed to double-strand DNA and incorporated into the host cell DNA within the nucleus. Therefore, the viral DNA is not detected by the host immune system. In the early life cycle of lentivirus, its DNA becomes exposed due to the release of the capsid and this exposed DNA is disrupted in a targeted manner by sequence-specific DNA nuclease. To prove this experimentally, a pseudo type of lentiviral DNA was designed in which GFP was used as the reporter. Then g RNA was designed against the GFP-coding region and noncoding LTRs. This g RNA that was designed had only viral specificity and had no specificity toward the human genome. To track the Cas9 expression inside the cells, a human codon-optimized Cas9 reporter was designed, which is equivalent to humanized Cas9 (hCas9) in reporter rescue assay using GFP. This g RNA together with hCas9 caused a significant decrease in the amount of expression of GFP as well as its fluorescence. Also, the fluorescence signal from virus-infected cells got decreased to the level of noninfected cells. The results also showed that GFP disruption is caused by insertion or deletion and viral DNA is permanently inhibited by the total disruption of the proviral genome using CRISPR CAS9. The study also showed that secondary delivery of CRISPR Cas9 leads to depletion of the viral expression, which suggests that low dose treatment is less cytotoxic and also can effectively eradicate the proviral genome for the long term. Finally, the results suggested that CRISPR CAS9 target both the preintegrated viral genome and integrated viral genome either in linear or circular double-stranded DNA (ds DNA) for disruption.

To examine the generality of CRISPR CAS9-mediated immunity in humans, both retrovirus and adenovirus have been used. The viral DNA was targeted by CRISPR CAS9 and subsequently the expression of the viruses decreased which suggests the application of CRISPR/CAS9 technology for antiviral therapy. Experimental data show that CRISPR Cas9 directly targets the viral coding or noncoding regions of the preintegrated or integrated viral genome and creates double-stranded breaks and thus the viral DNA is susceptible to degradation by exonuclease in the host cells. It has also been found that the transcriptional regulation of the lentiviral genome within the host cell during HIV infection is mediated by LTRs, which are critical elements for lentiviral expression within the cells. So, targeting the LTR sequence leads to a higher impact on HIV1 expression (Liao et al., 2015). LTR directed g RNA and CAS9 help in effective targeting and elimination of the HIV1 virus and also immunizes the cell with high specificity and efficiency (Hu et al., 2014). Although, CRISPR Cas9 has been used efficiently for HIV1 targeting, the NHEJ repair process may facilitate the viral escape sometimes that may lead to the targeting of nonviral sequences. The CAS9-associated NHEJ may lead to a reduction in the fitness of the replicating virus. Toward a more efficient treatment for HIV1, the combination of CRISPR and other therapy like antiviral drugs or other gene therapy have been reported (Wang et al.,

2016a–2016c). Researchers have successfully inhibited the expression of both transcriptionally active proviral DNA and latent integrated virus by the use of CRISPR/CAS9 technology. In this method, the CAS9 protein successfully creates a single-stranded DNA cut by its RuvC and HNH domain. To make it more efficient, the delivery system of CRISPR CAS9 into the cell must be improved. The off-target effects should be reduced before the use of Cas9 to render increased effectivity (Ebina et al., 2013).

10.5.5 THE IMPLICATION OF CRISPR IN CARDIAC DISEASE

Cardiovascular diseases constitute a major and increasing health problem in today's world. There are many challenges to gain deeper knowledge about the common and less common causes of cardiovascular mortality. Genetic testing and bioinformatic analyzes have helped to identify the susceptibility of subjects to particular cardiac diseases like coronary and peripheral artery disease (CAD and PAD), rheumatic, hypertensive and congenital heart disease, cerebrovascular disease (stroke), and arrhythmias. The innovative discovery of induced pluripotent stem cells revolutionized the field of genome editing and now the iPS cells are widely used in cardiomyogenesis. iPS cells derived from cardiomyocytes (CM) are used as a unique tool in the field, wherein several scientists have made efforts toward satisfactory in vitro maturation of iPS cells to generate a better model for cardiac pathologies. The iPS cell has many potential functions. For example, if an individual is affected with long QT interval then the iPS cells from this individual may be induced to generate functional CM, which may address the problem. In the case of individuals with structural cardiac defects such as dilated cardiomyopathy (DCM) and hypertrophic cardiomyopathy (HCM), the iPS cells show mutations. The iPS cells derived from DCM show a mutation in a gene that encodes for troponin which leads to abnormality in Ca^{2+} handling and also abnormal arrangement of sarcomeric actin that leads to a decrease in the contractility of the heart. A single missense mutation in the MYH7 gene of the iPS cells derived from HCM patients show unorganized sarcomere and decreased electrophysiological polarity that leads to serious problems. The CRISPR-CAS9 system has been experimentally used for gene knockin and knockout in human iPS cells. So this technology may be used to correct the genetic mutations in the iPS cell model. Barth syndrome is an X-linked genetic cardiac disease that may be eradicated by the combination of iPS cells and CAS9-mediated genome editing. Mutation in the tafazzin (TAZ) gene of iPS cells of a healthy donor by CAS9-mediated genome editing tools have helped to identify the relationship between TAZ gene mutations that cause the disease and mitochondrial phenotypes. The titin gene mutations in DCM have also been evaluated by the CRISPR-CAS9 system (Motta et al., 2017). The CRISPR-CAS9 system can also edit genes of somatic cells in vivo. It can disrupt the proprotein convertase subtilisin/Kexin type 9 (PCSK9) gene, which leads to the lowering of blood cholesterol resulting in the lowering of coronary heart disease. By rectifying the mutation in the calmodulin gene of iPS cells, we can overcome the problem of long QT syndrome. Long QT-associated calmodulin inactivation occurs due to the mutation in CALM1, CALM2 and CALM 3 gene. The long QT syndrome is mediated by the mutation in six calmodulin producing alleles and CRISPR interference technology can selectively rectify these mutated alleles. In this technique, the dCas9 and its associated g RNA bind to the promoter and inactivate RNA polymerase leading to transcriptional deactivation of the mutant alleles.

Organisms like Drosophila, zebrafish, and mouse have been used as a model to study cardiac diseases. Recently, a critical cardiovascular homeostasis pathway has been identified in drosophila by RNAi screening mechanism. CRISPR CAS9 has gained significant popularity to be used for mouse genome engineering. Experimentally, it has been proved that knockout of phospholamban and overexpression of calsequestrin leads to the improvement of cardiac function in the mouse. Recently, it was also shown that the application of CRISPR CAS9 genome editing in mouse zygote can correct the Duchenne muscular dystrophy by restoring the expression of in skeletal as well as in cardiac muscles. Zebrafish has recently been used to study cardiac development and to gain deeper knowledge about many inherited cardiac diseases. However, we should also remember that although CRISPR technology is very efficient in gene editing, its efficacy relies on the development of successful strategies that may vary from patient to patient (Motta et al., 2017).

10.5.6 IMPLICATIONS OF CRISPR IN NEUROSCIENCE

The manipulation of gene expression in physiological context is the principle way to identify the function of a particular gene. The manipulation can be done either by transgenesis or by RNAi technique, although RNAi has been more frequently used in manipulation. But it has some disadvantages like it often target transcripts rather than genomic locus and it sometimes also exerts off-target effects. The scientific literature shows the successful application of

CRISPR CAS9 in the field of neurosciences. It has been reported to disrupt the gene expression of the developmental neural stem cell of the mammalian brain (Kalebic et al., 2016). The CRISPR-CAS9 system has also been used to study the synaptic and circuit function, neural development, and several neural diseases. The conventional RNAi-mediated knockdown or transgenic knockout studies to understand human neural development has some limitations, which can be overcome by using the modern CRISPR CAS9 genome-editing tool (Kalebic et al., 2016). Recently, CRISPR/CAS9 technology has been used for the treatment of neurodegenerative diseases like Parkinson's disease or Huntington disease in animal models. Toward the generation of new models for neurodegenerative diseases, the CRISPR-Cas9 system can be used to directly target a particular gene or one or two alleles in the embryonic genome (Yang et al., 2014). The postmitotic neurons in adult brain cells do not undergo any division. These postmitotic neurons can be manipulated by using the CRISPR-CAS9 system. The homology dependent repair is mainly used to repair the double-stranded break in this case. Nonhomologous end joining is the predominant mechanism that is used to repair double-stranded breaks of postmitotic neurons. However, in an experiment where rat`s hippocampal cell culture was transfected with designed CAS9 and sg RNA expression vector by biolistic transfection method, the rate of e-mutation is almost 100% in CAS9 positive cells and the most frequent mutation is indel and frameshift type of mutations. This suggests that the mutagenesis efficacy in postmitotic neurons is almost the same as in normal dividing cells and nonhomologous end joining is the major repair mechanism (Ren-Jie et al., 2017). The complex structure and gene expression of the brain makes most of the brain diseases multifactorial and the causal link between an individual gene and central nervous system (CNS) diseases is largely unknown. Complex interactions of the multiple alleles which work with several environmental factors determine the genetic risk of the CNS disorders. Although new technologies have been developed for real-time tracking and visualization of the complex brain structure and neural activity, CRISPR/CAS9 technology provides an efficient way to disclose the complexity of CNS and its disease-causing factors. The invention of mouse models and the development of new novel models for human disease is now possible due to the advent of the CRISPR-CAS9 system. In vitro neural cells, ex vivo brain cells, and in vitro adult and embryonic mouse brain cells have been cultured and manipulated by CRISPR technology. It is now possible to create patient-specific mutation by using the unique gene-editing process of CRISPR. In this process, a panel of mouse models is created in which a mouse allele is modified to carry different mutations thus covering the complete allelic series. In the humanization project, the mouse models are used for CRISPR-genome editing to improve the accuracy of complex human traits. In vivo knockout of neurons mediated by CRISPR has been done using ultra electroporation of CAS9 and sg RNA directly to knockout the Grin1 gene thereby disrupting the NMDA receptor. To knockout the Huntingtin gene from the brain affected with Huntington disease, adeno-associated virus-mediated strategy has been successfully applied (Vesikansa, 2018). There is an overexpression of α-synuclein in the model transgenic animal with Parkinson's disease. Application of CRISPR CAS9, in this case, leads to depletion of the level of the mutated gene resulting in gene inactivation in dopaminergic neurons. Experimentally, it has also been proven that by microinjecting CAS9 and g RNA to the neural stem cell in an organotypic slice culture can disrupt certain gene expression (Kalebic, 2016). The causative landscape of most of the CNS disease is made up of ten to hundreds of different genes that carry many different mutations. By the application of CRISPR technology, multiple alleles can be simultaneously engineered causing three to five mutations in a single generation allowing a combinational study of several genes. The advances in the CRISPR technique provided an excellent way for modeling neurodegenerative diseases using mouse models, which results in less time consumption and also is less costly. The neuropsychiatric disorder modeling in mouse is challenging because many of the symptoms cannot be monitored in the mouse brain. The in vivo application of CRISPR has some drawbacks because the commonly used nuclease spCas large in size and incorporation of nuclease along with sg RNA expression (an important gene element) is limited due to the limited packing capacity of AAV. A nonviral in vivo platform for the expression of CRISPR has been developed recently where CAS9 sg RNA ribonucleoprotein is used to express CRISPR in several neural subtype cells. The RNPs penetrate the cell using the insertion pattern of the SV40 virus and it has a specific nuclear localization signal for CAS9. Gold nanoparticles have also been successfully used to deliver CAS9 along with sg RNA and template DNA that edit the genome by homologous end joining in an in vivo condition. By using CRISPR-mediated gene correction technique, humanized induced pluripotent stem cells or neurons derived from specific disease affected donors can be modified to produce an isogenic control cell line to treat a particular inherited neurological disease. HDR-independent target genome editing has also been developed recently to edit postmitotic neurons (Vesikansa, 2018).

Although there are many powerful applications of CRISPR CAS9 in the field of genome editing, there are also some limitations of CRISPR in the context of its application in neurological diseases. It has a low efficiency of transgene expression and the viral dependent transfection of CRISPR is not useful for neural delivery. So biolistic transfection, in utero electroporation, infusion of the recombinant protein by liposomal approaches has been used for neural systems (Savell and Day, 2017).

10.5.7 CRISPR-MEDIATED AGRICULTURAL ADVANCES AND IMPROVEMENT OF CROP PRODUCTIVITY

New crop varieties are developed for domestication and making major advances in feeding the world and society and for this different plant breeding techniques are successfully used (Borrelli et al., 2018). The conventional mutagenesis method in the case of plants like ethyl methyl sulfate mutagenesis and γ radiation for genome editing is now progressively being replaced by precise manipulation of specific genome sequences thus creating genetic variation in the plant crop for sustainable agriculture. CRISPR technology has been applied in many animals and model organisms. Likewise, this technology has also been used in many plant models like *Nicotiana tabacum*, *Arabidopsis*, wheat, and rice. High efficiency of mutation has been reported both in Arabidopsis and in Rice plants using this technology as the mutation frequency (the number of plants that contain the CRISPR CAS9 transgene) is very high. Regarding the mutagenesis of plants during early embryonic division, it may be heterozygous (contain mutation either one of the two alleles and repair occur on the same allele), homozygous (both the alleles contain mutation and repair occurs on the same allele), or biallelic (both the allele of same gene mutagenized but repair occur in different alleles). In many cases, mutations occur at the later stage of development at a specific cell or tissue as a result of which chimeric plant may form with normal wild-type traits. Multiple genome editing can be done by the expression of two or more sg RNA at a time, which is only possible by the use of CRISPR technology. As compared to the animal gene-editing process where off-target activity has a considerable effect on genome editing, in plants the off-target mutation is less problematic because that may be overcome by test crossing. It has also been observed that plants having homozygous, heterozygous, or biallelic mutations may not carry a further mutation in the next generations indicating that Cas9 may not be able to bind the target site once the target has been modified. The off-target effects may be reduced by using truncated g RNA, which is more sensitive to nucleotide mismatches (so that off-target mutation may escape) or by using the mutant variant of CAS9 that may reduce off-target mutation by 100-fold (Belhaj et al., 2015). The CRISPR/CAS9 technology has been used to knockout the polyploidy in many plant species. In an allohexaploid wheat variety, three selected genes were mutated using CRISPR technology and it was observed that the frequency and heritability of the mutations were different for the three different genes. By the application of CRISPR CAS9 in a single cell, a new fertile plant can be regenerated. The CRISPR efficiency is species-specific and genotype-specific and thus only lab varieties of different crops may be modified by the use of CRISPR. The introduction of the CRISPR technique in conventional transgenic methods like Agrobacterium-mediated gene transfer or biolistic method or direct transformation of protoplast is also beneficial. The biolistic and direct protoplast transformation facilitates the transfection of CAS9 and sg RNA along with the ribonucleoprotein. These two methods also have a higher repair rate of DNA. To increase the copy number, the biolistic method is used to insert a DNA template inside the host cell. Gene transfer in rice is mostly done by Agrobacterium-mediated gene transfer but the biolistic method is also used in many cases. The transient protoplast fusion is the most efficient way of transformation in Maize namely CRISPR, whereas the protoplast fusion in rice plants is a very difficult task (Borrelli et al., 2018). Experimentally, CAS9/sg RNA has been used for better transformation efficiency in rice. In this study, an *Agrobacterium*-based binary vector was designed, which contains CAS9 recombinant sequence along with many sg RNA sequences and this was used to transfect the callus cells of immature rice embryos (Zhou et al., 2014).

The global agricultural crop loss has increased from 20% to 40% and mainly attributed to a variety of stresses like viral stress, fungal stress, bacterial diseases, etc. CRISPR CAS9 is an effective tool to eradicate viral, fungal, and bacterial diseases. The *Polymyxa graminis virus* is a potent source of loss in the crop fields that lead to a major loss in agriculture. *Begomoviruses* are most dangerous in terms of the agricultural economy. Their genome consists of 220 bps and they replicate namely rolling circle model of replication. Recent studies reveal that CRISPR CAS9 introduction into the host plant nucleus could help to disrupt the vector DNA. Some mutated virus has also been constructed by CRISPR. Recent work has shown that when CAS9 and sg RNA is on the coding strand of the virus, the new variants of viruses may be created that are capable of replication. There are many challenges to rescue the crop plants from biotic stresses. For example, protection against RNA viruses was not obtained by the use of the

CRISPR technique as the RNA isolated from *Streptococcus pyrogens* only recognises ds DNA. However, a very recent study reveals that FnCas9 (a variety of normal CAS9) is capable to cut the RNA by its nuclease activity. The FnCAS9 expression can provide 20%–40% resistance that is stable and long term for up to the sixth generation. The mechanism of action of FnCas9 is based on CRISPR interference. The catalytic inactivation of FnCAS9 can limit the formation of viral mutant variants. FnCAS9 also has an advantage over normal Cas9 that is unlike CAS9 that requires nuclear localization signal to disrupt the ds DNA, FnCAS9 does not require any nuclear localization signal. It can directly disrupt the RNA in the cytoplasm. Another way to get rid of viral diseases is to modify the plant genome in such a way that it shows virus-resistant traits. A fungus is also able to create serious issues in the agricultural economy. Fungi that produce mycotoxins generate massive crop diseases like smut disease, mildew disease, rot diseases, etc. Nowadays, potential genes that provide disease resistance to fungi are targeted to be edited by the use of CRISPR CAS9. CAS9-based RNP technique also seems to be effective in editing a cluster of the genome in apple plants that may help to prevent blight disease. In perennial plants, genome editing has been done by protoplast technique as the stably integrated CRISPR CAS9 requires more time to segregate namely backcross. In annual plants, it only requires a few months of generation time. The RNP technique is more approachable as normal CRISPR CAS9 has an off-target mutation effect whereas RNP-mediated CRISPR CAS9 provides target specificity. CRISPR CAS9 has been widely used to make crop plants disease resistant as it can knockout a susceptible gene and provide the protection. However, it is challenging to make the CRISPR work in the field condition and reach the same effect which has been shown in laboratory conditions. The durability of disease resistance is also a serious concern and convincing general people about technology is also challenging (Borrelli et al., 2018). By the use of CRISPR editing, the agricultural productivity of rice can be improved. CRISPR CAS9 multi genome editing technique can also be used to improve quantitative traits such as yield and quality. These quantitative traits are controlled by multiple loci. In the case of rice, knocking out certain genes like GS3 and DEP1, etc. by the CRISPR system increases the grain size. To improve cold stress tolerance in rice, CRISPR technology has been applied recently. The gene TYFY1b and its analog TIFY1a have been identified to be the cold stress tolerance gene in rice and by the use of CRISPR editing the function of these genes may be improved. A recent study also shows that CRISPR Cpf1 has more efficient and accurate genome editing than CRISPR CAS9. The CRISPR Cpf1 can perform cleavage without the need of any tracer DNA.

Also, there are many advantages of CRISPR Cfp1 over the CRISPR-CAS9 system (Mishra et al., 2018).

10.5.8 USE OF CRISPR AS ANTIMICROBIALS

Antibiotics are known for their potential application to disrupt bacterial infection. In recent times, antimicrobials are threatened to undoing. The overuse and abuse of antimicrobials have led to the development of multi-drug-resistant microbes and also the beneficial amicrobial flora is destroyed (Beisel et al., 2015). As a result of this antimicrobial resistance, many common bacterial diseases are becoming increasingly difficult to cure. The development of modern antimicrobials without antimicrobial resistance requires more effort in the upcoming years. The CRISPR is an important part of bacterial immunity systems. Out of six types of the CRISPR system, type II CRISPR is structurally very simple and also it has a versatile application in genome editing and ecological engineering. Several years ago, it was shown that synthetic CRISPR may be used to eradicate certain microbes. Recent studies suggest that the CRISPR technique can be used as a potential tool for the disruption of multidrug-resistant bacterial strains. CRISPR/CAS9 technique is used to remove the antimicrobial-resistant gene from a selective strain of bacteria thus enabling the resensitization of particular bacteria to antimicrobials. To remove the specific strains from the mixed population of bacteria, CRISPR CAS9-mediated transformation has been done. For this purpose, CRISPR CAS9 has been incorporated into the host by phagemid-mediated delivery (Pursey et al., 2018). In this method, plasmids with selective phage packaging signals along with other components like *Streptococcus pyrogens* nuclease encoding sequence, artificially designed CRISPR RNA sequence and trans activating RNA sequence has been delivered into selective microbial strains. Here phage is used because phage is naturally evolved to inject their DNA into the host bacterial cell. After the packaging of phagemid and its delivery to the host, there is a rapid disruption of bacterial cells. These phagemids possess specific antimicrobial resistance plasmids, which result in the efficient removal of the plasmid (Beisel et al., 2014). Delivery of CRISPR CAS9 containing phagemids may cause sequence-specific killing of bacteria. In yet another study, it has been shown that the delivery of CRISPR CAS9 namely conjugative plasmid leads to the death of the bacteria that contain antimicrobial-resistant genes (Pursey et al., 2018). The traditional antibiotics have a poor range of application because most of the bacteria contain antimicrobial resistance genes.

This serious problem can now be overcome by CRISPR-based editing of an antimicrobial gene. Besides, CRISPR CAS9 cannot only provide resistance to the undesired gene but can also target mobile genetic elements like transposons to disrupt it thus ensuring genetic homeostasis. CRISPR CAS9 can also be used for vaccination by targeting invasive genetic elements. The acquisition of virulence trait by natural DNA transformation is prevented when CRISPR technology creates a potent barrier for bacteriophages (Barrangou, 2015).

Engineered CRISPR has a wide range of applications and it has been explored by scientists in recent research to design antimicrobial drugs and to modify antimicrobial resistance genes in many bacteria. The antimicrobials designed by CRISPR CAS9 have a broad spectrum of activity and target specificity for a DNA sequence. RNA-guided nucleases (RGN) are delivered to the microbial cells by using specific bacteriophage or bacteria carrying transmissible plasmids, which integrates into the genome namely conjugation. The RGN system is a recent approach by scientists where the undesirable gene or polymorphic sequences are targeted. This RGN system can also be used to target specific strains of bacteria by identifying their genetic signature and then knocking down the virulence or antimicrobial gene of the selected strains. It also reduces the off-target effect and remodels the genetic makeup of microbes. As compared to natural CRISPR, there is a simple modification in the spacer region of RGN CRISPR locus as a result of which it can cleave any DNA sequences. In the design of RGN, the only requirement is the NGG motif at the 3′ end of the cleavage site. With all the features, functional RGN is packed within a bacteriophage, which creates a functional lethal device that enables disruption of any undesired DNA sequences. Sometimes, bacteria have high copy numbers of plasmids that carry a multidrug-resistant gene, which is transmissible in the microbial population. To verify the action of RGN against the multiple copies of the drug-resistant genes, it has been tagged with GFP and subsequently, it was proved that RGN is sufficient to get rid of the high number of plasmids that have multidrug-resistant genes. There are multiple applications of RGN in antimicrobial designing. Although the current approach of antibiotics or probiotics is poorly described by the use of RGN, alternative paths for modulating complex microbial populations and reducing off-target effects should be achieved in near future (Citorik et al., 2014). Experimentally, it has also been shown that CRISPR interference can inhibit the transformation of noncapsulated strain to capsulated virulent strains (Bikard et al., 2012). There are many limitations of antimicrobials designing by the application of CRISPR. For example, a high range of host and low range of bacteriophage and we have little knowledge to expand the host range. Another limitation is that modification of surface receptors (which is required for the cell to be infected by bacteriophage) is very critical and is often risky. The delivery of CAS9 using nanoparticles has been attempted but it requires further explorations and standardizations. We must choose the target sequence or specific strains for efficient killing so that other beneficial strains may survive (Beisel et al., 2014).

Parasites cause the majority of the threatening diseases to livestock and also to humans and cause economic loss worldwide. After the advancement of CRISPR in biological research, CRISPR technology is now used for parasitological study to edit the genome of *Leishmania, Trypanosoma,* etc. A recent study has reported an effective way by which CAS9 manipulated the genome of *Leishmania* by knocking out one of the genes in a particular locus (paraflagellar rod 2 locus) and it was done by a single transfection in short time period. Loss of function, deletion, insertion mutation of particular sequences by CRISPR CAS9 has also been reported. In another parasite, *Trypanosoma cruzi,* the CRISPR-CAS9 system has been used to target selective 65 genes for knockout resulting in a decreased level of β-galactofuranosyl glycosyltransferase. CRISPR technology has been used to prevent many parasites like *Leishmania, Trypanosoma, toxoplasma,* etc. So this technique can be adapted for the future study of the structure and function of other parasites (Sastya et al., 2018).

10.6 ROLE OF CRISPR IN EVOLUTION

Evolution is a long-term process that can be studied by fossils or other past evidence. Sometimes, past evidence may be imprinted in DNA that provides a better way to understand the gene and its heredity. In the case of bacteria and viruses, CRISPR CAS9 provides a unique cue of coevolution that helps to maintain biodiversity. The study of coevolution is easy because CRISPR provides all the evidence of evolution in a complete DNA sequence of a single organism. The bacteria record their evolution as well as the pathogen's evolution in their CRISPR locus. The metagenomic study of coevolutionary patterns has been done either by observing the geographical pattern of adaptation of the host or by the direct temporal sampling of host and phage. The local pathogen selects their host mainly by antagonistic mechanism and that is the main method of coevolution in the host as well as in parasite. The geographical sample of two distant populations analyzed by CRISPR sampling results in very different outcomes as compared to normal bacterial genome sequence analysis. The study shows no similarity in spacer sequences in two different bacteria. This result suggests that strains are dispersed and genetically homogenous and thus there is almost no difference in their main genomic sequence. However, they have rapid divergence in their CRISPR loci because they are exposed to different phages in different environments (Vale and Little, 2010). In the study of the coevolutionary dynamics of CRISPR, it was found that highly diverse assemblage starts from a low diversity of initial condition. In a study, it was observed that coevolutionary dynamics can be controlled by three factors in the environment. Firstly, when a rare strain invades the bacteria, the acquired spacer from the invader strain will give more fitness to the host and reoccurrence of the rare strain will give more fitness advantages due to changes in genetic states. Secondly, invasion by a group of strains with identical immune phenotypes but different genotypes will provide more fitness. Also, it was suggested that only the first few spacers are providing coevolutionary forces because CRISPR immunity is dominated by the most recent acquired spacer. In a coevolutionary model, the ecological, molecular, and evolutionary components have to be considered. The ecological component mainly determines host and virus density and it depends on the reproduction of the host and the inactivation of the viral particle outside the host. Here there are two possible consequences, the first host dies because of viral infection or host disable the viral genome and modify its genome in a normal fashion. The molecular components determine whether the viral component leads to host lysis or there may be viral deactivation or spacer integration during infection. The evolutionary component introduces new host–virus interaction that can modify the genetic states of both the host and parasite (Childs et al., 2019). In the case of temporal assay or time-shift assay of coevolution dynamics, the host and parasite are compared for their past and future counterpart in a given time. In a host–parasite interaction, the parasite shows their host range as well as specifications to the host. This provides the basic idea of coevolution. The coevolutionary study of the host–parasite is particularly important because coevolutionary mechanisms can be influenced by many factors such as community structure diversity and population productivity (Betts et al., 2014).

Various experiments suggest that interaction between bacteria and phage leads to the coevolutionary changes in both phage and bacteria. Bacteria have a very limited reciprocal cycle of evolution (1–1.5). It has been reported that if the phage is unable to affect the evolution of the bacterial sequence then the coevolution stops. This challenge can be overcome by the experimental modification of the bacterial cell wall as a result of which the phage target may also be modified. Among the interactions of bacteria and phage, the best documented is that of *Pseudomonas fluorescens* SBW25 and the T7-like *podovirus* Φ2. In this interaction, it was shown that if the bacteria are cultured in rich media as a batch culture for a longer period then it may show the coevolutionary changes.

Coevolution is generally of different types. The type in which bacterial resistance and infectivity of phage evolve side by side is known as arms-race coevolution. In arms-race coevolution, phage evolves to infect the common bacterial genotype and thus the rare strain of bacteria gets resistant (Koskella and Brockhurst, 2014). In an experimental study of some particular CRISPR locus, it was found that there may be a spacer of transposons and other extrachromosomal plasmids. It also suggests that CRISPR loci may also be associated with other genetic records rather than only the main genomic memories (Koskella and Brockhurst, 2014).

For a proper understanding of CRISPR CAS9-associated coevolutionary studies, recently mathematical as well as computational methods have been used. The computational and mathematical study is based on a simple assumption about phage and bacterial interaction and CRISPR Cas9-based adaptive immune system. This study suggests that bacteria with CRISPR-CAS9 system can also be invaded by the phage and bacteria that do not have this CRISPR-based immune system can also prevent the phage infection. However, there are many limitations in mathematical and computational-based coevolutionary study and thus significant expertize is needed toward understanding the mathematical and computational coevolutionary studies (Levin et al., 2013).

10.7 CRISPR-MEDIATED NEW FINDINGS IN MODERN BIOLOGICAL ENGINEERING

The application of CRISPR is now becoming diversified in many branches of biology as researchers have started to explore this field in modern genetic engineering and genome editing. This CRISPR tool has a huge application in medical biology as it can target any DNA sequence with the help of g RNA. A new CAS9 enzyme has recently been developed that can target almost any DNA sequence and is structurally similar to normal CAS9. By using editing techniques, researchers have tried to find other varieties of CAS9 without compromising the accuracy of the system (Chatterjee et al., 2018). To use the CRISPR in a particular cell only, researchers have discovered a switching mechanism. It can be preferably used in cancer cells where CAS9 variant that is pro CAS9 is used as a sensor for viral infection where the viral infection is ultimately resulting in the formation of cancer. Pro CAS9 is functionally the same as the normal CAS9 but the only difference is that it carries a short protein sequence that can be cut out before it binds to the DNA. Depending on this mechanism, pro CAS9 can sense the type of cell in which it will act (Oakes et al., 2019). The off-target editing error that is the major hurdle in the field of CRISPR genome editing can now be overcome by simply changing 1400 amino acids of normal CAS9 of *Streptococcus pyrogens (*Slaymaker et al., 2015). CRISPR CAS9-mediated genome editing technique is now used to edit mammalian cell lines like in mice where female germline cells can be altered to drive the expression of white fur.

Recently, CRISPR Chip, a digital detection system that enables detection of DNA mutations without amplification, has been invented. It was designed by Kiana Aran of the Keck Graduate Institution who combined the power of CRISPR nucleic acid editing with ultra-sensitivity of graphene. This CRISPR chip is a biosensor for electronic detection of unamplified target DNA and it is also able to detect single point mutation. Recently, a new anti-CRISPR protein has been identified in the soil as well as the human gut. Many more findings are related to the use of this new genome-editing tool. Though there are many limitations to the application of CRISPR, still it may be used to create revolutionary findings in modern biology for future applications.

10.8 ADVANTAGE AND LIMITATIONS OF CRISPR

The CRISPR/CAS9 technology has many advantages over existing conventional editing technologies. CRISPR CAS9 can be used to rectify disease-causing mutations by removing a single base or large scale deletion of DNA bases. It can also be used to correct the altered function of a mutated protein. For restoring the normal protein function, CRISPR CAS9 along with the g RNA creates deletion in the DNA at the mutation site. The simple and efficient genome editing strategy of CRISPR CAS9 makes it a more advanced genetic tool. It can also be applied for genome editing of the embryo because it requires lesser time to modify target gene, whereas conventional genome editing of an embryonic stem cell requires much more time for targeting the genome (Flora and Welcker, 2017). The large sg RNA libraries that are designed with the help of the CRISPR/CAS9 technology make the target identification and functional screening very easy. Large populations and global warming create a huge food crisis worldwide. This problem may be overcome by the promising approach of CRISPR CAS9 mediate crop improvements (Omodamilola and Ibrahim, 2018). CRISPR-CAS9 system can also be used in somatic cell genome editing of animals in vivo to treat many genetic disorders (Savić and Schwank, 2016). Targeting multiple genes at a time to induce mutation in them by the help of multiple g RNA is the most recent use of the CRISPR-CAS9 system. It has also been used to screen noncoding DNA that fosters gaining more knowledge about the regulatory sequences, which play an important role in human genetic diseases.

One of the limitations of CRISPR is that it requires a PAM sequence for targeting and linking of the nuclease. However, this problem can be overcome by increasing the number of PAM sequences or by modifying the sequence of the PAM region. Another major problem related to the application of CRISPR CAS9 is in embryo mutagenesis. The off-target effect may lead to nonspecific targeting of a nontarget location of the genome. The off-target effect occurs due to the unwanted binding of sg RNA. By modifying the CAS9-binding sites or disrupting some of the CAS9-binding sites, the off-target effect may be reduced to some extent. Another major issue is that the efficacy of genome editing may vary from one cell to another especially *in vivo* conditions. The viability of editing efficacy is also a major issue related to the CRISPR CAS9-editing technique. The double-stranded break is repaired by either HDR or NHEJ. NHEJ has a certain frequency of introducing insertion or deletion. Thus, the usage of HDR can improve editing efficacy. The CRISPR CAS9 has also been reported to be used to find out the pluripotency transcription factor OCT4 in human embryogenesis in which the OCT4 encoding gene has compromised expression (Motta et al., 2017).

10.9 CONCLUDING REMARKS

The CRISPR-CAS9 system opens up a new era in the genome editing platform. It has transformed our knowledge of genome manipulation. Genome engineering by the use of CAS9 and sg RNA is now routinely used in many laboratories (Sternberg and Doudna, 2017). To make genome engineering more powerful, improvisation of the capabilities of CRISPR may also require. There is a lot to be known about the CRISPR system, which will be unraveled only by the regular practice of CRISPR tools in laboratories (Brokowski and Adli, 2018). Experiments to date have revealed the complex process of how CRISPR CAS9 can be used to target and cleave the DNA at a specific site. The targeting is done by PAM recognition and cleavage is done by CAS9. However, the exact mechanisms of off-target effects are yet to be understood (Jiang and Doudna, 2017). Gene therapy has been proposed to cure many genetic diseases but the rapid generation of modified nuclease for targeting the desired position of the genome is only possible by the use of CRISPR because it is less costly and user-friendly. In the future, to cure inherited mutation, CRISPR technology may be used to edit human embryonic genes. The selection of the gene to be edited depends on researchers and so there is an ethical and legal problem in using CRISPR as a clinical drug (Pandey et al., 2017). After the discovery of this bacterial adaptive immune system, its use is now extensive in the field of stem cell genome editing, which leads to many innovative ideas. The sequential editing of the genome of an organoid can also be done by the CRISPR CAS9-editing tool. It also has functions other than guide dependent genome editing. It can be used as a probe, altering the genomic environment and also affecting the stability in disease-causing genes (Freiermuth et al., 2017). CRISPR provides a huge range of advances as compared to other conventional genome-editing techniques. In the case of plants, the off-target mutation still needs to be solved systematically, whereas in animals the off-target effect has now been overcome. Taken together, CRISPR CAS9 can be used to revolutionize the fields of applied science and also in the case of human genetic disease and plant science (Belhaj et al., 2015).

KEYWORDS

- **CRISPR**
- **CAS9**
- **genome editing**
- **g RNA**
- **double-strand break**
- **stem cells**
- **cancer**

REFERENCES

Aguirre, A. J., Meyers, R. M., Weir, B. A., Vazquez, F., Zhang, C. Z., Ben-David, U., Cook, A., Ha, G., Harrington, W. F., Doshi, M. B., Kost-Alimova, M., Gill, S., Xu, H., Ali, L. D., Jiang, G., Pantel, S., Lee, Y., Goodale, A., Cherniack, A. D., Oh, C., Kryukov, G., Cowley, G. S., Garraway, L. A., Stegmaier, K., Roberts, C. W., Golub, T. R., Meyerson, M., Root,

D. E., Tsherniak, A., Hahn, W. C. Genomic copy number dictates a gene-independent cell response to CRISPR-Cas9 targeting. *Cancer Discov.* 2016, 6(8), 914–929.

Barrangou, R. Diversity of CRISPR-Cas immune systems and molecular machines. *Genome Biol.* 2015, 16, 247.

Barrangou, R., Fremaux, C., Deveau, H., Richards, M., Boyaval, P., Moineau, S., Romero, D. A., Horvath, P. CRISPR provides acquired resistance against viruses in prokaryotes. *Science* 2007, 315(5819), 1709–1712.

Bassuk, A. G., Zheng, A., Li, Y., Tsang, S. H., Mahajan, V. B. Precision medicine: genetic repair of retinitis pigmentosa in patient derived stem cells. *Sci. Rep.* 2016, 6, 19969.

Beisel, C. L., Gomaa, A. A., Barrangou, R. A CRISPR design for next-generation antimicrobials. *Genome Biol.* 2015, 15(11), 516.

Belhaj, K., Chaparro-Garcia, A., Kamoun, S., Patron, N. J., Nekrasov, V. Editing plant genomes with CRISPR-Cas9. *Curr. Opin. Biotechnol.* 2015, 32, 76–84.

Betts, A., Kaltz, O., Hochberg, M. E. Contrasted coevolutionary dynamics between a bacterial pathogen and its bacteriophages. *Proc. Natl. Acad. Sci. USA.* 2014, 111(30), 11109–11114.

Bikard, D., Hatoum-Aslan, A., Mucida, D., Marraffini, L. A. CRISPR interference can prevent natural transformation and virulence acquisition during in vivo bacterial infection. *Cell Host Microbe.* 2012, 12(2), 177–186.

Bjursell, M., Porritt, M. J., Ericson, E., Taheri-Ghahfarokhi, A., Clausen, M., Magnusson, L., Admyre, T., Nitsch, R., Mayr, L., Aasehaug, L., Seeliger, F., Maresca, M., Bohlooly-Y, M., Wiseman, J. Therapeutic genome editing with CRISPR-Cas9 in a humanized mouse model ameliorates α1-antitrypsin deficiency phenotype. *EBioMed.* 2018, 29, 104–111.

Borrelli, V. M. G.; Brambilla, V., Rogowsky, P., Marocco, A., Lanubile, A. The enhancement of plant disease resistance using CRISPR-Cas9 technology. *Front. Plant Sci.* 2018, 9, 1245.

Brokowski, C., Adli, M. CRISPR ethics: moral considerations for applications of a powerful tool. *J. Mol. Biol.* 2019, 431(1), 88–101.

Charpentier, E., Richter, H., van der Oost, J., White, M. F. Biogenesis pathways of RNA guides in archaeal and bacterial CRISPR-Cas adaptive immunity. *FEMS Microbiol. Rev.* 2015, 39(3), 428–441.

Chatterjee, P., Jakimo, N., Jacobson, J. M. Minimal PAM specificity of a highly similar SpCas9 ortholog. *Sci. Adv.* 2018, 4(10), eaau0766.

Chira, S., Gulei, D., Hajitou, A., Zimta, A. A., Cordelier, P., Berindan-Neagoe, I. CRISPR-Cas9: transcending the reality of genome editing. *Mol. Ther. Nucleic Acids* 2017, 7, 211–222.

Citorik, R. J., Mimee, M., Lu, T. K. Sequence-specific antimicrobials using efficiently delivered RNA-guided nucleases. *Nat. Biotechnol.* 2014, 32(11), 1141–1145.

Cong, L., Ran, F. A., Cox, D., Lin, S., Barretto, R., Habib, N., Hsu, P. D., Wu, X., Jiang, W., Marraffini, L. A., Zhang, F. Multiplex genome engineering using CRISPR-Cas systems. *Science* 2013, 339(6121), 819–823.

Demirci, Y., Zhang, B., Unver, T. CRISPR-Cas9: An RNA-guided highly precise synthetic tool for plant genome editing. *J. Cell Physiol.* 2018, 233(3), 1844–1859.

Doudna, J. A., Charpentier, E. Genome editing: the new frontier of genome engineering with CRISPR-Cas9. *Science* 2014, 346(6213), 1258096.

Ebina, H., Misawa, N., Kanemura, Y., Koyanagi, Y. Harnessing the CRISPR-Cas9 system to disrupt latent HIV-1 provirus. *Sci. Rep.* 2013, 3, 2510.

Feng, Y., Sassi, S., Shen, J. K., Yang, X., Gao, Y., Osaka, E., Zhang, J., Yang, S., Yang, C., Mankin, H. J., Hornicek, F. J., Duan, Z. Targeting Cdk11 in osteosarcoma cells using the CRISPR-Cas9 system. *J. Orthop. Res.* 2015, 33(2), 199–207.

Flora, A and Welker, J. CRISPR genome engineering: advantages and limitations. https://www.taconic.com/taconic-insights/model-generation-solutions/crispr-genome-engineering-advantages-limitations.html 2017.

Freiermuth, J. L., Powell-Castilla, I. J., Gallicano, G. I. Toward a CRISPR Picture: Use of CRISPR-Cas9 to model diseases in human stem cells in vitro. *J. Cell Biochem.* 2018, 119(1), 62–68.

Hartenian, E., Doench, J. G. Genetic screens and functional genomics using CRISPR-Cas9 technology. *FEBS J.* 2015, 282(8), 1383–1393.

Hille, F., Charpentier, E. CRISPR-Cas: biology, mechanisms and relevance. *Philos. Trans. R. Soc. Lond. B. Biol. Sci.* 2016, 371(1707), pii: 20150496.

Hsu, P. D., Lander, E. S., Zhang, F. Development and applications of CRISPR-Cas9 for genome engineering. *Cell* 2014, 157(6), 1262–1278.

Hu, W., Kaminski, R., Yang, F., Zhang, Y., Cosentino, L., Li, F., Luo, B., Alvarez-Carbonell, D., Garcia-Mesa, Y., Karn, J., Mo, X., Khalili, K. RNA-directed gene editing specifically eradicates latent and prevents new HIV-1 infection. *Proc. Natl. Acad. Sci. USA.* 2014, 111(31), 11461–11466.

Irion, U., Krauss, J., Nüsslein-Volhard, C. Precise and efficient genome editing in zebrafish using the CRISPR-Cas9 system. *Development.* 2014, 141(24), 4827–4830.

Jiang, F., Doudna, J. A. CRISPR–Cas9 structures and mechanisms. *Annu. Rev. Biophys.* 2017, 46, 505–529.

Kalebic, N., Taverna, E., Tavano, S., Wong, F. K., Suchold, D., Winkler, S., Huttner, W. B., Sarov, M. CRISPR-Cas9-induced disruption of gene expression in mouse embryonic brain and single neural stem cells in vivo. *EMBO Rep.* 2016, 17(3), 338–348.

Koonin, E. V. CRISPR: a new principle of genome engineering linked to conceptual shifts in evolutionary biology. *Biol. Philos.* 2019, 34(1), 9.

Koskella, B., Brockhurst, M. A. Bacteria–phage coevolution as a driver of ecological and evolutionary processes in microbial communities. *FEMS Microbiol. Rev.* 2014, 38(5), 916–931.

Levin, B. R., Moineau, S., Bushman, M., Barrangou, R. the population and evolutionary dynamics of phage and bacteria with CRISPR-mediated immunity. *PLoS Genet.* 2013, 9(3), e1003312.

Li, Y., Pan, S., Zhang, Y., Ren, M., Feng, M., Peng, N., Chen, L., Liang, Y. X., She, Q. Harnessing Type I and Type III CRISPR-Cas systems for genome editing. *Nucleic Acids Res.* 2016, 44(4), e34.

Liao, H. K., Gu, Y., Diaz, A., Marlett, J., Takahashi, Y., Li, M., Suzuki, K., Xu, R., Hishida, T., Chang, C. J., Esteban, C. R., Young, J., Izpisua Belmonte, J. C. Use of the CRISPR-Cas9 system as an intracellular defense against HIV-1 infection in human cells. *Nat. Commun.* 2015, 6, 6413.

Liu, Y., Qi, X., Zeng, Z., Wang, L., Wang, J., Zhang, T., Xu, Q., Shen, C., Zhou, G., Yang, S., Chen, X., Lu, F. CRISPR-Cas9-mediated p53 and Pten dual mutation accelerates hepatocarcinogenesis in adult hepatitis B virus transgenic mice. *Sci. Rep.* 2017a, 7(1), 2796.

Makarova, K. S., Koonin, E. V. Annotation and classification of CRISPR-cas systems. *Methods Mol. Biol.* 2015, 1311, 47–75.

Manica, A., Zebec, Z., Steinkellner, J., Schleper, C. Unexpectedly broad target recognition of the CRISPR-mediated virus defense system in the archaeon Sulfolobus solfataricus. *Nucleic Acids Res.* 2013, 41(22), 10509–10517.

Marraffini, L. A. CRISPR-Cas immunity in prokaryotes. *Nature* 2015, 526(7571), 55–61.

Martinez-Lage, M., Puig-Serra, P., Menendez, P., Torres-Ruiz, R., Rodriguez-Perales, S. CRISPR-Cas9 for cancer therapy: hopes and challenges. *Biomedicines.* 2018, 6(4), pii: E105.

Maxwell, K. L., Garcia, B., Bondy-Denomy, J., Bona, D., Hidalgo-Reyes, Y., Davidson, A. R. The solution structure of an anti-CRISPR protein. *Nat. Commun.* 2016, 7, 13134.

Mishra, R., Joshi, R. K., Zhao, K. Genome editing in rice: recent advances, challenges, and future implications. *Front. Plant Sci.* 2018, 9, 1361.

Motta, B. M., Pramstaller, P. P., Hicks, A. A., Rossini, A. The impact of CRISPR-Cas9 technology on cardiac research: from disease modelling to therapeutic approaches. *Stem Cells Int.* 2017, 2017, 8960236.

Mougiakos, I., Bosma, E. F., Ganguly, J., van der Oost, J., van Kranenburg, R. Hijacking CRISPR-Cas for high-throughput bacterial metabolic engineering: advances and prospects. *Curr. Opin. Biotechnol.* 2018, 50, 146–157.

Oakes, B. L., Fellmann, C., Rishi, H., Taylor, K. L., Ren, S. M., Nadler, D. C., Yokoo, R., Arkin, A. P., Doudna, J. A., Savage, D. F. CRISPR-Cas9 circular permutants as programmable scaffolds for genome modification. *Cell* 2019, 176(1-2), 254–267.

Omodamilola, O. I., Ibrahim, A. U. CRISPR technology:advantages, limitations, and future direction. *J. Biomed. Pharm. Sci.* 2018, 1, 115.

Özcan, A., Pausch, P., Linden, A., Wulf, A., Schühle, K., Heider, J., Urlaub, H., Heimerl, T., Bange, G., Randau, L. Type IV CRISPR RNA processing and effector complex formation in *Aromatoleum aromaticum*. *Nat. Microbiol.* 2019, 4, 89–96.

Pandey, V. K., Tripathi, A., Bhushan, R., Ali, A., Dubey, P. K. Application of CRISPR-Cas9 genome editing in genetic disorders: a systematic review up to date. *J. Genet. Syndr. Gene Ther.* 2017, 8(2), 1000321.

Pursey, E., Sünderhauf, D., Gaze, W. H., Westra, E. R., van Houte, S. CRISPR-Cas antimicrobials: Challenges and future prospects. *PLoS Pathog.* 2018, 14(6), e1006990.

Rath, D., Amlinger, L., Rath, A., Lundgren, M. The CRISPR-Cas immune system: biology, mechanisms, and applications. *Biochimie* 2015, 117, 119–128.

Sastya, S., Chouhan, M., Karunakaran, V., Jamra, N. Role of CRISPR-Cas9 gene editing tool in parasitology: a review. *J. Entomol. Zool. Stud.* 2018, 6(4), 243–246.

Savell, K. E., Day, J. J. Applications of CRISPR-Cas9 in the mammalian central nervous system. *Yale J. Biol. Med.* 2017, 90(4), 567–581.

Savić, N., Schwank, G. Advances in therapeutic CRISPR-Cas9 genome editing. *Transl. Res.* 2016, 168, 15–21.

Shin, J.W., Kim, K. H., Chao, M. J., Atwal, R. S., Gillis, T., MacDonald, M. E., Gusella, J. F., Lee, J. M. Permanent inactivation of Huntington's disease mutation by personalized allele-specific CRISPR-Cas9. *Hum. Mol. Genet.* 2016, 25(20), 4566–4576.

Slaymaker, I. M., Gao, L., Zetsche, B., Scott, D. A., Yan, W. X., Zhang, F. Rationally engineered Cas9 nucleases with improved specificity. *Science* 2016, 351(6268), 84–88.

Sternberg, S. H., Doudna, J. A. Expanding the Biologist's Toolkit with CRISPR-Cas9. *Mol. Cell* 2015, 58(4), 568–574.

Vale, P. F., Little, T. J. CRISPR-mediated phage resistance and the ghost of coevolution past. *Proc. Biol. Sci.* 2010, 277(1691), 2097–2103.

Vesikansa, A. Unraveling of central nervous system disease mechanisms using CRISPR genome manipulation. *J. Cent. Nerv. Syst. Dis.* 2018, 10, 1179573518787469.

Wang, X., Cao, C., Huang, J., Yao, J., Hai, T., Zheng, Q., Wang, X., Zhang, H., Qin, G., Cheng, J., Wang, Y., Yuan, Z., Zhou, Q., Wang, H., Zhao, J. One-step generation of triple gene targeted pigs using CRISPR-Cas9 system. *Sci. Rep*. 2016a, 6, 20620.

Wang, H., La Russa, M., Qi, L. S. CRISPR-Cas9 in genome editing and beyond. *Annu. Rev. Biochem*. 2016b, 85, 227–264.

Wang, G., Zhao, N., Berkhout, B., Das, A. T. CRISPR-Cas9 can inhibit HIV-1 replication but NHEJ repair facilitates virus escape. *Mol. Ther*. 2016c, 24(3), 522–526.

Wright, A. V., Nuñez, J. K., Doudna, J. A. Biology and applications of CRISPR systems: harnessing nature's toolbox for genome engineering. *Cell* 2016, 164(1), 29–44.

Xie, F., Ye, L., Chang, J. C., Beyer, A. I., Wang, J., Muench, M. O., Kan, Y. W. Seamless gene correction of β-thalassemia mutations in patient-specific iPSCs using CRISPR-Cas9 and piggyBac. *Genome Res*. 2014, 24(9), 1526–1533.

Yang, L., Grishin, D., Wang, G., Aach, J., Zhang, C. Z., Chari, R., Homsy, J., Cai, X., Zhao, Y., Fan, J. B., Seidman, C., Seidman, J., Pu, W., Church, G. Targeted and genome-wide sequencing reveal single nucleotide variations impacting specificity of Cas9 in human stem cells. *Nat. Commun*. 2014, 5, 5507.

Yin, H., Xue, W., Chen, S., Bogorad, R. L., Benedetti, E., Grompe, M., Koteliansky, V., Sharp, P. A., Jacks, T., Anderson, D. G. Genome editing with Cas9 in adult mice corrects a disease mutation and phenotype. *Nat. Biotechnol*. 2014, 32(6), 551–553.

Young, C. S., Hicks, M. R., Ermolova, N. V., Nakano, H., Jan, M., Younesi, S., Karumbayaram, S., Kumagai-Cresse, C., Wang, D., Zack, J. A., Kohn, D. B., Nakano, A., Nelson, S. F., Miceli, M. C., Spencer, M. J., Pyle, A. D. A single CRISPR-Cas9 deletion strategy that targets the majority of DMD patients restores dystrophin function in hiPSC-derived muscle cells. *Cell Stem Cell*. 2016, 18(4), 533–540.

Yuan, M., Webb, E., Lemoine, N. R., Wang, Y. CRISPR-Cas9 as a powerful tool for efficient creation of oncolytic viruses. *Viruses* 2016, 8(3),72.

Yuen, K. S., Chan, C. P., Wong, N. M., Ho, C. H., Ho, T. H., Lei, T., Deng, W.; Tsao, S. W., Chen, H., Kok, K. H., Jin, D. Y. CRISPR-Cas9-mediated genome editing of Epstein–Barr virus in human cells. *J. Gen. Virol*. 2015, 96(3), 626–636.

Zhan, T., Rindtorff, N., Betge, J., Ebert, M. P., Boutros, M. CRISPR-Cas9 for cancer research and therapy. *Semin. Cancer Biol*. 2019, 55, 106–119.

Zhou, H., Liu, B., Weeks, D. P., Spalding, M. H., Yang, B. Large chromosomal deletions and heritable small genetic changes induced by CRISPR-Cas9 in rice. *Nucleic Acids Res*. 2014, 42(17), 10903–10914.

CHAPTER 11

Revolutionary Approaches of Induced Stem Cells in Disease Prevention

STANZIN LADOL

Department of Zoology, Central University of Jammu, J&K, India,
E-mail: stladol22@gmail.com

ABSTRACT

Induced stem cells are derived from somatic cells by reprogramming methods. This technique was first introduced by a Japanese scientist in 2006 and has since been used to treat a wide range of diseases. Induced stem cells bring a revolution in stem cell research and a paradigm shift in stem cell therapy. Depending on the number of cell types they can generate, the nature of induced stem cells can be induced pluripotent, induced multipotent, or induced unipotent stem cells. The expression of transcription factors such as oct4, Sox2, c-myc, and klf4 is reported to be higher in embryonic stem cells. Hence, retro- or lentiviral vectors are used to deliver such transcription factors into somatic cells to generate induced stem cells. It is therefore evident that the roles of factors in the induction process is essential. Induced stem cells can be used as an alternative to embryonic stem cell therapies because the former bypass challenges such as immuno-incompatibility and ethical issues. Moreover, induced stem cells are indispensable for drug screening, drug discovery, preclinical toxicological assessment, and thus confer hope for personalized medicine. Induced stem cells are significant in the treatment of cancer, diabetes, cardiovascular, hematological, nervous system, Alzheimer's, and Parkinson's diseases. The diverse research prospects and applications of induced stem cells broaden the horizon in global health care. This chapter outlines the current state of knowledge about induced stem cells and highlights its role in disease prevention.

11.1 INTRODUCTION

Stem cell is the undifferentiated cell having the potential to differentiate into a spectrum of cell types. Stem cells act as a reservoir and their capacity lies in the self-renewal and repair mechanism. The ability of a cell to produce other cell types is known as potency. Depending on how many cell types they can generate they are termed unipotent, multipotent, or pluripotent stem cells. Pluripotent stem cells can generate all types of tissues in the body therefore, treatment strategies are significantly higher.

Cell isolation and culture techniques are not new; Hans Spemann performed the first experiment of somatic cell nuclear transfer in 1928. This breakthrough experiment revolutionized cloning technique, but the underlying mechanism was not known. Understanding the signaling mechanism in stem cells is important in order to know its architectural and differentiation ability. Later, Bongso et al. (1994) were successful in obtaining stem cells from human-derived inner cell mass of blastocysts (Thomson et al., 1998). Since then, this line of research is evolving very fast and provides a platform for regenerative medicine.

Pluripotent stem cells have a specific and complex system of signaling mechanisms and genomes. Transcription factors like Oct4, Sox2, and Nanog play an important role in pluripotency (Boyer et al., 2005). The anatomical source of stem cells plays a major role in the generation of tissue type. There are various sources of embryonic stem cells (Wagers and Weissman, 2004). Embryonic stem cells, due to its pluripotent nature, have greater potential for self-renewal and produce a wide variety and number of cells. Despite its great potential, embryonic stem cells are harder to isolate, characterize, and the self-renewal process is slow as compared to induced pluripotent stem cells. Three important characteristics of embryonic stem cells are; First, they express factors such as Oct4, Sox2, Tert, Utf1, and Rex (Carpenter and Bhatia, 2004). Second, they are unspecialized and can replenish, and third, under some physiological or experimental conditions in vitro they can be induced to generate specific cells such as cardiomyocytes, liver cells, nerve cell precursors, endothelial cells, hematopoietic cells, and insulin-secreting cells (Xu et al., 2002; Zhang et al., 2001). Contrary, adult stem cells show less proliferation and have a restricted range of differentiation capacities to generate cells. Some tissues have limited distinct subpopulation of stem cells, whereas other tissue types, such as brain tissue, prohibits and arduous access limits the use of adult stem cells. To date, the only adult-derived bone marrow and skin cell-based medical therapies have regenerative potentials. However, induced stem cells can differentiate and generate into various specific functional cell types in vitro, such as cardiac muscle cells, neuronal progenitors, hematopoietic stem cells (HSCs), endothelial cells, and bone cells (Lanza et al., 2004). Furthermore, they are ethically and legally less problematic, technically more feasible, and can be used directly in the emerging field of personalized medicine.

Induced pluripotent stem cells (iPSCs), which are remarkably similar to embryonic stem cells in genotypic and phenotypic properties, can be derived from human somatic tissues (Park et al., 2008b). Many comparative studies are reported that explained the molecular and functional similarities and differences between embryonic stem cells and iPSCs, for example Chin et al. (2009) compared three human embryonic stem cells lines and five iPSCs lines by microarray technique and identified hundreds of differentially expressed genes. They proposed that iPSCs are unique and may be considered as a subtype of pluripotent cells with potential remedial property (Chin et al., 2009). Deng et al. (2009) performed the targeted bisulfite sequencing of three human embryonic stem cell clones and four induced pluripotent stem cell lines and reported that there are differences in DNA methylation between these two types of cell lines. Subsequently, other studies also compared the gene expression between embryonic stem cells and iPSCs and found persistent donor cell gene expression in iPSCs (Ghosh et al., 2010; Marchetto et al., 2009). Later, other studies stated persistent donor cells epigenetic memories in human iPSCs (Kim et al., 2011) suggesting its application in disease model development, drug screening, drug discovery, and preclinical toxicological assessment.

11.2 IPSCS

iPSCs were generated by Takahashi and Yamanaka from specialized murine cells via a technique called reprogramming using a retroviral infection system in 2006. This scientific finding was later replicated in human cells by Takahashi et al. (2007) and Yu et al. (2007). This revolutionary work was conferred with the Nobel Prize in Physiology or Medicine in 2012. Since then many research groups have developed techniques to generate iPSCs from various somatic cell sources like fibroblasts, adipocyte stem cells, neural stem cells (NSCs), HSCs, and peripheral blood mononuclear cells (Hester et al., 2009; Liu et al., 2012). Various animal species, like mice, rats, rhesus monkeys, and humans, from which iPSCs were previously generated. Several methods like transfection by chemical reagents or viral infection and electroporation are efficient in delivering exogenous genes into human stem cells (Eiges et al., 2001; Zwaka and Thomson, 2003).

Many different effective delivery systems were introduced, such as episomal vectors, RNA-based transfection, protein transduction, and Sendai viral vectors have been used to generate cell lines from a wide array of adult somatic cell types. These are the preferred agents because they do not integrate into the host genome, whereas some retroviral vector leaves a residual effect on the host genome. These methods can be used as the basis for replacement therapy.

Different induction methods affect the efficiency and safety of the iPSCs for use in cell replacement therapies. It was observed that the gene expression of the transcription factor varies in different cell types. Depending on the cellular properties, various combinations of transcription factors can be used in the induction of pluripotent stem cells. For instance, mouse fibroblasts require OSKM to induce reprogramming process (Yoshida et al., 2007), while NSCs only require OCT4 for reprogramming (Kim et al., 2009a).

Some scientists proposed the minimal use of reprogramming factors to induce pluripotency in somatic cells that express appropriate levels of endogenous complementing factors. Adult mouse NSCs express high levels of Sox2 and c-Myc. However, exogenous Oct4 along with either Klf4 or c-Myc is adequate to generate iPSCs from NSCs (Jeong et al., 2008). Although NANOG is also a part of the pluripotency transcription complex, it is not crucial for the generation of iPSCs (Chambers et al., 2003; Mitsui et al., 2003; Takahashi and Yamanaka, 2006).

Normally, in reprogramming the transcription factors involved are KLF4, SOX2, OCT4, and c-MYC, whereas KLF4 and SOX2 can be replaced with other factors. While OCT4 is essential and c-MYC is superfluous and associated with a high risk of tumor development, but it can increase the reprogramming efficiency. In order to maintain increased reprogramming efficiency, c-MYC can be substituted by L-MYC, vitamin C, DNA methyltransferase inhibitors, and histone deacetylase inhibitors with a lower risk of tumorigenesis (Gunaseeli et al., 2010).

The use of nongenetic methods can replace the transduced transcription factors methods and improve reprogramming in the generation of iPSCs without major genetic manipulation. Three techniques to generate integration free iPSCs are; (1) the use of vectors that do not integrate into the host cell genome, (2) the use of integrating vectors that can be consecutively removed from the genome after the induction process is complete, and (3) without the use of vectors. Recently, successful reprogramming has been achieved without the use of viral or plasmid vectors. Explicitly, iPSCs have been generated from both murine and human fibroblasts by reprogramming vectors as purified recombinant proteins (Zhou et al., 2009) or as whole cell extrects, either from embryonic stem cells (Cho et al., 2010) or genetically engineered HEK293 cells (Kim et al., 2009b). The understanding of reprogramming technologies has helped elucidate the association between developmental stages and the induction process. And highlight the correlation between developmental biology and reprogramming technologies. Recent studies open new avenues for generating patient-specific iPSCs from a variety of diseases, including amyotrophic lateral sclerosis, muscular dystrophy, and Huntington's disease (HD) (Dimos et al., 2008). This chapter outlines the current state of knowledge about induced stem cells and highlights its role in disease prevention.

11.3 NEURODEGENERATIVE DISEASES

Many neurodegenerative diseases are complex and progressive, with a lack of clear mechanisms and effective treatment methods, as the differences between animal and human brain are wide and this remains one of the major challenges in animal-based models of human brain disease. Furthermore, animal-based models of neurodegenerative diseases are time consuming and resource intensive. However, iPSCs give us novel approaches to combating neuronal diseases. Many researchers have successfully developed induced pluripotent stem cell lines from patients with neurodegenerative diseases. (Alves et al., 2015) to study the etiology and mechanism of diseases. Neurogenesis has been found to occure in two neuronal niches in the adult brain, the subependymal zone lining the lateral ventricles and the subgranular zone of the dentate gyrus. This fascinating finding opened up the possibility of converting nonneurogenic astroglia into neurons when induced with an appropriate transcriptional factor (Alonso et al., 2012).

iPSCs derived dopaminergic neurons improved motor function in a rat model of Parkinson's disorder. Electrophysiology and morphology analysis confirmed the improvement in disease symptoms (Soldner et al., 2009). Human-derived iPSCs have been generated into dopaminergic neurons and successfully transplanted into the brains of the rat model of Parkinson's disorder (Hargus et al., 2010). Evolution in biomedical research enables the generation of induced stem cells and its use in diseases associated with cell death or impairment, such as Alzheimer's disease, Parkinson's disease, and many other neurodegenerative diseases.

11.4 NERVOUS SYSTEM DISEASES

Two separate studies reported disease specific induced pluripotent stem cell lines in 2008 (Dimos et al., 2008; Park et al., 2008a, 2008b), and the reconstruction of the disease state was first successfully carried out in vitro in the spinal muscular atrophy disorder (Ebert et al., 2009). Dopamine and motor neuron produced from human iPSCs revert damage in spinal cord injuries, Parkinson's disease, and amyotrophic lateral sclerosis (Karumbayaram et al., 2009). Neural tissue like motor neurons can be engineer using the combination of small molecules releasing puro microsphere and retinoic acid releasing microsphere. And it is used to manipulate human-derived induced pluripotent stem cell culture as it promotes cell differentiation. The engineered tissue expresses multiple neural markers such as β-tubulin III, TF olig2, HB9, ChaT, and generates specific cell lineages (Laura et al., 2018). This technique can be developed as an alternative in a biomaterial-based system by using drug-releasing microspheres as a tools for engineering neural tissue and treating spinal cord injuries (Lancaster et al., 2013). These studies show the prospect for the reproduction of disease-specific phenotypes by using patient-derived iPSCs and the potential applications of using these cells in pharmacology.

11.5 CARDIOVASCULAR DISEASES

miRNAs are small noncoding RNAs that are highly expressed in endothelial cells, and data suggest that they regulate angiogenesis and vascular development. Furthermore, it was found that miR-199b-5p significantly promotes proliferation, cell migration, and tube formation. miRNAs upregulation was observed during the differentiation of human iPSCs. Therefore, these findings have promoted an approach to treat ischemic injury through the mechanism of neovascularization (Liu et al., 2017; Wang et al., 2015). Moreover, there is data that iPSCs derived from cardiac cells protect cardiac cells from H_2O_2-induced oxidative stress by inhibiting caspase3/7 activation. These results taken together suggested the possibility that iPSCs derivatives might serve as a source of treatment for peripheral artery disease, especially atherosclerotic disease (Urbich et al., 2012).

Diabetic cardiomyopathy is a common complication of diabetes and affects millions of people worldwide. Cardiomyocytes derived from human iPSCs have more specific phenotypic and metabolic relevance to establishing a cardiovascular-related disease model (Cecilia et al., 2019). Recently, a study illustrated that the genome editing technique can be combined with human iPSCs derived from an asymptomatic patient and the resulting cell lines have the clinical implications for the development of personalized human iPSCs cardiomyocytes (Ma et al., 2018).

11.6 CANCER

iPSCs derived from cancer patients can be used as a novel strategy to study certain types of cancer. Thomson and his team in 1998 conducted an experiment in which they injected human embryonic stem cells into immunodeficient mice. And produce teratomas with three embryonic germ layers through differentiation assay technique, but the risk factors involved are tumor formation and immune rejection. Contrary, in cancer cell reprogramming, the cells may lose their carcinogenic properties and in turn acquire the properties of cancer stem cells and develop into cancer-iPSCs.

Induction protocols require minimal mutation in the cell genome, careful analysis of genomic and epigenomic integrity, and malignant transformation (the expression of Oct4, Sox2, Klf4, and c-Myc genes is associated to tumor developments) (Okita et al., 2007). Consequently, it is important to find other genes that could substitute c-Myc and Klf4 in induced pluripotent stem cell production. A study reported that in human somatic cell reprogramming, these genes can be successfully substituted by NANOG and LIN28 (Yu et al., 2007). Knowledge of the underlying carcinogenesis process can be elucidated using a cancer-induced pluripotent stem cell model. Moreover, cancer-induced pluripotent stem cell culture in vitro serves as a resource for pharmacology and toxicology as personal disease models (Hsiao and Edward, 2018).

11.7 BLOOD DISEASES

Sickle cell anemia is an inherited red blood cell disorder caused by a single point mutation in the β-globin gene of hemoglobin. The creation of induced pluripotent stem cell lines by gene targeting and replacement therapy repairs gene defects in hematopoietic progenitors. And the resulting progenitors produced normal red blood cells and cured the disease (Hanna et al., 2007). Cord blood is a suitable source of iPSCs due to its accessibility and minor genetic alterations (Giorgetti et al., 2009; Takenaka et al., 2009). Peripheral blood is another important source of iPSCs for instance; T-lymphocytes can be transformed into iPSCs using transient expression methods (Okita et al., 2013). This protocol can now be applied to many other diseases for which treatment or medicine.

11.8 GENETICS DISEASES

iPSCs derived from HD patients express significant amounts of mutant huntingtin (HTT) protein. Although mutant HTT aggregates accumulation is one of the indicators of the disease, however, HD-iPSCs do not exhibit increased cell death, sensitivity to cell stressors, or defects in GABAergic neuronal differentiation (Koyuncu et al., 2018), which indicates that the cells suppress the aggregation of mutant HTT aggregates (Koyuncu et al., 2018). Consequently, HD-iPSCs have been used to characterize antiaggregation mechanisms, which subsequently mimic differentiated neurons to suppress HTT aggregation and can essentially be differentiated into striatal neurons (Koyuncu et al., 2018), an indicator of cell viability and function. In vitro disease specific stem cells enable disease investigation and drug development in turn offers the opportunity to recapitulate pathological human tissue formation (Park et al., 2008a).

11.9 OTHER DISEASES

iPSCs in vitro pass through the same differentiation steps as in patients allow recapitulation and understanding of the molecular basis of pathology at an early stage. The autologus method based on iPSCs has simplified the lengthy skin biopsy and reduced the risk of complications and immunological rejection (Araki et al., 2013; Guha et al., 2013). Treating iPSCs cultures with extrinsic factors such as cytokines, and specific structural and microenvironment niches help to understand the developmental processes. This procedure can be used to generate retina, cerebral cortex, and pituitary gland tissues with suitable three-dimensional structures in vitro (Eiraku et al., 2008). In another example; in vivo injection of wild-type rat pluripotent stem cells into blastocysts of Pdx1deficient mice (which are incapable of forming a pancreas) resulted in the generation of functional pancreatic tissue in mice (Kobayashi et al., 2010).

Recently, a team led by ophthalmologist Kohji Nishida generated corneal cells by reprogramming methods using sheets of tissue from iPSCs (Hayashi et al., 2016). The Japanese committee has provisionally approved the use of reprogrammed stem cells to treat damaged corneas. However, the final approval is from the Japanese Ministry of Health is pending to test the treatment in humans.

Another breakthrough research carried out by a team of researcher led by Ravindra Gupta employs stem cell transplantation in a blood cancer patient by replacing the white blood cells with HIV resistant versions. The blood cancer patient with HIV was virus free after treatment (Gupta et al., 2019). Normally, the HIV binds to CCR5 receptors and attacks the cells, but a mutation with loss of function can prevent the virus from binding to the receptors and hence avert cell damage. This paves a way forward in the emerging scenario of increased resistance to treatments and the development of potential future treatments. A standard set of induced pluripotent stem cell lines could provide immunological matches for large groups of people (Taylor et al., 2011). However, the feasibility of such a strategy has yet to be assessed.

Animal models have contributed immensely to a better understanding of disease mechanisms. However, there are drawbacks to the accurate representation of human diseases. For instance, several drugs have been developed that have showed potential therapeutic utility in rodent models of amyotrophic lateral sclerosis. But sadly, they are found to be ineffective in human patients, underscoring the need for human cell disease models (Desnuelle et al., 2001). The advantage of this technique (iPSCs) is that defective gene can be repaired through homologous recombination and that the transplanted tissues are genetically matched to that of the recipient without immunorejection. Considerably, in the unforeseen future, an individual patient could have a convenient supply of their types of tissues to use. This paves the way for personalized medicine that will revolutionize pharmacology by putting the individual at the center of future health care.

11.10 CONCLUSION

In conclusion, this chapter summarized some of the potential applications of iPSCs and its associated techniques in stem cell research. These include applications in basic science, regenerative medicine, cell replacement therapy, and disease modeling. Combine with other budding technologies, such as epigenomic profiling, deep sequencing, and CRISPR technology the future of iPSCs technique is vivid from both a laboratory and the clinical perspective.

CONFLICT OF INTEREST

The authors declare that they have no conflict of interest.

KEYWORDS

- **induced pluripotent stem cells**
- **disease**
- **embryonic stem cells**
- **therapy**
- **treatments**
- **transcription factors**

REFERENCES

Alonso, M, Lepousez, G., Sebastien, W., Bardy, C., Gabellec, M. M., Torquet, N., Lledo, P.M. Activation of adult-born neurons facilitates learning and memory. *Nat. Neurosci.* 2012, 15, 897–904.

Alves, C.J., Dariolli, R., Jorge, F.M., Monteiro, M.R., Maximino, J.R., Martins, R.S., Strauss, B.E., Krieger, J.E., Callegaro, D., Chadi, G. Gene expression profiling for human iPS-derived motor neurons from sporadic ALS patients reveals a strong association between mitochondrial functions and neurodegeneration. *Front Cell Neurosci.* 2015, 9, 289.

Araki, R., Uda, M., Hoki, Y., Sunayama, M., Nakamura, M., Ando, S., Sugiura, M., Ideno, H., Shimada, A., Nifuji, A., Abe, M. Negligible immunogenicity of terminally differentiated cells derived from induced pluripotent or embryonic stem cells. *Nature* 2013, 494, 100–104.

Bongso, A., Fong, C. Y., Ng, S. C., Ratnam, S. Isolation and culture of inner cell mass cells from human blastocysts. *Hum. Reprod.* 1994, 9, 2110–2117.

Boyer, L.A., Lee, T.I., Cole, M.F., Johnstone, S.E., Levine, S.S., Zucker, J.P., Guenther, M.G., Kumar, R.M., Murray, H.L., Jenner, R.G., Gifford, D.K. Core transcriptional regulatory circuitry in human embryonic stem cells. *Cell* 2005, 122, 947–956.

Carpenter, M. K., Bhatia, M. *Characterization of Human Embryonic Stem Cells*. Amsterdam: Elsevier. 2004. vol. 1.

Cecilia, G., Ryan, H., Gabriella, B., Jane, S., Peter, S. Diabetic cardiomyopathy modelling using induced pluripotent stem cell derived cardiomyocytes: recent advances and emerging models. *Stem Cell Rev. Rep.* 2019, 15,13–22.

Chambers, I., Colby, D., Robertson, M., Nichols, J., Lee, S., Tweedie, S., Smith, A. Functional expression cloning of Nanog, a pluripotency sustaining factor in embryonic stem cells. *Cell* 2003, 113, 643–655.

Chin, M.H., Mason, M.J., Xie, W., Volinia, S., Singer, M., Peterson, C., Ambartsumyan, G., Aimiuwu, O., Richter, L., Zhang, J., Khvorostov, I. Induced pluripotent stem cells and embryonic stem cells are distinguished by gene expression signatures. *Cell Stem Cell* 2009, 5, 111–123.

Cho, H.J., Lee, C.S., Kwon, Y.W., Paek, J.S., Lee, S.H., Hur, J., Lee, E.J., Roh, T.Y., Chu, I.S., Leem, S.H., Kim, Y. Induction of pluripotent stem cells from adult somatic cells by protein-based reprogramming without genetic manipulation. *Blood* 2010, 116, 386–395.

Deng, J., Shoemaker, R., Xie, B., Gore, A., LeProust, E.M., Antosiewicz-Bourget, J., Egli, D., Maherali, N., Park, I.H., Yu, J., Daley, G.Q. Targeted bisulfite sequencing reveals changes in DNA methylation associated with nuclear reprogramming. *Nat. Biotech.* 2009, 27, 353–360.

Desnuelle, C., Dib, M., Garrel, C., Favier, A. A double-blind, placebo-controlled randomized clinical trial of alpha-tocopherol (vitamin E) in the treatment of amyotrophic lateral sclerosis. *Amyotroph. Lateral Scler. Other Motor Neuron Disord.* 2001, 2, 9–18.

Dimos, J.T., Rodolfa, K.T., Niakan, K.K., Weisenthal, L.M., Mitsumoto, H., Chung, W., Croft, G.F., Saphier, G., Leibel, R., Goland, R., Wichterle, H. Induced pluripotent stem cells generated from patients with ALS can be differentiated into motor neurons. *Science* 2008, 321, 1218–1221.

Ebert, A. D., Yu, J., Rose, F. F., Mattis, V. B., Lorson, C. L., Thomson, J. A., Svendsen, C. N. Induced pluripotent stem cells from a spinal muscular atrophy patient. *Nature* 2009, 457, 277–280.

Eiges, R., Schuldiner, M., Drukker, M., Yanuka, O., Itskovitz-Eldor, J., Benvenisty, N. Establishment of human embryonic stem cell-transfected clones carrying a marker for undifferentiated cells. *Curr. Biol.* 2001, 11, 514–518.

Eiraku, M., Watanabe, K., Matsuo-Takasaki, M., Kawada, M., Yonemura, S., Matsumura, M., Wataya, T., Nishiyama, A., Muguruma, K., Sasai, Y. Self-organized formation of polarized cortical tissues from ESCs and its active manipulation by extrinsic signals. *Cell Stem Cell* 2008, 3, 519–532.

Ghosh, Z., Wilson, K.D., Wu, Y., Hu, S., Quertermous, T., Wu, J.C. Persistent donor cell gene expression among human induced pluripotent stem cells contributes to differences with human embryonic stem cells. *PLoS One* 2010, 5, e8975.

Giorgetti, A., Montserrat, N., Aasen, T., Gonzalez, F., Rodríguez-Pizà, I., Vassena, R., Raya, A., Boué, S., Barrero, M. J., Corbella, B. A., Torrabadella, M. Generation of induced pluripotent stem cells from human cord blood using OCT4 and SOX2. *Cell Stem Cell* 2009, 5, 353–357.

Guha, P., Morgan, J. W., Mostoslavsky, G., Rodrigues, N. P., Boyd, A. S. Lack of immune response to differentiated cells derived from syngeneic induced pluripotent stem cells. *Cell Stem Cell* 2013, 12, 407–412.

Gunaseeli, I., Doss, M.X., Antzelevitch, C., Hescheler, J., Sachinidis, A. Induced pluripotent stem cells as a model for accelerated patient- and disease-specific drug discovery. *Curr Med Chem.* 2010, 17, 759–766.

Gupta, R.K., Abdul-Jawad, S., McCoy, L.E., Mok, H.P., Peppa, D., Salgado, M., Martinez-Picado, J., Nijhuis, M., Wensing, A.M., Lee, H., Grant, P. HIV-1 remission following CCR5Δ32/Δ32 haematopoietic stem-cell transplantation. *Nature* 2019, 568, 244.

Hanna, J., Wernig, M., Markoulaki, S., Sun, C.W., Meissner, A., Cassady, J.P., Beard, C., Brambrink, T., Wu, L.C., Townes, T.M., Jaenisch, R. Treatment of sickle cell anemia mouse model with iPS cells generated from autologous skin. *Science* 2007, 318, 1920–1923.

Hargus, G., Cooper, O., Deleidi, M., Levy, A., Lee, K., Marlow, E., Yow, A., Soldner, F., Hockemeyer, D., Hallett, P.J., Osborn, T., Jaenisch, R., Isacson, O. Differentiated Parkinson patient-derived induced pluripotent stem cells grow in the adult rodent brain and reduce motor asymmetry in Parkinsonian rats. *Proc. Natl. Acad. Sci. USA* 2010, 107, 15921–15926.

Hayashi, R., Ishikawa, Y., Sasamoto, Y., Katori, R., Nomura, N., Ichikawa, T., Araki, S., Soma, T., Kawasaki, S., Sekiguchi, K., Quantock, A.J., Tsujikawa, M., Nishida, K. Co-ordinated ocular development from human iPS cells and recovery of corneal function. *Nature* 2016, 531, 376–380.

Hester, M.E., Song, S., Miranda, C.J., Eagle, A., Schwartz, P.H., Kaspar, B.K. Two factor reprogramming of human neural stem cells into pluripotency. *PLoS One* 2009, 4, e7044.

Hsiao, M. C., Edward, C. Patient-derived induced pluripotent stem cells for models of cancer and cancer stem cell research. *J. Formos. Med. Assoc.* 2018, 117, 1046–1057.

Jeong, B. K., Holm, Z., Guangming, W., Luca, G., Kinarm, K., Vittorio, S., Marcos, J., Arau, B., David, R., Dong, W. H., Martin, Z., Hans, R. Pluripotent stem cells induced from adult neural stem cells by reprogramming with two factors. *Nature* 2008, 454, 646–650.

Karumbayaram, S., Novitch, B.G., Patterson, M., Umbach, J.A., Richter, L., Lindgren, A., Conway, A.E., Clark, A.T., Goldman, S.A., Plath, K., Wiedau-pazos, M. Directed differentiation of human-induced pluripotent stem cells generates active motor neurons. *Stem Cells* 2009, 27, 806–811.

Kim, J.B., Greber, B., Arauzo-Bravo, M.J., Meyer, J., Park, K.I., Zaehres, H., Schöler, H.R. Direct reprogramming of human neural stem cells by OCT4. *Nature* 2009a, 461, 649–653.

Kim, D., Kim, C.H., Moon, J.I., Chung, Y.G., Chang, M.Y., Han, B.S., Ko, S., Yang, E., Cha, K.Y., Lanza, R., Kim, K.S. Generation of human induced pluripotent stem cells by direct delivery of reprogramming proteins. *Cell Stem Cell* 2009b, 4, 472–476.

Kim, K., Zhao, R., Doi, A., Ng, K., Unternaehrer, J., Cahan, P., Huo, H., Loh, Y.H., Aryee, M.J., Lensch, M.W., Li, H. Donor cell type can influence the epigenome and differentiation potential of human induced pluripotent stem cells. *Nat. Biotech.* 2011, 29, 1117–1119.

Kobayashi, T., Yamaguchi, T., Hamanaka, S., Kato-Itoh, M., Yamazaki, Y., Ibata, M., Sato, H., Lee, Y. S., Usui, J., Knisely, A. S., Hirabayashi, M. Generation of rat pancreas in mouse by interspecific blastocyst injection of pluripotent stem cells. *Cell* 2010, 142, 787–799.

Koyuncu, S., Saez, I., Lee, H.J., Gutierrez, G. R., Pokrzywa, W., Fatima, A., Hoppe, T., Vilchez, D. The ubiquitin ligase UBR5 suppresses proteostasis collapse in pluripotent stem cells from Huntington's disease patients. *Nat. Commun.* 2018, 9, 2886.

Lancaster, M. A., Renner, M., Martin, C. A., Wenzel, D., Bicknell, L.S., Hurles, M.E., Homfray, T., Penninger, J.M., Jackson, A.P., Knoblich, J.A. Cerebral organoids model human brain development and microcephaly. *Nature* 2013, 501, 373.

Lanza, R., Gearhart, J., Hogan, B., Melton, D. W., Pedersen, R., Thomson, J., West, M. Handbook of Stem Cells. Amsterdam Elsevier. 2004, Vol. 1.

Laura, D. V., Karina, K., Stephanie, M. W. Engineering Neural Tissue from Human Pluripotent Stem Cells Using Novel Small Molecule Releasing Microspheres. *Adv. Biosys.* 2018, 2, 1800133.

Liu, X., Qing, L., Xin, N., Bin, H., Shengbao, C., Wenqi, S., Jian, D., Changqing, Z., Yang, W. Exosomes secreted from human-induced pluripotent stem cell-derived mesenchymal stem cells prevent osteonecrosis of the femoral head by promoting angiogenesis. *Int. J. Biol. Sci.* 2017, 13, 232–244.

Liu, T., Zou, G., Gao, Y., Zhao, X., Wang, H., Huang, Q., Jiang, L., Guo, L., Cheng, W. High efficiency of reprogramming CD34(þ) cells derived from human amniotic fluid into induced pluripotent stem cells with Oct4. *Stem Cells Dev.* 2012, 21, 2322–2332.

Ma, N., Zhang, J., Itzhaki, I., Zhang, S.L., Chen, H., Haddad, F., Kitani, T., Wilson, K.D., Tian, L., Shrestha, R., Wu, H. Determining the pathogenicity of a genomic variant of uncertain significance using CRISPR/Cas9 and human-induced pluripotent stem cells. *Circulation* 2018, 138, 2666–2681.

Marchetto, M.C., Yeo, G.W., Kainohana, O., Marsala, M., Gage, F.H., Muotri, A.R. Transcriptional signature and memory retention of human-induced pluripotent stem cells. *PLoS One* 2009, 4, e7076.

Mitsui, K., Tokuzawa, Y., Itoh, H., Segawa, K., Murakami, M., Takahashi, K., Maruyama, M., Maeda, M., Yamanaka, S. The homeoprotein Nanog is required for maintenance of pluripotency in mouse epiblast and ES cells. *Cell* 2003, 113, 631–642.

Okita, K., Ichisaka, T., Yamanaka, S. Generation of germline-competent induced pluripotent stem cells. *Nature* 2007, 448, 313–317.

Okita, K., Yamakawa, T., Matsumura, Y., Sato, Y., Amano, N., Watanabe, A., Goshima, N., Yamanaka, S. An efficient nonviral method to generate integration-free human-induced pluripotent stem cells from cord blood and peripheral blood cells. *Stem Cells* 2013, 31, 458–466.

Park, I. H., Arora, N., Huo, H., Maherali, N., Ahfeldt, T., Shimamura, A., Lensch, M. W., Cowan, C., Hochedlinger, K., Daley, G. Q. Disease-specific induced pluripotent stem cells. *Cell* 2008a, 134, 877–886.

Park, I.H., Zhao, R., West, J.A., Yabuuchi, A., Huo, H., Ince, T.A., Lerou, P.H., Lensch, M.W., Daley, G.Q. Reprogramming of human somatic cells to pluripotency with defined factors. *Nature* 2008b, 451, 141–146.

Soldner, F., Hockemeyer, D., Beard, C., Gao, Q., Bell, G.W., Cook, E.G., Hargus, G., Blak, A., Cooper, O., Mitalipova, M., Isacson, O., Jaenisch, R. Parkinson's disease patient-derived induced pluripotent stem cells free of viral reprogramming factors. *Cell* 2009, 136, 964–977.

Takenaka, C., Nishishita, N., Takada, N., Jakt, L. M., Kawamata, S. Effective generation of iPS cells from CD34(+). cord blood cells by inhibition of p53. *Exp. Hematol.* 2009, 38, 154–62.

Takahashi, K., Tanabe, K., Ohnuki, M., Narita, M., Ichisaka, T., Tomoda, K., Yamanaka, S. Induction of pluripotent stem cells from adult human fibroblasts by defined factors. *Cell* 2007, 131, 861–872.

Takahashi, K., Yamanaka, S. Induction of pluripotent stem cells from mouse embryonic and adult fibroblast cultures by defined factors. *Cell* 2006, 126, 663–676.

Taylor, C.J., Bolton, E.M., Bradley, J.A. Immunological considerations for embryonic and induced pluripotent stem cell banking. *Philos. Trans. R. Soc. Lond B. Biol. Sci.* 2011, 366, 2312–2322.

Thomson, J.A., Itskovitz-Eldor, J., Shapiro, S. S., Waknitz, M. A., Swiergiel, J. J., Marshall, V. S., Jones, J. M. Embryonic stem cell lines derived from human blastocysts. *Science* 1998, 282, 1145–1147.

Urbich,C., Kaluza,D., Frömel,T., Knau,A., Bennewitz,K., Boon,R.A., Bonauer,A., Doebele, C., Boeckel, J.N., Hergenreider, E., Zeiher, A.M., Kroll, J., Fleming, I., Dimmeler, S. MicroRNA-27a/b controls endothelial cell repulsion and angiogenesis by targeting semaphorin 6A. *Blood* 2012, 119, 1607–1616.

Wagers, A. J., Weissman, I. L. Plasticity of adult stem cells. *Cell* 2004, 116, 639–648.

Wang, Y., Zhang. L., Li, Y., Chen, L., Wang, X., Guo, W., Zhang, X., Qin, G., He, S.H., Zimmerman, A., Liu, Y., Kim, I.M., Weintraub, N.L., Tang, Y. Exosomes/microvesicles from induced pluripotent stem cells deliver cardioprotective miRNAs and prevent cardiomyocyte apoptosis in the ischemic myocardium. *Int. J. Cardiol.* 2015, 192, 61–69.

Xu, R. H., Chen, X., Li, D. S., Li, R., Addicks, G. C., Glennon, C., Zwaka, T. P., Thomson, J. A. BMP4 initiates human embryonic stem cell differentiation to trophoblast. *Nat. Biotech.* 2002, 20, 1261–1264.

Yoshida, Y., Yamanaka, S. Induced pluripotent stem cells 10 Years later: for cardiac applications. *Circ Res.* 2010, 120, 1958–1968.

Yu, J., Vodyanik, M.A., Smuga-Otto, K., Antosiewicz-Bourget, J., Frane, J.L., Tian, S., Nie, J., Jonsdottir, G.A., Ruotti, V., Stewart, R., Slukvin, I.I. Induced pluripotent stem cell lines derived from human somatic cells. *Science* 2007, 318, 1917–1920.

Zhang, S. C., Wernig, M., Duncan, I. D., Brustle, O., Thomson, J. A. In vitro differentiation of transplantable neural precursors from human embryonic stem cells. *Nat. Biotech.* 2001, 19, 1129–1133.

Zhou, W., Freed, C.R. Adenoviral gene delivery can reprogram human fibroblasts to induced pluripotent stem cells. *Stem Cells* 2009, 27, 2667–2674.

Zwaka, T. P., Thomson, J. A. Homologous recombination in human embryonic stem cells. *Nat. Biotech.* 2003, 21, 319–321.

CHAPTER 12

Stem Cell Biology: An Overview

SUMIT SIDDHARTH

Department of Oncology, Johns Hopkins University School of Medicine and the Sidney Kimmel Comprehensive Cancer Center at Johns Hopkins, Baltimore 21231 MD, USA, E-mail: siddharthsumit@gmail.com

ABSTRACT

Development is the sum of two distinct cellular processes; division and differentiation. Division is the progression of one cell to produce two daughter cells, while differentiation is the procedure by which a cell changes from one type to another and specializes to perform a specific function. Merriam-Webster dictionary defines stem cells as "an unspecialized cell that gives rise to differentiated cells" (https://www.merriam-webster.com/dictionary/stem%20cell). According to Encyclopedia Britannica, a stem cell is an undifferentiated cell that can divide to produce some offspring cells that continue as stem cells and some cells that are destined to differentiate (become specialized) (https://www.britannica.com/science/stem-cell). Stem cells are the cells with a potential to mature into numerous diverse cell types in the body. Stem cells act as the raw materials of the body, which produces all other cell types with specialized functions. Under specific conditions in vivo or in vitro, stem cells undergo cell division and form daughter cells, which can become new stem cells or differentiated cells with specific functions like liver cells (performing liver-related functions), blood cells (performing blood-related functions), heart cells (performing cardiac-related functions), bone cells (performing bone-related functions) etc. Stem cells are undifferentiated cells because they are not devoted to a particular developmental cascade to form a specific tissue or organ. Stem cells possess two important characteristics, which distinguishes them from other cell types and the two characteristics are (1) stem cells possess the potential of self-renewal through cell division even after long inactive periods and (2) stem cells can form tissue or organ-specific cells under the influence of the factors present in the milieu.

12.1 INTRODUCTION

A sperm fuses with an ovum to form a single-celled zygote that later forms an embryo. This division goes on till an entire organism is formed consisting of billions of specific cells arranged in an orderly fashion to perform specific tasks. All these specific cells arose from a single-celled zygote, a totipotent stem cell with the potential to form any kind of cells. The importance of stem cells lies in the fact that these are the only cells in the body with the ability to generate all other cells. Historical timeline of stem cell research can be dated back to 1962 when Sir John Gurdon successfully generated tadpoles from an enucleated frog egg cell transplanted with intestinal epithelial somatic cell nucleus without fertilization. This method was termed as somatic cell nuclear transfer (SCNT) (Gurdon, 1962). Using the same SCNT method, Sir Ian Wilmut cloned Dolly in 1997 (Wilmut et al., 1997). These two scientific endeavors proved the notion that a somatic cell nucleus contains all the essential information needed to generate the whole organism and the egg cell possesses enough factors necessary for the reprogramming of the somatic cell. Table 12.1 shows the historical timeline of stem cell research.

Stem cells can aid the researchers and doctors in:

1. Understanding the mechanism of disease occurrence.
2. Generating healthy cells to swap the diseased cells (regenerative medicine).
3. Screening new drugs.

12.2 EMBRYONIC STEM CELLS (ESCS)

As its name implies, ESCs are descendants of the embryo. Maximum of the ESCs was extracted from embryos obtained from oocytes fertilized in vitro in an in vitro-fertilization clinic and then contributed to research and development with donor's informed consent. ESCs are not obtained from the eggs of a fertilized woman (https://stemcells.nih.gov/info/basics/3.htm).

The practice of culturing and propagating cells in the research laboratory is called cell culture. Human embryonic stem cells (hESCs) are produced by taking cells out of an embryo into the preimplantation phase and transferring them to the cell culture dishes containing appropriate culture medium with growth factors. The cells are incubated at 37 °C under 5% CO_2 environmental conditions to divide and grow on the top of the cell culture flask/dish. The cell culture dishes are coated with a layer of mouse embryonic skin cells called a feeder layer. This feeder layer provides adherence to the ESCs. The cells composing feeder layer also secrete nutrients and growth factors in the environment (culture medium). But nowadays as an advancement of cell culture techniques, researchers have developed ways to grow and culture ESCs without the feeder layer. The use of the feeder layer always accompanied the threat of transmitting viruses and other associated macromolecules from mice to humans. When the cells grow, divide, and form colonies on the cell culture dish or flask, then, they are removed and replated in fresh culture dishes. This process of re-plating cells for subculturing continues many number of times. Each cycle of subculture is denoted as the passage. During this process of ESCs culture, many cells can be frozen in appropriate growth media, serum, and cryoprotectant or only in serum with cryoprotectant.

TABLE 12.1 Historical Timeline of the Development of Stem Cell Research

Year	Major Discoveries	Researcher
1962	Replacement of the cell nucleus of frog egg cell with mature cell from tadpole intestine	John Gurdon
1981	Generation of mouse embryonic stem cells	Martin Evans, Mathew Kaufman, Gail Martin
1997	Successful cloning of sheep (Dolly) using SCNT technology	Ian Wilmut
1998	Generation of human embryonic stem cells	James Thomson
2006	Successful generation of mouse iPSCs using Oct4, Sox2, Klf4 and cMyc	Shinya Yamanaka and Kazutoshi Takahashi
2007	Successful generation of human iPSCs with Oct4, Sox2, Klf4 and cMyc	Shinya Yamanaka
2007	Successful generation of human iPSCs with Oct4, Sox2, Nanog and Lin28	James Thomson
2012	Nobel Prize in Physiology and Medicine	Shinya Yamanaka and John Gurdon

At several points through the ESC generation process, scientists examine the cells for the fundamental properties of ESCs, which is called characterization of the cells. There are no gold standard tests to characterize the ESCs. There are numerous tests that laboratories perform to characterize the embryonic stem cell lines. Few of the techniques are mentioned below.

1. Allow the stem cells to grow and subculture for many months. This confirms that the cells can ensure long-term growth and self-renewal. Cells are observed microscopically to confirm that they maintain the healthy phenotype and remain undifferentiated.
2. To confirm that the cells are maintaining the ESCs like feature, the expression level of ESC markers are regularly checked by western blotting or RT-PCR. Few of the most common ESC markers are Nanog, Sox2, and Oct4.
3. Similarly, the protein level of several surface markers connected with undifferentiated cells is also detected by western blotting or flow cytometry.
4. Microscopic examination of the chromosomes is performed to detect for chromosomal damage or changes in chromosome number.
5. The frozen cells are revived to ensure that the frozen stocks are appropriate and can be subcultured whenever needed.
6. The pluripotency of human ESCs is checked by differentiating the cells spontaneously in cell culture; regulating the cells to differentiate into cells typical of the three germ layers; or implanting the cells into an immunocompromised mouse for benign tumor formation (teratoma). Because the immunocompromised mouse is not capable of rejecting the foreign hESCs, it becomes easier for the scientist to observe the growth and differentiation of these cells in the natural environment (https://stemcells.nih.gov/info/basics/3.htm).

If the ESCs grow under suitable culture conditions, they continue to be undifferentiated (nonspecialized). However, when cells form embryoid bodies by growing as a clump of cells, they spontaneously differentiate. They can form different specialized cells like muscle cells, nerve cells, etc. Although spontaneous differentiation certainly indicates that embryonic stem cell culture is healthy, it is not controlled and, therefore, is an incompetent strategy to generate certain cell-type cultures. To create a culture of the precise kind of differentiated cells (e.g., cardiac muscle cells, blood cells, or nerve cells), scientists are modulating the differentiation of ESCs. They achieve this by altering the culture medium, modifying the cell culture plate or the cells by introducing certain genes (cloning). Years of experimentation have helped scientists establish the basic protocols for the targeted differentiation of ESCs to specific cell types.

If scientists continue to maneuver ESC differentiation into the precise type of cells, these cells can be used to cure specific diseases like diabetes, muscular dystrophy, cardiac diseases, etc. in the future.

12.3 ADULT STEM CELLS (ASCS)

ASCs are considered as undifferentiated cells in the pool of differentiated cells in tissues or organs. They can self-renew as well as differentiate to produce certain or all the major specific cell types of a tissue or organ. The major role played by ASCs in living organisms is to sustain and repair their parent tissues. ASCs are also denoted as somatic stem cells where somatic means body cells and not germ cells (sperm or eggs) (https://stemcells.nih.gov/info/basics/4.htm).

Studies on ASCs have spawned great interest. The presence of these cells in many tissues has prompted scientists to query the usage and availability of ASCs for transplantation. Bone marrow stem cells (hematopoietic stem cells) have already been used for transplantation. Scientists have also found the presence of ASCs in the heart and brain, where stem cells were thought not to exist initially. Controlled differentiation of ASCs may lead to transplantation-based therapeutics.

ASC research began 60 years ago. During the 1950s, it was found that bone marrow has no less than two types of stem cells. One of the population is called hematopoietic cells, which form all kinds of blood cells. Another population is known as bone marrow stromal stem cells (mesenchymal stem cells or skeletal stem cells), which can form bones, cartilage, and fat cells for maintenance of blood and fibrous connective tissue, and it constitutes a small part of the bone marrow interstitial cell population.

During the 1960s, researchers discovered two brain regions in rat containing dividing cells, which eventually become neurons. In spite of these reports, researchers believe that the adult brain fails to create new neurons. By the 1990s, they agreed that stem cells contained in the adult brain could generate three major types of brain cells: nonneuronal cells like astrocytes and oligodendrocytes; and neurons (https://stemcells.nih.gov/info/basics/4.htm).

12.4 INDUCED PLURIPOTENT STEM CELLS (IPSCS)

iPSCs are the adult somatic cells, which are reprogrammed with the aid of reprogramming factors (Oct4, Sox2, Klf4, cMyc, Nanog, etc.) to attain the embryonic-like pluripotent state that allows them to grow and differentiate into any cell type. Differentiation is not a one-way road. It can be reprogrammed, and developmental efficacy can be returned in a somatic nucleus.

12.5 GENERATION OF IPSCS

iPSCs can be generated by introducing the "reprogramming factors" into a cell. These reprogramming factors are also known as "Yamanaka factors." They are Oct4, Sox2, cMyc, and Klf4. There can be other transcription factors, miRNAs, or other small molecules, which can also induce reprogramming into the adult cells. Shinya Yamanaka firstly generated iPSCs (Takahashi and Yamanaka, 2006). Yamanaka and colleagues hypothesized the genes that are required for ESCs may also reactivate an embryonic stage in well-differentiated adult cells under appropriate condition. This hypothesis led the beginning of a breakthrough, which earned Shinya Yamanaka Nobel prize in 2012 (along with John B Gurdon). Out of 24 genes chosen by the researchers initially, they finally identified four factors (Oct-4, Sox2, cMyc, and Klf4) responsible for inducing embryonic stem-like state in cells under selective condition.

In 2007, Shinya Yamanaka of Kyoto University and James Thomson from the University of Wisconsin-Madison reprogrammed human cells to iPSCs. Yamanaka and the group successfully transformed the human fibroblast into iPSCs using Oct4, Sox2, cMyc, and Klf4 by the help of the retrovirus system (Takahashi et al., 2007), while the group led by Thomson achieved it using different factors (Oct4, Sox2, Nanog, and Lin28) by the help of the lentiviral system in 2007, respectively (Yu et al., 2007). The next year, followed the development of iPSCs using human keratinocytes. Later in 2010, peripheral blood cells were used to generate iPSCs (Staerk et al., 2010) and in 2012, the iPSCs were generated from the urine using the renal epithelial cells (Zhou et al., 2012).

Ideally, iPSCs can be produced from any somatic cell using proper reprogramming elements. The whole process of generation of iPSCs involves three major steps:

1. Starting up the initial cell culture.
2. Activating the iPSCs.
3. Characterizing and expanding the iPSCs.

The first and the foremost, cells are extracted and grown followed by treatment with reprograming factors, so that these factors are introduced into the cultured cells, either by using an integrating system approach (retroviral, lentiviral, or inducible lentiviral) or nonintegrating system approach (sendaivirus, adenovirus, transgenes, recombinant proteins, plasmid DNA transfer). Transfected cells are then transferred onto the feeder layer of cells (fibroblasts, keratinocytes) with appropriate media for propagation. Once the reprogramming factors are expressed, then, the iPSCs are generated. Figure 12.1 shows the stepwise generation of iPSCs.

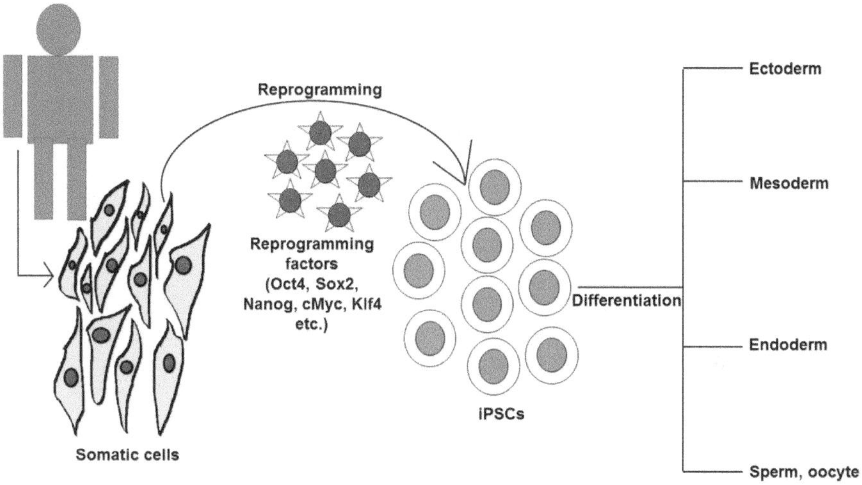

FIGURE 12.1 Generation of iPSCs. Somatic cells are isolated from an adult organism, grown and cultured under laboratory conditions. The reprogramming factors (Oct4, Sox2, Nanog, Klf4, cMyc, etc.) are introduced into the somatic cells and are converted into iPSCs. These iPSCs, then differentiate into multiple cell types.

These cultured iPSCs cells can be characterized phenotypically as well as by checking the cell surface proteins like SSEA-4 alkaline phosphatase and embryonic stem cell markers like Oct4, Sox2, Nanog, Klf4, cMyc, etc.

12.5.1 VARIOUS REPROGRAMMING FACTORS

Oct-3/4 (Pou5f1)

It represents a family of "Octamer transcription factors." Oct3/4 is a critical player in maintaining pluripotency. Other members of this family like Oct1 and Oct6 fail to induce stemness, which reveal the selectiveness of Oct3/4 in stemness induction.

Sox Family

In contrast to Oct3/4 that are pluripotent factors, members of Sox family are stemness associated factors for multipotent and unipotent cells. Sox2 is one of the members of the Yamanaka factors.

Myc Family

Though Myc family members are well-known proto-oncogenes, Yamanaka and others reported the role of cMyc in the human as well as mouse iPSCs generation. Despite being a Yamanaka factor, cMyc has faced challenges because one-fourth of mice bearing Myc transplantation activated iPSCs and developed lethal teratomas.

Klf Family

Another member of Yamanaka factor is Klf4. Yamanaka and colleagues identified Klf4 as a factor involved in human iPSCs development. But, Thomson et al. reported that Klf4 is not essential for stemness. Still Klf2 and Klf4 of the Klf family are used for iPSCs generation. Klf1 and Klf5 are also used for generating iPSCs but display lower efficiency.

Nanog

Thomson and group reported Nanog as a necessary transcription factor required to produce iPSCs. Nanog together with Sox2 and Oct3/4 is essential for pluripotency induction in ESCs.

Lin28

It is an mRNA binding protein (Ali et al., 2012) reported by Thomson as an essential factor for the iPSCs generation along with Oct4, Sox2, and Nanog (Yu et al., 2007).

12.6 APPLICATIONS OF IPSCS

Lack of knowledge of the mechanistic progression of many diseases makes it impossible for them to be treated. Hence, disease modeling becomes important to explore the mechanism of disease progression. Many disease testing models have been developed in the past and a few of them serve as a model system for representing the human cellular microenvironment. Animal models like rats, mice, dogs, monkeys, and primates have been utilized successfully for disease modeling. But, as the genetic constitution of animals varies compared to humans, it leads to variation in the results obtained from these studies. More importantly, as animals and humans are different species, so, will express different types of proteins, hence, no animal models exactly replicate the human microenvironment. Now, iPSCs come to the rescue of the researchers under these circumstances. As iPSCs are functional in vitro as well as in vivo, so, it serves as an alternative model for the study of disease modeling, regenerative medicine and drug discovery.

12.6.1 DISEASE MODELING

As iPSCs have the potential of self-renewal and differentiation in all types of human cells, they are utilized for the establishment of different disease models. The three-dimensional culture of iPSCs with extracellular matrix proteins mimic the in vivo microenvironment. Familia 1 Dysautonomia disease model was established by Lee et al. using iPSCs. Similarly, Moad et al. successfully formed iPSCs using prostate and urinary tract cells for exploring the mechanisms regulating the differentiation of these cells. This study revealed that iPSCs produced from prostate and urinary tract cells have a higher efficiency of differentiation into the parent cells (prostate and urinary tract cells) than the iPSCs generated from skin fibroblast cells. Hence, this study emphasized that organ of origin has a major role in imparting the differentiation efficiency (Moad et al., 2013). iPSCs research has been used for the study of many deficiency linked diseases.

Down's syndrome is caused by an extra copy of chromosome 21. Briggs et al. (2013) have utilized the iPSCs research to identify the molecular networks associated with the pathogenesis of Down's syndrome. iPSCs have been utilized in the field of neurogenerative disorders like Parkinson's disease (PD) in which there occurs a loss of dopaminergic neurons of substantia nigra. The treatment of PD had always been a difficult task because the neurons have already been lost by the time the PD gets clinically manifested. Due to this, the mechanism of this disease has not been explored. Nguyen et al. have used iPSCs and studied G2019S mutation in leucine-rich repeat of kinase 2 (LRRK2) gene, which has been reported in sporadic and familial PD cases. Hence, disease modeling using iPSCs research provides an improved understanding of the molecular mechanisms driving a disease.

12.6.1 REGENERATIVE MEDICINE

In regenerative medicine, damaged or degraded tissue is repaired by using iPSC to produce these tissues, which are then transplanted to damaged or denatured sites. Two major issues concern the use of gene therapy:

1. Tissue or organ availability
2. Immunorejection.

There is an increasing need of organs and tissues due to the ever-increasing number of accidents and degenerative diseases. But the population is unaware of the shortage of healthy donors, which leads to the nonavailability of organs. At the same time, the transplantation of cells, tissues, or organs in a patient can only happen from a disease-free person whose physiological characteristics are consistent with the physiological characteristics of the patient. In view of these risks, several tests are performed prior to implanting the tissue or organ into a patient. The use of iPSCs provides an excellent approach to these treatments, as cells that are implanted into the patient will be generated from the iPSCs produced by somatic cells in the patient's own body. iPSCs have been utilized to treat many pathologic and degenerative diseases. Injuries caused due to accidents or by natural disasters can also be treated with gene therapy using iPSCs.

Kazuki et al. corrected the gene defect in iPSCs in a patient with Duchenne muscular dystrophy (DMD). They utilized human artificial chromosome (HAC) to express the full sequence of dystrophin (DYS). They generated iPSCs using fibroblasts from DMD patient. The error of the DYS gene in iPSCs was corrected using HAC by transferring DYS-HAC (containing complete genomic sequence of the DYS in HAC) using microcellular-mediated chromosome transfer (Kazuki et al., 2010). Generating hepatocytes using fetal or adult progenitors face certain obstacles like loss of function of cultured hepatocytes or limited organ availability. Hepatocytes can also be generated using iPSCs.

Tissue Repair

Park et al. generated "vascular progenitor" cells from the human iPSCs that represented the vascular stem cells. They identified these cells using CD31 and CD146. They observed that the mouse retinal vascular cells repaired when the iPSCs were injected into the vitreous of the damaged retina (Tang et al., 2013).

Drug Discovery

Drug discovery is another important field where iPSCs research is utilized. Animals or cells are used in in vitro test systems, but are restricted due to their inability to reproduce physiological phenotypic properties in humans. It was believed that the results obtained in animals would replicate in humans also, but; it was found that the efficacy of drugs in animals varies from one animal to another and is not reciprocated in humans. Animal models are also not good models for drug toxicity studies because the drugs can be toxic to one species and nontoxic to another. Even carcinogenic agents have varying degrees of carcinogenicity depending upon the species, like formaldehyde is a better carcinogenic agent in rats than in mice (Kerns et al., 1983). All these evidences and reports suggest that a more reliable model is needed for the drug discovery studies compared to the ones that are existing. As these drugs will be used for human diseases, it becomes very important that the drug testing models should be closely related to humans.

There are several sequential steps that a chemical has to pass to be approved for human use. Chemical toxicity should be properly evaluated before administration to patients. Only a small percentage of drugs in the clinical trials reach the actual market. The expense of developing a drug is growing, and the estimated cost is around 1.2–1.5 billion US dollars per compound (Kaitin, 2008; Gunaseeli et al., 2010; Sollano et al., 2008). The progress of 30% of drugs was abandoned due to lack of efficiency and 30% because of associated safety concerns (cardiotoxicity, hepatotoxicity) (Laustriat et al., 2010). Drug development becomes more expensive and slower due to the lack of early diagnosis of drug toxicity models closely associated to humans. Better toxicity models are needed, which may provide the cardiotoxicity, hepatotoxicity, and other toxicity associated data before the drug reaches the clinical trials. This will also decrease the time consumed for the clinical trials of the drugs, which will eventually fail in the trials due to toxic effects. Many researchers worldwide, studied iPSCs and identified the compound toxicity. Using iPSCs proposes a better choice than conventional toxicological tests and provides a better chemical safety assessment as they offer an environment like human physiological conditions compared to traditional systems involving the guinea pigs or mice, etc.

Figure 12.2 summarizes the potential application of iPSCs research.

12.7 CHALLENGES FACED BY THE IPSCS RESEARCH

A major problem with iPSCs research is that the gene sets used for pluripotency also induces neoplastic development. For example, the Oct4 overexpression is a major factor for murine epithelial cell dysplaisa (Hochedlinger et al., 2005). Similarly, Sox2 activation is also related to colon cancer (Zhao et al., 2017). The altered cMyc expression is commonly seen in 70% of human carcinomas (Kuttler and Mai, 2006) and Klf4 expression is often elevated in breast cancer (Ghaleb et al., 2005). Low efficiency (0.02%) of reprogramming of human iPSCs from fibroblast also serves as a major disadvantage of the iPSCs research (McCall and Johnston, 2007). The probability of acquiring genetic mutations by iPSCs remains high during reprogramming. There is a necessity for the development and standardization of more stringent protocols for iPSCs generation, maintenance, and differentiation. As retroviruses are used for gene delivery, there exists a possibility that these retroviral vectors get randomly integrated into the host genome and results in genetic aberration (Howe et al., 2008).

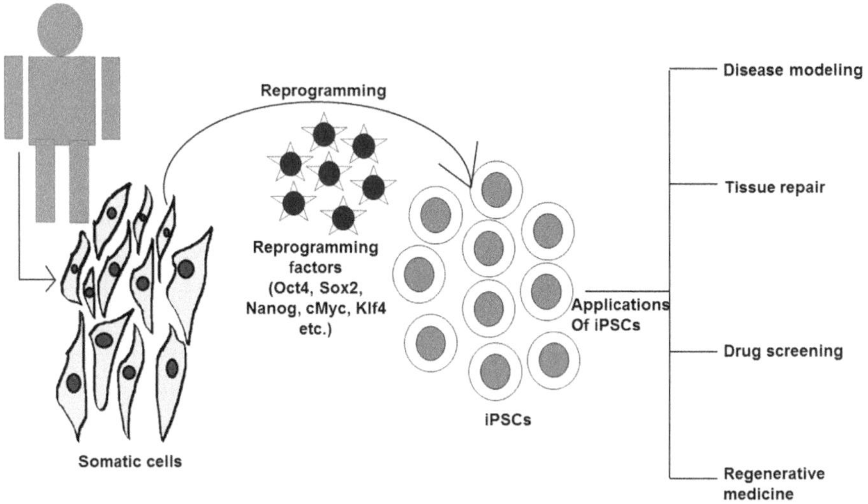

FIGURE 12.2 Potential applications of iPSCs. iPSCs are produced from the somatic cells by inducing them with reprogramming factors. These iPSCs can be used as an in vitro model system for analyzing various deadly diseases for mechanistic study as well as for drug screening. iPSCs are also used for the regenerative medicine and tissue repair.

12.8 CONCLUSION

Since its inception, stem cell research has revolutionized the field of medical science. The branch of iPSCs has considerably grown a lot in the last decade with the implications in disease modeling, drug screening, tissue repair, and regenerative medicine. But there are certain limitations like low efficiency of reprogramming, probability of acquiring genetic mutations, time and cost necessity for the development of patient-derived iPSCs. The need of the hour is to improvise the reprogramming strategies, reduce the probability of neoplastic development associated with reprogramming factors (ESC markers), and reduce the possibility of genetic aberrations associated with retroviral delivery, which will accelerate iPSCs research. iPSCs technology is a breakthrough toward personalized medicine.

KEYWORDS

- **division**
- **differentiation**
- **stem cells**
- **self-renewal**

REFERENCES

Ali, P. S., Ghoshdastider, U., Hoffmann, J., Brutschy, B., Filipek, S. Recognition of the let-7g miRNA precursor by human Lin28B. *FEBS Lett.* 2012, 586, 3986–3990.

Briggs, J. A., Mason, E. A., Ovchinnikov, D. A., Wells, C. A., Wolvetang, E. J. Concise review: new paradigms for down syndrome research using induced pluripotent stem cells: tackling complex human genetic disease. *Stem Cells Transl. Med.* 2013, 2, 175–184.

Ghaleb, A. M., Nandan, M. O., Chanchevalap, S., Dalton, W. B., Hisamuddin, I. M., Yang, V. W. kruppel-like factors 4 and 5: the yin and yang regulators of cellular proliferation. *Cell Res.* 2005, 15, 92–96.

Gunaseeli, I., Doss, M. X., Antzelevitch, C., Hescheler, J., Sachinidis, A. Induced pluripotent stem cells as a model for accelerated patient- and disease-specific drug discovery. *Curr. Med. Chem.* 2010, 17, 759–766.

Gurdon, J. B. The developmental capacity of nuclei taken from intestinal epithelium cells of feeding tadpoles. *J. Embryol. Exp. Morphol.* 1962, 10, 622–640.

Hochedlinger, K., Yamada, Y., Beard, C., Jaenisch, R. Ectopic expression of Oct-4 blocks progenitor-cell differentiation and causes dysplasia in epithelial tissues. *Cell* 2005, 121, 465–477.

Howe, S. J., Mansour, M. R., Schwarzwaelder, K., Bartholomae, C., Hubank, M., Kempski, H., Brugman, M. H., Pike-Overzet, K., Chatters, S. J., De Ridder, D., Gilmour, K. C., Adams, S., Thornhill, S. I., Parsley, K. L., Staal, F. J., Gale, R. E., Linch, D. C., Bayford, J., Brown, L., Quaye, M., Kinnon, C., Ancliff, P., Webb, D. K., Schmidt, M., Von Kalle, C., Gaspar, H. B., Thrasher, A. J. Insertional mutagenesis combined with acquired somatic mutations causes leukemogenesis following gene therapy of Scid-X1 patients. *J. Clin. Invest.* 2008, 118, 3143–3150.

Kaitin, K. I. Obstacles and opportunities in new drug development. *Clin. Pharmacol Ther.* 2008, 83, 210–212.

Kazuki, Y., Hiratsuka, M., Takiguchi, M., Osaki, M., Kajitani, N., Hoshiya, H., Hiramatsu, K., Yoshino, T., Kazuki, K., Ishihara, C., Takehara, S., Higaki, K., Nakagawa, M., Takahashi, K., Yamanaka, S., Oshimura, M. Complete genetic correction of Ips cells from duchenne muscular dystrophy. *Mol. Ther.* 2010, 18, 386–393.

Kerns, W. D., Pavkov, K. L., Donofrio, D. J., Gralla, E. J., Swenberg, J. A. carcinogenicity of formaldehyde in rats and mice after long-term inhalation exposure. *Cancer Res.* 1983, 43, 4382–4392.

Kuttler, F., Mai, S. C-Myc, genomic instability and disease. *Genome Dyn.* 2006, 1, 171–190.

Laustriat, D., Gide, J., Peschanski, M. Human pluripotent stem cells in drug discovery and predictive toxicology. *Biochem. Soc. Trans.* 2010, 38, 1051–1057.

Mccall, K., Johnston, B. Treatment options in end-of-life care: the role of palliative chemotherapy. *Int. J. Palliat. Nurs.* 2007, 13, 486–488.

Moad, M., Pal, D., Hepburn, A. C., Williamson, S. C., Wilson, L., Lako, M., Armstrong, L., Hayward, S. W., Franco, O. E., Cates, J. M., Fordham, S. E., Przyborski, S., Carr-Wilkinson, J., Robson, C. N., Heer, R. A novel model of urinary tract differentiation, tissue regeneration, and disease: reprogramming human prostate and bladder cells into induced pluripotent stem cells. *Europ. Urol.* 2013, 64, 753–761.

Sollano, J. A., Kirsch, J. M., Bala, M. V., Chambers, M. G., Harpole, L. H. The economics of drug discovery and the ultimate valuation of pharmacotherapies in the marketplace. *Clin. Pharmacol. Ther.* 2008, 84, 263–266.

Staerk, J., Dawlaty, M. M., Gao, Q., Maetzel, D., Hanna, J., Sommer, C. A., Mostoslavsky, G., Jaenisch, R. Reprogramming of human peripheral blood cells to induced pluripotent stem cells. *Cell Stem Cell* 2010, 7, 20–24.

Takahashi, K., Tanabe, K., Ohnuki, M., Narita, M., Ichisaka, T., Tomoda, K., Yamanaka, S. Induction of pluripotent stem cells from adult human fibroblasts by defined factors. *Cell* 2007, 131, 861–872.

Takahashi, K., Yamanaka, S. Induction of pluripotent stem cells from mouse embryonic and adult fibroblast cultures by defined factors. *Cell* 2006, 126, 663–76.

Tang, H., Sha, H., Sun, H., Wu, X., Xie, L., Wang, P., Xu, C., Larsen, C., Zhang, H. L., Gong, Y., Mao, Y., Chen, X., Zhou, L., Feng, X., Zhu, J. Tracking induced pluripotent stem cells-derived neural stem cells in the central nervous system of rats and monkeys. *Cell Reprogram.* 2013, 15, 435–442.

Wilmut, I., Schnieke, A. E., Mcwhir, J., Kind, A. J., Campbell, K. H. Viable offspring derived from fetal and adult mammalian cells. *Nature* 1997, 385, 810–813.

Yu, J., Vodyanik, M. A., Smuga-Otto, K., Antosiewicz-Bourget, J., Frane, J. L., Tian, S., Nie, J., Jonsdottir, G. A., Ruotti, V., Stewart, R., Slukvin, Ii, Thomson, J. A. Induced pluripotent stem cell lines derived from human somatic cells. *Science* 2007, 318, 1917–1920.

Zhao, X., Zhang, H. W., Zhang, Y., Li, S., Xu, R. X., Sun, J., Zhu, C. G., Wu, N. Q., Gao, Y., Guo, Y. L., Liu, G., Dong, Q., Li, J. J. Analysis of lipoprotein subfractions in 920 patients with and without type 2 diabetes. *Heart Lung Circ.* 2017, 26, 211–218.

Zhou, T., Benda, C., Dunzinger, S., Huang, Y., Ho, J. C., Yang, J., Wang, Y., Zhang, Y., Zhuang, Q., Li, Y., Bao, X., Tse, H. F., Grillari, J., Grillari-Voglauer, R., Pei, D., Esteban, M. A. generation of human induced pluripotent stem cells from urine samples. *Nat. Protoc.* 2012, 7, 2080–2089.

CHAPTER 13

Recent Advances in Imaging and Analysis of Cellular Dynamics in Real Time

CHANDRA BHAN, PANKAJ DIPANKAR, SHIBA PRASAD DASH, PAPIYA CHAKRABORTY, NIBEDITA DALPATI, and PRANITA P. SARANGI*

Department of Biotechnology, Indian Institute of Technology, Roorkee, Roorkee 247667, Uttarakhand, India

Corresponding author. E-mail: psarafbt@iitr.ac.in

ABSTRACT

Understanding the complex cellular and molecular processes inside a living cell are essential for proper decoding of the intricate network of information that drives the biology of a living system. Over the past few decades, because of limitation in the cell imaging techniques, it was difficult or rather impossible to study the kinetics, dynamics, and three-dimensional architecture of biomolecules that operate within the cell. However, a gradual development in the field of optics, high-quality electronics, biomedical devices, and experimental methods, had led to the advancement of imaging technologies. In recent years many noninvasive and living cell visualization techniques have been evolved such as confocal, multiphoton intravital microscopy with potential application in several fields in biomedical sciences. The aforementioned devices have given better insight into the dynamic behavior of living cells in terms of gene expression, protein–protein interaction, DNA–protein interaction, cell division, colocalization, and intracellular transport of biomolecules in a real-time manner. The introduction of advanced imaging techniques such as intravital *in vivo* imaging including very recent cryoelectron microscopy has opened avenues for in-depth analysis of the cellular process in a pathophysiological condition that can lead to the foundation of novel therapeutic approaches for many diseases. Indeed, real-time visualization of live cells, animals, plants including smaller entities such as viruses have taken the scientific development in life sciences to the next higher level. This chapter covers the entire range of latest imaging techniques, available to date to efficiently capture various aspects of cellular dynamics from single to multicellular levels.

13.1 INTRODUCTION

The existence of living beings involves proper communication of various organs and organ systems at cellular and molecular levels. To date, biologists have come up with numerous ways to understand complex cellular and molecular processes operating inside a living cell and decode nature's secrets that drive the biology of a living system. Amongst all the research techniques, microscopic imaging has served as an essential and indispensable tool for close observation and visualization of cells and tissues that have evolved from a simple and low-resolution bright field imaging to super high-resolution multiplexing with three-dimensional imaging technologies. Over the past few decades, the curiosity of the researchers has led to the invention of sophisticated optical instrumentations that have significantly surpassed the majority of the limitations involving traditional imaging systems. Modern imaging technologies have enabled the researchers to study the kinetics, dynamics, and three-dimensional architecture of biomolecules that operate within a cell and control many biological and physiological processes. Recent inventions and developments in optics, electronics, and biomedical fields have led to the advancement of image processing. In recent years, many noninvasive visualization techniques have evolved, such as confocal and intravital multiphoton microscopy (IVMPM) with potential applications in different fields of biomedical sciences. Furthermore, these techniques have also provided better insight into the dynamic behavior of living cells such as gene expression, protein–protein interaction, DNA–protein interaction, cell division, colocalization, and intracellular transport of biomolecules in real time. In addition, they have been very helpful in understanding the orchestrated behavior of immune cells under various environmental conditions. Similarly, for defining the pathophysiology of various infectious and noninfectious diseases including cancers it is imperative to understand the mechanisms governing the migration and crosstalk of various cell types in the vascular network and tissue spaces. The latest technologies such as IVMPM and cryoelectron microscopy (Cryo-EM) have opened many avenues for in-depth analysis of various cellular processes in different pathophysiological condition as a foundation for the development of novel therapeutic approaches. Similarly, recent developments in real-time visualization of animal and plant cells including smaller entities such as viruses have given the scientific development in life sciences a significant boost. In addition to the state of the art electronics and creative inclusion of elegantly engineered procedures such as photoacoustic *in vivo* imaging, and the use of near-infrared quantum dots imaging have contributed significantly to the evolution of imaging sciences. Taken together, this chapter will discuss various cellular and animal imaging techniques covering their basic working principles to the salient features of instrumentation and applications in biology.

13.2 EVOLUTION OF MICROSCOPY AND CELLULAR IMAGING

Visualization of a single cell was the first founding step, which pushed the scientific community into a new era of cell biology. This was possible due to the pioneering work of Leeuwenhoek (1632–1723) and Robert Hooke (1635–1703) when they first discovered the concept of looking at magnified objects. Subsequently, in the 19th century, the emergence of different disciplines such as biochemistry, molecular biology, and systems biology enabled an in-depth understanding of cellular physiology. Till 1950, the cellular and histological studies were performed exclusively with the help of the light microscopes, which had the limitation in both resolving power and magnification thereby creating hindrance in investigating useful information about cellular compositions and architecture (Breidenmoser et al., 2010). Consequently, researchers invented the electron microscope with electron beams, which significantly improved the magnification power to 10,000,000-fold, which facilitated a deeper understanding of molecular structures in a broad range of samples including microbial cells and tissue biopsy sample. In contrast to the light microscope, the electron microscope has two modes of visualization; the scanning electron microscopy (SEM) that helped in the visualization of the topology and transmission that gives an insight into the inner composition and complexity of specimens. However, the use of electron microscopy was limited to fixed and dehydrated dead cells and thus could not provide full information regarding cellular dynamics (Harris, 2015). Thus, with a need for better imaging tools with high-quality electronics, noninvasive visualization techniques such as fluorescence, confocal, multiphoton, and intravital microscopy (IVM) gradually made their way to the world of cellular imaging and visualization. Evolution of noninvasive microscopic techniques began with the introduction of fluorescence microscopy, which was based on the fluorescence signals from certain molecules when they absorb the light of certain wavelengths followed by emission of light with longer wavelengths.

Fluorescence microscopy was broadly applied in medicine, diagnostics, and in biological research (Marcu, 2012). In addition, it was used for studying biological events such as protein–protein interactions, monitoring of intracellular pH, and ion concentration (Hoppe et al., 2009). However, the greatest disadvantage with fluorescent microscopy was photobleaching (loss of fluorescence of fluorophores due to long exposure to intense light) and phototoxicity (exposure to intense light generates reactive chemicals that are toxic to the cells) and required experimental optimization (Sanderson et al., 2014). Subsequently, confocal microscopy emerged as the high-end fluorescence microscopy that provided sharper images with high contrast due to the use of laser as a source of light, detected from one single focal plane as shown in Figure 13.1. Confocal microscopy was used for multiple applications such as to study cellular components, the topology of the cell surface, examination of the tissue sample, and nuclear structures. With different controlled parameters such as temperature, humidity, and CO_2 levels, confocal microscopy also provided time-lapse imaging of living cells and their processes. However, the primary limitation of confocal microscopy was the inability to penetrate deeper into the tissues including photobleaching (St Croix et al., 2005). Consequently, researchers developed multiphoton microscopy, which allowed imaging of living tissues up to 1 mm in depth. Two-photon-excitation microscopy utilized near-infrared light that could excite different fluorophores. However, for each excitation, it absorbed two photons of infrared light. The use of infrared light minimized scattering in the tissues, reduced bleaching, and also concealed the background signals thus resulting in deeper penetration into the tissue (Konig, 2000). However, the aforementioned noninvasive imaging techniques were primarily limited to *in vitro* studies and were not adequate to investigate the physiological and biological events occurring inside the living organism in real time. Thus, the next development came in the form of IVM including intravital multiphoton microscopy (IVMPM). These advanced imaging technologies provided a platform for studying complex cellular processes in a live animal (*in vivo*) at high resolution. This technique has been extensively used by biomedical researchers in multiple fields such as studying cell migration, cellular interactions, drug delivery systems, the pharmacokinetics of drugs, etc. (Jensen, 2013). Currently, multiple instrumentation technologies have been launched by major research and imaging focused companies with basic to advanced features with upgradation facilities. More recently, Cryo-EM, the latest addition to the imaging field that has revolutionized the field of imaging and microscopy can image smallest entities such as viruses in their native state (Murata and Wolf, 2018). A detailed timeline for major inventions in the field of imaging and microscopy are listed in Table 13.1.

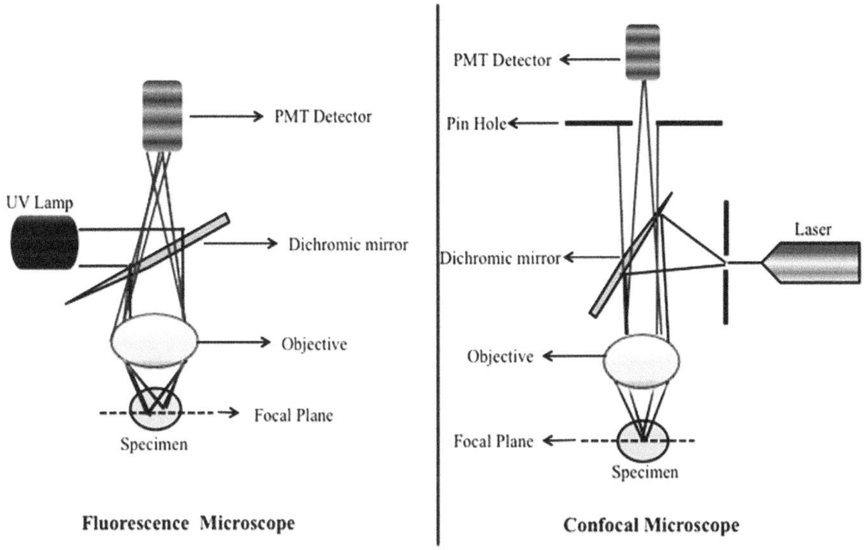

FIGURE 13.1 Schematic representation of fluorescence microscope & confocal microscope.

TABLE 13.1 Major Landmarks in the Field of Microscopy

Sr. No.	Time Period	Contribution
1.	14th century	The technique for grinding lenses was developed in Italy.
2.	1590	Hans and Zacharias Janssen built the first true microscope.
3.	1667	Robert Hooke modified the microscope and discovered the anatomy of cork cells, which was published in Micrographia.
4.	1675	Antoni van Leeuwenhoek used a microscope for observing tiny living organism like bacteria, etc.
5.	1830	Joseph Jackson Lister showed the use of multiple lenses at a certain distance can improve image quality.
6.	1878	Ernst Abbe had given a mathematical equation which correlates resolving power of the lens to the light intensity.
7.	1903	Richard Zsigmondy invented the ultra-microscope and in 1925, he received the Nobel Prize in Chemistry.
8.	1913	The first commercially luminescence bright-field microscope was launched by Zeiss.

TABLE 13.1 *(Continued)*

Sr. No.	Time Period	Contribution
9.	1929	First commercial intravital microscope launched by the firm Carl zeiss.
10.	1932	Frits Zernike developed the phase-contrast microscope that allowed the study of unstained material. in 1953, he was awarded the Nobel Prize in Physics.
11.	1931	Knoll and Ruska built the first transmission electron microscope. For this invention, in 1986, they received the Nobel Prize in Physics.
12.	1938	First SEM by Zworykin.
13.	1939	Siemens produced the first commercial TEM.
14.	1955	Marvin Lee Minsky filed patent for first "double focusing stage scanning microscope".
15.	1965	The first commercial SEM was developed by the Cambridge Scientific Instruments Mark I "Stereoscan".
16.	1981	Gerd Binnig and Heinrich Rohrer invented the scanning tunneling microscope that provided 3D- the image of a material at an atomic level and they won the Nobel Prize in Physics for the year 1986.
17.	1982	First commercially confocal scanning laser microscope was launched by Oxford Optoelectronics.
18.	1993–1996	Stefan Hell, Eric Betzig, and William Moerner developed the first super-resolution microscope, which allowed the visualization of matter smaller than 0.2 μm. For this work all the three scientists shared Nobel Prize in 2014.
19.	1990	Multiphoton microscope was developed by Winfried Denk and James Strickler at Cornell University.
20.	1990	Joachim Frank was the first to prepare the 3D image of a protein using cryo-electron microscope for which in 2017 he shared the Nobel prize in chemistry with Richard Henderson and Jacques Dubochet.

Subsequent contents of this chapter will describe basic working principles and potential applications, which have constantly evolved in high-end imaging technologies such as confocal laser-scanning microscopy, IVMPM that have significantly helped the scientific community to get a closer look of the living cells in real time.

13.3 CONFOCAL MICROSCOPY

13.3.1 OVERVIEW

After the invention of conventional fluorescence microscopy (also known as widefield fluorescence microscopy), it became an essential imaging tool in different streams of biological sciences. In 1955, Marvin Minsky introduced the concept of confocal microscopy to the scientific community, which was a specialized version of the existing imaging technology. Modification in the newly introduced technology consisted of an aperture in front of the detectors or photomultiplier tubes (PMTs) allowing the entry of emitted fluorescence from the focused optical plane within the specimen. This exact positioning of the optical components enabled the screening of the fluorescence coming from out-of-focus planes of the objective lens resulting in sharper image of a specimen with better contrast and spatial resolution (Vangindertael et al., 2018).

13.3.2 PRINCIPLE

The uniqueness of confocal microscopy involved focusing of the laser on a specific spot within the specimen rather than exposing the entire tissue thereby reducing photobleaching and phototoxicity. Secondly, the emitted fluorescence from each point is allowed to pass through a pinhole or aperture placed in front of the detectors allowing the entry of fluorescence signals coming from the illuminated spot in the specimen to enter into the detectors or PMTs. Confocal microscopy was designed to scan objects in a raster form and create pictures in a single optical plane. In other words, the confocal microscope prepares multiple optical sections of the specimen acting as an optical microtome allowing the preparation of a 3D model of scanned tissues. This special feature of stalking up the images from multiple optical planes on the *z*-axis was termed as Z-stacking. This feature allowed the researchers to image thicker sections of >30 μm (Fellers and Davidson, 2007). Thus, this technique was termed as confocal microscopy due to the specialized merging of two focal planes. Figure 13.2 shows the Z-stacking images of HeLa cells. The cells were labeled and Z slices were taken at 300 nm steps to visualize cellular structures in 3D.

13.3.3 INSTRUMENTATION AND TECHNICAL FEATURES

In a confocal microscope, high-intensity lasers are used as light sources to achieve clear and high contrast images instead of mercury and xenon lamp. Preferred laser sources for confocal imaging are gas, solid-state, and semiconductor lasers. A pinhole or aperture that limits the out-of-focus light rays is the final photosensitive detector that could amplify the low-intensity signal coming from the aperture into electrical signals (Sanderson et al., 2014). In addition, for timely collection of information from living cells two types of confocal microscopes with specialized features using different fluorochromes and fluorescent proteins with overlapping spectral signals were introduced that are described in Table 13.2.

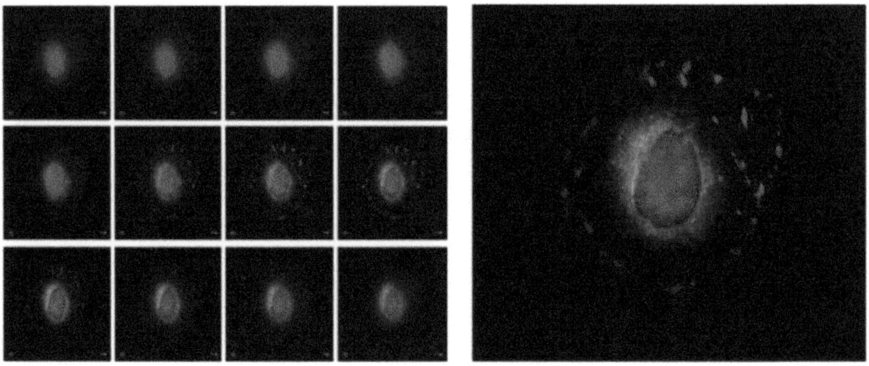

FIGURE 13.2 HeLa cell labeled using CellLight™ ER-GFP BacMam 2.0, CellLight™ Talin-RFP BacMam 2.0, and NucBlue™ Live ReadyProbes™ Reagent. Z slices were taken at 300 nm steps using an EVOS™ FL Auto 2 Imaging System equipped with an Olympus 60X Apochromat Oil objective and a GFP, RFP, and DAPI EVOS™ light cubes to visualize cellular structures in 3D. Images display a montage of individual slices (right) and a mean intensity projection of all 12 slices (left). (Courtesy of Thermo Fisher Scientific, USA.)

TABLE 13.2 Difference Between Spectral Scanning and Disk-Based Systems

Sr. No.	Attributes	Spectral Scanning	Disk based
1.	Source of light	Laser	Laser or arc
2.	Size of pin hole	Small	Large
3.	Method of scanning	Point-by-point scanning of specimen	One time scanning of entire specimen
4.	Type of detector	Photomultiplier tube (PMT) and assigning pseudo-color detector	Charge coupled devices (CCD) camera
5.	Image construction mode	Generation of 3D image	Used for Time Lapse imaging (4D-Image)
6.	Merit	Point-by-point imaging because of ability to generate optical section of an object	Less time consuming and higher transmission rate

13.3.4 TYPES OF CONFOCAL MICROSCOPY

13.3.4.1 SPECTRAL SCANNING

The discovery of different fluorochromes and fluorescent proteins in recent years have significantly helped the researchers to target multiple molecules in a cell simultaneously. However, their closely overlapping spectral signals could create problems in detecting the exact source of signals. Thus, spectral scanning confocal microscope was developed and equipped with an advanced spectral detection system allowing decisive segregation of overlapping emission spectra by capturing each spectrum of focused fluorescence light in the specific spot of the specimen. Currently, multiple technologies such as 8 and 32 spectral bins/scans are being used by leading manufacturers to improve signal separation and accurate detection (Oreopoulos et al., 2014).

13.3.4.2 DISK-BASED SYSTEMS

Confocal microscopy has provided an excellent scope to biological scientists for studying multiple parameters of cells and tissues with minimal noise. However, imaging biological processes in real time with millisecond timescale requires much faster scanning of fluorophores compared to point-by-point scanning methodology. This challenge in the acquisition of high-resolution images in living tissues was further resolved by using a spinning-disk system, which allowed the recording of information with high-speed and in a real-time manner (Oreopoulos et al., 2014). The basic difference between the spectral or point scanning and disc scanning microscope is that the former acquires the signal from each pixel and builds an image sequentially whilst later technique, parallel pinholes are used to illuminate the entire specimen to acquire multiple signals at a time. In addition, the introduction of an advanced camera system such as an electron-multiplying charge-coupled device (CCD) in the disc-based scanning also helps in taking better pictures in low intensity of light.

13.3.5 APPLICATION

Due to high resolution and 3D-imaging capabilities confocal microscopy serves as an important technique to study a variety of cellular processes. For instance, the microtubules dynamic can be studied in cortical neurons with neuronal dendritic marker MAP2 as shown in Figure 13.3(A). As described earlier, the discovery of different fluorophores and genetically engineered fluorescent proteins has allowed biologists to target and label multiple cellular components for imaging resulting in increasing the range of applications. In addition, confocal microscopy has enabled the users to capture multiple optical manipulations of excitation and emission spectrums of various fluorophores. For instance, optical systems such as fluorescence resonance energy transfer (FRET), fluorescence recovery after photobleaching (FRAP), fluorescence lifetime imaging (FLIM), fluorescence loss in photobleaching, fluorescence localization after photobleaching, fluorescence in situ hybridization (Dey, 2005; Jerome and Price, 2018) are regularly used in scientific research. Below are a few specific implications of confocal imaging technique.

Developmental Biology: Due to the clarity and higher spatial resolution, confocal microscopy serves as an essential tool for imaging elegant cellular processes in both plant and animal developmental biology. Confocal microscopy has facilitated the study of many development aspects of living organisms such as gastrulation, organization of microtubules in *Caenorhabditis elegans*, and embryonic development at the two-celled stage. This technique has its implication in the detailed study of stem cells and cardiovascular development as well (O'Connell and Golden, 2014).

Microbiology: Confocal microscopy has played an important role in obtaining information on multiple aspects of microbes such as bacteria and fungi including mobilization of cellular components in various environmental conditions, intracellular and intercellular communications, morphological changes, etc. (Frey-Klett et al., 2011). In addition to basic research, confocal microscopy has been successfully used in the early diagnosis of many infectious diseases. For example, *in vivo* confocal microscopy is a recent emerging noninvasive technique that has allowed rapid diagnosis of fungal keratitis with very high sensitivity and specificity (Kumar et al., 2010). Similarly, in the food industry, this technique has been used to produce 3D-image of living microbes that grow on food matrices as well as for the visualization and investigation of dynamics of antimicrobial compounds within the biofilms (Wilson et al., 2017).

Cell Biology and Translational Research: Confocal microscopy has been used as an essential component in studying cellular functions such as the natural killer cell-mediated killing of cancer cells as shown in Figure 13.3(C). The ability of this technique to scan through the cellular layers has immensely helped in studying cell localization, cellular subsets, cell behavior, and cellular processes under physiological, and inflammatory conditions (Rigby and Goldie, 1999; Ilie et al., 2019). Confocal microscopy has been used for evaluation of programmed cell death of various cells using fluorescent dye such as Annexin V that binds to phosphatidylserine on the cell membrane (Mukhopadhyay et al., 2007). Significant insight into the pathophysiology of human disease such as cancer has been achieved using confocal imaging of tissue sections isolated from experimental animals. Figure 13.3(B) shows the visualization of HeLa cell proliferation using the ki67 marker, which indicates the high proliferative state of these cancer cells. As a rapidly evolving technique, confocal microscopy-based imaging of live cells has indeed contributed immensely to biomedical research (Jensen, 2013).

A B C

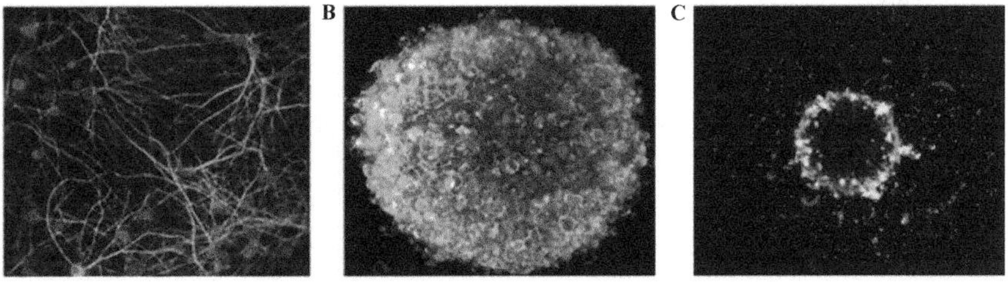

FIGURE 13.3 (A) Rat primary cortical neurons labeled with neuronal dendritic marker MAP2 antibody, neuronal cell body marker, Hu C/D followed by Alexa Fluor® 488 (cyan) and Alexa Fluor® 647 (pink) secondary antibodies, respectively, and Hoechst 33420 (blue). (B) HeLa spheroids labeled with the Ki67 antibody followed by Alexa Fluor® 647 (pink) and Alexa Fluor® 488 (green) phalloidin and Hoechst 34580 for nuclei (blue). (C) Natural killer cells labeled with CellTracker™ Deep Red attacking HeLa spheroids labeled with pHrodo Green AM Intracellular pH indicator (green). [Courtesy of (A) Bhaskar Mandavilli and Michael Derr (B) Bhaskar Mandavilli and Marcy Wickett (C) Christopher Langsdorf and Bhaskar Mandavilli, Thermo Fisher, USA].

13.3.6 ADVANTAGES

Confocal microscopy has several benefits over the traditional optical microscopy. For instance, confocal microscopy has bring down the background seen in the case of a thicker section via serial optical sections of Z-stacking. The important feature of confocal microscopy is the spatial filter that prevents the entry of out-of-focus light in thick tissue sections. More recently, confocal microscopy has been used for taking high-quality images in research and diagnostics of many diseases (Jonkman and Brown, 2015).

13.3.7 LIMITATIONS

There are certain limitations associated with confocal microscopy, which makes it difficult to use for delicate samples that include phototoxicity, photobleaching, including poor signal detection due to overlapping emission. In addition, the visualization of specimens is constrained to a certain depth (up to 100 μm) (Wnek and Bowlin, 2008).

13.4 ELECTRON MICROSCOPE

13.4.1 OVERVIEW

An electron microscope uses electron beams as a light source instead of visible light (Gordon, 2014). Electron beams allow the visualization of small size objects with very high resolution and magnification than a normal light microscope due to the short wavelength of the electron beam, which can be a hundred thousand times shorter compared to the visible light photons. Researchers and scientists have used it to examine metals, crystal structures, biological samples including microorganisms, cells, tissues, and medical biopsy samples, etc.

13.4.2 PRINCIPLE

Electrons, the subatomic particles orbiting around the nucleus, fly off from the atom when excited by heat energy. In the electron microscope, a beam of electron of around 0.05 Å wavelength is generated by an electron gun employing 60–80 kV electrons. Secondly, this technology uses electromagnetic lenses, which assist in the generation of the electromagnetic field, which is essential for guiding the beam of electron onto the specimen.

13.4.3 INSTRUMENTATION AND TECHNICAL FEATURES

An electron microscope uses multiple components such as electron gun, electromagnetic lenses, and image processing system. A heated tungsten filament, which works as the electron gun, that is responsible for generating electron microscopy. The condenser lens system focuses the electron beam to the specimen and depending upon its thickness the electrons are scattered and fall on the objective lens, which forms the intermediate magnified image. Finally, the third set of lens also called the projector lens that further magnifies the image. In an electron microscope, the images of the specimen are recorded and projected on a fluorescent screen for visualization.

13.4.4 TYPES OF ELECTRON MICROSCOPES

13.4.4.1 TRANSMISSION ELECTRON MICROSCOPE (TEM)

TEM is a microscopic technique in which the electron beam is passed through a very thin section (usually not more than 100 nm thick) to form an image. Usually, TEM creates two-dimensional and black and white images, which provide morphologic, compositional, and crystallographic information about the samples (Shaikjee et al., 2011). Compared to light microscopy, TEM requires special sample preparation methodology by which very thin enough sections are prepared to allow the electron beams to pass them. The sample preparation method is complex and includes dehydration, sputter coating of nonconductive materials, cryofixation, sectioning, and finally staining (Schrand et al., 2010).

13.4.4.2 SEM

In contrast to TEM, the SEM generates images of specimens that provide information about the architecture of the samples. SEM can achieve resolution better than 1 nm. The electron beam interacts with atoms in the specimen and generates various signals including secondary electrons, backscattered electrons, X-rays, and transmitted electrons that are used to generate the images. For SEM, the specimen must be dry because of the high-vacuum specimen chamber. Therefore, living cells and tissue need to be chemically fixed and dehydrated with an organic solvent before examination (Sokolova et al., 2011).

13.4.4.3 CRYO-EM

Cryo-EM is a special technique in which the samples are frozen to cryogenic temperatures by a process called vitrification, forming an amorphous solid, which does not cause any damage to the specimen (Razinkov et al., 2016). Unlike TEM and SEM, no fixing or dying is required to preserve the integrity and the samples are maintained in their native state. The development of this technique is an alternate approach to X-ray crystallography or NMR spectroscopy due to the determination of macromolecular structure without the need for crystallization. Cryo-EM gives high resolution, which solves the structure that is impossible for X-ray crystallography and NMR (Murata and Wolf, 2018). A schematic diagram of the working principle of cryo-EM is presented in Figure 13.4. In cryo-EM, the specimens are usually frozen (frozen-hydrated) by liquid ethane or liquid nitrogen before observation. Freezing is performed very quickly to prevent the formation of ice cubes around the specimen. Unlike conventional microscopy, no dye or stain is needed for visualization under the microscope (De Boer et al., 2015).

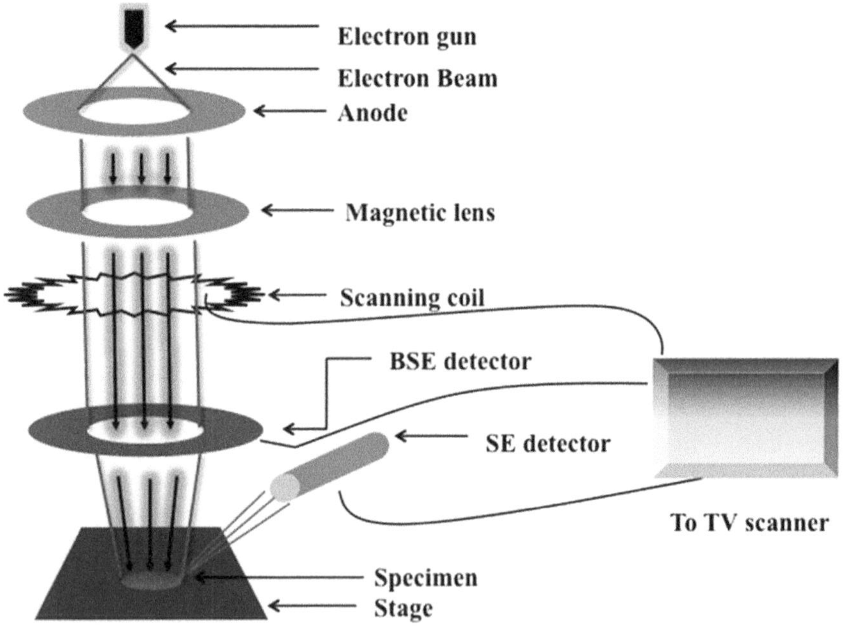

FIGURE 13.4 Schematic diagram of cryo-EM.

13.4.5 APPLICATIONS

Cryo-EM is the latest and most advanced addition to the field of high-resolution microscopy, which has helped the researchers to visualize microstructures such as viruses in their native conformation. For example, the introduction of cryo-EM enabled the researchers to understand the structure of the native HIV core and the capsid proteins (Zhao et al., 2013). In addition to HIV-I, cryo-EM has helped in solving three-dimensional structures of many viruses and their proteins, which have impact on the health worldwide. Currently, many emerging viral infections such as Chikungunya, Dengue, and Zika do not have any approved vaccines, which necessitate the development of effective antiviral drugs for keeping a check on outbreaks. Cryo-EM has played an important role in understanding the 3D structures or native and transient states of many viral proteins during entry and attachment on the cell surfaces, which have facilitated the process of antiviral drug designing and development (Walls et al., 2016). In addition to studying viruses, Cryo-EM has also helped in understanding many other pathogen-associated functional studies such as multidrug resistance mechanisms in bacteria (Belousoff et al., 2019). Similarly, study of complex events such as 3D-protein folding, the structure of amyloid fibrils associated with Alzheimer's, etc. has been possible due to the use of cryo-EM (Gremer et al., 2017).

13.4.6 ADVANTAGE OF CRYO-EM

Cryo-EM has the advantage of analyzing large, complex, and flexible structures, which makes it superior to conventional structural biology techniques like X-ray crystallography or NMR spectroscopy. Compared to other microscopy techniques, cryo-EM requires very small samples and still produces better images. Since the samples are rapidly frozen and are in a hydrated state without direct contact with other substances, it is assumed that their native conformation is perfectly preserved (Lyumkis, 2019).

13.4.7 LIMITATIONS OF CRYO-EM

In addition to the heavy cost of the instrument, one of the greatest disadvantages of the cryo-EM is the low signal to noise ratio. With time, further research on the instrumentation could improve the contrast and quality of images acquired by cryo-EM, which will help in answering many unsolved questions.

13.5 SINGLE AND IVMPM

13.5.1 OVERVIEW

As discussed earlier, invention of confocal and electron microscopy paved the path for high-end imaging devices with superior magnification, high resolution, and sensitivity, which dramatically improved the understanding of the biological phenomenon in various organisms (Sandoval and Molitoris, 2017). However, visualization and analysis of cells and tissues with these techniques were limited to the surface view mostly in nonliving cells and tissues and the resolution would reduce with an increase in the thickness of the specimens (Menger and Lehr, 1993). Consequently, it prompted the invention of IVM (Sandoval and Molitoris, 2017), which is an advanced microscopic technique that can be used to study molecular and cellular dynamics with highly improved spatial and temporal resolution including identification of single cells in a living tissue (Jain et al., 2002). There are various types of IVM techniques including widefield, laser-scanning confocal, spinning disk, and multiphoton microscopy (two or three photons) that could be used for detailed imaging of tissue architecture (Belperron et al., 2018).

The single-photon IVM imaging technique had few limitations where imaging of deeper architectures increased the chances of light getting scattered as well as photobleaching and loss of contrast (Guesmi et al., 2018), which restricted the use of IVM to *in vitro* and *ex vivo* and 2D image analysis (Hyun et al., 2012). Thus, multiphoton IVM microscopy was introduced that later became the most suitable procedure for studying deep living tissues at high resolution with three-dimensional visualization (Zipfel et al., 2003). As described in Figure 13.5, in IVMPM, two less energetic photons in the infrared spectrum are used to illuminate the target causing the least damage to the surrounding tissues, enabling the researchers to experiment for a longer period (Svoboda and Yasuda, 2006; Theer and Denk, 2006). For IVMPM, variety of signals are used for imaging including two-photon excited fluorescence (usually the primary signal), three-photon excited fluorescence, second and third-harmonic generation (SHG, THG) (Zipfel et al., 2003), which helps in imaging up to four types of channels such as blue, green, red, and far-red emitting fluorophores (Sandoval and Molitoris, 2017). Due to many advantages, IVMPM has been used in various biological fields such as neuroscience, tumor metastasis, angiogenesis, immune cell migration, and developmental biology (Zipfel et al., 2003).

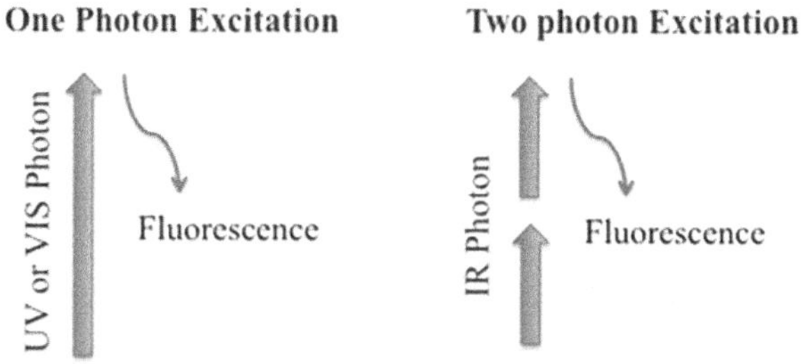

Parameters	One Photon Excitation	Two photon Excitation
Specimen	Thin tissue sections (Depth: 100 M)	Intact animals (skin, ear, vasculature etc) Thick sections (Depth: 1 mm)
Noise	More scattering	Less Scattering
Photo bleaching	More photo bleaching and toxicity	Less photo bleaching and toxicity
Application	Disease diagnosis, developmental biology, etc.	Solid tumor diagnosis, metabolic imaging, etc.

FIGURE 13.5 The flow-chart represents an overview of the differences between one-photon and two-photon microscopy. One-photon excitation uses visible or ultra-violet light as a source to illuminate the target, whereas two/multiphoton excitation uses infrared spectrum light source. The rapid pulses of infrared light in multiphoton microscopy can penetrate deep into the sample up to a depth of 1 mm, whereas conventional one-photon microscopes can penetrate up to 100 um, which limits its use for thick sections such as skin, ear, vasculature, etc. In addition, two/multiphoton imaging provides the facility to image a sample with minimum photobleaching and phototoxicity, along with less scattering as compared to a single-photon excitation. The above features facilitate the use of multiphoton imaging in a wide variety of samples and research areas.

13.5.2 PRINCIPLE

The basic principle of IVMPM includes the absorption of two or more photons that is infrared or near-infrared region (700–1000 nm range) simultaneously. The absorbed photons coordinate to produce the appropriate excitation energy for a certain fluorescent molecule (Figure 13.5) (So et al., 2001; Svoboda and Yasuda, 2006). The intensity is maximum near the area in focus, which reduces with increasing distance. Therefore, the excitation of fluorophores remains localized in a small focused region with the least diffraction. In addition, the nonlinear excitation ensures that the scattered photons are not strong enough to cause background fluorescence (Denk et al., 1994; Zipfel et al., 2003).

13.5.3 INSTRUMENTATION AND TECHNICAL FEATURES

The multiphoton microscopy is a combination of two-photon absorption with the use of a laser-scanning confocal. The laser is focused on a precise point in the sample at the desired plane, which excites the fluorescent molecules and generates photons detected by photodetectors (Tsai et al., 2002). They are further processed and presented in the form of the image in a data acquisition computer. Like the confocal microscope, IVMPM also includes four major components of the imaging system, namely lasers, scanners, objectives, and detectors. To generate adequate signal, usually, high intensity of excitation light is necessary. Therefore, lasers that produce rapid short pulses with constant average power and pulse repetition are preferred for efficient working of IVMPM (Svoboda and Yasuda, 2006). After its introduction in 1986 (Moulton, 1986), the mode-locked titanium-doped sapphire laser has been the best choice for IVMPM over dye lasers due to its ability to emit ultra-short optical pulses in the range of femtosecond interval and tunable excitation wavelength ranging from <700 nm to >1000 nm covering the excitation wavelength of commonly used important fluorophores (Zipfel et al., 2003).

Along with the above lasers, rotating polygonal mirrors, or resonant galvanometers are used as scanners in IVMPM to perform high-speed scanning in IVMPM. In addition, acousto-optic deflectors may be used for better and improved scanning (So et al., 2001; Svoboda and Yasuda, 2006). In IVMPM for obtaining maximum excitation efficiency and high resolution, objectives with high numerical aperture are used. Importantly, detectors such as PMTs, CCD cameras, and photodiodes are used in IVMPM. Widely used detectors for multiphoton microscopes are PMTs due to their higher sensitivity (20%–40% quantum efficiency) in the blue-green spectral region, low cost, and robust nature (Batchelor and Goulian, 2006; Zinter and Levene, 2011).

For better visualization of cellular components, the selection of appropriate fluorochromes and reporter molecules plays a very important role in obtaining high-resolution images with IVMPM. The probability of two-photon absorption is expressed quantitatively by measuring two-photon cross-section, which is represented in units of Goeppert–Mayer (GM) where

1GM is 10^{-50} cm^4 s (Kauert et al., 2006). Two-photon spectra are broader with a single laser simultaneously exciting more than one fluorescent molecule. Enhanced green fluorescent protein (GFP), cyan fluorescent protein, yellow fluorescent protein, and red fluorescence proteins have good cross-sections (w100 GM) and therefore are regularly used for IVM imaging (Spiess et al., 2005). While using multiplex imaging, dichroic mirrors are also used on the detection side for achieving better separation of colors. Similarly, to visualize fast-moving cells in the blood without much interference from endogenous fluorescence, recently semiconductor nanocrystals (quantum dots, QDots), near infrared quantum dots are also proposed for use in IVM (Austin et al., 2017; Pons et al., 2019).

13.5.4 APPLICATION

IVMPM have significantly helped in the progress of biomedical research and taken the understanding of scientists about the living cell biology to the next level. The following are the few examples of its application in research and diagnosis.

As a Diagnostic Method for Solid Tumors: For diagnostics, IVMPM is capable of imaging specimens up to 1000 µm of depth through an intact fibrous tumor capsule without requiring intravenous contrast agents. Therefore, this method is used to examine solid tumors keeping the resection margins intact. Similarly, this technique provides a solution to repetitive and invasive tissue analysis, while analyzing the details of cancerous tissues and reduces the number of experimental animals for specific experiments (Muensterer et al., 2017). Furthermore, MPM can be used in understanding the dynamics of abnormal vasculature in tumors (Gabriel et al., 2018).

Visualization of Organelles and Cellular Processes in Living Tissues: IVMPM is used to view cell organelles such as Golgi apparatus in live cells and deep inside living tissue with very low cytotoxicity (Xu et al., 1996). This technique is also used in monitoring alterations in the metabolism by investigating the relative amounts of reduced NADH and FAD in normal tissues compared to cancer (Skala et al., 2007).

Visualization of Infections and Immunological Responses in Live Animals: The dynamics of immune cell movement in and out of the blood vessel with response to inflammatory, cancerous, and infectious stimulations have been investigated using IVMPM. This technique has enabled the researchers to understand and study the exact movement pattern and localization in the tissues. For example, Figure 13.11 shows the extravasation of neutrophils in a cremaster muscle microvasculature when the muscle layer was exposed to f-MLP (*N*-formyl-methionine-leucyl-phenylalanine) in live animals. Use of IVMPM in immunology have led to the publication of numerous mechanistic studies related to cellular interactions, antigen-specific immune cell behaviors, role of adhesion receptors in immune cell trafficking, which have revolutionized our understanding of host immunity to infectious and noninfectious antigens (Hyun et al., 2012; Lim et al., 2015; Boldock et al., 2018). Similarly, in one more study, the mechanism of B-cell selection in the germinal center and the role of anatomically different light zone and dark zone were elucidated using IVMPM in combination with flow cytometry to study specific B cells and mechanism of germinal enter selection (Victora et al., 2010). IVMPM has been used to directly observe viral pathogens in different tissues including observation of trans-infection events of viruses *in vivo* (Sewald, 2018).

13.5.5 ADVANTAGES

Multiphoton microscopy is highly advanced imaging technique with multiple benefits, such as penetration up to 500 mm within a tissue block and reduction in the loss of light because of near-infrared-based excitation (Tauer, 2002). In addition, localized excitation at the focal plane and longer wavelengths in IVMPM prevents photobleaching and phototoxicity in the layers above and below the focal plane enabling imaging for a longer duration (Denk et al., 1990).

13.5.6 LIMITATIONS

In spite of so many advantages and benefits of IVMPM, it also has certain disadvantages, which requires attention while using the technique. For instance, although the photodamage above and below the focal plane is minimum in IVMPM, the damaging effects could be greater in certain specimens containing pigments such as hemoglobin, melanin, or chlorophyll due to their photo absorbing properties. This could result in morphological and biophysical damages to the cells and tissues and also cause significant damages to cellular components such as protein and DNA (Drummond et al., 2002).

13.6 LIVE-CELL IMAGING

13.6.1 OVERVIEW

Two different approaches have been followed to study biological processes over time. The first approach is to capture images of a series of fixed cells or tissues and the second is to capture the images of processes in a living state, which is technically more strenuous. But, with the evolution of imaging sciences and improvement in the range and sensitive fluorescent proteins, quantum dots, and synthetic fluorophore technology, there is an increase in the number of researchers worldwide who are currently using live-cell imaging as a reliable technique, for a detailed understanding of the fundamental processes of cellular and subcellular activities. Currently, live-cell imaging has become an important and routine analytical technique in most of the biological research laboratories and is widely being used in a variety of biomedical research areas such as immunology, pharmacology, cancer biology, and developmental biology. The observation of dynamic changes by live-cell imaging provides critical insight into the cellular processes rather than a snapshot provided by fixed cell imaging. Moreover, live-cell imaging is less prone to experimental artifacts thereby providing significant and reliable data compared to the microscopy of fixed cells or tissues (Sung and McNally, 2011).

13.6.2 PRINCIPLE

Various microscopic techniques can be used to perform live-cell imaging experiments. The conventional approach of live-cell imaging includes common contrast enhancement techniques such as bright field, differential interference contrast (DIC), phase contrast, Hoffman modulation contrast, and widefield fluorescence microscopy. The advanced and modern high-end imaging techniques such as fluorescence, confocal microscopy, and multiphoton microscopy also offer the facility for imaging living cells in their physiological tissue environments. Hence, the principle of individual microscopic technique applies to live-cell imaging depending upon the type of technical system is being used for recording data. The unique and vital technical challenge, as well as the requirement of live-cell imaging, is the maintenance of the cells in a healthy and functional state during the entire period of imaging process, which is essential for a successful live-cell imaging experiment.

13.6.3 INSTRUMENTATION AND TECHNICAL FEATURES

13.6.3.1 MICROSCOPY TECHNIQUES AND LABELING USED FOR LIVE-CELL IMAGING

Different microscopic techniques could be used for imaging live cells depending upon the biological questions. For conventional live-cell imaging, different transmission light microscopy techniques such as phase contrast, bright field, and DIC could be used. In bright field microscopy, the samples are usually not stained but common stains such as toluidine blue could be used to enhance the contrast of the specimen. The phase-contrast microscopy is generally used to visualize giant structures such as nucleus or organelles. But for fluorescently labeled specimen, an imaging device equipped with fluorescent microscopy along with facilities such as maintenance of temperature and humidity is used. In such devices, a bright light source like LED or mercury lamp is used to excite the fluorophores. Similarly, as shown in Figure 13.6, fluorescent multiplexing could also be combined with live-cell imaging to visualize cellular and subcellular dynamics in a single experiment. As discussed earlier, for faster image capture, laser scanning, and spinning-disk confocal microscopy are widely used in live-cell imaging, with the latter being less time consuming. With the recent development in microscopy and fluorescent technology over the past few decades, advanced microscopic techniques such as multiphoton microscopy, TIRF - Total Internal Reflection Fluorescence, FRET, FLIM, BRET - Bioluminescence Resonance Energy Transfer, FRAP, and PALM - Photoactivated Localization Microscopy are widely used to capture image sequences and videos of live cells (De Los Santos et al., 2015).

13.6.3.2 LIVE-CELL IMAGING CHAMBERS

Specimen chambers are an integral part of live-cell imaging providing excellent optical conditions, while maintaining a healthy environment for the specimens during the experiment. Both open and closed chambers are commercially available for live-cell imaging. While open chambers provide quick access to the researcher to add supplements, media, drugs, or perform microinjections, closed chambers keep the environment sealed and thus prevent evaporation of media and gases such as CO_2. Usually, the closed chambers insulate the specimen from the outer environment and are designed with additional ports to supplement medium or drugs during the experiment without affecting the imaging procedure. Applications of wide varieties of chambers are published to be used during live-cell imaging ranging from simplest coverslips to much-advanced perfusion chambers. For short-term experiments, coverslips sealed with different kinds of sealant such as agarose, a mixture of vaseline, lanolin, paraffin are used. In contrast, specially designed chambers equipped with advanced options and features are commercially available for long-term experiments. In addition, the type of specimen under examination, time duration, and purpose of the experiment decide the variety of culture chambers to be used for individual experiments.

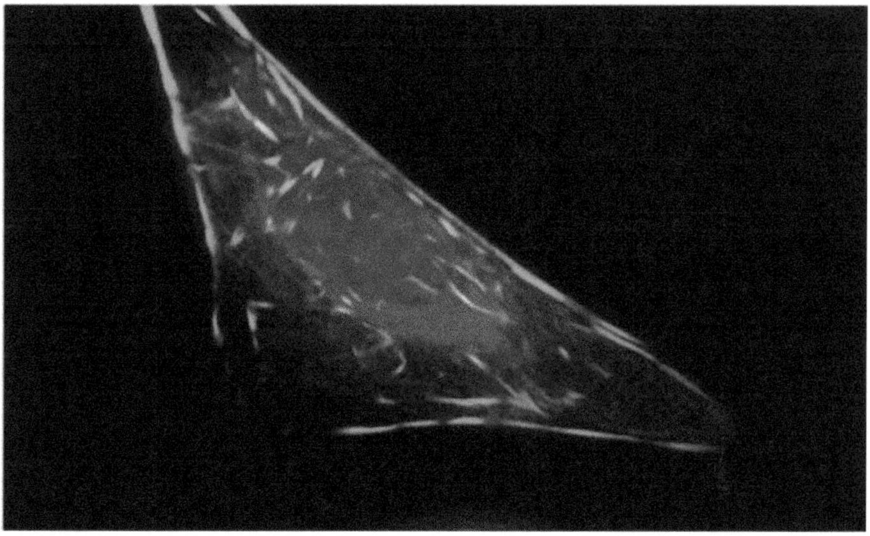

FIGURE 13.6 U2OS cells labeled using NucBlue™ Live ReadyProbes™ Reagent, CellLight™ Actin-GFP BacMam 2.0, CellLight™ Mitochondria-RFP BacMam 2.0, and Tubulin Tracker Deep Red show superb multiplexing capability and staining specificity. Cells were imaged in Gibco™ HBSS buffer containing calcium and magnesium, supplemented with 1X Probenecid solution. Images were generated using an EVOS™ FL Auto 2 Imaging System with an Olympus 60X Apochromat Oil objective using DAPI, GFP, RFP, and Cy™5 EVOS™ light cubes. (Courtesy of Thermo Fisher Scientific, USA.)

13.6.3.3 MAINTAINING THE HEALTH OF LIVING CELLS DURING IMAGING

For visualization of biological properties of a living cell and the detailed dynamics of biological and physiological processes in real time, it is essential to provide the cells an optimal physiological environment during a live-cell imaging experiment. Stringent control of the culture environment is an important factor in acquiring information during live-cell imaging and the accuracy of data collected depends largely on the maintenance of cells on the microscope stage. Health of a cell is comparatively less critical during short-term experiments, while for long-term experiments it is imperative to watch out all the possible factors such as (1) pH, (2) oxygenation, (3) temperature, and (4) osmolarity, which could alter metabolic function of the specimen (Daniels, 2012; Cole, 2014).

1. **pH:** Most of the mammalian cell line show optimal growth in a pH range of 7.2–7.4 and maintenance of CO_2 concentration in the chamber is crucial to regulate the physiological pH during the entire imaging process. Cells usually have an optimal growth condition when supplemented with 10% fetal bovine serum and 5% of CO_2 concentration. HEPES buffer at a concentration of 10–20 mM is often used as a buffering agent due to its cost-effectiveness for a short period of experiments. But, for long-term experiments, which last usually more than 10–12 h, HEPES buffer is not recommended (Daniels, 2012).
2. **Oxygenation:** Oxygenation is not a concern for short-term live-cell imaging techniques, but it is a critical factor for cell health during longer experiments. To manage the oxygenation during long-term experiments, the media could be changed frequently or a larger quantity of media can be used at the beginning of the experiment. In addition, imaging chambers could be connected with inlets for required gases such as oxygen and carbon dioxide.
3. **Temperature:** Temperature serves as the most critical factor that decides the outcome of the experiment. The cellular physiology will be disturbed even with a smaller variation in temperature parameter. To maintain the optimum temperature, either a small stage-top incubator can be used to maintain sample temperature, or large incubators could be used to heat the entire microscope. The objective can act as a heat sink during live-cell imaging experiment. Hence, objective heaters are designed to overcome this situation. Metal foil blanket with Velcro anchor, copper tubing water jacket, and proportionally controlled closed-loop heaters are some of the designs used to thermally control the objectives (Daniels, 2012).
4. **Osmolarity:** Osmolarity is maintained in a range of 260–320 mOsm/kg during live-cell imaging, which is achieved by averting the evaporation of culture media by using properly sealed culture chambers. Alternatively, a humidified environment can be maintained to achieve the required osmolarity.

13.6.4 APPLICATIONS

The recent advances over the last two decades have revolutionized the biomedical research in terms of innovative applications of live-cell imaging techniques. Kinetic imaging of live cells and animals has enabled the researcher to characterize dynamic cellular processes. It has enabled researchers to study the cellular and subcellular processes in real time. Figure 13.7 shows a real-time image of a cell undergoing mitosis and with the use of appropriate staining, it is possible to monitor the assembly of microtubules into a mitotic spindle in real time. Fine structures like astral and kinetochore microtubules as well as cortical actin and condensed DNA could be seen in the picture. The live-cell imaging technique can capture pictures with a rate lesser than a frame per second enabling quantification of brisk cellular and subcellular events. With the help of live-cell imaging techniques longer processes like cell migration, proliferation can be studied in real time. Figure 13.8A shows time-lapse imaging of human dermal fibroblast wound healing. It shows the migration of the cells into the wound to close it up for 25 h. Similarly, Figure 13.8B represents the time-lapse images obtained from the video of contracting cardiomyocytes showing the flux of calcium moving across the cells over time. Figure 13.9 shows a 5 h time-lapse montage of HeLa cell lines. Even, time-lapse imaging of live cells can be compiled to make movies of the dynamic cellular processes. Live-cell imaging has a wide variety of applications in biomedical research including proliferation assay, migration or chemotaxis assays, phagocytosis, immune cell killing, apoptosis, wound healing assays, NETosis, and angiogenesis including the effect of drugs on specific cell types (Christakou et al., 2013; Hoppenbrouwers et al., 2017; Matsushita et al., 2017; Sarangi et al., 2018; Huuskes et al., 2019). The facility for reading multiple wells at a time has helped in recording and screening living cell data with a variety of conditions in less time. Few recent applications of the live-cell imaging are discussed below.

Long-Term Study of Neuronal Stem Cells (NSCs): The NSCs are highly mobile, photosensitive, and can grow at confluent densities. These properties of NSCs limit the long-term study of these cells. Pillti et al. described a method of live-cell time-lapse imaging to study the cell cycle and migration dynamics, along with lineage selection in *in vitro* cultured multipotent stem cells (both human and mouse) using single-cell tracking. The published time-lapse imaging experiment lasted for a long term for up to 7–14 days. The long-term studies of NSCs will help to investigate some fundamental processes involved in cell maintenance and lineage-specific differentiation. This will help the researchers to investigate the dynamics

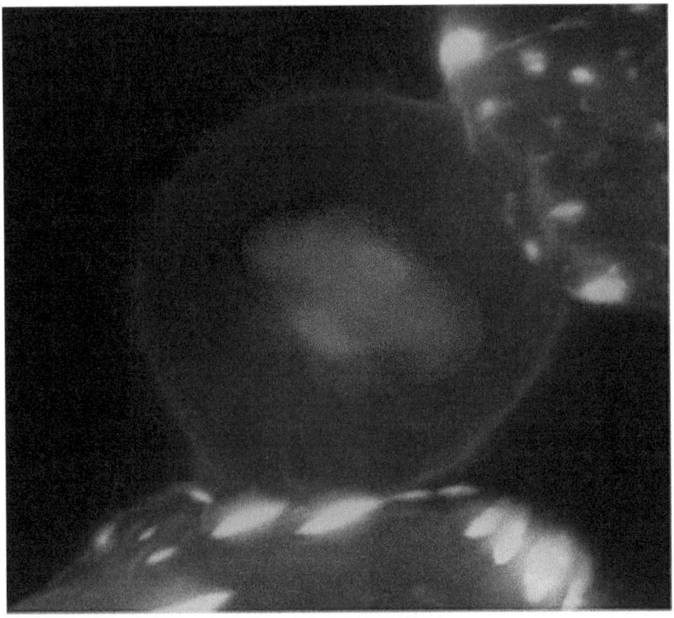

FIGURE 13.7 HeLa cells labeled using NucBlue™ Live ReadyProbes™ Reagent, CellLight™ Talin-GFP BacMam 2.0, and Tubulin Tracker Deep Red. Inset shows a mitotic cell with microtubules assembled into a mitotic spindle with visible astral and kinetochore microtubules, as well as cortical actin and condensed DNA. Cells were imaged in Gibco™ HBSS buffer containing calcium and magnesium, supplemented with 1X Probenecid solution. Images were generated using an EVOS™ FL Auto 2 Imaging System with an Olympus 100X Super Apochromat Oil objective using DAPI, GFP, and Cy™5 EVOS™ light cubes. (Courtesy of Thermo Fisher Scientific, USA.)

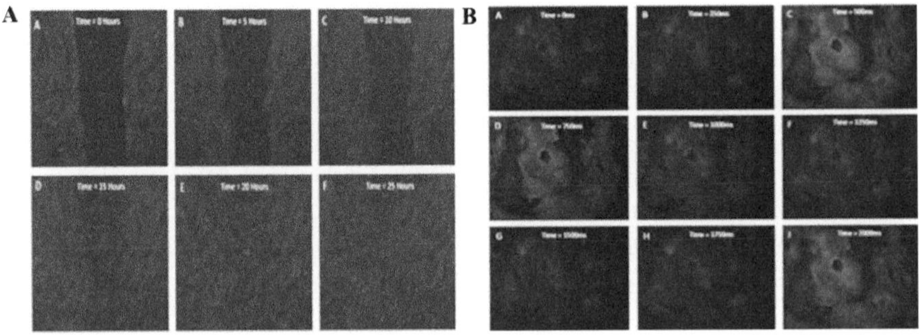

FIGURE 13.8 (A) Human dermal fibroblasts were labeled with Cell Tracker Deep Red. Time-lapse imaging of human dermal fibroblast wound healing shows the cell migrating into the wound to close it up over 25 hours (B) Time-lapse images obtained from the video of contracting cardiomyocytes show the flux of calcium moving across the cells over time. (Courtesy of Thermo Fisher Scientific, USA.)

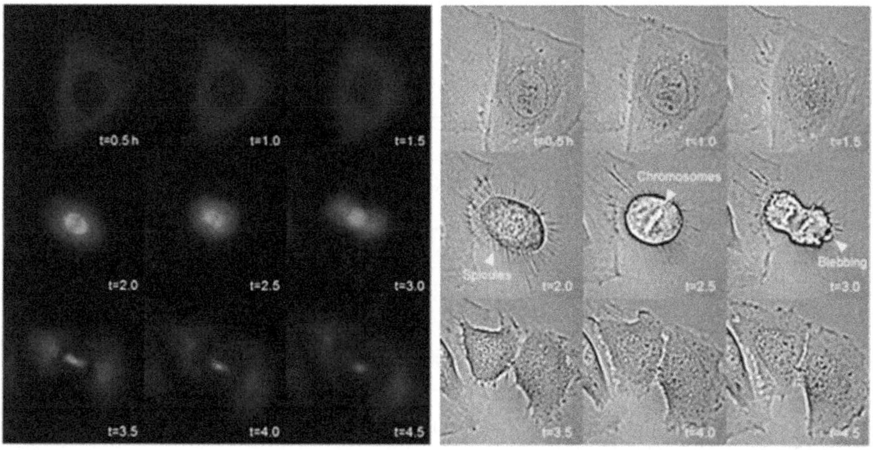

FIGURE 13.9 HeLa cells labeled using CellLight™ GFP-Tubulin BacMam 2.0. Cells were imaged in live cell imaging solution using an EVOS™ FL Auto 2 Imaging System equipped with an Onstage Incubator to allow long-term time-lapse imaging. Images were acquired at 30 minute sampling intervals using an Olympus 60X Apochromat Oil objective and a GFP EVOS™ light cube to visualize microtubules and transmitted light to visualize cellular morphology. (Left: GFP-Tubulin, Right: Bright field). (Courtesy of Thermo Fisher Scientific, USA.)

of the cell cycle, their migration, and the possibility of the use of NSCs to treat brain injuries and degenerative diseases could be understood (Piltti et al., 2018).

Visualization of mRNA in Live Cells: mRNAs are the intermediate molecules containing the genetic information, which are translated into proteins involved in different cellular events. Visualization of specific mRNAs at cellular level could be achieved using various probes such as nano probes, fluroscent proteins, flurogenic RNA aptamers for providing deeper insight into the distribution and dynamics of mRNA expression of a particular gene inside living cells (Wang et al., 2019) For instance, Bratu et al. (2003) used nuclear-resistant molecular beacons to visualize transport and localization of *oskar* mRNA in oocytes of *Drosophila melanogaster.* In addition, colocalization of mRNA with mitochondria in live cells has been visualized using molecular beacons, which has the potential to provide innovative insights into the transport, dynamics, and function of specific mRNA (Santangelo et al., 2005). Recently, aptamer-initiated fluroscence complementation was introduced for real-time tracking and imaging of mRNAs (Wang et al., 2018).

Detection of *Trypanosoma cruzi:* *T. cruzi,* a parasitic euglenoid protozoans is responsible for Chagas disease or American trypanosomiasis

(Chagas, 1909) Recently, Ferreira et al. described a live-cell imaging methodology for observing *T. cruzi* inside host tissue. For the very first time, individual amastigotes, intermediate forms, and motile trypomastigotes were visualized inside the tissues of the mammalian hosts (Ferreira et al., 2016).

13.7 LIVE ANIMAL IMAGING

13.7.1 OVERVIEW

Data obtained from experimental animal models are critical for validating the results obtained from *in vitro* and *ex vivo* studies. Although experimental animals have been an integral part of biomedical research, it has recently become possible to study biological processes inside the living organism due to live-cell imaging technique. In the modern era of biomedical research, advanced imaging techniques have become an indispensable part of most of the biological and translational research laboratories. Live animal or *in vivo* imaging is an elegant noninvasive imaging technique used for real-time study of biological events at the cellular and molecular level inside living systems. To date, live animal imaging and multiple animals were used for longitudinal studies, where each animal represents a single time point but *in vivo* imaging has succeeded in using a single animal to study dynamic processes in longitudinal studies. This technique provides deeper insight into dynamic biological processes at the anatomical and molecular level in physiological conditions and has broad applications in biological research ranging from studying basic biological processes to advanced biomedical and preclinical investigations.

13.7.2 PRINCIPLE

In live animal/*in vivo* imaging procedure, the experimental animal is anaesthetized first, followed by injecting different substrates locally or systemically wherever necessary and imaged using different imaging modalities such as bioluminescence imaging, fluorescence imaging, IVMPM, magnetic resonance imaging (MRI), computed tomography (CT), and ultrasonography/ultrasound (US) (Willmann et al., 2008). Furthermore, these imaging technologies are sometimes combined to obtain a better understanding of the detailed cellular and subcellular events inside the living system in real time. The working principle of different imaging modalities used in *in vivo* imaging is further discussed in the section below.

13.7.3 INSTRUMENTATION AND TECHNICAL FEATURES

The instruments used for in vivo imaging techniques include the facilities for anesthesia, monitoring the health of the experimental animal, and different imaging modalities.

Anesthesia: Both inhalation and injectable form of anesthetics are used during *in vivo* imaging procedures (Tremoleda et al., 2012). The two injectable anesthetic combinations used routinely in preclinical imaging are ketamine and fentanyl-based agents (Tremoleda et al., 2012). These are administered intraperitoneally or intravenously before imaging procedure. For example, isoflurane and halothane are mostly used as inhalation anesthetics during *in vivo* imaging procedures in combination with oxygen (Tremoleda et al., 2012). The *in vivo* imaging device often includes the instrumentation for inhalation anesthesia where inlet and outlet ports are provided for proper anesthesia during the experimental procedure. Inhalation anesthetic agents lower depression of cardiac functions compared to the injectable agents, while it reduces peripheral resistance resulting in reduced blood pressure in the experimental animal (Kersten et al., 1996; Kober et al., 2004).

Monitoring the Health of the Experimental Animal: It is imperative to monitor the health of the anesthetized experimental animal during the imaging procedure. Respiratory, central and peripheral nervous cardiovascular, and musculoskeletal systems are generally included in the monitoring procedure (Tremoleda et al., 2012). The parameters taken into consideration include: respiratory rate, body temperature, heart rate, rhythm and muscle tone, and reflexes of the central nervous system (Tremoleda et al., 2012).

Imaging Modalities: To date, several different imaging modalities have been used for live animal or *in vivo* imaging with their advantages and limitations. Broadly they are divided into two categories, anatomical or morphological imaging or primarily molecular imaging technique (Willmann et al., 2008). Primarily anatomical/morphological imaging includes the US, CT, and MRI, whereas primarily molecular imaging technique includes optical (fluorescence and bioluminescence) imaging techniques, single-photon emission computed tomography (SPECT), and positron emission tomography (PET) techniques. Amongst various techniques, multiphoton optical imaging techniques are regularly used in studying the dynamic biological processes including cell behaviors in living animals. Optical imaging of live animals includes both fluorescence and bioluminescence imaging, which are sensitive and cost-effective. In fluorescence imaging, fluorochromes such as GFP and RFP are used, while bioluminescence *in vivo* imaging is based on the chemiluminescent reaction between an enzyme and substrate. A thermoelectrically cooled CCD camera cooled to −120 °C to −150 °C is used to capture high-quality images and the cooling procedure makes the camera impeccably sensitive even to very weak signals (Willmann et al., 2008).

13.7.4 APPLICATION

Live animal imaging has a variety of applications in various biomedical research disciplines including immunology, oncology, pharmacology, cardiology, developmental biology, neurology, and drug development. Its application ranges from the study of basic cellular dynamics to preclinical trials and drug testing. More recently, different combinations of imaging modalities are used to gain deeper insight into the dynamic cellular and subcellular processes. For instance, Rompolas et al. recently used a noninvasive, two-photon live imaging approach to study physiological hair-follicle regeneration in live mice (Rompolas et al., 2012). Similarly, Dunn et. al. (2007) studied different renal function by administering fluorescent probes into live animals. Furthermore, bioluminescent imaging has been used to study bacterial pathogenesis in plants and animals in living state with bacteria having lux operon (Kassem et al., 2014). Imaging of inflammation, neovascularization, and identification of active vascular calcification is achieved using ^{18}F-Sodium fluoride-PET and ^{18}F-Flurodeoxy glucose-PET, respectively (Irkle et al., 2015; Osborn and Jaffer, 2015). Xu et al. (2014) developed a strategy to control T-cell trafficking optically by using photoactivable-chemokine motif receptor-4 and used FRET-based technology to study optogenetic control of CXCR-4 signaling and T-cell migration. Few of the other advanced applications of *in vivo* imaging are discussed below.

Photoacoustic Imaging (PAI): This is a recently introduced hybrid imaging modality, which combines the specificity and sensitivity of optical imaging with high spatial resolution of US imaging (Beard, 2011; Wang and Yao, 2016). It is an emerging live animal imaging modality having a wide range of applications in biomedical research. PAI has been used in multiple *in vivo* imaging systems such as angiogenic vasculature detection, to evaluate the heterogeneity of receptor expression in tumor cells and to visualize protease activity (Weber et al., 2016). Recently, PAI has been used to detect EGFR-expressing tumors in a mouse model of orthotopic pancreatic cancer (Hudson et al., 2014). Jathoul et al. (2015) have used a tyrosine-based genetic reporter for deep *in vivo* PAI of mammalian tissues. More recently, Hariri et al. (2019) has used LED-based PAI for molecular imaging of oxidative stress sensing in ovarian cancer.

Drug Development: The advances in the molecular imaging technique have enabled researchers to better understand the disease biology and activities of drugs during the preclinical and clinical phases of drug development. Molecular imaging techniques such as PET, SPECT, and optical imaging are widely being used to investigate the effect of drug candidates in tumor metabolism, proliferation, angiogenesis, apoptosis, and to identify critical biomarkers (Gambhir, 2002; Willmann et al., 2008). In addition, it plays a significant role in cardiovascular drug discovery and development (Lindner and Link, 2018).

Cell Migration and Function *In Vivo*: Live animal imaging with different anatomical and molecular imaging techniques such as MRI, PET, and optical imaging are used to study cell migration and distribution in real time (Kircher et al., 2011). Figure 13.10 shows the activity of myeloperoxidase enzyme activity in lipopolysaccharide or phosphate buffer saline injected endotoxemic animals via IVIS imaging system where luminol-based bioluminescence signal is used to detect the enzyme activity in live animals. Recently, in a separate study, researchers have also used MRI-based cell tracking method to study the adipose-derived mesenchymal stem cells using Manganese oxide-based nanoparticles (Kim et al., 2011). IVMPM imaging can also be used to study immune cell migration, distribution, and extravasation in live tissue. Figure 13.11 shows the extravasation of neutrophils in a cremaster muscle microvasculature when the muscle layer was exposed to *N-Formylmethionyl-leucyl-phenylalanine*, using IVMPM imaging. In an interesting *in vivo* FRET experiment, Hyun et al. have investigated a distinct role of adhesion molecules such as Mac-1 and LFA-1 in the process of neutrophil extravasation by using LFA-1 FRET and Mac-1 FRET transgenic mice (Hyun et al., 2019).

Detection of Therapeutic T cells in Glioma Patients: Glioblastoma multiforme (GM) is the most common brain tumor in patients with a survival time of 12 months (Krex et al., 2007). One of the proposed therapy for improving survival in GM patients is the adoptive cellular gene immunotherapy. Yaghouti et al. for the first time used PET imaging to detect therapeutic cytolytic T cells in a glioma patient by infusion *ex vivo* expanded genetically engineered autologous cytolytic T cells (Yaghoubi et al., 2009). Recently, PET-based imaging technique was also used to assess the efficacy of cytotoxic T-cell targeted immunotherapy in glioma subjects (Keu et al., 2017).

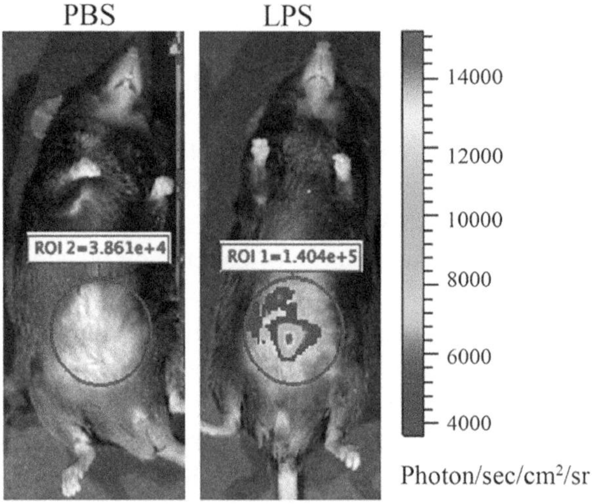

FIGURE 13.10 Intravital live animal imaging showing myeloperoxidase enzyme activity in the peritoneal cavity of phosphate buffer saline and lipopolysaccharide injected live C57BL/6 animals using an IVIS imaging system. (Courtesy of the Minsoo Kim Lab at University of Rochester Medical Center Rochester, NY; https://www.urmc.rochester.edu/labs/minsoo-kim.aspx)

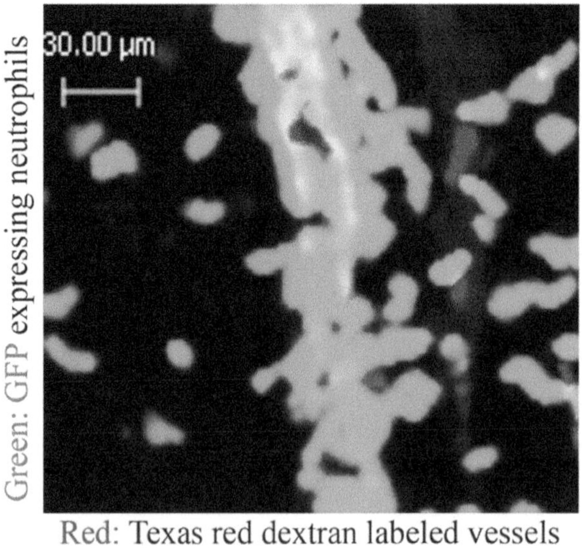

FIGURE 13.11 Intravital multiphoton imaging showing the extravasation of neutrophils in a cremaster muscle microvasculature when the muscle layer was exposed to (N-Formylmethionine-leucyl-phenylalanine). (Courtesy of the Minsoo Kim Lab at University of Rochester Medical Center Rochester, NY; https://www.urmc.rochester.edu/labs/minsoo-kim.aspx)

13.7.5 ADVANTAGES

The advantage of live animal imaging techniques over conventional methods is that these techniques can be performed in live intact animals with an adequate spatial and temporal resolution to study dynamic cellular and subcellular events in real time (Willmann et al., 2008). Live animal imaging techniques have enabled researchers to use the same organism for longitudinal studies. Thus, the same organism becomes the control of itself keeping the uniformity in the experimental procedure and reduces the need for a higher number of experimental animals thereby reducing the cost compared to conventional histochemical methods. In preclinical studies, it plays a crucial role in which the prospective drug or any biological molecule can be tracked and different biological processes can be simultaneously examined in real time. Anatomical imaging techniques like CT, MRI, and US provide high spatial resolution while the molecular imaging techniques such as ECT, SPECT, and optical imaging have the potential to detect small cellular and subcellular changes.

13.7.6 LIMITATIONS

Although the *in vivo* imaging procedure is a noninvasive state of the art imaging technology, the instrumentation is very expensive, thereby reducing the accessibility and availability. The imaging modalities equipped have certain limitations such as the primarily anatomical/morphological imaging techniques like CT, US, and MRI have the limitation of detecting smaller changes. Similarly, the primary molecular imaging techniques such as optical imaging, ECT, and SPECT have the limitation of poor spatial resolution.

13.8 CONCLUSION

Multidimensional images of living structures have been able to trigger numerous extraordinary biological questions in the minds of researchers. Seeing the unfolding of biological events has helped in correlating the findings from other data collection research and development platforms. With an increase in curiosity and the need for better visualization, the field of microscopy has evolved from simple lenses to ultramodern IVMPM, cryo-EM, and other technically challenging and expensive microscopic tools and methodologies. Along with the development of engineered marker proteins and other reagents such as Q-dots nanocrystals and nanoparticle-based biosensors have broadened the scope of imaging science. Similarly, a combination of interdisciplinary approaches such as the use of microfluidics, 4D technologies along with super-resolution microscopy could bring in holistic progress in the field of research and diagnosis. Biological imaging has helped the researchers in both clinical and basic sciences in understanding key events in living organisms and their contribution to normal physiology and disease biology. Still, our knowledge is limited in many fundamental aspects of cell biology, which may be improved with the refinement of existing techniques and the addition of newer processing and analysis methodologies.

ACKNOWLEDGMENT

We would like to thank the Thermo fisher scientific for providing the high-quality picture of confocal, Z-stacking, and time-lapse imaging. The authors also thank Dr. Minsoo Kim of the University of Rochester, New York, USA, for giving his consent to use the IVMPM imaging cremaster muscle and IVIS bioluminescence imaging. This work is funded by the DBT, Govt. of India BT/010/IYBA/2017/04 (P.P.S.). UGC, Govt. of India fellowship to CB and SPD, ICMR fellowship to PD, and, MHRD, Govt. of India fellowship to PC & CSIR-fellowship to ND. The authors would like to thank Mr. Puneet Kumar for critical reading of the manuscript.

AUTHOR CONTRIBUTION

CB and PD contributed to the preparation of overview and confocal microscopy. SPD compiled the information for live cell and animal imaging. PC wrote the IVMPM and ND contributed to electron microscopy. PPS supervised and organized the entire chapter.

KEYWORDS

- **confocal microscopy**
- **cryoelectron microscopy (Cryo-EM)**
- **intravital multiphoton imaging**
- **live-cell imaging**
- **live animal imaging**

REFERENCES

Austin, A., Lietman, T., Rose-Nussbaumer, J. Update on the management of infectious Keratitis. *Ophthalmology.* 2017, 124, 1678–1689.

Batchelor, E., Goulian, M. Imaging OmpR localization in Escherichia coli. *Mol. Microbiol.* 2006, 59, 1767–1778.

Beard, P. Biomedical photoacoustic imaging. *Interf. Focus.* 2011, 1, 602–631.

Belousoff, M. J., Venugopal, H., Wright, A., Seoner, S., Stuart, I., Stubenrauch, C., Bamert, R. S., Lupton, D. W., Lithgow, T. cryoEM-guided development of antibiotics for drug resistant bacteria. *Chem. Med. Chem.* 2019, 14, 527–531.

Belperron, A. A., Mao, J., Bockenstedt, L. K. Two photon intravital microscopy of Lyme Borrelia in mice. *Methods. Mol. Biol.* 2018, 1690, 279–290.

Boldock, E., Surewaard, B. G., Shamarina, D., Na, M., Fei, Y., Ali, A., Williams, A., Pollitt, E. J., Szkuta, P., Morris, P. Human skin commensals augment *Staphylococcus aureus* pathogenesis. *Nat. Microbiol.* 2018, 1, 881–890.

Bratu, D. P., Cha, B.-J., Mhlanga, M. M., Kramer, F. R., Tyagi, S. Visualizing the distribution and transport of mRNAs in living cells. *Proc. Natl. Acad. Sci.* 2003, 100, 13308–13313.

Breidenmoser, T., Engler, F. O., Jirikowski, G., Pohl, M., Weiss, D. G. *Transformation of Scientific Knowledge in Biology, Changes in our Understanding of the Living Cell Through Microscopic Imaging.* Max Planck Institute for the History of Science. 2010. pp 7–81.

Chagas, C. New human trypanosomiasis. Morphology and life cycle of Schyzotrypanum cruzi, the cause of a new human disease. *Mem. Inst. Oswaldo Cruz.* 1909, 1, 159–218.

Christakou, A. E., Ohlin, M., Vanherberghen, B., Khorshidi, M. A., Kadri, N., Frisk, T., Wiklund, M., Önfelt, B. Live cell imaging in a micro-array of acoustic traps facilitates the quantification of natural killer cell heterogeneity. *Int. Biol.* 2013, 5, 712–719.

Cole, R. Live-cell imaging, The cell's perspective. *Cell. Adh. Migr.* 2014, 8, 452–459.

Daniels, J. W. *Live Cell Imaging Methods Review*. Labome. United States. 2012. pp 3–81.

De Boer, P., Hoogenboom, J. P., Giepmans, B. N. Correlated light and electron microscopy, ultrastructure lights up! *Nat. Methods* 2015, 12, 503.

De Los Santos, C., Chang, C. W., Mycek, M. A., Cardullo, R. A. FRAP, FLIM, and FRET, detection and analysis of cellular dynamics on a molecular scale using fluorescence microscopy. *Mol. Reprod. Dev.* 2015, 82, 587–604.

Denk, W., Delaney, K. R., Gelperin, A., Kleinfeld, D., Strowbridge, B. W., Tank, D. W., Yuste, R. Anatomical and functional imaging of neurons using 2-photon laser scanning microscopy. *J. Neurosci. Methods* 1994, 54, 151–162.

Denk, W., Strickler, J. H., Webb, W. W. Two-photon laser scanning fluorescence microscopy. *Sci.* 1990, 248, 73–76.

Dey, V. R. N. *The Basics of Confocal Microscopy*. Springer-Verlag New York Inc. 2005.

Drummond, D., Carter, N., Cross, R. Multiphoton versus confocal high-resolution z sectioning of enhanced green fluorescent microtubules, increased multiphoton photobleaching within the focal plane can be compensated using a Pockels cell and dual widefield detectors. *J. Microsc.* 2002, 206, 161–169.

Dunn, K. W., Sutton, T. A., Sandoval, R. M. Liven Animal imaging of renal function by multiphoton microscopy. *Curr. Protoc. Cytom.* 2007, 41, 12.9. 1–12.9. 18.

Fellers, T. J., Davidson, M. W. *Introduction to Confocal Microscopy*. Olympus Fluoview Resource Center. 2007.

Ferreira, B. L., Orikaza, C. M., Cordero, E. M., Mortara, R. A. Trypanosoma cruzi, single cell live imaging inside infected tissues. *Cell. Microbiol.* 2016, 18, 779–783.

Frey-Klett, P., Burlinson, P., Deveau, A., Barret, M., Tarkka, M., Sarniguet, A. Bacterial-fungal interactions, hyphens between agricultural, clinical, environmental, and food microbiologists. *Microbiol. Mol. Biol. Rev.* 2011, 75, 583–609.

Gabriel, E. M., Fisher, D. T., Evans, S., Takabe, K., Skitzki, J. J. Intravital microscopy in the study of the tumor microenvironment, from bench to human application. *Oncotarget.* 2018, 9, 20165.

Gambhir, S. S. Molecular imaging of cancer with positron emission tomography. *Nat. Rev. Cancer*. 2002, 2, 683.

Gordon, R. E. Electron microscopy, a brief history and review of current clinical application. *Methods Mol. Biol.* 2014, 1180, 119–135.

Gremer, L., Scholzel, D., Schenk, C., Reinartz, E., Labahn, J., Ravelli, R. B. G., Tusche, M., Lopez-Iglesias, C., Hoyer, W., Heise, H., Willbold, D., Schroder, G. F. Fibril structure of amyloid-beta(1-42) by cryo-electron microscopy. *Science.* 2017, 358, 116–119.

Guesmi, K., Abdeladim, L., Tozer, S., Mahou, P. Dual-color deep-tissue three-photon microscopy with a multiband infrared laser. *Light Sci. Appl.* 2018, 7, 12.

Hariri, A., Jeevarathinam, A. S., Zhao, E., Jokerst, J. V. Molecular imaging of oxidative stress sensing using LED-based photoacoustic imaging (Conference Presentation). In: *Photons Plus Ultrasound, Imaging and Sensing 2019.* International Society for Optics and Photonics. 2019. 108782C.

Harris, J. R. Transmission electron microscopy in molecular structural biology, a historical survey. *Arch. Biochem. Biophys.* 2015, 581, 3–18.

Hoppe, A. D., Seveau, S., Swanson, J. A. Live cell fluorescence microscopy to study microbial pathogenesis. *Cell. Microbiol.* 2009, 11, 540–550.

Hoppenbrouwers, T., Autar, A. S., Sultan, A. R., Abraham, T. E., Van Cappellen, W. A., Houtsmuller, A. B., Van Wamel, W. J., Van Beusekom, H. M., Van Neck, J. W., De Maat, M. P. In vitro induction of NETosis, Comprehensive live imaging comparison and systematic review. *PLoS One* 2017, 12, e0176472.

Hudson, S. V., Huang, J. S., Yin, W., Albeituni, S., Rush, J., Khanal, A., Yan, J., Ceresa, B. P., Frieboes, H. B., Mcnally, L. R. Targeted noninvasive imaging of EGFR-expressing orthotopic pancreatic cancer using multispectral optoacoustic tomography. *Cancer Res.* 2014, 74, 6271–6279.

Huuskes, B. M., Debuque, R., Kerr, P. G., Samuel, C. S., Ricardo, S. D. The use of live cell imaging and automated image analysis to assist with determining optimal parameters for angiogenic assay in vitro. *Front. Cell Dev. Biol.* 2019, 7, 45.

Hyun, Y.-M., Choe, Y. H., Park, S. A., Kim, M. LFA-1 (CD11a/CD18) and Mac-1 (CD11b/CD18) distinctly regulate neutrophil extravasation through hotspots I and II. *Exp. Mol. Med.* 2019, 51, 39.

Hyun, Y.-M., Sumagin, R., Sarangi, P. P., Lomakina, E., Overstreet, M. G., Baker, C. M., Fowell, D. J., Waugh, R. E., Sarelius, I. H., Kim, M. Uropod elongation is a common final step in leukocyte extravasation through inflamed vessels. *J. Exp. Med.* 2012, 209, 1349–1362.

Ilie, M. A., Caruntu, C., Lixandru, D., Georgescu, S. R., Constantin, M. M., Constantin, C., Neagu, M., Zurac, S. A., Boda, D. In vivo confocal laser scanning microscopy imaging of skin inflammation, Clinical applications and research directions. *Exp. Ther. Med.* 2019, 17, 1004–1011.

Irkle, A., Vesey, A. T., Lewis, D. Y., Skepper, J. N., Bird, J. L., Dweck, M. R., Joshi, F. R., G allagher, F. A., Warburton, E. A., Bennett, M. R. Identifying active vascular microcalcification by 18 F-sodium fluoride positron emission tomography. *Nat. Commun.* 2015, 6, 7495.

Jain, R. K., Munn, L. L., Fukumura, D. Dissecting tumour pathophysiology using intravital microscopy. *Nat. Rev. Cancer* 2002, 2, 266–276.

Jathoul, A. P., Laufer, J., Ogunlade, O., Treeby, B., Cox, B., Zhang, E., Johnson, P., Pizzey, A. R., Philip, B., Marafioti, T. Deep in vivo photoacoustic imaging of mammalian tissues using a tyrosinase-based genetic reporter. *Nat. Photonics* 2015, 9, 239.

Jensen, E. C. Overview of live-cell imaging, requirements and methods used. *Anat Rec* (Hoboken). 2013, 296, 1–8.

Jerome, W. G., Price, R. L. *Basic Confocal Microscopy*. Springer. 2018.

Jonkman, J., Brown, C. M. Any way you slice it—a comparison of confocal microscopy techniques. *J. Biomol. Tech.* 2015, 26, 54.

Kassem, I. I., Splitter, G. A., Miller, S., Rajashekara, G. Let there be light! bioluminescent imaging to study bacterial pathogenesis in live animals and plants. *Bioluminescence, Fundamentals and Applications in Biotechnology-Volume 3.* Springer, 2014.

Kauert, M., Stoller, P. C., Frenz, M., Rička, J. Absolute measurement of molecular two-photon absorption cross-sections using a fluorescence saturation technique. *Opt. Express* 2006, 14, 8434–8447.

Kersten, J. R., Schmeling, T. J., Hettrick, D. A., Pagel, P. S., Gross, G. J., Warltier, D. C. Mechanism of myocardial protection by isoflurane role of adenosine triphosphate-regulated potassium (KATP) channels. *Anesthesiol. J. Am. Soc. Anesthesiol.* 1996, 85, 794–807.

Keu, K. V., Witney, T. H., Yaghoubi, S., Rosenberg, J., Kurien, A., Magnusson, R., Williams, J., Habte, F., Wagner, J. R., Forman, S., Brown, C., Allen-Auerbach, M., Czernin, J., Tang, W., Jensen, M. C., Badie, B., Gambhir, S. S. Reporter gene imaging of targeted T cell immunotherapy in recurrent glioma. *Sci. Transl. Med.* 2017, 9, eaag2196.

Kim, T., Momin, E., Choi, J., Yuan, K., Zaidi, H., Kim, J., Park, M., Lee, N., Mcmahon, M. T., Quinones-Hinojosa, A., Bulte, J. W., Hyeon, T., Gilad, A. A. Mesoporous silica-coated hollow manganese oxide nanoparticles as positive T1 contrast agents for labeling and MRI tracking of adipose-derived mesenchymal stem cells. *J. Am. Chem. Soc.* 2011, 133, 2955–2961.

Kircher, M. F., Gambhir, S. S., Grimm, J. Noninvasive cell-tracking methods. *Nat. Rev. Clin. Oncol.* 2011, 8, 677.

Kober, F., Iltis, I., Cozzone, P., Bernard, M. Cine-MRI assessment of cardiac function in mice anesthetized with ketamine/xylazine and isoflurane. Magnetic Resonance Materials in Physics, *Bio. Med.* 2004, 17, 157–161.

Konig, K. Multiphoton microscopy in life sciences. *J. Microsc.* 2000, 200, 83–104.

Krex, D., Klink, B., Hartmann, C., Von Deimling, A., Pietsch, T., Simon, M., Sabel, M., Steinbach, J. P., Heese, O., Reifenberger, G., Weller, M., Schackert, G. Long-term survival with glioblastoma multiforme. *Brain* 2007, 130, 2596–2606.

Kumar, R. L., Cruzat, A., Hamrah, P. Current state of in vivo confocal microscopy in management of microbial keratitis. *Semin Ophthalmol* 2010, 25, 166–170.

Lim, K., Hyun, Y.-M., Lambert-Emo, K., Capece, T., Bae, S., Miller, R., Topham, D. J., Kim, M. Neutrophil trails guide influenza-specific CD8+ T cells in the airways. *Sci.* 2015, 349, aaa4352.

Lindner, J. R., Link, J. Molecular imaging in drug discovery and development. *Circ. Cardiovasc. Imaging* 2018, 11, e005355.

Lyumkis, D. Challenges and opportunities in cryo-EM single-particle analysis. *J. Bio.Chem.* 2019, 294, 5181–5197.
Marcu, L. Fluorescence lifetime techniques in medical applications. *Ann. Biomed. Eng.* 2012, 40, 304–331.
Matsushita, J., Inagaki, S., Nishie, T., Sakasai, T., Tanaka, J., Watanabe, C., Mizutani, K.-I., Miwa, Y., Matsumoto, K., Takara, K. Fluorescence and bioluminescence imaging of angiogenesis in flk1-nano-lantern transgenic mice. *Sci. Rep.* 2017, 7, 46597.
Menger, M. D., Lehr, H. A. Scope and perspectives of intravital microscopy--bridge over from in vitro to in vivo. *Immunol. Today* 1993, 14, 519–522.
Moulton, P. F. Spectroscopic and laser characteristics of Ti, Al_2O_3. *JOSA B,* 1986, 3, 125–133.
Muensterer, O. J., Waldron, S., Boo, Y. J., Ries, C., Sehls, L., Simon, F., Seidmann, L., Birkenstock, J., Gödeke, J. Multiphoton microscopy, a novel diagnostic method for solid tumors in a prospective pediatric oncologic cohort, an experimental study. *Int. J. Surgery* 2017, 48, 128–133.
Mukhopadhyay, P., Rajesh, M., Hasko, G., Hawkins, B. J., Madesh, M., Pacher, P. Simultaneous detection of apoptosis and mitochondrial superoxide production in live cells by flow cytometry and confocal microscopy. *Nat. Protoc.* 2007, 2, 2295–2301.
Murata, K., Wolf, M. Cryo-electron microscopy for structural analysis of dynamic biological macromolecules. *BBA-Gen. Subjects* 2018, 1862, 324–334.
O'connell, K. F., Golden, A. Confocal imaging of the microtubule cytoskeleton in C. elegans embryos and germ cells. *Methods Mol. Biol.* 2014, 1075, 257–272.
Oreopoulos, J., Berman, R., Browne, M. Spinning-disk confocal microscopy, present technology and future trends. *Methods In Cell Biol.* 2014, 123, 153–175.
Osborn, E. A., Jaffer, F. A. Imaging inflammation and neovascularization in atherosclerosis, clinical and translational molecular and structural imaging targets. *Curr. Opin. Cardiol.* 2015, 30, 671.
Piltti, K. M., Cummings, B. J., Carta, K., Manughian-Peter, A., Worne, C. L., Singh, K., Ong, D., Maksymyuk, Y., Khine, M., Anderson, A. J. Live-cell time-lapse imaging and single-cell tracking of in vitro cultured neural stem cells–tools for analyzing dynamics of cell cycle, migration, and lineage selection. *Methods* 2018, 133, 81–90.
Pons, T., Bouccara, S., Loriette, V., Lequeux, N., Pezet, S., Fragola, A. In Vivo Imaging of Single Tumor Cells in Fast-Flowing Bloodstream Using Near Infrared Quantum Dots and Time-Gated Imaging. *ACS nano.* 2019, 13(3), 3125–3131.
Razinkov, I., Dandey, V. P., Wei, H., Zhang, Z., Melnekoff, D., Rice, W. J., Wigge, C., Potter, C. S., Carragher, B. A new method for vitrifying samples for cryoEM. *J. Struct. Biol.* 2016, 195, 190–198.
Rigby, P. J., Goldie, R. G. Confocal microscopy in biomedical research. *Croat Med. J.* 1999, 40, 346–52.
Rompolas, P., Deschene, E. R., Zito, G., Gonzalez, D. G., Saotome, I., Haberman, A. M., Greco, V. Live imaging of stem cell and progeny behaviour in physiological hair-follicle regeneration. *Nat.* 2012, 487, 496.
Sanderson, M. J., Smith, I., Parker, I., Bootman, M. D. Fluorescence microscopy. *Cold Spring Harb Protoc.* 2014, 071795.
Sandoval, R. M., Molitoris, B. A. Intravital multiphoton microscopy as a tool for studying renal physiology and pathophysiology. *Methods.* 2017, 128, 20–32.
Santangelo, P. J., Nitin, N., Bao, G. Direct visualization of mRNA colocalization with mitochondria in living cells using molecular beacons. *J. Biomed. Opt.* 2005, 10, 044025.

Sarangi, P. P., Chakraborty, P., Dash, S. P., Ikeuchi, T., De Vega, S., Ambatipudi, K., Wahl, L., Yamada, Y. Cell adhesion protein fibulin-7 and its C-terminal fragment negatively regulate monocyte and macrophage migration and functions in vitro and in vivo. *Faseb J.* 2018, 32, 4889–4898.

Schrand, A. M., Schlager, J. J., Dai, L., Hussain, S. M. Preparation of cells for assessing ultrastructural localization of nanoparticles with transmission electron microscopy. *Nat. Protoc.* 2010, 5, 744.

Sewald, X. Visualizing viral infection in vivo by multi-photon intravital microscopy. *Viruses* 2018, 10, 337.

Shaikjee, A., Franklyn, P. J., Coville, N. J. The use of transmission electron microscopy tomography to correlate copper catalyst particle morphology with carbon fiber morphology. *Carbon* 2011, 49, 2950–2959.

Skala, M. C., Riching, K. M., Gendron-Fitzpatrick, A., Eickhoff, J., Eliceiri, K. W., White, J. G., Ramanujam, N. In vivo multiphoton microscopy of NADH and FAD redox states, fluorescence lifetimes, and cellular morphology in precancerous epithelia. *Proc. Natl. Acad. Sci.* 2007, 104, 19494–19499.

So, P. T. C., Kim, K. H., Buehler, C., Masters, B. R., Hsu, L., Dong, C.-Y. Basic Principles of Multiphoton Excitation Microscopy. In: PERIASAMY, A. (ed.) *Methods in Cellular Imaging.* New York, NY, Springer New York, 2001. pp. 201–211

Sokolova, V., Ludwig, A.-K., Hornung, S., Rotan, O., Horn, P. A., Epple, M., Giebel, B. Characterisation of exosomes derived from human cells by nanoparticle tracking analysis and scanning electron microscopy. *Colloids Surf. B Biointerfaces* 2011, 87, 146–150.

Spiess, E., Bestvater, F., Heckel-Pompey, A., Toth, K., Hacker, M., Stobrawa, G., Feurer, T., Wotzlaw, C., Berchner-Pfannschmidt, U., Porwol, T. Two-photon excitation and emission spectra of the green fluorescent protein variants ECFP, EGFP and EYFP. *J. Microsc.* 2005, 217, 200–204.

St Croix, C. M., Shand, S. H., Watkins, S. C. Confocal microscopy, comparisons, applications, and problems. *Biotechniques* 2005, 39, S2–5.

Sung, M. H. & Mcnally, J. G. Live cell imaging and systems biology. *WIRES. Syst. Biol. Med.* 2011, 3, 167–182.

Svoboda, K., Yasuda, R. Principles of two-photon excitation microscopy and its applications to neuroscience. *Neuron* 2006, 50, 823–839.

Tauer, U. Advantages and risks of multiphoton microscopy in physiology. *Exp. Physiol.* 2002, 87, 709–714.

Theer, P., Denk, W. On the fundamental imaging-depth limit in two-photon microscopy. *J. Opt. Soc. Am. A. Opt. Image Sci. Vis.* 2006, 23, 3139–3149.

Tremoleda, J. L., Kerton, A., Gsell, W. Anaesthesia and physiological monitoring during in vivo imaging of laboratory rodents, considerations on experimental outcomes and animal welfare. *EJNMMI Res.* 2012, 2, 44.

Tsai, P. S., Nishimura, N., Yoder, E. J., Dolnick, E. M., White, G. A., Kleinfeld, D. Principles, design, and construction of a two-photon laser-scanning microscope for in vitro and in vivo brain imaging. In: *In Vivo Optical Imaging of Brain Function.* CRC Press. 2002.

Vangindertael, J., Camacho, R., Sempels, W., Mizuno, H., Dedecker, P., Janssen, K. An introduction to optical super-resolution microscopy for the adventurous biologist. *Methods Appl. Fluores.* 2018, 6, 022003.

Victora, G. D., Schwickert, T. A., Fooksman, D. R., Kamphorst, A. O., Meyer-Hermann, M., Dustin, M. L., Nussenzweig, M. C. Germinal center dynamics revealed by multiphoton microscopy with a photoactivatable fluorescent reporter. *Cell* 2010, 143, 592–605.

Walls, A. C., Tortorici, M. A., Bosch, B.-J., Frenz, B., Rottier, P. J., Dimaio, F., Rey, F. A., Veesler, D. Cryo-electron microscopy structure of a coronavirus spike glycoprotein trimer. *Nat.* 2016, 531, 114.

Wang, Z., Liu, W., Fan, C., Chen, N. Visualizing mRNA in live mammalian cells. *Methods* 2019, 161, 16–23.

Wang, Z., Luo, Y., Xie, X., Hu, X., Song, H., Zhao, Y., Shi, J., Wang, L., Glinsky, G., Chen, N. In Situ spatial complementation of aptamer mediated recognition enables live cell imaging of native RNA transcripts in real time. *Angew. Chem.* 2018, 130, 984–988.Wang, L. V., Yao, J. A practical guide to photoacoustic tomography in the life sciences. *Nat. Methods* 2016, 13, 627.

Weber, J., Beard, P. C., Bohndiek, S. E. Contrast agents for molecular photoacoustic imaging. *Nat. Methods* 2016, 13, 639.

Willmann, J. K., Van Bruggen, N., Dinkelborg, L. M., Gambhir, S. S. Molecular imaging in drug development. *Nat. Rev. Drug Discov.* 2008, 7, 591.

Wilson, C., Lukowicz, R., Merchant, S., Valquier-Flynn, H., Caballero, J., Sandoval, J., Okuom, M., Huber, C., Brooks, T. D., Wilson, E. Quantitative and qualitative assessment methods for biofilm growth, A mini-review. *Res. Rev. J. Eng. Technol.* 2017, 6.

Wnek, G. E., Bowlin, G. L. Confocal Microscopy/Denis Semwogerere, Eric R. Weeks. *Encyclopedia of Biomaterials and Biomedical Engineering.* CRC Press, 2008.

Xu, Y., Hyun, Y.-M., Lim, K., Lee, H., Cummings, R. J., Gerber, S. A., Bae, S., Cho, T. Y., Lord, E. M., Kim, M. Optogenetic control of chemokine receptor signal and T-cell migration. *Proc. Natl. Acad. Sci.* 2014, 111, 6371.

Xu, C., Zipfel, W., Shear, J. B., Williams, R. M., Webb, W. W. Multiphoton fluorescence excitation, new spectral windows for biological nonlinear microscopy. *Proc. Natl. Acad. Sci.* 1996, 93, 10763–10768.

Yaghoubi, S. S., Jensen, M. C., Satyamurthy, N., Budhiraja, S., Paik, D., Czernin, J., Gambhir, S. S. Noninvasive detection of therapeutic cytolytic T cells with 18F-FHBG PET in a patient with glioma. *Nat. Clin. Pract. Oncol.* 2009, 6, 53–58.

Zhao, G., Perilla, J. R., Yufenyuy, E. L., Meng, X., Chen, B., Ning, J., Ahn, J., Gronenborn, A. M., Schulten, K., Aiken, C., Zhang, P. Mature HIV-1 capsid structure by cryo-electron microscopy and all-atom molecular dynamics. *Nat.* 2013, 497, 643–646.

Zinter, J. P., Levene, M. J. Maximizing fluorescence collection efficiency in multiphoton microscopy. *Opt. Express* 2011, 19, 15348–15362.

Zipfel, W. R., Williams, R. M., Webb, W. W. Nonlinear magic, multiphoton microscopy in the biosciences. *Nat. Biotechnol.* 2003, 21, 1369–1377.

CHAPTER 14

Integrated Omics Technology for Basic and Clinical Research

KULDEEP GIRI, VINOD SINGH BISHT, SUDIPA MAITY, and KIRAN AMBATIPUDI*

Department of Biotechnology, Indian Institute of Technology Roorkee, Roorkee, Uttarakhand 247667, India

Corresponding author. E-mail: kiran.ambatipudi@bt.iitr.ac.in

ABSTRACT

In the era of early disease diagnosis, omics-based approaches which include genomics, transcriptomics, proteomics and metabolomics have gained significant momentum to characterize and analyse huge data in a nontargeted and nonbiased way. Due to the recent technological advancement, scientists have emphasized to avoid unnecessary investigations to help overcome the disease burden. Multiomics coupled with bioinformatics has been recently involved in the detection of diagnostic and prognostic biomarkers in diseases (e.g, diabetes, breast and ovarian cancer) and their therapy prediction including monitoring treatment and therapeutic intervention. However, the reliability of the current potential biomarker identification approaches is associated with other different predictive tests for their accuracy and efficacy. Nevertheless, a paradigm shift was observed in the fields of precision medicine, agricultural sciences, and food microbiology due to steady progress made in multiomics technology. The immense investigation of precise effects including analysis of tissues and multi-organ data has been enabled with the successive improvement of machine learning methods. Although integrative omics techniques are spreading rapidly in heterogeneous research areas, it is limited by its high cost including the need for specific bioinformatics and biostatics tools for integrating large datasets obtained at each level. However, with careful planning and investment of human resources, the multiomics approach has the promising potential to unravel hidden features or trace molecular components in multiscale spatial organization. Thus, a successful integrative omics approach calls for a prominent need to establish a global infrastructure for achieving high-output to translate basic research clinical applications from bench to bedside.

14.1 INTRODUCTION

In the era of early disease diagnosis for therapeutic intervention, omics-based approaches have gained significant momentum due to its holistic approach to characterize and analyze huge data in a nontargeted and non-biased way. The cascade of omics (Figure 14.1), which is a hypothesis-driven or reductionist approach, includes genomics (gene), transcriptomics (mRNA), proteomics (peptides/proteins) and metabolomics (metabolites) to answer the complexity of molecules, part of a cell/tissue, or organism affected by genetic and/or epigenetic factors (Horgan and Kenny, 2011). Due to the recent advancement made in different omics fields, an early disease diagnosis and therapeutic intervention are pertinent to avoid unnecessary investigations or tests and help in personalized care and management to overcome disease burden.

Multiomics approach in conjunction with bioinformatics has been recently involved in the finding of diagnostic and prognostic biomarkers, therapy prediction including monitoring treatment, recurrence and therapeutic intervention in different biological systems such as human, animal and plant. For example, panels of biomarkers have been assessed for early diagnoses of diseases such as diabetes (e.g., HbA1c) (Lyons and Basu, 2012), ovarian (e.g., CA125) (Bast et al., 1998), breast (e.g., CA15-3) (Zaleski et al., 2018), and autoimmune disease such as primary Sjögren's syndrome (pSS) (e.g., miRNAs) (Chen et al., 2015). However, the reliability of the current biomarker identification approaches is dependent on other diagnostic/predictive tests for their accuracy and efficacy.

Recently, advanced multiomics technology has contributed immensely in the disciplines of precision medicine, agricultural sciences, and food microbiology. Although integrative omics techniques are spreading rapidly in heterogeneous research areas, it lacks cost-effectiveness and requires specific bioinformatics and biostatics tools to integrate large datasets obtained at different levels. However, with careful planning and investment of time and human resources, the multiomics approach has the promising potential to unravel hidden features or trace molecular components in multiscale spatial organization. Thus a successful integrative omics approach calls for a prominent need to establish a global infrastructure for achieving high-output to translate basic research to clinical applications from bench to bedside. Due to the advantages of omics technologies with a wide/whole range of applications in different fields, this book chapter will discuss newer omics technologies with a particular emphasis on the transfer of application from bench-to-bedside. Furthermore, we will discuss a few examples of integrated omics approach together with bioinformatics and statistics in different fields of biology such as microbiology, biomedicine, pharmacology and their clinical validation.

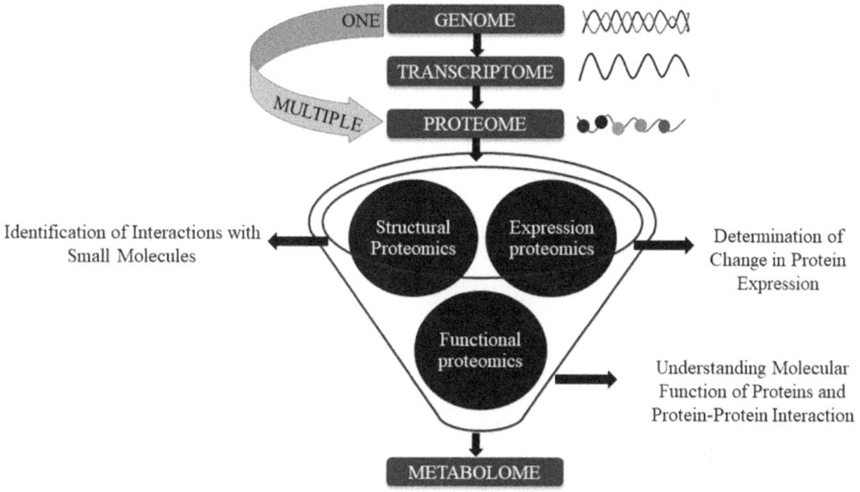

FIGURE 14.1 Approaches to characterise and analyse data. The cascade of omics includes genomics (gene), transcriptomics (mRNA), proteomics (peptides/proteins) and metabolomics (metabolites) to answer the complexity of molecules, part of a cell or tissue or organism. Proteomics can be classified as 1) structural proteomics which deals with identification of 3D structures of biomolecules (e.g. protein and protein complex), 2) functional proteomics studies molecular function of proteins, protein-protein interactions and, 3) expression proteomics is associated with quantitative analysis of proteome and change in expression/ abundance between different samples (e.g. health vs. disease) including identification of biomarker discovery for diagnostic, prognostic and predictive analysis.

14.2 OVERVIEW OF HISTORICAL PROGRESS OF OMICS

The term "Omics" is a well-informed English-language neologism informally addressing the various activities in biology ending in the suffix-omics like genomics and transcriptomics, referring to the study of genes and their transcripts, respectively. In contrast, the comprehensive study of protein expression and metabolites is known as proteomics and metabolomics, respectively. The term genomics was proposed by Tom H. Roderick in 1986, a geneticist from the United States, in a beer bar during a discussion on starting a new scientific journal—now known as *Genomics* (Kuska, 1998). Similarly, subsequent terms with suffix-omics, for instance, proteomics was proposed much later by Marc Wilkins, a doctoral student in Keith Williams's laboratory at Macquarie University, Sydney, Australia, in 1995 to describe the whole protein complement of an organism's genome (Yadav 2007). Taken together, omics field is an old concept in the modern biological world with general consideration that the center of this field is "genome" that was initially proposed by Hans Winkler, professor of Botany, University of Hamburg, Germany, in 1920 (Noguera-Solano et al., 2013). There are thousands of words in different fields ending in omics that may come in the future because it has become a trend after 2003 to end any biological field with omes or omics. The first acceptance of the ome- and omics was by the Cambridge Healthtech Institute group and later by the Nature group that started an Omics web portal.

The science of data generation and analysis started in the 1930s and progressed with the advancement in technologies to study genes, transcripts, and proteins in the system and informatics biology. In the 1940s–1950s, graph theory, a mathematical tool, was proposed to analyze biological networks including an idea to use a computer for practical general purpose (to perform the most common computing task). Concurrently, mathematical and physical theories including laws were available for different applications in biological systems (Manzoni et al., 2018). The evolution in the study methods of biomolecules (gene, transcript, and proteins) related to informatics started when the holistic general system theory was applied with formulas to systems across different areas of science (Manzoni et al., 2018). However, the roots of omics evolution started in the 1950s with genomics and transcriptomics used to analyze and predict the genes and their structures. For instance, in 1958, characterization of DNA molecule was discovered by Watson and Crick and in 1953, first high-resolution crystal structure (by X-ray crystallography) of myoglobin was reported by John Kendrew for which he shared Nobel prized with Max Perutz. In subsequent years, informatics and proteomic methods were progressive with International Business Machines (1975) and Apple (1976) revolutionizing the field of informatics and Protein Data Bank by creating a database of 3D structures. Furthermore, the invention of mass spectrometry (MS) in 1989 to analyze proteins and metabolites further accelerated the omics field to characterize and quantify complex biomolecules. The Human Genome Project (HGP), which was aimed to determine the entire sequence of the human genome and map all gene, was a collective effort of Department of Energy and National Institutes of Health (NIH) for 13 years (1986 to 2003) (Nusbaum et al., 2001). Subsequently, the aim was to develop technologies to determine the data carried by genes in static DNA stretch. Consequently, newer technologies were developed to study functional analysis of genome ranging from simple traditional polymerase chain reaction (PCR) to modern analytical tools like Next Generation Sequencing (NGS), MS, and nuclear magnetic resonance (NMR) (Moorthie et al., 2011). Similarly, databases such as Uniprot (2003) and Protein Exchange Consortium (2006) were developed to collect sequence and functional information such as subcellular location, a pattern of expression, and protein–protein interaction of various biomolecules. The progressive development in the applied field of (1) genomics, (2) transcriptomics, (3) proteomics, and (4) metabolomics are listed in Table 14.1.

TABLE 14.1 History of Genomics, Transcriptomics, Proteomics and Metabolomics

Discovery	Year of Discovery	Investigators
Genomics		
DNA as a genetic material	1944	Oswald T. Avery, Colin M. MacLeod Maclyn McCarty
chain-termination and chemical methods of DNA sequencing	1977	Frederick Sanger and Walter Gilbert
GenBank	1982	Walter G
PCR procedure	1983	Kary Mullis
First automated DNA sequencer	1986	Leroy Hood Lloyd Smth Michael Hunkapiller Tim Hunkapiller
Genomics	1986	Tom H. Roderick
BLAST	1990	Altshul et al.
Microarray	1995	Pat Brown Ron Davis
Next generation sequencing (Illumina/Solexa sequencing)	1998	Shankar Balasubramanian and David Klenerman
Cluster analysis for Microarray data	1998	Eisen et al.
Oligonucleotide microarray	1998	Affymetrix
Solid sequencing	2006	Life technologies
Ion semiconductor sequencing	2010	Ion Torrent Systems, Inc.

TABLE 14.1 *(Continued)*

Discovery	Year of Discovery	Investigators
Transcriptomics		
Northern blot	1977	James Alwine, David Kemp, and George Stark
RNA sequencing	2005	Margulies
Expressed sequence tag	1991	Adams and co-workers
Serial analysis of gene expression (SAGE)	1995	Victor Velculescu
Cap analysis gene expression (CAGE)	2003	Shiraki et al.
Proteomics		
Complete amino acid sequence of insulin determined	1952	F. Sanger
Atlas of Protein Sequence and Structure	1965	Margaret Dayhoff
Brookhaven protein database	1973	Brookhaven National Laboratory
Two-dimensional electrophoresis	1975	P. O'Farrell
NMR for proteins	1980	K. Wuthrich
SWISS-PROT	1986	SWISS-PROT
Clustal multiple sequence alignment	1988	H.D. Higgins P.M. Sharp
Proteomics	1995	Marc Wilkins
Mass spectrometry imaging	1990	Richard Caprioli
Human proteome organization	2001	Under HUPO services
Metabolomics		
METLIN (Metabolite and Tandem MS Database)	2005	The Scripps Research Institute
Human metabolic project	2005	University of Alberta
Human metabolic database	2007	University of Alberta

14.3 SIGNIFICANCE OF BIOLOGICAL DATABASES

Biological research databases are powerful research tools providing students with an original collection of journals. For instance, journals such as *Nucleic Acids Research*, *Scientific Reports*, and *Bioinformatics* publish special issues on database research. These databases rigorously update and provide opportunities for big data storage, processing, exchange, and curation.

Biological databases are categorized according to the types of data they managed. They are divided into (1) DNA, (2) RNA, (3) protein, (4) expression, (5) pathway, (6) disease, (7) nomenclature, (8) literature and, (9) standard and ontology. For example, GenBank, a DNA database provides DNA sequences that are publicly available. Similarly, protein databases such as UniProt is a resourceful media for protein information. Another such database, known as Proteomics Identifications (Vizcaíno et al., 2016) database collects and stores experimentally validated MS-based identified proteins. Additionally, a database like ProGlycProt is a collection of experimentally characterized glycoproteins and glycosyltransferases of bacteria and archaea. Taken together, biological databases efficiently provide curated sources of information that could analyze massive data sets into big discovery (Zou et al., 2015).

14.4 TYPES OF OMICS-BASED DATA

Many fields of biological activity integrated with systems biology have a range of applications and are classified under omics. The global analyses of biological molecules within a cell/tissue or organism in a high-throughput way to provide a comprehensive detail about complex systems are called omics. Different types of omics data can be obtained depending on the field of study; however, it is not practical to provide a category-wise list of omics data. However, few omics disciplines like genomics, proteomics, and metabolomics are mentioned below. Types of omics data and their repositories are listed in Table 14.2.

14.4.1 GENOMICS

The uniqueness of an organism relies on tiny variations in the DNA. All the genes with necessary structural and functional information are designated as the genome of an organism. The single genome of an organism is constant over time, however; mutations and different rearrangement of chromosomes (DNA bundles) provide an essential variation (Iafrate et al., 2004; Redon et al., 2006). The comprehensive analysis (characterization/quantification of a gene) of an organism's genome is referred to as genomics. Genomic data is categorized into three main streams (Figure 14.2): (1) genotyping aims to study genome sequence including the structural and physiological function of genes along with identification of genes specifically susceptible

TABLE 14.2 Databases of OMICS Techniques

Database Name	Location	Database Specification
Genomic Database		
DNA databank of Japan (DDBJ)	National Institute of Genetics Mishima, Japan	DNA sequences
European Molecular Biology Laboratory (EMBL)	Heidelberg, Germany	Cell biology, developmental biology, genome biology, structural and computational biology
GenBank	National Centre for Biological Information	DNA sequences
Online Mendelian Inheritance in Man (OMIM)	Johns Hopkins University, USA	Human genes and its phenotypes
The Human Gene Mutation Database (HGMD)	Institute of Medical Genetics, UK	Database of human gene mutations
Gene card	Weizmann Institute of Science, Israel	Information related to genomic, transcriptomic, proteomic and clinical function of a gene
Saccharomyces Genome Database (SGD)	Stanford University	Yeast genes and proteins repository
GenAtlas	University Paris Descartes, Paris	Gene function, structure, expression, mutation and diseases
Transcriptome Database		
Gene Expression Omnibus (GEO)	NCBI, NIH USA	Functional and expression genomic database
DNA databank of Japan (DDBJ)	National Institute of Genetics Mishima, Japan	Gene expressed sequences
Transcriptome Shotgun Assembly Sequence Database (TSA)	NCBI, USA	Transcript sequences from EST and next generation sequencing
PdumBase	Iowa State University, United States	developmental transcriptome database of *Platynereis dumerilii*
Brain Transcriptome Database (BrainTx)	RIKEN Center for Brain Science, Japan	Transcriptome related to brain function, development and dysfunction of brain

TABLE 14.2 *(Continued)*

Database Name	Location	Database Specification
Proteomic Database		
Worldwide protein Data bank (wwPDB)	RCSB PDB, BMRB, PDBe and	Macro or micro molecule 3D structures
Protein databank of Japan (PDBJ)	Institute for Protein Research, Osaka University, Japan	macromolecular structures and bioinformatics tools
Protein Data Bank in Europe (PDBe)	EMBL-EBI	Biological macromolecular structure and sequences
Universal Protein Resource (UniProt)	Joint collaboration of EMBL-EBI, Swiss Institute of Bioinformatics and Protein Information Resource	Bioinformatics tools, protein sequences and their biological function
Protein Information Resources (PIR)	University of Delaware and Georgetown University Medical Center	Protein database and bioinformatics tools
OWL	The University of Manchester	non-redundant protein sequence database
Protein Research Foundation (PRF)	Osaka, Japan	Peptide, protein and synthetic compounds
Biological Magnetic Resonance Data Bank (BMRB)	University of Wisconsin, Madison	NMR data for protein, peptide, nucleic acid and other macromolecules
Swiss prot	EMBL-EBI	Protein sequences, its structure, function and post-translational modification
RCSB PDB	Rutgers, The State University of New Jerse and San Diego Supercomputer Center and the Skaggs School of Pharmacy and Pharmaceutical Sciences, University of California, San Diego	3D shape of protein, nucleic acid and other molecules
Metabolomics Database		
Human Metabolome database	University of Alberta, Alberta	human metabolites function, structure and related disease

TABLE 14.2 *(Continued)*

Database Name	Location	Database Specification
METLIN (Metabolite and Tandem MS Database)	The Scripps Research Institute, California	MS/MS data of metabolites
MetaboLights	EMBL-EBI, United Kingdom	Metabolite structure, spectra, their location and fuunction
Chemical Entities of Biological Interest (ChEBI)	EMBL-EBI, United Kingdom	Molecular entities of chemical compounds, their structure, formula, charge, average mass,
Reactome	ELIXIR, United Kingdom	Human biological pathways and reactions

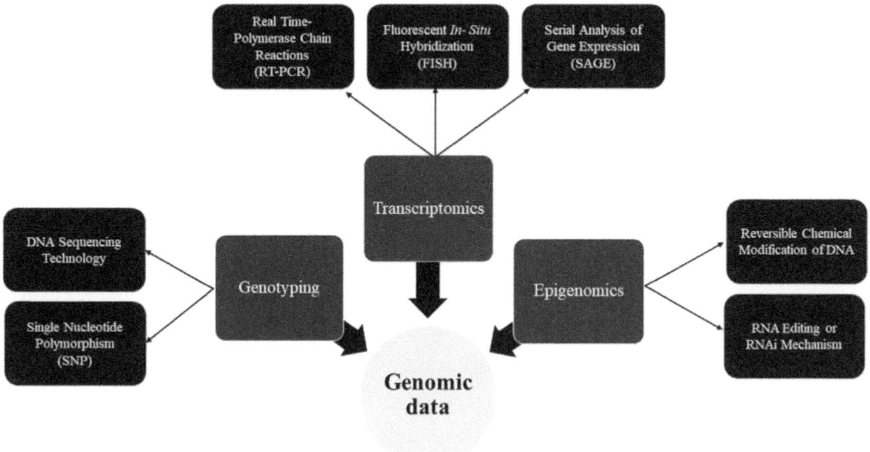

FIGURE 14.2 Types of genomic data. Genomic data is categorised into three main streams. 1) Genotyping aims to study genome sequence including structural and physiological function of genes and their susceptible to disease. Two basic approaches for genotyping are a) Single Nucleotide Polymorphism (SNP) and b) DNA sequencing technology for complete or partial DNA sequence identification and associated phenotypes. 2) Transcriptomics (Genome-to-Gene) focuses on studying complete set of RNA transcripts including coding transcriptome (e.g. mRNA, rRNA, tRNA and miRNA) and non-coding transcriptome (e.g. ncRNA). Modern technologies which can analyse mRNA expression or differential mRNA expression are Fluorescent in- situ hybridization (FISH), Real time- polymerase chain reactions (RT-PCR), and Serial Analysis of Gene Expression (SAGE) 3) Epigenomics focuses on understanding the epigenetic processes which are persistent for generations without changing the basic sequence of a DNA. The epigenetic processes could be tissue/cell type specific and affected by environmental changes or through disease progression. The main epigenetic processes are revisable chemical modification of DNA (e.g. hyper/hypo methylation in CpG islands), histone modification and RNA editing or RNAi mechanism.

to diseases. The two basic approaches for genotyping are (a) single nucleotide polymorphism (SNP) including Tag-SNP and fine mapping (Gasperskaja and Kučinskas, 2017) and (b) DNA sequencing technology for complete or partial DNA sequence identification and associated phenotypes. The more advanced genotyping approach is arrays-SNP with thousands of nucleotides with known sequence designed to bind specifically to variants within DNA sequence (Anderson and Schrijver, 2010); (2) Transcriptomics (genome to the gene) focuses on studying the whole set of RNA including coding transcriptome (mRNA, rRNA, tRNA, and miRNA) and noncoding transcriptome (ncRNA). The expression of a set of genes in the form of different transcripts is studied either by microarray (nucleotide probe-based) or RNA sequencing technology. The microarray can measure the extent of expression of several genes in a single mRNA corresponding to existing genomic sequencing information. RNA sequencing technology with the ability to quantify unknown transcripts can be used for newly transcriptome (lacking reference genome) and for differential expression analysis. The altered expression of a gene can be correlated with the abundance of a particular transcript within a mixture of transcripts. Of note, modern technologies to analyze mRNA expression or differential mRNA expression are reporter gene, fluorescent in situ hybridization, Serial Analysis of Gene Expression (SAGE), real-time polymerase chain reactions, Northern blotting, and tiling arrays (Henry et al., 2014); and (3) epigenomics focuses on understanding the epigenetic processes that are persistent for generations without changing the basic sequence of a DNA. The epigenetic processes could be tissue/cell type-specific and affected by environmental changes or through disease progression. The three main epigenetic processes are (1) revisable chemical modification of DNA (e.g., hyper/hypomethylation in CpG islands), (2) histone modification (e.g., acetylation/deacetylations) of chromosome largely affecting gene transcription or gene silencing process, and (3) RNA editing or RNAi mechanism, which have recently gained considerable momentum to investigate gene expression and/or their regulations (Abi Khalil, 2014). Few techniques to analyze genomic data including their advantage and limitations are provided in Table 14.3.

14.4.2 PROTEOMICS

The proteome—a pool of proteins—expressed within a biological entity like cell, organ, and organism by a genome at a given time (Wilkins et al., 1996). The protein primary structure (linear chain of amino acids) determines the secondary and tertiary structures (functional unit) including different protein functions. The complexity of proteins in comparison to static gene sequence of a DNA is deeper, as proteome is neither dynamic nor static as genome (Mirza and Olivier, 2008). The proteomics experimental analysis in a large scale (proteomic analysis) is important from a functional aspect (alteration in the level of gene expression) as few studies claimed that genome or their transcripts (with or without modifications) analysis is substantially unchanged but affect the phenotype (Godovac-Zimmermann and Brown, 2001). The proteome of a cell performs diverse roles including signal transduction and metabolism of the cell, which are essential for the survival of an organism. Similarly, protein–protein interactions (physical contact and their interacting partners) to form dimers (reverse transcriptase) and multiprotein

TABLE 14.3 Some OMICS-based Technologies, Associated Upside and Downside

Techniques	Upside	Downside	Authors
Genomics			
Fluorescence *in situ* Hybridization (FISH)	Clinical diagnostic and preventive medicine, with high sensitivity and specificity	Applicable only for specific probe-based labelling	Henry et al., 2014
Next Generation Sequencing (NGS) like illumine and sequencing by ligation	Low cost, simple and highly scalable, require very less amount of biomolecules	Highly expensive and slow	Rao et al., 2014
Sanger sequencing	Produce high quality result for the unknown sequence with high reproducibility	For the larger sequence it produce slow results	Kircher and Kelso, 2010
Chip based assays	For the high quality and fast GWA	Based on the large scale antibody characterization	Shendure et al., 2004
Transcriptomics			
RNA-Seq	direct analysis of unknown genomic sequence and alternative splice variants	Discrepancies in the sequencing of isoforms variants	Shendure et al., 2004
fluorescence-based reporter gene	Detection of gene expression though direct fluorescence	Difficulties in detection of regulatory genes throughout the genome	Masto et al., 2006
cDNA microarray	Quantitative-High throughput techniques	Produce complex data	Pedrotty et al., 2012
Proteomics			
MS (Microscopy)	High throughput and able to detect medium and low abundant proteins, require lowest amount of sample, high sensitivity	Require highest quality samples with desalting procedure	Bassim et al., 2012; Ambatipudi et al., 2009

TABLE 14.3 *(Continued)*

Techniques	Upside	Downside	Authors
Yeast two hybrid	Simplest approach to identify and quantify the protein-protein interactions	Require large scale characterization of antibody and associated with the specific sensitivity with the target proteins	Stynen et al., 2012
ELISA	Quantitatively detect specific target with high sensitivity and specificity	Limited to only available antibody, time consuming	Ambatipudi et al., 2009
Metabolomics			
Gas chromatography GC-MS	High through put, fast, less time and low sample requirement, high separation efficiency	Require well maintained large database	Scalbert et al., 2009

complexes (proteasome degradation complexes) or with nucleic acid affect several molecular pathways and may act as regulatory element (prokaryotic/eukaryotic transcription factors) for gene or transcript expression (Gonzalez and Kann, 2012).

Currently, the two important approaches for comprehensive quantification and characterization of proteins are (1) antibody-based protein detection (immunoassays) and (2) MS-based protein identification and quantification (Figure 14.3). Enzyme-linked immunosorbent assay (ELISA) (Figure 14.4A) or western blotting (Figure 14.4B) are traditional and most common strategies for protein quantification with high specificity and sensitivity. However, these techniques are limited by the availability of sensitive/specific antibodies including the cost and time for the development of multiplexing immunoassays (Ambatipudi et al., 2012). Thus the use of MS-based techniques such as single/multiple reaction monitoring (SRM/MRM) or accurate inclusion mass screening gives impetus to the confirmation of differential expression of marker proteins (Parker and Borchers, 2014). Additionally, it is also critical to identify interacting proteins that could play a significant role in various cell–cell communications, metabolic, signal transduction

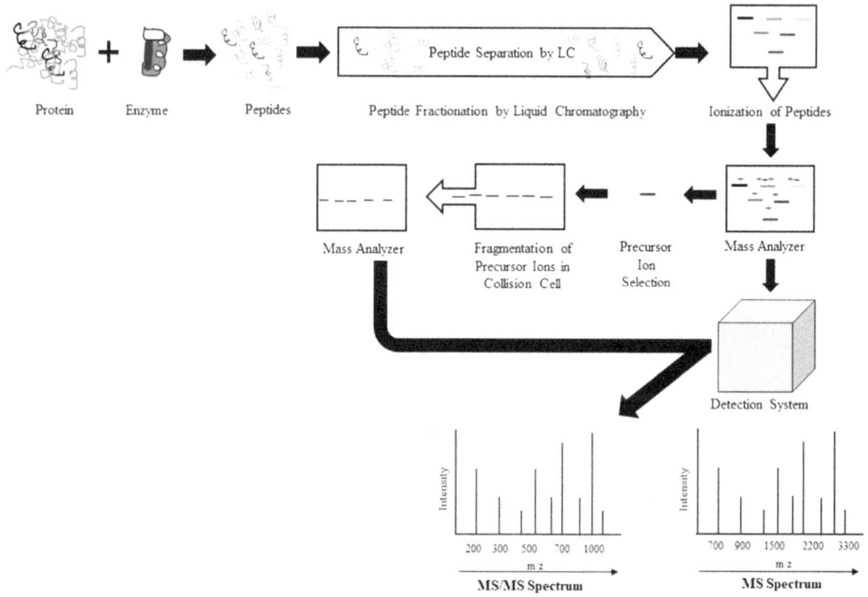

FIGURE 14.3 Peptides derived from enzymatic digestion (i.e. trypsin) of selected protein are separated based on their hydrophobicity by reversed-phase liquid chromatography followed by their characterization using mass spectrometry.

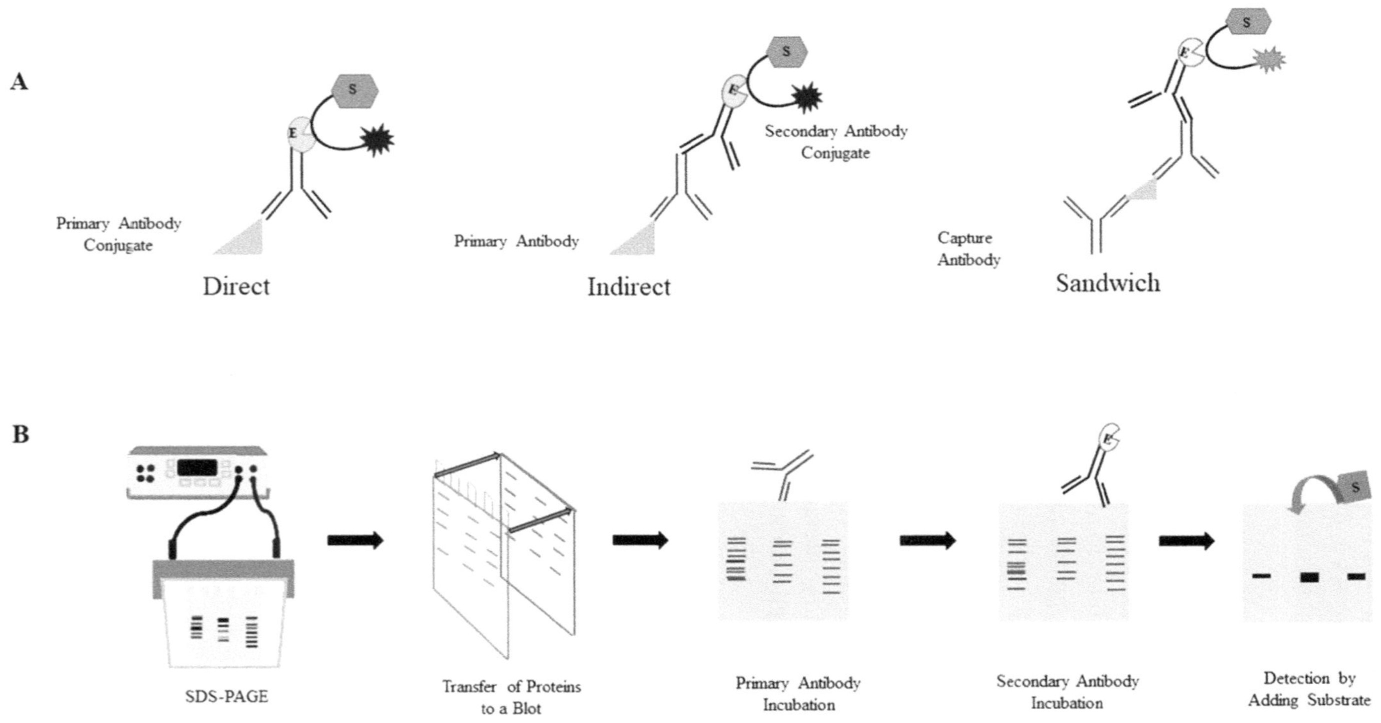

FIGURE 14.4 The comprehensive characterisation and quantitation of proteins. The traditional and most common strategies for characterisation and quantification of proteins with high specificity and sensitivity are antibody-based protein detection (immunoassays) approaches. A) Enzyme Linked Immunosorbent Assays (ELISA)-which includes three basic types-Direct, Indirect and Sandwich ELISA and, B) Sodium dodecyl sulphate-Polyacrylamide gel electrophoresis (SDS-PAGE) with Western blotting (WB) identifies proteins on the basis of specific antigen-antibody complex formation followed by enzymatic reaction on a substrate visible under UV light.

pathways, and development. The field that deals with the recognition and evaluation of protein interaction is referred as interactomics. For example, yeast two-hybrid (Stynen et al., 2012) system is the most frequent system to study protein interactions in living cells followed by NGS or microarrays. However, these in vitro methods have a limitation with high false-positive annotation of molecules (Rao et al., 2014).

14.4.3 METABOLOMICS

The evolution of metabolomics was initiated in 1970 by Arthur Robinson. Subsequently, development of omics or functional genomics opened a new path to analyze the effect or application of various mixtures in medicine. The human metabolome had its first blueprint completed in 2007 by concerted scientists' efforts from Alberta and Calgary Universities who successfully characterized over 2000 metabolic substances, 1000 drugs, and 3000 food components in a human body (Wishart, 2007). The biochemical activity of a cell or an organism in its current state is represented by metabolome or metabolites consisting of a mixture of small biomolecules or end products of many biochemical activities such as sugars, lipids, and hormones. The metabolites of a cell largely affected by environmental stress including exposure to chemicals, drugs, and disease state can be specifically monitored by analyzing the final products of cells or an organism. The study associated with metabolome is referred to as metabolomics (metabonomics) or the systemic approach that enables the study of distinctive chemical fingerprints of different cellular processes. The component of metabolome can be specifically characterized by different techniques such as chromatography, MS, and NMR (Weckwerth, 2003; Nicholson and Lindon, 2008). Subsequently, results from these techniques are compared to computerized libraries of MS to facilitate compound identification. The major applications of the metabolomics include assessment of toxicity in urine, blood, and plasma samples; phenotypic changes due to deletion or insertion in an organism's genome; and to predict gene function of an unknown gene by comparison with metabolic perturbations caused by gene alterations (Weckwerth, 2003; Shulaev, 2006).

14.5 OMICS APPROACHES IN DISEASE DIAGNOSIS AND THERAPY

Since the evolution in informative machinery, omics technologies were aimed to globally detect genes, transcripts, and downstream molecules such as proteins and metabolites. Disease predictive tests based on omics technologies including genomics and proteomics pledge to address relevant clinical methods for early disease diagnosis and therapy management. Omics technologies such as nucleotide sequencing, RNA microarray, and high throughput MS have applications to establish the design for clinical trials and patient stratification for suitable treatment regime and evaluation of therapy outcomes. Additionally, omics-based personalized and precision medicine are gaining momentum since the US government in 2012 notified and emphasized the *Big Data Research* and *Development Initiative*, patient-powered research, which would be hugely beneficial in predicting the right therapy to the right subject at a right time. Thus the focus in the section below will be to discuss and emphasize the key role of omics data in creating windows for disease identification and therapy prediction.

14.6 GENOMICS IN DISEASE AND MEDICINE

Omics profiling studies and revolution in omics technologies have significantly broadened the knowledge on elucidating disease and disease mechanism at the molecular level that has substantially benefited disease-oriented studies and health care (Chen et al., 2012). Similarly, the beneficial outcomes of genomics and genetics in the last couple of decades in human health have revolutionized the medical sector. For instance, the association of breast cancer gene1/2 (BRCA1/2) mutation to breast cancer and ovarian cancers, mapping of the gene responsible for cystic fibrosis transmembrane conductance regulator, and Tay–Sachs disease predict the vulnerability of the disease to individuals with a family history. However, it is essential to acknowledge that both disciplines are different but complementary. Genetics covers the details of the inheritance pattern of genes from one generation to another, whereas genomics provide insights on genes, genes' products, and functions as well as gene–gene interaction in an individual. The top-of-the -line technologies in genomics and genetics contribute to generating huge data for disease management. The application of genomics and genetics in predicting tests for the disease and targeting medicine gained momentum pertaining to the HGP completion. The premise to elucidate the reference normal human genome sequence and any variation (polymorphism) (HGP) was to address the abnormal gene that could be responsible for disease resulting in effective disease management (Garay and Gray, 2012). Genomics including molecular phenotyping and genetics employed in the clinical laboratory has the ability to alter current clinical practices and drug development. Genomics shows promise in understanding the relation of some diseases with their root cause. Due to the progress made in health care, genomics in conjunction with various sequencing and bioinformatics techniques will succeed in detecting different disease-associated abnormalities and mechanisms for effective therapeutic management (Bell, 2004).

The alliance of genomics with human health has enlightened the common objective of disease management. In the postgenomic era, two projects namely Genome-Wide Association Studies (GWAS) and haplotype map (HapMap) studies were initiated to identify common genetic variants to a particular trait or disease. The GWAS particularly emphasized the identification of any association between SNPs and common traits or common human disease that has enabled to assess approximately 500,000 SNPs and copy number variants (CNV) in DNA using DNA microarray in a large population with approximately 100,000 participants (Rabbani et al., 2016). The study included a range of diseases, for example, breast and prostate cancers, to complex metabolic diseases like schizophrenia, diabetes, and Crohn's diseases, an inflammatory bowel disease. Similarly, the HapMap project created a genome-wide database that categorized common patterns of variation in human sequence within individuals and across populations. The HapMap project focused on the "common disease common variants" hypothesis and triggered the identification of common occurring complex disease (non-Mendelian) in individuals caused by genetic variants in a subset of the population. The HapMap hypothesis concluded that the existence of a common pattern of genetic or allelic variants causing the specific phenotype affects more than 5% of a population for a common trait (Bloss et al., 2011). Although, both the GWAS and HapMap projects contributed significantly in predicting risk variants and implementation of drugs for effective treatment, these projects had limitations in identifying gene–gene interaction, nonallelic interaction, and influence of environmental factors on gene and disease (Korte and Farlow, 2013).

14.7 NEXT-GENERATION SEQUENCING IN DISEASE DIAGNOSIS AND PERSONALIZED MEDICINE

The technologies used for sequencing are pilots for genomic medicine. The whole-genome sequencing (WGS) or whole-exome sequencing (WES), collectively known as NGS, analyzes a single nucleotide base of a genome/exome, has gained momentum over the GWAS studies. The NGS approach parallelizes the process of sequencing by producing a huge (millions) sequence in a very rapid and profitable way. It allows the analysis of the whole genome and genetic bases for diseases in unprecedented ways (Behjati and Tarpey, 2013). By NGS approach massive production of short sequences with multiple DNA (mtDNA) fragments is generated in sufficient quantity to redundantly represent every base in the target genome. Researchers use WGS to identify CNVs, disease genes, small insertions and/or deletions, single nucleotide variants, and diseases related to structural chromosomal anomalies with higher sensitivity (McCarthy et al., 2013). The applications and workflow of NGS are shown in Figure 14.5.

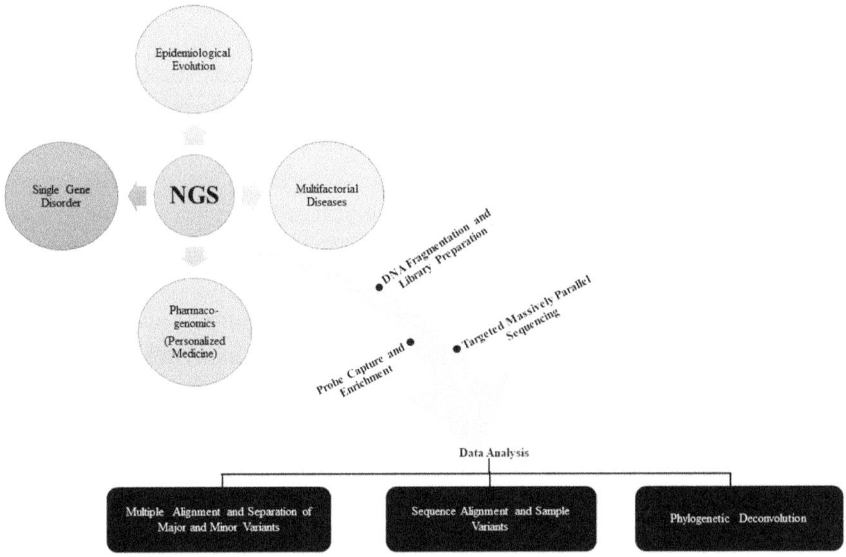

FIGURE 14.5 Applications of next generation sequencing (NGS) and a brief workflow. Various applications of next generation sequencing includes identification of single gene disorder (SGD) or Mendelian disorder, multifactorial diseases, epidemiological evolutions and pharmacogenomics in personalized medicine. The workflow of NGS starts with DNA fragmentation and library preparation followed by probe capture and enrichment. This approach leads to sufficient quantity of short sequences of multiple DNA (Mtdna) fragments to redundantly represent every base in the target genome. The data analysis is based on (i) multiple alignments of major and minor variants, (ii) sequence alignment and sample variants and (iii) phylogenetic deconvolution.

WGS differs from gene sequencing (determining the order of nucleotide in a DNA sequence) as well as exome sequencing (determining protein-coding regions genome) is used to map one's unique DNA. In fact, WGS has an advantage over exome sequencing by performing one test analysis for comparison to single-gene sequencing or genotyping to identify total variants in an array of known mutations of a genome (Welch et al., 2011). By this approach, WGS ensures the identification of any mutated gene in a large volume of DNA responsible for the variation in individuals. The mutations or alteration in any region of a genome may increase the risk or cause of the disease affected due to the mutated region, however, not all genetic changes in a genome cause the disease. Through the WGS approach, researchers and/or clinicians could confidently identify the responsible gene for genetic disorder and ensure the selection of appropriate treatment at the right time. For instance, a large cohort studied by Nagasaki et al. (2015) identified a large specific dataset of DNA variants that could be helpful in understanding the epidemiological evolution. Additionally, WGS has been useful in identifying the molecular ground of human disorders like intellectual disability and adenomatous polyposis in a family. The outcomes of WGS can be categorized into (1) single gene disorder or Mendelian disorder, which is responsible for the occurrence of genomic disease based on the mutation(s) in a single gene of DNA such as fragile X syndrome, muscular dystrophy, Huntington disease, cystic fibrosis, and sickle cell anemia; (2) multifactorial diseases that occur by cause of more than one genetic changes in DNA, for example, diabetes or heart disease, whose onset is generally accounted for at least two factors (multifactor) including several genes, lifestyle, and gene interaction with the environment; and (3) pharmacogenomics (PGx) that predicts the response of a drug based on genetic code of an individual for the prescribed amount of the drug (Bell, 2004). For example, Mizzi et al. (2017) identified pharmacogene-related variants that were involved in a given individual's response to treatment. Although the diagnostic applications of WGS are still at a nascent stage and not so cost-effective, new bioinformatics pipelines for data handling might make WGS a better strategy of choice to identify alterations in DNA sequence (Precone et al., 2015).

In 2009, the WES technique was first developed to selectively map DNA of the coding region of a genome to identify a rare or common associated variant with significant abnormalities or disorders. One percent of the human genome comprises the total exome and covers approximately 80%–85% mutations in the exome region responsible for different diseases. Thus it is imperative to study the exome variants in DNA for many diseases such as Mendelian (predisposing SNP) and rare-Mendelian diseases. To date, WES has successfully identified 30,000 variants including common and rare variants with the identification of 200 rare variants potentially responsible for diseases such as Huntington's due to Huntingtin (HTT) gene and Duchenne muscular dystrophy (DMD) due to Dystrophin (DMK) (Robinson, 2010). The basic concept behind the WES is to develop a strategy for the elucidation of disease variants. For example, (1) exome trio strategy—for the recognition of de novo alteration in the patients (not inherited by either parent) and (2) shared phenotype strategy—for the identification/analysis of same or different genetic variants in a gene of individuals of the same but rare syndrome (Gilissen et al., 2012; Bamshad et al., 2011). In the current decade, WES has been successful in identifying rare variants responsible for many heterogeneous diseases like hair loss, types of monogenic diabetes, a disorder associated with the lysosome, and nonsyndromic mental retardation. This technique has been a choice for clinical practitioners to address the Barter syndrome, a recessive renal tubular disorder (Choi et al., 2009) and parental aneuploidies to characterize the trisomy in chromosomes 21 and 18 compared to other screening systems. One area that has hugely benefited from NGS or WES is cancer, which is primarily due to the mutation or alteration in the gene for effective characterization. To understand the cancer genetics, WES has been proven useful for characterizing circulating tumor cells (CTCs) in prostate (metastatic), gastric, breast cancer, lymphomas, and melanomas (Precone et al., 2015).

In WES, to sequence the exomes, selected target enrichment approaches are currently used, namely, molecular inversion probe, multiplex PCR, array-based technologies, hybridization and in solution hybridization based on capture approaches such as Illumina TruSeq, Agilent Sure Select, and Nimble Gen Seq Cap EZ, and selector probe (HaloPlexand). Each of the above approaches of WES has different capture efficiency and researchers are using this technology for diagnostic application based on the type of choice and need.

Both the NGS sequencing methods, WGS (for the coding or noncoding regions) and WES (only for the coding regions), are being used in the clinic as well as by researchers. However, both these approaches are associated with high cost and percentage of coverage of sequence. Nevertheless, a detailed study of different DNA fragments through exome and genome sequencing would help to determine new genetic (rare or common) variants that are associated with patients' health conditions and in the future could facilitate by diagnosing more diseases (Manzoni et al., 2018).

14.8 PHARMACOGENOMICS IN PRECISION/PERSONALIZED MEDICINE

PGx, also known as drug-gene testing, helps to understand an individual's genetic code (genetic makeup) to predict responsiveness to drugs (adverse or favorable), efficacy, and metabolism based on gene polymorphism. The main goal of PGx is to understand a drug's pharmacokinetics and pharmacodynamics. Pharmacokinetics surrounds four essential processes, that is, absorption, distribution, metabolism, and excretion from the patient's body. Similarly, pharmacodynamics deals with the drug's molecular action on receptors, ion channels, enzymes, and immune systems (Adams, 2008). Although in early stage, PGx has brought a revolution in the health sector by aiming to provide a tailored-made medication based on an individual's specific genetic construction. The joint effort of pharmacology that deals with drug design and genomics has offered a comprehensive evaluation of genes and related functions leading to the evolution of PGx. The term PGx is sometimes interchangeably used as pharmacogenetics, with the latter focusing on single gene-drug response, whereas the former focusing on more than one gene-drug response including epigenetics and polymorphism within the gene.

Tailored drugs or personalized medicines have been in constant demand since adverse drug response was responsible for the hospitalization and death of more than hundreds of thousands of patients in developed countries such as in the United States (Vogenberg et al., 2010). Consequently, conscientious efforts have been shaped through PGx to improve drug development, dose selection, and its interaction with the target site. The PGx has application at numerous stages of the drug development process including drug target identification, different levels of clinical trials including inclusion/exclusion criteria (patients selection), dose range selection and modification, interpretation of clinical trial results, regulatory issues, and personalized therapy (Adithan, et al., 2008). To improve patient safety, predict drug response, and guide the safe dosage, gene-testing affects many health areas such as anesthesiology, psychiatry, cancer adjuvant therapy, neurology, medicine related to heart disease, dermatology, hematology and associated disorders, gastroenterology, human immunodeficiency virus (HIV), rheumatology, and other infectious diseases including mortality risk due to *Staphylococcus aureus* infection (Rieder et al., 2005; Mangravite et al., 2006; Jensen and Lyon, 2009).

The first time PGx was applied to patients treated with warfarin (Coumadin, Jantoven), a drug commonly used to treat blood clotting. For the safe use of warfarin, a genetic sensitivity test was carried out to check for genetic alteration in CYP2C9, an enzyme required in drug-metabolism and VKORC1, responsible for vitamin K activation. Due to considerable success achieved by this therapy, the FDA has recommended its usage in all patients' genotype data before anticlot treatment by warfarin (Rieder et al., 2005). Furthermore, recent evidence suggests that genetic risk assessment test for warfarin, although still in their early phase, has been employed successfully in improving patients' health by avoiding serious risk and fatal adverse effects.

14.8.1 SUCCESS OF GENOMICS IN PRECISION/PERSONALIZED MEDICINE

The "one-for-all" or "one-size-fits-all" strategy-based drugs are currently used; however, the response of these drugs is still unclear. For example, on an average, percentage of patients who are unresponsive to drugs treated for antidepressants, asthma, diabetic, arthritis, Alzheimer, and cancer is 38%, 40%, 43%, 53%, 70%, 75%, respectively (Spear et al., 2001). Personalized medicine has its ability to effectively benefit the patient ensuring that the "drug is only for you" and the health system. To date, biomarker targets for personalized drugs for different diseases have been legalized by the FDA and the numbers are continuously rising with a new genetic testing approach (Weinshilboum and Wang, 2017). Similarly, few theranostic approaches that are currently in use for the successful treatment outcomes of patients affected by cancer and HIV are (1) Dako's Hercep genotyping test for Trastuzumab, (2) Myriad's test for BACA1/2 to assess breast and/or ovarian cancer risk, (3) Bayer's TruGene genotyping test for HIV, (4) the Roche's AmpliChip CYP 450 Test, and (5) Monogram's Trofile coreceptor tropism assay for HIV infection to predict the response for various therapies prescribed to patients. In the case of breast cancer, two prognostic complex tests, Oncotype DX and Mamma Print test, are employed for the prognosis of breast cancer using prognostic biomarker (Vogenberg et al., 2010). The Oncotype Dx helps clinicians to predict the benefit of chemotherapy to women affected with a certain type of breast cancer, while Mamma Print tests the susceptibility of patients for distant recurrence. Many drugs have been tailored-made based on molecular analysis for patient's treatment, which is discussed in the subsequent section (Schmidt 2017).

Oncology is a branch of science that deals with tumors/cancers and based on gene expression pattern breast cancer tumor can be categorized into five subtypes, that is, luminal A and B, triple-negative/basal-like, and Her-2 enriched breast cancer. Owing to the heterogeneity of the disease, clinical outcomes may vary among affected women. For example, breast cancer arises due to the mutation in the breast cancer susceptibility gene (BRCA1/BRCA2) could be handled differently than other breast cancer subtypes (Dai et al., 2015). Similarly, drugs such as Tamoxifen targeting EEBR2 biomarker and Trastuzumab targeting ESR1, ESR2, PGR, F5, F2, and CYP2D6 biomarker have been used for treating other subtypes of breast cancer (e.g., estrogen/progesterone positive cases). The molecular diagnostic test for HER2 overexpression helps in identifying patients who could be benefited from drugs such as Lapatinib. Similarly, for melanoma and nonsmall cell lung cancer treatment, targeting genetic variants such as proto-oncogene BRAF, ALK, and EGFR reduced the cost and time in the trial and error process. The Dabrafenib and Crizotinib have already been legalized for melanoma and lung cancer treatment. Approximately, 73% melanomas, 56% thyroids, 51% colorectal, 43% endometrial, 41% lung and pancreatic, 32% breast, and 21% ovarian and head and neck cancer have been targeted with specific drugs.

Cardiovascular diseases, a general term used to specify the ailment of heart vessels and blood pumping effort, include angina and heart stroke—causing significant mortality in developed as well as developing countries. Certain drugs such as carvedilol, isosorbide-dinitrate and hydralazine (Bidil), metoprolol, and warfarin have been the choice of drugs for effective treatment by clinicians. Similarly, treatment of infectious diseases such as Hepatitis, HIV, leprosy, and tuberculosis, drugs such as Abacavir, Boceprevir, Dapsone, and Isoniazid respectively, have been successfully implemented to target certain biomarkers (e.g., HLA-B57:01, IFNL3, G6PD, and NAT1; NAT2) (Weinshilboum and Wang, 2017).

Although PGx has the potential application in drug designing and development, there are certain limitation associated with it: (1) PGx considers only one test for one drug at a time, that is, multiple tests are needed for multiple drugs, (2) PGx tests are not available for all the drugs, for example, there is no test for drug aspirin and other pain reliever drugs, and (3) PGx test are associated with ethical issues such as discrimination of health insurance companies in issuing certain policies based on one's genetic data. Furthermore, false-positive or inaccurate findings for diseases could have a huge burden on healthcare expenses (Schmidt, 2017).

14.8.2 CHALLENGES AND OPPORTUNITIES IN GENOMIC MEDICINE

The assimilation of the genomic environment with therapeutics is an emerging field of clinical science. Clinical science data of the human genome created a transforming landmark by fueling the genomic marker-based detection of disease variation, early disease diagnosis, disease risk and susceptibility, disease recurrence, and drug response. For accurate disease management, the National Academy of Sciences suggested adopting genomic-based data to maximize the health benefit and reduce the potential risk and cost including checking the circumstances in which genome editing should be allowed. In addition, the Division of Genomic Health of NIH has developed several electronic programs and networks for the promotion and development of genomic-based medicines. Of note, the US FDA has approved several genome-based tests for metabolic disorders, cardiac transplant, cystic fibrosis, severe combined immunodeficiency, cancer, and microbial genome-based pathogen screening. For example, recently the US FDA has authorized the first saliva-based genomic screening for drug metabolism associated with genetic variations. Similarly, certain drugs have also been legalized by the US FDA for the clinical trial. For instance, Larotrectinib, which is used for treating neurotrophic receptor tyrosine kinase gene fusion-positive solid tumors, has 75% drug response rate.

In the form of personalized and precision medicine, genomic medicine translated itself into clinical practice. Consequently, personalized medicine shifted from bench-to-bedside and increased our understanding of complex diseases, diseases with mutant variation, and target-based drug development. However, there are few technical limitations related to data storage, processing, and subsequent clinical data interpretation from genomic calculations. The extraction of valid data from complex genomic information for its integration with the current existing health system is another challenge for genomics. The direct implication of genomic-based therapeutics in health management is a complicated step and linked with ethical, social, and legal issues with human subjects or tissues. Patient heterogeneity and risk of false-positive results are other important considerations for genomic-based medicines. The selection of appropriate genetic markers and progression in developing cost-effective therapeutics is a significant challenge for genomics. Genomic medicine regulation and guidelines along with trained clinical staff appointments is another important challenge for the implication of genomics in medicine (Schmidt 2017). The difference in the approach of traditional medicine for a population, that is, "one size fit all" and precision medicine (tailored) for selected-population are shown in Figure 14.6.

14.9 TRANSCRIPTOMICS

Another promising technique for early disease diagnosis is the transcriptomics-based interpretation of gene functions and their expression profiling under different conditions. Regulation of genes that expressed differentially by internal (e.g., growth factors, hormones, signal molecules) or external (e.g., environmental stimuli, infection, or disease conditions) factors make a transcriptome more heterogeneous, informative, and dynamic compared to genomic data. A transcriptome that represents a very small proportion of genetic code is transcribed to all RNAs is approximately 5% of the human genome (Frith et al., 2005). Nevertheless, the study of the transcriptome is more diverse and complex because one gene can produce more than one RNA transcript due to the presence of different RNA editing sites, splicing sites, and alternative initiation and termination sites for transcription. The study of transcriptome that refers to all transcripts including protein-coding transcript (mRNA) and nonprotein coding transcript (ncRNA) of a DNA in a cell was initiated in 1990. Subsequently, the emergence of hybridization and sequencing-based techniques has made transcriptomics a universal approach for gene profiling analysis (Adams, 2008).

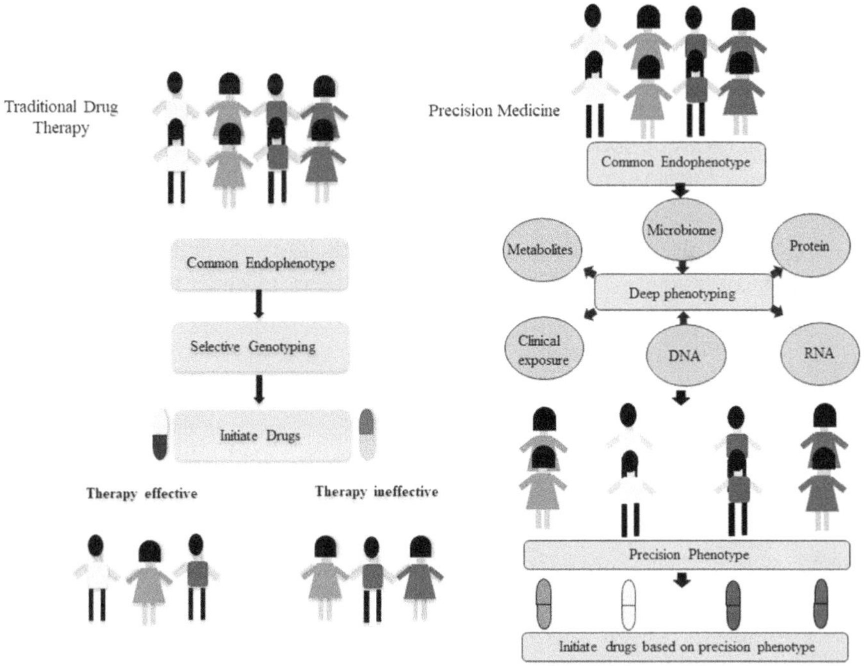

FIGURE 14.6 Precision medicine approach compared to traditional drug therapy. Individuals with similar endophenotype could be susceptible to a disease but they may be biologically distinct with different disease profiles. Under precision medicine, an individual undergoes deep phenotyping, an integrative approach that considers an individual's genetics, lifestyle and exposure to environments to predict and prognosticate health, detect biomarkers and improve an individual's health.

Across all the cells, 99% DNA is similar but according to surrounding cellular microenvironment or factors that directly affect the biophysical or biochemical properties of a cell, the transcribed RNA differs across cell types. RNA sequencing and microarray have revolutionized the transcriptome sequencing and quantification for biomedical applications. Transcriptome-based studies have also improved disease diagnosis and therapeutics by uncovering patient heterogeneity at the differential gene transcription level (Pedrotty et al., 2012). Additionally, it has been helpful in unraveling the puzzle of "gene and diseases" for gene-based disease development, prognosis, and target-based drug development.

Hybridization and sequencing-based techniques have outspread the transcriptome application to the agriculture sector (e.g., plant breeding, plant stress management, phytohormone response and developmental study, disease and pest resistant crop development), food science, medicine, and applied fields such as microbiome and evolutionary studies. Expressed sequence tags, RT-qPCR, SAGE, and microarray-based advancement enabled screening of short transcript sequences (Casamassimi et al., 2017). Next-generation or massively parallel sequencing allows screening of several pathogenic or mutant loci of disease-causing agents. Currently, identification of the transcriptome of an individual cell or multiple cells for biomedical purposes is an achievable target. Whole transcriptome sequencing and cDNA microarray allow the study of transcriptome of multiple cells, while single-cell RNA sequencing enables the study of transcriptome from a single cell.

14.9.1 TRANSCRIPTOME-BASED DISEASE MARKER DISCOVERY

Transcriptomics comprises functional genomics, describing gene function and differential genes expression to relate the variable response of cells under different conditions. Transcriptomics reliability, feasibility, and less biased gene transcription enabled the use of transcriptome in clinical studies. Previously, serological-based macromolecular markers (e.g., lipid, sugar, and protein), microbial staining, and growth-based pathogenesis were used for disease diagnosis, prognosis, and treatment; but molecular marker-based techniques have broadened our current understanding for early detection and effective disease management. Transcriptome-based disease marker study offers screening of changes in gene transcription levels during healthy and disease condition including response to a drug. For instance, higher expression of NPC2 in tuberculosis patient's peripheral blood is directly linked with tuberculosis prognosis (de Araujo et al., 2016). Similarly, transcriptome-based study in pancreatic cancer showed that HNF1A acts as a tumor suppressor gene (Luo et al., 2015). Furthermore, overexpression of interferon (IFN)-stimulated genes was detected in insulitis pancreatic islets of type 1 diabetic patients (Lundberg et al., 2016).

Transcriptome-based therapy prediction or drug targeting also balances the abnormal physiological characters, for example, restoration of transcriptome abnormality re-establish/return the physiological dyslipidemia characters (Wagner et al., 2015). Similarly, Desai et al. (2017) correlated circulating transcriptome markers (e.g., SELENBP1, SLC4A1, EPB42, NELL2, GOLGA8B) to access the patient's death. Furthermore, population-based study in Haiti identified altered expressed gene transcript signatures (e.g., ANXA3, CACNA1E, CR1, CREB5, DYSF, ENTPD1, GK, GYG1, LIMK2, MAPK 14, NAIP, OSM, PFKFB3, PGLYRP1, PHTF1, RAB20, SIPA1L2, SOC3, SRPK1, WDFY3) for validating sputum *Mycobacterium tuberculosis* load and antituberculosis treatment response (Dupnik et al., 2018). Similarly, researchers in Indian tuberculosis patients identified 10 signature sequences (e.g., HK3, RAB13, FCGR1A, RBBP8, BCL6, IFI44L, TIMM10, SMARCD3, SLPI, and CYP4F3) to discriminate healthy and pulmonary tuberculosis patients whose differential expression decreased after treatment (Sambarey et al., 2017). Yu et al (2010) detected adenocarcinoma in lung cancer patient's sputum using miRNA-based transcript that showed 80.6% sensitivity and specificity of 91.7%. Similarly, of 12 peripheral blood gene transcript signatures early response of 3 (IER3) genes was significantly enhanced in Huntington's disease (Runne et al., 2007). Recently, US FDA approved several gene expression-based diagnostic tests such as Corus cad for obstructive coronary artery diseases, Presage (soluble ST2 receptor-based) gene score for chronic heart failure prognosis (Siemelink and Zeller, 2014), and AlloMap gene-based test score to detect rejection of transplanted heart (Yamani et al., 2007).

14.9.2 CURRENT SCENARIO OF TRANSCRIPTOME-BASED THERAPEUTICS

The genomic variation between the same population of patients and the further gap between genomic and phenotype characters led to transcriptomics-based precision and personalize therapeutics development. In the beginning of 1990, mRNA expression in the eukaryotic cytoplasm allowed mRNA as a therapeutic molecule resulting in the development of the mRNA-based vaccines

(Pardi et al., 2018). Transcriptome-based drug development emerged first time during the early 1980s with the revelation of catalytic RNA (Li et al., 2016), and subsequently several new small RNA molecules (e.g., siRNA, ribozymes, oligonucleotides, aptamers) have been used as therapeutics. Currently, for drug development RNA is being used as mRNA inhibitor (e.g., antisense RNA), catalysis (e.g., ribozyme), ligand (e.g., aptamer), and RNA interference (e.g., RNAi) (Burnett and Rossi, 2012). Similarly, siRNA-based (e.g., Bcr-Abl siRNA, iPsiRNA, siG12D LODER) (Burnett and Rossi, 2012) and mRNA-based (e.g., RNA-lipoplex, Lipo-MERIT) (Grabbe et al., 2016) vaccines are under clinical trial phase II and III for cancer (lung cancer, pancreatic cancer) and against infectious pathogens including HIV, influenza, Zika, Ebola, rabies, hepatitis C virus, human papillomavirus, *Streptococcus* spp., and *Toxoplasma gondii.*

Self-replicating and self-nonreplicating are two types of mRNA-based vaccines that are currently in the developmental stage (Pardi et al., 2018). In August 2018, the USA and European FDA approved Patisiran (double-stranded transthyretin siRNA enclosed lipid nanoparticle) for the cure of hereditary transthyretin (TTR)-mediated amyloidosis (hATTR) (Hoy 2018). Similarly, a drug named Inotersen (an antisense oligonucleotide inhibitor of transthyretin mRNA) is approved by both US FDA and EMA for the curing hereditary transthyretin amyloidosis (Keam 2018). Whole transcriptome RNA sequencing study revealed that mutation and downregulation of EthA promoter along with the toxin-antitoxin module (e.g., mazF5 mazE5) and tRNAs (e.g., leuX and thrU) are responsible for extensive drug resistance to ethionamide in *M. tuberculosis* (de Welzen et al., 2017). Verbist et al. (2015) used transcriptome-based data for drug discovery (oncology, metabolic, neurology, and virology) projects for decision (go/no-go decisions) making during the early stage of drug development for reducing drug failure during clinical trial stage III and IV.

Aptamer and ribozymes are RNA-based therapeutics of which certain drugs of this group have been legalized for clinical uses, whereas few drugs are in developmental stage. Ribonucleic acid-based aptamer (pegaptanib 28-bp aptamer) initially approved by US FDA in 2004 to treat age-related macular degeneration, which was later approved in Canada, European Union, Australia, Switzerland, and Brazil as a selective $VEGF_{165}$ isoform antagonist (Vinores, 2006). Ribozyme-based angiozyme, L-TR/Tat-neo, MY-2, and RRz1 had completed their clinical trials (Burnett and Rossi, 2012), whereas several aptamers- based therapeutics such as REG1 and ARC1905 are under clinical trials.

14.9.3 LIMITATION AND CHALLENGES OF TRANSCRIPTOMICS

Transcriptome-based disease diagnosis and therapeutics have facilitated better management of complex diseases, for example, cancer and autoimmune diseases. It has enabled comprehensive recognition and evaluation of genomic expression within a cell or a pool of cells. Currently, transcriptomics-based diagnosis and therapeutic techniques are subjected to clinical trial studies. In spite of the progress made through transcriptomics, differential gene expression and transcript splicing mediated cellular heterogeneity are a bottleneck with the translational implication of transcript-based therapeutics. Gene transcript studies generate complex, noisy, prefiltered, and heterogeneous large size data, analyzed through hypothesis-driven computational programs. Similarly, a conclusive unbiased integration of hypothetical data with biological studies is another challenge with transcript studies (Lundstrom 2018). Although transcriptome-based markers are developed in laboratories for different therapeutic applications, their integration, standardization, and validation with clinical studies are major hurdles for transcriptome-based diagnosis and therapy prediction (Starobova et al., 2018). Furthermore, the cell RNase that affects the RNA stability limits the transcriptome-based therapeutics for translational application. Similarly, developing a suitable comprehensive delivery system with transcript stability is another challenge for RNA-based drug development (Guo et al., 2012).

14.10 PROTEOMICS

The proteome of a bio-system is complex and dynamic as opposed to a static genomic entity. The variability (both temporal and spatial) in the proteome of a cell is important for clinical diagnosis as it emphasizes the pathophysiological complexity and contrast among two biosystems (Andersen and Mann, 2006). However, technological advancement has enabled a comprehensive analysis of the proteomes in any biological entity. Proteomics involves analysis of changes in the proteome (decrease or increase in protein expression level) in response to the surrounding environmental changes or disease including overall gradual assessment of the changes in the proteome of a cell (Petricoin et al., 2002). Thus proteomics is imperative for the characterization of proteomes of a cell along with linking proteins with their possible known structures and functions.

The fruitful execution of the HGP resulted in a conclusive identification of 35,000 genes. Subsequently, the major concern was to determine the function of the genes with approximately 500,000 estimated proteins derived from the coding region of the genome in humans (Stein, 2004). By employing genomic as approaches, for example, genome sequencing, it is highly improbable to predict the exact function of the gene(s) as transcription happens at mRNA level including the possibility of splicing and modifications. The protein array techniques can be applied for performing proteomics analysis in a similar way of measurement of RNA by using gene chip. However, immobilization of long length proteins in the native state and performing the expression analysis of thousands of proteins simultaneously seems to be difficult. The high throughput MS-based approaches like LC-MS, multidimensional protein identification technology (MudPIT), and shotgun approach of proteomics revolutionized the proteomics approach through the capability of high-fragmentation and resolution to detect low abundant proteins (Mann and Kelleher, 2008). Thus proteomics-based techniques could complement the finding of genomics and transcriptomics in its true sense at the protein level by defining the structure–function relation with the genes. The general approach of proteomics is represented in Figure 14.7.

Currently, proteomics and associated technologies are impacting various research fields ranging from diagnosis to validation of biomarkers in complex diseases, predicting the pathophysiology of disease and related mechanism, and evaluation of vaccine production program (Aslam et al., 2017). Different proteomic technologies such as sodium dodecyl sulfate polyacrylamide gel electrophoresis, 2D- differential gel electrophoresis, and matrix-assisted laser desorption and ionization (MALDI)/surface-enhanced laser desorption and ionization (SELDI), time of flight (TOF), liquid chromatography coupled with tandem MS (LC-MS/MS) have been employed to determine various aspects of proteins such as sequence, posttranslational modifications, structure, and possible cross-talks. Different proteomic tools in various fields are discussed in subsequent sections.

14.10.1 PROTEOMICS: GENERAL CLASSIFICATION

The study of proteomics can be classified as (1) structural proteomics deals with the identification of 3D structures of biomolecules (protein and protein complex), subproteome isolation and organelle composition using tool such as NMR spectroscopy and X-ray crystallography, (2) functional proteomics is associated with the understanding of the molecular function of the protein,

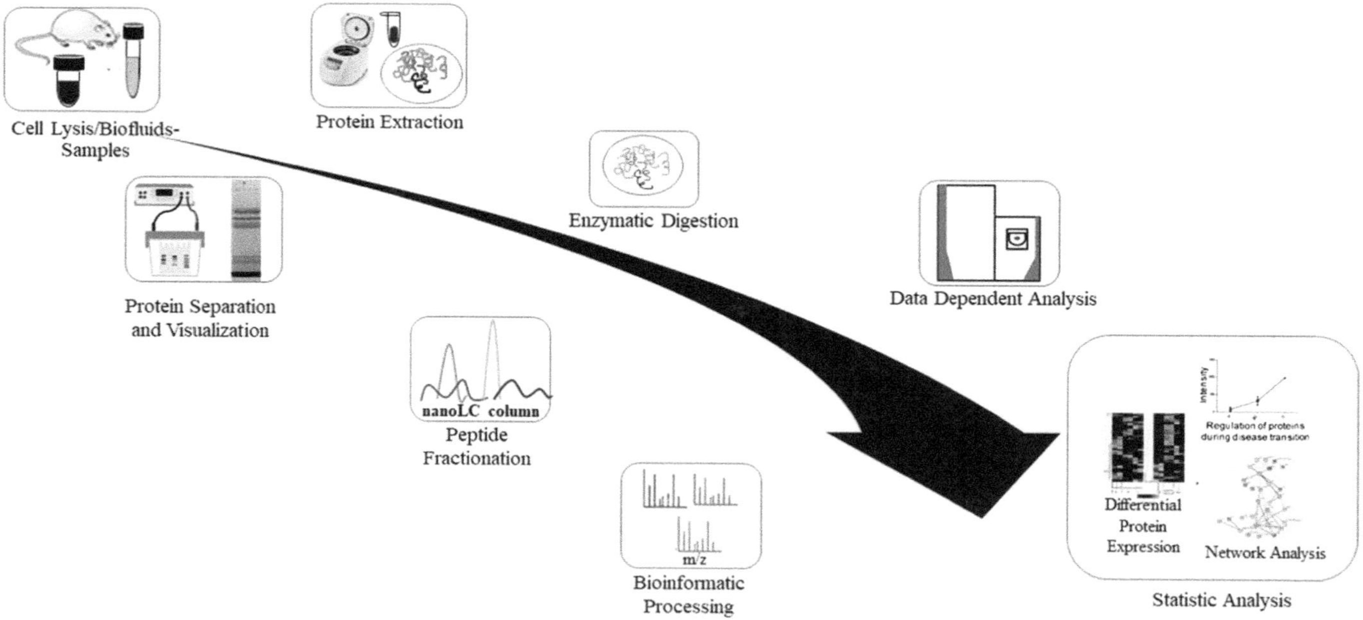

FIGURE 14.7 Protein characterization by mass spectrometry. The general approach for protein characterization involves (i) extraction of proteins from different sources (e.g. cell or tissue or biofluids), (ii) protein separation and visualization through electrophoresis, (iii) enzymatic digestion of protein to peptides followed by their separation using nano liquid chromatography column and, (iv) Identification of proteins by mass spectrometry followed analysis by bioinformatics analysis.

protein–protein interaction, elucidation of extended network of signaling pathways, cellular localization and modifications (post-translational) such as glycosylation, phosphorylation, and proteolysis using MALDI/SELDI and tissue microarrays. The characterization of protein interaction is important to address the pathways of protein complex formation. Proteome interaction can be studied using important techniques such as yeast-two-hybrid system, affinity purification, and MS, and (3) expression proteomics is associated with quantitative analysis of proteome and change in expression/abundance between samples such as disease and health. Additionally, expression proteomics is associated with biomarker (diagnostic, prognostic and predictive) discovery through differential proteomic analysis. Expression-based proteomics studies majorly use techniques such as 1D or 2D SDS PAGE coupled with the MALDI-TOF, SELDI, liquid chromatography-electrospray ionization (ESI) MS/MS and protein microarray. The protein expression profiling further includes subareas such as medical microbiology and disease mechanism. Similarly, proteome mining includes drug discovery, target identification/validation and differential display (Domon and Aebersold, 2006; Aslam et al., 2017). A schematic diagram (Figure 14.1) represents a different application of proteomics.

14.10.2 PROTEOMICS TECHNIQUES IN CURRENT CLINICAL DIAGNOSIS

Currently, immunological-techniques based on antibodies are the preferred method to detect the antigen in a biosystem (e.g., blood). The cruciality of this technique are good sensitivity and ease of experimental handling. To date, numerous commercialized ELISA kits have been exploited to detect allergens, diagnose autoimmune defects like thyroid, Grave's and Hashimoto's disease, anemia, and infectious diseases such as HIV/acute immunodeficiency syndrome (AIDS), Ebola, and Lyme disease (www.omegadiagnostic.com/products). However, the limited availability of broad ranges of antibody and presence (amount) of antigen in the sample affect the overall application of ELISA. In contrast, 2-DE approach, although low throughput technique, has identified several diagnostic biomarkers such as CTCs including CD45, EpCAM, and cytokeratin's 8, 18+, 19+ from whole blood to monitor the progression of breast cancer, Pro2PSA from serum for benign cancer detection and ROMA (HE4+CA125) from serum to predict ovarian cancer progression (Füzéry et al., 2013). Interestingly, clinical proteomics accelerated with newer innovative technologies such as MALDI-TOF and MS-based techniques. These techniques changed the lookout of the current diagnostic approach of proteomics, for example, MALDI has successfully been applied to diagnosis disease such as gastrointestinal, prostate cancer, renal cancer, and colorectal cancers (Rodrigo et al., 2014), Alzheimer's disease (Kitamura et al., 2017) and renal amyloidosis disease (Casadonte et al., 2015). Similarly, ESI-MS and LC-MS have gained their usage due to higher sensitivity and specificity as well as ease of sample preparation. Furthermore, selected reaction monitoring (SRM)/MRM are newer quantitative approaches based on MS-triple quadrupole (QQQ) to selectively isolate the precursor ions corresponding to a mass of the targeted peptide and determine the peptide-specific fragment ions for targeted quantification and accurate identification of a proposed set of proteins from a mixture of complex background. The advantage of the SRM/MRM relies on its high sensitivity that makes it a method of choice in clinical diagnosis; however, per sample cost is higher that limits its usage in a clinical setup, especially in developing countries (Picotti et al., 2010).

In today's date, widely used methodologies to characterize peptides/proteins are bottom-up and top-down approaches. In the bottom-up approach, which is more specific and sensitive, includes liquid chromatographic-based separation of peptides derived from enzymatic digestion (i.e., trypsin) of selected protein and their subsequent identification through MS. However, posttranslation modification such as glycosylation and phosphorylation is identified by performing a top-down approach that involves intact protein identification through the MS (Barbosa et al., 2012). Similarly, for labeled quantification of proteins, Isotope Coded Affinity Tag, is a sensitive technique used for relative quantitative analysis by tagging cysteine residues. Similarly, isobaric tags for relative and absolute quantification (iTRAQ) in protein extract and stable isotope labeling with amino acid in cell culture (SILAC) is used for quantitative comparison of experimentally tagged proteins to discriminate their abundances by MS in different samples (healthy vs. disease state) (Ong et al., 2002; Ross et al., 2004). High throughput MS technique such as MudPIT which is a two dimensional chromatography technique, using high-performance liquid chromatography (incorporate strong cation exchange and reversed-phase columns) coupled with tandem MS identification and quantifies protein from a complex mixture and determination of protein-protein interaction (Washburn et al., 2001). The label-free and labeled (SILAC and iTRAQ) approach for proteomes' quantitation and subsequent analysis is represented in Figure 14.8.

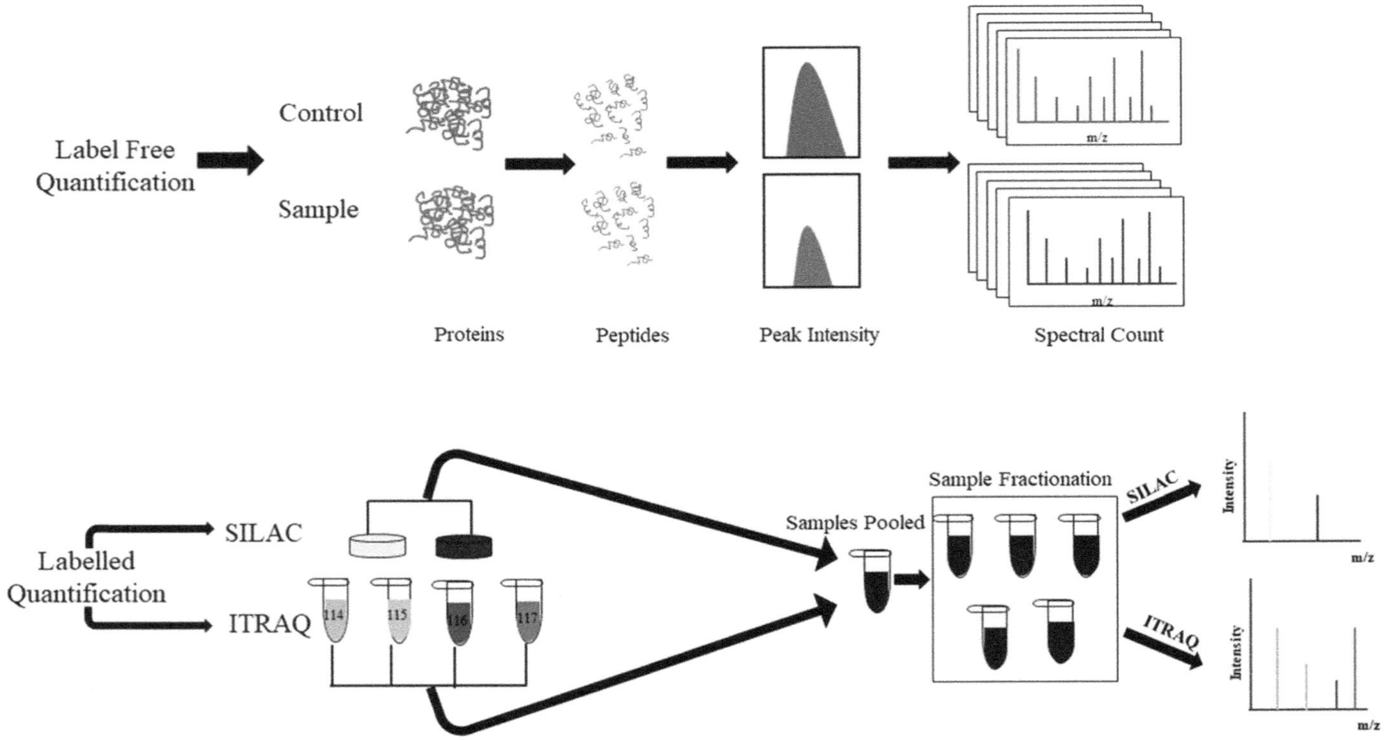

FIGURE 14.8 Identification and quantitation of proteins are performed by label-free and labelled approach. Label-free approach is based on correlation of protein abundance with area under the curve (peak intensity) of precursor ion spectra or tandem mass spectrometry (MS/MS) of peptides obtained from a specific protein. Labelled approach involves tagging the samples with isotopes followed by quantitation of peak intensities of the tags to observe differential expression of proteins by mass spectrometry.

14.10.3 PROTEOMICS APPLICATION IN DISEASE DIAGNOSIS

Different types of alterations in proteome at their structure, function and protein abundance level are useful in the prediction of significant abnormalities during the asymptomatic stage of the disease or before the onset of the diagnostic characteristics during the clinical trials (Xiao et al., 2005). The continuous advancement in the proteomic tools has enabled proteomics to study different research areas including human biology and medicine. For instance, proteomics has the potential for initial screening of human diseases and the prediction of novel biomarkers related to disease diagnosis, prognosis and therapeutic intervention (Azad et al., 2006). To date, proteomics has found its relevance in application in cell, tissue and biofluids for the identification and characterization of potential biomarkers. Amongst different matrices analyzed, proteomics of biofluids (e.g., saliva, urine, milk) has recently gained popularity over cell/tissue proteomics due to ease and noninvasive collection procedure including longitudinal monitoring (Giri et al., 2019). For example, screening tests through OVA1 for ovarian cancer using SELDI-TOF, prostate-specific antigen indicator of prostate cancer illness and the receptor tyrosine kinase CD340 for breast cancer illness have been few successful examples of proteomic analysis of biofluids (Schiess et al., 2009). Similarly, different biofluids, for example, saliva, urine, and blood, have also been characterized for the clinical outcomes. For instance, saliva has been characterized for the recognition of breast cancer metastasis (Giri et al., 2019), urinary proteomics has been characterized the proteins for the Anderson-Fabry disease (Vojtová et al., 2010), cerebrospinal fluid has been characterized by the proteomics for the Alzheimer disease (Wallin et al., 2010), multiple and amyotrophic lateral sclerosis (Ottervald et al., 2010; Zhou et al., 2010), and Parkinson disease (van Dijk et al., 2010). Similarly, proteomics of biofluid includes the nipple aspirate test for breast cancer (Pawlik et al., 2005), tear (Lema et al., 2010), and synovial (Gobezie et al., 2007) fluid for osteoarthritis and keratoconus, respectively. Subsequently, proteomics has been implemented for the identification of proteins from serum and plasma. A significant amount of proteomics data from patients have been obtained for different cancers such as head and neck (Freed et al., 2008), bladder (Minami et al., 2010), colon (Ransohoff, 2008), and breast (Hamrita et al., 2009). Similarly, proteomics-based identification of markers have been carried out for diseases such as diabetes (Isabel Padrão et al., 2012), autoimmune diseases (Ambatipudi et al., 2012; Bassim et al., 2012), heart and infectious diseases (He, 2003; Kiernan et al., 2006), and gestational Down syndrome (Kolla et al., 2010).

14.10.4 PROTEOMICS IN DIAGNOSIS OF DISEASES RELATED TO BLOOD

Proteomics has been applied for the detection of blood-related cancers such as acute leukemia including myeloid and lymphoblastic leukemia. Fujii et al. (2005) performed proteomics analysis of blood helped in successfully identifying 37 inflammatory and 26 cell communication proteins. Subsequently, a hypothesis was proposed on multinucleate cells evading the cytotoxic T-cells and triggering inflammation and reactive proliferation. Similarly, a combination of 2-DE and MALDI-TOF has been employed for the discovery of differentially expressed proteins in leukemic conditions compared to healthy controls. Furthermore, researchers were successful in identifying proteins involved in glycolysis, apoptosis, angiogenesis, and metastasis (López-Pedrera et al., 2006). The response of the valosin-containing protein (VCP) to glucocorticoid therapy has been analyzed by 2-DE and SELDI-TOF MS that detected the high expression of VCP indicating poor prognosis of the acute lymphoblastic leukemia disease (Lauten et al., 2006).

14.10.5 DIAGNOSTIC APPLICATION OF PROTEOMICS IN INFECTIOUS DISEASE

Proteomics is potentially exploited to diagnose infectious diseases, for example, mastitis, malaria, measles, tuberculosis, HIV/AIDS, and many respiratory diseases. Below are a few examples related to the diagnosis of tuberculosis and HIV/AIDS.

Tuberculosis: Tuberculosis is a global burden caused by *M. tuberculosis*. Agranoff (2006) employed SELDI-TOF on diseased serum samples and healthy controls to successfully identify reliable markers like serum amyloid protein and transthyretin for early diagnosis of tuberculosis.

HIV/AIDS: To demonstrate HIV infected T-cells surface proteins, MALDI-TOF was employed. The Bruton's tyrosine kinase and X-linked inhibitor of apoptosis have been characterized and could possibly serve as potential diagnostic markers for HIV if validated in a large patient cohort (Guendel et al., 2015; Berro et al., 2007). Furthermore, proteomics has rigorously found its use to demonstrate the interaction of HIV with mononuclear phagocytic cells (Bone marrow and alveolar derived macrophages, histiocytes, and Kupffer cells) and the mechanism by which the immunomodulatory properties of the cells get altered (Ciborowski and Gendelman, 2006).

14.10.6 PROTEOMICS IN TARGETED THERAPEUTICS

Disease mechanism could be predicted through different interconnected networks of proteins, structures, and allied functions. Extraction of disease-associated information could be helpful in recognizing possible causes, associated consequences, and mechanism (Petricoin et al., 2002). The comprehensive knowledge about a disease can help in targeting newly selected therapeutics, response to medications, and continuation of therapy (Lee et al., 2011). For the effective management of the disease, newer therapeutics are imperative for precision and personalized medicine. Thus it is worthwhile to have profiling data of proteins using different strategies of clinical proteomics with higher precision that can auger potential disease biomarker identification and enhance the development processes of targeted therapeutic (Zhou et al., 2016). Earlier the substantial omics approach—genomics and transcriptomics were the only approaches for accurate prediction of therapy and its response. However, proteomics advancement threw some lights on molecular medicine that is heading toward the use of MS-based proteomics approaches. Thus modern high throughput proteomic technologies with subproteomics (e.g., clinical proteomics) promise research to be translated from bench to bedside. Below are a few proteomics-based drugs and therapies currently in use.

In 1970, US FDA approved tamoxifen as a first target-based anticancer therapy for estrogen receptor (ER) positive breast cancer patients, acting as estrogens' antagonist for ERs. Target-based anticancer drugs such as denosumab for bone metastasis, lapatinib for breast cancer, pazopanib and sorafenib tosylate for renal cell carcinoma have been approved by US FDA for the precise treatment of cancer (Yan et al., 2011). For the effective management of arthritis, tumor necrosis factor (TNF)-based anti-TNF therapy has been used to treat severe complications of the disease. Furthermore, infliximab has been used to treat the autoimmune disorder and was the first chimeric monoclonal antibody has been approved by US FDA as an inhibitor for TNFα to treat arthritis (Perdriger 2009). Similarly, etanercept and cardiac troponin T is another drug, approved by the US FDA, as a TNFα inhibitor for the arthritis therapeutics (Azevedo et al., 2015) and as a safety marker for cardiotoxicity and acute myocardial infarction (McRae et al., 2019). Protein-based markers are useful for the evaluation of drug-induced toxicity. For instance, renal papillary antigen 1 is a US FDA approved protein marker to screen out drug-induced nephrotoxicity (Rouse et al., 2011). Thus taken together protein-based markers have revolutionized the applied biomedical science by improving the accuracy of disease profiling and identifying markers for drug-induced toxicological screening and precise target-based inhibitors.

14.10.7 PROTEOMIC CHALLENGES AND OPPORTUNITIES

Protein-based disease prediction offers both structural and functional information about proteins responsible for simple to complex pathological conditions. Proteomics due to novelty, multidisciplinary, indispensable, unbiasedness and universality approach provided a dispensable choice for different biomedical applications globally. In comparison to other omics techniques, proteomics offers precise identification of disease prognosis and marker-based drug development (Kentsis, 2011). Post 19th and early 20th century, major development and up-gradation occurred in MS-based proteomics techniques (e.g., SELDI/TOF, MALDI-TOF, and ESI) resulting in the analysis of complex molecules including quantitation of analyte in different matrices Additionally, in 2002 Human Proteome Organization (HUPO) was established and initiated the Human Proteome Project, which was as a major milestone in human protein-based studies (Kavallaris and Marshall, 2005).

Currently, there are many protein-based diagnostic and therapeutic available for clinical applications. Along with continuous development in the technology, there are structural, technical, and data-related challenges in proteomics. For instance, tissue samples and body fluids generally used for pathological screening are complex in protein content and possess several challenges. For instance, protein isoforms and different secondary and tertiary structures result in protein variability including posttranslational modifications adding further heterogeneity to proteins. Similarly, a high abundance of proteins creates difficulties during isolation of medium and low copy number of proteins from healthy and pathological samples (Ambatipudi et al., 2009). Furthermore, complex instrumentation and analytical algorithms make proteomics more challenging for translational and clinical application including the cost of instrumentation. The necessity of a huge cohort of clinical samples for unbiased clinical validation and the progression in clinically significant output is another major challenge. Similarly, integrating proteomic results with biological specimens is another challenge in proteomics (Schubert et al., 2017). Although there are many advanced tools to study large proteins, there are calls for desperate measures in the advancement of large protein isolation, solubility and MS-based studies including complex data analysis algorithms (Gregorich and Ge, 2014).

14.11 METABOLOMICS

Metabolomics comprises functional genomics which addresses the journey of a gene-to -proteins-to-phenotypic representation. In 1998, metabolomics was introduced into omics field with an aim to qualitative and quantitative study the metabolome-products in an organism. Although genomic-based studies can explain the initial and subsequent stages of a biological process; however, metabolome expresses the final phenotypic effect of a gene in an organism (Griffin, 2006). The continuous advancement and integration with other omics fields like genomics, transcriptomics, and proteomics have broadened our knowledge to understand the mechanism of complex disease onset and progression (i.e., diabetes, Wilson, Alzheimer, and Parkinson diseases). Integrating multiomics-based clinical data, databases like MS/MS-based metabolite database METLIN (Guijas et al., 2018) and Human Metabolome Database were programed in 2005 and 2007, respectively. With the advancement in a variety of sensitive metabolic analytical tools, data processing and availability of metabolic databases have facilitated the precise screening of various metabolites. Currently, NMR, gas chromatography-mass spectrometry, and LC-MS are the three main techniques currently used to analyze the metabolome or gene products (Tan et al., 2016). The multidisciplinary and phenotypic oriented approach of metabolomics makes it suitable for the precise development of medicine and preventive healthcare practices. The clinical metabolomics involves an array of pathological samples like urine, serum, tissue biopsy, saliva, stool for the mapping of disease development and target prediction for the development of both endogenous (e.g., host) and exogenous (e.g., pathogen) metabolite-based therapies (Hocher and Adamski, 2017). In addition, metabolites control developmental and physiological processes (e.g., Retinoic acid), cell stem characters, DNA replication, transcription and translation (e.g., ATP, GTP, S-adenosyl methionine, acetyl CoA), cell interaction, and signaling pathways. By metabolic-based studies, we can easily predict the alteration in a biological sample by observing alteration in phenotypic characters (Johnson et al., 2016). Metabolites quantitation by MS is shown in Figure 14.9.

14.11.1 METABOLIC SIGNATURE-BASED DISEASE DIAGNOSIS

Metabolite-based disease identification is not a new approach as Ayurveda and traditional Chinese medicine have used ants to identify sugar in urine to detect diabetes. Physicians of ancient Egypt and Arab have also described

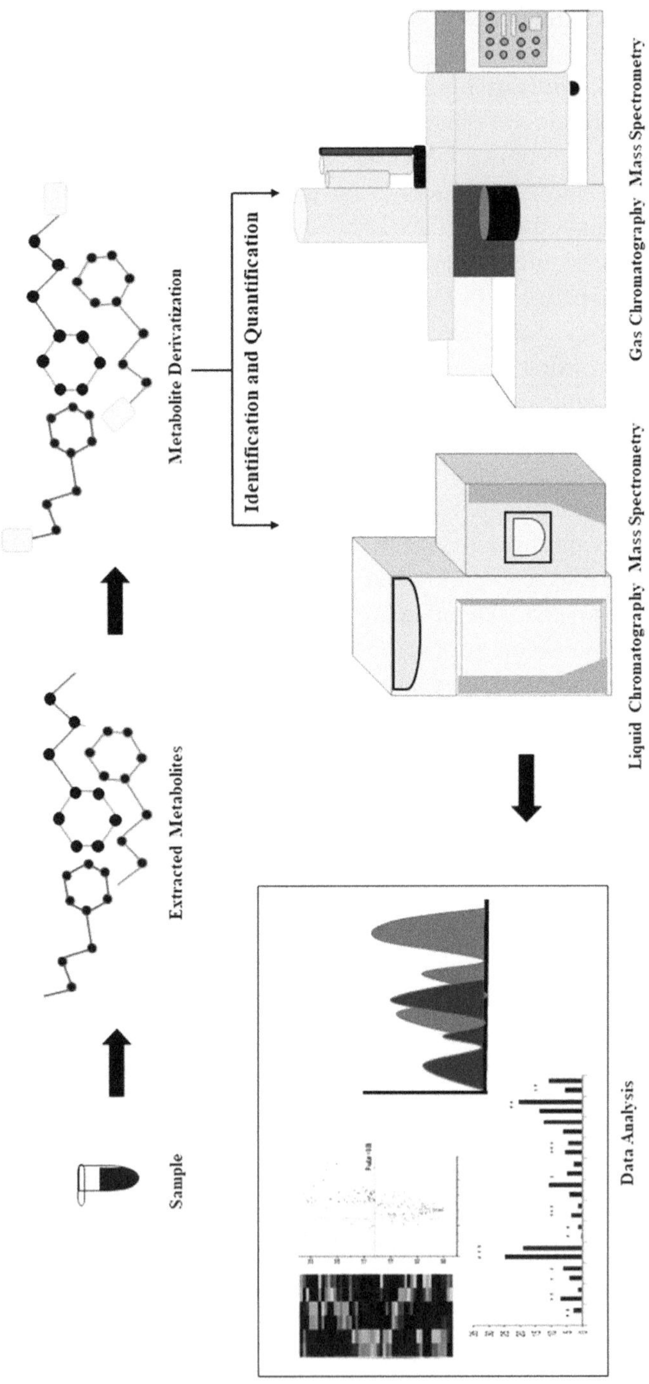

FIGURE 14.9 A workflow of metabolomics. Extracted metabolites can be identified and quantitated either by gas- and/or liquid chromatography coupled tandem mass spectrometry followed by data analysis to identify metabolites involved in different pathways.

diabetes based on physiological markers such as excessive thirst and urination (Karamanou et al., 2016). Metabolic markers are directly linked with the physiologic state of an organism, for instance, to discriminate between healthy and patients. In established arthritis patients, along with C-reactive proteins, 3-hydroxybutyrate concentration is elevated compared to healthy controls. Higher numbers of lactate and lipid metabolites have been identified in the same population of arthritis patients (Young et al., 2013). Similarly, 2-hydroxybiphenyl is a diabetic metabolic marker associated with high risk of diabetes development, whereas phosphatidylcholine, (3S)-7-hydroxy-2',3',4',5',8-pentamethoxyisoflavan, and tetrapeptides (MEIR and LDYR) are associated with lower risk of diabetes prognosis in American Indians (Zhao et al., 2015).

Metabolic markers are clinically useful as a diagnostic tool for disease identification, risk, and prognosis. Their sensitivity and selectivity enables the detection of alterations in metabolic pathways during disease prognosis and to predict drug response. Biochemical disturbance in methionine, tyrosine, tryptophan, and purine pathways leads to elevation in 5-hydroxy indole acetic acid, glutathione, vanillylmandelic acid, methionine, xanthosine, and glutathione in Alzheimer's patient's cerebrospinal fluid, whereas in mild cognitive impairment patients the cerebrospinal fluid has elevated levels of hypoxanthine, 5-hydroxy indole acetic acid, and methionine (Kaddurah-Daouk et al., 2013).

Metabolites are genome associated metabotypes (metabolic phenotype), which is broadly employed for disease identification and classification of disease severity. Based on the metabolic markers, asthmatics are classified as a mild, moderate, and severe form of pathogenesis. For instance, the serological study based on healthy and asthmatics patients showed an increase in levels of oleoylethanolamide, sphingosine-1-phosphate, and *N*-palmitoyltaurine in moderate and severe asthmatics (Reinke et al., 2017). Metabolite-based precise markers provide specificity and sensitivity to detect multifactorial disorders. Alteration in the creatine-to-creatinine ratio in DMD disease patient's serum samples was found to be associated with progression in disease development (Boca et al., 2016). Continued technical modernization in LC-MS has enabled rapid identification, validation, and standardization of metabolic markers for disease diagnosis, mapping of host-pathogen interaction, and host physiological changes during microbial infection.

Metabolomics along with genomics and proteomics can diagnose and prognoses cancer (Prostate, breast and ovarian cancer) depending on host metabolic system and analyzes altered metabolites (Zhang et al., 2013;

Turkoglu et al., 2016; Hadi et al., 2017). Tan et al. (2013) identified 39 differential serum metabolites in healthy subjects and 40 metabolites (e.g., gut flora, urea cycle, nucleotide synthesis, TCA cycle, fatty acid, phenylalanine, amino acid, and carbohydrate metabolism) in colorectal cancer patients. Cardiovascular diseases (e.g., hypertension, myocardial ischemia, acute coronary syndrome, and heart failure) are silent killers; their risk assessment is still a complex puzzle in clinical studies (Wang et al., 2017). Gao et al. (2017) identified 42 biomarkers to capture the physiological changes during the early stage of Coronary Atherosclerosis patients versus healthy controls. Of 42 biomarkers, lysophosphatidylcholine, elaidic acid, L-fucose, monoglyceride, diglyceride, indoxylsulfuric acid, prasterone sulfate, and phosphatidylglycerol were uniquely identified in the early stage of coronary atherosclerosis patients. These results indicate a significant role of metabolic markers in the prediction of early onset and diagnosis of coronary atherosclerosis, cardiovascular diseases, and many cancers.

14.11.2 METABOLOMICS-BASED THERAPEUTICS

Several metabolic products (e.g., insulin, growth hormone, steroids, vitamins, and minerals) are being used as therapeutics to suppress interconnected metabolic cycles. Similarly, many metabolic products are also being treated as drug targets. Recently in 2017, US FDA approved the first metabolic based anticancer drug for isocitrate dehydrogenase-2 (ID2) mutated acute myelogenous leukemia, which targets the oncometabolite 2-hydroxyglutarte produced by mutant ID2 and inhibits its activity (Dugan and Pollyea, 2018).

Based on clinical parameters, currently, physicians use step-up or step-down dose selection method for chronic disease management. A higher dose or unpredicted dose of any drug can show harmful effects and is possible that the prescribed dose is not sufficient to achieve clinically useful results. Unfortunately, such conditions have led to unpredictable drug-induced adverse effects. Consequently, cellular metabolic markers are slowly emerging to predict therapeutic response. For example, methotrexate used as a chemotherapy drug blocks the dihydrofolate reductase enzyme; however, it shows toxic effects when used continuously and exposed to a higher concentration. Both in vivo and in vitro results showed enhanced histidine catabolism in cells linked with methotrexate sensitivity (Kanarek et al., 2018). Similarly, pharmacometabolomics-based markers can be used for personalized medicine development for individual patients during acute and chronic conditions (e.g., cardiovascular, arthritis, diabetes, and obesity). In HepG2 cell line based study for liver toxicity, that is, hepatotoxins altered approximately 90 metabolites and disrupted the urea cycle, protein metabolism, ammonia recycling, glycolysis, nucleotide metabolism, fatty acid metabolism, phospholipids synthesis, and mitochondrial disturbance (García-Cañaveras et al., 2015).

14.11.3 METABOLOMICS CHALLENGES AND LIMITATIONS

Metabolomics-based disease diagnosis and therapeutic approaches are continuously evolving translational clinical science. Since ancient times, there are several success stories of metabolomics and its role in disease diagnosis and therapeutics. Coupling metabolomics with genomics and proteomics has allowed a comprehensive evaluation of anomalies in physiological pathways linked to complex diseases. However, like other diagnostic approaches, it has few limitations, for instance, different metabolites of interconnected metabolic pathways regulate the activity of other metabolites through direct or feedback mechanisms (Scalbert et al., 2009). There are several interconnected regulations that affect the accuracy in the prediction of a single or multiple metabolites or gene products that are responsible for the defects in the regular mechanism. The necessity of comprehensive coverage is another major problem with metabolomics as its concentration varies from one organism to another organism, including distribution of gut microbiome, diet, environment, daily life, sex, age, and body weight. Lack of absolute metabolite concentration and variation in experimental data between researchers makes metabolic data more diverse (Sas et al., 2015). Furthermore, MS and NMR-based metabolite study generate complex and noisy data narrowing its translational application in clinical studies. Additionally, calibration and specific standard development for each metabolite is another major challenge (Gowda and Djukovic, 2014). Thus to capture physiological changes during disease development, further technological up-gradation is mandatory. Nevertheless, metabolomics undoubtedly has disseminated more realistic and reliable data facilitating better decision making and policy intervention that could accelerate the clinical translational research.

14.12 SUMMARY

Since the evolution in informative machinery, omics-based technologies were aimed for the comprehensive detection of the genes, transcripts, and downstream molecules like proteins and metabolites. This chapter has substantially covered the few different applications of omics-based approaches to evaluate complex disease diagnosis, allied mechanism, and concerned therapy for effective management of diseases such as cancers, cardiovascular, and infectious diseases. For instance, omics-based technologies such as WGS, WES, NGS-based approach for gene and transcript and MS-based approach for proteins and metabolites mark the discovery of different biomarkers related to diagnosis, prognosis, recurrence, and therapeutic intervention. The omics technologies such as genomics and transcriptomics—nucleotide sequencing, RNA microarray, and proteomics and metabolomics—high-throughput MS, have applications to establish the design for clinical trials and patient stratification for suitable treatment regime and evaluation of therapeutic outcomes. More importantly, omics-based tailored medicine are gaining momentum as they are associated with empowering the patients by predicting the right therapy for the right individual at the right time. Furthermore, different omics databases and repositories for clinical data submission have accelerated new findings that could serve different health care approaches for biomarker discovery and its translation for better therapeutics. Finally, with technological advancement including newer bioinformatic and statistical programs, omics-based approaches hold great promise for translating research from the bench-to-bedside.

ACKNOWLEDGMENT

This work is supported by the Department of Biotechnology (DBT), Government of India; Grant no. BT/PR12721/AAQ/1/618/2015 to K.A. Ph.D. support fellowship was provided to K.G. and V.S.B by Council of Scientific and Industrial Research (CSIR) and S.M. by Ministry of Human Resource and Development (MHRD).

KEYWORDS

- **multiomics**
- **biomarkers**
- **disease diagnosis**
- **therapeutics**
- **personalized medicines**

REFERENCES

Adams, J. U. Building the Bridge from Bench to Bedside. *Nat. Rev. Drug Discov.* **2008**, *7*(6), 463–464.

Adithan, C.; Surendiran, A.; Pradhan, S. C. Role of Pharmacogenomics in Drug Discovery and Development. *Indian J. Pharmacol.* **2008**, *40*(4), 137-143.

Agranoff, D.; Fernandez-Reyes, D.; Papadopoulos, M. C.; Rojas S. A.; Herbster, M.; Loosemore, A.; Tarelli, E.; Sheldon, J.; Schwenk, A.; pollok, R.; Rayner, C. F.; Krishna, S. Identification of Diagnostic Markers for Tuberculosis by Proteomic Fingerprinting of Serum. *Lancet.* **2006**, *368*(9540), 1012–1021.

Ambatipudi, K. S.; Lu, B.; Hagen, F. K.; Melvin J. E.; Yates, J. R. Quantitative Analysis of Age Specific Variation in the Abundance of Human Female Parotid Salivary Proteins. *J. Proteome Res.* **2009**, *8*(11), 5093–5102.

Ambatipudi, K. S.; Swatkoski, S.; Moresco J. J.; Tu, P. G.; Coca, A.; Anolik, H. J.; Gucek, M.; Sanz, I.; Yates, J. R.; Melvin, J. E. Quantitative Proteomics of Parotid Saliva in Primary Sjögren's Syndrome. *Proteomics.* **2012**, *12*(19–20), 3113–3120.

Andersen, J. S.; Mann, M. Organellar Proteomics: Turning Inventories into Insights. *EMBO Rep.* **2006**, *7*(9), 874–879.

Anderson, M. W.; Schrijver, I. Next Generation DNA Sequencing and the Future of Genomic Medicine. *Genes.* **2010**, *1*(1), 38–69.

de Araujo, L. S.; Vaas, L. A.; Ribeiro-Alves, M.; Geffers, R.; Mello, F. C.; de Almeida, A. S.; Moreira, A. D.; Kritski, A. L.; Lapa E Silva, J. R.; Moraes, M. O.; Pessler, F.; Saad M. H. Transcriptomic Biomarkers for Tuberculosis: Evaluation of DOCK9. EPHA4, and NPC2 MRNA Expression in Peripheral Blood. *Front. Microbiol.* **2016**, *7*, 1586.

Aslam, B.; Basit, M.; Nisar, M. A.; Khurshid, M.; Rasool, M. H. Proteomics: Technologies and Their Applications. *J. Chromatogr. Sci.* **2017**, *55*(2), 182–196.

Azad, N. S.; Rasool, N.; Annunziata, C. M.; Minasian, L.; Whiteley, G.; Kohn, E. C. Proteomics in Clinical Trials and Practice: Present Uses and Future Promise. *Mol. Cell Proteom.* **2006**, *5*(10), 1819–1829.

Azevedo, V. F.; Galli, N.; Kleinfelder, A.; D'Ippolito, J.; Urbano, P. C. Etanercept Biosimilars. *Rheumatol. Int.* **2015**, *35*(2), 197–209.

Bamshad, M. J.; Ng, S. B.; Bigham, A. W.; Tabor, H. K.; Emond, M. J.; Nickerson, D. A.; Shendure, J. Exome Sequencing as a Tool for Mendelian Disease Gene Discovery. *Nat. Rev. Genet.* **2011**, *12*(11), 745–755.

Barbosa, E. B.; Vidotto, A.; Polachini, G. M.; Henrique, T.; Marqui, A. B.; Tajara, E. H. Proteomics: Methodologies and Applications to the Study of Human Diseases. *Rev. Assoc. Med. Bras.* (1992). **2012**, *58*(3), 366–375.

Bassim, C. W.; Ambatipudi, K. S.; Mays, J. W.; Edwards, D. A.; Swatkoski, S.; Fassil, H.; Baird, K.; Gucek, M.; Melvin, J. E.; Pavletic, S. Z. Quantitative Salivary Proteomic Differences in Oral Chronic Graft-versus-Host Disease. *J. Clin. Immunol.* **2012**, *32*(6), 1390–1399.

Bast, R. C.; Xu, F. J.; Yu, Y. H.; Barnhill, S.; Zhang, Z.; Mills, G. B. CA 125: The Past and the Future. *Int. J. Biol. Markers.* **1998**, *13*(4), 179–187.

Behjati, S.; Tarpey, P. S. What Is next Generation Sequencing? *Arch. Dis. Child. Educ. Pract. Ed..* **2013**, *98*(6), 236–238.

Bell, J. Predicting Disease Using Genomics. *Nature.* **2004**, *429*(6990), 453–456.

Berro, R.; de la Fuente, C.; Klase, Z.; Kehn, K.; Parvin, L.; Pumfery, A.; Agbottah, E.; Vertes, A.; Nekhai, S.; Kashanchi, F. Identifying the Membrane Proteome of HIV-1 Latently Infected Cells. *J. Biol. Chem.* **2007**, *282*(11), 8207–8218.

Bloss, C. S.; Jeste, D. V.; Schork, N. J. Genomics for Disease Treatment and Prevention. *Psychiatr. Clin. North Am.* **2011**, *34*(1), 147–166.

Boca, S. M.; Nishida, M.; Harris, M.; Rao, S.; Cheema, A. K.; Gill, K.; Wang, D.; An, L.; Gauba, R.; Seol, H.; Morgenroth, L. P.; Henricson, E.; McDonald, C.; Mah, J. K.; Clemens, P. R.; Hoffman, E. P.; Hathout, Y.; Madhavan, S. Correction: Discovery of Metabolic Biomarkers for Duchenne Muscular Dystrophy within a Natural History Study. *PLoS One.* **2016**, *11*(7), e0159895.

Burnett, J. C.; Rossi, J. J. RNA-Based Therapeutics: Current Progress and Future Prospects. *Chem. Biol.* **2012**, *19*(1), 60–71.

Casadonte, R.; Kriegsmann, M.; Deininger, S. O.; Amann, K.; Paape, R.; Belau, E.; Suckau, D.; Fuchser, J.; Beckmann, J.; Becker, M.; Kriegsmann, J. Imaging Mass Spectrometry Analysis of Renal Amyloidosis Biopsies Reveals Protein Co-Localization with Amyloid Deposits. *Anal. Bioanal. Chem.* **2015**, *407*(18), 5323–5331.

Casamassimi, A.; Federico, A.; Rienzo, M.; Esposito, S.; Ciccodicola, A. Transcriptome Profiling in Human Diseases: New Advances and Perspectives. *Int. J. Mol. Sci..* **2017**, *18*(8), 1652.

Chen, R.; Mias, G. I.; Li-Pook-Than, J.; Jiang, L.; Lam, H.Y.; Chen, R.; Miriami, E.; Karczewski, K. J.; Hariharan, M.; Dewey, F. E.; Cheng, Y.; Clark. M. J.; Im, H.; Habegger, L.; Balasubramanian, S.; O'Huallachain, M.; Dudley, J. T.; Hillenmeyer, S.; Haraksingh, R.; Sharon, D.; Euskirchen, G.; Lacroute, P.; Bettinger, K.; Boyle, A. P.; Kasowski, M.; Grubert, F.; Seki, S.; Garcia, M.; Whirl-Carrillo, M.; Gallardo, M.; Blasco, M. A.; Greenberg, P. L.; Snyder, P.; Klein, T. E.; Altman, R. B.; Butte, A. J.; Ashley, E. A.; Gerstein, M.; Nadeau, K. C.; Tang, H.; Snyder, M. Personal Omics Profiling Reveals Dynamic Molecular and Medical Phenotypes. *Cell.* **2012**, *148*(6), 1293–1307.

Chen, W.; Cao, H.; Lin, J.; Olsen, N. Zheng SG. Biomarkers for Primary Sjögren's Syndrome. *Genom. Proteom. Bioinform.* **2015**, *13*(4), 219–223.

Choi, M.; Scholl, U. I.; Ji, W.; Liu, T.; Tikhonova, I. R.; Zumbo, P.; Nayir, A.; Bakkaloğlu, A.; Ozen, S.; Sanjad, S.; Nelson-Williams, C.; Farhi, A.; Mane, S.; Lifton, R. P. Genetic Diagnosis by Whole Exome Capture and Massively Parallel DNA Sequencing. *Proc. Natl. Acad. Sci.* **2009**, *106*(45), 19096–190101.

Ciborowski, P.; Gendelman, H. E. Human Immunodeficiency Virus-Mononuclear Phagocyte Interactions: Emerging Avenues of Biomarker Discovery, Modes of Viral Persistence and Disease Pathogenesis. *Curr. HIV Res.* **2006**, *4*(3), 279–291.

Dai, X.; Li, T.; Bai, Z.; Yang, Y.; Liu, X.; Zhan, J.; Shi, B. Breast Cancer Intrinsic Subtype Classification, Clinical Use and Future Trends. *Am. J. Cancer Res.* **2015**, *5*(10), 2929–2943.

Desai, A. A.; Lei, Z.; Bahroos, N.; Maienschein-Cline, M.; Saraf, S. L.; Zhang, X.; Shah, B. N.; Nouraie, S. M.; Abbasi, T.; Patel, A. R.; Lang, R. M.; Lussier, Y.; Garcia, J. G. N.; Gordeuk, V. R.; Machado, R. F. Association of Circulating Transcriptomic Profiles with Mortality in Sickle Cell Disease. *Blood.* **2017**, *129*(22), 3009–3016.

van Dijk, K. D.; Teunissen, C. E.; Drukarch, B.; Jimenez, C. R.; Groenewegen, H. J.; Berendse, H. W.; van de Berg, W. D. Diagnostic Cerebrospinal Fluid Biomarkers for Parkinson's Disease: A Pathogenetically Based Approach. *Neurobiol. Dis.* **2010**, *39*(3), 229–41.

Domon, B.; Aebersold, R. Mass Spectrometry and Protein Analysis. *Science.* **2006**, *312*(5771), 212–217.

Dugan, Jamem.; Pollyea, D. Enasidenib for the Treatment of Acute Myeloid Leukemia. *Expert Rev. Clin. Pharmacol.* **2018**, *11*(8), 755–760.

Dupnik, K. M.; Bean, J. M.; Lee, M. H.; Jean Juste, M. A.; Skrabanek, L.; Rivera, V.; Vorkas, C. K.; Pape, J. W.; Fitzgerald, D. W.; Glickman, M. Blood Transcriptomic Markers of Mycobacterium Tuberculosis Load in Sputum. *Int. J. Tuberc. Lung Dis.* **2018**, *22*(8), 950–958.

Freed, G. L.; Cazares, L. H.; Fichandler, C. E.; Fuller, T. W.; Sawyer, C. A.; Stack, B. C.; Schraff, S.; Semmes, O. J.; Wadsworth, J. T.; Drake, R. R. Differential Capture of Serum Proteins for Expression Profiling and Biomarker Discovery in Pre- and Posttreatment Head and Neck Cancer Samples. *Laryngoscope*. **2008**, *118*(1), 61–68.

Frith, M. C.; Pheasant, M.; Mattick, J. S. The Amazing Complexity of the Human Transcriptome. *Eur. J. Hum. Genet.* **2005**,*13*(8), 894–897.

Fujii, S.; Miyata, A.; Takeuchi, M.; Yoshino, T. Acute Lymphoblastic Leukemia (L3) with t(2;3)(P12;Q27), t(14;18)(Q32;Q21), and t(8;22)(Q24;Q11). *Jap. J. Clin. Hematol.* **2005**, *46*(2), 134–140.

Füzéry, A. K.; Levin, J.; Chan, M. M.; Chan, D. W. Translation of Proteomic Biomarkers into FDA Approved Cancer Diagnostics: Issues and Challenges. *Clin. Proteom*. **2013**, *10*(1), 13.

Gao, X.; Ke, C.; Liu, H.; Liu, W.; Li, K.; Yu, B.; Sun, M. Large-Scale Metabolomic Analysis Reveals Potential Biomarkers for Early Stage Coronary Atherosclerosis. *Sci. Rep.* **2017**, *7*(1), 11817.

Garay, J. P.; Gray, J. W. Omics and Therapy—A Basis for Precision Medicine. *Mol. Oncol.* **2012**, *6*(2): 128–139.

García-Cañaveras, J. C.; Jiménez, N.; Gómez-Lechón, M. J.; Castell, J. V.; Donato, M. T.; Lahoz, A. LC-MS Untargeted Metabolomic Analysis of Drug-Induced Hepatotoxicity in HepG2 Cells. *Electrophoresis*. **2015**, *36*(18), 2294–2302.

Gasperskaja, E.; Kučinskas, V. The Most Common Technologies and Tools for Functional Genome Analysis. *Acta Med. Litu.* **2017**, *24*(1), 1–11.

Gilissen, C.; Hoischen, A.; Brunner, H. G.; Veltman, J. A. Disease Gene Identification Strategies for Exome Sequencing. *Eur. J. Hum. Genet.* **2012**, *20*(5), 490–497.

Giri, K.; Mehta, A.; Ambatipudi, K. In Search of the Altering Salivary Proteome in Metastatic Breast and Ovarian Cancers. *FASEB Bioadv.* **2019**, *1*(3), 191–207.

Gobezie, R.; Kho, A.; Krastins, B.; Sarracino, D. A.; Thornhill, T. S.; Chase, M.; Millett, P. J.; Lee, D. M. High Abundance Synovial Fluid Proteome: Distinct Profiles in Health and Osteoarthritis. *Arthritis Res. Ther.* **2007**, *9*(2), R36.

Godovac-Zimmermann, J.; Brown, L. R.. Perspectives for Mass Spectrometry and Functional Proteomics. *Mass Spectrom. Rev.* **2001**, *20*(1), 1–57.

Gonzalez, M. W.; Kann, M. G. Chapter 4: Protein Interactions and Disease. *PLoS Comput. Biol.* **2012**, *8*(12), e1002819.

Gowda, G. A.; Djukovic, D. Overview of Mass Spectrometry-Based Metabolomics: Opportunities and Challenges. *Methods in Mol. Biol.* **2014**, *1198*, 3–12.

Grabbe, S.; Haas, H.; Diken, M.; Kranz, L. M.; Langguth, P.; Sahin, U. Translating Nanoparticulate-Personalized Cancer Vaccines into Clinical Applications: Case Study with RNA-Lipoplexes for the Treatment of Melanoma. *Nanomedicine*. **2016**, *11*(20), 2723–34.

Gregorich, Z. R.; Ge, Y. Top-down Proteomics in Health and Disease: Challenges and Opportunities. *PROTEOMICS*. **2014**, *14*(10), 1195–1210.

Griffin, J. L. The Cinderella Story of Metabolic Profiling: Does Metabolomics Get to Go to the Functional Genomics Ball? *Philos. Trans. R. Soc. Lond. Biol. Sci.* **2006**, *361*(1465), 147–161.

Guendel, I.; Iordanskiy, S.; Sampey, G. C.; Van Duyne, R.; Calvert, V.; Petricoin, E.; Saifuddin, M.; Kehn-Hall, K.; Kashanchi, F. Role of Bruton's Tyrosine Kinase Inhibitors in HIV-1-Infected Cells. *J. NeuroVirol.* **2015**, *21*(3), 257–275.

Guijas, C.; Montenegro-Burke, J. R.; Domingo-Almenara, X.; Palermo, A.; Warth, B.; Hermann, G.; Koellensperger, G.; Huan, T.; Uritboonthai, W.; Aisporna, A. E.; Wolan, D. W.; Spilker, M. E.; Benton, H. P.; Siuzdak, G. METLIN: A Technology Platform for Identifying Knowns and Unknowns. *Anal. Chem.* **2018**, *90*(5), 3156–64.

Guo, P.; Haque, F.; Hallahan, B.; Reif, R.; Li, H. *Nucleic Acid Ther.* **2012**, *22*(4), 226–245.

Hadi, N. I.; Jamal, Q.; Iqbal, A.; Shaikh, F.; Somroo, S.; Musharraf, S. G. Serum Metabolomic Profiles for Breast Cancer Diagnosis, Grading and Staging by Gas Chromatography-Mass Spectrometry. *Sci. Rep.* **2017**, *7*(1), 1715.

Hamrita, B.; Chahed, K.; Trimeche, M.; Guillier, C. L.; Hammann, P.; Chaïeb, A.; Korbi, S.; Chouchane, L. Proteomics-Based Identification of Alpha1-Antitrypsin and Haptoglobin Precursors as Novel Serum Markers in Infiltrating Ductal Breast Carcinomas. *Clin. Chim. Acta.* **2009**, *404*(2),111–118.

He, Q. Y. Rethinking about the Prevalence, Prevention and Treatment of SARS. *Zhonghua Yi Xue Za Zhi.* **2003**, *83*(12), 1012–1013.

Henry, V. J.; Bandrowski, A. E.; Pepin, A. S.; Gonzalez, B. J.; Desfeux, A. OMICtools: An Informative Directory for Multi-Omic Data Analysis. *Database (Oxford).* **2014**, (0), bau069–bau069.

Hocher, B.; Adamski, J. Metabolomics for Clinical Use and Research in Chronic Kidney Disease. *Nat. Rev. Nephrol.* **2017**, *13*(5), 269–284.

Horgan, R. P.; Kenny, L. C. 'Omic' Technologies: Genomics, Transcriptomics, Proteomics and Metabolomics Author Details. *Obstetr. Gynaecol.* **2011**, *13*(3), 189–195.

Hoy, S. M. Patisiran: First Global Approval. *Drugs.* **2018**, *78*(15), 1625–1631.

Iafrate, A. J.; Feuk, L.; Rivera, M. N.; Listewnik, M. L.; Donahoe, P. K.; Qi, Y.; Scherer, S. W.; Lee, C. Detection of Large-Scale Variation in the Human Genome. *Nat. Genet.* **2004**, *36*(9), 949–951.

Lander, E. S.; Linton, L. M.; Birren, B.; Nusbaum, C.; Zody, M. C.; Baldwin, J.; Devon, K.; Dewar, K.; Doyle, M.; FitzHugh, W.; Funke, R.; Gage, D.; Harris, K.; Heaford, A.; Howland, J.; Kann, L.; Lehoczky, J.; LeVine, R.; McEwan, P.; McKernan, K.; Meldrim, J.; Mesirov, J. P.; Miranda, C.; Morris, W.; Naylor, J.; Raymond, C.; Rosetti, M.; Santos, R.; Sheridan, A.; Sougnez, C.; Stange-Thomann, Y.; Stojanovic, N.; Subramanian, A.; Wyman, D.; Rogers, J.; Sulston, J.; Ainscough, R.; Beck, S.; Bentley, D.; Burton, J.; Clee, C.; Carter, N.; Coulson, A.; Deadman, R.; Deloukas, P.; Dunham, A.; Dunham, I.; Durbin, R.; French, L.; Grafham, D.; Gregory, S.; Hubbard, T.; Humphray, S.; Hunt, A.; Jones, M.; Lloyd, C.; McMurray, A.; Matthews, L.; Mercer, S.; Milne, S.; Mullikin, J. C.; Mungall, A.; Plumb, R.; Ross, M.; Shownkeen, R.; Sims, S.; Waterston, R. H.; Wilson, R. K.; Hillier, L. W.; McPherson, J. D.; Marra, M. A.; Mardis, E. R.; Fulton, L. A.; Chinwalla, A. T.; Pepin, K. H.; Gish, W. R.; Chissoe, S. L.; Wendl, M. C.; Delehaunty, K. D.; Miner, T. L.; Delehaunty, A.; Kramer, J. B.; Cook, L. L.; Fulton, R. S.; Johnson, D. L.; Minx, P. J.; Clifton, S. W.; Hawkins, T.; Branscomb, E.; Predki, P.; Richardson, P.; Wenning, S.; Slezak, T.; Doggett, N.; Cheng, J. F.; Olsen, A.; Lucas, S.; Elkin, C.; Uberbacher, E.; Frazier, M.; Gibbs, R. A.; Muzny, D. M.; Scherer, S. E.; Bouck, J. B.; Sodergren, E. J.; Worley, K. C.; Rives, C. M.;

Gorrell, J. H.; Metzker, M. L.; Naylor, S. L.; Kucherlapati, R. S.; Nelson, D. L.; Weinstock, G. M.; Sakaki, Y.; Fujiyama, A.; Hattori, M.; Yada, T.; Toyoda, A.; Itoh, T.; Kawagoe, C.; Watanabe, H.; Totoki, Y.; Taylor, T.; Weissenbach, J.; Heilig, R.; Saurin, W.; Artiguenave, F.; Brottier, P.; Bruls, T.; Pelletier, E.; Robert, C.; Wincker, P.; Smith, D. R.; Doucette-Stamm, L.; Rubenfield, M.; Weinstock, K.; Lee, H. M.; Dubois, J.; Rosenthal, A.; Platzer, M.; Nyakatura, G.; Taudien, S.; Rump, A.; Yang, H.; Yu, J.; Wang, J.; Huang, G.; Gu, J.; Hood, L.; Rowen, L.; Madan, A.; Qin, S.; Davis, R. W.; Federspiel, N. A.; Abola, A. P.; Proctor, M. J.; Myers, R. M.; Schmutz, J.; Dickson, M.; Grimwood, J.; Cox, D. R.; Olson, M. V.; Kaul, R.; Raymond, C.; Shimizu, N.; Kawasaki, K.; Minoshima, S.; Evans, G. A.; Athanasiou, M.; Schultz, R.; Roe, B. A.; Chen, F.; Pan, H.; Ramser, J.; Lehrach, H.; Reinhardt, R.; McCombie, W. R.; de la Bastide, M.; Dedhia, N.; Blöcker, H.; Hornischer, K.; Nordsiek, G.; Agarwala, R.; Aravind, L.; Bailey, J. A.; Bateman, A.; Batzoglou, S.; Birney, E.; Bork, P.; Brown, D. G.; Burge, C. B.; Cerutti, L.; Chen, H. C.; Church, D.; Clamp, M.; Copley, R. R.; Doerks, T.; Eddy, S.; Eichler, E. E.; Furey, T. S.; Galagan, J.; Gilbert, J. G.; Harmon, C.; Hayashizaki, Y.; Haussler, D.; Hermjakob, H.; Hokamp, K.; Jang, W.; Johnson, L. S.; Jones, T. A.; Kasif, S.; Kaspryzk, A.; Kennedy, S.; Kent, W. J.; Kitts, P.; Koonin, E.; Korf, I.; Kulp, D.; Lancet, D.; Lowe, T. M.; McLysaght, A., Mikkelsen, T.; Moran, J. V.; Mulder, N.; Pollara, V. J.; Ponting, C. P.; Schuler, G.; Schultz, J.; Slater, G.; Smit, A. F.; Stupka, E.; Szustakowki, J.; Thierry-Mieg, D.; Thierry-Mieg, J.; Wagner, L.; Wallis, J.; Wheeler, R.; Williams, A.; Wolf, Y. I.; Wolfe, K. H.; Yang, S. P.; Yeh, R. F.; Collins, F.; Guyer, M. S.; Peterson, J.; Felsenfeld, A.; Wetterstrand, K. A.; Patrinos, A.; Morgan, M. J.; de Jong, P.; Catanese, J. J.; Osoegawa, K.; Shizuya, H.; Choi, S.; Chen, Y. J.; Szustakowki, J. Initial Sequencing and Analysis of the Human Genome. *Nature* **2001**, *409*(6822), 860–921.

Isabel Padrão, A.; Ferreira, R.; Vitorino, R.; Amado, F. Proteome-Base Biomarkers in Diabetes Mellitus: Progress on Biofluids' Protein Profiling Using Mass Spectrometry. *Proteomics – Clin. Appl.* **2012**, *6*(9–10), 447–466.

Jensen, S. O.; Lyon, B. R. Genetics of Antimicrobial Resistance in *Staphylococcus aureus*. *Future Microbiol.* **2009**, *4*(5), 565–582.

Johnson, C. H.; Ivanisevic, J.; Siuzdak, G. Metabolomics: Beyond Biomarkers and towards Mechanisms. *Nat. Rev. Mol. Cell Biol.* **2016**, *17*(7), 451–459.

Kaddurah-Daouk, R.; Zhu, H.; Sharma, S.; Bogdanov, M.; Rozen, S.; Matson, W.; Oki, N. O.; Motsinger-Reif, A. A.; Churchill, E.; Lei, Z.; Appleby, D.; Kling, M. A.; Trojanowski, J. Q.; Doraiswamy, P. M.; Arnold, S. E. Alterations in Metabolic Pathways and Networks in Alzheimer's Disease. *Transl. Psychiatry.* **2013**, *3*(4), e244–e244.

Kanarek, N.; Keys, H. R.; Cantor, J. R.; Lewis, C. A.; Chan, S. H.; Kunchok, T.; Abu-Remaileh, M.; Freinkman, E.; Schweitzer, L. D.; Sabatini, D. M. Histidine Catabolism Is a Major Determinant of Methotrexate Sensitivity. *Nature.* **2018**, *559*(7715), 632–636.

Karamanou, M.; Protogerou, A.; Tsoucalas, G.; Androutsos, G.; Poulakou-Rebelakou, E. Milestones in the History of Diabetes Mellitus: The Main Contributors. *World J. Diabetes.* **2016**, *7*(1), 1.

Kavallaris, M.; Marshall, G. M. Proteomics and Disease: Opportunities and Challenges. *Med. J. Aust.* 2005, *182*(11), 575–579.

Keam, S. J. Inotersen: First Global Approval. *Drugs.* **2018**, *78*(13), 1371–1376.

Kentsis, A. Challenges and Opportunities for Discovery of Disease Biomarkers Using Urine Proteomics. *Pediatr. Int.* **2011**, *53*(1), 1–6.

Kiernan, U. A.; Nedelkov, D.; Nelson, R. W. Multiplexed Mass Spectrometric Immunoassay in Biomarker Research: A Novel Approach to the Determination of a Myocardial Infarct. *J. Proteome Res.* **2006**, *5*(11), 2928–2934.

Kircher, M.; Kelso, J. High-throughput DNA sequencing--concepts and limitations. *Bioessays.* **2010**, *32*(6), 524-36.

Kitamura, Y.; Usami, R.; Ichihara, S.; Kida, H.; Satoh, M.; Tomimoto, H.; Murata, M.; Oikawa, S. Plasma Protein Profiling for Potential Biomarkers in the Early Diagnosis of Alzheimer's Disease. *Neurol. Res.* **2017**, *39*(3), 231–238.

Kolla, V.; Jenö, P.; Moes, S.; Tercanli, S.; Lapaire, O.; Choolani, M.; Hahn, S. Quantitative Proteomics Analysis of Maternal Plasma in Down Syndrome Pregnancies Using Isobaric Tagging Reagent (ITRAQ). *J. Biomed. Biotechnol.* **2010**, *2010*, 1–10.

Korte, A.; Farlow, A. The Advantages and Limitations of Trait Analysis with GWAS: A Review. *Plant Methods.* **2013**, *9*(1), 29.

Kuska, B. Beer, Bethesda, and Biology: How "Genomics" Came into Being. *J. Natl. Cancer Inst.* **1998**, *90*(2), 93.

Lauten, M.; Schrauder, A.; Kardinal, C.; Harbott, J.; Welte, K.; Schlegelberger, B.; Schrappe, M.; von Neuhoff, N. Unsupervised Proteome Analysis of Human Leukaemia Cells Identifies the Valosin-Containing Protein as a Putative Marker for Glucocorticoid Resistance. *Leukemia.* **2006**, *20*(5), 820–826.

Lee, J. M.; Han, J. J.; Altwerger, G.; Kohn, E. C. Proteomics and Biomarkers in Clinical Trials for Drug Development. *J. Proteomics.* **2011**, *74*(12), 2632–2641.

Lema, I.; Brea, D.; Rodríguez-González, R.; Díez-Feijoo, E.; Sobrino, T. Proteomic Analysis of the Tear Film in Patients with Keratoconus. *Mol. Vis.* **2010**, *16*, 2055–2061.

Li, S.; Todor, A.; Luo, R. Blood Transcriptomics and Metabolomics for Personalized Medicine. *Comput. Struct. Biotechnol. J.* **2015**, *14*, 1–7.

López-Pedrera, C.; Villalba, J. M.; Siendones, E.; Barbarroja, N.; Gómez-Díaz, C.; Rodríguez-Ariza, A.; Buendía, P.; Torres, A.; Velasco, F. Proteomic Analysis of Acute Myeloid Leukemia: Identification of Potential Early Biomarkers and Therapeutic Targets. *Proteomics.* **2006**, *6*(S1), S293–S299.

Lundberg, M.; Krogvold, L.; Kuric, E.; Dahl-Jørgensen, K.; Skog, O. Expression of Interferon-Stimulated Genes in Insulitic Pancreatic Islets of Patients Recently Diagnosed With Type 1 Diabetes. *Diabetes.* **2016**, *65*(10), 3104–3110.

Lundstrom, K. Latest Development on RNA-Based Drugs and Vaccines. *Future Sci. OA.* **2018**, *4*(5): FSO300.

Luo, Z.; Li, Y.; Wang, H.; Fleming, J.; Li, M.; Kang, Y.; Zhang, R.; Li, D. Hepatocyte Nuclear Factor 1A (HNF1A) as a Possible Tumor Suppressor in Pancreatic Cancer. Edited by Jose G. Trevino. *PLoS One.* **2010**, *10*(3), e0121082.

Lyons, T. J.; Basu, A. Biomarkers in Diabetes: Hemoglobin A1c, Vascular and Tissue Markers. *Transl. Res.* **2012**, *159*(4), 303–312.

Mangravite, L. M.; Thorn, C. F.; Krauss, R. M. Clinical Implications of Pharmacogenomics of Statin Treatment. *Pharmacogenomics J.* **2006**, *6*(6), 360–374.

Mann, M.; Kelleher, N. L. Precision Proteomics: The Case for High Resolution and High Mass Accuracy. *Proc. Natl. Academy Sci. U S A.* **2008**, *105*(47), 18132–18138.

Manzoni, C.; Kia, D. A.; Vandrovcova, J.; Hardy, J.; Wood, N. W.; Lewis, P. A.; Ferrari, R. Genome, Transcriptome and Proteome: The Rise of Omics Data and Their Integration in Biomedical Sciences. *Brief Bioinform.* **2018**, *19*(2), 286–302.

Maston, G. A.; Evans, S. K.; Green, M. R. Transcriptional regulatory elements in the human genome. *Annu. Rev. Genomics. Hum. Genet.* **2006**, *7, 29-59.*

McCarthy, J. J.; McLeod, H. L.; Ginsburg, G. S. Genomic Medicine: A Decade of Successes, Challenges, and Opportunities. *Sci. Transl. Med.* **2013**, *5*(189), 189sr4.

McRae, A.; Graham, M.; Abedin, T.; Ji, Y.; Yang, H.; Wang, D.; Southern, D.; Andruchow, J.; Lang, E.; Innes, G.; Seiden-Long, I.; DeKoning, L.; Kavsak, P. Sex-Specific, High-Sensitivity Cardiac Troponin T Cut-off Concentrations for Ruling out Acute Myocardial Infarction with a Single Measurement. *CJEM.* **2019**, *21*(1), 26–33.

Minami, S.; Sato, Y.; Matsumoto, T.; Kageyama, T.; Kawashima, Y.; Yoshio, K.; Ishii, J.; Matsumoto, K.; Nagashio, R.; Okayasu, I. Proteomic Study of Sera from Patients with Bladder Cancer: Usefulness of S100A8 and S100A9 Proteins. *Cancer Genomics & Proteomics.* **2010**, *7*(4), 181–189.

Mirza, S. P.; Olivier, M. Methods and Approaches for the Comprehensive Characterization and Quantification of Cellular Proteomes Using Mass Spectrometry. *Physiol. Genomics.* **2008**, *33*(1), 3–11.

Mizzi, C.; Dalabira, E.; Kumuthini, J.; Dzimiri, N.; Balogh, I.; Başak, N.; Böhm, R.; Borg, J.; Borgiani, P.; Bozina, N.; Bruckmueller, H.; Burzynska, B.; Carracedo, A.; Cascorbi, I.; Deltas, C.; Dolzan, V.; Fenech, A.; Grech, G.; Kasiulevicius, V.; Kádaši, Ľ.; Kučinskas, V.; Khusnutdinova, E.; Loukas, Y. L.; Macek, M.; Makukh, H.; Mathijssen, R.; Mitropoulos, K.; Mitropoulou, C.; Novelli, G.; Papantoni, I.; Pavlovic, S.; Saglio, G.; Sertić, J.; Stojiljkovic, M.; Stubbs, A. P.; Squassina, A.; Torres, M.; Turnovec, M.; van Schaik, R. H, Voskarides, K.; Wakil, S. M.; Werk, A.; Del Zompo, M.; Zukic, B.; Katsila, T.; Lee, M. T.; Motsinger-Rief, A.; Mc Leod, H. L.; van der Spek, P. J.; Patrinos, G. P. Correction: A European Spectrum of Pharmacogenomic Biomarkers: Implications for Clinical Pharmacogenomics. *PLoS One.* **2017**, *12*(2), e0172595.

Moorthie, S.; Mattocks, C. J.; Wright, C. F. Review of Massively Parallel DNA Sequencing Technologies. *HUGO J.* **2011**, *5*(1–4), 1–12.

Nagasaki, M.; Yasuda, J.; Katsuoka, F.; Nariai, N.; Kojima, K.; Kawai, Y.; Yamaguchi-Kabata, Y.; Yokozawa, J.; Danjoh, I.; Saito, S.; Sato, Y.; Mimori, T.; Tsuda, K.; Saito, R.; Pan, X.; Nishikawa, S.; Ito, S.; Kuroki, Y.; Tanabe, O.; Fuse, N.; Kuriyama, S.; Kiyomoto, H.; Hozawa, A.; Minegishi, N.; Douglas Engel, J.; Kinoshita, K.; Kure, S.; Yaegashi, N.; ToMMo Japanese Reference Panel Project, Yamamoto, M. Rare Variant Discovery by Deep Whole-Genome Sequencing of 1,070 Japanese Individuals. *Nat. Commun.* **2015**, *6*(1), 8018.

Nicholson, J. K.; Lindon, J. C. Systems Biology: Metabonomics. *Nature.* **2008**, *455*(7216), 1054–1056.

Noguera-Solano, R.; Ruiz-Gutierrez, R.; Rodriguez-Caso, J. M. Genome: Twisting Stories with DNA. *Endeavour.* **2013**, *37*(4), 213–219.

Ong, S. E.; Blagoev, B.; Kratchmarova, I.; Kristensen, D. B.; Steen, H.; Pandey, A.; Mann, M. Stable Isotope Labeling by Amino Acids in Cell Culture, SILAC, as a Simple and Accurate Approach to Expression Proteomics. *Mol. Cell. Proteomics.* **2002**, *1*(5), 376–386.

Ottervald, J.; Franzén, B.; Nilsson, K.; Andersson, L. I.; Khademi, M.; Eriksson, B.; Kjellström, S.; Marko-Varga, G.; Végvári, A.; Harris, R. A.; Laurell, T.; Miliotis, T.; Matusevicius, D.; Salter, H.; Ferm, M.; Olsson, T. Multiple Sclerosis: Identification and Clinical Evaluation of Novel CSF Biomarkers. *J. Proteomics.* **2010**, *73*(6), 1117–1132.

Pardi, N.; Hogan, M. J.; Porter, F. W.; Weissman, D. mRNA Vaccines - a New Era in Vaccinology. *Nat. Rev. Drug Discov.* **2018**, *17*(4), 261–279.

Parker, C. E.; Borchers, C. H. Mass Spectrometry Based Biomarker Discovery, Verification, and Validation--Quality Assurance and Control of Protein Biomarker Assays. *Mol. Oncol.* **2014**, *8*(4), 840–858.

Pawlik, T. M.; Fritsche, H.; Coombes, K. R.; Xiao, L.; Krishnamurthy, S.; Hunt, K. K.; Pusztai, L.; Chen, J. N.; Clarke, C. H.; Arun, B.; Hung, M. C.; Kuerer, H. M. Significant Differences in Nipple Aspirate Fluid Protein Expression between Healthy Women and Those with Breast Cancer Demonstrated by Time-of-Flight Mass Spectrometry. *Breast Cancer Res. Treat.* **2005**, *89*(2), 149–157.

Pedrotty, D. M.; Morley, M. P.; Cappola, T. P. Transcriptomic Biomarkers of Cardiovascular Disease. *Prog. Cardiovasc Dis.* **2012**, *55*(1), 64–69.

Perdriger, A. Infliximab in the Treatment of Rheumatoid Arthritis. *Biologics.* **2009**. *3*, 183–191.

Petricoin, E. F.; Zoon, K. C.; Kohn, E. C.; Barrett, J. C.; Liotta, L. A. Clinical Proteomics: Translating Benchside Promise into Bedside Reality. *Nat. Rev. Drug Discov.* **2002**, *1*(9), 683–695.

Picotti, P.; Rinner, O.; Stallmach, R.; Dautel, F.; Farrah, T.; Domon, B.; Wenschuh, H.; Aebersold, R. High-Throughput Generation of Selected Reaction-Monitoring Assays for Proteins and Proteomes. *Nat. Methods.* **2010**, *7*(1), 43–46.

Precone, V.; Del Monaco, V.; Esposito, M. V.; De Palma, F. D.; Ruocco, A.; Salvatore, F.; D'Argenio, V. Cracking the Code of Human Diseases Using Next-Generation Sequencing: Applications, Challenges, and Perspectives. *Biomed Res. Int.* **2015**, 161648.

Rabbani, B.; Nakaoka, H.; Akhondzadeh, S.; Tekin, M.; Mahdieh, N. Next Generation Sequencing: Implications in Personalized Medicine and Pharmacogenomics. *Mol. Biosyst.* **2016**, *12*(6), 1818–1830.

Ransohoff, D. F. The Process to Discover and Develop Biomarkers for Cancer: A Work in Progress. *J. Natl. Cancer Ins.* **2008**, *100* (20): 1419–1420.

Rao, V. S.; Srinivas, K.; Sujini, G. N.; Kumar, G. N. Protein-Protein Interaction Detection: Methods and Analysis. *Int. J. Proteomics.* **2014**, 147648.

Redon, R.; Ishikawa, S.; Fitch, K. R.; Feuk, L.; Perry, G. H.; Andrews, T. D.; Fiegler, H.; Shapero, M. H.; Carson, A. R.; Chen, W.; Cho, E. K.; Dallaire, S.; Freeman, J. L.; González, J. R.; Gratacòs, M.; Huang, J.; Kalaitzopoulos, D.; Komura, D.; MacDonald, J. R.; Marshall, C. R.; Mei, R.; Montgomery, L.; Nishimura, K.; Okamura, K.; Shen, F.; Somerville, M. J.; Tchinda, J.; Valsesia, A.; Woodwark, C.; Yang, F.; Zhang, J.; Zerjal, T.; Zhang, J.; Armengol, L.; Conrad, D. F.; Estivill, X.; Tyler-Smith, C.; Carter, N. P.; Aburatani, H.; Lee, C.; Jones, K. W.; Scherer, S. W.; Hurles, M. E. Global Variation in Copy Number in the Human Genome. *Nature*. **2006**, *444*(7118), 444–454.

Reinke, S. N.; Gallart-Ayala, H.; Gómez, C.; Checa, A.; Fauland, A.; Naz, S.; Kamleh, M. A.; Djukanović, R.; Hinks, T. S.; Wheelock, C. E. Metabolomics Analysis Identifies Different Metabotypes of Asthma Severity. *Eur. Respir. J.* **2017**, *49*(3), 1601740.

Rieder, M. J.; Reiner, A. P.; Gage, B. F.; Nickerson, D. A.; Eby, C. S.; McLeod, H. L.; Blough, D. K.; Thummel, K. E.; Veenstra, D. L.; Rettie, A. E. Effect of VKORC1 Haplotypes on Transcriptional Regulation and Warfarin Dose. *N. Engl. J. Med.* **2005**. *352*(22), 2285–2293.

Robinson, P. N. 2010. Whole-Exome Sequencing for Finding de Novo Mutations in Sporadic Mental Retardation. *Genome Biol.* **2010**, *11*(12), 144.

Rodrigo, M. A.; Zitka, O.; Krizkova, S.; Moulick, A.; Adam, V.; Kizek, R. MALDI-TOF MS as Evolving Cancer Diagnostic Tool: A Review. *J. Pharm. Biomed. Anal.* **2014,** *95*, 245–255.

Ross, P. L.; Huang, Y. N.; Marchese, J. N.; Williamson, B.; Parker, K.; Hattan, S.; Khainovski, N.; Pillai, S.; Dey, S.; Daniels, S.; Purkayastha, S.; Juhasz, P.; Martin, S.; Bartlet-Jones, M.; He, F.; Jacobson, A.; Pappin, D. J. Multiplexed Protein Quantitation in Saccharomyces

Cerevisiae Using Amine-Reactive Isobaric Tagging Reagents. *Mol. Cell. Proteomics.* **2004**, *3*(12), 1154–1169.

Rouse, R. L.; Zhang, J.; Stewart, S. R.; Rosenzweig, B. A.; Espandiari, P.; Sadrieh, N. K. Comparative Profile of Commercially Available Urinary Biomarkers in Preclinical Drug-Induced Kidney Injury and Recovery in Rats. *Kidney Int.* **2011**, *79*(11), 1186–1197.

Runne, H.; Kuhn, A.; Wild, E. J.; Pratyaksha, W.; Kristiansen, M.; Isaacs, J. D.; Régulier, E.; Delorenzi, M.; Tabrizi, S. J.; Luthi-Carter, R. Analysis of Potential Transcriptomic Biomarkers for Huntington's Disease in Peripheral Blood. *Proc. Natl. Acad. Sci. U S A.* **2007**, *104*(36), 14424–14429.

Sambarey, A.;, Devaprasad, A.; Mohan, A.; Ahmed, A.; Nayak, S.; Swaminathan, S.; D'Souza, G.; Jesura, J. A.; Dhar, C.; Babu, S.; Vyakarnam, A.; Chandra, N. Unbiased Identification of Blood-Based Biomarkers for Pulmonary Tuberculosis by Modeling and Mining Molecular Interaction Networks. *EBioMedicine.* **2017**, *15*, 112–126.

Sas, K. M.; Karnovsky, A.; Michailidis, G.; Pennathur, S. Metabolomics and Diabetes: Analytical and Computational Approaches. *Diabetes.* **2015**, *64*(3), 718–732.

Scalbert, A.; Brennan, L.; Fiehn, O.; Hankemeier, T.; Kristal, B. S.; van Ommen, B.; Pujos-Guillot, E.; Verheij, E.; Wishart, D.; Wopereis, S. Mass-Spectrometry-Based Metabolomics: Limitations and Recommendations for Future Progress with Particular Focus on Nutrition Research. *Metabolomics.* **2009**, *5*(4), 435–458.

Schiess, R.; Wollscheid, B.; Aebersold, R. Targeted Proteomic Strategy for Clinical Biomarker Discovery. *Mol. Oncol.* **2009**, *3*(1), 33–44.

Schmidt, S. The Personalized Medicine Report. *PMC.* 2017. *1*, 1–64.

Schubert, O. T.; Röst, H. L.; Collins, B. C.; Rosenberger, G. Aebersold RQuantitative Proteomics: Challenges and Opportunities in Basic and Applied Research. *Na. Protoc.* **2017**, *12*(7), 1289–1294.

Shendure, J.; Mitra, R. D.; Varma, C.; Church, G. M. Advanced sequencing technologies: methods and goals. *Nat. Rev. Genet.* **2004**, *5*(5), 335–344.

Shulaev, V. Metabolomics Technology and Bioinformatics. *Brief. Bioinform.* **2006**, *7*(2): 128–139.

Siemelink, M. A.; Zeller, T. Biomarkers of Coronary Artery Disease: The Promise of the Transcriptome. *Curr. Cardiol. Rep.* **2014**, *16*(8), 513.

Soriano, V. Hot News: Hepatitis B Gene Therapy Coming to Age. *AIDS Rev.* **2018**, *20*(2), 125–127.

Spear, B. B.; Heath-Chiozzi, M.; Huff, J. Clinical Application of Pharmacogenetics. *Trends Mol. Med.* **2001**, *7*(5), 201–4.

Starobova, H.; S. W. A. H.; Lewis, R, J.; Vetter, I. Transcriptomics in Pain Research: Insights from New and Old Technologies. *Mol. Omics.* **2018**, *14*(6), 389–404.

Stein, L. D. End of the Beginning. *Nature.* **2004**, *431*(7011): 915–916.

Stynen, B.; Tournu, H.; Tavernier, J.; Van Dijck, P. Diversity in Genetic in Vivo Methods for Protein-Protein Interaction Studies: From the Yeast Two-Hybrid System to the Mammalian Split-Luciferase System. *Microbiol. Mol. Biol. Rev.* **2012**, *76*(2), 331–382.

Tan, B.; Qiu, Y.; Zou, X.; Chen, T.; Xie, G.; Cheng, Y.; Dong, T.; Zhao, L.; Feng, B.; Hu, X.; Xu, L. X.; Zhao, A.; Zhang, M.; Cai, G.; Cai, S.; Zhou, Z.; Zheng, M.; Zhang, Y.; Jia, W. Metabonomics Identifies Serum Metabolite Markers of Colorectal Cancer. *J. Proteome Res.* **2013**,*12*(6), 3000–3009.

Tan, S. Z.; Begley, P.; Mullard, G.; Hollywood, K. A.; Bishop, P. N. Introduction to Metabolomics and Its Applications in Ophthalmology. *Eye (Lond).* **2016**, *30*(6): 773–783.

Turkoglu, O.; Zeb, A.; Graham, S.; Szyperski, T.; Szender, J. B.; Odunsi, K.; Bahado-Singh, R. Metabolomics of Biomarker Discovery in Ovarian Cancer: A Systematic Review of the Current Literature. *Metabolomics*. **2016**, *12*(4), 60.

Verbist, B.; Klambauer, G.; Vervoort, L.; Talloen, W.; QSTAR Consortium, Shkedy, Z.; Thas, O.; Bender, A,; Göhlmann, H. W.; Hochreiter, S. Using Transcriptomics to Guide Lead Optimization in Drug Discovery Projects: Lessons Learned from the QSTAR Project. *Drug Discov. Today.* **2015**, *20*(5), 505–513.

Vinores, S. A. Pegaptanib in the Treatment of Wet, Age-Related Macular Degeneration. *Int. J. Nanomedicine*. **2006**, *1*(3), 263–268.

Vizcaíno, J. A.; Csordas, A.; Del-Toro, N.; Dianes, J. A; Griss, J.; Lavidas, I.; Mayer, G.; Perez-Riverol, Y.; Reisinger, F.; Ternent, T.; Xu, Q. W.; Wang, R.; Hermjakob, H. Update of the PRIDE Database and Its Related Tools. *Nucleic Acids Res.* **2016,** *44*(D1): D447–456.

Vogenberg, F. R.; Isaacson, Barash C.; Pursel, M. Personalized Medicine: Part 1: Evolution and Development into Theranostics. P T. **2010,** *35*(10), 560–576.

Vojtová, L.; Zima, T.; Tesař, V.; Michalová, J.; Přikryl, P.; Dostálová, G.; Linhart, A. Study of Urinary Proteomes in Anderson-Fabry Disease. *Ren. Fail.* **2010,** *32*(10), 1202–1209.

Wagner, A.; Cohen, N.; Kelder, T.; Amit, U.; Liebman, E.; Steinberg, D. M.; Radonjic, M.; Ruppin, E. Drugs That Reverse Disease Transcriptomic Signatures Are More Effective in a Mouse Model of Dyslipidemia. *Mol. Syst. Biol.* **2015,** *11*(3), 791.

Wallin, A. K.; Blennow, K.; Zetterberg, H.; Londos, E.; Minthon, L.; Hansson, O. CSF Biomarkers Predict a More Malignant Outcome in Alzheimer Disease. *Neurology.* **2010,** *74*(19), 1531–1537.

Wang, J.; Tan, G.; Han, L. N.; Bai, Y. Y.; He, M; Liu, H. B. *J. Geriatr. Cardiol.* **2017**, *14*(2), 135–150.

Washburn, M. P.; Wolters, D.; Yates, J. R 3rd. Large-Scale Analysis of the Yeast Proteome by Multidimensional Protein Identification Technology. *Nat. Biotechnol.* **2001,** *19*(3), 242–247.

Weckwerth, W. Metabolomics in systems biology. *Annu. Rev. Plant Biol.* **2003**, *54*(1), 669–89.

Weinshilboum, R. M.; Wang, L. Pharmacogenomics: Precision Medicine and Drug Response. *Mayo Clin. Proc.* **2017,** *92*(11), 1711–1722.

Welch, J. S.; Westervelt, P.; Ding, L.; Larson, D. E.; Klco, J. M.; Kulkarni, S.; Wallis, J.; Chen, K.; Payton, J. E.; Fulton, R. S.; Veizer, J.; Schmidt, H.; Vickery, T. L.; Heath, S.; Watson, M. A.; Tomasson, M. H.; Link, D. C.; Graubert, T. A.; DiPersio, J. F.; Mardis, E. R.; Ley, T. J.; Wilson, R. K. Use of Whole-Genome Sequencing to Diagnose a Cryptic Fusion Oncogene. *JAMA.* **2011,** *305*(15), 1577.

de Welzen, L.; Eldholm, V.; Maharaj, K.; Manson, A. L.; Earl, A. M.; Pym, A. S. Whole-Transcriptome and -Genome Analysis of Extensively Drug-Resistant Mycobacterium Tuberculosis Clinical Isolates Identifies Downregulation of EthA as a Mechanism of Ethionamide Resistance. *Antimicrob. Agents Chemother.* **2017,** *61*(12), e01461-17.

Wilkins, M. R.; Pasquali, C.; Appel, R. D.; Ou, K.; Golaz, O.; Sanchez, J. C.; Yan, J. X.; Gooley, A. A.; Hughes, G.; Humphery-Smith, I.; Williams, K. L.; Hochstrasser, D. F. From Proteins to Proteomes: Large Scale Protein Identification by Two-Dimensional Electrophoresis and Amino Acid Analysis. *Biotechnology (N Y).* **1996,** *14*(1): 61–65.

Wishart, D. S. Proteomics and the Human Metabolome Project. *Expert Rev Proteomics.* **2007**, *4*(3), 333–335.

Xiao, Z.; Prieto, D.; Conrads, T. P.; Veenstra, T. D; Issaq, H. J. Proteomic Patterns: Their Potential for Disease Diagnosis. *Mol. Cell. Endocrinol.* **2005**, *230*(1–2), 95–106.

Yadav, S. P. The Wholeness in Suffix -Omics, -Omes, and the Word Om. *J Biomol Tech.* **2007,** *18*(5), 277.

Yamani, M. H.; Taylor, D. O.; Rodriguez, E. R.; Cook, D. J.; Zhou, L.; Smedira, N.; Starling, R. C. Transplant Vasculopathy Is Associated With Increased AlloMap Gene Expression Score. *J. Heart Lung Transplant.* **2007**, *26*(4), 403–406.

Yan, L.; Rosen, N.; Arteaga, C. Targeted Cancer Therapies. *Chin. J. Cancer.* **2011**, *30*(1), 1–4.

Young, S. P.; Kapoor, S. R.; Viant, M. R.; Byrne, J. J.; Filer, A.; Buckley, C. D.; Kitas, G. D.; Raza, K. The Impact of Inflammation on Metabolomic Profiles in Patients With Arthritis. *Arthritis Rheum.* **2013**, *65*(8), 2015–2023.

Yu, L.; Todd, N. W.; Xing, L.; Xie, Y.; Zhang, H.; Liu, Z.; Fang, H.; Zhang, J.; Katz, R. L.; Jiang, F. Early Detection of Lung Adenocarcinoma in Sputum by a Panel of MicroRNA Markers. *Int. J. Cancer.* **2010**, *127*(12), 2870–2878.

Zaleski, M.; Kobilay, M.; Schroeder, L.; Debald, M.; Semaan, A.; Hettwer, K.; Uhlig, S.; Kuhn, W.; Hartmann, G.; Holdenrieder, S. Improved Sensitivity for Detection of Breast Cancer by Combination of MiR-34a and Tumor Markers CA 15-3 or CEA. *Oncotarget.* **2018**, *9*(32), 22523–22536.

Zhang, T.; Watson, D. G.; Wang, L.; Abbas, M.; Murdoch, L.; Bashford, L.; Ahmad, I.; Lam, N. Y.; Ng, A. C.; Leung, H. Y. Application of Holistic Liquid Chromatography-High Resolution Mass Spectrometry Based Urinary Metabolomics for Prostate Cancer Detection and Biomarker Discovery. Edited by Irina U. Agoulnik. *PLoS One.* **2013**, *8*(6), e65880.

Zhao, J.; Zhu, Y.; Hyun, N.; Zeng, D.; Uppal, K.; Tran, V. T.; Yu, T.; Jones, D.; He, J.; Lee, E. T.; Howard, B. V. Novel Metabolic Markers for the Risk of Diabetes Development in American Indians. *Diabetes Care.* **2015,** *38*(2), 220–227.

Zhou, J. Y.; Afjehi-Sadat, L.; Asress, S.; Duong, D. M.; Cudkowicz, M.; Glass, J. D.; Peng, J. Galectin-3 Is a Candidate Biomarker for Amyotrophic Lateral Sclerosis: Discovery by a Proteomics Approach. *J. Proteome Res.* **2010**, *9*(10), 5133–5141.

Zhou, L.; Wang, K.; Li, Q.; Nice, E. C.; Zhang, H.; Huang, C. Clinical Proteomics-Driven Precision Medicine for Targeted Cancer Therapy: Current Overview and Future Perspectives. *Expert Rev. Proteomics.* **2016**, (*4*), 367–381.

Zou, D.; Ma, L.; Yu, J.; Zhang, Z. Biological Databases for Human Research. *Genomics Proteomics Bioinformatics.* **2015**, *13*(1), 55–63.

CHAPTER 15

Current State of Malaria Diagnosis: Conventional, Rapid, and Safety Diagnostic Methods

BARSA BAISALINI PANDA* and RUPENANGSHU KUMAR HAZRA

ICMR-Regional Medical Research Centre, Chandrasekharpur, Bhubaneswar 751023, Odisha, India

Corresponding author. E-mail: barsapanda007@gmail.com

ABSTRACT

Malaria is a severe disease and endemic in most countries in the world. Despite all efforts, it remains a serious public health hazard. For malaria control, effective diagnosis and treatment are important but in most countries, malaria diagnosis remains a challenge. The present methodologies and approaches for malaria diagnosis are practically helpful for the laboratory technician and for the medical practitioner concerned. In comparison with traditional methods, many technologies now available provide new and additional features that can be more effective in diagnosing malaria. Almost all novel technologies for malaria identification involve nucleic acid sequences, immune chromatographic capture, and conjugated monoclonal antibodies, which provide the signs of infection. Quick recognition and prompt treatment of malarial parasite infection is the most significant aim of disease management. The importance and urgency of obtaining results rapidly from the blood samples of patients showing symptoms offer a number of sensitive approaches to the diagnosis of malaria, which are inappropriate for daily laboratory use. New technologies must be fit into the concept of successful vector control strategies, which acquire a much greater meaning today by bringing all the available technologies within a folder.

15.1 INTRODUCTION

Malaria is a severe disease and endemic in most countries in the world (WHO, 2018). India contributes for around 49% of *Plasmodium falciparum* (*Pf*) and 25% of *Plasmodium vivax* (*Pv*) malaria cases of the approximate 145 million total malaria cases confirmed in Southeast Asia (World Malaria Report (WMR), 2017). Despite all control measures, it remains a serious health hazard in the world. Worldwide, malaria cases increasing due to the increase in transmission risk in regions where proper malaria control mechanisms have not been put in place. Environmental factors, socio-economic burden, population movement, international travel, and rising trends of drug-resistant parasite strains also amplify transmission risks.

For malaria control, WHO has two key strategies. The first strategy is vector control, which has two components: long-lasting insecticide-treated mosquito nets and another one is Indoor residual spraying (Michael, 2013). But recently, malaria vectors are seen to be developing resistance to almost all insecticides. Hence, proper operational insecticide resistance management implementation strategies should be developed in order to control malaria transmission. According to WHO's World Malaria Report 2011, over the last few years, malaria transmission has declined as a result of vector control efforts. For the treatment of patients, effective malaria diagnosis is important but certain constraints have to be recognized here: diagnosis becomes difficult when patients have fever but do not show any other symptoms. Without proper diagnosis, patients should not use the antimalarial drug. This might lead to developing drug resistance later. Patients are slowly developing resistance to Chloroquine-based therapy. So, it is not safe to resort to this. Thus, artemisinin combination therapy must be used in this case (Michael, 2013). In most of the countries, malaria diagnosis presents a challenge to laboratories and most of the cases are found in those countries where difficulties are faced in the malaria diagnosis and training of the workforce. Surveillance and identification of malaria are difficult in an inaccessible area, where facilities for keeping and handling of samples do not exist. Effective diagnosis can reduce both transmission and death of malaria. So to diminish malaria transmission burden conventional, rapid, safe, and successful diagnosis is essential. Different types of diagnosis methods are available to detect malaria infections which include microscopic and nonmicroscopic tests. Microscopy procedure is time-consuming, requires proper training and expertise. Advanced technology used in microscopy provides an alternative to light microscopy. The nonmicroscopic test includes Empiric/

syndromic diagnosis, diagnosis through nucleic acid sequences, malaria diagnosis from dried blood spots (DBSs), rapid diagnostic test (RDTs), immunochromatographic test and aldolase enzymes for malaria diagnosis. Those methods are described below.

15.2 MICROSCOPIC SLIDE EXAMINATION

For the detection of parasites in endemic areas, microscopic slide examination is treated as a gold standard and is the most widely used method. Depending upon the type of parasite, environmental factors, and other factors sensitivity of microscopy varies, but, in general, diagnostic sensitivity is not higher than 75%. For patients who have a low level of parasitemia or partial immunity, diagnostic sensitivity is less than 75%. In microscopy, thick and thin films are made from the same blood sample to detect the parasite. A different number of stains are used to detect characteristics of malaria but not all the stains (e.g., Schüffner dots). Take 5 μL of blood on a glass slide to make thick films and allow the blood to dry and then lyse the blood (usually with water) before staining. Giemsa, field's or diluted wright's stain is used in thick blood film to stained red blood cells. During the staining, it is a challenge to identify the parasite within the platelets and WBC due to lysis of the RBC, and experience is required in finding the parasites. Minimum 100 fields should be read before announcing that no malaria parasites are observed. If parasites are found, scan extra 100 fields to increase the chance of recognizing mixed infections. The thin film should always be observed to confirm the same malaria parasite. Thick film procedure is more subtle for parasite detection but it does not differentiate the type of parasites. But it is possible in thin films, because of the RBC fixed monolayer which is available in it. In thin films, methanol fixed and diluted Giemsa stain, buffer water (pH 7.2) is used to highlight the parasite insertions into the RBC. This procedure is preferred all the time because the organisms are easier to count.

Though the microscopy technology is very simple and treated as a gold standard, for slide preparation and identification of malaria smears adequate training and expert technician are required. The advantages of using a microscope are that it (1) identifies mixed infections; (2) determines enormity of parasitemia; (3) requires less laboratory infrastructure; (4) is comparatively less cost-effective. Microscopy also has some drawbacks: (1) it fails to detect low parasitemia; (2) with very high and low parasitemia, errors in interpretation can occur; (3) an expert is required.

15.3 FLUORESCENCE MICROSCOPY DIAGNOSIS

To improve malaria parasite detection in blood films, different types of methods have been used. Particular fluorescent dyes are used to identify parasite nucleic acid, which attaches to the parasite's nucleic acid. The nucleus will strongly fluoresce while the UV light is excited at an appropriate wavelength. The most frequently used fluorochromes are Acridine orange (AO) and Benzothio carboxypurine (BCP). At 490 nm level, these fluorochromes are excited and release green (apple green) and yellow fluorescence. AO staining is used either directly or combined with the concentration method, namely, thick blood film (Traore et al., 1996; Gay et al., 1996).

Using centrifugation, QBC II combines with AO coated capillary tube and an internal float. Then it separates a layer of WBC and plates. In RBC upper layer or in the layer of WBC and platelets, parasites appear. By using a long focal length objective in a fluorescence microscope, parasites can be observed over the capillary tube (Figure 15.1a; Gaye et al., 1999). Kawamoto technique is used as a first excitation filter equestrian in the path of the transmitted light beam, and to pass the stained film this technique permits the excitation wavelength of AO from 470 to 490 nm. For seeing the fluorescing parasite which is stained with AO, a second filter (510 nm) is positioned in the ocular. Strong sunlight or a quartz halogen basis is used as an exciting wavelength source (Kawamoto, 1991).

AO techniques have been employed both in field and laboratory conditions. The QBC fluorescence method is costly and required specialized types of equipment like fluorescence microscope with high-intensity mercury vapor to provide the excitation wavelength and centrifuge machine to separate the cell layers. The microscopist using AO must be able to distinguish fluorescence stained parasites. AO staining reactivity for identification of malaria parasites has been ranging between 41% and 93% which infected with parasite and level of the parasite is <100 parasites/μL (0.002% parasitemia). The specificity of *P. falciparum* infection is more than 93% (Gaye, 1999). AO staining specificity for identification of later stages of development of *Pf* and *Pv* infections is very lower (52%).

BCP is an alternative fluorochrome procedure; it is directly applied to a lysed blood suspension to unfixed thick blood film and stains the DNA of *P. falciparum* parasites (Cooke, 1992). Leukocyte nuclei are badly stained in this procedure (Figure 15.1b). This method has some disadvantages; it needs rapid detection to prevent the disappearance or precipitation of the dye. In comparison with Giemsa staining, its sensitivity and specificity are higher (>95% for *P. falciparum*).

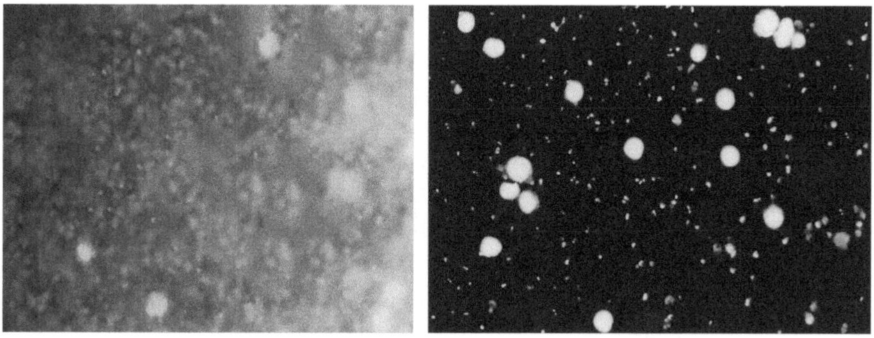

(a) (b)

FIGURE 15.1 a. *P. falciparum* trophozoites stained with AO in the QBC UV fluorescence method. b. *P. falciparum* trophozoites stained with BCP in the fluorescence method.

Fluorochrome compounds more accurately and rapidly detect parasites. However, an important disadvantage of this method is that both AO stain and BCP stains are not able to differentiate among *Plasmodium* species easily. Fluorescence methods for malaria identification need costly equipment and special training (Delacollett and Van der Stuyft, 1994). This method is viable and second to Romanowsky staining. In a remote area, where microscopes especially fluorescence or proper training are not available, we can use nonmicroscopic diagnosis for potential benefits (Antony, 2002).

15.4 EMPIRIC/SYNDROMIC DIAGNOSIS

This method is the most extensively used diagnosis used on the basis of signs, symptoms, and clinical history. In many remote areas, patients with fever are treated for malaria disease without any diagnostic confirmation. Many disadvantages associated with this approach. (1) Remarkable clinical overlap among febrile illnesses may be there and alone fever is unspecific to make any particular diagnosis; (2) if any mix infections are there, then treatment for one does not treat others; (3) asymptomatic malaria is also there which is not treated with this method (Bell et al., 2006; Michael, 2013).

15.5 DIAGNOSIS THROUGH NUCLEIC-ACID SEQUENCES

Multiplex and nested polymerase chain reaction (PCR) methods can be used to detect malaria parasites when morphological identification of parasites (less/μL of blood) is difficult. Different types of PCR assays (like single or multiplex PCR) have been developed to detect malaria parasite DNA from whole blood (Barker et al., 1992; Snounou et al., 1993). These assays are additionally sensitive than other assays for initial diagnosis. A different study shows different PCR results which may be due to different collection techniques, storage of the samples, DNA extraction method and selection of primers, amplification conditions, and analysis of the amplified product. Further approaches in PCR technology have advantages to detect viable or nonviable parasites, stain variation, mutation, and drug resistance strains involved in parasite genes. Improvement in DNA isolation methods and development in a thermocycler may validate the amplification of malaria parasite DNA to be performed within a time frame that is clinically applicable for acute diagnosis in both laboratory and field conditions.

Multiplex, nested, and reverse transcriptase PCR are used to identify four *Plasmodium* species using circumsporozoite and 18s rRNA genes as target genes in *Plasmodium* spp. Among *Plasmodium* spp., the most conserved gene is 18s rRNA gene and for identification of *P. vivax* species-specific region, circumsporozoite gene has been used and combined with specific radiolabeled or fluorescein probes. PCR can detect a low level of parasitemia in the patients' blood and identify it in the parasite level, which are the major advantages of using a PCR-based technique. Infection with five *Plasmodium* spp. or less/μL parasites can be identified with equal specificity and 100% sensitivity (Kawamoto et al., 1996; Snounou et al., 1993). Positive results from sub patent infections are obtained due to the additional sensitivity of PCR.

15.6 MALARIA DIAGNOSIS FROM DBSs

When performing epidemiological or diagnostic surveys in an inaccessible area, the collection, storage, and transportation of samples (Wheat, 2001) present difficulties. By using DBSs, the new method can solve all the problems (Mastronardi et al., 2015). DBSs are one of the most successful ways to collect and store samples from infected persons who live in inaccessible areas. DBSs serve as a very cheap and good source of collection of samples and it can be easily transported to the laboratory. In the DBS method, blood samples from infected humans or persons with symptoms of the disease are collected on filter papers and can use to extract DNA using different techniques. It can be assayed for a large collection of data that include sporozoite detection. Depending upon the pore size and thickness, different types of filter paper brands composed of 100% cellulose are available.

Generally, two filter papers Whatman no 1 and Whatman 3 MM (Whatman, Maidstone, UK) are used widely (Mei et al., 2010). Many techniques and procedures are available to extract DNA from DBSs like tris-EDTA buffer, phosphate buffer saline, methonal, and Chelex-100 method.

In TE-buffer method, firstly TE buffer incubated is at 55 °C for a minute (1 mL of 1 M Tris base [pH 8.0] and 0.2 mL EDTA [0.5 M] for 100 mL solution). Then the punches of DBS were cut from the collected sample and soaked in preheated TE buffer. Then it is incubated at 97 °C and centrifuged at 12,000 rpm at room temperature for 2 minutes. Collected supernatant is kept at 4 °C for later use and −20 °C for longer period use. This DNA extraction method requires less manpower and it takes 45 minutes approximately to complete and Tris buffer is a good source to store DNAs for a longer period in a pH stable state. It requires incubation at 55 °C and 97 °C followed by one successive centrifugation hence a very cheap and effective way to isolate DNA in resource-poor areas.

In methanol extraction method, DBS punches are soaked in 100% methanol. After incubating the sample at room temperature for a minute the methanol is removed and the sample is dried. Methanol can easily dissolve polar solvents and elutes DNA. The samples are dried and autoclaved water is added. The samples are mashed and incubated at 97 °C for 10 min for DNA elution. The samples are then transferred and stored at 4 °C. Methanol is very cheap. The entire process takes approximately 30 minutes and is incubated at 97 °C without the use of centrifugation.

In phosphate buffer saline (PBS) 3 days method, DBS cuts are placed with Lysis buffer (containing KCl, $MgCl_2$, Tris Cl, and Tween-20) containing Proteinase K overnight. Lysis buffer components degrade the cellular integrity and lysed the cell. On day 2, by centrifugation, the cellular debris is removed and in the supernatant equal volume of phenol is mixed. The nonpolar phenol part mixes well with protein and organic portion while DNAs remain in the aqueous phase. After centrifugation the phenol part is discarded and then phenol: Chloroform is added into the mixture in a 1:1 ratio. The density of nonpolar portion is increased by this, which helps in better separation of the aqueous and phenol-chloroform portion. After the separation of these two phases, Chloroform:Isoamyl alcohol in a 24:1 ratio are mixed to remove any foam produced by the proteins and the aqueous phase is separated through successive centrifugations at 10,000 rpm. This whole procedure is carried out at 4 °C to prevent the naked DNA present in an aqueous medium from degrading. On the third day, naked DNA is eluted by applying cold ethanol and isolated from the aqueous phase by centrifugation at 12,000 rpm. Then the ethanol is removed and the samples are again centrifuged with 70% ethanol for best results. The samples are dried and stored by adding autoclaved water at 4 °C. This whole procedure is accomplished at 0 °C to confirm the reliability of naked DNA and to preserve and prevent it from degrading. This procedure is carried out in 3 days at three different temperatures. First, the sample is crushed and washed with PBS. PBS is an isotonic detergent with pH slightly alkaline ranging between 7.4 and 8.0. Being an isotonic solvent it does not interrupt the integrity of the cell and the sample is easily washed because of its detergent ability. This method of DNA extraction, either from DBS or from whole mosquito results in a good yield of DNA, while being advantageous at storing the DNA sample for as long as 6 months after extraction. Extraction of DNA from a DBSs sample is very fast and cost-effective when Chelex-100 molecular grade resins were used. They are very effective at binding DNA which can be separated simply by centrifugation. This technique approximately takes 30 min to complete the DNA extraction procedure from the Chelex beads which are comparatively cheaper than other DNA isolation kits or chemicals and the whole procedure needs only heating and incubation at 99 °C followed by two successive centrifugations at room temperature. Chelex protects the sample from DNases. DNA extracted using Chelex-100 Resin is suitable for PCR. Its sensitivity is higher than 93% (Panda et al., 2019).

15.7 RAPID DIAGNOSTIC TESTS (RDTS)

This test is used in malaria diagnosis. It detects the malaria parasites that are present in blood. RDTs are always used as an alternative to microscopy, particularly in remote inaccessible areas where microscopy service is not provided (Kakkilaya, 2003). From the infected blood, RDTs detect particular antigens produced by malaria parasites. Depending upon the varieties of RDTs, it can detect one species (*P. falciparum*) or other three species (*P. vivax*, *P. malariae*, and *P. ovale*). Finger pricks blood is used for the test. Many manufacturers used prelabeled antibody RDTs kit to detect specific antigens by adding a washing buffer. If the blood contains the target antigen, then the antigen–antibody complex is made and it migrates up to the test strip to bind with a coated antibody. A washing buffer is added to wash the hemoglobin and permits visualization of colored lines formed by the immobilized Ag-Ab complexes. The parasite lactate dehydrogenase (pLDH) test is performed to identify a parasitemia of more than 100–200 parasites/μL and PfHRP2 tests identify asexual parasitemia of more than 40 parasites/μL in the asexual cycle. RDTs may take 5–30 min time depending upon the kit. The cost of RDTs differs from nation to nation and test to test, ranging from \$1.20 to \$13.50 per test. RDTs have cross-reactivity with autoantibodies, namely, rheumatoid factor which gives false-positive results for malaria, this is the main disadvantage of RDTs. Those patients who have +ve rheumatoid factors have shown the false +ve reactions and this false +ve reaction is greater with the PfHRP2 tests using IgG capture antibody linked to the PfHRP2 tests using IgM antibodies, and the pLDH test. PfHRP2 tests using IgG capture antibody, the PfHRP2 tests using IgM antibodies, and the pLDH test have shown 16.5%–83%, 6.6% and 3.3% false-positive reaction, respectively. The sensitivity of RDTs for density >100 parasites/μL blood is >90%. Some studies show >95% sensitivity achieved at parasitemia is nearly equal to 500 parasites/μL.

15.8 IMMUNOCHROMATOGRAPHIC TESTS

The principle of immunochromatography is established on the movement of liquid over the surface of a nitrocellulose membrane. These tests use either polyclonal or monoclonal antibodies to detect antigens of parasites from the outermost blood. Currently, this test targets the *P. falciparum* histidine-rich protein 2 (a pan-malarial Plasmodium aldolase) and the specific lactate dehydrogenase of the parasite. These experiments can be done without using any special equipment, a laboratory, and electricity. Histidine-rich protein 2 is produced from the asexual and gametocytes stages of *Pf,* which is a soluble protein. When antimalarial therapy is started, it is expressed on the surface of the red blood cell membrane and is shown up to 28 days. Several RDTs which are directing PfHRP2 have been also developed. From both the stages that is, sexual and asexual stages of the live parasites, an enzyme called pLDH is produced, which is released from the infected erythrocytes of the parasite and it is a solvable glycolytic enzyme. It has been developed in *P. falciparum, P. vivax, P. ovalae, and P. malariae* species, and for each species, different type of isomers of pLDH exist. A qualitative immunochromatographic dipstick and a quantitative immunocapture assay using monoclonal antibodies have been developed with pLDH as the target. An immunodot assay and a dipstick assay have been also developed using polyclonal antibodies with pLDH as the target (Kakkilaya, 2003).

15.9 ALDOLASE ENZYMES FOR MALARIA DIAGNOSIS

For quick malaria diagnostic tests, aldolase enzymes are considered as targets which are recognized inside the glycolytic pathway of the malaria parasite.

Citric acid cycle is not present in human malaria blood-stage metabolism; therefore, ATP generation fully depends on the glycolytic cycle. In this pathway, aldolase plays an important role as a key enzyme. In *P. berghei* that is, rodent malaria parasite, two glycolytic aldolase enzymes (aldo-1 and aldo-2) are used to examine the particular expression of aldolase isoenzymes. The aldo-1 was alike *P. falciparum* aldolase, while 13% sequence diversity was seen in aldo-2 (Meier et al., 1992). In the blood, asexual stages and sporozoite of malaria parasites were detected by aldo-1 and aldo-2 by using specific antibodies. Monoclonal antibodies created in contradiction of *Plasmodium* aldolase, which is pan-specific and with the combination of HRP-2 it has been used to identify 2 *Plasmodium* species that is, *Pf* and *Pv* in blood. The combined *Pf/Pv* immunochromatographic test (ICT Pf/Pv) was evaluated by Tjitra et al. (1999) and these were designed with capture stripes for both aldolase and HRP-2. They have reported that the negative predictive value and specificity were 98.2% and 94.8%, respectively, for the analysis of *P. vivax* but the positive predictive value of 50% and the overall sensitivity of 75% for *P. vivax* malaria were less than advisable in their study. The sensitivity was 96% for more than 500 parasites/μL (0.01% parasitemia), but for values lower than 500 parasites/μL was only 29%. In a study at Brisbane, Australia, 13 patients were detected as ill with malaria by the ICT *Pf/Pv* test. Eisen and Saul (2000) also displayed less sensitivity for the recognition of *P. vivax* when the parasite count was less than 500 parasites/μL (0.01% parasitemia). They explained that unlike the HRP2 line with *Pf* or *Pv* infection during treatment if the intensity of pan-specific line is declined, it means it became negative though it is positive for a considerable period.

15.10 CONCLUSIONS

Microscopic examination is the gold standard method for the routine laboratory diagnosis of malaria. A trained person can count 50 parasites/μL (0.001% parasitemia) and identify 98% of the species level. This procedure is recognized as laborious, time-consuming, and requiring proper training to achieve the required skills. In the last few years, efforts to replace the traditional but laborious reading of blood films have led to the development of new techniques for the detection of malarial parasites. The sensitivity of new technology is better than microscopy. Methods using fluorescence microscopy have helped improve the sensitivity but not the specificity. For the diagnosis of four malaria parasites at one time, PCR has been established as a sensitive method. In PCR we can detect <5 parasites/μL whose sensitivity and specificity make it an excellent technique in comparison with the nonmicroscopic method. However, PCR takes more time and better infrastructure for the identification of malaria, which is an impractical standard against acute malaria diagnosis. For the surveillance and identification of infectious diseases, filter paper has played an important role in the last 50 years. DBS also plays a key role in the identification of parasitic infections. Extraction of DNA from DBS is very easy and takes less time. DBS is one of the most successful ways to collect and store samples from infected persons who live in inaccessible areas. DBS serves as a very cheap and good means of collection of samples and it can be easily transported to a laboratory.

The immunochromatographic method is easy to handle and more rapid than other nonmicroscopic methods for the diagnosis of malaria. However, WHO pays attention to a number of issues while considering the current status of development of the RDT for malaria. In nonimmune populations, the sensitivity of RDT remains a great problem. For confident diagnosis, parasite densities should be between 5000 parasites/µL (0.1% parasitemia) and 50 parasites/µL (0.001% parasitemia). For *P. vivax* diagnosis, OptiMAL tests have 90%–96% sensitivity and ICT *Pf/Pv* tests have 75%–95% sensitivity. ICT *Pf/Pv* immunochromatographic test with pan malaria antibody failed to identify *P. malariae* symptomatic infection with a parasitemia of 4080 parasites/µL in Papua New Guinea and East Timor (Tijitra et al., 1999). There is no information on experience with this test format for *P. ovale*. The clinical and epidemiological significance of dipsticks recognizing gametocytes of *Plasmodium* is very important. In comparison to pLDH, HRP-2 rapidly detects young forms of *P. falciparum* from sexual stages but not so rapidly in later ones, which is noticed to be lively in young forms. So, a negative RDT result should be confirmed through microscopic examination. The latest generation of RDT provides practical opportunities for diagnosing malaria away from the laboratory and the patient in endemic areas who are unable to come to hospitals. *Pf* diagnosis for both enzyme-based assays and HRP-2 with sensitivity more than 100 parasites/µL (0.002% parasitemia) is fairly good as the maximum number of clinical laboratory staff in nonprofessional laboratories try to diagnose the disease microscopically with restricted experience of malaria cases (Mills et al., 1999). This procedure is useful for diagnosing most of the *non-falciparum* malaria positive cases and act as major supportive procedures for its detection. There are many points to be taken into considerations while studying the techniques for laboratory detection of malaria (Table 15.1). The present discussion on the overview of new malaria tests based on innovative technologies indicates new directions. However, it does not lose sight of the essentiality of studying accurately

TABLE 15.1 Comparison of methods for diagnosing Plasmodium infection in blood. (Modified from Anthony, 2002)

Parameter	Microscopic slide examination	PCR based method	PCR using Dried Blood Spots method	Fluorescence microscopy diagnosis	Rapid Dipstick HRP-2 antigen capture assay	Dipstick pLDH, ICT pf/pv
Sensitivity (parasite/ μL)	Not higher than 75	>100	>93	>50	97-99 (High malaria season) 85.8 (Low malaria season)	97-99 (High malaria season) 85.8 (Low malaria season)
Specificity	5 *Plasmodium* species	5 *Plasmodium* species	5 *Plasmodium* species	*P. falciparum* infection is more than 93% (Gaye et al., 1999)	*P. falciparum* only	Specificity of *P. falciparum*, *P. ovalae* and *P. malariae* is good
Parasitemia	Yes	Yes	Yes	No	Approximate estimation	Approximate estimation
Skill level	Superior	Superior	Superior	Moderate	Low	Low
Time required for result	30-60 mins	2-4 hr	2-4hr	30-60 mins	15 mins	20 mins
Equipment	Microscope	PCR apparatus	PCR apparatus	QBC apparatus or direct fluorescence microscope	Kit	Kit
Cost/test	Low	Moderate	Moderate	Moderate/ low	Moderate	Moderate
Safety	Only personnel trained and approved by the principal investigator should operate the microscope	Biosafety cabinet can be used	Biosafety cabinet can be used	Biosafety regulation should be followed	Non-experienced person can also handle.	Non-experienced person can also handle.

stained thick and thin blood films as the gold standard or standard operating procedure, when malaria is replacing an existing training program for the identification of the different *Plasmodium* species and for recognition of parasitemia lower than the existing edge value of exposure by RDT.

KEYWORDS

- **malaria diagnosis**
- **microscopy**
- **polymerase chain reaction**
- **dried blood spots**
- **rapid diagnostic test**
- **immunochromatographic test**

REFERENCES

Bell, D.; Wongsrichanalai, C.; Barnwell, J. W. Ensuring quality and access for malaria diagnosis, how can it be achieved? *Nat. Rev. Microbiol.* **2006**, *4*, 682–695.

Barker, R. H.; Banchongaksorm, N.T.; Courval, M. M.; Suwonkerd, W.; Rimwungtragoon, K.; Wirth, D. R. A simple method to detect *Plasmodium falciparum* infection in human patients, a comparison of the DNA probe method to microscopic diagnosis. *Am. J. Trop. Med. Hyg.* **1992**, *41*, 266–272.

Cooke, A. H.; Moody, A. H.; Lemon, K.; Chiodini, P. L.; Horton, J. Use of the fluorochrome benzothiocarboxypurine in malaria diagnosis. *Trans. Royal Soc. Trop. Med. Hyg.* **1992**, *87*, 549.

Delacollett, D.; Van der Stuyft, P. Direct acridine orange staining is not a 'miracle' solution to the problems of malaria diagnosis in the field. *Trans. Royal Soc. Trop. Med. Hyg.* **1994**, *88*, 187–188.

Eisen, D. P.; Saul, A. Disappearance of pan-malarial antigen reactivity using the ICT Malaria Pf/Pv kit parallels decline of patent parasitemia as shown by microscopy. *Trans. Royal Soc. Trop. Med. Hyg.* **2000**, *94*, 169–170.

Gay, F.; Traore, B.; Zanoni, J.; Danis, M.; Gentilini, M. Direct acridine orange fluorescence examination of blood slides compared to current techniques for malaria diagnosis. *Trans. Royal Soc. Trop. Med. Hyg.* **1996**, *90*, 516–518.

Gaye, O.; Diouf, M.; Diallo, S. A comparison of thick films, QBC malaria, PCR and PATH *falciparum* malaria test strip in *Plasmodium falciparum* diagnosis. *Parasite.* **1999**, *6*, 273–275.

Kawamoto, F. Rapid diagnosis of malaria by fluorescence microscopy. *Lancet*, **1991**, 624–625.

Kakkilaya, B. S. Rapid diagnosis of Malaria. *Lab Med.* **2003**, *8*, 602–608.

Kawamoto, F.; Miyake, H.; Kaneko, O.; Kimura, M.; Nguyen, T. D.; Lui, Q.; Zhou, M.; Le, D. D.; Kawai, S.; Isomura, S.; Wataya, Y. Sequence variation in the 18S rRNA gene, a target for PCR-based malaria diagnosis, in *Plasmodium ovale* from southern Vietnam. *J. Clin. Microbiol.* **1996**, *34*, 2287–2289.

Mastronardi, C. A.; Whittle, B.; Tunningley, R.; Neeman, T.; Giberto, P. F. The use of dried blood spots sampling for the measurement of HbAIC, a cross sectional study. *BMC Clin. Pathol.* **2015,** *15*, 13.

Michael, L. W. Laboratory diagnosis of malaria, conventional and rapid diagnostic methods. *Arc. Pathol. Lab. Med.* **2013**, *137*, 805–811.

Mei, J. V.; Zobel, S. D.; Hall, E. M.; DeJesus, V. R.; Adam, B. W.; Hannon, W. H. Performance properties of filter paper devices for whole blood collection. *Bioanalysis* **2010**, *2*, 1397–1403.

Meier, B.; Dobeli, H.; Certa, U. Stage-specific expression of aldolase isoenzymes in the rodent malaria parasite *Plasmodium bergei. Mol. Biochem. Lab.* **1992**, *52*, 15–27.

Mills, C. D.; Burgess, D. C. H.; Taylor, H.; Kain, K. C. Evaluation of a rapid and inexpensive dipstick assay for the diagnosis of *Plasmodium falciparum* malaria. *Bull. WHO.* **1999**, *77*, 553–559.

Moody, A. Rapid diagnostic tests for malaria parasites. *Clin. Microbio.Rev.* **2002**, *15*, 66–78.

Panda, B. B.; Meher, A. S.; Hazra, R. K. Comparison between different methods of DNA isolation from dried blood spots for determination of malaria to determine specificity and cost effectiveness. *J. Parasit. Dis.* **2019**. https,//doi.org/10.1007/s12639-019-01136-0.

Snounou, G.; Viriyakosol, S.; Jarra, W.; Thaithong, S.; Brown, K. N. Identification of the four human malarial species in field samples by the polymerase chain reaction and detection of a high prevalence of mixed infections. *Mol. Biochem. Parasitol.* **1993**, *58*, 283–292.

Tjitra, E.; Suprianto, S.; Dyer, M.; Currie, B. J.; Anstey, N. M. Field evaluation of the ICT malaria Pf/Pv immunochromatographic test for detection of *Plasmodium falciparum* and *Plasmodium vivax* in patients with presumptive clinical diagnosis of malaria in eastern Indonesia. *J. Clin. Microbiol.* **1999**, *37*, 2412–2417.

Wheat, P.F. History and development of antimicrobial susceptibility testing methodology. *J. Antimicrobiol. Chemother.* **2001**, *48(1)*, 1–4.

PART III

Nanotechnological Intervention in Life Sciences

CHAPTER 16

Current Perspective of Biofunctionalized Nanomaterials in Biology and Medicine

NAMITA BHOI[1] and ISWAR BAITHARU[2*]

[1]Nano Research Centre, School of Chemistry, Sambalpur University, Burla, Odisha, India

[2]P.G. Department of Environmental Sciences, Sambalpur University, Burla, Odisha, India

[*]Corresponding author. E-mail: iswarbaitharu@suniv.ac.in

ABSTRACT

Nanomaterials have asserted their position in almost every field due to their wide applicability. While many nanomaterials find direct application in the physical and chemical industry, their applications in pharmacology and medicine require proper modifications to make them biocompatible through a process called biofunctionalization. Biofunctionalization is a fundamental technique for permanent or temporary modification of materials by changing the properties or modification of the surface to have a biologically compatible function. Synthesis and subsequent biofunctionalization of a nanomaterial for a specific biomedical application basically depend on physicochemical and biological properties of nanomaterial. Biofunctionalization can be done using the top-down or bottom-up approach. The top-down approaches such as grinding, ball millings, and heating are the most popular physical processing methods for large scale industrial production of biofunctionalized nanomaterials that are further engineered for biomedical uses. In the bottom-up approach, physical processes like laser ablation, physical vapor deposition, and chemical processes like chemical vapor deposition, self-assembled monolayer formation are involved to make them biocompatible. Functionalization of nanomaterials surfaces implicates liposomes, polymer drug conjugates, dendrimer, polymeric nanoparticle, nucleic acid-based nanoparticles, and quantum dots for targeted drug delivery. To engineer surfaces of these nanoparticles, various biocompatible targeting ligands categorized as organic nanocarriers such as liposomes, dendrimer, polymeric nanoparticle, peptides, aptamers, and inorganic nanoparticles like metal nanoparticles are primarily used. Surface functionalization of nanomaterials is based on the basic principle of noncovalent and covalent interactions. The noncovalent approaches include the adsorption phenomenon in which the targeting ligand is adsorbed on the surface of the nanoparticle through noncovalent forces like the electrostatic force of attraction, hydrophobic interactions, and hydrogen bonding. This chapter intends to delineate the processes and mechanisms involved in the biofunctionalization of nanomaterials and highlight their potential applications in medicine and pharmacology.

16.1 INTRODUCTION

Nanotechnology has asserted its position in almost every major field due to its wide applicability. Bio-nanotechnology is an interdisciplinary science that utilizes biological starting materials as well as biological design or fabrication principles in biotechnology and medicine (Fakruddin et al., 2012). For the targeted delivery of therapeutic agents, the development of nanoparticles has introduced new opportunities for applications in pharmaceuticals and biomedical sciences. Recent advancement in the functionalization of nanomaterials has resulted in remarkable progress in the field of detection, imaging, and the targeted drug delivery and also in the treatment of various diseases. Several multifunctional devices developed using new imaging agents are now capable of dealing with various difficulties encountered during targeted delivery of drugs to infectious cells or tissues. However, very often physiochemical methods used in functionalization process lead to the production of nanomaterials with reduced biocompatibility. Biofunctionalization is a fundamental technique that involves permanent or temporary modification of materials by changing their surface properties to have biologically compatible functions. The biofunctionalization can furnish the nanomaterials with enhanced biocompatibility and high affinity toward receptors or proteins important for biological recognition. Synthesis and subsequent biofunctionalization of a nanomaterial for a specific biomedical application basically depend on the physicochemical and biological properties of nanomaterials. The properties of nanomaterials such as small size and large surface area make them different from their large counterparts and offer excellent platforms for designing biosensing devices. This chapter narrates the current advancement on the biofunctionalization of nanomaterials with appropriate biological agents to develop greater biocompatibility for various biomedical applications.

16.2 MECHANISM OF BIOFUNCTIONALIZATION OF NANOMATERIALS

For the biofunctionalization of nanomaterials, their active surfaces need to be engineered prior to various therapeutic applications. Numerous stabilizing agents are used for the purpose of surface functionalization of nanoparticles. Such agents are classified as (1) monomeric stabilizers, for example, thiol group, (2) inorganic materials, for example, silica, gold, and (3) organic polymer such as polyethylene glycol (PEG), polyvinyl alcohol, DNA molecules, peptides, polysaccharides (Kettler et al., 2014). The biofunctionalization of nanoparticle (NP) surfaces generally depends on the biocompatibility of the cell surfaces (Pelaz et al., 2015). The bioengineering of surfaces of nanomaterials results in several advantages such as (1) targeted delivery of therapeutics, (2) selective binding to the surface required receptors, (3) controlled release of drug, (4) endocytosis, (5) biodistribution, (6) prolonged retention time in blood, (7) maintenance of the monodispersity of colloid (Raliya et al., 2016).

Surface functionalization of nanomaterials is based on the basic principle of noncovalent and covalent interactions.

Noncovalent coupling: The noncovalent approaches include an adsorption phenomenon (physisorption) in which the targeting ligand is adsorbed on the surface of nanoparticle through noncovalent forces like the electrostatic force of attraction, hydrophobic interactions, and hydrogen bonding. Noncovalent approaches involve a process named steric stabilization (refers to the stabilization of colloidal nanoparticles).

Covalent coupling: On the other hand, ligands possessing different functional groups are adsorbed on the surface of nanomaterials on the basis of covalent bond interaction through a process called chemisorption forming a self-assembled monolayer.

16.3 BIOCONJUGATION OF NANOPARTICLE FOR APPLICATION IN DIAGNOSTICS

Over time, there are several nanoparticle platforms that have been studied so far enabling them to be used in various medical fields. There are various nanoparticle platforms that involve polymeric nanoparticles liposomes, nanoshells, polymer–drug conjugates, dendrimer, etc. Nanoparticle platforms such as liposomes and the polymer–drug conjugates are responsible for more than 80% of the available nanoparticles as therapeutics in biomedical use.

16.3.1 LIPOSOMES

These are nano- to micro-sized spherical vesicles that contain a bilayer membrane structure composed of natural or synthetic amphiphilic lipid molecules (Figure 16.1(A); Sercombe et al., 2015). Liposomes have been used widely as pharmaceutical carriers in the field of clinical research because of their unique advantages like biocompatibility and biodegradable composition. Liposomes are used to encapsulate both hydrophobic and hydrophilic drugs and are utilized widely as excellent therapeutic carriers. To extend its circulation half-life, polymers like PEG can be used to cover the liposomes to minimize the harmful effects on living cells. Further, the liposomes can also be coated with a polymer that enables them to carry a functional group for ligand-targeted drug delivery to affected tissue or organ system.

16.3.2 POLYMER–DRUG CONJUGATES

Polymer–drug conjugates have been extensively studied as one of the important nanoparticle-mediated drug delivery platforms. The polymer in these complexes is usually bound to a targeting ligand which is successively used as an excellent nanocarrier for targeted delivery of therapeutics. Compounds like PEG are popularly used as drug conjugates, known to enhance the serum stability and solubility of drugs whereas it reduces the cell-mediated immune response. Several other linear polymers used as a carrier of therapeutics include polyglutamic acid, *N*-(2-hydroxypropyl) methacrylamide, polysaccharide, and poly(allylamine hydrochloride) (Davis et al., 2015). The advantages of polymer drug conjugate are to extend the circulation half-life of the drug molecules for longer durations and also reduce the process of endocytosis (Lv et al., 2016). However, there are major challenges such as induction of cell-mediated immune response, less drug-carrying ability, uncontrolled release of the drug, challenges in the synthesis of these conjugates, and also the polymer can cause harmful effects on the human system which need special attention.

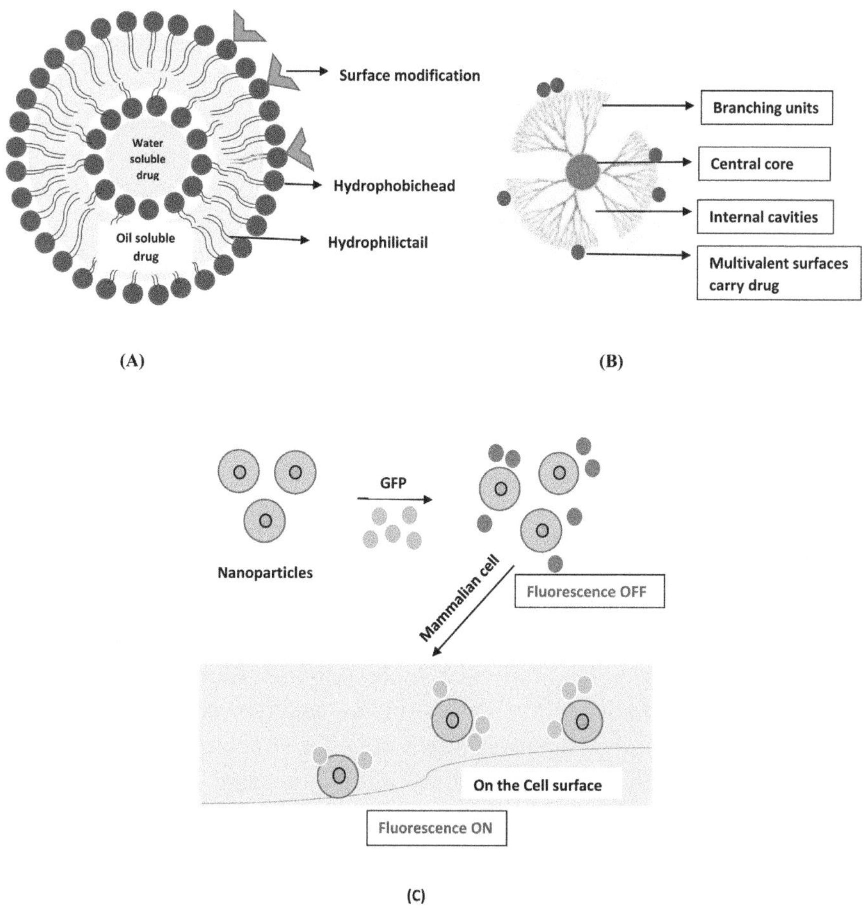

FIGURE 16.1 (A) Liposome; (B) Dendrimer; (C) Represents a complex formed by binding of a protein with a (quenched) GFP-NP (Green Fluorescent Protein-Nanoparticle) showing the presence or absence of fluorescence in a mammalian cell surface. (Bajaj et al. 2010)

16.3.3 DENDRIMER

Dendrimers are repetitively branched, globular, and macromolecular structures having size ranges from 2.5 to 10 nm. As shown in Figure 16.1 (B), these are characterized by three major structural components that include the central core, the branching units whose multivalent surfaces carry the drug, and the internal cavities composed of several natural or synthetic building blocks such as sugars, nucleic acids, amino acids, peptides, etc. Dendrimers are found with wider application because of their properties like monodispersed size, large surface area, the ability of surface functionalization, and long term stability. In biology and medicine, polyamidoamine and polypropylene imine are commonly utilized dendrimers. Dendrimer can either be loaded with drugs within the internal cavities or in their central core through hydrophobic interaction, hydrogen bonding, or by a chemical bond. The branching units of the dendrimer can carry drugs through covalent conjugation or electrostatic adsorption. The dendrimer surfaces can also be readily modified by insertion of new chemical functional groups or molecular targeting groups which make them suitable to be used as detecting and imaging agents, and in therapeutic attachment sites (Mintzer et al., 2011). It is also possible to make dendrimer water soluble, by functionalizing their outer shell with charged species or other hydrophilic groups.

16.3.4 BIOCONJUGATION OF POLYMERIC NANOPARTICLE

The biodegradable polymeric nanoparticles have been used as an excellent therapeutic carrier. These are generally synthesized by forming self-assembly of copolymers consisting of two or more polymer blocks with different hydrophobicity (Khalid et al., 2017). In an aqueous environment, these copolymers are rapidly organized into a micellar structure having a central core. Polymeric nanoparticles have the ability to carry biomolecules like proteins, nucleic acids macromolecules inside their cavity and also act as a carrier of both water-soluble and insoluble drugs (Torchilin et al., 2007). To functionalize the polymeric micelle surface, PEG is the most commonly used polymer. Diseases like tuberculosis can be treated by using antibiotics coated with PLGA NPs (polylactide-*co*-glycolide) (O'Hara et al., 2000).

16.3.5 NANOSHELL

These are usually spherical NPs consisting of a dielectric core surrounded by a layer of thin metallic shell, mostly gold. The metal layer absorbs a specific wavelength of radiation and subsequently produces heat that destroys the tumor cells without causing any harm to the normal cells (Singh, 2018). The gold nanoshells having silica core use near-infrared radiation which can be used in the diagnosis and treatment of cancer, which also enhances optical phenomena like light absorption, scattering, fluorescence, refraction, etc. (Huang et al., 2007).

16.3.6 NUCLEIC ACID-BASED NANOPARTICLES

These include multivalent DNA and RNA molecules that can be used as a base in designing nanocarriers for therapeutic approaches. Multivalent surface of the DNA macromolecules can be used as a therapeutic nanocarrier for targeted delivery of drugs and genes, and also for bioimaging (Xiao et al., 2019). Similarly, RNA macromolecules (25–40 nm) with a trivalent central core, aptamers, and siRNA (small interfering RNA) in targeted drug delivery applications are now gaining more attention in biomedicine research.

16.3.7 BIOCONJUGATION OF QUANTUM DOTS

Quantum dots (QDs) are semiconducting colloidal nanocrystals having a diameter within 1–20 nm. These often show unique size-dependent electro-optic characteristics. Due to the small size of QDs, it is useful for various imaging as well as sensing application in biology and medicine (Abbasi et al., 2016). QDs usually have longer life spans, a higher ability to produce bright color light, and are resistant to photobleaching. QDs can be functionalized by binding several biomolecules either by attachment on the surface of QDs directly or by a surface that stabilizes it by forming a crosslinker between the biomolecules and the QDs' surfaces. The biomolecules should bind either through covalent or noncovalent coupling. Covalent coupling involves the use of thiol coupling chemistry that works by replacing thiol acids with thiolated biomolecules that reside on the surface of the QDs. These strategies can be used to attach proteins such as antibodies (Wang et al., 2002), oligonucleotides such as DNA (Mitchell et al., 1999) on the surface of QD. On the other hand, noncovalent ligand-QD coupling uses electrostatic or hydrophobic forces such as electrostatic coupling. It is also used to bind the positively charged protein on the surface of oppositely charged alkyl-carboxyl-capped QDs. (Lin et al., 2003).

16.4 THE PROCESS INVOLVED IN THE USE OF BIOFUNCTIONALIZED NANOMATERIALS IN MEDICAL DIAGNOSIS

Many unique properties of NPs provide advantages over several current techniques in the field of biomedical research. The development of various targeting ligands brought about progress in the field of drug discovery and targeted delivery of therapeutics. The various in vivo processes involved in molecular targeting of biofunctionalized NPs include active targeting and passive targeting. Targeted drug delivery is known to improve efficiency by reducing the toxic side effect in the field of biology and medicine.

16.4.1 PASSIVE TARGETED DRUG DELIVERY

Nanomaterials can be successfully utilized in the diagnosis of cancer. Tumor tissue vasculature is different from normal tissue in its arrangement, that is, it has an abnormal disorganized structure. Also, the tumor microenvironment is characterized by high vascular density, high permeability, and impaired lymphatic drainage. Passive targeting takes into account the advantages of the enhanced permeability and retention (EPR) effect that result in an increased concentration of NPs near the tumor tissue as compared to normal tissue. The passive targeting process is mediated by properties such as the NPs particle's composition, size, shape, and surface characteristics. Thus, by optimizing the physicochemical characteristics of NPs, it can be delivered to the targeted cells or tissues without causing any harm to normal tissue.

16.4.2 ACTIVE TARGETED DRUG DELIVERY

In case of active targeting, ligand should be used for enhancing the targeted delivery of NPs in a more specific manner. There are several methods of active targeting; one such method includes the bioconjugation of specific targeting ligands to NPs that are complementary to the receptors present on the cell surfaces (Glavin et al. 2011). This can be achieved through both covalent and noncovalent interactions. Other methods involve pH-responsive drug delivery, redox potential based mechanism. Targeting ligands for active targeted drug delivery include small molecules, peptides, monoclonal antibodies and their fragments, and nucleic acids such as aptamers.

16.4.2.1 MONOCLONAL ANTIBODIES

Monoclonal antibodies have the ability to bind only one epitopic site of an antigen with high affinity and specificity. For molecular targeting purposes, artificially engineered monoclonal Antibodies are most commonly used.

They generally elude the immune system. Several engineered monoclonal antibodies used as therapeutics include rituximab, trastuzumab, cetuximab, and bevacizumabuse to target various disease processes. Monoclonal antibodies have several limitations because of their large size, poor permeability, potential immunogenicity, and need surface modifications to be functional, have high production cost and decreased efficiency with successive use make them a less preferred option in drug delivery (Weinberg et al., 2005).

16.4.2.2 APTAMERS

These are short oligonucleotide ligands; nucleic acid such as DNA or RNA strands that specifically bind to target molecules such as proteins, peptides, carbohydrates, and also living cells. Aptamers have the ability to adopt various shapes by folding through intramolecular interaction and thus can bind to a variety of ligands. Using a process called SELEX (Systemic Evolution of Ligands by a chemical process called Exponential Enrichment), aptamers that can specifically bind to the target molecule can be selected for various biomedical applications. Aptamers are widely used as targeting ligands due to their unique advantages like small size (~15 kD) and hence large surface area, nonimmunogenic, good biocompatibility, and also better solubility and permeability. More than 200 aptamers have been isolated so far. Conversely, the major disadvantages of aptamers include their high manufacturing cost and their plasma instability.

16.4.2.3 SMALL MOLECULES

Because of the small size, good permeability, low-cost production, and easy modification, small molecules have an increasing interest as a class of targeting molecules. Folic acid (folate) is one of the most extensively studied small molecules targeting moieties in targeted drug delivery. In cancer cells, folate receptors are commonly overexpressed, so these ligands are often used for targeting cancer cells (Marchetti et al., 2014). The folates have a strong binding affinity toward the folate receptors (KD = $\sim 10^{-9}$ M); forming different conjugates that allow therapeutic drugs to reach the cancer cells without causing any damage to the unaffected cells. It has been extensively used as a targeting ligand by combining with several carriers of therapeutics such as liposomes, nanoshells, polymer NPs, dendrimer, nucleic acid-based NPs, etc.

16.4.2.4 PEPTIDES

Alternative targeting molecules such as peptides are extensively used as compared to the conventional antibodies due to their moderate sizes, large surface to volume ratio, low immune response (immunogenicity), high affinity, stability, and they can be easily synthesized. Peptides are identified as targeting ligands by using peptide phage libraries (~10^{11} different peptide sequences), bacterial peptide display library, plasmid peptide library, and several other chemical synthetic peptide library technologies. Peptides serve as a more effective source for targeted drug delivery. For instance, a cyclic peptide called cilengitide which binds to integrins is now in Phase II clinical trials for the treatment of lung cancer and pancreatic cancer. The only drawback associated with peptides is their vulnerability to proteolysis.

16.5 APPLICATIONS OF BIOFUNCTIONALIZED NANOMATERIALS IN BIOLOGY AND MEDICINE

16.5.1 AS SENSOR

NPs are often used in sensor applications including detection of analytes at very low concentrations, detection of pathogens, detection and separation of various cells, and also in the detection of biological functions. Besides, these are also useful in determining the toxic content in water, drug residues in food, blood glucose levels in diabetes patients.

16.5.1.1 DETECTION OF ANALYTES

Biofunctionalization of NPs with various targeting ligands is useful in sensing and detecting analytes with high specificity. By using various sensing methods, specific binding of biological components such as proteins, antibodies, nucleic acids, DNA, RNA with the analytes can be detected even at very low concentrations. NPs due to their unique size-dependent properties as well as surface properties can be useful for more efficient and rapid detection of these analytes (Saha et al., 2012). Gold NPs are now widely used for the detection of various analytes. Conjugated polymer NPs are formed when the fluorescent protein is bound with the gold NP. Binding of gold NP to fluorescence protein decreases the intensity of emitted light representing fluorescence OFF but when it is detached from the NP surface, it again shows fluorescence response that represents fluorescence ON (Figure 16.1C). In these systems, according to their different fluorescence responses, the target protein is detected which binds with gold NPs on the cell surface (De et al., 2009). For the detection and identification of biomolecules such as glucose, hemoglobin and also bacteria, and viruses a process known as SERS (surface-enhanced Raman scattering) involving gold NPs can be used. Nanosensors carry the advantages of various magnetic NPs such as superparamagnetic iron oxide NPs (SPIONs). Most of the biological samples show a small susceptibility toward magnetic properties. Biofunctionalization of SPIONs with amino groups allows its multifunctional surface to carry sulfhydryl-bearing molecules for targeted delivery of therapeutics. Relaxation sensors having magnetic NPs can be used to detect various biomolecules such as DNA and proteins (Haun et al., 2010).

16.5.1.2 DETECTION AND SEPARATION OF PATHOGEN

For the detection of bacteria, one of the most commonly used methods involves the use of magnetic nanoparticle as biosensor. Nanoparticles are functionalized with antibodies in contrast to antigens on the bacterial surfaces. The Weissleder group has reported the use of NPs coated with antibodies in a microfluidic device for the detection of bacteria. Recent report by Liong et al. (2013) focus on strategies based on the principle of magnetic barcoding without any antibodies to detect mutations in single-gene (Liong et al., 2013). To label various bacteria, NP functionalized with small molecules can be effectively used. *E. coli* bacterial strains developed by using magnetic glyco-NPs remove 88% of bacteria from the sample by utilizing the bacterial interaction with carbohydrates on mammalian cell surfaces (Pan et al., 2012). Kell et al. (2008) described that antibiotics such as vancomycin-coated SPIONs can be used to detect as well as to effectively separate various Gram-positive and Gram-negative bacteria through magnetic capture strategy. QDs can also be used as pathogen sensors to detect *E. coli* bacterial species from a diverse bacterial species. Also, it can separate as little as 10 bacteria/mL by using streptavidin-coated QDs from the experimental samples (Gilmartin et al., 2012).

16.5.1.3 DETECTION AND SEPARATION OF VARIOUS CELLS

Biofunctionalization of NPs plays an important role to effectively separate and isolate the required cells from the mixture of different types of cells.

Circulating tumor cells can be identified and detected by magnetic NP immunomagnetic techniques which involve a ligand-receptor-based mechanism. It has been reported that biofunctionalized iron Oxide NPs with tumor markers such as anti-HER2 (human epidermal growth receptor 2) can be used for the separation of 73.6% of HER2 over-expressed cancer cells (Xu et al., 2011). Targeted polymer-coated gold NPs can be used to quantify circulating tumor cells along with WBC count (Dreaden et al., 2012). A technique called biomimetic nanotechnology can also be used for this purpose.

16.5.2 AS IMAGING AGENT

In molecular imaging, NPs have been developed as unique contrast agents and biomarkers. It also induces cellular reactions accompanied by various diseases like cancer and cardiovascular diseases. Due to its small size and large surface area, it can carry a huge number of imaging agents for more efficient imaging. Targeted imaging includes the entry of NPs passively into tumor cells through the EPR effect or it can lead to the accumulation of contrast agents more near the targeting tumor tissues as compared to the normal tissues and use it as amplifiers for targeted imaging in vivo.

Dye-doped nanoparticles are formed by incorporating organic or inorganic dyes into a polymer or silica-based particle. They can be either incorporated into the nanoparticles or can be bound to the surface of it by covalent or noncovalent interactions and used for imaging applications. Dyes can be protected from oxygen which in turn enhances the photostability of nanoparticles. For example, as compared to an ordinary dye, incorporation of pyrene dyes into polystyrene particles increases the emission intensity by 40 times at identical concentrations using a normal microemulsion technique (Gao et al., 2002). Inorganic NPs include QDs (fluorescent) are used in targeted imaging of cells because they can release specific wavelengths of light in which autofluorescence of tissue is minimized and penetration depth of the excited light is increased. It also includes various applications such as molecular imaging of lymph nodes, embryonic stem cells, as well as tumor cells. Gold NPs functionalized with small organic molecules can be widely used for in vivo cancer imaging. Gold NPs conjugated with anti-HER2 antibodies using cyanine dye can be used to detect HER2 positive breast cancer (Maiti et al., 2012). Magnetic NPs can be used as favorable contrast agents in magnetic resonance imaging (MRI), a biomedical technique that involves interaction between the magnetic nuclei. Also, SPIONs functionalized with dextran or carboxydextran can be used as a contrast agent in MRI for the detection of tumor (Unterweger et al., 2018).

16.5.3 INVESTIGATION OF BIOLOGICAL PROCESSES

To overcome the disadvantages of conventional methods, various nonconventional processes use biofunctionalized NPs to initiate cell signaling to promote protein productions. The QDs semiconducting properties offer very good tunable and photophysical characteristics. Some important biological phenomena that use QD include fluorescence, immunostaining of biomolecules such as proteins and cytoskeletal filaments, and also used to display the dynamics of individual molecules inside a living cell (Wang et al., 2016). Biofunctionalized NPs with target-specific ligands can be developed by coating the surface with a biocompatible active layer using a magnetic field (Etoc et al., 2013). Magnetic NPs complexed with protein are introduced into the cells, which can significantly bind to target proteins on the cell surfaces and help in explaining the factors that regulate the cell morphology by stimulating various signal transduction pathways. Also, IONPs doped with zinc can be conjugated with antibodies for the development of magnetic switches (Cho et al., 2012). The three-dimensional arrangement of cells plays an essential role in cell culture for basic biological study. In vitro conventional 2D cell cultures are of low performance as they are not able to mimic the in vivo conditions, therefore 3D cell cultures are now developed to obtain more appropriate physical and biochemical microenvironment. Magnetic levitation method is a unique approach for growing 3D tissue by the introduction of cells followed by the treatment with a group of magnetic NP such as iron oxide NPs, gold NPs, and also polymers (Souza et al., 2010).

16.5.4 IN THERAPEUTICS

In therapeutics, the drug delivery approach provides an essential mechanism to enhance the effectiveness of drugs or medicine through improved pharmacokinetics and biodistribution of nanoparticles. Metal NPs can be widely used as carriers that can carry genes and drugs either through a passive or active pathway or by direct routes. Emphasis on the active targeting approach continues to yield results along with the structures and functionalization of metal NPs through different coatings. Herceptin conjugated porous Fe_3O_4 NPs can be released into the breast cancer cells under the low pH conditions inside the affected cells which lead to further opening of the pores so that cisplatin can be released into the cells in a controlled manner (Cheng et al., 2009).

16.5.5 EFFECTS OF NPS ON BIOLOGICAL PROCESSES

Biological systems interact with various NPs which are affected by the various bulk as well as surface characteristics of NPs including their shape, size, surface area, functional groups, charge on the surface, and also arrangement both in vitro and in vivo conditions. Due to the EPR effect and on the basis of the small size of NP, molecules of specific sizes are more prone toward the tumor tissue as compare to the normal tissues. NPs can be delivered actively or passively to the tumor cells. However, the in vitro cellular uptake of NPs can be affected by the various phases of the cell cycle. It is also found that cellular uptake of various NP and distribution inside the cell occurs by fragmentation between the daughter cells when the cell splits.

Here is the list of some important NPs and their biofunctionalization using various targeting ligands for their use in biomedical applications given in Table 16.1.

16.6 BIOFUNCTIONALIZED NANOMATERIALS AND ITS LIMITATIONS

The major challenges of biofunctionalization are (1) maintenance of monodispersity of nanoparticles by reducing their aggregation, (2) reduction of unspecified adsorption onto the exterior of the nanoparticle, and (3) reduction of cytotoxicity. To attain these criteria, the surface properties of nanoparticles need to be enhanced by a protective coating.

16.6.1 COLLOIDAL STABILITY

The balance between attractive van der Waals and repulsive electrostatic forces gives the measure of colloidal stability. If the magnitude of attractive van der Waals force is less than that of the repulsive electrostatic force then there will be no nanoparticle aggregation (the particle remains stable). But in a biological environment, nanoparticles forming aggregate are not stable, but it can be shifted in favor of van der Waals forces if oppositely

TABLE 16.1 List of some important nanoparticles and their biofunctionalization using various targeting ligands for their use in biomedical applications.

Some important nanoparticles	Targeting ligand for biofunctionalization	Application	Reference
Gold NPs	Green fluorescent protein	Fluorescence sensor	De et al., 2009
SPIONs(Super Paramagnetic Iron oxide NPs)	Antibiotics (vancomycin)	Separation of Gram-positive and Gram-negative bacteria	Kell et al., 2008
Quantum dots	Streptavidin	Pathogen sensor	Gilmartin et al., 2012
Iron oxide NPs (IONPs)	Anti-HER2	Identification of circulating tumor cells (CTCs)	Xu et al., 2011
Polymer (polystyrene)/Silica based particle	Pyrene dye	Enhancement of photostability of NPs	Gao et al., 2002
SPIONS	Carbohydrates such as Dextran/ carboxy dextran	MRI contrast agent	Unterweger et al., 2018
Iron oxide NPs	Zinc metal	Magnetic switch	Cho et al., 2012
Porous Fe_3O_4	Herceptin	Treatment of breast cancer	Cheng et al., 2009
Liposomes	Folate	Target leukemia cells (Cancer treatment)	Pan et al., 2002
Gold NPs	Aptamers (RNA)	in vitro targeted therapy for epidermal growth factor receptor (EGFR)	Li et al., 2010

charged ions surround the charges that are present on the surface of the nanoparticle.

16.6.2 BIOCOMPATIBLE SURFACES

To design "bioinert" surfaces, it is necessary to reduce the unspecified attachment of all the biomolecules on the surfaces of nanomaterials. The biomolecules like proteins can interact with the surfaces because they have the ability to adsorb on the surface at the interface area primarily by noncovalent interactions such as van der Waals interactions, hydrophobic bonds, and hydrogen bonds. These forces usually arise due to the "surface inhomogeneity," that are commonly marked by spots on the surface that may be either hydrophilic or hydrophobic, charged or neutral (Andrade et al., 1992). Accordingly, the process, rate, and magnitude of protein adsorption are also affected by the surface properties of the biomaterials.

16.6.3 CYTOTOXICITY

Several phenomena cause toxicity to cells. Biofunctionalization of nanoparticle sometime exacerbates the toxic effect on mammalian cells by adsorbing on the cell surfaces. Toxicity is caused because biofunctionalization of nanoparticles can induce slow release inside the cells. Depending on the chemical composition of nanoparticles, toxicity to different cell types varies. Metal nanoparticles show marked toxicity because of their greater biocompatibility. There is a necessity to reduce toxicity by, if all the possible sources of cytotoxicity are prevented then the nanoparticles show a very narrow range of cytotoxicity.

16.7 FUTURE PERSPECTIVE

There is great enthusiasm in both academia and industry for the development of biofunctionalized nanoparticles as therapeutic agents. In preclinical studies, numbers of nanoparticles have already shown exciting results explaining their use as excellent carriers of therapeutics. Various surface-functionalized nanoparticles both organic and inorganic are presently used in various sensing and imaging applications, such as SPIONs in MRI and tissue imaging as contrast agents and QDs as pathogen sensors. Several nanomaterials are seen as the future of targeted drug delivery in medicine in improving human health. Nanomaterials must be made "stimuli-responsive" which allows them to function by using various internal or external stimuli such as heat, light sources, magnetic field, and pH. These smart materials found huge applications in targeted drug delivery, controlled release of drugs, medical implants, tissue scaffolding, and wound dressing. However, there are several process limitations and drawbacks in biofunctionalization of nanoparticle demanding extensive research and innovation to overcome them.

The increasing use of biofunctionalized nanomaterials in therapeutics has massive advantages although there are risks that increase concerns on the adverse side-effects on human health. Toxicity of these biofunctionalized nanomaterials is now an emerging matter of concern and research. The drug delivery pathways may also cause the introduction of several toxic nanomaterials into the human body. Though biofunctionalization of nanoparticles has numerous prospective applications in medicine and in various other fields, there are necessities of rigorous research and innovation in optimizing the fictionalization process itself. Further the potential application of biofunctionalized nanomaterials and their impact on human health also needs to be studied extensively.

KEYWORDS

- **biofunctionalization**
- **nanomaterials**
- **medicine**
- **surface modification**

REFERENCES

Abbasi, E.; Kafshdooz, T.; Bakhtiary, M.; Nikzamir N.; Nikzamir, M.; Mohammadian, M.; Akbarzadeh, A. Biomedical and biological applications of quantum dots. *Artif. Cells Nanomed. Biotech.* **2016**, *44*, 885–891.

Andrade, J. D.; Hlady, V.; Wei, A. P. Adsorption of complex proteins at interfaces. *Pure Appl. Chem.* **1992**, *64*, 1777–1781.

Bajaj, A.; Rana, S.; Miranda, O. R.; Yawe, J. C.; Jerry, D. J.; Bunz, U. H. F.; Rotello, V. M. Cell surface-based differentiation of cell types and cancer states using a gold nanoparticle-GFP based sensing array. *Chem. Sci.* **2010**, *1*, 134–138.

Cheng, K.; Peng, S.; Xu, C.; Sun, S. Porous hollow Fe_3O_4 nanoparticles for targeted delivery and controlled release of cis-platin. *J. Am. Chem. Soc.* **2009**, *131*, 10637–10644.

Cho, M. H.; Lee, E. J.; Son, M.; Lee, J. H.; Yoo, D.; Kim, J. W.; Park, S. W.; Shin, J. S.; Cheon, J. A magnetic switch for the control of cell death signaling in *invitro* and *invivo* systems. *Nat. Mat.* **2012**, *11*, 1038–1043.

Davis F. F. The origin of pegnology. *Adv. Drug Deliv. Rev.* **2002**, *54*, 457–458.

De, M.; Rana, S.; Akpinar, H.; Miranda, O. R.; Arvizo, R. R.; Bunz, U. H.; Rotello, V. M. Sensing of proteins in human serum using conjugates of nanoparticles and green fluorescent protein. *Nat. Chem.* **2009**, *1*, 461–465.

Dreaden, E. C.; Alkilany, A. M.; Huang, X.; Murphy, C. J.; El-Sayed, M. A. The golden age, gold nanoparticles for biomedicine. *Chem. Soc. Rev.* **2012**, *41*, 2740–2779.

Etoc, F.; Lisse, D.; Bellaiche, Y.; Piehler, J.; Coppey, M.; Dahan, M. Subcellular control of Rac-GTPase signaling by magnetogenetic manipulation inside living cells. *Nat. Nanotech.* **2013**, *8*, 193–198.

Fakruddin, M.; Hossain, Z.; Afroz, H. Prospects and applications of nanobiotechnology, a medical perspective. *J. Nanobiotech.* **2012**, *10*, 31.

Gao, H.; Zhao, Y.; Fu, S.; Li, B.; Li, M. Preparation of novel polymeric fluorescent nanoparticles. *Coll. Polymer Sci.* **2002**, *280*, 653–660.

Gilmartin, N.; O'Kennedy, R. Nanobiotechnologies for the detection and reduction of pathogens. *Enz. Micro. Tech.* **2012**, *50*, 87–95.

Glavin, P.; Thompson, D.; Rayan, K. B.; Mccarthy, A.; Moore, A. C.; Burke, C. S.; Dyson, M.; Maccaraith, B. D.; Gun'Ko, Y. K.; Byrne, M. T.; Volkov, Y.; Keely, C.; Keehan, E.; How, M., Duffy, C.; Macloughlin, R. Nanoparticle-based drug delivery, case studies for cancer and cardiovascular applications. *Cell. Mol. Life Sci.* **2011**, *69*, 389–404.

Haun, J. B.; Yoon, T. J.; Lee, H.; Weissleder, R. Magnetic nanoparticle biosensors. *Interdis. Rev., Nanomed. Nanobiotech.* **2010**, *2*, 291–304.

Huang, X.; Jain, P. K.; El-Sayed, I. H.; El-Sayed, M. A. Gold nanoparticles, interesting optical properties and recent applications in cancer diagnostics and therapy. *Nanomedicine* **2007**, *2*, 681–93.

Kell, A. J.; Stewart, G.; Ryan, S.; Peytavi, R.; Boissinot, M.; Huletsky, A.; Bergeron, M. G.; Simard, B. Vancomycin-modified nanoparticles for efficient targeting and preconcentration of gram-positive and gram-negative bacteria. *ACS Nano.*, **2008**, *2*, 1777–88.

Kettler, K.; Veltman, K.; Vande Meent, D.; Van Wezel, A.; Hendriks, A. J. Cellular uptake of nanoparticles as determined by particle properties, experimental conditions, and cell type. *Env. Toxico. Chem.* **2014**, *33*, 481–492.

Khalid, M.; Hossam, S. Polymeric nanoparticles, Promising platform for drug delivery. *Int. J. Pharm.* **2017**, *528*, 675–69.

Li, N.; Larson, T.; Nguyen, H. H.; Sokolov, K.V.; Ellington, A. D. Directed delivery to Cells. *Chem. Commu.* **2010**, *46*, 392.

Lin, Z.; Cui, S.; Zhang, H.; Chen, Q.; Yang, B.; Su, X.; Zhang, J.; Jin, Q. Studies on quantum dots synthesized in aqueous solution for biological labeling via electrostatic interaction. *Analy. Biochem.* **2003**, *319*, 239–243.

Liong, M.; Hoang, A. N.; Chung, J.; Gural, N.; Ford, C. B.; Min, C.; Shah, R. R.; Ahmad, R.; Fernandez-Suarez, M.; Fortune, S. M.; Toner, M.; Lee, H.; Weissleder, R. Magnetic barcode assay for genetic detection of pathogens. *Nat. Commu.* **2013**, *4*, 1752.

Lv, F.; Qiu, T.; Liu, L.; Ying, J.; Wang, S. Recent advances in conjugated polymer materials for disease diagnosis. *Small.* **2016**, *12*, 696–705.

Maiti, K. K.; Dinish, U. S.; Samanta, A.; Vendrell, M.; Soh, K. S.; Park, S. J.; Olivo, M.; Chang, Y. T. Multiplex targeted in vivo cancer detection using sensitive near-infrared SERS nanotags. *Nano Today*. **2012**, *7*, 85–93.

Marchetti, C.; Palaia, I.; Giorgini, M.; De Medici C.; Iadarola R.; Vertechy L.; Di Donato V.; Tomao F.; Muzzi L.; Benedetti, P. Targeted drug delivery via folate receptors in recurrent ovarian cancer, a review. *OncoTargets. Ther*. **2014**, *7*, 1223–36.

Mintzer, M. A.; Grinstaff, M. W. Biomedical applications of dendrimers, a tutorial. *Chem. Soc. Rev*. **2011**, *40*, 173–90.

Mitchell, G. P.; Mirkin, C. A.; Letsinger, R. L. Programmed assembly of DNA functionalized quantum dots. *J. Am. Chem. Soc*. **1999**, *121*, 8122–8123.

O'Hara, P.; Hickey, A. J. Respirable PLGA microspheres containing rifampicin for the treatment of tuberculosis, manufacture and characterization. *Pharma. Res*. **2000**, *17*, 955–61.

Pan, X. Q.; Zheng, X.; Shi, G.; Wang, H.; Ratnam, M.; Lee, R. J. Strategy for the treatment of acute myelogenous leukemia based on folate receptor beta-targeted liposomal doxorubicin combined with receptor induction using all-trans retinoic acid. *Blood*. **2002**, *100*, 594–602.

Pan, Y.; Du, X.; Zhao, F.; Xu, B. Magnetic nanoparticles for the manipulation of proteins and cells. *Chem. Soc. Rev*. **2012**, 41, 2912–2942.

Pelaz, B.; Pino, P.; Maffre, P.; Hartmann, R.; Gallego, M.; Rivera-Fernández, S.; Fuente, J. M.; Nienhaus, G. U.; Parak, W. J. Surface functionalization of nanoparticles with polyethylene glycol, effects on protein adsorption and cellular uptake. *ACS Nano*. **2015**, *9*, 6996–7008.

Raliya, R.; Singh, C. T.; Haddad, K.; Biswas, P. Perspective on nanoparticle technology for biomedical use. *Cur. Pharm. Des*. **2016**, *22*, 2481–90.

Saha, K.; Agasti, S. S.; Kim, C.; Li, X.; Rotello, V. M. Gold nanoparticles in chemical and biological sensing. Chem. Rev. **2012**, *112*, 2739–2779.

Sercombe, L.; Veerati, T.; Moheimani, F.; Wu, S. Y.; Sood, A. K.; Hua, S. Advances and challenges of liposome assisted drug delivery. *Front. Pharm*. **2015**, *6*, 286.

Singh, M. R. Application of Metallic Nanomaterials in Nanomedicine. In, Adhikari, R., Thapa, S. (eds) Infectious Diseases and Nanomedicine III. *Adv. Exp. Med. Bio*. **2018**, 1052.

Souza, G. R.; Molina, J. R.; Raphael, R. M.; Ozawa, M. G.; Stark, D. J.; Levin, C. S.; Bronk, L. F.; Ananta, J. S.; Mandelin, J.; Georgescu, M. M.; Bankson, J. A.; Gelovani, J. G.; Killian, T. C.; Arap, W.; Pasqualini, R. Three-dimensional tissue culture based on magnetic cell levitation. *Nat. Nanotech*. **2010**, *5*, 291–296.

Torchilin, V. P. Micellar nanocarriers, pharmaceutical perspectives. *Pharmaceutical Res*. **2007**, *24*, 1–16.

Unterweger, H.; Dezsi, L.; Matuszak J.; Janko C.; Poettler M.; Jordan J.; Bauerle T.; Szebeni J.; Fey, T.; Boccaccini, A.; Alexiou, C.; Cicha, I. Dextran-coated superparamagnetic iron oxide nanoparticles for magnetic resonance imaging, evaluation of size-dependent imaging properties, storage stability and safety. *Int. J. Nanomed*. **2018**, *13*, 1899–1915.

Wang, J.; Wang, F.; Li, F.; Zhang W.; Shen Y.; Zhou D.; Gou, S. Multifunctional poly(curcumin) nanomedicine for dual-modal targeted delivery, intracellular responsive release, dual-drug treatment and imaging of multidrug-resistant cancer cells. *J. Mater. Chem*. **2016**, *4*, 2954–2962.

Wang, S. P.; Mamedova, N.; Kotov, N. A.; Chen, W.; Studer, J., Antigen/antibody immunocomplex from CdTe nanoparticle bioconjugates. *Nano Lett*. **2002**, *2*, 817–822.

Weinberg, W. C.; Frazier-Jessen, M. R.; Wu, W. J.; Weir, A.; Hartsough, M.; Keegan, P.; Fuchs, C. Development and regulation of monoclonal antibody products, challenges and opportunities. *Cancer. Metastasis Rev.* **2005**, *24*, 569–84.

Xiao, Y.; Shi, K.; Qu, K.; Chu, B.; Qian, Z. Engineering nanoparticles for targeted delivery of nucleic acid therapeutics in tumor. *Mol. Thera. Meth. Clin. Develop.* **2019**, *12*, 1–18.

Xu, H.; Aguilar, Z. P.; Yang, L.; Kuang, M.; Duan, H.; Xiong, Y.; Wei, H.; Wang, A. Antibody conjugated magnetic iron oxide nanoparticles for cancer cell separation in fresh whole blood. *Biomaterials* **2011**, *32*, 9758–9765.

CHAPTER 17

Nano-System as Therapeutic Means

ANANYA GHOSH[1] and ANIRUDDHA MUKHERJEE[2*]

[1]*P.G. Department of Biotechnology, Utkal University, Vani Vihar, Odisha, India*

[2]*Department of Microbiology, Gurudas College, Kolkata, West Bengal, India*

[*]*Corresponding author. E-mail: tupulmicro@gmail.com*

ABSTRACT

The use of nanoparticles in therapeutics offers wide-ranging opportunities for exploration and intensive research as a carrier of a drug and has gained a front seat in recent years, due to the advantages, which it offers over its predecessor, micelles. These benefits include the relatively simple methods for the preparation of nanoparticles, the use of easily available solvents, better stability, and more targeted therapy. Among other diseases, cancer has had the spotlight in nanoparticle-based therapy for the last few years. The application of nanoparticles in medicine was initially popularized to reduce the toxic side-effects of drugs. Much later, the fact that they may stimulate toxicity came into light and the field of nanotoxicology was established and thus biocompatibility remains an important factor for any material to be used on a living system. Nanoparticles can be prepared from various materials as per convenience and depending upon the drug of choice, the mode of action, and the disease, which is being targeted. Nanoparticles made from silver, chitosan, silicon dioxide, have shown positive results in terms of compatibility and have successfully dodged hypersensitive response in most cases. Here, discussions on the applications of nanoparticles being used in therapeutics for the delivery of drugs have been provided along with an insight into the contemporary research trends. This chapter covers the various types of nanoparticles, their preparation, selecting the drug of choice and its loading, characterization of these nanoparticles, their application and mode of action, and the various risks associated with their use and nanotoxicants.

17.1 INTRODUCTION

The advent of nanoparticles and the idea of nanotechnology in modern science in the second half of the 20th century paved the way for a revolution in almost all spheres of science. This change was brought in due to the extensive use of nanoparticles in technology and application-based science. The field of medicine was not an exception to this either. Nano-sized particles were being used widely in the treatment of multiple ailments as a medium for drug delivery. Very soon, research in this field saw immense growth with almost all the developed and developing countries doing extensive research on this property of nanoparticles. Another objective for its rapid popularity was that it provided a number of benefits over the methods, which were in use at that time, like micelles and injections.

Before long, one of the deadliest maladies in the world, cancer or malignancy, as it is called associated its name with the field of nanotechnology. Cancer is a condition of undifferentiated and uncontrolled growth of cells, which is typically characterized by a formation of tumor or neoplasm. The malignant form of the primary tumor can migrate from its place of origin to different parts of the organism's body (a phenomenon designated as metastasis) resulting in formation of more tumors. Laboratories around the world were testing nanoparticles loaded with their drug of choice on neoplasms for their anticancer properties. Other diseases that came in forefront for research using nanoparticulate matters as delivery agents are psychological disorders like depression, schizophrenia, Alzheimer's disease, etc. Here, we have restricted the primary focus on cancer, tuberculosis, AIDS/HIV, and central nervous system-related disorders. There are, however, many other infections, cardiovascular diseases, metabolic, and psychotic disorders that are also being extensively researched to find their cure with these ultrafine molecules; we have summarized these findings in Table 17.1.

Nanoparticles used for drug delivery medium can be made of various materials and sizes depending upon their utilities and the drug that they entrap. Nanoparticles made of gold, hydrogels, liposomes, biopolymers like chitosan, starch, gelatin, alginate, etc. have gained attention among others as suitable agents for drug delivery in living systems. Manufacturing nanoparticles in an ideal laboratory condition can employ several diverse methods and reagents. Generally, particles that are used for such therapeutic purposes should have a diameter varying between 10 and 100 nm to allow maximum transport within living cells. Any nanoparticle should have at least two components, one being the carrier and the other a compound of therapeutic importance to combat a disease it is being made against.

TABLE 17.1 Various Types of Nanocarrier Systems Under Investigation for Delivery of Drugs, Their Advantages Over the Naked Drug and the Diseases They Target

NP Material	Encapsulated/ Conjugated Drugs	Indication(s)	Advantage	Reference
Chitosan (in native and in modified forms)	Insulin	Diabetes mellitus	Oral intake, protection against acid degradation in GI tract	Wong et al. (2017)
Alginate based	Insulin	Diabetes mellitus	Increased stability of insulin	
Alginate, dextran sulfate, chitosan and albumin (multilayered)	Insulin	Diabetes mellitus	Reduced basal blood sugar level, withstands against acidic pH	
Dextran-Vitamin B12	Insulin	Diabetes mellitus	Reduce the basal blood sugar level	
Phospholipid-conjugated Poly (lactic-*co*-glycolic acid) (PLGA)	Insulin	Diabetes mellitus	Increased biodegradability and biocompatibility	
Polyallylamine (PAA) based	Insulin	Diabetes mellitus	Protection against trypsin and pepsin, high entrapment efficiency (78%–93%)	Thompson et al. (2009, 2010)
Perfluorocarbon	Doxorubicin or paclitaxel	Restenosis	inhibit the proliferation of vascular smooth muscle cells	Lanza et al. (2002)
Paramagnetic perfluorocarbon	Fumagillin	Angiogenesis	Disrupts plaque formation, noninvasive, lower doses of drug required	Winter et al. (2006)
Polyethylene glycol (PEG)-coated prednisolone phosphate liposome	Glucocorticoids	Rheumatoid arthritis	Longer duration of circulation	Metselaar et al. (2003)
Betamethasone phosphate, flavinmononucleotide and zirconium oxide (hybrid)	Glucocorticoids (GC)	Multiple sclerosis	Increased cell type-specificity and reduces the side-effects of free GC	Montes-Cobos et al. (2016)

TABLE 17.1 *(Continued)*

NP Material	Encapsulated/ Conjugated Drugs	Indication(s)	Advantage	Reference
Silver (via green synthesis)	Glycyrrhizin	Gastric ulcer	Cost-efficient and highly biocompatible	Sreelakshmy et al. (2016)
Ethyl cellulose-based	Piroxicam (PX)	Inflammation, analgesic, and anti-pyretic	Greatly reduces the GI tract irritation caused by free PX	El-Habashy et al. (2016)
Zinc oxide	Ketoprofen	Inflammation, analgesic, and anti-pyretic	Improved gastric activity and anti-inflammatory properties than standard zinc oxide	Olbert et al. (2017)
Ureido conjugated chitosan-tripolyphosphate (TPP)	Amoxicillin	*Helicobacter pylori* causing gastric ulcers and related infections	Tolerant to acidic pH in GI tract and ensures a targeted delivery	Jing et al. (2016)
Concanavalin-A (Con-A)-conjugated gastro-retentive PLGA	Acetohydroxamic acid and clarithromycin	*Helicobacter pylori* causing gastric ulcers and related infections	Maximum activity against *H. pylori* as confirmed by *in vitro* data	Jain et al. (2016)
Silver	-	Fungal infection by *Trichosporon asahii*	Antimicrobial activity of silver along with elevated biocompatibility	Xia et al. (2016)
Solid lipid nanoparticles (SLNs)	Retinoic acid	Acne vulgaris	Prevents skin drying	Goyal et al. (2016)
PLGA	Azelaic acid	Acne vulgaris	Improved efficacy, tolerability, compliance, and cosmetic acceptability	
Chitosan	Peptide-based vaccines/ microbicides	Venereal diseases affecting vagina	Heightened bioavailability, biocompatibility and reduced toxicity	Marciello et al. (2017)

TABLE 17.1 *(Continued)*

NP Material	Encapsulated/ Conjugated Drugs	Indication(s)	Advantage	Reference
SLNs	Aripiprazole	Schizophrenia	Enhance in vivo efficacy and oral bioavailability of aripiprazole	Silki and Sinha (2018)
Poly-ε-caprolactone	Haloperidol	Schizophrenia	Efficacy to the amount of drug crossing the BBB	Sun et al. (2016)
Calixarene	Chlorpromazine	Psychosis	Improved the entrapment and drug loading efficiency	
PLGA	Clozapine	Schizophrenia	Reduces fatal side effects	
Chitosan and TPP	Olanzapine	Psychosis	Improves bioavailability	
Poly(caprolactone), Poloxamer 188 and 407	Aripiprazole	Psychosis	Enhanced brain targeting and rapid action	Sawant et al. (2016)

After a stable nanoparticle has been prepared with proven pharmaceutical importance, there are quite a several steps in its journey from the lab to the patient's bedside. Characterization of nanoparticles and an elaborate toxicity study of it in various in vitro and in vivo systems constitute a major part of this expedition. The ladders of characterization as important as the preparatory phase or the clinical trials because this is where it is found out if a particle system has been properly assembled, its stability and other properties like size, uniformity, drug loading efficiency, the kinetics of drug release, etc. For studying the toxicity of the drug-loaded nanoparticles, usually stable cell lines for in vitro representations and murine model and/or model invertebrate organism like *Artemia* spp. are favored.

Even though research in this sector follows a diverse range of approaches and employs different materials and methods yet, the basic goals of research remain unaffected. They require attributes like increased stability, maximum bioavailability, rapid action of the clinical drug, and faster recovery from the diseased state, and reduce the risk of bio-incompatibility and nanotoxicity.

Ironically, the first introduction of nanoparticles and their applications in therapeutic medicine gained momentum because nanoparticles are way more biocompatible than other mediums which were then in use and largely removed the risk of toxic side-effects. But soon medical practitioners and clinical researchers came to realize that nanoparticles are also accountable for generating toxic reactions as well as for hypersensitive reactions in the body of certain individuals. This lead to a major shift in approaches to using nanoparticles in therapeutic medicine and gave rise to the field of nanotoxicology.

17.2 NANOPARTICLES: A QUICK RECAP

"There's plenty of room at the bottom," the notion of nanoparticles was first articulated in modern science by celebrated Nobel laureate Dr Richard Feynman in his famous lecture dated in 1959 at the California Institute of Technology in Pasadena, USA. But, the tale of nanoparticles begins centuries before that. Scientists presume that the earliest record of using nanoparticles in medicine dates back to ancient India when the Ayurveda Rasayana Shastra was documented. The modern nanoparticles were somewhat similar to *bhasmas* used for ailments in those days. Strikingly, even the latest technologies that are used in the synthesis of these particles bear a resemblance to those age-old methods of preparing *bhasmas*. Use of such ultrafine particles has witnessed many applications in industries, since its inception but its use in therapeutics and agriculture have always raised questions of ethics and safety, restricting its usage as a result (Chaudhury, 2011).

17.3 APPLICATIONS AND ADVANTAGES FOR NANOPARTICLE-BASED DRUG DELIVERY

While discussing nanoparticle-mediated drug delivery methods, the most elementary question that one has to confront is why the necessity for this novel mode of therapy arose. This query has been brought up numerous numbers of times and every time answers became less and less subtle. The benefits of these ultrafine particles are enormous which makes them appropriate to be used in therapeutics. Earlier, the function of nanoparticles was performed by liposomes and micelles. At times, these membrane-based systems are destabilized even during their administration in the body, via oral or injectable means. Biomolecules like nucleic acids or proteins require a medium of high efficacy for protection from unwanted enzymatic degradation (nucleases and proteases). They might also invoke the body's defence mechanism, which is capable of mounting an immune response, followed by clearance of drug nanocarriers (Vo et al., 2012). Simple membrane-bound drug carriers might not promise such defence and are too vulnerable for such significant therapeutic roles.

Polymeric nanoparticles can be yielded from both natural and synthetic polymers that are easily biodegradable. Due to their reliable stability and a minimal size ranging from 10 to 200 nm, nanoparticles have the advantage of crossing blood-brain barriers (BBB), gain entry to the respiratory system, and also easy absorption through tight junctions (TJs) of endothelial skin cells (Betzer et al., 2017). This enables therapeutic drugs loaded in it to reach obscure locations of the body, which was almost impossible earlier with liposomes.

Another prerequisite of its small size and larger surface area is that nanoparticles have improved solubility in body fluids, amplifying the bioavailability of the compounds they encase. Medicinal drugs, which were previously rejected for having low solubility and bioavailability, were again introduced into research in therapeutic nanotechnology. Nanoparticle-based therapy has also demonstrated targeted delivery of the drugs they are carrying and they can be concentrated at tumor region or sites of inflammation, and antigen sampling sites as a result of enhanced permeability and retention (EPR) effect of the vasculature. Once hoarded at the target site, nonpolar, biodegradable polymeric nanoparticles can act as a local storehouse of the drug depending on the components of the carrier, supplying an incessant source of encapsulated therapeutic compound(s) at diseased sites. While medications delivered in lipid membrane-bound systems are susceptible to even enzymes like lipases, nanoparticle-based therapy has been successful in evading the endogenous enzymes from degrading the drugs (Rizvi et al., 2018). This has paved way for a novel branch of pharmaceuticals, wholly dependent on the research of nanoparticle-mediated therapies.

17.4 CHARACTERISTICS OF THERAPEUTIC NANOPARTICLES

In light of the ongoing discussion, before we focus on how drug-loaded nanoparticles are synthesized or what mechanism they abide by in combating disorders, we must study what are the fundamental characters of nanoparticles, which are currently in use for medicinal purposes. This will give us a basic understanding of the properties of these ultrafine entities and will also provide us with a set of approximate guidelines to be followed while synthesizing any new nanoparticulate system for pharmaceutical utilization.

17.4.1 SIZE OF PARTICLE

The size of a nanoparticle is of utmost significance when used as a therapeutic in any biotic system. Currently, the fastest and the quintessential method is by photon-correlation spectroscopy or dynamic light scattering. With the dimensions of nanocarrier systems, varies in their in vivo distribution, toxicity, biological fate, stability, and drug loading and drug-releasing abilities. Ideally, the diameter of the particles should be under 200 nm, but it may vary to some extent depending upon the drug it has been loaded with, location of target organ, etc. Experiments demonstrate that particles with diameter 100 nm location of have shown 2.5-fold and 6-fold greater uptake than particles of 1 um and 10 um, respectively (Desai et al., 1997). In a similar study, it was seen that nano-sized particles could infiltrate through the submucosal layers in a murine intestinal model, while micro-sized particles were only able to navigate the epithelial lining. The entire purpose of using nanoparticles in therapy is to confirm that the drug that it bears must be released and should be released efficiently. Smaller particles have a higher surface area to volume ratio, which implies that the quantity of the drug in proximity to the surface is higher. This will enhance the chances of a quicker and more efficient and targeted release of the drug (Buzea et al., 2007). Smaller size of these nano-sized vectors also enable higher uptake by a variety of cellular and intracellular targets.

Another important factor, which is to be considered, is the immune reaction that nanoparticles might be confronted with in the body. The vascular and the lymph systems which transport the particles are also sites for clearing out foreign matters or immunogens. Studies show that particles with dimensions higher than 200 nm evoke responses from the lymphatic system quicker, thus getting discarded shortly after administration (Prokop et al., 2008). From the above points, a conclusion can be hence drawn that therapeutic nanoparticles should be big enough to avert their leakage from blood capillaries, yet, it should be minuscule enough to not trigger the body's defence cells (Sykes et al., 2016).

17.4.2 SURFACE PROPERTIES

It has already been recognized how size plays a pivotal function in nanoparticle-based drug formulations. Amendment in the surface properties can also have a direct impact on the behavior and effectiveness of therapeutic complexes (Bantz et al., 2014). To produce an optimum nanoparticle-based drug delivery system, surface curvature, incorporation of suitable targeting ligands, and reactivity in an in vivo model are of extreme significance (Khanbabaie and Jahanshahi, 2012). These factors directly govern the inhibition of aggregation, increased stability, receptor-binding properties of particles which ultimately culminate in determining the pharmaceutical competence of a drug coupled nanoparticulate system.

As mentioned before, particles in the nanometer range are detectable by the lymphatic cells as a part of body's immune mechanism and thus the surface of nanoparticles should be modified in such a way so that this inevitable elimination can at least be delayed. Hydrophobicity of the surface endorses clearance of the ultrafine entities by allowing blood components to bind with it; thus evidently, increase in hydrophilic molecules on the exterior side of nanoparticles will allow them to last for an extended duration in the circulation (Kou, 2013). For this reason, nanoparticles are coated with polymers or surfactants or by synthesis of copolymers like polyethene glycol (PEG; prevents hepatic and splenic localization), polyethene oxide, poloxamine, poloxamer, and polysorbate 80 (Tween 80) has been proven useful, which has even been successful in crossing BBB (Huang et al., 2016). PEG is hydrophilic and almost inert which when combined at the surface of nanoparticles prevents plasma proteins from binding, thus antagonizing opsonization by mononuclear phagocyte system (MPS). For this attribute, PEGylated nanoparticles are termed as "stealth" nanoparticles as they remain practically undetectable by the reticuloendothelial system (Angra et al., 2011; Li and Huang, 2010).

Aggregation is another issue to be addressed while synthesizing nanoparticles. Especially particles in the nanometer range forming micelles, quantum dots, and dendrimers are more prone to clumping due to their large surface area. Particles are often coated with capping agents to avoid aggregation (Li and Kaner, 2006).

Zeta potential of nanoparticles is used with the purpose of characterization of its surface charge property. It reveals the electric potential of drug-delivering particles and depends upon their composition and the medium in which they are dispersed. Nanoparticles having a value of zeta potential ±30 mV or above prevent aggregation and are considered stable for therapeutic uses. Zeta potential also can determine if the encapsulated drug or active compound possesses any charge or not (Bhattacharjee, 2016) and hence can be a major contributing factor for the determination of surface properties of the nanoparticular assemblies.

17.4.3 DRUG LOADING AND RELEASE

Undoubtedly, a vital aspect of therapeutic nanoparticles is the loading of the medicinal drug, its targeted delivery, and its subsequent release. The whole purpose of this mode of therapy is defeated if the nanoparticulate system fails to efficiently entrap the drug during synthesis or is unsuccessful in releasing it efficiently or releasing it at all. This attribute might be dependent on a lot of factors including temperature, pH, solubility of drug, deabsorption of surface-bound or absorbed drug, diffusion of drug through the nanoparticle matrix, swelling of nanoparticle matrix, erosion, and combination of diffusion and erosion processes (Mura et al., 2013; Siepmann and Göpferich, 2001; Son et al., 2017).

The release of the drug also differs depending upon what polymer is being used for nanoparticle preparation, but in all cases, incorporation of the drug of choice takes place during the formulation of nano-sized carriers. The synthesized polymeric nanoparticles, based on their composition, can be termed as nanocapsules or nanospheres. Nanospheres are homogeneous entities in which the drug is physically as well as uniformly distributed in the inner matrix of the nanoparticle-forming polymer. Alternatively, nanocapsules are heterogeneous systems containing an oil-like compound in its inner reservoir which encapsulates a medicinal compound and the polymer membrane surrounds the cavity.

In the matrix of the nanosphere, the drug is substantially and uniformly dispersed and is released by erosion of the matrix. There is a rapid burst to release the drug that is related to weakly bound to the large surface area of the nanoparticle followed by a sustained release (Lee and Yeo, 2015). On the other hand, when nanocapsules are deployed, the release of the drug is meticulously regulated by diffusion of the drug through the polymeric layer. Hence, drug diffusibility through the polymer forming nanocapsules is unquestionably a decisive factor of its ability to be delivered. Ionic interactions between polymer-forming material and the drug, complexes will be formed which shall hinder the release of the drug. This might be evaded by the addition of other auxiliary agents like polyethene oxide-propylene oxide and will interfere with the interactions between the drug and capsule matrix, thus permitting the better release of drug to target tissues (Rani et al., 2017; Calvo et al., 1997).

17.4.4 TARGETED DRUG DELIVERY

After ensuring proper loading of the drug and its sustained release, the subsequent step, which follows, is the targeted delivery of the drug. Nanoparticles can breach infected or inflamed tissue via larger epithelial junctions that are formed near the damaged region. Vector-mediated penetrations of a medicinal drug can take place actively or passively. The designation of active targeting is applicable if the drug carrier system is coupled to a cell or tissue-specific ligand; alternatively, passive targeting involves nanoparticles reaching target tissue or organ through leaky junctions or the EPR effect (Varshosaz and Farzan, 2015).

An ideal polymeric nanoparticulate system should be competent to reach, recognize, interact, and unload its encapsulated drug to specific tissues which are exhibiting symptoms of pathogenicity or damage. It is also of equal importance that the drug circumvents healthy tissues and cells by not causing any damage to them. This attribute is achieved in many cases by the fact that nanoparticles cannot invade healthy cells due to their intact membranes or junctions, while neoplastic cells do not have that and are prone to have leakages in cell junctions. But, in many cases, to assure specificity, the nanocarrier surface is coated with specific ligands (Figure 17.1) which can be peptides, small molecules, designed proteins, antibodies, and nucleic acid aptamers, etc. (Liu et al., 2009).

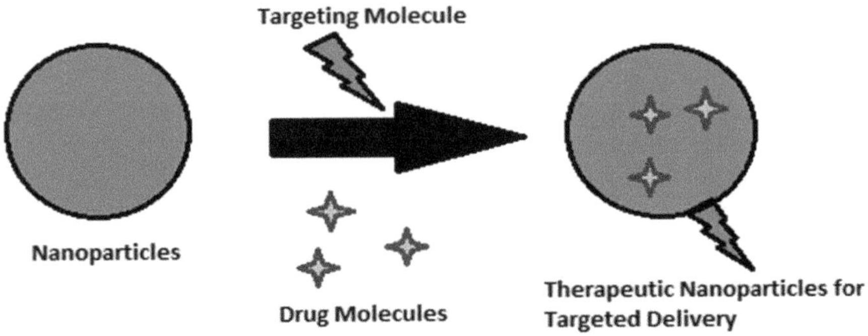

FIGURE 17.1 A schematic representation of nanoassemblies with encapsulated drug for targeted delivery.

Another incentive of using nanoparticles is that due to their minuscule dimensions ranging in nanometers, it can permeate through several biological membranes. For instance, the actions of many antineoplastic, antiviral drugs have been hindered for their inability to navigate through the BBB. Such compounds, if encapsulated or coupled with a nanocarrier system, can be delivered to the target site for treatment. It has been reported that nanovectors can pass through the BBB resulting from the opening of TJs by hyperosmotic mannitol, which can also provide sustained delivery of therapeutic agents for difficult-to-treat diseases for example, brain tumors (Rizvi et al., 2018; Singh et al., 2009).

17.5 NANOPARTICLE-BASED DRUG DELIVERY IN CANCER

"The Emperor of Maladies," as it is rightly called cancer tolls the second-highest cases of mortality worldwide, which if remain unchecked, is destined to increase by 24% toward the end of this current decade. Regrettably, conventional methods of treatment like chemotherapy and radiation therapy, though effective in most cases, could be inadequate to combat some cases of malignancy. Therapies such as these are not only extremely expensive, but they come with side effects (Baudino, 2015) which cannot be overlooked in those scenarios. Therefore, the necessity for an alternative means to tackle cancer has triggered extensive biomedical research, especially in nano-based approaches involving drug delivery. Increased bioavailability, biocompatibility, immunocompatibility, efficient excretion, targeted delivery, and controlled release established their impending potential in applied science.

Cancer can be of hundred different types and are typically categorized by the formation of tumor or neoplasms. The mass of undifferentiated cells forming neoplasm is due to the uncontrolled and unregulated division of cells without their execution via apoptosis. Tumor-forming cells also possess an attribute of disintegrating from their native site (metastasis) and are mobilized to other body parts where they exhibit novel symptoms of malignancy. Statistics speak that of the millions of cancer cases diagnosed worldwide, some 90%–95% cases are due to genetic mutation, triggered by lifestyle and environmental factors while the remaining 5%–10% patients are affected by inherited genetics. Common environmental factors and lifestyle practices that lead to cancer deaths include tobacco (25%–30%), unhealthy eating habits and obesity (30%–35%), radiation (both ionizing and nonionizing, up to 10%), infections (15%–20%), stress, lack of exercise and pollution (Anand et al., 2008).

The affected genes that cause the improper cell division in cancer are divided into two distinct categories. Oncogenes are those who promote cell growth and reproduction. Tumor suppressor genes act to inhibit cell division and survival (Knudson, 2001).

As mentioned before, there is more than one option for combating malignancy but these methods might not be effective in all patients or for all variations of cancer. Radiotherapy, chemotherapy, immunotherapy, anticancer drugs no doubt save many lives, but there is still immense scope and extensive need for improvements. Nanoparticle-based delivery of anticancer drugs have been showing encouraging results in the laboratory and some of them have also been approved for therapeutic uses on humans while many more being in trial stages. Discussing that, here are a few nano carrier-based approaches that have gained considerable recognition and are anticipated to have a promising future in the battle against cancer.

17.5.1 HYDROGELS

Cancer is a complex disease that demands high-throughput research. The dynamic tumor microenvironment undergoes changes in extracellular matrix properties, such as alterations in proteins and hormones, and activation of hypoxia-inducible factors, over a long period.

Delivering therapeutic agents to cancerous tissues is a challenge to physicians and researchers due to both the complexity and dynamic nature of the tumor microenvironment and the instability and off-target effect of associated drugs. However, in recent researches, numerous hydrogel-based reconstructive tumor models have shown encouraging results. Hydrogel-based drug delivery system (DDS) has been shown to improve the efficacy of chemotherapy by increasing drug half-life, facilitating controlled and adjustable drug release, and subsequently decreasing nontargeted exposure of the drug to minimize risks of damaging healthy cells. Hydrogels have given researchers plenty of possibilities to regulate cancer DDSs.

Curcumin, a pleiotropic medicinal compound present in *Curcuma longa*, has been extensively acknowledged in cancer research specifically in pancreatic and breast cancer, hepatocellular carcinoma, skin cell sarcoma, etc. It has been demonstrated to hinder the proliferation of cancer cells and to induce apoptosis through down-regulation of Bcl2 and Survivin and can regulate the cell cycle to prevent uninhibited cell division. Curcumin can also affect the regulation of microRNAs (e.g., miR-15a, miR-16, miR-196, miR-22, and miR-34a) which have been anticipated in recent studies to be potent biomarkers for breast cancer (Bahrami et al., 2017).

However, the clinical applications of curcumin were restricted due to its low solubility and minimal systemic bioavailability. This adversity was surmounted by encapsulation of the compound in a polymeric nanoparticle, forming "nano curcumin." The hydrogel-based nano curcumin demonstrated inhibition of NF-κB activation and IL-6 synthesis as well as reduction of pro-inflammatory cytokines to combat pancreatic cancer (Tajbakhsh et al., 2017). Hydrogels-based DDS can also deliver multiple therapeutic agents concurrently. A drug, alone, might not be efficient enough against all cancer cells owing to the heterogenic nature of tumors; hence, concurrent delivery of multiple drugs with different molecular targets could answer to drug resistance and can minimize the chances of metastasis. To study the synergistic effects of Lapatinib (LAPA) and Paclitaxel (PTX) on breast cancer caused by overexpression of human epidermal growth factor receptor 2, the drugs were combined into a thermosensitive hydrogel system. LAPA, an oral inhibitor of human epidermal growth factor receptor 2 tyrosine kinase, when administered alone had low bioavailability and thus, the high dosage was prescribed for tangible results. On the other hand, the use of PTX was associated with several side effects and demanded a more specific and targeted delivery to avoid causing any harm to healthy cells. The combined therapy, however, showed significantly steady accumulation of LAPA in the tumors with an acceptable level of safety with consequent antineoplastic effects (Hu et al., 2015, Sepantafar et al., 2017).

17.5.2 LIPOSOMES AND MICELLES

Micelles are defined as super-molecular assemblies (spherical) composed of the amphiphilic copolymer. The core of micelles is capable of housing hydrophobic compounds, and the outer shell is a hydrophilic brush-like corona which enhances its polarity, thereby allowing the delivery of contents with low solubility. 20-(S)-Camptothecin, extracted from the bark of *Camptotheca acuminate* (Chinese happy tree), is a natural DNA topoisomerase 1 inhibitor, and its interference with the cytoskeleton (microtubules and actin filaments) makes it a potent agent for antineoplastic therapies. Nevertheless, its use is delimited due to low solubility, toxicity, and instability of the drug (Wang et al., 2016; Martino et al., 2017). If the surface of the micelle is further PEGylated, it escalates the ability of the nanocarrier to navigate via fenestrated vascular sites in tumors and inflamed tissue regions through passive transport, thus contributing to higher drug concentration in tumors. Several polymeric micelles containing antineoplasmic drugs, such as NK911, NK105, NC-6004, NK012, and SP1049C are under clinical trials (Oerlemans et al., 2010) while Genexol-PM (PTX), a drug with antimalignant property has been approved for usage against breast cancer (Zhang et al., 2014).

17.5.3 DENDRIMERS

Dendrimers (*dendrons* meaning trees in Greek) are defined as extensively, yet perfectly branched macromolecules to which drugs, targeting, and imaging agents can adhere via the modification of its numerous functional groups. Various structural features govern the properties of the compounds of medicinal and diagnostic importance, such as absorption, distribution, metabolism, and subsequent elimination (ADME profile) from the living system. The syntheses of dendrimers can either be Tomalia-type divergent synthesis (dendrimer is formed in a stepwise manner from the center to the periphery) or Fréchet-type convergent synthesis (dendrons are synthesized first following which they are anchored to a multifunctional core). Most of the commercially available dendrimers follow the former method of production.

Dendrimers-based drug delivery has proven to be beneficial over other chemotherapeutic methods due to their optimum biodistribution of the drug and its controlled release at the tumor site. The high aqueous solubility, low toxicity, compact globular shape, and controlled surface functionalities make further facilitates their usage as the treatment of choice.

Camptothecin, a naturally derived alkaloid, as mentioned previously, had limited usage in anticancer treatment due to low water-solubility and imposing risks of harm to healthy tissues, even leading to bladder inflammation was administered (Somani and Dufes, 2014).

17.5.4 CARBON NANOTUBES

Double- and single-walled carbon nanotubes are a very recent approach involving an allotropic form of carbon with a cylindrical framework. The nanotubes are synthesized by deepening the number of sheets in the concentric cylinders. The core of this cylindrical structure is hydrophobic thus favoring the storage of water-insoluble drugs in it. The large outer surface area allows its functionalization and can be coupled with tumor receptors to ensure specificity. The additional rewards of using CNTs include elevated EPR effects, uncomplicated procedures for leading of drugs (hydrophilic molecules bind to the outer surface which the inner core houses the hydrophobic compounds) and targeted intracellular accumulation of the nanotubes to cause minimal systemic damage. The needle-like shape also favors transmembrane penetration by enabling the nano-vectors to transport its contents to obscure locations in the body that were previously inaccessible by the drug alone. A very popular anticancer agent, Doxorubicin (DOX), has been conjugated with CNTs and both in vivo and in vitro tests in that study confirm that DOX with CNT showed a more targeted delivery and were more efficient in the destruction of tumor cells in comparison to much higher doses of free DOX. Moreover, the release of DOX was seen to be elicited at lower pH, which is a typical characteristic of tumor microenvironments (Dinesh et al., 2016; Sanginario et al., 2017).

17.5.5 BUCKMINSTERFULLERENE C60

The spherical molecule and its derivatives have shown tremendous impending possibilities not only in the field of cancer research and pharmacology. From particular research, reports show that polyethyleneimine fullerene (C60-PEI), which, through an amide linker, was amended with folic acid, followed by the encapsulation of the antitumor compound, docetaxel onto the C60-PEI-FA nanoparticles. This C60-based DDS exhibited an improved antitumor efficacy while dodging the evident ill-effects to normal organs owing to its prolonged presence in circulation and at the same time enhanced the docetaxel uptake of tumor cells by 7.5-fold. It is also capable of binding up to six electrons and can hence function as a potent scavenger of reactive oxygen species (Shi et al., 2016).

17.6 NANOPARTICLE-BASED THERAPEUTIC ADVANCEMENTS FOR COMBATING TUBERCULOSIS (TB)

The air-borne communicable disease, spreading via sneeze droplets from infected personnel, records as one of the primary causes of death all around the globe. WHO reports show that TB has affected over 10 million people in 2017, claiming the life of 1.6 million. The causative organism, *Mycobacterium tuberculosis*, mainly affects the lungs and the infection is typically symptomized by chest pain and the presence of blood in sputum. The bacterium survives and thrives within the phagocytic cells like macrophages by the subversion of their effector functions, thus crippling an important component of innate immunity. Generally, rifampicin and isoniazid (INH) are the drugs of choice against TB and recommended as first-line anti-TB medications. However, improper or mismanaged usage of these might lead to the advancement of the disease due to the bacterium becoming resistant to the drugs (multidrug resistance or MDR TB). MDR TB might take up to 2 years to be completely cured and requires the administration of second-line drugs along with the previous ones.

Various researches are being carried out to upgrade the quality of treatment for TB by reducing the adverse effects of the drugs used and the outcomes have been quite rewarding. The rate of TB has steadily declined by 2% annually; however, the desired target is to achieve at least 4%–5% reduction of the diseased cases per year to reach the 2020 milestone of the End TB Strategy. Here we discuss how the latest nanotechnological approaches have aided in accomplishing this goal.

An important second regimen drug of choice for MDR TB has been ethionamide (ETH), a prodrug that is converted to its active form by enzyme EthA and functions by the disruption of the mycolic acid. The mono-oxygenase enzyme EthA is controlled by the EthR gene of *M. tuberculosis*, which is a transcriptional repressor. Despite being a potent drug for MDR, the utilization of ETH has been delimited due to the various detrimental side effects it accompanies. Nausea, vomiting, and related gastrointestinal disorders are frequently reported by users. Other major drawbacks of this medication include hepatotoxicity and psychiatric disturbances among patients like depression and instances of encephalopathy and peripheral neuropathy. Hence, the drug is almost impermissible for individuals with a history of liver dysfunction, even for healthy individuals careful monitoring and regular liver function tests are needed during the treatment. Another major obstacle of using this medication is that ETH has a short in vivo half-life and large doses (at least 500 mg) are imperative for substantial results.

Recently, molecules have improved the relatively low level of ETH bioactivation by EthA, which induces conformational changes in, promoting the inhibition of its repressor function. These molecules were found to substantially augment the activity of ETH and were designated as "ETH booster" or "boosters." Nanoparticles made of biodegradable poly D,L-lactic-co-glycolic acid or PLGA, and polylactic acid and of polymeric β-cyclodextrins was prepared and ETH-booster pair was co-encapsulated in them, overcoming low water solubility of the drug and ETH's tendency to crystallize in an aqueous environment. Among the synthesized nanocarrier systems, polymeric β-cyclodextrins not only yielded the best antimicrobial response in both in vivo and in vitro studies, but they were also shown not to cause any adversity in healthy lung cells, even after repeated doses of administration (Costa-Gouveia et al., 2017).

Another recent approach, which addressed the problem of systemic exposure to anti-TB drugs, aimed to develop a new model of therapy that will localize the effects of the chemotherapy at the site of infection and diminish risks of damaging healthy cells. The procedure involved the incorporation of rifabutin into solid lipid nanoparticles (SLNs). Rifabutin functions by inhibition of bacterial DNA-dependent RNA polymerases in the affected cells of the MPS. They are efficiently engulfed by the infected macrophages and have exhibited complete release of the encapsulated drug. The SLNs are synthesized from lipids that are in a solid-state at both room and body temperature and biocompatible surfactants are used to prevent its aggregation. The ultrafine entities were then aerosolized and were delivered via inhalation in the diseased animals (Gasper et al., 2016).

Silica nanoparticles have also emerged as a suitable alternative to traditional antibiotics for MDR TB. Studies carried out with beta (β)-glucan-conjugated silica nanoparticles containing INH, one of the most acknowledged first-line treatments, produced substantial results to indicate toward the development of a novel mode of drug administration. INH, just like ETH, is a prodrug which requires activation by a bacterial catalase-peroxidase enzyme in *Mycobacterium tuberculosis* called KatG. KatG catalyzes the synthesis of isonicotinic acyl radical, which spontaneously pairs with NADH to form a nicotinoyl-NAD adduct. This complex is capable of strong interaction with enoyl-acyl carrier protein reductase InhA, hence restricting the natural enoyl-AcpM substrate and the function of fatty acid synthase.

This progression of reactions interferes with the production of mycolic acids, an integral component of the mycobacterial cell wall. Another mechanism of action of INH is the production of nitric oxide, produced during KatG activation, which was shown to aid the activity of another antimycobacterial prodrug, pretomanid (Jangsun et al., 2018; Suarez et al., 2009).

17.7 NANOPARTICLE-MEDIATED DRUG DELIVERY FOR COMBATING AIDS

Another menace to the modern society, Acquired Immunodeficiency Syndromes or AIDS, as it is commonly abbreviated, has been accountable for deaths of approximately 1.3 million globally, according to the Joint United Nations Programme on HIV/AIDS (UNAIDS) report while another 36.9 million lives are thriving in a vulnerable condition due to it. The retrovirus accountable for initiating AIDS, Human Immunodeficiency Virus (HIV) attacks the cytotoxic T-lymphocytes, hence crippling an important constituent of the immune system and rendering the affected individual susceptible to secondary infections. The disease typically spreads via the exchange of body fluids and thus, unprotected sexual intercourse, blood transfusion without proper medical tests of the donor, reuse of contaminated needles, all can be considered as responsible routes of viral entry and subsequent infection.

After the first case of AIDS being discovered in the United States in the 1980s and understanding what a grim threat it is to our civilization, physicians, and medical researchers have been working tirelessly to develop effective anti-HIV therapies and medications. Even though the efforts conceived fruitful results and many medicinal drugs were discovered to suppress the virus to undetectable levels in the patient's blood, yet it was quite clear that they were not capable of curing the disease. The current combination of antiretroviral therapies using antiretroviral drugs is a long-term commitment. Disruption or discontinuation of the therapy often leads to a rebound of the virus and the progression of AIDS. Hence, there rose an urgent requirement for extensive research, which will provide knowledge of the source of the residual virus, the underlying virologic, physiologic, and pharmacologic mechanisms behind the prognosis of the disease, and viral persistence to develop targeted drug therapies to ensure the removal of residual virus in AIDS/HIV patients.

A single infectious virus is sufficient for the onset of the disease. The $CD4^+$ T-lymphocytes serve as its major host and is the site for replication and pathogenesis. Other host cell receptors include CCR5, CXCR4, etc. while other phagocytic cells like macrophages and mononuclear phagocytes might also harbor the causative microbe. The viral RNA, upon entry into the host cell, is transcribed to DNA by the HIV reverse transcriptase and is integrated into the host genome, thereby taking the cell hostage for the reproduction of viral progeny.

Of the various drugs, whose therapeutic roles are being scrutinized as encapsulated within various nanoparticles, some have discoursed in this section. Zidovudine has been tested in rodent models by delivering liposomes loaded with zidovudine myristate (a prodrug of zidovudine). The intravenously injectable particles have exhibited enhanced entrapment efficiency and a longer half-life in blood plasma. Various other tactics are also being adapted to ensure more targeted delivery of the medications and minimize the risks of systemic damage. Another drug, which is of great importance as one of the WHO recommended drugs for combination of antiretroviral therapies, is Efavirenz which has been trialed separately with three different classes of NPs: polymeric micelles, SLNs, and polymeric dendrimers. The pharmacokinetics and distribution in in vitro models are also being investigated for another primary antiretroviral compound named tenofovir by encapsulating it with nanofibres as well as traditional liposomal nanostructure (Shao et al., 2016).

17.8 THERAPEUTIC APPROACHES OF NANOTECHNOLOGY IN CNS-RELATED DISORDERS

The fact that nanoparticles are capable of crossing the BBB was a boom for medical researchers and physicians, which indicated the impending potential of them in the treatment of neurodegenerative disorders. The CNS or central nervous system, comprising of the brain and the spinal cord, acts as the governing center for all life processes. BBS is the most selective and stringent interface separating the periphery from the CNS, and being mainly made up of endothelial cells connected by TJs and adherens junctions is nearly impermeable to microbes and foreign molecules. Pathological conditions or neurodegenerative disorders of the CNS alter the integrity of BBB, thereby allowing the entry of pathogen and immunogens and rendering the most significant part of the body in a vulnerable state.

Here, we focus our attention to some of the most widely affecting neural disorders which not have recorded grim figures of mortality rates worldwide. Disorders like stroke, Alzheimer's, and Parkinson's disease (PD) have had a huge impact on society and public health. Even though effective treatment options are available to increase the lifespans of the affected, they are accompanied by various side effects and ultimately deteriorate the general quality of life.

17.8.1 STROKE

A widely affecting, causing long-term disability in adults all across the globe, affecting almost 800,000 per annum, is the medical condition of stroke. It is caused by the deprivation of the brain of the blood supply as a result of a bleeding vessel (hemorrhagic stroke) or by an obstruction of a vessel due to a blood clot (ischemic stroke). An ischemic stroke involves the opening of BBB for a short period (minutes to hours), before a refractory interval and then, reopening of the barrier for an extended period (hours to days). The restitution of the blood supply or reperfusion is crucial to minimize cerebral injury, but it can also deteriorate damage, which is termed reperfusion injury contributing to the latter opening. Ischemic stroke mainly involves the loss and the disruption of the TJs, which are degraded by matrix material proteinases (MMP). The extracellular matrix degradation of type IV collagen and further promoting BBB permeability can also be attributed to the MMPs. Delivery of drugs via a nanocarrier system must consider the disintegration of TJs and the opening of the BBB because of it and can use it to its advantage. The leaky BBB indeed surges the probability of NPs to navigate to their target sites and exhibit their therapeutic potentials largely.

Neuroprotective drugs, that were initially incompetent in in vivo models due to restricted permeability via the BBB, low solubility, and rapid clearance, could be coupled with NPs as a potential form of therapy. Chitosan nanoparticle, for instance, was coupled with transferrin receptor (TfR) antibody and loaded with a specific caspase-3 inhibitor peptide (*N*-benzyloxycarbonyl-Asp(OMe)-Glu(OMe)-Val-Asp(OMe)-fluoromethyl ketone or Z-DEVD-FMK) was shown to diminish the neurological deficits caused by an episode of ischemic stroke. The monoclonal TfR antibody was employed to recognize the TfR type 1 which is highly concentrated in brain capillary endothelium and probably triggers receptor-mediated transport across the BBB via transcytosis. Caspase-3 was reported to be activated from its preform following cleavage in ischemic stroke and its regional distribution corresponded to the region destined to infarct, hence proving its direct involvement in cell death. The results of this study showed tangible evidence that the NPs, loaded with drugs and conjugated to the antibody, could successfully reduce the caspase-3 activity and the infarct volume in the mice model. The motor functions, previously interrupted by stroke, were also restored to a great extent after the therapy (Karatas et al., 2009). Tanshinone IIA, a derivative of phenanthrenequinone, has also displayed effective actions against oxidative stress caused during neurological disorders. In spite of its potential role against ischemic injuries, its applications, like many other drugs, were restricted by its shorter half-life, poor solubility, and incompetency to permeate through the BBB. Cationic bovine serum albumin-conjugated tanshinone IIA PEGylated NPs delivered a suitable answer to this problem. In rodents, this approach recorded in ameliorated infarct volume due to its reduction approximately by 70%, decreased neurological impairments caused by the stroke, neutrophil infiltration, and apoptosis of neuronal cells after an intravenous injection during the period of reperfusion. Furthermore, these NPs also exhibited neuroprotective functions via modulation of inflammatory processes and neuronal signaling pathways by down-regulation pro-inflammatory cytokines for example, IL-8 and TNF-α, and up-regulating anti-inflammatory cytokines, like IL-10 and transforming growth factor-β1 (TGF-β1) in murine ischemic brain. Another molecule of therapeutic importance for stroke was revealed to adenosine when its conjugation with lipid squalene improved the drug's permeability via BBB and resulted in the reduction of stroke-induced infarct area and improvement of neurological deficit scores (Gaudin et al., 2014).

17.8.2 ALZHEIMER'S DISEASE (AD)

With almost 44 million people worldwide, AD is a neurodegenerative disease typically symptomized and associated with dementia and memory loss. Cerebral atrophy, accumulation of amyloid-beta (Aβ) peptide, hyperphosphorylated tau protein filaments, and related cerebrovascular changes leading to cerebral amyloid angiopathy the prime causes behind this chronic ailment. Research for AD can be particularly complicated as none of the animal models, currently, in use, are a true representation of the neuropathology spectrum of the disease. However, transgenic animal models altered by mutation can provide a better insight into the prognosis of it. P-glycoprotein (P-gp) and low-density lipoprotein receptor-related protein, are responsible for Aβ clearance in healthy organisms, while the Aβ influx to the brain is controlled by the receptor for advanced glycation end products (RAGE). Significantly, amplified amounts of Aβ in the AD affected induce upregulation of RAGE expression, hence spawning a positive feedback loop for subsequent exacerbation of Aβ accumulation in the brain alongside activating several inflammatory and oxidative cascades. Aβ-RAGE interaction also prompts TJ disruption via intracellular Ca^{2+}-calcineurin signaling and secretion of MMP-2 and -9.

Neuroprotective peptides may be a good investment in AD therapeutics, which can function by eroding Aβ plaque formation, degrading Aβ toxic peptides, and modulating some enzymes like secretases. For example, poly(ethylene glycol)-poly(lactic acid) block copolymer NPs modified with B6 peptide and coupled with the neuroprotective peptide NAPVSIPQ (NAP) were able to accumulate in the brain of mice in higher levels than those NPs without B6. This form of treatment meliorated spatial learning deficit and enhanced cholinergic function possibly via a decrease of acetylcholinesterase (acetylcholine degrading enzyme) and a surge of choline acetyltransferase (acetylcholine synthesizing enzyme) activity. Nerve growth factors (NGF) are another class of molecules, which is indicating an encouraging perspective in AD therapeutics. NGF-loaded poly(butyl cyanoacrylate) nanoassemblies were witnessed by researchers to reach the brain within 45 minutes of administration in a significantly higher amount than free NGF. Its effector functions in neurodegenerative animals, by reversion of amnesia and improved recognition and memory, established its importance in combating AD. Coenzyme Q, prevalent it is coenzyme Q form in humans as a powerful antioxidant in mitochondrial oxidative phosphorylation cascade, was also demonstrated to have therapeutic properties in AD. The coenzyme-loaded PLGA NPs, modified with trimethylated chitosan showed improved cognitive and spatial memory performances in transgenic mouse models with AD.

17.8.3 PARKINSON'S DISEASE

With almost 7–10 million cases globally, PD records as the second-most prevalent neurodegenerative disorder in seniors after Alzheimer's. The malady is typically caused by the selective degradation of dopaminergic neurons in the substantia nigra (SN) that in turn downregulates the dopamine present in the striatum and Lewy bodies in the neurons. An initial assumption regarding PD was that unlike other neurodegenerative disorders, it does not cause any alterations in the BBB. Later, studies showed that drugs like verapamil and benserazide, which were previously incapable of passing the BBB in healthy animals, were found in abundance in the brains of PD affected animal models. Lewy bodies, which are α-synuclein and protein inclusions in neurons, were also associated with PD, causing the disintegration of the BBB. Additionally, metals, such as iron and manganese were found in profusion in brain lesions of PD patients and animal models. Lactoferrin receptor levels were also found increased in SN dopaminergic neurons of PD affected and implicated that neuronal iron uptake probably endorses dopaminergic neuronal degeneration.

Dopamine replacement therapies, despite having yielded positive results for PD with improved motor symptoms, still raise a question when it comes to restoration of behavior and cognition. Hence, there rose the urgency for an alternative mode of medication, which not only will have the rewards of dopamine replacement but should also exhibit tangible results in improving the quality of life. Researchers have therefore loaded PLGA nanoparticles with dopamine confirming the potential to improve behavior in PD animal models without accompanying any abrupt changes in the cardiac functions. The nanoparticles were able to use the disintegration of the BBB to its advantage by crossing the interface in the SN and striatum, which are regions altered by the onset of PD. A slow and controlled release of the hormonal neurotransmitter from NPs ensures its presence in the plasma for a longer period. Thus, dopamine-loaded NPs can be a preferred alternative to circumvent the toxicity caused by bulk dopamine, when administered alone (Pahuja et al., 2015). As mentioned before, upregulation of lactoferrin and lactoferrin receptors have been linked with brain lesions caused in PD, but this information can be used for combating the same. NPs, tagged with lactoferrin, are spontaneously taken up by the lesion cells having the receptors specific for it. Moreover, lactoferrin can act as an excellent chelator of the iron, which peaks at the SN and striatum of PD patients, thus serving dual anti-PD roles. Another molecule of interest, urocortin, is a corticotrophin-releasing hormone-related peptide, found to protect dopaminergic neurons. This peptide, just like many other drugs, is incapable of crossing the BBB when administered unaccompanied but, PEG-PLGA NPs loaded with urocortin and conjugated with lactoferrins were detected in cortex, SN, and striatum after 1 hour after injection. However, NPs with lactoferrins could barely reach that target site (Saraiva et al., 2016).

17.9 CLEARANCE OF NPS FROM THE SYSTEM AND TOXICITY CONCERNS

Despite the advancement in nanocarrier-mediated delivery of drugs is emerging as a novel and optimistic branch of therapeutics, an issue, which is of major concern among all clinical researchers, is the question of clearance of the nanoparticles from the living system and the question of toxicity associated with it. Although both organic and inorganic nanoparticles have been proven to possess impending applications as an alternate medicament for surmounting the discrepancies of the conventional forms of therapies, the use of inorganic molecules in pharmaceuticals has been greatly compromised owing to this. Even though all the materials from which nanoparticles are made have been well characterized in their bulk state, yet in the nanoscale, they exhibit additional properties, which might result in their toxicity in the living systems. Moreover, traditional cytotoxicity assays are also not competent enough for the characterization of all nanoparticles since the ultrafine entities react with the dyes used for the assays by reducing them and hence, yield unreliable data. Most of the toxicity studies that have been performed for nanoparticles are in vitro, and in vivo studies are mandatorily required to support that. Some of the studies in animal models have detected the presence of inorganic NPs, made of metals like zinc and cadmium, to be present in the blood even after 90 days of administering.

The materials constituting the ultrafine particles are also accountable for toxicity studies. For instance, iron oxide nanoparticles have shown great potential as drug delivery vessels as well as imaging agents for diagnostic purposes. Iron oxide nanoparticles have exhibited low toxicity since they are biologically degraded to form iron ions, which is a vital trace element in humans. Another such element, which is present in scarce amounts in the human body, is silicon and intact silicon nanoparticles were found to be clearable by the reticuloendothelial system. However, if degraded to water-soluble silicic acid, the traces of porous silicon nanoparticles were quite noticeable even after 1 week of administration. Sophisticated regulation of particle size and its surface properties is vital since it is these properties that decide the pharmacokinetics, biodegradation, and clearance properties of these nanoparticles. Toxicity concern still and will remain a major hurdle on the way of clinical translation of nanoparticles. Extensive research work is demanded in the preclinical phases for determining the most suited method for the characterization and reduction of toxic effects of these ultrafine assemblies.

KEYWORDS

- **nanoparticles**
- **drug delivery**
- **therapeutics**
- **biocompatibility**
- **cancer**
- **chitosan**

REFERENCES

Anand, P.; Kunnumakkara, A. B.; Sundaram, C.; Harikumar, K. B.; Tharakan, S. T.; Lai, O. S.; Sung, B.; Aggarwal, B. B. Cancer is a preventable disease that requires major lifestyle changes. *Pharma. Res.* **2008**, *25(9)*, 2097–116.

Angra, P. K.; Rizvi, S. A. A.; Oettinger, C. W.; D'Souzaa, M. J. Novel approach for preparing nontoxic stealth microspheres for drug delivery. *Eur. J. Chem.* **2011**, *2*, 125–129.

Bahrami, A.; Aledavood, A.; Anvari, K.; Hassanian, S. M.; Maftouh, M.; Yaghobzade, A.; Avan, A. The prognostic and therapeutic application of microRNAs in breast cancer, tissue and circulating microRNAs. *J. Cell Physio.* **2017**, https,//doi.org/10.1002/jcp.25813.

Bantz, C.; Koshkina, O.; Lang, T.; Galla, H.; Kirkpatrick, C. J.; Stauber, R. H.; Maskos, M. The surface properties of nanoparticles determine the agglomeration state and the size of the particles under physiological conditions. Zellner, R., ed. Beilstein *J. Nanotechnol.* **2014**, *5*, 1774–1786.

Baudino, T. A. Targeted cancer therapy, the next generation of cancer treatment. *Curr. Drug Discov. Technol.* **2015**, *12 (1)*, 3–20.

Betzer, O.; Shilo, M.; Opochinsky, R.; Barnoy, E.; Motiei, M.; Okun, E.; Yadid, G; Popovtzer, R. The effect of nanoparticle size on the ability to cross the blood–brain barrier, an *in vivo* study. *Nanomedicine* **2017**, *12*, 1533.

Bhattacharjee, S. DLS and zeta potential—what they are and what they are not? *J. Control Release* **2016**, *235*, 337–351.

Buzea, C.; Pacheco, I. I.; Robbie, K. Nanomaterials and nanoparticles, sources and toxicity. *Biointerphases.* **2007**, *2(4)*,17–71.

Calvo, P.; Remuñan-López, C.; Vila-Jato, J. L; Alonso, M. J. Chitosan and chitosan/ethylene oxidepropylene oxide block copolymer nanoparticles as novel carriers for proteins and vaccines. *Pharm. Res.* **1997**, *14*, 1431–6143.

Chaudhary, A. Ayurvedic Bhasma, Nanomedicine of ancient India—its global contemporary perspective. *J. Biomed. Nanotechnol.* **2011**, *7*, 68–79.

Costa-Gouveia, J.; Pancani, E.; Jouny, S.; Machelart, A.; Delorme, V.; Salzano, G.; Iantomasi, R.; Piveteau, C.; Queval, C. J. J.; Song, O. R.; Flipo, M.; Deprez, B.; Saint-André J. P.; Hureaux J.; Majlessi, L.; Willand, N.; Baulard, A.; Brodin, P.; Gref, R. Combination therapy for tuberculosis treatment, pulmonary administration of ethionamide and booster co-loaded nanoparticles. *Sci. Rep.* **2017**, *7*, 5390.

Gaspara, D. P.; Fariaa, V.; Gonçalvesa, L. M. D.; Taboada, P., Remuñán-López, C.; Almeidaa, A. J. Rifabutin-loaded solid lipid nanoparticles for inhaled antitubercular therapy, physicochemical and *in vitro* studies. *Int. J. Pharm.* **2016**, *497 (1-2)*, 199–209.

Desai, M. P.; Labhasetwar, V.; Walter, E.; Levy, R. J.; Amidon, G. L. The mechanism of uptake of biodegradable microparticles in Caco-2 cells is size dependent. *Pharm. Res.* **1997**, *14*, 1568–1573.

Dinesh, B.; Bianco, A.; Ménard-Moyon, C. Designing multimodal carbon nanotubes by covalent multi-functionalization. *Nanoscale.* **2016**, *8(44)*, 18596–18611.

El-Habashy, S. E.; Allam, A. N.; El-Kamel, A. H. Ethyl cellulose nanoparticles as a platform to decrease ulcerogenic potential of piroxicam, formulation and in vitro/in vivo evaluation. *Int. J. Nanomed.* **2016**, *26(11)*, 2369–80.

Gaudin, A.; Yemisci, M.; Eroglu, H.; Lepetre-Mouelhi, S.; Turkoglu, O. F.; Donmez-Demir, B.; Caban, S.; Sargon, M. F.; Garcia-Argote, S.; Pieters, G.; Loreau, O.; Rousseau, B.; Tagit, O.; Hildebrandt, N.; Le Dantec, Y.; Mougin, J.; Valetti, S.; Chacun, H.; Nicolas, V.;

Desmaële, D.; Andrieux, K.; Capan, Y.; Dalkara, T.; Couvreur, P. Squalenoyl adenosine nanoparticles provide neuroprotection after stroke and spinal cord injury. *Nat. Nanotech.* **2014**, *9*, 1054–1062, http,//dx.doi.org/10.1038/nnano.2014.274.

Goyal, R.; Macri, L. K.; Kaplan, H. M.; Kohn, J. Nanoparticles and nanofibers for topical drug delivery. *J. Controlled Release*. **2015**. doi 10.1016/j.jconrel.2015.10.04.

Hu, H.; Lin, Z.; He, B.; Dai, W.; Wang, X.; Wang, J.; Zhang, X.; Zhang, H.; Zhang, Q. A novel localized co-delivery system with lapatinib microparticles and paclitaxel nanoparticles in a peritumorally injectable in situ hydrogel. *J. Control Release.* **2015**, *220*, 189–200.

Huang, Y.; Zhang, B.; Xie, S.; Yang, B.; Xu, Q.; Tan, J. Superparamagnetic iron oxide nanoparticles modified with Tween 80 pass through the intact blood–brain barrier in rats under magnetic field. *ACS Appl. Mater. Interfaces*. **2016**, *8*, 11336–11341.

Hwang, J.; Son, J.; Seo, Y.; Jo, Y.; Lee, K.; Lee, D.; Khan, M. S.; Chavan, S.; Park, C.; Sharma, A.; Gilad, A. A.; Choi, J. Functional silica nanoparticles conjugated with beta-glucan to deliver anti-tuberculosis drug molecules. *J. Ind.. Eng. Chem.* **2018**, *58*, 376–385.

Jain, S. K.; Haider, T.; Kumar, A.; Jain, A. Lectin-conjugated clarithromycin and acetohydroxamic acid-loaded plga nanoparticles, a novel approach for effective treatment of *H. pylori*. *AAPS Pharm. Sci. Tech.* **2016**, *17*, 1131.

Jing, Z.W.; Jia, Y.Y.; Wan, N.; Luo, M.; Huan, M.L.; Kang T.B.; Zhou, S.Y.; Zhang, B. L. Design and evaluation of novel pH-sensitive ureido-conjugated chitosan/TPP nanoparticles targeted to *Helicobacter pylori*. *Biomaterials* doi, 10.1016/j.biomaterials.2016.01.045.

Karatas, H.; Aktas, Y.; Gursoy-Ozdemir, Y.; Bodur, E.; Yemisci, M.; Caban, S.; Vural, A.; Pinarbasli, O.; Capan, Y.; Fernandez-Megia, E.; Novoa-Carballal, R.; Riguera, R.; Andrieux, K.; Couvreur, P.; Dalkara, T. A nanomedicine transports a peptide caspase-3 inhibitor across the blood–brain barrier and provides neuroprotection. *J. Neurosci.* **2009**, *29*, 13761–13769, http,//dx.doi.org/10.1523/JNEUROSCI.4246-09.2009.

Khanbabaie, R.; Jahanshahi, M. Revolutionary impact of nanodrug delivery on neuroscience. *Curr. Neuropharm.* **2012**,*10(4)*, 370–392.

Knudson, A. G. Two genetic hits (more or less) to cancer. *Nat. Rev. Cancer.* **2001**, *1(2)*, 157–162.

Lanza, G. M.; Yu, X.; Winter, P. M.; Abendschein, D. R.; Karukstis, K. K.; Scott, M. J.; Chinen, L. K.; Fuhrhop, R. W.; Scherrer, D. E.; Wickline, S. A. Targeted antiproliferative drug delivery to vascular smooth muscle cells with a magnetic resonance imaging nanoparticle contrast agent, implications for rational therapy of restenosis. *Circulation.* **2002**, *106*, 2842–2847.

Lee, J. H.; Yeo, Y. Controlled drug release from pharmaceutical nanocarriers. *Chem. Eng. Sci.* **2015**, *125*, 75–84.

Li, D.; Kaner, R. B. Shape and aggregation control of nanoparticles, not shaken. Not Stirred. *J. Am. Chem. Soc.* **2006**, *128(3)*, 968–975.

Li, S. D.; Huang, L. Stealth nanoparticles, high density but sheddable PEG is a key for tumor targeting. *J. Control Release* **2010**, *145(3)*, 178–181.

Liu, R.; Kay, B. K.; Jiang, S.; Chen, S. Nanoparticle delivery, targeting and nonspecific binding. *MRS Bull.* **2009**, *34(6)*, 432–440.

Lu, L. M.; Wang, X.; Marin-Muller, C.; Wang, H.; Lin, P. H.; Yao, Q.; Chen, C. Current advances in research and clinical applications of PLGA-based nanotechnology. *Expert. Rev. Mol. Diagn.* **2009**, *9*, 325-341.

Martino, E.; Volpe, D. S.; Terribile, E.; Benetti, E.; Sakaj, M.; Centamore, A.; Sala, A.; Collina, S. The long story of camptothecin, From traditional medicine to drugs. *Bioorg. Med. Chem. Lett.* **2017**, *27*, 701–707.

Marzia, M.; Silvia, R.; Carla, C.; Carmen, R. Freeze-dried cylinders carrying chitosan nanoparticles for vaginal peptide delivery. *Carbo. Polym*. 2017. http,//dx.doi.org/10.1016/j.carbpol.2017.04.051.

Metselaar, J. M.; Wauben, M. H.; Wagenaar-Hilbers, J. P. A.; Boerman, D. C.; Storm, G. Complete remission of experimental arthritis by joint targeting of glucocorticoids with long-circulating liposomes. *Arthritis Rheum*. **2003**, *48*, 2059–2066.

Montes-Cobos, E.; Ring, S.; Fischer, H. J.; Heck, J.; Strau, B. J.; Schwaninger, M.; Reichardt, S. D.; Feldmann, C.; Lühder, F.; Reichardt, H. M. Targeted delivery of glucocorticoids to macrophages in a mouse model of multiple sclerosis using inorganic-organic hybrid nanoparticles. *J. Control Release*. **2016**. doi,10.1016/j.jconrel.2016.12.00.

Mura, S.; Nicolas, J.; Couvreur, P. Stimuli-responsive nanocarriers for drug delivery. *Nat. Mater*. **2013**, *12*, 991–1003.

Oerlemans, C.; Bult, W.; Bos, M.; Storm, G.; Nijsen, J. F. W.; Hennink, W. E. Polymeric micelles in anticancer therapy, targeting, imaging and triggered release. *Pharm. Res*. **2010**, *27*, 2569–2589.

Olbert, M.; Gdula-Argasinska, J.; Nowak, G.; Librowski, T. Beneficial effect of nanoparticles over standard form of zinc oxide in enhancing the anti-inflammatory activity of ketoprofen in rats. *Pharma. Rep*. **2017**, *69(4)*, 679–682.

Pahuja, R.; Seth, K.; Shukla, A.; Shukla, R. K.; Bhatnagar, P.; Chauhan, L. K. S.; Saxena, P. N.; Arun, J.; Chaudhari, B. P.; Patel, D. K.; Singh, S. P.; Shukla, R.; Khanna, V. K.; Kumar, P.; Chaturvedi, R. K.; Gupta, K. C. Transblood-brain barrier delivery of dopamine loaded nanoparticles reverses functional deficits in parkinsonian rats. *ACS Nano*. **2015**. http,//dx.doi.org/10.1021/nn506408v (150331165747002).

Palkhiwala, S.; Bakshi, S. R. Engineered nanoparticles, Revisiting safety concerns in light of ethno medicine. *Ayu*. **2014**, *35*, 237–242.

Prokop, A.; Davidson, J. M. Nanovehicular intracellular delivery systems. *J. Pharm. Sci*. **2008**, *97(9)*, 3518–3590.

Rizvi, S. A.; Saleh, A. M. Applications of nanoparticle systems in drug delivery technology. *Saudi Pharm. J*. **2017**, 89–97.

Sanginario, A.; Miccoli, B.; Demarchi, D. Carbon nanotubes as an effective opportunity for cancer diagnosis and treatment. *Biosensors* **2017**, *7(1)*, 9.

Saraiva, C.; Praça, C.; Ferreira, R.; Santos, T.; Ferreira, L.; Bernardino, L. Nanoparticle-mediated brain drug delivery, overcoming blood–brain barrier to treat neurodegenerative diseases. *J Control Release*. **2016**, *235*, 34–47.

Sawant, K.; Pandey, A.; Patel, S. Aripiprazole loaded poly (caprolactone) nanoparticles, optimization and *in vivo* pharmacokinetics. *Mater. Sci. Eng*. **2016**, *66*, 230–243.

Sepantafar, M.; Maheronnaghsh, R.; Mohammadi, H.; Radmanesh, F; Baharvand, H. *Trends Biotech*. **2017**, *35*, 1074.

Shao, J.; Kraft, J. C.; Li, B.; Yu, J.; Freeling, J.; Koehn, J.; Ho, R. J. Y. Nanodrug formulations to enhance HIV drug exposure in lymphoid tissues and cells, clinical significance and potential impact on treatment and eradication of HIV/AIDS. *Nanomedicine*. **2016**, *11*, 545–64.

Shi, J.; Wang, B.; Wang, L.; Lu, T.; Fu, Y.; Zhang, H.; Zhang, Z. Fullerene (C60)-based tumor-targeting nanoparticles with "off–on" state for enhanced treatment of cancer. *J. Control Release* **2016**, *235*, 245–258.

Siepmann, J.; Göpferich, A. Mathematical modeling of bioerodible, polymeric drug delivery systems. *Adv. Drug Deliv. Rev*. **2001**, *48(2–3)*, 229–247.

Silki, S.; Sinha, V. R. Enhancement of In Vivo Efficacy and Oral Bioavailability of Aripiprazole With Solid Lipid Nanoparticles. *AAPS PharmSciTech*. 2018, *19*, 1264–1273.

Singh, R.; Lillard, J. W. Nanoparticle-based targeted drug delivery. *Exp. Mol. Pathol.* **2009**, *86*, 215–223.

Somani, S.; Dufes, C. Applications of dendrimers for brain delivery and cancer therapy. *Nanomedicine*. **2014**, *9(15)*, 2403–2414.

Son, G. H.; Lee, B. J.; Cho, C. W. Mechanisms of drug release from advanced drug formulations such as polymeric-based drug-delivery systems and lipidnanoparticles. *J. Pharm. Invest.* **2017**. https,//doi.org/10.1007/s40005-017-0320-1.

Sreelakshmy, V.; Deepa, M. K.; Mridula, P. Green synthesis of silver nanoparticles from *Glycyrrhiza glabra* root extract for the treatment of gastric ulcer. *J. Develop. Drugs*. **2016**, *5*, 152.

Suarez, J.; Ranguelova, K.; Jarzecki, A. A.; Manzerova, J.; Krymov, V.; Zhao, X.; Yu, S.; Metlitsky, L.; Gerfen, G. J.; Magliozzo, R. S. An oxyferrous heme/protein-based radical intermediate is catalytically competent in the catalase reaction of *Mycobacterium tuberculosis* catalase-peroxidase (KatG). *J. Biol. Chem.* **2009**, *284(11)*, 7017–7029.

Sun, Y.; Kang, C.; Liu, F.; Song, L. Delivery of antipsychotics with nanoparticles. *Drug Develop. Res.* **2016**, *77(7)*, 393–399.

Sykes, E. A.; Dai, Q.; Sarsons, C. D.; Chen, J.; Rocheleau, J. V.; Hwang, D. M.; Zheng, G.; Cramb, D. T.; Rinker, K. D.; Chan, W. C. W. Tailoring nanoparticle designs to target cancer based on tumor pathophysiology. *Proc. Nat. Acad. Sci.* **2016**, *113*, E1142–E1151.

Tajbakhsh, A.; Mokhtari-Zaer A.; Rezaee, M.; Afzaljavan, F.; Rivandi, M.; Hassanian, S. M.; Avan A. Therapeutic potentials of BDNF/TrkB in Breast cancer; current status and perspectives. *J. Cell. Biochem.* **2017**, *118(9)*, 2502–2515.

Thompson, C. J.; Tetley, L.; Cheng, W. P. The influence of polymer architecture on the protective effect of novel comb shaped amphiphilic poly(allylamine) against in vitro enzymatic degradation of insulin-towards oral insulin delivery. *Int. J. Pharm.* **2017**, *383*, 216–227.

Thompson, C. J.; Tetley, L.; Uchegbu, I. F.; Cheng, W. P. The complexation between novel comb shaped amphiphilic polyallylamine and insulin-towards oral insulin delivery. *Int. J. Pharm.* **2009**, *376*, 46–55.

Varshosaz, J.; Farzan, M. Nanoparticles for targeted delivery of therapeutics and small interfering RNAs in hepatocellular carcinoma. *World. J. Gastroenterol.* **2015**, *21(42)*, 12022–12041.

Vo, T. N.; Kasper, F. K.; Mikos, A. G. Strategies for controlled delivery of growth factors and cells for bone regeneration. *Adv. Drug Deliv. Rev.* **2012**, *64(12)*, 1292–1309.

Wang, X.; Tanaka, M.; Krstin, S.; Peixoto, H. S.; Moura, C. C.; Wink, M. Cytoskeletal interference-A new mode of action for the anticancer drugs camptothecin and topotecan. *Eur. J. Pharmacol.* **2016**, *789*, 265–274.

Winter, P. M.; Morawski, A. M.; Caruthers, S. D.; Harris, T. D.; Fuhrhop, R. W.; Zhang, H. Y.; Allen, J. S.; Lacy, E. K.; Williams, T. A.; Wickline, S. A., Lanza, G. M. Antiangiogenic therapy of early atherosclerosis with paramagnetic alpha(v)beta(3)-integrin-targeted fumagillin nanoparticles. *J. Am. Coll. Cardiol.* **2004**, *43*, 322A–323A.

Wong, C. Y.; Al-Salami, H.; Dass C. R. Potential of insulin nanoparticle formulations for oral delivery and diabetes treatment. *J. Control Rel.* **2017**, *264*, 247–275.

Xia, Z. K.; Ma, Q. H.; Li, S. Y.; Zhang, D. Q.; Cong, L.; Tian, Y. L.; Yang, R. Y. The antifungal effect of silver nanoparticles on *Trichosporon asahii*. *J. Microbiol. Immunol. Infect.* **2016**, *49(2)*, 182–188.

Zhang, X.; Huang, Y.; Li, S. Nanomicellar carriers for targeted delivery of anticancer agents. *Ther. Deliv.* **2014**, *5(1)*, 53–68.

Zhou, Q.; Zhang, L.; Wu, H. Nanomaterials for cancer therapies. *Nanotech. Rev.* **2014**, *6(5)*, 473–496.

CHAPTER 18

Recent Developments in Nanoparticulate-Mediated Drug Delivery in Therapeutic Approaches

JANMEJAYA BAG#, SWETAPADMA SAHU#, and MONALISA MISHRA*

Neural Developmental Biology Lab, Department of Life Science, NIT Rourkela, Rourkela 769008, Odisha, India

**Corresponding author. E-mail: mishramo@nitrkl.ac.in*

#These two authors contributed equally to this study.

ABSTRACT

In the modern decade, the development of nanoparticles (NPs) has emerged as a potential target to study applied science and technology. NP enhances the efficacy of drug delivery to many folds in various target organs. NPs are classified as inorganic or organic, carbon or polymeric or ceramic-based, and hard or soft. Factors like size, shapes, and properties affect the efficiency of NPs. Thus, all these properties are taken into account while selecting the NPs for drug delivery. NPs can be transported to the organ of interest through many routes that is, oral, transdermal, buccal, intravenous, rectal, and so forth. Various nanoshells, nanotubes, nanogels, metallic, and biological NPs are synthesized to upsurge the effectiveness of drug delivery. This chapter elaborates on different types of NPs used as a drug carrier and various factors that can surge the drug delivery efficiently.

18.1 INTRODUCTION

Nanoparticles (NPs) are comprehensively used in drug delivery as a carrier agent. It provides a more reliable and susceptible way to discharge the drug competently to the desired area. Its smaller size allows it to dock to the membrane of the target organ easily and release the drug more conveniently. NPs of various materials, shapes, and sizes are used for drug delivery. Doping, encapsulation, or coating of the NP by polymers allows us to deliver the drug through the membrane simply via diffusion (Zhang et al., 2014). The drug-releasing time depends on the ionic strength and interaction between the axillary ingredient with the drugs (Mudshinge et al., 2011). Thus, both the drug and the nanocarriers are optimized based on the biological, physical, and chemical properties, to enhance the process of drug delivery. NPs like solid-lipid NPs, polymers NPs, liposomes NPs, dendrimers NPs, carbon or silicon materials NPs, and magnetic NPs are used for drug delivery. NPs are encapsulated or doped covalently with the drugs. These covalent bonds break after getting the physical stimuli of the body (like enzymatic activity, pH, temperature, osmolality) and release the drug (Wilczewska et al., 2012). Since the nanocarriers interact with the blood vascular system of the body, they should be nontoxic and biocompatible (Barik and Mishra, 2018). Furthermore, the NP toxicity depends upon the shape, size, concentration, amount, administration route, the response of the host immune system, and the time the NPs stay in the bloodstream (Wilczewska et al., 2012).

NPs can be delivered at their target site orally, the direct injection method, and via inhalation. Once it enters into the body, it goes to the circulatory system and interacts with the proteins. This is the first step before it affects other organs (Mu et al., 2014). Next, the NPs are absorbed via the blood capillaries. The lymphatic systems detect the NPs as a foreign substance and try to filter these NPs from the blood. Thus, the host immune system is the main challenging element toward the drug delivery (Park et al., 2016; Alexis et al., 2008). Here, in this chapter we are discussing numerous parameters that can affect drug delivery.

18.2 VARIOUS FACTORS THAT CAN IMPROVE NP MEDIATED DRUG DELIVERY

Factors like size of NPs, temperature, pH, enzymes, antigens, electroporation, and ultrasound can affect the NP mediated drug delivery (Figure 18.1). Below, we are describing all the factors individually.

18.2.1 SIZE OF NANOPARTICLES

NP size has a significant role in delivering drugs to various target sites. Hence, during making formulation of NPs tagged with drugs, various shapes and sizes of NPs are used. Some organs can accept the larger-sized NPs whereas others cannot. Blood brain barrier (BBB) of each organ decides whether the NPs will be accepted or not. Thus, depending upon the characteristic of the BBB, NPs are being designed. The small-sized NPs are more competent than the larger ones (Buseck and Adachi, 2008; Tang et al., 2012; Sung et al., 2007). The NPs less than 100 nm are more competent and can uptake drugs 2.5 fold greater than the NPs of 1 μm (Desai et al., 1997) in size. Thus, the smaller the NPs better the capacity to deliver drugs inside cells (Buzea et al., 2007). Furthermore, the large sized NPs are quickly eliminated via the lymphatic system (van der Zande et al., 2012). Henceforth, the retention and assimilation capability of the drug decreases, and also the time of residing on the targeted site have also been reduced (Chen et al., 2019; Wray and Klaine, 2015).

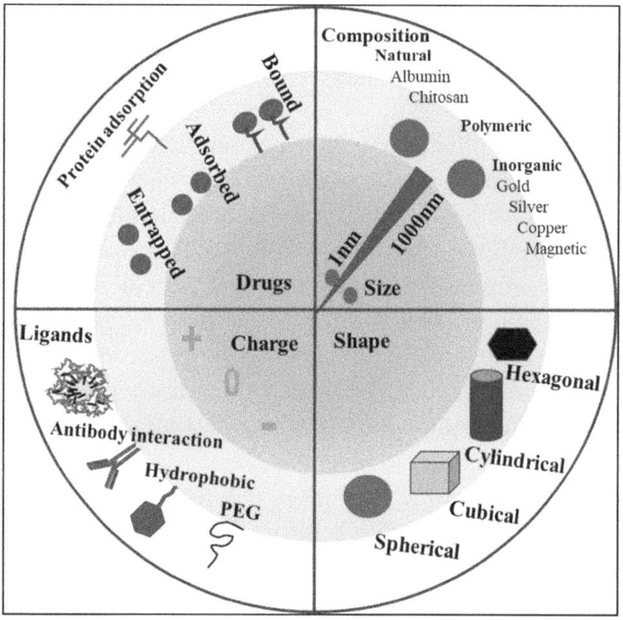

FIGURE 18.1 Structure of a nanoparticle with different compositions, charge, ligand, drugs, shape, and size.

18.2.2 TEMPERATURE

Temperature is a decisive factor in the therapeutic application (Figure 18.2). Cancer cells can be treated using thermal therapeutics. In thermal treatment, a specific region of the body is treated with an elevated temperature to optimize the treatment time (Jaque et al., 2014). Temperature determines the viability and dynamics of the living system, ranging from cells to tissues (Hildebrandt et al., 2002; Chicheł et al., 2007; Wust et al., 2002). In humans, the rise of temperature that is, >37 °C is deliberated as hyperthermia (Mackowiak, 1991; O'grady et al., 1998). Thermal therapy has been used since the 19th century for treating cancer (Habash et al., 2006). Uterine cervix cancer was successfully treated by heating circulating water (Hobohm, 2014; Wunderlich, 1871). Temperature is used to treat (1) irreversible injury treatments (Johannsen et al., 2005), (2) hypothermia treatments (Diederich, 2005), (3) diathermia treatments (Armar and Lachelin, 1993), (4) nanotechnology-based thermal and magneto thermal therapy (Hergt et al., 2006), (5) NPs in photo thermal therapy. NP-based thermal therapy has an upper hand as compared to the other therapies because: (1) NPs are smaller than 100 nm, so it has been incorporated into the cells faster and decrease the circulation time of drugs, (2) easily diffusible in biocompatible liquids, (3) it produces efficient heat when excited externally (Hervault and Thanh, 2014; Chen and Cai, 2015).

Temperature regulative magnetic nanoparticles (MNPs) generate a magnetic field by altering the temperature. So, they are more convenient than the other NPs. A hand full of reports is available to date on temperature-responsive MNPs (Horak et al., 2007; Zhang et al., 2007; Lai et al., 2007). Poly(N-isopropylacrylamide) at a low critical temperature (32 °C) can undergo a transition from the coil to the globular stage. Poly(ethylene oxide, PEO)-poly(propylene oxide, PPO)-poly(ethylene oxide, PEO) block polymers are (Wanka et al., 1994) also used for temperature-dependent drug delivery. A narrow range of temperature variation can induce the block co-polymers to self-reassemble and form vesicles as well as micelles. The mesostructured biocompatible NPs are implemented as drug carriers *in vivo* (Nakashima and Bahadur, 2006). Therefore, by regulating the temperature of NPs, a better targeted drug delivery system can be established.

18.2.3 pH

pH is a physiochemical metabolomics regulator of the living being. The pH of healthy human blood varies within 7.35–7.45. Our stomach is acidic (<pH 3.5) to facilitate easy digestion. The pH of our digestive tract varies from region to region. Therefore, we need a required pH value for the drugs to target the digestive tract. The ideal drug delivery system requires

the zero-release until it reaches the specific site, so that, the drugs will be dispersed in the appropriate buffer and act effectively near the area of interest (Neri and Supuran, 2011; Gao et al., 2010). Doxorubicin (DOX) is an anti-cancer drug, which activates at pH 5.2 to 2.0. At pH 7.4, a small amount of DOX is released, showing a good capping efficiency to the NP (e.g., silica) to avoid the undesirable leaching problem (Zhang et al., 2013). So, before the administration of the drugs into the body system orally or intravenously; pH of the blood, stomach, or the site of treatment must be analyzed using *in vivo* model (Barik and Mishra, 2018).

18.2.4 ENZYMES

Enzymes mediate the delivery of drugs at the specific site of action, which is taken through the oral route. Human alimentary canal possesses various enzymes throughout the digestive tract. Enzymes are catalytic in nature and hence can break the drug before reaching the targeted site (Han and Amidon, 2000; Mura et al., 2013). Therefore, to prevent its degradation from the enzymatic environment before being released to the target site, now the drugs are encapsulated. Cytochrome p-450 3A4 (CYP3A4) enzyme is found in humans (Table 18.1). The P-glycoprotein or multidrug resistance pumps are present in villi of the small intestine, which facilitate absorption. These factors (CYP3A4 and P-glycoprotein) assist in the absorbance of the drugs through villi (Benet et al., 1999; Kamath and Park, 1993; Veronese and Pasut, 2005). These classes of proteins are induced by some substrates and inhibited by specific inhibitors. Cyclosporine and tacrolimus are tested in humans, whereas a cysteine protease inhibitor is tested in the rat model (Aksungur et al., 2011; Pople and Singh, 2010). The inhibitor and inducer of P-glycoproteins and CYP3A4 are inhibited by proteases of cysteine (Paolini et al., 2017). The midazolam, indinavir, and rifabutin are substrate drugs and act upon CYP3A4 enzymes and affect intestinal metabolism (Benet et al., 1999; Sharma et al., 2016; Feleni et al., 2017). Therefore, it is essential to check the enzyme activity and properties before administration of a drug, through the oral or buccal route.

18.2.5 ANTIGEN

Antigen and adjuvants help in drug delivery by inducing and modifying the immune response. Most of the antigens have a particular property, this property decides whether the delivery of antigen will be effective to induce the immune system or not (Storni et al., 2005). Primarily, the antigen-mediated drug delivery depends on the chemical composition of the vaccines (Storni et al., 2005). The peptide vaccine is more difficult for stability and delivery, therefore the response rate toward cancer patients was very low (Ma et al., 2012). To suppress tumor growth, poly(D,L-lactide-*co*-glycolide) (PLGA)-NPs are employed (Ma et al., 2012). Using dendritic cells of humans, artificial antigen presenting cells are generated and loaded with PLGA-NPs to enhance drug delivery. Tumor antigenic peptides are encapsulated by these NPs, and checked for their efficiency (Ma et al., 2012). Chitosan NPs are developed in response to antigenic efficiency using sodium-sulfate precipitation method, to deliver the protein-based antigens (Nagamoto et al., 2004). Proteins as well as the peptides are internalized in a solution by antigen presenting cells through micropinocytosis, hence poorly recognized by MHC-I complex. When these are encapsulated with polymeric NPs, they are recognized easily and show more efficient phagocytosis. So, the polymeric NPs are extra competent than other methods and they can induce T-cell immunity strongly (Hanes et al., 1997; Elamanchili et al., 2004; Kwon et al., 2005).

TABLE 18.1 Nanocarrier Loaded With Drugs and Its Therapeutic Uses

Drugs	Nanocarriers	Therapeutic Activity	Ref.
Doxorubicin, Dox	PLGA	Breast cancer Tumors Antineoplastic agent	Prados et al. (2012) Park et al. (2009)
Cyclosporin	SLN	Dry eye syndrome	Aksungur et al. (2011)
Docetaxel, CYP3A4	PLGA	Hepatic metabolism	Paolini et al. (2017)
Tacrolimus	NLCs	Atopic dermatitis	Pople and Singh (2010), Pople and Singh (2012)
Midazolam	PLGA	Epilepsy diseases	Sharma et al. (2016)
Indinavir	SQD	Antiretroviral drugs	Feleni et al. (2017)

18.2.6 ELECTROPORATION

In electroporation, by applying electric current pores are created in the membrane (Neumann, 1989). Application of electroporation enhances the transdermal transport of drug and blood by analytical extraction method. The use of electric field has two advantages: (1) provide the extra force for the transportation of the molecules which are charged across the skin and transdermal site, (2) the produced ionic motion induces the convective flow across the skin, called as electro-osmosis. These two mechanisms are crucial in the transdermal transport of the molecules during iontophoresis. Here, mainly AC or DC current density has been applied which is more than 0 to 1 mA/cm^2. A constant voltage is applied since the membrane resistance changes with time (Kost et al., 2000). Electric stimuli were used to conduct the polymeric bulk material and to improve the release of the molecules via an implantable electronic device. Poly(3,4-ethylene dioxythiophene) was prepared and coated with poly(L-lactide, PLLA) nanofibers with dexamethasone (Dex). This complex is being encapsulated into the electronic device. After the degradation of the PLLA, the conducting nanofiber provides a precise release of Dex (Abidian et al., 2006). Again, the electrode was coated by polypyrrole/Dex film (Ge et al., 2011), so when the voltage is applied, the electricity triggers the release of the Dex (Wadhwa et al., 2006). Thus, with an increased surface area of NPs the drug loading efficiency increases upon applying electric field (Ge et al., 2011).

18.2.7 ULTRASOUND

Ultrasound is an audio sound, which can travel by air and water, but it has a higher frequency (i.e., 20 kHz) than the hearing frequency of humans. Ultrasound can be reflected, focused, refracted, or absorbed. Ultrasound waves vary physically with a variation of pressure and the movement of molecules occurs due to contraction and expansion (Pitt et al., 2004). Ultrasound drug delivery method is also known as sonophoresis drug delivery system (Figure 18.2). For enhancing drug delivery, ultrasound is incorporated with electroporation, iontophoresis, mechanical pressure, or electric field (Kost et al., 2000). For diagnostic purposes, ultrasound is used in therapeutics applications (Shapiro et al., 2016; Bouakaz et al., 2016).

It helps to deliver the drugs to a tumor with minimal systemic toxicity (Wang et al., 2013; Chowdhury et al., 2016). In liver tumors, ultrasound therapy guides the drug to deliver at the tumor area (Chowdhury et al., 2016). By combining ultrasound, sonoporation, and microbubble (USMB) techniques, pores are formed on the wall of the blood vessels and thus enhances extravascular delivery of therapeutics at the interest site of action (Delalande et al., 2013; Chowdhury et al., 2017). It is safe and allows the drugs to deliver without negotiating with physiological barriers and defense mechanism of tissue in therapeutics (Chen and Hwang, 2013).

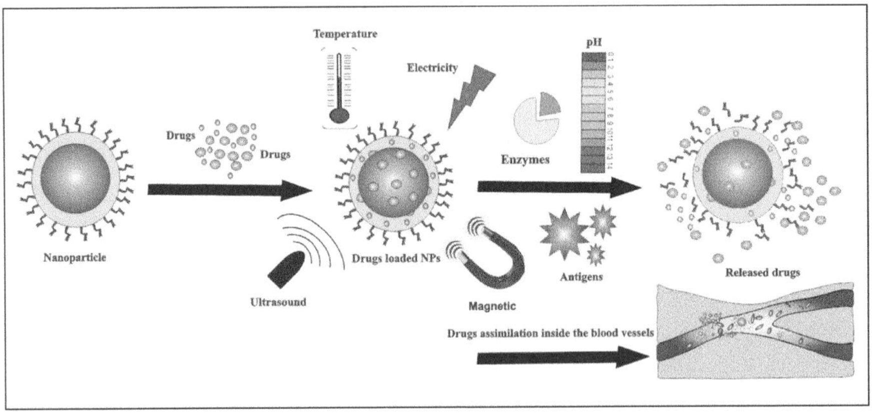

FIGURE 18.2 Various factors that can improve NP-mediated drug delivery system.

18.3 VARIOUS ROUTES OF ADMINISTRATION

Drugs are delivered according to the disease state of the patients. Most of the drugs are delivered via oral and prenatal route. Also, the drugs are administered via buccal, nasal, subcutaneous, and rectal routes. Factors affecting the various route of drug delivery have been discussed below.

18.3.1 ORAL OR BUCCAL

In recent years, macromolecular drugs are delivered using NP carriers. These macromolecules cannot be directly administered to the body via oral, buccal, or by the oromucosal route (Figure 18.3A and B). This route has been successfully administered for systematic and local drug delivery of macromolecules with a low molecular weight (Cario, 2012). Oromucosal route of drug delivery can also be controlled by setting the required parameters. This is the best route for administration because:

1. easy for self-administration
2. high passage permeability
3. moderate pH value
4. absence of keratinized region
5. avoidance of hepatic pass metabolism
6. avoidance of degradation of drug within the gastrointestinal tract
7. lack of enzymatic barrier (Trastullo et al., 2016).

The sublingual region of the oral mucosae is densely packed with dendritic cells and lymphatic vessels via which the antigen added dendritic cells move toward the lymph nodes (Shojaei, 1998; Hovav, 2014). Hence, the sublingual region and the oral buccal cavity are the potentially favorable site for the tolerance of allergen and antigen by inducing a specific immune response. As the mucosal surface is the main route for pathogen entry, therefore it is a challenge for the researchers to administer the drugs safely and effectively. To avoid this condition, buccal mucosal route is used. The mucosal layer can clear the administered drug and antigen by the process of mucosal clearance mechanism. However, now researchers developed a new technique to overcome this problem by developing mucus-penetrating NPs which can tag to mucoadhesive drugs. It can quickly dissolve the mucous films, liquid formulation, sprays, and absorption affinity of the NPs (Masek et al., 2017; Trastullo et al., 2016). For buccal and oral delivery, lipid-based polymeric NPs are used (Tang et al., 2009). For vaccination, nucleic acids, proteins, and peptide NPs are also used (De La Fuente et al., 2008). Polymer like chitosan is commonly used for the formulation of NPs. It can be dissolved in an acidic medium and contributes a positive charge to it. The positive charge interacts with a negatively charged molecule like tripolyphosphate to yield NP for drug encapsulation. Propranolol hydrochloride (PNL), used to treat heart diseases (Newton et al., 2015) is taken orally many times because of its short half-life inside the GI tract. Furthermore, its bioavailability is low due to high first-pass metabolism (Al-Kassas et al., 2016). Therefore, PNL loaded NPs are loaded into mucoadhesive drugs for delivery through the buccal mucosal route, which helps the PNL to overcome this problem (Kraisit et al., 2018). A drug like Zolpidem is taken orally, which has a low affinity toward the site of action. Owing to its small life period, its duration of action is less. To enhance the life span of the Zolpidem on site of action, the researchers developed Zolpidem nanospheres with PLGA encapsulation and incorporate them into the buccoadhesive film.

18.3.2 NASAL

The nasal mucosal membrane is a useful site for drug delivery at a greater extent in combination with decreased first-pass effect with enhancing patient acceptability (Giri et al., 2012). The drugs to be administered through the nose loaded with liquid NPs are effective to treat disorders associated with the central nervous system. It allows direct drug delivery through nose-to-brain by lipid NPs (Cunha et al., 2017). The intranasal drug administration is effective against cardiovascular diseases, infection, pain, and menopausal syndrome (Cunha et al., 2017). The nasal mucosa is highly permeable and easy to access the drug absorption site (Romeo et al., 1998) (Figure 18.3C). However, this path is limited to insulin drug delivery. So, to overcome the problem, many absorption enhancers, solutions of bioadhesive polymers, surfactant, a protease inhibitor, or bioadhesive macromolecules have been proposed (Lee, 1990; Hirai et al., 1978). A new type of chitosan NP formulated using the ionotropic gelation method, which has an excellent capacity for the association of drugs against diphtheria toxin and insulin (Calvo et al., 1997; Fernandez-Urrusuno et al., 1999). Chitosan NPs are more effective than chitosan solutions while absorbing insulin via the nose (Fernandez-Urrusuno et al., 1999). The delivery of drugs from nose to brain routes play a greater value for the large molecular weight of biologically administered solution (Ong et al., 2014). But the toxicity issue is there using nasal mucosa to safely transport the NPs to the brain.

18.3.3 SUBCUTANEOUS

Many drugs are delivered via intravenous injection, as protein-based drugs are incompatible with the oral route of delivery (Figure 18.3D). Hence, many drugs are administered via the subcutaneous route (Tetteh and Morris, 2014) (Figure 18.3E). During administration, the volume and the formulation of drug constituents need to be considered. In the infusion method, the slow administration for a long period of time may be inconvenient, but more than 10 mL can be delivered via this method. However, during delivery through subcutaneous way 1–2 mL of insulin can be delivered (Rejinold et al., 2016). Thus, the subcutaneous method is the only viable alternative to vein infusion (Rejinold et al., 2016). By this method, drugs are infused with the lymphatic system and via a subclavian vein, they go to the systemic circulation (Trevaskis et al., 2015). The starling force generated by the myriad factors are present in lymphatic vessels, help to complete the circulation of drugs from the interstitial matrix to lymphatic capillaries (Berteau et al., 2015; Collins et al., 2017). In this delivery, different properties of drugs like charge, isoelectric point, and the molecular mass are deliberated (Collins et al., 2017). By this drug delivery system, side effects are observed when injected in large volumes (>20 mL). The side effects include bleb formation, edema, swelling, pressure build-up, and erythema (Kurtin et al., 2012). However, these side effects are seen only when the drug administered is more than 20 mL (Shapiro, 2010). As an advancement, the NPs are introduced as a drug carrier to reduce the volume of the drug and enhance the efficacy of drug delivery (Yang et al., 2003; Schoellhammer et al., 2015).

18.3.4 RECTAL

Rectal route is an effective and localized drug delivery method for systemic treatment (Figure 18.3F). This is applicable for children, comatose, and elderly patients (Jannin et al., 2014). This method is banned in some countries due to privacy concerns. The rectal drug delivery is done in post-treatment when patients can not swallow, show regular vomiting or hesitant to parenteral administration (Batchelor, 2014). Advantages of the rectal route are (1) no enzyme in rectum, (2) easy drug delivery for the patients suffering from vomiting, pediatric, geriatric, and difficulty in swallowing (Ban and Kim, 2013; Jadhav et al., 2009), (3) easy for patients suffering from constipation, (4) protects the gastric mucosa from exposing toward irritant drugs. With modern pharmaceutics, rectal drug delivery systems increase the bioavailability and thus control the drug delivery. Various formulation, integration, and principles are applied to modify the rectal drug delivery systems (Purohit et al., 2018; Maisel et al., 2015; Basavaraj et al., 2013). Rectal administration is the shortest route to reach the targeting colon site, but sometimes the drug is not easy to access at a specific site with a specific time period. Rectal administration is sometimes uncomfortable to the patients and may be less effective. The drug delivery via rectal administration may be via foam, suspension, solution, and suppository. The corticosteroid like prednisolone is administrated via rectum for the ulcerative colitis treatment. However, this drug is absorbed by the large bowel due to mainly its topical applications. The amount and the concentration of drugs reaching the colon depend upon the formulation factor, retention time, and efficiency of drug release (Shendge and Sayyad, 2013; Watts and Lllum, 1997). Polymers that are pH sensitive can be coated to drugs and help in targeting the colon (Gupta et al., 2001),

Moreover, rectal drug administration is also applied in microbial infection/disease treatment. Microbicides are developed to fight against the AIDS/HIV retrovirus. Polymers of nanocarriers have been used widely for treating viral infection through rectal/colon administration.

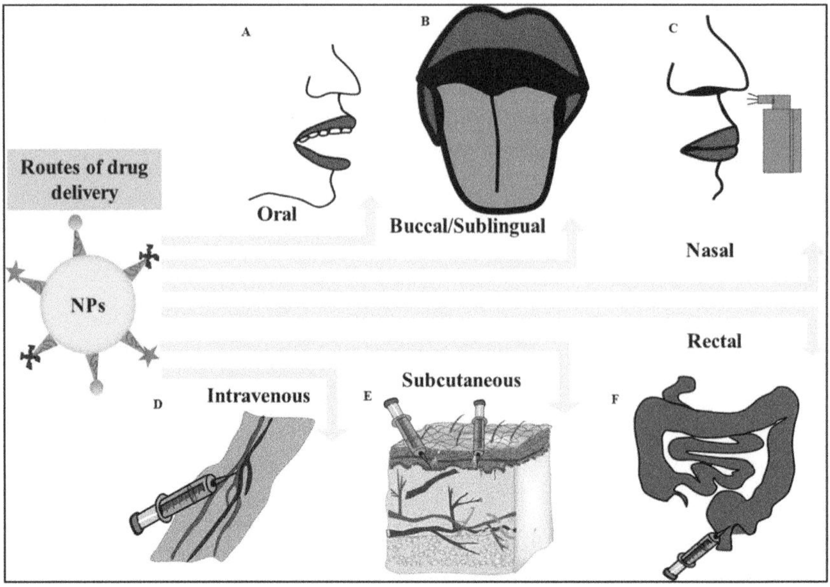

FIGURE 18.3 Various routes of administration of NP-loaded drugs.

18.4 TYPES OF NANOPARTICLES USED FOR THERAPY

NPs are the drug delivery carriers in therapeutics applications. Many types of NPs and modified NPs are nowadays developed for enhancing the drug loading capacity. Thus, biological NPs (nanoghost, nanoneedle, exosome, nanoclew, etc.), inorganic (gold, silver, copper, magnetic NPs, etc.), hybrid, and polymeric NPs are developed to enhance the drug delivery.

18.4.1 LIPOSOMES

Liposomes are spherical in shape with hydrophobic phospholipid bilayer outside and central hydrophilic compartments and its size is <400 nm. The presence of phospholipids, make it highly assessable to the membrane. The liposomal bilayers can be made up of natural or either synthetic phospholipids. Liposomal based drug loading depends upon the constituent of the phospholipid, steric hindrance, permeability charge, and density of the membrane (Figure 18.4). The lipid bilayer merges itself into the interaction between the hydrophobic phosphate group and water molecules to form liposomes (Qiu et al., 2008). The liposome formation is spontaneous due to the amphiphilic property of the bilayer. From the bloodstream, they go to the blood to the target site by the mechanism of extravasation through passive and active strategies into the interstitial space. The drug loading via liposome can be enhanced by altering the pH for loading lipophilic or saturated soluble drugs. After drug loading, the mononuclear phagocyte system can easily remove the liposomal vesicles from the body. This opsonization (it is an immune process wherever foreign particles are directed for damage by an immune cell called as a phagocyte) process is a hindrance during targeted drug delivery by using liposomes. Thus, PEGylation (polyethene glycol) coated with liposomes enhances the drug's half-life. Now the liposomes can be escaped from the clearance mechanism by the immune cells (Malam et al., 2009). These PEG-liposomes are the dose-independent molecules (Allen and Hansen, 1991). At very low dosage they are prone to clearance by mononuclear phagocytes. The PEG-liposome formulations were employed during the study of Kaposi's sarcoma in HIV protein and multiple myeloma, which is used with DOX (Ning et al., 2007; James et al., 1994). Doxil®, a drug used for treating breast cancers is in association with liposomes. Liposomal-Doxil has a lower toxicity than the free DOX. Liposomal drug delivery is an effective treatment over the conventional chemotherapeutics drugs and it is cost-effective. Liposomal drugs are administrated via the ocular, transdermal, intravenous, or parenteral way (Bochot and Fattal, 2012; Schroeter et al., 2010; Gomez et al., 2012). Liposome is not used for oral drug delivery routes since the gastrointestinal enzymes can degrade the liposomal products and thus the drugs have a poor effect on targeted cells. Retro convection can enhance the potentiality of drugs to target the brain (Krauze et al., 2007). Marqibo® is the most recent FDA approved liposomal drug, which helps to treat acute lymphoblastic leukemia (Allen and Cullis, 2013). Now new formulations on liposomal drug delivery are emerged to be used as anticancer drugs and to improve from side effects of chemotherapy (Table 18.2).

18.4.2 NIOSOMES

Niosomes can carry hydrophilic or hydrophobic or multiple drugs at a time. Niosomes are formed by lipids, nonionic surfactants, and a hydration medium. Formerly, they are used in the cosmetic industries, and nowadays it has been implicated for drug delivery (Keservani et al., 2011; Khandare et al., 1994) since it does not have any side effects or toxicity

(Yoshioka et al., 1994). Niosomes can be assembled by the cholesterol and a nonionic, nontoxic surfactant like polyglycerol alkyl ethers, tweens, and span, triton, or glucosyl dialkyl ethers (Table 18.2). They can be of multilamellar or unilamellar vesicles. In water medium, it is enwrapped in the ordered bilayer of a nonionic surfactant and behaves similarly to that of liposomes. For drugs like acyclovir sodium and vancomycin hydrochloride, niosomes are the applicable carriers (Mohamed et al., 2017). High entrapment of niosomes is associated with an equimolar concentration of the cholesterol to the drug and nonionic surfactant. Drugs are entrapped by the carriers for a longer time which is achieved by increasing the concentration of cholesterol in niosomes. To increase the entrapment volume, cholesterol is needed. Due to its charge, it can enhance the interlamellar distance and decrease the size of the vesicles, and increase in membrane curvature (Schreier and Bouwstra, 1994). Niosomes can entrap the drugs/solute more efficiently than liposome. They are osmotically active, enhance drug stability, improve oral bioavailability, enhance skin penetration, easy to store and handle the surfactant, possess both hydrophilic as well as hydrophobic moieties together, biodegradable, nonimmunogenic, biocompatible, make the drug to reach the targeted site by parenteral, oral or topical routes. All these features make niosomes an ideal drug carrier than liposomes (Namdeo and Jain, 1996; Moghassemi and Hadjizadeh, 2014). They can protect the drugs from the cellular environment and delay the clearance from mononuclear phagocyte system by circulation and enhance therapeutic performance.

18.4.3 POLYMERIC NANOPARTICLE

Recently, polymeric nanoparticles are highly used as carriers. Polymeric NPs are biocompatible and biodegradable (Marin et al., 2013), but sometimes they are a little immunogenic due to different properties and characteristics. Synthetic polymeric NPs are well-characterized polymers so that it can be handled easily. These polymers are used for the drug carrier with sugar, proteins, and various macromolecules which are biocompatible and pharmaceutically approved (Mansour et al., 2010). Considering the physical and chemical properties of drugs, the polymeric nanoparticles are prepared and utilised (Siddiqui et al., 2020; Bag et al., 2020; Mishra et al., 2019) (Figure 18.4).

18.4.4 NANOFIBER

Nanofibers are prepared by the electrospinning method (electrospun). In this method, high voltage is being applied to make a polymeric solution. In the polymeric solution, the stable drop shape is maintained due to the surface tension generated after applying voltage. When the electric potential reaches a critical point, it makes the liquid to a cone-shaped structure known as Taylor cone. By applying an electric current, the solvent evaporates and the solution gets solidified with a nanofibrous structure. Thus, the nanofibrous mesh gets synthesized. Electrospinning method depends upon various properties of the nanofibrous drugs: they include (1) molecular structure and weight; (2) solution properties of polymer-like charge, viscosity, dielectric constant, surface tension; (3) processing electric potential, rate of flow of concentration; and (4) temperature, velocity, and humidity of electrospinning chamber (Yoo et al., 2009; Sill and von Recum, 2008; Meng et al., 2011). The biomimicking nanofibrous meshes (collagen fibers) are in demand due to their structural similarity with the extracellular matrix of a human (Yoo et al., 2009). Protein, carbohydrate, and nucleic acids are added as a bioactive molecule in the electrospinning method. It has been modified to immobilized form and incorporated to the surface of nanofibers to increase the bio potential drug delivery (Bhattarai et al., 2006). Paclitaxel, a biodegradable hydrophobic anticancer drug is directly electrospun to have a drug-releasing nanofiber mesh (Xie and Wang, 2006). Proteins besides nucleic acids have hydrophobic charged macromolecules, which are immobilized covalently as well as physically, with the surface of a nanofibrous mesh for moderating various cellular functions. These electrospun nanofibers mesh have extremely interconnected exposed nanoporous sheets along with a specified surface area for local and sustained drug delivery (Yoo et al., 2009).

18.4.5 NANOPORE

Among various new nanomaterials, the nanopore and nanotube contribute a lot in the drug delivery system. They exhibit high surface area, high surface chemistry, low-cost fabrication, rigidity, chemical, and mechanical resistivity (Kayser et al., 2005; Hughes, 2017; Lavan et al., 2003; Vallet-Regi et al., 2007). Mesoporous silica NP is used for drug delivery systems from the last two decades. It is synthesized, organically by an electrochemical process with mesoporous silicon (Anglin et al., 2008; Salonen et al., 2008). However, now the researchers have focused on developing nanopores and nanotubes

(Ghicov and Schmuki, 2009). Nanoporous silica, alumina oxides, and titania nanotubes are the best examples of the nanoporous carrier to be used for drug delivery in therapeutics (Grimes, 2007; Schmid, 2002).

18.4.6 NANOTUBE

Delivery of bioactive molecules for cancer therapy is a great challenge due to the poor pharmacokinetics, low permeability, and less liver/kidney clearance. This can be overcome by protecting the nanomaterials by encapsulation, delivering more than one type of drug, targeting specific tumor, controlling the physical and chemical property of the drug by pharmacokinetics. The Apo2L and TRAIL are the pro-apoptotic receptors, which induce apoptosis in cancerous cells. So, PLGA (Kim et al., 2013), PEG (Lim et al., 2011; Park et al., 2009), hydrogels (Kim et al., 2010), and human serum albumin NPs (Bae et al., 2012) can be made by the combinations of Apo2L and TRAIL which can act as a carrier for targeting the cancerous cells. These studies show the improved quality and sustained delivery of Apo2L and TRAIL in cancer therapy. Carbon nanotubes, due to their hollow cylindrical structure, high surface area, and capacity to cross the barrier are widely used as nanocarriers (Kam et al., 2006). Carbon nanotube (CNT) is a one-dimensional large cylindrical molecule, having a hexagonal arrangement of carbon at the center with multilayered graphene sheets in its wall (Figure 18.4). CNT exhibit in two forms, such as single-walled CNT and multiple walled CNT. Single-walled CNT is more compatible and stable than the multiwalled CNT. Therefore, single-walled CNT is used for (1) targeted drug delivery to the lymph (gemcitabine, polyacrylic acid-magnetic NP) (Liu et al., 2009), (2) tumor (paclitaxel, chemophore) (Liu et al., 2008), and (3) central nervous system (DOX) (Liu et al., 2009) (Table 18.2). However, instead of many applications of CNT in drug delivery as a carrier, many studies have also reported its biosafety and toxic effects on tissue and cells *in vitro* in addition to *in vivo* (Warheit et al., 2004).

18.4.7 DENDRIMER

"Dendrimer" is a Greek word which means branches of the tree. A dendrimer is a polymer of multiple monomers, which have a property to self-reunite (Warheit et al., 2004). Dendrimer NPs have encapsulation properties (Turnbull and Stoddart, 2002) and thus have a highly ordered structure (Lee et al., 2006)

(Figure 18.4). Many polymers are encapsulated with the dendrimer and thus upsurges the potential for drug delivery. Poly L-glutamic, polyamidoamine, melamine, polyethyleneimine, polypropylineimine, chitin, and poly-ethylene glycol are incorporated with the dendrimers (Table 18.2). These NPs are used in cancer therapeutics as targeting molecules (Majoros et al., 2005), gene therapy, antisense property, and in magnetic resonance imaging (MRI). Dendrimer drug delivery has been focused on drug encapsulation which regulates the discharge of the drugs incorporated with the dendrimer. Later, dendronized polymers were developed by combining dendrimer and polymer, which possess the linear series of a polymer having dendrimer in each repeating unit. Therefore, it enhances drug delivery. DOX is conjugated with the dendrimer and optimized with the blood circulation time according to their molecular composition and size (Lee et al., 2006). DOX dendrimer is applied in multiple drug delivery systems as incorporated with the PEGylation and drugs can release through hydrazine dendrimer linkage (Singh and Lillard Jr, 2009). DOX can also be applied on mice model suffering from colon carcinoma and tumors, resulting in complete regression of tumor cells. Furthermore, the mice can recover 100% after 2 months of treatment. DOX dendrimers are more efficient and not as much toxic than free DOX and increase their strength by nine-fold than the free DOX (Singh and Lillard Jr, 2009).

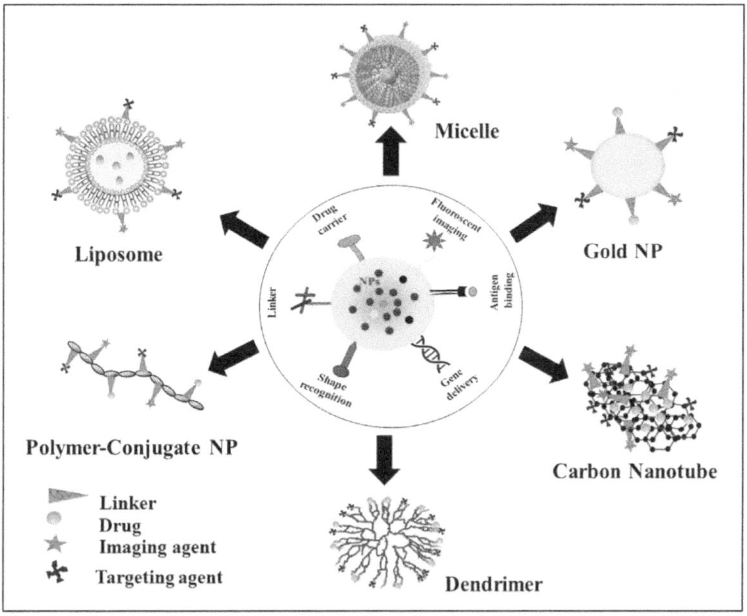

FIGURE 18.4 Various nanoparticles used in drug delivery as a therapeutic agent.

18.4.8 XPCLAD® NANOPARTICLE

NPs are sparingly soluble in the aqueous solution. It offers greater bioavailability to be absorbed easily. Recently, researchers have developed an NP which has more retention capacity and soluble in the aqueous drug solution. These NPs are named as XPclad® NPs. These are formulated by using vibratory ball milling and planetary ball milling techniques. In the ball milling procedure, it grinds the powder by maximum velocity, while in the case of vibratory milling, the particle is broken by grinding with respect to motion. Moreover, the planetary ball mill is being utilized in many industrial sectors and vibratory ball mill procedure has been applied in medical industries. In general, XPclad® NPs are of 5–30 nm and can be modified up to 10–60 μm by controlling its size, speed of ball mill, cycle, and centrifugal forces. It is more convenient than the other NPs due to its uniform size, efficiency to load hydrophilic in addition to hydrophobic drug, oral, cutaneous, and systemic routes of administration, easy to protect, controlled, and targeted release. Thus, XPclad® NPs are made up of biopolymers like starch, alginate, collagen, and so forth (Table 18.2). XPclad® NPs are employed in cancer drugs, therapeutic proteins, and vaccines (Koo et al., 2006). Novel XPclad® NPs are used in tumor therapy due to their less toxicity and capability to kill prostate cells tumor and Treg cells (Singh and Lillard Jr, 2009). The XPclad® NP's surface is modified either with PCL or PEG. By the combination of folate and antibodies, surface molecules can be modified whereas the inner core can be amended to the hydrophobic and hydrophilic core by drugs, proteins, nucleic acids, metal ions, peptides, a fluorophore, etc., to deliver the active encapsulated agents. For example, XPclad® NPs coated with folic acids are used to treat mice with PC3 tumors. XPclad® NPs encapsulate therapeutic proteins for oral delivery. Monoclonal antibodies can act as antiprotective antigens, which can also be incorporated with Clad® NPs to enhance passive immunity against anthrax toxin (Singh and Lillard, Jr 2009) (Table 18.2).

18.4.9 QUANTUM DOTS (QDS)

Quantum dots (QDs) have size less than 10 nm. It fluoresces brightly in the presence of the laser light source. The fluorescence of QD differs depending on the size of their crystals (Dey and Rao, 2011). It has a narrow range of spectral band with narrow excitation spectra. It holds photonics, electronics and thus, has wide applications in the arena of electronics and photonics. The luminescent properties of these semiconductor nanocrystals are sensitive to the environment and surface preparation of nanocrystals. The nanocrystals are enclosed in a geometric core-shell of the wider semiconductor bandgap, which resulted in the increased quantum efficiencies and photochemical stability (Jain et al., 2007; Alivisatos et al., 2005). The QDs range from 500 nm produce the light-emitting diode in the blue range and it has been used in solid-state semiconductor to generate luminescent light. It is used in cellular and biomolecular imaging. The unique properties of QDs include (1) resistance against photobleaching, (2) improved signal strength, (3) size-dependent tunable light emission, and (4) simultaneous multiple color excitation fluorescence. The quantum dot size determines its emission frequency. Its size includes 680 nm and 620 nm (emits maple red-orange), 600 nm (fort orange), and 520 nm (green color).

QDs have many unique properties such as (1) it can be changed to various shapes and combined with a variety of biomolecules, (2) the designer atoms offer several optical and electrical properties, (3) QD luminescence under UV light, has red and bright green color, (5) these fluorescent QDs belong to group II to VI and III to V of periodic table, that is, Te, Se, Zn, Ag, Cd, Ln, P, and Pb, etc., (6) the wavelength will be shorter with a decrease in size of the QDs, (7) the excitation range is wide, (8) it has precise emission wavelength, thus multiple fluorescence emission spectra do not overlap (Rathore et al., 2006).

QD has numerous advantages: (1) highly resistant to degradation as compared to other imaging probes and can track the cells for a longer time, (2) have size-tunable emission from UV to IR, (3) gives good contrast in imaging due to the presence of nanocrystals, (4) small size allows it to be incorporated with many materials like polymer matrices, fabrics, or liquid mixtures (Rathore et al., 2006).

All these properties make QD an ideal nanomaterial for various clinical applications. Luminescent and stable QDs are used to visualize the cancer cells. Under the fluorescent microscope, they help to detect the damaged cells at higher magnification. In breast cancer detection, QDs are labeled with the polyacrylate cap that covalently encapsulates with the antibodies for the immunodetection of breast cancer marker *Her2*. It helps to detect viral diagnosis such as respiratory syncytial virus (Zhong et al., 2003). In pharmaceutical fields, QD detects the drugs and carriers at a time by incorporating within it (such as in polymers, liposomes, and so forth) (Sonvico et al., 2005). To detect the metallic NPs, MRI helps a lot. Nontargeted near IR emitting QD core t2-MP EviTags is tested in mice models having a tumor (Sonvico et al., 2005). In an immune-based assay, it is used for detecting the antigen and antibody by labeling with the fluorophore. ZnS coated with

CdSe QDs are used to detect antibodies (Sun et al., 2001). QDs are used to distinguish tumor in brain cells, cell death, DNA labeling, cell tracking. NP probes having multiple functions based on QDs are utilized for detecting and targeting cancers in living animals and evaluating multiple biomarkers in them (Wang et al., 2007; Stroh et al., 2005; Vu et al., 2005). The application of QD as a carrier to both cells and small animals are mentioned in the table (Table 18.2).

18.4.10 CERAMIC NANOPARTICLE

Ceramic NPs are porous, high heat resistant, and their size is less than 50 nm. Hence, they are ideal for drug delivery. They are formulated with phosphate, oxides, carbides, carbonates of metals on metalloids such as silicon, titanium, calcium, zirconia, alumina, hydroxyapatite, etc (Gautam et al., 2019; Gautam et al., 2021; Ekka et al., 2021; Nayak et al., 2021). They are used as a drug carrier against glaucoma, bacterial infection, and cancer. They can easily regulate the discharge of the drugs. They can carry the high load and easily incorporated into the hydrophilic as well as hydrophobic system. So, they are an excellent drug carrier for proteins, genes, drugs, and imaging agents. Bioceramic NPs are used for tissue regeneration therapy and in bone engineering. These bioceramic NPs interact with the tissues and cells to boost the formation of mineralization and osteogenesis (Tautzenberger et al., 2012). The small size allows it to penetrate the cell membrane and deliver the drug (Langer and Peppas, 2003; Sokolova et al., 2012). Metal incorporated ceramic NPs are used for various applications. (1) aluminum ceramic NPs in the bone scaffold (Webster et al., 2001), (2) calcium phosphate ceramic NPs in transfection (Jordan et al., 1996), gene silencing (Sokolova et al., 2012), and bone regeneration (Dorozhkin and Epple, 2002), (3) silica ceramic NPs in drug delivery gene silencing, and transfection (Hom et al., 2010; del Campo et al., 2005), (4) gold ceramic NPs in gene silencing, anticancer agent, imaging, and in transfection (del Campo et al., 2005; Rosi et al., 2006), (5) CNT-ceramic NPs in gene silencing and transfection (Balani et al., 2007) (Table 18.2).

18.4.11 NANOSHELLS

Nanoshells are also called core shells with a thickness of 1–20 nm. These are spherical and have a core concentric particle surrounded by the shells of other sheets (Jeong et al., 2005). Nanoshells are used in therapeutic applications (Loo et al., 2004) and in biomedical imaging. They have versatile features which have the optical property and reduced susceptibility to temperature and chemicals.

18.4.12 FULLERENES

Fullerenes are composed of carbon molecules, forming the hollow sphere, tubes, and ellipsoid structures. They share a similar structure with the graphite. Nanotubes are examples of cylindrical fullerenes and have carbon constituents only. Its size ranges from a very small nanometer to less than a 100 nm size. Fullerenes have both an open end and a closed-end, which make the fullerenes to act as an ideal medicinal drug delivery system (Holister et al., 2003). Due to this property, fullerenes have a high binding capacity for specific antibiotics to resist the bacteria or any other organism. It is also used as an antimicrobial mediator that can be activated by light and used in targeting the cancer cells (Tegos et al., 2005).

18.4.13 PROTEIN NANOPARTICLES

Protein-based NPs are categorized under the colloidal NPs. A colloidal NP has more advantages than the other drug carrier system. Proteins NPs are biocompatible, biodegradable, and nonantigenic. Furthermore, they are metabolizable and can easily modify the surface and covalently incorporated with the drugs and ligands. Proteins NPs have many structures as primary, secondary, and tertiary structures, it can easily alter the surface and modification of covalent drug attachment. Formulation of protein-based NPs is made using albumin, gliadin, and gelatin. Protein-NPs are used for protein drug delivery to pulmonary structures or be incorporated within the polymer of the ionosphere or microsphere for controlled release. Gelatin protein NPs can be achieved by the hydrolysis of fibrous, collagen, and insoluble proteins, which are found abundantly in skin and bones. Albumin protein is used for the macromolecular carrier and to prepare nanocapsules and nanospheres. In albumin NPs, ligands are easily attached to the covalent linkage. It is used as the drug carrier for antitumor drugs (Takakura et al., 1990) (Table 18.2).

TABLE 18.2 Nanoparticles with Coated Drugs, Route of Administration, and Its Therapeutic Applications

Nanoparticle	Coated With	Drugs	Route of Administration	Therapeutic Applications
Liposomes	PEGylation.	Doxorubicin, Marqibo.	Ocular, transdermal or subcutaneous, intravenous or parenteral	Multiple myelomas, breast cancer
Niosomes	Polyglycerol alkyl ether, cholesterol, Glucosyl dialkyl ether	Acyclovir sodium, Vancomycin hydrochloride	Oral, ocular, transdermal or subcutaneous, intravenous	Anticancer drugs development
Polymeric	Sugar, proteins	Dactinomycin and methotrexate	Oral, parenteral	Antitumor drugs development
Nanofiber	Carbohydrate, proteins, nucleic acids	Paclitaxel	Transdermal or subcutaneous	Anticancer drug development, antibiotics
Nanotubes	PLGA, PEG, hydrogel, HSAN	Gemcitabine, paclitaxel, doxorubicin	Intravenous, subcutaneous, nasal	Cancer therapy, imaging
Dendrimer	Poly L-glutamic, polyamidoamine, melamine, chitin, PEG, PEGylation	Doxorubicin	Oral, rectal, transdermal	MRI, cancer therapy, gene therapy, tumor therapy
XPclad®	Cellulose, alginate, starch, collagen, PCL-PEG	Folate, cisplatin	Cutaneous, oral	Tumor therapy, cancer drugs, therapeutic protein and vaccines, anti-anthrax toxin
Quantum dots	Polyacrylate	t2-MP EviTags	Intravenous	Electronics and photonic, biomolecular and cellular imaging, immunodetection
Ceramic	Phosphate, carbides, metalloids	Paclitaxel, cisplatin	Oral, nasal	Cancer, glaucoma, bacterial infection, gene silencing, bone scaffold
Protein	Albumin, gliadin, gelatin, PEG	Insulin, tetanus, calcitonin	Oral/buccal, nasal, cutaneous	Antitumor drugs

18.4.14 INORGANIC NANOPARTICLES

18.4.14.1 GOLD

Gold NPs are extensively used, because of their unique structure, dimension, a surface modification which allows releasing the drug in a controlled manner (Han et al., 2007) (Figure 18.4). The gold NP has distinctive physicochemical properties for unloading and transporting the drugs. Being a novel metal, it is inert and less toxic as reported from former studies (Connor et al., 2005). Besides that, it can be synthesized easily at the desired size (1–150 nm) (Ghosh et al., 2008). All these characters and properties make gold NP a biocompatible drug delivery system. Gold NPs are used for cancer treatment and immunodiagnostic purposes. Moreover, the colloidal gold NP has been implicated in a drug delivery system by liposomal and biodegradable polymers carrier (Muller et al., 2000). Colloidal gold NPs are applied to TNF for tumor-targeted drug delivery (Paciotti et al., 2004). Methotrexate is also an anticancer drug, which is loaded with 13 nm gold NPs and successfully administrated to treat cancer (Chen et al., 2007). Drug delivery success to the target site enhances the opportunity to treat various tumors.

18.4.14.2 SILVER

Silver is used in the preparation of NPs. They have antibacterial, antifungal, antiviral, and antioxidant activity. It can further enhance the thermal, electrical, optical, and catalytic properties (Ivanova et al., 2018; Phull et al., 2016; Le Ouay and Stellacci, 2015). Depending upon the anticancer activity, silver NPs can suppress the tumor cell growth and arrest the cell cycle on the human cell line (Patra et al., 2015). Silver NPs are synthesized by physical, chemical, and biological methods. For biological method plant sources, from leaves (Yu et al., 2007; Sudha et al., 2017; Gomathi et al., 2017), root (Suman et al., 2013; Singh et al., 2016), and barks are used (Bar et al., 2009; Johnson and Prabu, 2015). The laser ablation method (Tsuji et al., 2002) and the laser irradiation method are also used for the preparation of silver NPs. Silver NPs prevent and treat diseases like cauterization, skin wound healing, and pleurodesis (Klasen, 2000).

18.4.14.3 COPPER

Copper sulfide (CuS) NPs are developed by photothermal ablation method to treat cancer. In photothermal ablation, a laser light source focuses on tumor cells, and the absorbed light energy is transferred into heat energy and kills cancer cells. The copper sulfide NPs shows electrical, catalytical, and optical properties with an application of photo degradation of pollutants (Wang et al., 2009), DNA detection (Zhang et al., 2008), biological labeling (Dobson et al., 2001), laser light monitoring, and eyes protection (Yu et al., 2007). They react with near-infrared light and can generate heat, which could harness the photothermal activity of cancer cells (Li et al., 2010). Its small size, good photostability, water-solubility, narrow emission bandwidth, and enhanced surface area make CuS NPs a good carrier to deliver drugs. CuS NPs are also used in the detection of cellular components for selective imaging and therapy (Brannon-Peppas and Blanchette, 2012; Brigger et al., 2012). Hollow CuS NPs (HCuS) can also be made by engineering the particles possessing a photothermal coupling effect. HCuS NPs are delivered with the drug by the transdermal route. The CuS NPs unable to penetrate the dermal layer of skin efficiently, due to the presence of many layers on the transdermal that is lipid reach stratum corneum barrier that cannot pass the macromolecules easily. The HCuS NPs can penetrate the cutaneous barrier and easily reach the target site (Ramadan et al., 2012).

18.4.14.4 MAGNETIC IRON OXIDE NANOPARTICLES

MNPs are used in MRI (Dobson, 2006; Corot et al., 2006), magnetic cell sorting and immunopathological assay. The magnetic field treats cancer cells by decreasing body temperature (Pankhurst et al., 2003). Magnetic NP carriers can effectively reach the tumor cells and deliver the drug thereby minimizing the chances of systemic infection (Dobson, 2006). The efficiency of MNPs can be increased by reducing the size and generating nonfauling polymers on it. This enhances the stealthiest and increases the circulation time within the blood (Moghimi et al., 2001). The loaded drug complex carrier is then injected intravenously or via intra-arterial route. Here, the high gradient earth magnetic external field has been generated which guides and locates the drugs at the tumor site. The MNP carrier after reaching the target site *in vivo*, the drug therapeutics agents are released from the magnetic carrier by enzymatic activity or by changing the temperature, pH, and osmolarity of the targeted site of the tumor (Alexiou et al., 2000). Advantages of MNPs are (1) image visualized by superparamagnetic NPs, (2) heating of the magnetic field triggers the drug release, (3) guided or held by the magnetic field, and (4) generated heat produced hyperthermal ablation of tissues.

18.4.15 BIOLOGICAL NANOPARTICLES

18.4.15.1 EXOSOME

Exosomes are the tiny biological NPs which transfer the signal or information between the cells and offer the potential to treat and diagnose disease. The diameter of the exosomes varies from 40 to 200 nm (Figure 18.5). This offers a unique feature to cell-specific tropism toward the originated cells. These are utilized as a drug strategy to transport cargo consisting of proteins, drugs, and miRNA. Exosomes are biologically originated NPs, contain natural lipid bilayers and immunogenicity. It can clear various drugs from the body and easily pass the BBB, which is beneficial for designing personalized therapeutic drugs (Suri et al., 2007).

18.4.15.2 NANOBUBBLE

Nanobubbles are spherical and nano-sized structures filled with gas molecules and stabilized by the lipid shells (Figure 18.5). Nanobubbles combined with ultrasound, thermal, acoustic, or magnetic sensors are used for imaging and drug delivery. They are stable and showed a longer time of residence in the systemic circulation. Nanobubbles are introduced to plasmonic nanobubbles and applied in the cell theranostic process by optical scattering. This targets damaged portions of the cell by a mechanical and nonthermal way (Suri et al., 2007).

18.4.15.3 NANOCLUSTER

Nanoclusters are the self-assembled NPs formed of either polymers or organic molecules of smaller size. These polymers are cross-linked with the silver, gold, or magnetic particles (Diez and Ras, 2011; Xie et al., 2009; Ditsch et al., 2005) (Figure 18.5). Nanoclusters show both molecular and fluorescence properties, which makes them a perfect carrier for drug delivery, bioimaging, and biosensing.

18.4.15.4 NANONEEDLE

Nanomaterials enhance the bioavailability of medicines and drugs, without any harmful effects (Shende et al., 2018). In nanoneedle, the drugs are encapsulated to increase the penetration through the transdermal route and to promote the controlled release of drugs (Yum et al., 2010; Shende et al., 2018) (Figure 18.5). Different nanoneedles have been developed rendering to the administration route.

1. **Solid nanoneedle:** These nanoneedles are utilized for the pretreatment of skin, by increasing the porousness of the skin. These are fabricated by the polymer solutions mixed homogenously with drugs (Han et al., 2005). The loading of drugs delivered to a wide range of bioactive molecules like nucleic acids which are impermeable to cell and nuclear membrane. By this loading, nanoneedle can expose radially to the drug solution causing its rapid release (Chiappini et al., 2015).
2. **Porous nanoneedle:** The large surface area and pore volume provide a good pool to load the drugs and improves the payload density. In the electrostatic loading method on nanoneedle, the porous structure can be achieved by loading of drugs in the form of powder, which gives the higher loading concentration with the amorphous phase. The mesoporous needle is used for better penetration and well positioning for loading of nanoneedle. Various NPs have been implicated successfully within porous nanoneedle to enhance the therapeutic properties. Liposome loaded with siRNA into porous nanoneedle. Target genes for silencing for over 21 days: gold NPs are loaded for enhancing the MRI contrast potential, and gold nanoshells are improved for photothermal effects (Chiappini et al., 2015).
3. **Hollow nanoneedle:** These nanoneedles are placed along with the drug reservoir to the cytosols of the cell. In hollow nanoneedles, drugs are not loaded, but it only acts as a channel for drug delivery system (Peer et al., 2012). These are employed for the drug infusion to the skin. The drugs are targeted inside the needle in a cavity.

Porous silicon nanoneedle is made up of cathepsin B sensor. This sensor is operated for uncovering normal from tumor cells (Kawamura et al., 2016). Due to the transdermal delivery of the nanoneedle carriers' drugs have been delivered to the target site. It is useful in the delivery of toxic drugs since this delivery route enables the drug to reach on targeted sites. Injection of nanoneedle provides pain-free drug delivery without affecting the nerve endings. These are safe and easy to deliver the drug in the particulate form (Shende et al., 2018; Kolhar et al., 2011).

18.4.15.5 NANOCLEW

Nanoclew is a DNA-related biocompatible carrier. It can assemble itself to look like a cocoon or clew by rolling circle amplification (Suri et al., 2007). Here, the ssDNA makes the whole cocoon (Figure 18.5). Nanoclew is used to treat cancer cells. DNA nanoclew is used to shuttle CRISPR-Cas9 gene-editing tools into cells.

18.4.15.6 NANOGHOST

Nanoghost technology nowadays proposes a novel system for drug delivery. It is derived from the nano-vesicles and is made up of mesenchymal stem cell membrane by maintaining their unique surface. It is created by the scalable and reproducible hypotonic treatment and homogenization, to remove the cell cytoplasm to form the ghost cells and reduce the cell size to produce nanoghost (Figure 18.5). Nanoghost is associated with capabilities, which make allogeneic mesenchymal cells immune evasive, highly tolerated, biocompatible, and can be used in multiple administrations. Nanoghost loaded with pDNA efficiently targeted and transfected to different cancer cells.

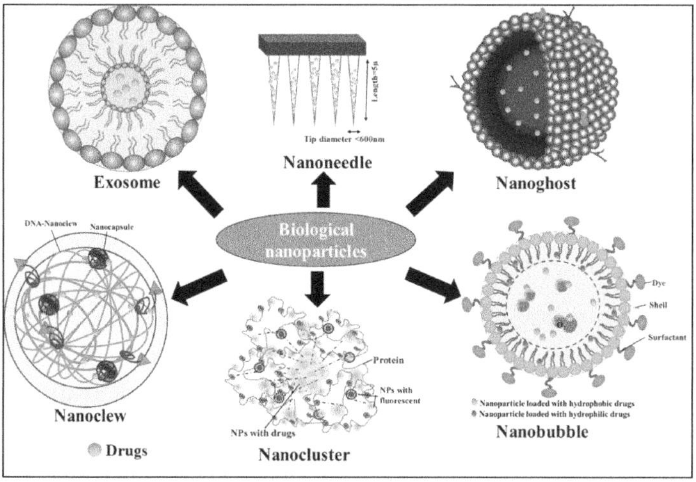

FIGURE 18.5 Biological nanoparticles used as a drug carrier.

18.4.16 HYBRID NANOPARTICLES

18.4.16.1 NANO-TERMINATOR

Nano-terminators are the biodegradable liquid metals that are the drug-loaded nanodroplets. It has a core containing liquid phase eutectic-gallium-indium and it is subsequently doped with thiolated polymeric shell and hyaluronic acid to attack the tumor/cancer cells. These are injected into the bloodstream directly and are absorbed into the tumor cells. Once the liquid metals get dissolved, the other carrier drugs are released due to the presence of highly acidic tumor surrounding. DOX an anticancer drug (Prados et al., 2012) is used in this nanodroplet liquid solution and form nano-Dox which can be injected into the bloodstream (Lu et al., 2015).

18.4.16.2 INJECTABLE NP GENERATOR

Injectable nanoparticle generator (iNPG) is micrometer in size and loaded as chemotherapeutic drugs. iNPG is loaded with DOX to poly(L-glutamic acid), which is also sensitive to pH. It is linked with a cleavable linker and loaded with a polymeric drug that is pDox to assemble and form iNPG-pDox (Bae et al., 2003). iNPG consists of various nonporous particles like silica NP (Xu et al., 2016; Tasciotti et al., 2008). Whenever pDox is released from iNPG, it forms a nanosized particle spontaneously in the presence of a water medium. When this iNPG-pDox is injected intravenously at the tumor sites, it gets accumulated and taken up by tumor cells. pDox NPs are trafficked closer to the nucleus of the cell, and then cleaved into DOX. Therefore, it avoids the excretion of the drug via efflux pumps. Thus, iNPG loaded NPs are used to deliver drugs to breast cancer patients at the metastatic stage (Xu et al., 2016).

18.5 CONCLUSION

NPs are used as a carrier to deliver drugs in the therapeutic drug delivery system. Besides its useful applications, some drawbacks are also seen. Innumerable reports are available for the toxicity of NPs toward various cells, tissues, and model organisms (Mishtra et al., 2021; Barik et al., 2018). To overcome this problem, researchers developed many techniques and equipment for the convenient use of the NPs as a carrier to deliver drugs. With all these nano techniques NP will become a boon for mankind.

ACKNOWLEDGMENTS

JB is thankful to BT/PR21857/NNT/28/1238/2017 for financial support. SS is thankful to DST-INSPIRE for financial support. MM lab is supported by Grant No. BT/PR21857/NNT/28/1238/2017, EMR/2017/003054, Odisha DBT 3325/ST(BIO)-02/2017.

KEYWORDS

- **fullerenes**
- **nanoclew**
- **nanoghost**
- **niosomes**
- **quantum dots**

REFERENCES

Abidian, M.R.; Kim, D.H.; Martin, D.C. Conducting polymer nanotubes for controlled drug release. *Adv. Mater.* **2006**, *18(4)*, 405–9.

Aksungur, P. M.; Demirbilek, E.B.; Denkbas, J.; Ludwig, V. A.; Unlu, N. Development and characterization of Cyclosporine A loaded nanoparticles for ocular drug delivery, Cellular toxicity, uptake, and kinetic studies. *J Control Release.* **2011**, *151(3)*, 286–94.

Al-Kassas, R.; Wen, J.; Cheng, A.E.-M.; Kim, A.M.-J.; Liu, S.S.M.; Yu, J. Transdermal delivery of propranolol hydrochloride through chitosan nanoparticles dispersed in mucoadhesive gel. *Carbohydr. Polym.* **2016**, *153*, 176–86.

Alexiou, C.; Arnold, W.; Klein, R. J.; Parak, F. G.; Hulin, P.; Bergemann, C.; Erhardt, W.; Wagenpfeil S.; Luebbe, A.S. Locoregional cancer treatment with magnetic drug targeting. *Cancer Res.* **2000**, *60(23)*, 6641–8.

Alexis, F.; Pridgen, E.; Molnar L. K.; Farokhzad, O. C. Factors affecting the clearance and biodistribution of polymeric nanoparticles. *Mol. Pharm.* **2008**, *5(4)*, 505–15.

Alivisatos, A. P.; Gu W.; Larabell, C. Quantum dots as cellular probes. *Annu. Rev. Biomed. Eng.* **2005**, *7*, 55–76.

Allen, T.; Hansen, C. Pharmacokinetics of stealth versus conventional liposomes, effect of dose. *BBA-Biomembranes.* **1991**, *1068(2)*, 133–41.

Allen, T. M.; Cullis, P. R. Liposomal drug delivery systems, from concept to clinical applications. *Adv. Drug Delivery Rev.* **2013**, *65(1)*, 36–48.

Anglin, E. J.; Cheng, L.; Freeman W. R.; Sailor, M. J. Porous silicon in drug delivery devices and materials. *Adv. Drug Deliv. Rev.* **2008**, *60(11)*, 1266–77.

Armar, N.; Lachelin, G. C. Laparoscopic ovarian diathermy, an effective treatment for anti-oestrogen resistant anovulatory infertility in women with the polycystic ovary syndrome. *An International J. Obstet. Gynaecol. Res.* **1993**, *100(2),* 161–4.

Bae, S.; Ma, K.; Kim, T. H.; Lee, E. S.; Oh, K. T.; Park, E.-S.; Lee K. C.; Youn, Y. S. Doxorubicin-loaded human serum albumin nanoparticles surface-modified with TNF-related apoptosis-inducing ligand and transferrin for targeting multiple tumor types. *Biomaterials.* **2012**, *33(5)*, 1536–46.

Bae, Y.; Fukushima, S.; Harada A.; Kataoka, K. Design of environment sensitive supramolecular assemblies for intracellular drug delivery, polymeric micelles that are responsive to intracellular pH change. *Angew. Chem. Int. Ed.* **2003**, *42(38)*, 4640–3.

Bag, J., Mukherjee, S., Ghosh, S.K., Das, A., Mukherjee, A., Sahoo, J.K., Tung, K.S., Sahoo, H. and Mishra, M. Fe_3O_4 coated guargum nanoparticles as non-genotoxic materials for biological application. *International Journal of Biological Macromolecules*. **2020,** 165, pp.333-345.

Balani, K.; Anderson, R.; Laha, T.; Andara, M.; Tercero, J.; Crumpler E.; Agarwal, A. Plasma-sprayed carbon nanotube reinforced hydroxyapatite coatings and their interaction with human osteoblasts in vitro. *Biomaterials.* **2007**, *28(4),* 618–24.

Ban, E.; Kim, C.-K. Design and evaluation of ondansetron liquid suppository for the treatment of emesis. *Arch. Pharm. Res.* **2013**, *36(5)*, 586–92.

Bar, H.; Bhui, D. K.; Sahoo, G. P.; Sarkar, P.; De S. P.; Misra, A. Green synthesis of silver nanoparticles using latex of Jatropha curcas. *Colloids Surf. A Physicochem. Eng. Asp.* **2009**, *339(1–3),* 134–9.

Barik, B. K.; Mishra, M. Nanoparticles as a potential teratogen, a lesson learnt from fruit fly. *Nanotoxicol.* **2018**, 1–27.

Basavaraj, B.V.; Devi, S.; Bharath, S.; Deveswaran R.; Madhavan, V. Design and evaluation of sustained release propranolol hydrochloride suppositories. *Int. J. Pharm. Sci. Rev. Res.* **2013**, *25*, 5–12.

Batchelor, H. Rectal drug delivery, in *Pediatric Formulations* Springer, **2014**, 303–10.

Benet, L. Z.; Izumi, T.; Zhang, Y.; Silverman J. A.; Wacher, V. J. Intestinal MDR transport proteins and P-450 enzymes as barriers to oral drug delivery. *J Control Rel.* **1999**, *62(1–2),* 25–31.

Berteau, C.; Filipe-Santos, O.; Wang, T.; Rojas, H. E.; Granger C.; Schwarzenbach, F. Evaluation of the impact of viscosity, injection volume, and injection flow rate on subcutaneous injection tolerance. *Med. Devices (Auckland, N.Z.),* **2015**, *8,* 473.

Bhattarai, N.; Li, Z.; Edmondson D.; Zhang, M. Alginate-based nanofibrous scaffolds, Structural, mechanical, and biological properties. *Adv. Mater.* **2006**, *18(11),* 1463–7.

Bochot, A.; Fattal, E. Liposomes for intravitreal drug delivery, a state of the art. *J Control Release.* **2012**, *161(2)*, 628–34.

Bouakaz, A.; Zeghimi A.; Doinikov, A. A. Sonoporation, concept and mechanisms, in *Therapeutic Ultrasound* Springer, **2016**, 175–89.

Brannon-Peppas, L.; Blanchette, J. O. Nanoparticle and targeted systems for cancer therapy. *Adv. Drug Deliv. Rev.* **2012**, *64*, 206–12.

Brigger, I.; Dubernet C.; Couvreur, P. Nanoparticles in cancer therapy and diagnosis. *Adv. Drug Deliv. Rev.* 2012, *64*, 24–36.

Buseck, P. R.; Adachi, K. Nanoparticles in the atmosphere. *Elements.* **2008**, *4(6)*, 389–94.

Buzea, C.; Pacheco I. I.; Robbie, K. Nanomaterials and nanoparticles, sources and toxicity. *Biointerphases.* **2007**, *2(4)*, MR17-MR71.

Calvo, P.; Remunan-Lopez, C.; Vila-Jato J.L.; Alonso, M. Novel hydrophilic chitosan-polyethylene oxide nanoparticles as protein carriers. *J. Appl. Polym. Sci.* **1997**, *63(1)*, 125–32.
Cario, E. Nanotechnology-based drug delivery in mucosal immune diseases, hype or hope? *Mucosal Immunol.* **2012**, *5(1)*, 2.
Chen, F.; Cai, W. Nanomedicine for targeted photothermal cancer therapy, where are we now? *Nanomedicine,* **2015**, *10(1)*, 1–3.
Chen, H.; Hwang, J. H. Ultrasound-targeted microbubble destruction for chemotherapeutic drug delivery to solid tumors. *J. Ther. Ultrasound.* **2013**, *1(1)*, 10.
Chen, P.; Wang, H.; He, M.; Chen, B.; Yang B.; Hu, B. Size-dependent cytotoxicity study of ZnO nanoparticles in HepG2 cells. *Ecotox. Environ. Safe.* **2019**,*171*, 337–46.
Chen, Y.-H.; Tsai, C.-Y.; Huang, P.-Y.; Chang, M.-Y.; Cheng, P.-C.; Chou, C.-H.; Chen, D.-H.; Wang, C.-R.; Shiau A.-L.; Wu, C.-L. Methotrexate conjugated to gold nanoparticles inhibits tumor growth in a syngeneic lung tumor model. *Mol. Pharm.* **2007**, *4(5)*, 713–22.
Chiappini, C.; De Rosa, E.; Martinez, J.; Liu, X.; Steele, J.; Stevens M.; Tasciotti, E. Biodegradable silicon nanoneedles delivering nucleic acids intracellularly induce localized in vivo neovascularization. *Nat. Mater.* **2015**, *14(5)*, 532.
Chicheł, A.; Skowronek, J.; Kubaszewska M.; Kanikowski, M. Hyperthermia–description of a method and a review of clinical applications. *Rep. Pract. Oncol. Radiother.* **2007**, *12(5)*, 267–75.
Chowdhury, S. M.; Lee, T.; Willmann, J. K. Ultrasound-guided drug delivery in cancer. *Ultrasonography.* **2017**, *36(3)*, 171.
Chowdhury, S. M.; Wang, T. Y.; Bachawal, S.; Devulapally, R.; Choe, J. W.; Elkacem, L. A.; Yakub, B. K.; Wang, D. S.; Tian, L.; Paulmurugan, R. Ultrasound-guided therapeutic modulation of hepatocellular carcinoma using complementary microRNAs. *J. Control. Rel.* **2016**, *238*, 272–80.
Collins, D.; Kourtis, L.; Thyagarajapuram, N.; Sirkar, R.; Kapur, S.; Harrison, M.; Bryan, D.; Jones, G.; Wright, J. Optimizing the bioavailability of subcutaneously administered biotherapeutics through mechanochemical drivers. *Pharm. Res.* **2017**, *34(10)*, 2000–11.
Connor, E. E.; Mwamuka, J.; Gole, A.; Murphy, C. J.; Wyatt, M. D. Gold nanoparticles are taken up by human cells but do not cause acute cytotoxicity. *Small.* **2005**, *1(3)*, 325–7.
Corot, C; Robert, P.; Idée, J. M.; Port, M. Recent advances in iron oxide nanocrystal technology for medical imaging. *Adv. Drug Delivery Rev.* **2006**, *58(14)*, 1471–504.
Cunha, S.; Amaral, M. H.; Lobo, J. S.; Silva, A. C. Lipid nanoparticles for nasal/intranasal drug delivery. *Criti. Rev. Thera. Drug Carrier Syst.* **2017**, 3*4(3)*, 257–282.
De La Fuente, M.; Csaba, N.; Garcia-Fuentes, M.; Alonso, M. J. Nanoparticles as protein and gene carriers to mucosal surfaces. *Nanomed.* **2008**, *3(6).*
Del Campo, A.; Sen, T.; Lellouche, J. P.; Bruce, I. J. Multifunctional magnetite and silica–magnetite nanoparticles, synthesis, surface activation and applications in life sciences. *J. Magn. Magn. Mater.* **2005**, *293(1)*, 33–40.
Delalande, A.; Kotopoulis, S.; Postema, M.; Midoux, P.; Pichon, C. Sonoporation, mechanistic insights and ongoing challenges for gene transfer. *Gene.* **2013**, *525(2)*, 191–9.
Desai, M. P.; Labhasetwar, V.; Walter, E.; Levy, R. J.; Amidon, G. L. The mechanism of uptake of biodegradable microparticles in Caco-2 cells is size dependent. *Pharm. Res.* **1997**, *14(11)*, 1568–73.
Dey, N. S.; Rao, M. B. Quantum dot, novel carrier for drug delivery. *Int. J. Res. Pharm. Biomed. Sci.* **2011**, *2*, 448–58.

Diederich, C. J. Thermal ablation and high-temperature thermal therapy, overview of technology and clinical implementation. *Int. J. Hypertherm.* **2005**, *21(8)*, 745–53.

Diez, I.; Ras, R. H. Fluorescent silver nanoclusters. *Nanoscale*, **2011**, *3(5)*, 1963–70.

Ditsch, A.; Lindenmann, S.; Laibinis, P. E; Wang, D. I.; Hatton, T. A. High-gradient magnetic separation of magnetic nanoclusters. *Ind. Eng. Chem. Res.* **2005**, *17*, 6824–36.

Dobson, J. Magnetic nanoparticles for drug delivery. *Drug Development Res.* **2006**, *67(1)*, 55–60.

Dobson, K. D.; Visoly-Fisher, I.; Hodes, G.; Cahen, D. Stabilizing CdTe/CdS solar cells with Cu containing contacts to pdCdTe. *Adv. Mater.* **2001**, *13(19)*, 1495–9.

Dorozhkin, S. V.; Epple, M. Biological and medical significance of calcium phosphates. *Angewandte Chemie Int. Ed.* **2002**, *41(17)*, 3130–46.

Ekka, B., Dhar, G., Sahu, S., Mishra, M., Dash, P. and Patel, R.K. Removal of Cr (VI) by silica-titania core-shell nanocomposites: In vivo toxicity assessment of the adsorbent by Drosophila melanogaster. *Ceramics International* **2021**, 47(13), pp.19079–19089.

Elamanchili, P.; Diwan, M.; Cao, M.; Samuel, J. Characterization of poly (D, L-lactic-co-glycolic acid) based nanoparticulate system for enhanced delivery of antigens to dendritic cells. *Vaccine.* **2004**, *22(19)*, 2406–12.

Feleni, U.; Ajayi, R. F.; Jijana, A.; Sidwaba, U.; Douman, S.; Baker, P.; Iwuoha, E. Tin selenide quantum dots electrochemical biotransducer for the determination of Indinavir-A protease inhibitor anti-retroviral drug. *J. Nano Res.* **2017**, 12–24.

Fernandez-Urrusuno, R.; Calvo, P.; Remunan-Lopez, C.; Vila-Jato, J. L.; Alonso, M. J. Enhancement of nasal absorption of insulin using chitosan nanoparticles. *Pharm. Res.* **1999**, *16(10)*, 1576–81.

Gao, W.; Chan, J. M.; Farokhzad, O. C. pH-responsive nanoparticles for drug delivery. *Mol. Pharm.* **2010**, *7(6)*, 1913–20.

Gautam, A., Gautam, C., Mishra, M., Mishra, V.K., Hussain, A., Sahu, S., Nanda, R., Kisan, B., Biradar, S. and Gautam, R.K. Enhanced mechanical properties of hBN–ZrO 2 composites and their biological activities on Drosophila melanogaster: synthesis and characterization. *RSC Advances* **2019**,9(70), pp.40977–40996.

Gautam, A., Gautam, C., Mishra, M., Sahu, S., Nanda, R., Kisan, B., Gautam, R.K., Prakash, R., Sharma, K., Singh, D. and Gautam, S.S. Synthesis, structural, mechanical, and biological properties of HAp-ZrO2-hBN biocomposites for bone regeneration applications. *Ceramics International* **2021**, 47(21), pp.30203-30220.

Ge, J.; Neofytou, E.; Cahill, T. J.; Beygui, R. E.; Zare, R. N. Drug release from electric-field-responsive nanoparticles. *ACS Nano.* **2011**, *6(1)*, 227–33.

Ghicov, A.; Schmuki, P. Self-ordering electrochemistry, a review on growth and functionality of TiO 2 nanotubes and other self-aligned MO x structures. *Chem. Commun.* **2009**, *20*, 2791–808.

Ghosh, P.; Han, G.; De, M.; Kim, C. K.; Rotello, V. M. Gold nanoparticles in delivery applications. *Adv. Drug Delivery Rev.* **2008**, *60(11)*, 1307–15.

Giri, T. K.; Thakur, A.; Alexander, A.; Badwaik, H.; Tripathi, D. K. Modified chitosan hydrogels as drug delivery and tissue engineering systems, present status and applications. *Acta Pharm. Sinica B.* **2012**, *2(5)*, 439–49.

Gomathi, M.; Rajkumar, P.; Prakasam, A.; Ravichandran, K. Green synthesis of silver nanoparticles using Datura stramonium leaf extract and assessment of their antibacterial activity. *Res. Efficient Tech.* **2017**, *3(3)*, 280–4.

Gomez, C.; Benito, M.; Teijon, J. M.; Blanco, M. D. Novel methods and devices to enhance transdermal drug delivery, the importance of laser radiation in transdermal drug delivery. *Thera. Delivery* **2012**, *3(3)*, 373–88.

Grimes, C.A. Synthesis and application of highly ordered arrays of TiO 2 nanotubes. *J. Mater. Chem.* **2007**, *17(15)*, 1451–7.

Gupta, V. K.; Beckert, T. E.; Price, J. C. A novel pH-and time-based multi-unit potential colonic drug delivery system. I. Development. *Int. J. Pharm.* **2001**, *213(1–2)*, 83–91.

Habash, R. W.; Bansal, R.; Krewski, D.; Alhafid, H. T. Thermal therapy, part 2, hyperthermia techniques. *Criti. Rev. Biomed. Eng.* **2006**, *34(6)*, 491–542.

Han, G.; Ghosh, P.; Rotello, V. M. Functionalized gold nanoparticles for drug delivery. *Nanomed.* **2007**, *2(1)*, 113–123.

Han, H. K.; Amidon, G. L. Targeted prodrug design to optimize drug delivery. *Am Asso. Pharm. Sci.* **2000**, *2(1)*, 48–58.

Han, S. W.; Nakamura, C.; Obataya, I.; Nakamura, N.; Miyake, J. A molecular delivery system by using AFM and nanoneedle. *Biosens. Bioelectr.* **2005**, *20(10)*, 2120–5.

Hanes, J.; Cleland, J. L.; Langer, R. New advances in microsphere-based single-dose vaccines. *Adv. Drug Delivery Rev.* **1997**, *28(1)*, 97–119.

Hergt, R.; Dutz, S.; Müller, R.; Zeisberger, M. Magnetic particle hyperthermia, nanoparticle magnetism and materials development for cancer therapy. *J. Phys. Condens. Mater.* **2006**, *18(38)*, S2919.

Hervault, A.; Thanh, N. T. K. Magnetic nanoparticle-based therapeutic agents for thermo-chemotherapy treatment of cancer. *Nanoscale.* **2014**, *6(20)*, 11553–73.

Hildebrandt, B.; Wust, P.; Ahlers, O.; Dieing, A.; Sreenivasa, G.; Kerner, T.; Felix, R.; Riess, H. The cellular and molecular basis of hyperthermia. *Criti. Rev. Oncol./Hematol.* **2002**, *43(1)*, 33–56.

Hirai, S.; Ikenaga, T.; Matsuzawa, T. Nasal absorption of insulin in dogs. *Diabetes.* **1978**, *27(3)*, 296–9.

Hobohm, H.U. 2014. *Healing heat, An essay on cancer immune defence*, BoD–Books on Demand.

Holister, P.; Weener, J. W.; Román, C.; Harper, T. Nanoparticles. *Tech. White Papers.* **2003**, *3*, 1–11.

Hom, C.; Lu, J.; Liong, M.; Luo, H.; Li, Z.; Zink, J. I., Tamanoi, F. Mesoporous silica nanoparticles facilitate delivery of siRNA to shutdown signaling pathways in mammalian cells. *Small.* **2010**, *6(11)*, 1185–90.

Horak, D.; Babič, M.; Mackova, H.; Beneš, M. J. Preparation and properties of magnetic nano and microsized particles for biological and environmental separations. *J. Separat. Sci.* **2007**, *30(11)*, 1751–72.

Hovav, A. Dendritic cells of the oral mucosa. *Mucosal Immunol.* **2014**, *7(1)*, 27.

Hughes, G.A. Nanostructure-mediated drug delivery. *Nanomed. Cancer*. **2017**, 47–72.

Ivanova, N.; Gugleva, V.; Dobreva, M.; Stefanov, S.; Pehlivanov, I.; Andonova, V. Silver nanoparticles as multy-functional drug delivery systems. *Nanomed.* **2018**, 71–92.

Jadhav, U.; Dias, R.; Mali, K.; Havaldar, V. Development of in situ-gelling and mucoadhesive liquid suppository of ondansetron. *Int. J. ChemTech Res.* **2009**, *1(4)*, 953–61.

Jain, S.; Shukla, K.; Jain, V.; Saraf, S.; Saraf, S. Nanoparticles, emerging carriers for delivery of bioactive agents. *Pharma Times,* **2007**, *39(1)*, 30–5.

James, N.; Coker, R.; Tomlinson, D.; Harris, J.; Gompels, M.; Pinching, A.; Stewart, J. Liposomal doxorubicin (Doxil), an effective new treatment for Kaposi's sarcoma in AIDS. *Clin. Oncol.* **1994**, *6(5)*, 294–6.

Jannin, V.; Lemagnen, G.; Gueroult, P.; Larrouture, D.; Tuleu, C. Rectal route in the 21st Century to treat children. *Adv. Drug Del. Rev.* **2014**, *73*, 34–49.

Jaque, D.; Maestro, B.; Del Rosal, P.; Haro-Gonzalez, A.; Benayas, J.; Plaza, E. M.; Rodriguez, D.; Sole, J. G. Nanoparticles for photothermal therapies. *Nanoscale.* **2014**, *6(16)*, 9494–530.

Jeong, U.; Wang, Y.; Ibisate M.; Xia, Y. Some new developments in the synthesis, functionalization, and utilization of monodisperse colloidal spheres. *Adv. Funct. Mater.* **2005**, *15(12)*, 1907–21.

Johannsen, M.; Gneveckow, U.; Eckelt, L; Feussner, A; Waldöfner, N.; Scholz, R.; Deger, S.; Wust, P.; Loening, S.; Jordan, A. Clinical hyperthermia of prostate cancer using magnetic nanoparticles, presentation of a new interstitial technique. *Int. J. Hyperther.* **2005**, *21(7)*, 637–47.

Johnson, I.; Prabu, H. J. Green synthesis and characterization of silver nanoparticles by leaf extracts of Cycas circinalis, Ficus amplissima, Commelina benghalensis and Lippia nodiflora. *Int. Nano Lett.* **2015**, *5(1)*, 43–51.

Jordan, M.; Schallhorn, A.; Wurm, F. M. Transfecting mammalian cells, optimization of critical parameters affecting calcium-phosphate precipitate formation. *Nucleic Acids Res.* **1996**, *24(4)*, 596–601.

Kam, N. W. S.; Liu Z.; Dai, H. Carbon nanotubes as intracellular transporters for proteins and DNA, an investigation of the uptake mechanism and pathway. *Angew. Chem. Int. Ed.* **2006**, *45(4)*, 577–81.

Kamath, K. R.; Park, K. Biodegradable hydrogels in drug delivery. *Adv. Drug Deliv Rev.* **1993**, *11(1–2)*, 59–84.

Kawamura, R.; Shimizu, K.; Matsumoto, Y.; Yamagishi, A.; Silberberg, Y. R.; Iijima, M.; Kuroda, K. Fukazawa, S. I.; Ishihara K.; Nakamura, C. High efficiency penetration of antibody-immobilized nanoneedle thorough plasma membrane for in situ detection of cytoskeletal proteins in living cells. *J. Nanobiotechnol.* **2016**, *14(1)*, 74.

Kayser, O.; Lemke A.; Hernandez-Trejo, N. The impact of nanobiotechnology on the development of new drug delivery systems. *Curr. Pharm. Biotechnol.* **2005**, *6(1)*, 3–5.

Keservani, R. K.; Sharma, A. K.; Ayaz M.; Kesharwani, R. K. Novel drug delivery system for the vesicular delivery of drug by the niosomes. *Int. J. Res. Control Release*. **2011**, *1(1)*, 1–8.

Khandare, J.; Madhavi G.; Tamhankar, B. Niosomes-novel drug delivery system. *Eastern Pharm.* **1994**, *37*, 61–67.

Kim, I.; Byeon, H. J.; Kim, T. H.; Lee, E. S.; Oh, K. T.; Shin, B. S.; Lee K. C.; Youn, Y. S. Doxorubicin-loaded porous PLGA microparticles with surface attached TRAIL for the inhalation treatment of metastatic lung cancer. *Biomaterials.* **2013**, 34(27), 6444–53.

Kim, Y.-J.; Chae, S. Y.; Jin, C.-H.; Sivasubramanian, M.; Son, S.; Choi, K. Y.; Jo, D.-G.; Kim, K.; Kwon I. C.; Lee, K. C. Ionic complex systems based on hyaluronic acid and PEGylated TNF-related apoptosis-inducing ligand for treatment of rheumatoid arthritis. *Biomaterials.* **2010**, 31(34), 9057–64.

Klasen, H. A historical review of the use of silver in the treatment of burns. II. Renewed interest for silver. *Burns*. **2000**, 26(2), 131–8.

Kolhar, P.; Doshi N.; Mitragotri, S. Polymer nanoneedle-mediated intracellular drug delivery. *Small.* **2011**, 7(14), 2094–100.

Koo, O. M.; Rubinstein I.; Onyuksel, H. Camptothecin in sterically stabilized phospholipid nano-micelles, a novel solvent pH change solubilization method. *J. Nanosci. Nanotechnol.* **2006**, 6(9–10), 2996–3000.

Kost, J.; Pliquett, U.; Mitragotri, S. S.; Langer R. S.; Weaver, J. C. Effect of electric field and ultrasound for transdermal drug delivery, **2000**, Google Patents.
Kraisit, P.; Limmatvapirat, S.; Luangtana-Anan, M. Sriamornsak, P. Buccal administration of mucoadhesive blend films saturated with propranolol loaded nanoparticles. *Asian J. Pharm. Sci.* **2018**, 13(1), 34–43.
Krauze, M. T.; Noble, C. O.; Kawaguchi, T.; Drummond, D.; Kirpotin, D. B.; Yamashita, Y.; Kullberg, E.; Forsayeth, J.; Park, J. W.; Bankiewicz, K. S. Convection-enhanced delivery of nanoliposomal CPT-11 (irinotecan) and PEGylated liposomal doxorubicin (Doxil) in rodent intracranial brain tumor xenografts. *Neuro-Oncol.* **2007**, 9(4), 393–403.
Kurtin, S.; Knop, C. S.; Milliron, T. Subcutaneous administration of bortezomib, strategies to reduce injection site reactions. *J. Adv. Pract. Oncol.* **2012**, 3(6), 406.
Kwon, Y. J.; Standley, S. M.; Goh, S. L.; Fréchet, J. M. Enhanced antigen presentation and immunostimulation of dendritic cells using acid-degradable cationic nanoparticles. *J. Control Release.* **2005**, 105(3), 199–212.
Lai, J. J.; Hoffman, J. M.; Ebara, M.; Hoffman, A. S.; Estournès, C.; Wattiaux, A.; Stayton, P. S. Dual magnetic-/temperature-responsive nanoparticles for microfluidic separations and assays. *Langmuir.* **2007**; 23(13), 7385–91.
Langer, R.; Peppas, N. A. Advances in biomaterials, drug delivery, and bionanotechnology. *AIChE J.* **2003**, 49(12), 2990–3006.
Lavan, D. A.; McGuire, T.; Langer, R. Small-scale systems for in vivo drug delivery. *Nat. Biotechnol.* **2003**, 21(10), 1184.
Le Ouay, B.; Stellacci, F. Antibacterial activity of silver nanoparticles, a surface science insight. *Nano Today.* **2015**, 10(3), 339–54.
Lee, C. C.; Gillies, E. R.; Fox, M. E.; Guillaudeu, S. J.; Frechet, J. M.; Dy, E. E.; Szoka, F. C. A single dose of doxorubicin-functionalized bow-tie dendrimer cures mice bearing C-26 colon carcinomas. *Proc. Natl. Acad. Sci.* **2006**, 103(45), 16649–54.
Lee, V.H. Protease inhibitors and penetration enhancers as approaches to modify peptide absorption. *J. Control. Release*. **1990**, 13(2–3), 213–23.
Li, Y.; Lu, W.; Huang, Q.; Li, C.; Chen, W. Copper sulfide nanoparticles for photothermal ablation of tumor cells. *Nanomed.* **2010**, 5(8), 1161–71.
Lim, S. M.; Kim, T. H.; Jiang, H. H.; Park, C. W.; Lee, S.; Chen, X.; Lee, K. C. Improved biological half-life and anti-tumor activity of TNF-related apoptosis-inducing ligand (TRAIL) using PEG-exposed nanoparticles. *Biomaterials* **2011**, 32(13), 3538–46.
Liu, Z.; Chen, K.; Davis, C.; Sherlock, S.; Cao, Q.; Chen, X.; Dai, H. Drug delivery with carbon nanotubes for in vivo cancer treatment. *Cancer Res.* **2008**, 68(16), 6652–60.
Liu, Z.; Fan, A. C.; Rakhra, K.; Sherlock, S.; Goodwin, A.; Chen, X.; Yang, Q.; Felsher, D. W.; Dai, H. Supramolecular stacking of doxorubicin on carbon nanotubes for in vivo cancer therapy. *Angewandte Chemie Int. Eds.* **2009**, 48(41), 7668–72.
Loo, C.; Lin, A.; Hirsch, L.; Lee, M. H.; Barton, J.; Halas, N.; West, J.; Drezek, R. Nanoshell-enabled photonics-based imaging and therapy of cancer. *Tech. Cancer Res. Treatment.* **2004**, 3(1), 33–40.
Lu, Y.; Hu, Q.; Lin, Y.; Pacardo, D. B.; Wang, C.; Sun, W.; Ligler, F. S.; Dickey, M. D.; Gu, Z. Transformable liquid-metal nanomedicine. *Nat. Commun.* **2015**, 6, 10066.
Ma, W.; Chen, M.; Kaushal, S.; McElroy, M.; Zhang, Y.; Ozkan, C.; Bouvet, M.; Kruse, C.; Grotjahn, D.; Ichim, T. PLGA nanoparticle-mediated delivery of tumor antigenic peptides elicits effective immune responses. *Int. J. Nanomed.* **2012**, 7, 1475.
Mackowiak, P. A. Fever. *Basic Mech. Manage.* **1991**, 300–1.

Maisel, K.; Ensign, L.; Reddy, M.; Cone, R.; Hanes, J. Effect of surface chemistry on nanoparticle interaction with gastrointestinal mucus and distribution in the gastrointestinal tract following oral and rectal administration in the mouse. *J. Control. Release.* **2015**, 197, 48–57.

Majoros, I. J.; Thomas, T. P.; Mehta, C. B.; Baker, J. R. Poly (amidoamine) dendrimer-based multifunctional engineered nanodevice for cancer therapy. *J. Med. Chem.* **2005**, 48(19), 5892–9.

Malam, Y.; Loizidou, M.; Seifalian, A. M. Liposomes and nanoparticles, nanosized vehicles for drug delivery in cancer. *Trends. Pharm. Sci.* **2009**, 30(11), 592–9.

Mansour, H. M.; Sohn, M.; Al-Ghananeem, A.; DeLuca, P. P. Materials for pharmaceutical dosage forms, molecular pharmaceutics and controlled release drug delivery aspects. *Int. J. Mol. Sci.* **2010**, 11(9), 3298–322.

Marin, E.; Briceno, M. I.; Caballero-George, C. Critical evaluation of biodegradable polymers used in nanodrugs. *Int. J. Nanomed.* **2013**, 8, 3071.

Masek, J.; Lubasova, D.; Lukac, R.; Turanek-Knotigova, P.; Kulich, P.; Plockova, J.; Maskova, E.; Prochazka, L.; Koudelka, S.; Sasithorn, N. Multi-layered nanofibrous mucoadhesive films for buccal and sublingual administration of drug-delivery and vaccination nanoparticles-important step towards effective mucosal vaccines. *J. Control. Release.* **2017**, 249, 183–95.

Meng, Z.; Xu, X.; Zheng, W.; Zhou, H.; Li, L.; Zheng, Y.; Lou, X. Preparation and characterization of electrospun PLGA/gelatin nanofibers as a potential drug delivery system. *Coll. Surf. B, Biointer.* **2011**, 84(1), 97–102.

Mishra, P.K., Ekielski, A., Mukherjee, S., Sahu, S., Chowdhury, S., Mishra, M., Talegaonkar, S., Siddiqui, L. and Mishra, H., 2019. Wood-based cellulose nanofibrils: haemocompatibility and impact on the development and behaviour of Drosophila melanogaster. *Biomolecules.* **2019**, 9(8), p.363.

Mishra, M. and Panda, M. Reactive oxygen species: the root cause of nanoparticle-induced toxicity in Drosophila melanogaster. *Free Radical Research.* **2021**, pp.1–17.

Moghassemi, S.; Hadjizadeh, A. Nano-niosomes as nanoscale drug delivery systems, an illustrated review. *J. Control. Release.* **2014**, 185, 22–36.

Moghimi, S. M.; Hunter, A. C.; Murray, J. C. Long-circulating and target-specific nanoparticles, theory to practice. *Pharm. Rev.* **2001**, 53(2), 283–318.

Mohamed, H. B.; El-Shanawany, S. M.; Hamad, M. A.; Elsabahy, M. Niosomes, a strategy toward prevention of clinically significant drug incompatibilities. *Sci. Rep.* **2017**, 7(1), 6340.

Mu, Q.; Jiang, G.; Chen, L.; Zhou, H.; Fourches, D.; Tropsha, A.; Yan, B. Chemical basis of interactions between engineered nanoparticles and biological systems. *Chem. Rev.* **2014**, 114(15), 7740–81.

Mudshinge, S. R.; Deore, A. B.; Patil, S.; Bhalgat, C. M. Nanoparticles, merging carriers for drug delivery. *Saudi Pharm. J.* 2011, 19(3), 129–41.

Muller, R. H.; Mader, K.; Gohla, S. Solid lipid nanoparticles (SLN) for controlled drug delivery–a review of the state of the art. *Euro. J. Pharm. Biopharma.* **2000**, 50(1), 161–77.

Mura, S.; Nicolas, J.; Couvreur, P. Stimuli-responsive nanocarriers for drug delivery. *Nat. Mater.* **2013**, 12(11), 991.

Nagamoto, T.; Hattori, Y.; Takayama, K.; Maitani, Y. Novel chitosan particles and chitosan-coated emulsions inducing immune response via intranasal vaccine delivery. *Pharm. Res.* **2004**, 21(4), 671–4.

Nakashima, K.; Bahadur, P. Aggregation of water-soluble block copolymers in aqueous solutions, recent trends. *Adv. Coll. Interface Sci.* **2006**, 123, 75–96.

Namdeo, A.; Jain, N. Niosomes as drug carriers. *Ind. J. Pharm. Sci.* **1996**, 58(2), 41.

Nayak, N., Basha, S.A., Tripathi, S.K., Biswal, B.K., Mishra, M. and Sarkar, D. Non-cytotoxic and non-genotoxic wear debris of strontium oxide doped (Zirconia Toughened Alumina) (SrO-ZTA) implant for hip prosthesis. *Materials Chemistry and Physics* **2021**, 274, p.125187.

Neri, D.; Supuran, C. T. Interfering with pH regulation in tumours as a therapeutic strategy. *Nat. Rev. Drug Dis.* **2011**, 10(10), 767.

Neumann, E. The relaxation hysteresis of membrane electroporation. *Electroporation Electrofusion Cell Bio.* **1989**, 61–82.

Newton, A.; Indana, V.; Kumar, J. Chronotherapeutic drug delivery of Tamarind gum, Chitosan and Okra gum controlled release colon targeted directly compressed Propranolol HCl matrix tablets and in-vitro evaluation. *Int. J. Bio. Macromol.* **2015**, 79, 290–9.

Ning, Y.; He, M. K.; Dagher, R.; Sridhara, R.; Farrell, A.; Justice, R.; Pazdur, R. Liposomal doxorubicin in combination with bortezomib for relapsed or refractory multiple myeloma. *Oncology.* **2007**, 21(12).

O'grady, N. P.; Barie, P. S.; Bartlett, J. G.; Bleck, T.; Garvey, G.; Jacobi, J.; Linden, P.; Maki, D. G.; Nam, M.; Pasculle, W. Practice guidelines for evaluating new fever in critically ill adult patients. *Rev. Infect. Dis.* **1998**, 26(5), 1042–59.

Ong, W. Y.; Shalini, S. M.; Costantino, L. Nose-to-brain drug delivery by nanoparticles in the treatment of neurological disorders. *Cur. Med. Chem.* **2014**, 21(37), 4247–56.

Paciotti, G. F.; Myer, L.; Weinreich, D.; Goia, D.; Pavel, N.; McLaughlin, R. E.; Tamarkin, L. Colloidal gold, a novel nanoparticle vector for tumor directed drug delivery. *Drug Del.* **2004**, 11(3), 169–83.

Pankhurst, Q. A.; Connolly, J.; Jones, S.; Dobson, J. Applications of magnetic nanoparticles in biomedicine. *J. Appl. Phy.* **2003**, 36(13), R167.

Paolini, M.; Poul, L.; Berjaud, C.; Germain, M.; Darmon, A.; Bergere, M.; Pottier, A.; Levy, L.; Vibert, E. Nano-sized cytochrome P450 3A4 inhibitors to block hepatic metabolism of docetaxel. *Int. J. Nanomed.* **2017**, 12, 5537.

Park, H. S.; Nam, S. H.; Kim, J.; Shin, H. S.; Suh, Y. D.; Hong, K. S. Clear-cut observation of clearance of sustainable upconverting nanoparticles from lymphatic system of small living mice. *Sci. Rep.* **2016**, 6, 27407.

Park, J.; Fong, P. M.; Lu, J.; Russell, K. S.; Booth, C. J.; Saltzman, W. M.; Fahmy, T. M. PEGylated PLGA nanoparticles for the improved delivery of doxorubicin. *Nanomed. Nanotech. Bio. Med.* **2009**, 5(4), 410–8.

Patra, S.; Mukherjee, S.; Barui, A. K.; Ganguly, A.; Sreedhar, B.; Patra, C. R. Green synthesis, characterization of gold and silver nanoparticles and their potential application for cancer therapeutics. *Mater. Sci. Eng.* **2015**, 53, 298–309.

Peer, E.; Artzy-Schnirman, A.; Gepstein, L.; Sivan, U. Hollow nanoneedle array and its utilization for repeated administration of biomolecules to the same cells. *ACS Nano.* **2012**, 6(6), 4940–6.

Phull, A. R.; Abbas, Q.; Ali, A.; Raza, H.; Zia, M.; Haq, I. U. Antioxidant, cytotoxic and antimicrobial activities of green synthesized silver nanoparticles from crude extract of Bergenia ciliata. *Future J. Pharm. Sci.* **2016**, 2(1), 31–6.

Pitt, W.G.; Husseini, G.A.; Staples, B.J. Ultrasonic drug delivery–a general review. *Expert Opin. Drug Deliv.* **2004**, 1(1), 37–56.

Pople, P. V.; Singh, K. K. Targeting tacrolimus to deeper layers of skin with improved safety for treatment of atopic dermatitis. *Int. J. Pharm.* **2010**, 398(1–2), 165–78.

Pople, P. V.; Singh, K. K. Targeting tacrolimus to deeper layers of skin with improved safety for treatment of atopic dermatitis—Part II, In vivo assessment of dermatopharmacokinetics, biodistribution and efficacy. *Int. J. Pharm.* **2012**, 434(1–2), 70–9.

Prados, J.; Melguizo, C.; Ortiz, R.; Velez, C.; Alvarez, P. J.; Arias, J. L.; Ruiz, M. A.; Gallardo, V.; Aranega, A. Doxorubicin-loaded nanoparticles, new advances in breast cancer therapy. *Anti-Cancer Agents Med. Chem.* **2012**, 12(9), 1058–70.

Purohit, T. J.; Hanning, S. M.; Wu, Z. Advances in rectal drug delivery systems. *Pharm. Develop. Tech.* **2018**, 23(10), 942–52.

Qiu, L.; Jing, N.; Jin, Y. Preparation and in vitro evaluation of liposomal chloroquine diphosphate loaded by a transmembrane pH-gradient method. *Int. J. Pharm.* **2008**, 361(1–2), 56–63.

Ramadan, S.; Guo, L.; Li, Y.; Yan, B.; Lu, W. Hollow copper sulfide nanoparticle-mediated transdermal drug delivery. *Small.* **2012**, 8(20), 3143–50.

Rathore, K.; Lowalekar, R.; Nema, R.; Jain, C. Quantum dots, a future drug delivery system. *Pharm. Rev.* **2006**, 4, 30–2.

Rejinold, N. S.; Shin, J. H.; Seok, H. Y.; Kim, Y. C. Biomedical applications of microneedles in therapeutics, recent advancements and implications in drug delivery. *Exp. Opin. Drug Delivery* **2016**, 13(1), 109–31.

Romeo, V.; Gries, W.; Xia, W.; Sileno, A.; Pimplaskar, H.; Behl, C. Optimization of systemic nasal drug delivery with pharmaceutical excipients. *Adv. Drug Delivery Rev.* **1998**, 29(1–2), 117–33.

Rosi, N. L.; Giljohann, D. A.; Thaxton, C. S.; Lytton-Jean, A. K.; Han, M. S.; Mirkin, C. A. Oligonucleotide-modified gold nanoparticles for intracellular gene regulation. *Science.* **2006**, 312(5776), 1027–30.

Salonen, J.; Kaukonen, A. M.; Hirvonen, J.; Lehto, V. P. Mesoporous silicon in drug delivery applications. *J. Pharm. Sci.* **2008**, 97(2), 632–53.

Schmid, G. Materials in nanoporous alumina. *J. Mater. Chem.* **2002**, 12(5), 1231–8.

Schoellhammer, C. M.; Schroeder, A.; Maa, R.; Lauwers, G. Y.; Swiston, A.; Zervas, M.; Barman, R.; DiCiccio, A. M.; Brugge, W. R.; Anderson, D. G. Ultrasound-mediated gastrointestinal drug delivery. *Sci. Transl. Med.* **2015**, 7(310), 310ra168–310ra168.

Schreier, H.; Bouwstra, J. Liposomes and niosomes as topical drug carriers, dermal and transdermal drug delivery. *J. Control. Release.* **1994**, 30(1), 1–15.

Schroeter, A.; Engelbrecht, T.; Neubert, R. H.; Goebel, A. S. New nanosized technologies for dermal and transdermal drug delivery. A review. *J. Biomed. Nanotech.* **2010**, 6(5), 511–28.

Shapiro, G.; Wong, A. W.; Bez, M.; Yang, F.; Tam, S.; Even, L.; Sheyn, D.; Ben-David, S.; Tawackoli, W.; Pelled, G. Multiparameter evaluation of in vivo gene delivery using ultrasound-guided, microbubble-enhanced sonoporation. *J. Control Release.* **2016**, 223, 157–64.

Shapiro, R. Subcutaneous immunoglobulin therapy by rapid push is preferred to infusion by pump, a retrospective analysis. *J. Clin. Immuno.* **2010**, 30(2), 301–7.

Sharma, D.; Kumar Sharma, R.; Bhatnagar, A.; Nishad, D. K.; Singh, T.; Gabrani, R.; Sharma, S. K.; Ali, J.; Dang, S. Nose to brain delivery of midazolam loaded PLGA nanoparticles, in vitro and in vivo investigations. *Curr. Drug Delivery* **2016**, 13(4), 557–64.

Shende, P.; Sardesai, M.; Gaud, R. Micro to nanoneedles, a trend of modernized transepidermal drug delivery system. *Artif. Cells Nanomed. Biotech.* **2018**, 46(1), 19–25.

Shendge, R. S.; Sayyad, F. J. Formulation development and evaluation of colonic drug delivery system of budesonide microspheres by using spray drying technique. *J. Pharm. Res.* **2013**, 6(4), 456–61.

Shojaei, A. H. Buccal mucosa as a route for systemic drug delivery, a review. *J. Pharm. Pharm. Sci.* **1998**, 1(1), 15–30.

Siddiqui, L., Bag, J., Mittal, D., Leekha, A., Mishra, H., Mishra, M., Verma, A.K., Mishra, P.K., Ekielski, A., Iqbal, Z. and Talegaonkar, S. Assessing the potential of lignin nanoparticles as drug carrier: Synthesis, cytotoxicity and genotoxicity studies. *International Journal of Biological Macromolecules*. **2020,** 152, pp.786–802.

Sill, T. J.; Von Recum, H. A. Electrospinning, applications in drug delivery and tissue engineering. *Biomaterials* **2008**, 29(13), 1989–2006.

Singh, P.; Kim, Y. J.; Wang, C.; Mathiyalagan, R.; Yang, D. C. The development of a green approach for the biosynthesis of silver and gold nanoparticles by using Panax ginseng root extract, and their biological applications. *Artif. Cells Nanomed. Biotech.* **2016**, 44(4), 1150–7.

Singh, R.; Lillard, J. W. Nanoparticle-based targeted drug delivery. *Exper. Mol. Pathog.* **2009**, 86(3), 215–23.

Sokolova, V., Rotan, O.; Klesing, J.; Nalbant, P.; Buer, J.; Knuschke, T.; Westendorf, A. M.; Epple, M. Calcium phosphate nanoparticles as versatile carrier for small and large molecules across cell membranes. *J. Nanopart. Res.* **2012**, 14(6), 910.

Sonvico, F.; Dubernet, C.; Colombo, P.; Couvreur, P. Metallic colloid nanotechnology, applications in diagnosis and therapeutics. *Curr. Pharm. Des.* **2005**, 11(16), 2091–105.

Storni, T.; Kundig, T. M.; Senti, G.; Johansen, P. Immunity in response to particulate antigen-delivery systems. *Adv. Drug Delivery Rev.* **2005**, 57(3), 333–55.

Stroh, M.; Zimmer, J. P.; Duda, D. G.; Levchenko, T. S.; Cohen, K. S.; Brown, E. B.; Scadden, D. T.; Torchilin, V. P.; Bawendi, M. G.; Fukumura, D. Quantum dots spectrally distinguish multiple species within the tumor milieu in vivo. *Nat. Med.* **2005**, 11(6), 678.

Sudha, A.; Jeyakanthan, J.; Srinivasan, P. Green synthesis of silver nanoparticles using Lippia nodiflora aerial extract and evaluation of their antioxidant, antibacterial and cytotoxic effects. *Res. Efficient Tech.* **2017**, 3(4), 506–15.

Suman, T.; Rajasree, S. R.; Kanchana, A.; Elizabeth, S. B. Biosynthesis, characterization and cytotoxic effect of plant mediated silver nanoparticles using Morinda citrifolia root extract. *Coll. Surf. B Biointerf.* **2013**, 106, 74–8.

Sun, B.; Xie, W.; Yi, G.; Chen, D.; Zhou, Y.; Cheng, J. Microminiaturized immunoassays using quantum dots as fluorescent label by laser confocal scanning fluorescence detection. *J. ImmunoMeth.* **2001**, 249(1–2), 85–9.

Sung, J. C.; Pulliam, B. L.; Edwards, D. A. Nanoparticles for drug delivery to the lungs. *Trends Biotech.* **2007**, 25(12), 563–70.

Suri, S. S.; Fenniri, H.; Singh, B. Nanotechnology-based drug delivery systems. *J. Occup. Med. Toxicol.* **2007**, 2(1), 16.

Takakura, Y.; Fujita, T.; Hashida, M.; Sezaki, H. Disposition characteristics of macromolecules in tumor-bearing mice. *Pharm. Res.* **1990**, 7(4), 339–46.

Tang, B. C.; Dawson, M.; Lai, S. K.; Wang, Y. Y.; Suk, J. S.; Yang, M.; Zeitlin, P.; Boyle, M. P.; Fu, J.; Hanes, J. Biodegradable polymer nanoparticles that rapidly penetrate the human mucus barrier. *Proc. Nat. Acad. Sci.* **2009**, 106(46), 19268–73.

Tang, F.; Li, L.; Chen, D. Mesoporous silica nanoparticles, synthesis, biocompatibility and drug delivery. *Adv. Mater.* **2012**, 24(12), 1504–34.

Tasciotti, E.; Liu, X.; Bhavane, R.; Plant, K.; Leonard, A. D.; Price, B. K.; Cheng, M. M. C.; Decuzzi, P.; Tour, J. M.; Robertson, F. Mesoporous silicon particles as a multistage delivery system for imaging and therapeutic applications. *Nat. Nanotech.* **2008**, 3(3), 151.

Tautzenberger, A.; Kovtun, A.; Ignatius, A. Nanoparticles and their potential for application in bone. *Int. J. Nanomed.* **2012**, 7, 4545.

Tegos, G. P.; Demidova, T. N.; Arcila-Lopez, D.; Lee, H.; Wharton, T.; Gali, H.; Hamblin, M. R. Cationic fullerenes are effective and selective antimicrobial photosensitizers. *Chem. Bio.* **2005**, 12(10), 1127–35.

Tetteh, E. K.; Morris, S. Evaluating the administration costs of biologic drugs, development of a cost algorithm. *Health Eco. Rev.* **2014**, 4(1), 26.

Trastullo, R.; Abruzzo, A.; Saladini, B.; Gallucci, M. C.; Cerchiara, T.; Luppi, B.; Bigucci, F. Design and evaluation of buccal films as paediatric dosage form for transmucosal delivery of ondansetron. *Euro. J. Pharm. Biopharm.* **2016**, 105, 115–21.

Trevaskis, N.L.; Kaminskas, L.M.; Porter, C.J. From sewer to saviour—targeting the lymphatic system to promote drug exposure and activity. *Nat. Rev. Drug Dis.* **2015**, *14(11),* 781.

Tsuji, T.; Iryo, K.; Watanabe, N.; Tsuji, M. Preparation of silver nanoparticles by laser ablation in solution, influence of laser wavelength on particle size. *Appl. Surf. Sci.* **2002**, *202(1–2)*, 80–5.

Turnbull, W. B.; Stoddart, J. F. Design and synthesis of glycodendrimers. *Rev. Mol. Biotech.* **2002**, *90(3–4)*, 231–55.

Vallet-Regi, M.; Balas, F.; Arcos, D. Mesoporous materials for drug delivery. *Angewandte Chemie Int. Eds.* **2007**, *46(40)*, 7548–58.

Van der Zande, M.; Vandebriel, R. J.; Van Doren, E.; Kramer, E.; Herrera Rivera, Z.; Serrano-Rojero, C. S.; Gremmer, E. R.; Mast, J.; Peters, R. J.; Hollman, P. C. Distribution, elimination, and toxicity of silver nanoparticles and silver ions in rats after 28-day oral exposure. *ACS Nano.* **2012**, *6(8)*, 7427–42.

Veronese, F. M.; Pasut, G. PEGylation, successful approach to drug delivery. *Drug Dis. Today.* **2005**, *10(21)*, 1451–8.

Vu, T. Q.; Maddipati, R.; Blute, T. A.; Nehilla, B. J.; Nusblat, L.; Desai, T. A. Peptide-conjugated quantum dots activate neuronal receptors and initiate downstream signaling of neurite growth. *Nano Lett.* **2005**, *5(4),* 603–7.

Wadhwa, R.; Lagenaur, C. F.; Cui, X. T. Electrochemically controlled release of dexamethasone from conducting polymer polypyrrole coated electrode. *J. Control. Rel.* **2006**, *110(3)*, 531–41.

Wang, J.; Yong, W. H.; Sun, Y. H.; Vernier, P. T.; Koeffler, H. P.; Gundersen, M. A.; Marcu, L. Receptor-targeted quantum dots, fluorescent probes for brain tumor diagnosis. *J. Biomed. Opt.* **2007**, *12(4)*, 044021.

Wang, T. Y.; Wilson, K. E.; Machtaler, S.; Willmann, J. K. Ultrasound and microbubble guided drug delivery, mechanistic understanding and clinical implications. *Curr. Pharm. Biotech.* **2013**, *14(8)*, 743–52.

Wang, X. Y.; Fang, Z.; Lin, X. Copper sulfide nanotubes, facile, large-scale synthesis, and application in photodegradation. *J. Nanoparticle Res.* **2009**, *11(3)*, 731–6.

Wanka, G.; Hoffmann, H.; Ulbricht, W. Phase diagrams and aggregation behavior of poly (oxyethylene)-poly (oxypropylene)-poly (oxyethylene) triblock copolymers in aqueous solutions. *Macromolecules* **1994**, *27(15)*, 4145–59.

Warheit, D. B.; Laurence, B. R.; Reed, K. L.; Roach, D. H.; Reynolds, G. A.; Webb, T. R. Comparative pulmonary toxicity assessment of single-wall carbon nanotubes in rats. *Toxico. Sci.* **2004**, *77(1)*, 117–25.

Watts, P. J.; Lllum, L. Colonic drug delivery. *Drug Develop. Industr. Pharm.* **1997**, *23(9)*, 893–913.

Webster, T. J.; Siegel, R. W.; Bizios, R. Nanoceramic surface roughness enhances osteoblast and osteoclast functions for improved orthopaedic/dental implant efficacy. *Scripta Mater.* **2001**, *44(8–9)*, 1639–42.

Wilczewska, A. Z.; Niemirowicz, K.; Markiewicz, K. H.; Car, H. Nanoparticles as drug delivery systems. *Pharm. Rep.* **2012**, *64(5)*, 1020–37.
Wray, A. T.; Klaine, S. J. Modeling the influence of physicochemical properties on gold nanoparticle uptake and elimination by Daphnia magna. *Env. Toxico. Chem.* **2015**, *34(4)*, 860–72.
Wunderlich, C. A. On the Temperature in Diseases, A Manual of Medical Thermometry, New Sydenham Society. 1871.
Wust, P.; Hildebrandt, B.; Sreenivasa, G.; Rau, B.; Gellermann, J.; Riess, H.; Felix, R.; Schlag, P. Hyperthermia in combined treatment o f cancer. *Lancet Oncol.* **2002**, *3(8)*, 487–97.
Xie, J.; Wang, C. H. Electrospun micro-and nanofibers for sustained delivery of paclitaxel to treat C6 glioma in vitro. *Pharm. Res.* **2006**, *23(8)*, 1817.
Xie, J.; Zheng, Y.; Ying, J. Y. Protein-directed synthesis of highly fluorescent gold nanoclusters. *J. Am. Chem. Soc.* **2009**, *131(3)*, 888–9.
Xu, R.; Zhang, G.; Mai, J.; Deng, X.; Segura-Ibarra, V.; Wu, S.; Shen, J.; Liu, H.; Hu, Z.; Chen, L. An injectable nanoparticle generator enhances delivery of cancer therapeutics. *Nat. Biotech.* **2016**, *34(4)*, 414.
Yang, M. X.; Shenoy, B.; Disttler, M.; Patel, R.; McGrath, M.; Pechenov, S.; Margolin, A. L. Crystalline monoclonal antibodies for subcutaneous delivery. *Proc. Nat. Acad. Sci.* **2003**, *100(12)*, 6934–9.
Yoo, H. S.; Kim, T. G.; Park, T. G. Surface-functionalized electrospun nanofibers for tissue engineering and drug delivery. *Adv. Drug Delivery Rev.* **2009**, *61(12)*, 1033–42.
Yoshioka, T.; Sternberg, B.; Florence, A. T. Preparation and properties of vesicles (niosomes) of sorbitan monoesters (Span 20, 40, 60 and 80) and a sorbitan triester (Span 85). *Int. J. Pharm.* **1994**, *105(1)*, 1–6.
Yu, X. L.; Cao, C. B.; Zhu, H. S.; Li, Q. S.; Liu, C. L.; Gong, Q. H. Nanometer sized copper sulfide hollow spheres with strong optical limiting properties. *Adv. Func. Mater.* **2007**, *17(8)*, 1397–401.
Yum, K.; Wang, N.; Yu, M. F. Nanoneedle, a multifunctional tool for biological studies in living cells. *Nanoscale.* **2010**, *2(3)*, 363–72.
Zhang, H.; Liu, D.; Shahbazi, M.A.; Makila, E.; Herranz-Blanco, B.; Salonen, J.; Hirvonen, J.; Santos, H.A. Fabrication of a multifunctional nano in micro drug delivery platform by microfluidic templated encapsulation of porous silicon in polymer matrix. *Adv. Mater.* **2014**, *26(26)*, 4497–503.
Zhang, J.; Srivastava, R.; Misra, R. Core–shell magnetite nanoparticles surface encapsulated with smart stimuli-responsive polymer, synthesis, characterization, and LCST of viable drug-targeting delivery system. *Langmuir.* **2007**, *23(11)*, 6342–51.
Zhang, J.; Yuan, Z. F.; Wang, Y.; Chen, W. H.; Luo, G. F.; Cheng, S. X.; Zhuo, R. X.; Zhang, X. Z. Multifunctional envelope-type mesoporous silica nanoparticles for tumor-triggered targeting drug delivery. *J. Am. Chem. Soc.* **2013**, *135(13)*, 5068–73.
Zhong, X.; Feng, Y.; Knoll, W.; Han, M. Alloyed Zn x Cd1-x S nanocrystals with highly narrow luminescence spectral width. *J. Am. Chem. Soc.* **2003**, *125(44)*, 13559–63.
Zhang, S.; Zhong, H.; Ding, C. Ultrasensitive flow injection chemiluminescence detection of DNA hybridization using signal DNA probe modified with Au and CuS nanoparticles. *Analy. Chem.* **2008**, *80(19)*, 7206–12.

CHAPTER 19

Beneficial Utility and Perspective of Nanomaterials Toward Biosensing

RAVINDRA PRATAP SINGH* and KSHITIJ RB SINGH

Department of Biotechnology, Indira Gandhi National Tribal University, Anuppur, Amarkantak 484887, Madhya Pradesh, India

**Corresponding author. E-mail: rpsnpl69@gmail.com*

ABSTRACT

Nanomaterial-based biosensor technology provides a promise to unravel the biocompatibility and enhance applications for the detection of toxicants/pollutants/carcinogens/adulterants and microbes derived from plants, animals, foods, soil, air, and water. The biomolecules are using like enzymes, antibodies, aptamer, receptors, ds/ssDNA, organelles, and cells in the fabrication of biosensor. The biosensor utilizes biomolecules like enzymes (enzymosensor), antibodies (immunosensors), DNA (genosensor), cell/tissue, microbes, aptamers, and receptors; these biosensors have revealed their potential applications in medical, environmental, and agricultural domains. The utilization of biosensor is simple, quick, and it offers low-price detection of the desired analyte, which has increased its demand globally. Nanomaterial development and utilization toward technologies have created a vast potential contribution to the life sciences particularly biotechnology exploitation metal and metal oxide nanoparticles, semiconductor nanoparticles, magnetic nanoparticles, dendrimers, graphene, fullerene, carbon nanotubes, nanocomposite, and quantum dots that have found their places toward the biosensor/nanobiosensor field. The prospects and challenges of NMs pertaining to potential applications of biosensor development and its miniaturization have focused noteworthy influence on the event of recent ultrasensitive biosensing devices (nanobiosensor) to resolve the many environmental issues that affect adversely different numerous comforts of living entities. Nanobiosensors are incredibly specific, selective, sensitive, and reusable and independent of pH and temperature. The useful utility of nanomaterial primarily based biosensors is in the food business to envision quality and safety of food, in bioscience early stage and fast detection of the analyte of interest are vital aspects particularly in drug discovery and lifestyle diseases, in defense, and for marine applications. Cell- and a tissue-based biosensor with immobilized enzyme detect hormones, drugs, or toxins, whereas DNA (genosensor) detect toxicants, toxic metals, and the pathogen is the potential goals to be highlighted in this work. The nanomaterial-based biosensor is ultrabiosensor that have engaging prospects and nice challenges toward clinical diagnosing, food analysis, process management, and environmental observance that utilizes the structure and potentiality of nanomaterials and biomolecules to style single-molecule multifunctional nanocomposites, nanofilms, and nanoelectrodes that remains complicated and difficult. Utilization of nanomaterials enhances magnetic, optical, and electrical properties of biosensors/nanobiosensors that are still terribly difficult in regard to fabrication, characterization, detection, and biosensing approach for the potential applications in medical/clinical/health, environmental, and agriculture particularly food industries. The useful utility and perspective of nanomaterials toward biosensor technology would be showed increased sensitivity, detection time, price, and automation. In this chapter, a number of the few recent and previous advances in biosensor technology are mentioned that highlight a comprehensive perspective to the readers.

19.1 INTRODUCTION

Nanomaterials at the nanoscale dimension can develop processes and products. They have powerfully created their presence in various areas like health care, implants, prostheses, textiles, energy, space, solar, defense, security, terrorism, and surveillance (Sailer et al., 2005). Nanobiotechnology has emerged as a brand-new exciting field of nanotechnology, which is recognized as a replacement knowledge base frontier within the field of natural and material science. The prospects and challenges of the nanomaterials-based biochip, nanomotors, nanocomposite, nanobiomaterials, nanobiosensors, and nanodrug-delivery systems in the industry, defense, energy, space, and health care are very potential. Biomolecules as bionanomatrials may assume to play the vital significance in nanobiotechnology, for instance, peptide nucleic acid in place of DNA acts as a probe in the genetic science, medicine, pharmaceutics/clinical/cytogenetics applications (Singh et al., 2010). Biosensor with biosensing recognition material and an electrical device, that is, the transducer will be used for the detection of biological and chemical agents as shown in Figure 19.1 (Singh, 2011). Singh et al. (2011) reported the fast determination of not only H_2O_2 but also conjointly compound in biological samples using CAT/PANi/ITO thin film as bioelectrode and suggested that this economical film is used for the immobilization of not only solely enzyme but also conjointly different enzymes and bioactive molecules (Singh, 2011; Singh et al., 2009). Biosensors became an emerging space of knowledge-based analysis for the detection of varied styles of targeted biomolecules (Singh, 2011; Singh et al., 2010, 2009; Kim et al., 2009).

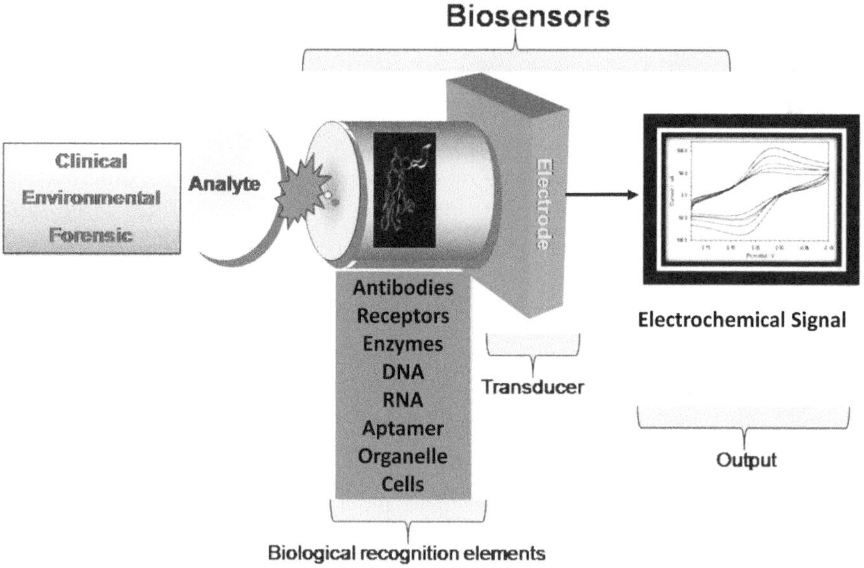

FIGURE 19.1 A simplified matrix that can lead to a variety of combinations of molecular recognition elements and transducers to produce biosensors.

In addition, the nanobiosensor could be integrated with lab-on-a-chip toward biochemical analysis and clinical diagnostics or biomedical diagnostics and can have huge implications for the good thing about society and human health. The biosensors do utilize nanomaterials known as ultrasensitive biosensors and have potential applications in agriculture, clinical healthcare, and environment for real-time detection of toxins, pesticides, antibiotics, pathogens, and foodstuff. The rising trends toward biosensing development in bioelectronics likely to be addressed and resolved the challenges of pollution issues regarding not only human health but all living entities (Singh et al.,

2012; Jianrong et al., 2004; Tombelli et al., 2005; Seeman1999). Singh and Choi (2010) reported single biomolecule based probe detection technology and opened new R&D in NS & NT. Nanomaterial-based nanobiosensor exhibited high surface area, electronic properties, electrocatalytic activity, and in addition pretty much as good biocompatibility and accustomed to achieve direct immobilization of enzymes to the conductor surface, that is, electrode, to push chemical reaction, and to amplify a proof of biorecognition events (Xiao and Chang 2010; Pumera et al., 2007). Few potential nanobiosensors have been reported using carbon nanotube (CNT) modified electrodes for NADH detection, gold nanoparticles for electrochemical immunosensors, nanoparticles-based biosensor, nanowire-based biosensor and nanowires (NWs) act as an electrode or conductor or interconnectors in nanoelectronic devices (Agui et al., 2008; Gooding 2005; He et al., 2006; Kumar et al., 2008; Rivas et al., 2007; Wang 2005; Umasankar and Chen 2008; Merkoci et al., 2005; Pingarron et al., 2008; Yogeswaran and Chen 2008). Chen et al. (2008) reported charged Ni–Al–OH nanosheets to immobilize horseradish peroxidase (HRP) onto GC Electrode for the reduction of H_2O_2 and trichloroacetic acid. Various researchers have been reported metal and metal oxides using in biosensing for biosensors (Lu et al., 2008; Tang et al., 2005; Vamvakaki et al., 2008; Zhou et al., 2008).

The biosensing platform is the need for the time that requires miniaturization, cost-effectiveness, and either single and multiple analytes of interest detection. Sassolas et al. (2008) reported genosensors and gene microarrays for the specific binding of complementary base sequences. Hahn et al. (2005) reported DNA microarrays fabrication on solid supports of 10–100 μm reaction sites for the immobilization of gene sequences. Electrochemical transducers have used for detective work of DNA conjugation and thanks to their high sensitivity, small size, and low cost. There are variously labeled DNA probe, enzyme, ferrocene, an intercalator for the label-free genosensors (Daniel et al., 2007; Arora et al., 2006). Kumar and Dill (2005) reported a DNA microarray with numerous microelectrodes having RAM, whereas Davies et al. (2005) reported random assemblies of microdisks with carbon wires embedded epoxy resin. Ordeig and Banks (2006) reported microfluidic environments for the biosensing tasks to require fewer amounts of probe molecules and target analyte onto boron-doped diamond microelectrode disks made up by the electrodeposition of Cu, Ag, and Au (Fletcher and Harne 1999; Wang 2006). Microfluidic technology is a single-chip device to detect viruses and bacteria in biological specimens. The device is very small made up of arrays of microelectrodes controlled by complementary metal-oxide semiconductor electronic equipment. The genetic analysis, disease identification, and detection of biological warfare agents were possible by this technology based on impedance, potentiometry, voltammetry, and ion-sensitive assays (Liu 2004; Simm et al., 2005; Hassibi and Lee (2006).

Nanomaterials-based biosensors have vast utility in the food sector for monitoring food quality and safety; further, the nanomaterial-based biosensors in bioscience have application toward early and fast detection of an analyte of interest for combating lifestyle diseases. Cell and tissue-based biosensor detects hormones, drugs, or toxins, whereas genosensor detects infectious pathogens, toxicants are the potential goals to be highlighted. The nanomaterial-based biosensor may be ultrabiosensor that have engaging prospects and nice challenges toward clinical diagnosing, food analysis, process management, and environmental observance that utilizes the structure and potentiality of nanomaterials and biomolecules to style single-molecule multifunctional nanocomposites, nanofilms, and nanoelectrodes that remains complicated and difficult. Utilization of nanomaterials enhances magnetic, optical, and electrical properties of biosensors/nanobiosensors that are still terribly difficult in regard to fabrication, characterization, detection, and biosensing approach for the potential applications in medical/clinical/health, environmental, and agriculture particularly food industries. The useful utility and perspective of nanomaterials toward biosensor technology would be showed increased sensitivity, detection time, price, and automation. In this chapter, a number of the few recent and previous advances in biosensor technology are mentioned that highlight a comprehensive perspective to the readers.

19.2 NANOSTRUCTURED MATERIALS AND BIOSENSING

The metal and metal oxide nanoparticles are derived from respective salts and used in biosensing applications. Nanomaterials (1–100 nm) show unique and different properties due to their tiny size at the nanoscale (Cai et al., 2002, 2004; Kerman et al., 2004). The improvement in oxidation–reduction properties of AuNPs not to mention silver has led to their immense application toward biosensing development with exceptional sensitivity (Wang et al., 2003; Ozsoz et al., 2003). K'Owino et al. (2003) reported silver monolayer for the enhancing metal electrochemical sensing of anticancer drugs. Ngundi (2003) reported the detection of DNA–protein, antigen–antibody, and DNA–RNA reactions. The aptamer is an artificial nucleic acid matter consisting of fiber DNA/RNA sequences used as a recognition component for biosensing applications to detect varied analytes of interest to reinforce sensitivity and property exploitation NPs and QDs labels (Lee et al., 2008; Song et al., 2008; Tombelli et al., 2007; Willner and Zayats 2007; Hansen et al., 2006; He et al., 2007; Numnuam et al., 2008).

Semiconductor or conducting chemical compound NWs are very enticing for coming up with high-density macromolecule arrays, thanks to their high surface-to-volume area and electron transport phenomenon on biomolecules. Hahm and Lieber (2004) reported the nanowire functionalized probe DNA detection based on DNA hybridization events. Patolsky et al. (2004) reported single-molecule SiNW functionalized with either peptide nucleic acid probes or antibodies for the detection of influenza. Zheng et al. (2005) reported an antibody functionalized silicon nanowire sensing element array to detect cancer markers in the biological fluid samples. Tang et al. (2007) reported a label-free electrochemical immunoassay-based electrode to detect carcino-embryonic antigen. Yang et al. (2004) reported biosensor to detect microorganisms impede metrically. Yang and Bashir (2008) reported electrochemical electric resistance for the fast detection of foodborne infective microorganism using interdigitated array microelectrodes-based microchips. Cretich et al. (2008) reported immobilizing DNA fragments for the bacterial genotyping and food pathogen identification using PDMS functionalized biomolecules in microarrays, biosensors, and cell culture. Liao et al. (2006) reported a microfabricated electrochemical sensor array to detect infectious pathogens in biological fluids. Karasinski et al. (2007) reported an autonomous electrochemical biosensor technique to detect microbes.

Clark and Lyons (1962) reported for the first time a glucose enzyme electrode-based biosensor. Wang et al. (2008) reported glucose biosensors for diabetes control. Liu et al. (2005) reported AuNPs or CNTs based biosensor for the detection of glucose using glucose oxidase enzyme. Patolsky et al. (2004) reported SWCNT using enzyme glucose oxidase to detect glucose blood samples. Lu et al. (2007) reported enzyme-functionalized Au NWs to detect glucose. Qu et al. (2007) reported CNTs and PtNW to detect glucose. Ramanavicius et al. (2005) reported PPy coated GOx NPs for biosensing. Li et al. (2000) reported conductivity-based glucose nanobiosensor based on conducting-polymer-based nanogap. Mediator less reagent in ultrabiosensor is the main advantage leading to high selectivity. Jing and Yang (2006) reported a boron-doped diamond electrode with the mediator-free reagent to detect glucose. Wilson (2005) reported continuous glucose watching for the therapeutic interventions, whereas Wilson and Gifford (2005) reported multi-electrode array to monitor glucose. Jung et al. (2004) reported two working electrodes for the enzyme and nonenzymic electrodes for measuring glucose concentration and background current for diabetic people. Heller (2005) reported glucose biosensors for regulating blood glucose patient monitoring. Nitric oxide is a useful substance of thrombocyte and microorganism adhesion. Gifford et al. (2005) reported glucose biosensors. The nanomaterials are also targeted contact for the oxidation–reduction in between GOx and conductor supports. Andreescu and Luck (2008) reported genetically built periplasmic aldohexose entities on AuNPs for the glucose biosensor. Berg et al. (2002) reported painless or noninvasive in vitro testing of glucose using immobilized artificial receptors for aldohexose within a biocompatible membrane.

19.3 CARBON NANOMATERIALS AND BIOSENSING

Curl, Kroto, and Smalley discovered fullerenes in 1985 and got the noble prize in 1996. Fullerene is an elongated sphere of carbon atoms formed by interconnecting six-member rings and 12 isolated five-member rings forming hexagonal and pentagonal faces (Kroto et al., 1985, Zheng et al., 2000). CNTs are hollow cylinders of carbon atoms, rolled tubes of graphite specified their walls are hexangular carbon rings and sometimes fashioned in massive bundles that were discovered by Iijima (1991). The CNTs are two types namely SWNTs and MWNTs as shown in Figure 19.2 (Zheng et al., 2002; Sawamura et al., 2002; Lee et al., 2002; Margadonna et al., 2002; Schon et al., 2001; Iijima, 1991). CNTs showed rigidity, flexibility, toughness, and conductors or semiconductors (Overney et al., 1993, Mintmire et al., 1992, Hamada et al., 1992, Saito et al., 1992) and used as electrodes, power cables, fibers, composites, actuators, sensors, biosensors, in field emission-based flat panel displays, and computers. Figure 19.3 shows the most commonly used carbon nanomaterials like fullerene, CNTs (sinle and multiwalled), and graphene. Apart from these, unique properties of CNTs like little size, electrochemical activity, physical properties, low density, and biocompatibility are very useful and having immense potential applications. CNT nanofibers are used in biosensing applications for example in implantable applications for continuous monitoring of glucose, lactate, antibodies, and antigens. CNTs toxicity could not be ruled out (Tombler et al., 2000). Maiti et al. (2001) reported the toxic and hazardous nature of carbon-based nanomaterials.

Li et al. (2005) reported chitosan doped with CNT sensing to detect salmon sperm DNA using DNA indicator. So et al. (2005) reported SWNT-field-effect transistor (FET)-based biosensors using DNA aptamers to detect

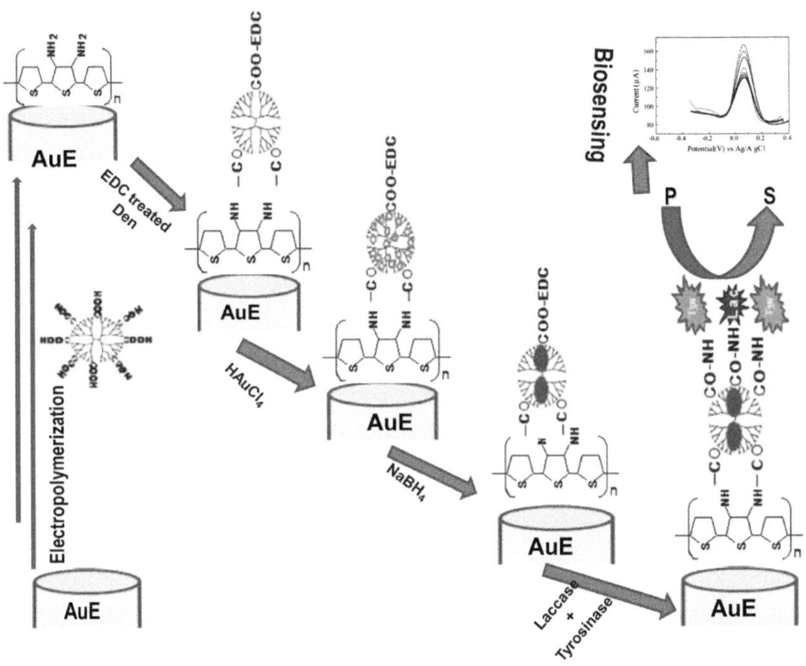

FIGURE 19.2 Representative nanobioelectrode fabrication for ultrabiosensing detection using dendrimer.

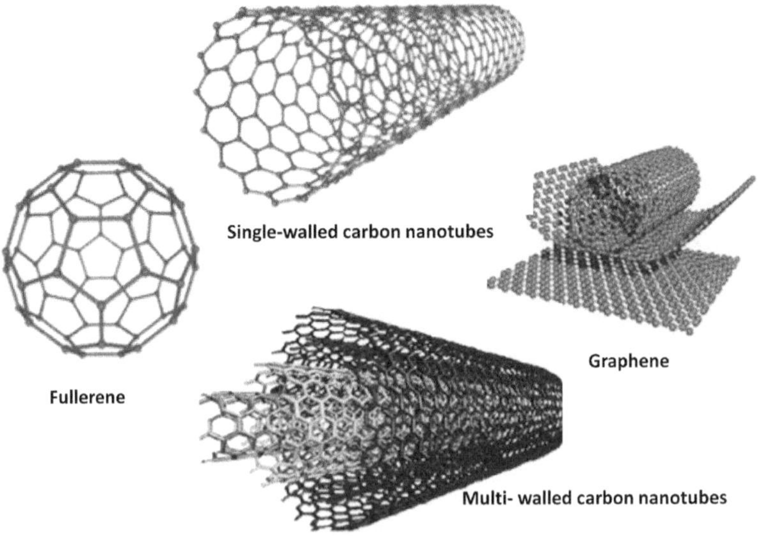

FIGURE 19.3 Carbon nanomaterials in zero-dimensional (0D) fullerene, 1D carbon nanotubes, and 2D graphene.

protein. Wang (2005) reported the nanomaterial fabricated bioelectronic devices. Singh et al. (2005) reported the fabrication of the functionalized CNTs molecular biosensors. Wang et al. (2004a) reported genosensor based on self-assembled MWCNT using DNA oligonucleotides as a probe and methylene blue as an indicator or intercalator. Furthermore, Wang et al. (2004b) reported CNT-modified glassy carbon transducers for assay of DNA crossing. Katz and Willner (2004) reported CNTs metallic NWs immobilized proteins and DNA hybrid systems as active FETs or biosensor devices (enzyme electrodes, immunosensors, or DNA sensors). Jung et al. (2004) reported DNA oligonucleotides on SWNT multilayer films as a future biosensor. Cai et al. (2003) reported a unique genosensor utilized MWCNTs–COOH) to monitor by DPV using electroactive intercalator daunomycin. Gao et al. (2011) reported an immunosensor to detect carcinoembryonic antigen (CEA). Meng et al. (2011) reported an immunosensor to detect cancer biomarkers (AFP). Cao et al. (2011) reported an immunosensor to detect casein in cheese samples based on AuNPs/poly (L-Arginine)/MWCNT nanocomposite film. Piao et al. (2011) reported an immunosensor based on CNTs coated with tyrosinase and magnetic nanoparticles to detect phenol, o-quinone, and catechol. Zhuo et al. (2011) reported an immunosensor to detect internal organ biomarkers using gold and CNTs composite as a fascinating platform. Zhuo et al. (2010) reported a replacement strategy for the antigen-antibody sensing process by functionalizing SiO2NPs labeled secondary antibodies supported a sandwich bioassay. Tang et al. (2010) reported a replacement chemical science bioassay protocol to detect alpha-fetoprotein. Li et al. (2010) reported the strategy of immunosensors to detect glucose. Sánchez et al. (2010) reported an immunosensor using CNT/polysulfone/RIgG transducer. Munge et al. (2010) reported an immunosensor to detect metalloproteinase-3, a cancer biomarker protein.

Graphene is marvel material that was discovered by Andre Geim and Konstantin Novoselov in 2004. It is a flexible material with 2D planar structure and showed potential applications in electronic devices, photonics and optoelectronics, medical medicine and drug delivery, sensors, versatile electronics, energy storage, nanocomposites, spintronics, etc. Graphene is used as a nanomaterial for biosensing for the early diagnosis of lifestyle diseases like diabetes, cancer, heart diseases, brain diseases, asthma, allergy, urethral stones, etc. Furthermore, in another way graphene is structurally one-atom-thick tabular sheets of sp2-bonded carbon atoms. Differently, graphene is single sheets of graphite, 2D honeycomb lattice with wonderful properties at room temperature. Graphene is Sp2 hybridized carbon nanomaterials typically in zero-dimensional (0D) fullerene, 1D CNTs, and 2D graphene

(Novoselov et al., 2004, Geim 2009). Graphene has attention-grabbing electrical, optical, mechanical, and chemical properties. Graphene is useful in creating biosensors, DNA sequencing, drug carrier in nanomedicine. Feng & Liu (2011) reported disease identification and treatment using graphene oxide, graphene platelets, and graphene nanoflakes. Zhao et al. (2011) reported graphene oxide-DNAzyme to detect Pb^{2+}. Tang et al. (2011) reported graphene nanosheet in bionanotechnology. He et al. (2011) reported reduced graphene oxide (rGO) thin-film transistors to detect fibronectin and avidin. Liu et al. (2011) reported fluorescent detection of single-nucleotide polymorphism (SNP) supported polymer ligase reaction and π-stacking. Xu et al. (2011) reported electrogenerated chemiluminescence of luminol for biosensing applications. Li et al. (2011) reported a graphene oxide-based platform using GO-peptide-QDs nanoprobes to detect protease activity. Liu et al. (2011) reported an immunoassay using gold nanoflower-labeled antibodies to detect p-nitrophenol. Lu et al. (2011) reported protein biosensor to detect aldohexose using metal nanoparticles-(PdNPs-) functioned graphene (nafion-graphene). Zhang et al. (2011) reported graphene-based biosensing platform-based peptides as probe biomolecules and dye-labeled peptides. Choi et al. (2011) reported SPR imaging biosensors used a graphene-on-silver substrate to increase the imaging. Feng et al. (2011) reported electrochemical devices that may understand label-free neoplastic cell detection. Zhang et al. (2011) reported a flexible molecular beacon-(MB) on the self-assembled ssDNA-graphene compound (ssDNA-GO) design to detect DNA, thrombin, Ag^{+}, Hg^{2+}, and cysteine. Song et al. (2011) reported an economical enzyme-based screen-printed electrode (SPE) for monitoring catechol a phenolic compound. Lu et al. (2011) reported conducting channel on a back-gated FET of r-GO). Ohno et al. (2010) reported an immunosensor using aptamer-modified graphene field-effect transistor to detect IgE. Li et al. (2010) reported a graphene-Nafion-modified glassy carbon (GN/GC) electrode to determine the anticancer herbal drug. Huang et al. (2011) reported COOH functionalized graphene to detect adenine and guanine by CV and DPV. Goh and Pumera (2011) reported graphene nanoribbons to detect explosives 2,4,6-trinitrotoluene (TNT) in saltwater. Wang et al. (2011) reported graphene-CdS (G-CdS) nanocomposite immobilization glucose oxidase (GOD) biosensors. Wan et al. (2011) reported an immunosensor doped with reduced graphene sheets to detect marine pathogenic sulfate-reducing bacteria. Li et al. (2010) reported MnO_2/graphene oxide hybrid nanostructure to detect H_2O_2 in an alkaline medium for catalyst biosensor and energy conversion applications. Liu et al. (2010) reported one-step fluoroimmunoassay using graphene sheets with CdTe quantum dots for biomarker supermolecule. Wang et al. (2010) reported graphene, chitosan, hemoglobin, and ionic liquid on a glassy carbon electrode to detect nitromethane. Li et al. (2010) reported graphene-based molecular beacon using water-soluble graphene oxide to hairpin and dsDNA to detect DNA. Wu et al. (2010) reported graphene-modified SPR biosensor to detect the refractive index to suggest the optical property of graphene. Yang and Gong (2010) reported an immunosensor-based graphene film to detect prostate-specific antigen, a cancer biomarker. Zhang et al. (2010) reported HRP/ss-DNA/GP/GC bioelectrode for H_2O_2 detection. He et al. (2010) reported rGO films onto terephthalate (PET) films to detect catecholamine. Wu et al. (2010) reported graphene/laccase/2,2-azino-bis(3-ethylbenzothiazoline-6-sulfonic acid) to detect oxygen released from human erythrocytes. Lightcap et al. (2010) reported a composite of rGO with metal NPs at separate sites for chemical and biological sensing. Pisana et al. (2010) reported graphene-based extraordinary magnetoresistance devices. Shan et al. (2010a) reported NADH and ethanol detection using an ionic liquid-functionalized graphene-modified electrode and ADH/IL-graphene/chitosan modified electrode respectively. Furthermore, Shan et al. (2010b) reported glucose biosensor consisted of graphene/AuNPs/GOD/chitosan composite-modified gold electrode for the electrochemical detection of glucose. Wu et al. (2009) reported the bionanocomposite film of GOx/Pt/graphene sheets/chitosan to detect glucose and provide a new chance for clinical diagnosing and point-of-care applications. Shan et al. (2009a) reported graphene sheets functionalized with poly-L-lysine in biological applications. Zhou et al. (2009) reported chemically reduced graphene-oxide-modified glassy carbon electrode (GCE) to detect free bases of DNA, hydrogen peroxide, dopamine, ascorbic acid, uric acid, and acetaminophen. Mohanty and Berry (2008) reported graphene nanostructured devices for diagnostics and therapeutics, for example, single-bacterium biodevice, and label-free DNA sensor. Lu et al. (2008) reported amperometric glucose biosensor used exfoliated carbon nanoplatelets and Pd NPs and nafion.

19.4 METALS AND METAL-OXIDE NANOSTRUCTURED MATERIALS AND BIOSENSING

The optical absorption is the property of light related to the SPR of noble metals. Absorption and scattering due to metal nanoparticles are very useful for biosensing and diagnosis. Metal oxide nanoparticles are useful to medical devices. The few most common examples are Fe_3O_4 NPs for therapeutic and diagnostic, ZnONPs for biomedical applications and TiO_2NPs have biomedical applications (Singh, 2016). Synthesis of nanomaterials is possible by bottom-up approach means process self-assembly of molecular species, with manageable chemical reactions, very economical and versatile, well-controlled form, surface properties, and developed porous structures. The engineered nanomaterials (0, 1, 2, and 3D) have attracted and emerging toward biosensing based on nanoelctrochemistry (Singh et al 2012). Hranisavljevic et al. (2002) reported hybrid inorganic–organic composites, emerging materials that hold important promising applications, for example, silicon with organic polymers known as polysilsesquioxanes, which is utilized in electronics, photonics, and others technologies. Karian (1999) reported morphology of inorganic/organic hybrids nanocomposites that showed physical, chemical, and size-dependent properties and also exhibited improvement in properties like solubility, thermal, and thermomechanical stability, toughness, optical transparency, gas permeability, dielectric constant, and fire retardancy, for example, beryllium silsesquioxane, poly(hydridosilsesquioxane), and polyhedral oligomeric silsesquioxane. Sperling and Parak (2010) reported inorganic colloidal nanoparticles to possess electron density, optical absorption, fluorescence (CdSe or CdTe), phosphorescence (doped ZnO), and magnetic moment (Fe_3O_4/CoNPs). Barbadillo et al. (2009) reported organic–inorganic hybrid nanocomposite for glucose biosensing. Shen and Shi (2010) reported dendrimer/organic/inorganic nanocomposite for biomedical applications like molecular diagnosis and targeted molecular imaging of cancer. Kerman et al. (2007) reported kinase peptide-modified electrochemical biosensors. Chen et al. (2008) reported 2D nanostructures in biochemical sensing and biological imaging. Zhou et al. (2009) reported bioassay of supermolecule cancer biomarkers by mistreatment of peroxidase-mimicking DNAzyme functionalized AuNPs as a catalytic probe. Liang et al. (2009) reported conductometric immunochemical assay using organic/inorganic hybrid membrane, which offers biosecurity application.

19.5 DENDRIMERS AND BIOSENSING

Dendrimers are branched molecules means dendron or tree or arborols or cascade molecules. It involves conjugating due to multivalent nature with chemical species onto the surface and detects like detecting agents, that is, affinity ligands, targeting molecule, imaging agent, drug active molecule, and solubilizing agent. They have potential applications ability in biosensing (Singh, 2011, 2012; Singh et al., 2019). Zhao et al. (1998) reported dendrimers as the organic nanoparticles, hyperbranched, high surface area with functional group, and water soluble. They are showing low systemic toxicity and maximally utilized in biomedical, catalysts, targeted drug delivery, MRI contrast agents, and bioadhesives (Garg et al., 2011, Wilbur et al., 1998). Dendrimers are manmade, nanomaterials with unique property due to polyvalency or multiple active groups on their surface. Dendrimer structure has a core, interior, and periphery unique size, shape, flexibility, and reactivity that makes it one of the smartest materials (Zhu et al., 2006; Bosman et al., 1999; Tully and Frechet, 2001; Patri et al., 2002; Lee et al., 2001; Balogh and Tomalia, 1998). The dendrimers showed biocompatibility and functional groups for biosensing to perform specific electrochemical reactions (Ledesma-García et al., 2003). Singh (2011) reported tyrosinase enzyme immobilized on an Au nanoparticle encapsulated-dendrimer bonded conducting polymer (CP) on a glassy carbon electrode for the estimation of catechol. Figure 19.3 shows the nanobiosensor development and construction strategy for biosensing and its biosensing performance in schematic presentation onto the nanobioelectrode using dendrimer. Tsukruk et al. (1997, 1998) reported an assembled film of dendrimers in monolayers or multilayers on a solid surface. Svobodova et al. (2002) reported poly(amidoamine) (PAMAM) dendrimers on an Au electrode to detect glucose. Snejdarkova et al. (2003) reported biosensor based on acetylcholinesterase and choline oxidase, using PAMAM dendrimers mixed with 1-hexadecanethiol. Koda et al. (2008) reported PAMAM dendrimers onto solid surface for the use in applied and basic sciences.

19.6 LIPOSOMES AND BIOSENSING

Liposomes are artificial lipid-bilayer spherical vesicles with different properties depending upon their composition, surface charge, and size. Akbarzadeh et al. (2013) classified liposome according to the number of lipid bilayers. Liu and Chen (2015) reported liposomes utility in drug delivery, gene therapy, and vaccines. Simao et al. (2015) reported the liposomal systems as carriers for bioactive compounds such as antioxidants, antimicrobials, and antigenic proteins. Ghalandarlaki et al. (2014) reported that liposomes are atoxic and degradable under physiological conditions, and nonimmunogenic. However, proteoliposomes are protein–lipid interactions that affect the surface of the membrane model to understand biochemical and biophysical phenomena and to evaluate biotechnological applications (Dong et al., 2016; Elkhodiry et al., 2016).

19.7 CONJUGATED CONDUCTING POLYMERS AND BIOSENSING

CPs are π-conjugated structure with sp2-hybridization that exhibit electrical conductivity known as conjugated polymers. Polyacetylene (PA) was first CP known in the 1970s. The aromatic CPs, like PPy, PANi, polythiophene (PT), and poly (3,4-ethyelenedioxythiophene) (PEDOT) are few most important examples that show thermal stability and conductivity. Table 19.1 shows few most commonly used aromatic CPs for biosensing use. The processes of polymerization and doping are important aspects. The charges in the CPs are possible due to doping either p-doping (oxidized) or n-doping (reduced) or neutrality as charged polarons (radical ions) or bipolarons (dications or dianions) and generate electrical conductivity (Singh, 2011, 2012; Jang, 2006; Li et al., 2012; Lai et al., 2016; Snook et al., 2011).

TABLE 19.1 Commonly Used Aromatic Conducting Polymers Used in Biosensing

Conducting Polymers	Chemical Structure	Applications
Polyaniline	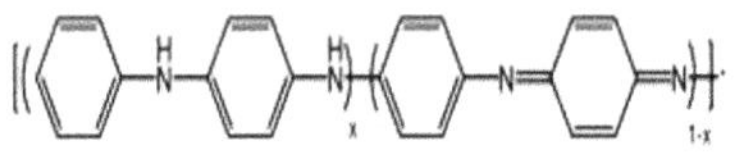	Biosensing, drug delivery
Polypyrrole	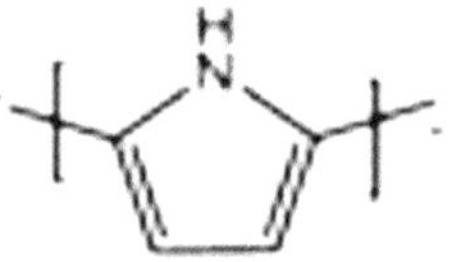	Biosensing, microsurgical tools
Polythiophene	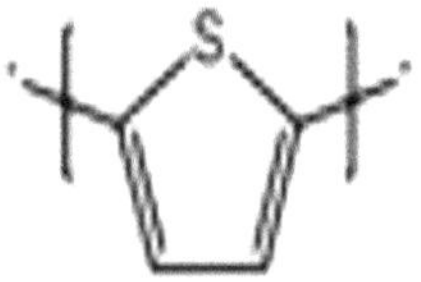	Biosensing, tissue engineering
PEDOT (Poly(3,4 ethylenedioxythiophene)	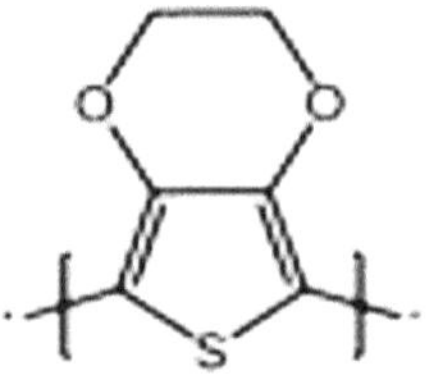	Biosensing, nerve grafts, heart muscles patches

CPs are used as transducers in biosensors, namely, PPy, PANi, PCz, POPD, and polyfuran in biomedical and environmental monitoring, for example, glucose biosensor, and cholesterol biosensors in clinical diagnostics (Newman and Turner, 2005; Wang, 1998; Adeloju and Moline, 2001; Das et al., 2016; Kumar et al., 2001). In a humidness sensing element, films of CPs, such as PEDOT/polystyrene sulfonate (PSS) and poly(anilinesulfonic acid) have been shown to be capable of detecting a change in humidity (Nohria et al., 2006; Wei et al., 2018). Gao et al. (2003) reported coaxial nanowire composites, that is, PPy/CNTs, which is flexible, lightweight, and showed electrical conductivity due to the electroactive surface area. CPs are broadly utilized in biomedical applications. CPs have an optical and electrical property with controlled morphology, high stability, and biocompatibility. The few CPs are PPy, PANI, PT, and PEDOT. They act as a transducer in biosensor, scaffold in tissue engineering for the growth, proliferation and differentiation of cells, and actuators for artificial muscles due to their volume change properties. CPs has limitations like lack of biodegradability, low hydrophilicity, low charge-carrier mobility, and a small loading capacity of biomolecules. However, possibilities for resolving these issues CPs should be blended with biodegradable polymers to enhance biodegradability and hydrophilicity via a surface modification and functionalization. CPs hybridization with CNTs and nanofibers has proven low charge-carrier mobility of the CPs. CP nanomaterials have utilized in healthcare, environmental, and biomedical applications. The conventional CP-based biosensors can be possible by incorporation of nanomaterials into biosensing, which is known as nanobiosensors (Nguyen and Yoon, 2016; Park et al., 2014; Oh et al., 2013; Yoon and Jang, 2009). Nanomaterial PPy is polymerized product of pyrrole in terms of NPs, nanotubes, core-shell, and hollow nanospheres; highly utilized in electronic devices. Kim et al. (2009) reported bioelectronic nose to detect an odorant mixture using CNT modified olfactory receptors (hORs). Jang and Yoon (2003) reported PPy NTs using reverse microemulsion polymerization in a polar solvent for the B-nose. Yoon et al. (2009) reported hORs-modified CPNTs FET to detect an odorant. Lee et al. (2012) reported nanobioelectronic nose to detect a gaseous odorant. Song et al. (2012) reported bioelectronic tongue-based FET to detect taste.

Aptamers are either oligonucleotide or peptide molecules act as a bioprobe to detect desired biomolecules. Kwon et al. (2012) reported electronic transistor aptasensors to detect antivascular epithelium protein. Huang et al. (2011) reported PPy-NW-grown gold electrodes to detect MUC1 aptamer biomarker. Sun et al. (2015) reported electrochemical aptasensor using PPy-Au nanocomposite (NC) to detect myoglobin. Immunoassays are an analytical tool for the diagnosis of diseases. Tabrizi et al. (2016) reported an immunosensor using antiH-IgG modified overoxidized PPy/AuNPsNPs screen-printed electrode to detect HIgG. Mishra et al. (2012) reported ZnS

NC/PPy film electrode to detect the C-reactive protein. Wang et al. (2015) reported multifunctionalized rGO-doped PPy/pyrrolepropylic acid NCs as an immunosensor to detect aflatoxin B1 (AFB1). Kwon et al. (2013) reported FET/HIV immunoassay devices using carboxylated PPyNPs/graphene to detect HIV antibody. Hydrogen peroxide, a reactive oxygen species, which is responsible for several diseases, like atherosclerosis, cancer, and Alzheimer's disease, is also important for cell growth, migration in healthcare, and biological monitoring. Mahmoudian et al. (2014) reported PPy/Ag nanostrip to detect H_2O_2. Nia et al. (2016) reported PPy-CuNPs to detect H_2O_2. Park et al. (2014) reported PPy/rGO transducer to detect H_2O_2 in biological fluids. Tao et al. (2014) reported PPy NPs function like peroxidase-like activity. Ghadimi et al. (2015) reported ultraPPy/PtNPs-GCE to detect dopamine. Weng et al. (2014) reported PPyNPs/HRP-GoD screen-printed carbon electrodes to detect glucose. PANi is a CP with good conductivity, stability, and showed diverse color changes and highly used in biosensors, drug delivery, and tissue engineering (Humpolicek et al., 2012). CP-based nanocomposite material like PANi is very important. Zhai et al. (2013) reported GOx/PANi and GOx/GNPs/PANi nanocomposites to detect glucose.

Feneley et al. (2000) incontestable the appliance of magneto-optical magnetic NPs (MNPs)-PANi in imaging victimization NPs functionalized targeting prostate-specific membrane matter, a particular biomarker expressed in prostate growth animal tissue cells for not only imaging and therapeutic interventions but also in applications in electronics, bioimaging, sensing, coating, and smart textiles. Li et al. (2008) reported photoelectrochemical biosensor using PANi. Zhu et al. (2016) reported TiO_2NT-PANI-AuNPs|LDH|NAD^+-ITO SPR biosensor to detect L-lactate.

A nanomaterial, poly(3,4-ethylenedioxythiophene) was derived from PT, transparent in thin film and exhibits excellent stability in the oxidized state. The limitation of PEDOT is solubility and stability but it overcome by adding water-soluble polyelectrolyte, poly (styrene sulfonic acid) and showed excellent physical properties for film-formation, high conductivity, transmittance, and physical stability (Jonas and Schrader, 1991, Heywang and Jonas, 1992). Khan and Armes (1999) reported PEDOT-coated polystyrene latex using seed polymerization. Heller (1999) reported glucose biosensors to detect glucose level of diabetic patients using the glucose oxidase enzyme on the surface of the PEDOT transistor. Zhu et al. (2004) reported organic electrochemical transistor (OECT) of PEDOT: PSS to detect glucose. Furthermore, Macaya et al. (2007) reported OECTs composed of PEDOT: PSS and a Pt gate electrode to detect glucose in human saliva. Liao et al. (2013) reported OECTs composed of GOx/graphene or/rGO gate electrodes to detect glucose. Wang et al. (2013) reported BPEI-Fc/PEDOT: PSS/GOx/SPCE bioelectrode to detect glucose. Yang et al. (2014) reported GOx/CP nanofibers/Pt microelectrode to detect glucose. Dopamine is a neurotransmitter in blood. Its deficiency may cause neurological disorders like PD disease, AD disease, and schizophrenia due to abnormal metabolism of dopamine. Vasantha et al. (2006) reported PEDOT film electrodes to detect dopamine and also ascorbic acid. Tang et al. (2011) reported OECTs to detect dopamine. Liao et al. (2013) reported nafion or chitosan on the graphene/Pt or rGO/Pt gate electrodes to detect dopamine.

19.8 NANOCOMPOSITES AND BIOSENSING

Nanocomposites became a central domain to assist humanity. The advances in nanostructured composites and their trends, organic process applications in numerous fields together with in water treatment, green energy generation, anticorrosive, arduous coatings, antiballistic, optoelectronic devices, solar cells, biosensors, and nanodevices, has been reportable by many investigators enormously (Singh, 2017). Nanocomposite materials are mixtures of two or a lot of elements; it generally consists of a couple of matrix containing one or additional fillers created from particles, sheets, and fibers with dimensions between 1–100 nm, which have better surface-to-volume quantitative relation. The vital property of nanocomposites is that they are less porous than regular plastics, creating them ideal to use within the packaging of foods and drinks, vacuum packs, and to safeguard medical instruments, film, and different product from outside contamination. These developments are attainable because of their distinctive properties of nanostructures composites at nanoscale level that might amendment dramatically by a reduction in size, shape, dimension and exhibit newer properties together with reactivity, electrical conduction, insulating behavior, elasticity, strength, and color (Singh et al., 2011, 2014; Singh, 2016). Multifunctional nanocomposite is fast-evolving field with higher surface assimilation capability, property, and stability. They have potential for detection of toxic metal, pollutants, toxicants, and adulterants from the contaminated water. The nanocomposites with additives as filler counterparts showed substantial property enhancements like mechanical properties, diminished porosity to hydrocarbons, thermal stability, chemical resistance, surface look, electrical conduction, and optical property (Zhu et al., 2001; Organ et al., 2002; Bourbigot et al., 2002).

Nanocomposite materials are used in general automotive and industrial applications and revolutionizing the world of materials. It has high impact in developing a replacement generation of composites with increased practicality and a good vary of applications. Nanostructured nanocomposites are used in nanobiosensors to detect the analyte of interest in biological, chemical, and environmental nature of the sample (Singh et al., 2012a, 2012b, 2012c, 2010, 2014; Tiwari et al., 2012a, 2012b, 2012c; Yadav et al., 2012). Arya et al. (2006) reported ChOx/FNAB/ODT/Au nanocomposite film to estimate cholesterol by SPR technique. Nowadays, monitoring of toxicants, contaminants, or pollutants in the different source of the sample is very important to save human health and ecosystems from the risk posed to them. In these contexts, the development and application of biosensor arrays using aptamers for environmental detection are highlighted (Singh et al., 2008). Singh et al. (2009a) reported Au/APTES/GDA/GST nanocomposite film to detect captan in contaminated water. Captan is a known harmful chemical and potential carcinogen to a water ecosystem. Singh and Choi (2009) reported biosensors based on polyaniline, polypyrrole, PT when blend with metal, ceramic to form nanocomposites for wide applications to detect toxicants in environmental samples. Singh et al. (2009b) reported CAT/PANi/ITO nanocomposite film to detect H_2O_2 and azide.

Research and development toward nanobiosensor utilizing nanocomposites to detect pesticides, antibiotics, pathogen, toxins, and biomolecules in food, soil, and water are very important and challenging for various applications to be focused not only on biosensor development but also on nanobiosensor (Singh, 2011). All these growing areas can have a noteworthy influence on the event of latest immoderate biosensing devices (nanobiosensor) and resolve the issues, challenges of the human health, and other living entities (Singh et al., 2012). Sabherwal et al. (2016) reported biofunctionalized carbon nanocomposites to generate sensitive detection systems in clinical/environmental monitoring. Song et al. (2015) reported porous Co nanobeads/rGO nanocomposites for glucose sensing. Wang et al. (2015) reported luminol electrochemiluminescence strategy using TiO_2/CNTs nanocomposites for the detection of glucose. Mohammed et al. (2016) reported NiS_2–CNT NCs nanocomposite for the toxic chemical 4-methoxyphenol for efficient phenolic sensor development in environment and healthcare. Shen et al. (2016) reported Au–Fe_3O_4 modified with 1-(2-mercaptoethyl)-1,2,3,4,5,6-hexanhydro-s-triazine-2,4,6-trione for the bimodal detection of melamine. Murugadoss et al. (2016) reported PbO/CdO/ZnO and PbS/CdS/ZnS nanocomposites for optical and electrochemical biosensing applications. Zhang et al. (2016) reported Prussian blue and Au modified polystyrene nanocomposites as peroxidase-like and catalase-like catalytic activity for glucose detection.

19.9 CONCLUSIONS

The main focus of this chapter is on the potentialities of nanomaterials for the biosensing applications. Apart from this, some other importance in areas of cancer diagnostics, detection of pathogenic organisms, food safety, agricultural, environmental, and clinical applications are included elsewhere. The key issues pertaining to biosensing in real-time samples is minimal cost in the analysis of complex clinical and environmental samples. Also, this chapter highlights the R&D activities on biosensing. It is backed up with the rapid development of nanotechnology; these pieces of research are making a progress in multidisciplinary research, which are beginning to make in the market place. Focusing on smart nanomaterials, device structures, and functionalities, the chapter highlights milestones achieved and explain further prospects and challenges in various domains.

ACKNOWLEDGMENTS

Ravindra P. Singh is thankful to IGN Tribal University, Amarkantak, Madhya Pradesh, India for providing facilities to prepare this chapter.

KEYWORDS

- **biosensor**
- **nanobiosensor**
- **genosensor**
- **immunosensor**
- **enzymosensor**
- **medical**
- **environmental**
- **agriculture**
- **food analysis**
- **nanomaterial**

REFERENCES

Amirabbas, N.; Abbaspour, S.; Masood, M.; Mirsattari, S.N.; Vahedi, A.; Mackenzie, K.J.D. Photocatalytic properties of mesoporous TiO_2 nanocomposites modified with carbon nanotubes and copper. *Ceram. Int.* **2016**, 42(10), 11901–11906.

Andrews, R.; Weisenberger, M.C. Carbon nanotube polymer composites. *Curr. Opin. Solid State Mater. Sci.* **2004**, 8(1), 31–37.

Arora, K.; Chaubey, A.; Singhal, R.; Singh, R.P.; Pandey, M.K.; Samanta, S.B. et al., Application of electrochemically prepared polypyrrole-polyvinylsulphonate films to DNA biosensor. *Biosens. Bioelectron.* **2006**, 21(9), 1777–1783.

Arya, S.K.; Solanki, P.R.; Singh, R.P.; Pandey, M.K.; Datta, M.; Malhotra, B.D. Application of octadecanethiol self assembled monolayer to cholesterol biosensor based on surface Plasmon Resonance technique. *Talanta.* **2006**, 69(4), 918–926.

Adeloju, S.B.; Moline, A.N. Fabrication of ultra-thin polypyrrole-glucose oxidase film from supporting electrolyte-free monomer solution for potentiometric biosensing of glucose. *Biosens. Bioelectron.* **2001**, 16, 133–139.

Andreescu, S.; Luck, L. A. Studies of the binding and signaling of surface-immobilized periplasmic glucose receptors on gold nanoparticles, a glucose biosensor application. *Anal. Biochem.* **2008**, 375(2), 282–290.

Arora, K.; Chaubey, A.; Singhal, R.; Singh, R.P.; Pandey, M.K.; Samanta, S.B.; Malhotra, B.D. Application of electrochemically prepared polypyrrole–polyvinyl sulphonate films to DNA biosensor. *Biosensors. Bioelectron.* **2006**, 21 (9), 1777–1783.

Agui, L.; Yanez-Sedeno, P.; Pingarron, J. M. Role of carbon nanotubes in electroanalytical chemistry. A review. *Anal. Chim. Acta.* **2008**, 622(1–2), 11–47.

Akbarzadeh, A.; Rezaei-sadabady, R.; Davaran S et al., Liposome, classification, preparation, and applications. *Nanoscale Res. Lett.* **2013**, 8(102), https://doi.org/10.1186/1556-276X-8-102.

Balogh, L.; Tomalia, D. A. Poly(amidoamine) dendrimer-templated nanocomposites. 1. Synthesis of zerovalent copper nanoclusters. *J. Am. Chem. Soc.* **1998**, 120, 29, 7355–7356.

Bosman, A.W.; Janssen, H. M.; Meijer, E. W. About dendrimers, structure, physical properties, and applications. *Chem. Rev.* **1999**, 99, 7, 1665–1688.

Barbadillo, M.; Casero, E.; Petit-Dominguez, M. D.; Vazquez, L.; Pariente, F.; Lorenzo, E. Gold nanoparticles-induced enhancement of the analytical response of an electrochemical biosensor based on an organic-inorganic hybrid composite material. *Talanta.* **2009**, 80, 2, 797–802.

Berg, J. M.; Tymoczko, J. L.; Stryer, L. Biochemistry, W. H. Freeman & Co., New York, NY, USA, 5th edition, **2002**.

Bourbigot, S.; Devaux, E.; Flambard, X. Flammability of polyamide-6/clay hybrid nanocomposite textiles. *Polym. Degrad. Stab.* **2002**, 75, 397–402.

Chen, X.; Fu, C.; Wang, Y.; Yang, W.; Evans, D. G. Direct electrochemistry and electrocatalysis based on a film of horseradish peroxidase intercalated into Ni–Al layered double hydroxide nanosheets. *Biosensors. Bioelectron.* **2008**, 24(3), 356–336.

Choi, S. H.; Kim, Y. L.; Byun, K. M. Graphene-on-silver substrates for sensitive surface plasmon resonance imaging biosensors. *Opt. Exp.* **2011**, 19(2), 458–466.

Cao, Q.; Zhao, H.; Yang, Y. et al., Electrochemical immunosensor for casein based on gold nanoparticles and poly (l-Arginine)/multi-walled carbon nanotubes composite film functionalized interface. *Biosens. Bioelectron.* **2011**, 26, 8, 3469–3474.

Cai, H.; Cao, X.; Jiang, Y.; He, P.; Fang, Y. Carbon nanotube-enhanced electrochemical DNA biosensor for DNA hybridization detection. *Anal. Bioanal. Chem.* **2003**, 375(2), 287–293.

Clark, L.C.; Lyons, C. Electrode systems for continuous monitoring in cardiovascular surgery. *Ann. NY Acad. Sci.* **1962**, 102, 29–45.

Cretich, M.; Sedini, V.; Damin, F.; Di Carlo, G.; Oldani, C.; Chiari, M. Functionalization of poly (dimethylsiloxane) by chemisorption of copolymers, DNA microarrays for pathogen detection. *Sens. Actuators B.* **2008**, 132(1), 258–264.

Cai, W.; Peck, J. R.; Van Der Weide, D. W.; Hamers, R. J. Direct electrical detection of hybridization at DNA-modified silicon surfaces. *Biosens. Bioelectron.* **2004**, 19(9), 1013–1019.

Cai, H.; Wang, Y.; He, P.; Fang, Y. Electrochemical detection of DNA hybridization based on silver-enhanced gold nanoparticle label. *Anal. Chim. Acta.* **2002**, 469(2), 165–172.

De Brabander, E. M. M.; Nijenhuis, A. G.; Borggreve, R.; Put. J. Star polycondensates, large scale synthesis, rheology and material properties. *Polym. News.* **1997**, 22, 1, 6–12.

Das, P.; Das, M.; Chinnadayyala, S.R.; Singha, I.M.; Goswami, P. Recent advances on developing 3rd generation enzyme electrode for biosensor applications. *Biosens. Bioelectron.* **2016**, 79, 386–397.

Daniel, S.; Rao, T. P.; Rao, K.S.; et al., A review of DNA functionalized/grafted carbon nanotubes and their characterization. *Sensor. Actuators B Chem.* **2007**, 122(2), 672–682.

Davies, T. J.; Ward, J. S.; Banks, C. E.; et al., The cyclic and linear sweep voltammetry of regular arraysof microdisc electrodes, fitting of experimental data. *J. Electroanal. Chem.* **2005**, 585, 1–62.

Dong, Y.; Yang, J.; Zhang, J.; Zhang, X. Nano-delivery vehicles/adjuvants for DNA vaccination against HIV. *J Nanosci. Nanotechnol.* **2016**, 16, 2126–2133.

Elkhodiry, M.A.; Momah, C.C.; Suwaidi, S.R.; et al., Synergistic nanomedicine, passive, active, and ultrasound-triggered drug delivery in cancer treatment. *J. Nanosci. Nanotechnol.* **2016**, 16,1–18.

Feneley, M.; Jan, H.; Granowska, M.; Mather, S.; Ellison, D.; Glass, J.; Coptcoat, M.; Kirby, R.; Ogden, C.; Oliver, R. Imaging with prostate-specific membrane antigen (PSMA) in prostate cancer. *Prostate Cancer Prost. Dis.* **2000**, 3, 47–52.

Feng, L.; Chen, Y.; Ren, J.; Qu. X. A graphene functionalized electrochemical aptasensor for selective label-free detection of cancer cells. *Biomaterials* **2011**, 32, 11, 2930–2937.

Feng, L.; Liu, Z. Graphene in biomedicine, opportunities and challenges. *Nanomedicine* **2011**, 6, 2, 317–324.

Fletcher, S.; Horne, M.D. Random assemblies of microelectrodes (RAMy electrodes) for electrochemical studies. *Electrochem. Commun.* **1999**, 1(10), 502–512.

Gooding, J. J. Nanostructuring electrodes with carbon nanotubes, a review on electrochemistry and applications for sensing. *Electrochim. Acta.* **2005**, 50(15), 3049–3060.

Gifford, R.; Batchelor, M. M.; Lee, Y.; Gokulrangan, G.; Meyerhoff, M. E.; Wilson, G. S. Mediation of in vivo glucose sensor inflammatory response via nitric oxide release. *J. Biomed. Mater. Res. A* **2005**, 75(4), 755–766.

Gao, X.; Zhang, Y.; Chen, H.; Chen, Z.; Lin, X. Amperometric immunosensor for carcinoembryonic antigen detection with carbon nanotube-based film decorated with gold nanoclusters. *Anal. Biochem.* **2011**, 414, 1, 70–76.

Geim, A. K. Graphene: status and prospects. *Science.* **2009**, 324 (5934), 1530–1534.

Goh, M. S.; Pumera, M. Graphene-based electrochemical sensor for detection of 2,4,6-trinitrotoluene (TNT) in seawater, the comparison of single, few, and multilayer graphene nanoribbons and graphite microparticles. *Anal. Bioanal. Chem.* **2011**, 399, 1, 127–131.

Garg, T.; Singh, O.; Arora, S.; Murthy, R. S. R. Dendrmer—a novel scaffold for drug delivery. *Int. J. Pharm. Sci. Rev. Res.* **2011**, 7, 2, 211–220.

Ghalandarlaki, N.; Alizadeh, A.M.; Ashkani-Esfahani, S. Nanotechnology-applied curcumin for different diseases therapy. *Biomed. Res. Int.* **2014**, 2014, 394264.

Gao, M.; Dai, L.; Wallace, G.G. Biosensors based on aligned carbon nanotubes coated with inherently conducting polymers. *Electroanalysis*. **2003**, 15, 1089–1094.

Gao, M.; Huang, S.; Dai, L.; Wallace, G.; Gao, R.; Wang, Z. Aligned coaxial nanowires of carbon nanotubes sheathed with conducting polymers. *Angew. Chem. Int. Ed.* **2000**, 39, 3664–3667.

Ghadimi, H.; Mahmoudian, M.; Basirun, W.J. A sensitive dopamine biosensor based on ultra-thin polypyrrole nanosheets decorated with Pt nanoparticles. *RSC Adv*. **2015**, 5, 39366–39374.

He, Q.; Wu, S.; Gao, S.; et al., Transparent, flexible, all-reduced graphene oxide thin film transistors. *ACS Nano*. **2011**, 5, 6, 5038–5044.

Heller, A. Integrated medical feedback systems for drug delivery. *AIChE J.* **2005**, 51(4), 1054–1066.

Hahm, J. I.; Lieber, C. M. Direct ultrasensitive electrical detection of DNA and DNA sequence variations using nanowire nanosensors. *Nano Lett*. **2004**, 4(1), 51–54.

Hahn, S.; Mergenthaler, S.; Zimmermann, B.; Holzgreve, W. Nucleic acid based biosensors, the desires of the user. *Bioelectrochemistry*. **2005**, 67(2), 151–154.

Hassibi, A.; Lee, T. H. A programmable 0.18 μ CMOS electrochemical sensor microarray for biomolecular detection. *IEEE Sens. J.* **2006**, 6(6), 1380–1388.

Hansen, J. A.; Wang, J.; Kawde, A. N.; Xiang, Y.; Gothelf, K. V.; Collins, G. Quantum-dot/aptamer-based ultrasensitive multi-analyte electrochemical biosensor. *J. Am. Chem. Soc.* **2006**, 128(7), 2228–2229.

Hamada, N.; Sawada, S. I.; Oshiyama. A. New one-dimensional conductors, graphitic microtubules. *Phys. Rev. Lett*. **1992**, 68, 10, 1579–1581.

He, P.; Shen, L.; Cao, Y.; Li, D. Ultrasensitive electrochemical detection of proteins by amplification of aptamer-nanoparticle bio bar codes. *Anal. Chem*. **2007**, 79(21), 8024–8029.

He, P.; Xu, Y.; Fang, Y. Applications of carbon nanotubes in electrochemical DNA biosensors. *Microchim. Acta*. **2006**, 152(3–4), 175–186.

Huang, K. J.; Niu, D. J.; Sun, J. Y.; et al., Novel electrochemical sensor based on functionalized graphene for simultaneous determination of adenine and guanine in DNA. *Colloid Surf. B*. **2011**, 82, 2, 543–549.

He, Q.; Sudibya, H. G.; Yin, Z.; et al., Centimeter-long and large-scale micropatterns of reduced graphene oxide films, fabrication and sensing applications. *ACS Nano*. **2010**, 4, 6, 3201–3208.

Hranisavljevic, J.; Dimitrijevic, N. M.; Wurtz, G. A.; Wiederrecht, G. P. Photoinduced charge separation reactions of J-aggregates coated on silver nanoparticles. *J. Am. Chem. Soc.* **2002**, 124, 17, 4536–4537.

Huang, J.; Luo, X.; Lee, I.; Hu, Y.; Cui, X.T.; Yun, M. Rapid real-time electrical detection of proteins using single conducting polymer nanowire-based microfluidic aptasensor. *Biosens. Bioelectron*. **2011**, 30, 306–309.

Humpolicek, P.; Kasparkova, V.; Saha, P.; Stejskal, J. Biocompatibility of polyaniline. *Synth. Met*. **2012**, 162, 722–727.

Heywang, G.; Jonas, F. Poly(alkylenedioxythiophene)s-New, very stable conducting polymers. *Adv. Mater*. **1992**, 4, 116–118.

Heller, A. Implanted electrochemical glucose sensors for the management of diabetes. *Annu. Rev. Biomed. Eng*. **1999**, 1, 153–175.

Iijima, S. Helical microtubules of graphitic carbon. *Nature*. 1991, 354, 6348, 56–58.

Jing, W.; Yang, Q. Mediator-free amperometric determination of glucose based on direct electron transfer between glucose oxidase and an oxidized boron-doped diamond electrode. *Anal. Bioanal. Chem.* **2006**, 385, 7, 1330–1335.

Jung, M. W.; Kim, D. W.; Jeong, R. A.; Kim, H. C. Engineering in medicine and biology society. In Proceedings of the 26th Annual International Conference of the IEEE (*IEMBS '04*) **2004**, 1, 1987–1989.

Jianrong, C.; Yuqing, M.; Nongyue, H.; Xiaohua, W.; Sijiao, L. Nanotechnology and biosensors. *Biotechnol. Adv.* **2004**, 22 (7), 505–518.

Jung, D.H.; Kim, B.H.; Ko, Y.K.; et al., Covalent attachment and hybridization of DNA oligonucleotides on patterned single-walled carbon nanotube films. *Langmuir* **2004**, 20, 20, 8886–8891.

Jonas, F.; Schrader, L. Conductive modifications of polymers with polypyrroles and polythiophenes. *Synth. Met.* **1991**, 41, 831–836.

Jang, J., Yoon, H. Facile fabrication of polypyrrole nanotubes using reverse microemulsion polymerization. *Chem. Commun.* **2003**, 720–721.

Jang, J. Conducting Polymer Nanomaterials and their applications. In Emissive Materials Nanomaterials; Eds. Abe, A.; Dusek, K.; Kobayashi, S. Springer, Berlin/Heidelberg, Germany, **2006**, 189260.

Kim, T.H.; Lee, S.H.; Lee, J.; Song, H.S.; Oh, E.H.; Park, T.H.; Hong, S. Single-carbon-atomic-resolution detection of odorant molecules using a human olfactory receptor-based bioelectronic nose. *Adv. Mater.* **2009**, 21, 91–94.

Kwon, O.S.; Park, S.J.; Hong, J. Y.; Han, A. R.; Lee, J.S.; Lee, J.S.; Oh, J.H.; Jang, J. Flexible FET-type VEGF aptasensor based on nitrogen-doped graphene converted from conducting polymer. *ACS Nano.* **2012**, 6, 1486–1493.

Rajesh, A.K.; Chaubey, A.; Grover, S.K.; Malhotra, B.D. Immobilization of cholesterol oxidase and potassium ferricyanide on dodecylbenzene sulfonate ion-doped polypyrrole film. *J. Appl. Polym. Sci.* **2001**, 2001, 82, 3486–3491.

Kwon, O.S.; Lee, S.H.; Park, S.J.; An, J.H.; Song, H.S.; Kim, T.; Oh, J.H.; Bae, J.; Yoon, H.; Park, T.H. Large-scale graphene micropattern nano-biohybrids, high-performance transducers for FET-type flexible fluidic HIV immunoassays. *Adv. Mater.* **2013**, 25, 4177–4185.

Khan, M.; Armes, S. Synthesis and characterization of micrometer-sized poly(3,4-ethylenedioxythiophene)- coated polystyrene latexes. *Langmuir* **1999**, 15, 3469–3475.

Koda, S.; Inoue, Y.; Iwata, H. Gene transfection into adherent cells using electroporation on a dendrimer-modified gold electrode. *Langmuir* **2008**, 24 (23), 13525–1353.

Kerman, K.; Chikae, M.; Yamamura, S.; Tamiya, E. Gold nanoparticle-based electrochemical detection of protein phosphorylation. *Anal. Chim. Acta.* **2007**, 588, 1, 26–33.

Karian, H. G. Handbook of Polypropylene and Polypropylene Composites, Eds. Dekker, M. NewYork, NY, USA, **1999**.

Kim, H.; Kang, D. Y.; Goh, H. J. Analysis of direct immobilized recombinant protein G on a gold surface. *Ultramicro.* **2009**, 108 (10), 1152–1156.

Kumar, S. A.; Chen, S. M. Nanostructured zinc oxide particles in chemically modified electrodes for biosensor applications. *Anal. Lett.* **2008**, 41(2), 141–158.

K'Owino, I. O.; Agarwal, R.; Sadik, O. A. Novel electrochemical detection scheme for DNA binding interactions using monodispersed reactivity of silver ions. *Langmuir* **2003**, 19(10), 4344–4350.

Kumar, A.; Dill, K. *IVD Technol.* **2005**, 11, 35–40.

Kerman, K.; Morita, Y.; Takamura, Y.; Ozsoz, M.; Tamiya, E. Modification of Escherichia coli single-stranded DNA binding protein with gold nanoparticles for electrochemical detection of DNA hybridization. *Anal. Chim. Acta.* **2004**, 510(2), 169–174.

Karasinski, J.; White, L.; Zhang, Y.; et al., Detection and identification of bacteria using antibiotic susceptibility and a multi-array electrochemical sensor with pattern recognition. *Biosens. Bioelectron.* **2007**, 22(11), 2643–2649.

Katz, E.; Willner, I. Biomolecule-functionalized carbon nanotubes, applications in nanobioelectronics. *ChemPhysChem.* **2004**, 5, 8, 1084–1104.

Kroto, H.W.; Heath, J.R.; O'Brien, S.C.; Curl, R.F.; Smalley. R.E. C60, Buckminsterfullerene. *Nature*. **1985**, 318, 6042, 162–163.

Liao, C.; Zhang, M.; Niu, L.; Zheng, Z.; Yan, F. Highly selective and sensitive glucose sensors based on organic electrochemical transistors with graphene-modified gate electrodes. *J. Mater. Chem. B* **2013**, 1, 3820–3829.

Liao, C; Zhang, M.; Niu, L.; Zheng, Z.; Yan, F. Organic electrochemical transistors with graphene-modified gate electrodes for highly sensitive and selective dopamine sensors. *J. Mater. Chem. B* **2013**, 2, 191–200.

Li, Q.; Wu, J.; Tang, Q.; Lan, Z.; Li, P.; Lin, J.; Fan, L. Application of microporous polyaniline counter electrode for dye-sensitized solar cells. *Electrochem. Commun.* **2008**, 10, 1299–1302.

Lee, S.H.; Kwon, O.S.; Song, H.S.; Park, S.J.; Sung, J.H.; Jang, J.; Park, T.H. Mimicking the human smell sensing mechanism with an artificial nose platform. *Biomaterials* **2012**, 33, 1722–1729.

Li, X.; Wang, Y.; Yang, X.; Chen, J.; Fu, H.; Cheng, T.; Wang, Y. Conducting polymers in environmental analysis. *Trends Anal. Chem.* **2012**, 39, 163–179.

Lai, J.; Yi, Y.; Zhu, P.; Shen, J.; Wu, K.; Zhang, L.; Liu, J. Polyaniline-based glucose biosensor, A review. *J. Electroanal. Chem.* **2016**, 782, 138–153.

Ledesma-García, J.; Manríquez, J.; Gutiérrez-Granados, S.; Godínez, L.A. Dendrimer modified thiolated gold surfaces as sensor devices for halogenated alkyl-carboxylic acids in aqueous medium. A promising new type of surfaces for electroanalytical applications. *Electroanalysis* **2003**, 15, 7, 659–666.

Liu, Y.; Chen, C. Role of nanotechnology in HIV/AIDS vaccine development. *Adv. Drug Deliv. Rev.* **2015**, 103, 76–89.

Lu, J.; Do, I.; Drzal, L.T.; Worden, R.M.; Lee, I. Nanometal-decorated exfoliated graphite nanoplatelet based glucose biosensors with high sensitivity and fast response. *ACS Nano* **2008**, 2, 9, 1825–1832.

Liang, K. Z.; Qi, J. S.; Mu, W. J.; Liu, Z. X. Conductometric immunoassay for interleukin-6 in human serum based on organic/inorganic hybrid membrane-functionalized interface. *Bioprocess Biosyst. Eng.* **2009**, 32, 3, 353–359.

Lu, L.M.; Li, H.B.; Qu, F.; Zhang, X.B.; Shen, G.L.; Yu, R.Q. In situ synthesis of palladium nanoparticle-graphene nanohybrids and their application in nonenzymatic glucose biosensors. *Biosens. Bioelectron.* **2011**, 26, 8, 3500–3504.

Li, L.; Du, Z.; Liu S.; et al., A novel nonenzymatic hydrogen peroxide sensor based on MnO_2/graphene oxide nanocomposite. *Talanta* **2010**, 82, 5, 1637–1641.

Liu, M.; Zhao, H.; Quan, X.; Chen, S.; Fan, X. Distance-independent quenching of quantum dots by nanoscale-graphene in self-assembled sandwich immunoassay. *Chem. Commun.* **2010**, 46, 42, 7909–7911.

Li, F.; Huang, Y.; Yang, Q.; et al., A graphene-enhanced molecular beacon for homogeneous DNA detection. *Nanoscale* **2010**, 2, 6, 1021–1026.

Lightcap, I.V.; Kosel, T.H.; Kamat, P.V. Anchoring semiconductor and metal nanoparticles on a two-dimensional catalyst mat storing and shuttling electrons with reduced graphene oxide. *Nano Lett.* **2010**, 10, 2, 577–583.

Li, W.; Yuan, R.; Chai, Y.; Chen, S. Reagentless amperometric cancer antigen 15–3 immunosensor based on enzyme-mediated direct electrochemistry. *Biosens. Bioelectron.* **2010**, 25, 11, 2548–2552.

Liu, M.; Zhao, H.; Chen, S.; Yu, H.; Zhang, Y.; Quan, X. A graphene-based platform for single nucleotide polymorphism (SNP) genotyping. *Biosens. Bioelectron.* **2011**, 26, 10, 4213–4216.

Li, J.; Lu, C.-H.; Yao, Q.-H.; et al., A graphene oxide platform for energy transfer-based detection of protease activity. *Biosens. Bioelectron.* **2011**, 26, 9, 3894–3899.

Liu, B.; Tang, D.; Tang, J.; Su, B.; Li, Q.; Chen, G. A graphene-based Au(111) platform for electrochemical biosensing based catalytic recycling of products on gold nanoflowers. *Analyst* **2011**, 136, 11, 2218–2220.

Lu, G.; Park, S.; Yu, K.; et al., Toward practical gas sensing with highly reduced graphene oxide, a new signal processing method to circumvent run-to-run and device-to-device variations. *ACS Nano* **2011**, 5, 2, 1154–1164.

Li, J.; Chen, J.; Zhang, X. L.; Lu, C. H.; Yang, H. H. A novel sensitive detection platform for antitumor herbal drug aloe-emodin based on the graphene modified electrode. *Talanta* **2010**, 83, 2, 553–558.

Lee, K.; Song, H.; Kim, B.; Park, J.T.; Park, S.; Choi, M.G. The first fullerene-metal sandwich complex, an unusually strong electronic communication between two C60 cages. *J. Am. Chem. Soc.* **2002**, 124, 12, 2872–2873.

Li, J.; Liu, Q.; Liu, Y.; Liu, S.; Yao, S. DNA biosensor based on chitosan film doped with carbon nanotubes. *Anal. Biochem.* **2005**, 346, 1, 107–114.

Lee, S.C.; Parthasarathy, R.; Duffin, T.D.; et al., Recognition properties of antibodies to PAMAM dendrimers and their use in immune detection of dendrimers. *Biomed. Microdevices* **2001**, 3, 1, 53–59.

Liu, R. Integrated microfluidic biochips for immunoassay and DNA bioassays. *Conf. Proc. IEEE Eng. Med. Biol. Soc.* **2004**, 7, 5394.

Liao, J.C.; Mastali, M.; Gau, V.; et al., Use of electrochemical DNA biosensors for rapid molecular identification of uropathogens in clinical urine specimens. *J. Clin. Microbiol.* **2006**, 44(2), 561–570.

Liu, J.; Chou, A.; Rahmat, W.; Paddon-Row, M.N.; Gooding, J.J. Achieving direct electrical connection to glucose oxidase using aligned single walled carbon nanotube arrays. *Electroanalysis* **2005**, 17(1), 38–46.

Lee, J. O.; So, H.M.; Jeon, E.K.; Chang, H.; Won, K.; Kim, Y.H. Aptamers as molecular recognition elements for electrical nanobiosensors. *Anal. Bioanal. Chem.* **2008**, 390(4), 1023–1032.

Lu, Y.H.; Yang, M.; Qu, F.; Shen, G.; Yu, R. Enzyme-functionalized gold nanowires for the fabrication of biosensors. *Bioelectrochemistry.* **2007**, 71(2), 211–216.

Li, C.Z.; He, H.X.; Tao, N.J. Quantized tunneling current in the metallic nanogaps formed by electrodeposition and etching. *Appl. Phys. Lett.* **2000**, 77(24), 3995–3997.

Lu, X.; Zhang, H.; Ni, Y.; Zhang, Q.; Chen, J. Porous nanosheet-based ZnO microspheres for the construction of direct electrochemical biosensors. *Biosens. Bioelectron.* 2008, 24, 1, 93–98.

Merkoci, A.; Aldavert, M.; Marín, S.; Alegret, S. New materials for electrochemical sensing V, nanoparticles for DNA labeling. *Trends Anal. Chem.* **2005**, 24(4), 341–349.

Mohanty, N.; Berry, V. Graphene-based single-bacterium resolution biodevice and DNA transistor, interfacing graphene derivatives with nanoscale and microscale biocomponents. *Nano Lett.* **2008**, 8(12), 4469–4476.

Munge, B.S.; Fisher, J.; Millord, L.N.; Krause, C.E.; Dowd, R.S.; Rusling, J.F. Sensitive electrochemical immunosensor for matrix metalloproteinase-3 based on single-wall carbon nanotubes. *Analyst* **2010**, 135, 6, 1345–1350.

Meng, L.; Gan, N.; Li, T.; Cao, Y.; Hu, F.; Zheng, L. A three-dimensional, magnetic and electroactive nanoprobe for amperometric determination of tumor biomarkers. *Int. J. Mol. Sci.* **2011**, 12, 1, 362–375.

Maiti, A.; Andzelm, J.; Tanpipat, N.; Von Allmen, P. Effect of adsorbates on field emission from carbon nanotubes. *Phys. Rev. Lett.* **2001**, 87, 15, Article ID 155502, 1–4.

Mintmire, J.W.; Dunlap, B.I.; White, C.T. Are fullerene tubules metallic? *Phys. Rev. Lett.* **1992**, 68, 5, 631–634.

Margadonna, S.; Aslanis, E.; Prassides, K. Ammoniated alkali fullerides (ND3) xNaA2C60, ammonia specific effects and superconductivity. *J. Am. Chem. Soc.* **2002**, 124, 34, 10146–10156.

Mohammed, M.R.; Ahmed, J.; Asiri, A.M.; Siddiquey, I.A.; Hasnat, M.A. Development of 4-methoxyphenol chemicalsensor based on NiS2-CNT nanocomposites. *J. Taiw. Inst. Chem. Engin.* **2016**, 64, 157–165.

Murugadoss, G.; Ramasamy, J.; Rangasamy, T.; Manavalan, R.K. PbO/CdO/ZnO and PbS/CdS/ZnS nanocomposites, studies on optical, electrochemical and thermal properties. *J. Lumin.* **2016**, 170, 78–89.

Muhammad, N.; Bergmann, C.; Geshev, J.; Quijada, R.; Galland, G.B. An efficient approach to the preparation of polyethylene magnetic nanocomposites. *Polymers*. **2016**, 97, 131–137.

Macaya, D.J.; Nikolou, M.; Takamatsu, S.; Mabeck, J.T.; Owens, R.M.; Malliaras, G.G. Simple glucose sensors with micromolar sensitivity based on organic electrochemical transistors. *Sens. Actuators B Chem.* **2007**, 123, 374–378.

Mahmoudian, M.; Alias, Y.; Basirun, W.; Woi, P.M.; Yousefi, R. Synthesis of polypyrrole coated silver nanostrip bundles and their application for detection of hydrogen peroxide. *J. Electrochem. Soc.* **2014**, 161, H487–H492.

Mishra, S.K.; Pasricha, R.; Biradar, A.M. Zns-nanocrystals/polypyrrole nanocomposite film based immunosensor. *Appl. Phys. Lett.* **2012**, 100, 053701.

Nia, P.M.; Meng, W.P.; Alias, Y. One-step electrodeposition of polypyrrole-copper nano particles for H_2O_2 detection. *J. Electrochem. Soc.* **2016**, 163, B8–B14.

Nohria, R.; Khillan, R.K.; Su, Y.; Dikshit, R.; Lvov, Y.; Varahramyan, K. Humidity sensor based on ultrathin polyaniline film deposited using layer-by-layer nano-assembly. *Sens. Actuators B*. **2006**,114, 218–222.

Nguyen, D.N.; Yoon, H. Recent advances in nanostructured conducting polymers, from synthesis to practical applications. *Polymers*. **2016**, 8, 118.

Numnuam, A.; Chumbimuni-Torres, K. Y.; Xiang, Y. et al.2008. Aptamer-based potentiometric measurements of proteins using ion-selective microelectrodes. *Anal. Chem.* **2008**, 80(3), 707–712.

Novoselov, K. S.; Geim, A. K.; Morozov, S. V.; et al., 2004. Field in atomically thin carbon films. *Sci.* **2004**, 306, 5696, 666–669.

Frost, M.; Meyerhoff, M. E. *In vivo* chemical sensors, tackling biocompatibility. *Anal. Chem.* **2006**, 78(21), 7370–7377.

Newkome, G. R.; Moorefield, C. N.; Vogtle. F. Dendritic Molecules, Concepts, Syntheses and Perspectives, VCH, Weinheim, Germany. **1996**.

Ngundi, M. Rational Design of Chemical and Biochemical Sensors, Chemistry Department, Binghamton University, Binghamon, NY, USA, **2003**.

Ordeig, O.; Banks, C. E.; Davies, T. J. et al., 2006. Regular arrays of microdisc electrodes, simulation quantifies the fraction of dead electrodes. *Analyst*. **2006**, 131, 3, 440–445.

Overney, G.; Zhong, W.; Tománek. D. Structural rigidity and low frequency vibrational modes of long carbon tubules. *Z. Phys. D*. **1993**, 27, 1, 93–96.

Ozsoz, M.; Erdem, A.; Kerman, K. et al., 2003. Electrochemical genosensor based on colloidal gold nanoparticles for the detection of Factor V Leiden mutation using disposable pencil graphite electrodes. *Anal. Chem.* **2003**, 75(9), 2181–2187.

Ohno, Y.; Maehashi, K.; Matsumoto, K. Label-free biosensors based on aptamer-modified graphene field-effect transistors. *J. Am. Chem. Soc.* **2010**, 132, 51, 18012–18013.

Oh, W.K.; Kwon, O.S.; Jang, J. Conducting polymer nanomaterials for biomedical applications, Cellular interfacing and biosensing. *Polym. Rev.* **2013**, 53, 407–442.

Park, S.J.; Kwon, O.S.; Lee, J.E.; Jang, J.; Yoon, H. Conducting polymer-based nanohybrid transducers, A potential route to high sensitivity and selectivity sensors. *Sensors* **2014**, 14, 3604–3630.

Park, S.J.; Song, H.S.; Kwon, O.S.; Chung, J.H.; Lee, S.H.; An, J.H.; Ahn, S.R.; Lee, J.E.; Yoon, H.; Park, T.H. Human dopamine receptor nanovesicles for gate-potential modulators in high-performance field-effect transistor biosensors. *Sci. Rep.* **2014**, 4, 1–8.

Patolsky, F.; Weizmann, Y.; Willner, I. Long-range electrical contacting of redox enzymes by SWCNT connectors. *Angew. Chem.* **2004**, 43(16), 2113–2117.

Piao, Y.; Jin, Z.; Lee, Z. et al., 2011. Sensitive and high-fidelity electrochemical immunoassay using carbon nanotubescoated with enzymes and magnetic nanoparticles. *Biosens. Bioelectron.* **2011**, 26, 3192–3199.

Pisana, S.; Braganca, P. M.; Marinero, E. E.; Gurney, B. A. Tunable nanoscale graphene magnetometers. *Nano Lett.* **2010**, 10, 1, 341–346.

Patri, A. K.; Majoros, I. J.; Baker, J. R. Dendritic polymer macromolecular carriers for drug delivery. *Curr. Opin. Chem. Biol.* **2002**, 6, 4, 466–471.

Patolsky, F.; Zheng, G.; Hayden, O.; Lakadamyali, M.; Zhuang, X.; Lieber, C. M. Electrical detection of single viruses. *Proc. Natl. Acad. Sci. USA*. **2004**, 101(39), 14017–14022.

Pumera, M.; Sánchez, S.; Ichinose, I.; Tang, J. Electrochemical nanobiosensors. *Sens. Actuators B*. **2007**, 123 (2), 1195–1205.

Pingarron, J.M.; Yanez-Sedeno, P.; Gonzalez-Cortes, A. Gold nanoparticle-based electrochemical biosensors. *Electrochim. Acta*. **2008**, 53(19), 5848–5866.

Qu, F.; Yang, M. H.; Shen, G. L.; Yu, R. Q. Electrochemical biosensing utilizing synergic action of carbon nanotubes and platinum nanowires prepared by template synthesis. *Biosens. Bioelectron.* **2007**, 22(8), 1749–1755.

Ramanavicius, A.; Kausaite, A.; Ramanaviciene, A. Polypyrrole-coated glucose oxidase nanoparticles for biosensor design. *Sens. Actuators B*. **2005**, 111–112, 532–539.

Rivas, G. A.; Rubianes, M. D.; Rodriguez, M. C. et al., 2007. Carbon nanotubes for electrochemical biosensing. *Talanta*. **2007**, 74(3), 291–307.

Shan, C.; Yang, H.; Han, D.; Zhang, Q.; Ivaska, A.; Niu. L. Electrochemical determination of NADH and ethanol based on ionic liquid-functionalized grapheme. *Biosens. Bioelectron.* **2010**, 25, 6, 1504–1508.

Shan, C.; Yang, H.; Han, D.; Zhang, Q.; Ivaska, A.; Niu. L. Graphene/AuNPs/chitosan nanocomposites film for glucose biosensing. *Biosens. Bioelectron.* **2010**, 25, 5, 1070–1074.

Shea, K. J.; Loy, D. A. Bridged polysilsesquioxanes. Molecular-engineered hybrid organic-inorganic materials. *Chem. Mat.* **2001**, 13, 10, 3306–3319.

Shan, C.; Yang, D.; Han, H.; Zhang, Q.; Ivaska, A.; Niu. L. Water-soluble graphene covalently functionalized by biocompatible poly-L-lysine. *Langmuir*. **2009**, 25, 20, 12030–12033.

Shen, M.; Shi, X. Dendrimer-based organic/inorganic hybrid nanoparticles in biomedical applications. *Nanoscale*. **2010**, 2, 9, 1596–1610.

Sanchez, S.; Fabregas, E.; Pumera, M. Detection of biomarkers with carbon nanotube-based immunosensors. *Methods Mol. Biol.* **2010**, 625, 227–237.

Simao, A.M.S.; Bolean, M.; Cury, T.A.C. et al., 2015. Liposomal systems as carriers for bioactive compounds. *Biophys. Rev.* **2015**, 7,391–397.

Singh, R.; Pantarotto, D.; McCarthy, D. et al., Binding and condensation of plasmid DNA onto functionalized carbon nanotubes, toward the construction of nanotube-based gene delivery vectors. *J. Am. Chem. Soc.* **2005**, 127, 12, 4388–4396.

Singh R.P.. A catechol biosensor based on a gold nanoparticles encapsulated-dendrimer. *Analyst*. **2011**,136, 1216.

Singh, R.P.; Kumar, K.K.; Rai, R.; Choi, J. W.; Tiwari, A.; Pandey, A. Synthesis, characterization of metal-Oxide nanomaterials for biosensors. *Synth. Charact. Appl. Smart Mat.* **2012**, 225–237.

Singh, R.P.; Tiwari, A.; Choi, J.W.; Pandey, A.C. Smart nanomaterials for biosensors, biochips and molecular bioelectronics. In Smart Nanomaterials for Sensor Application; Eds. Songjun, L., Yi, G., and He, L., **2012**, 3–41.

Singh, R. P.; Choi, J. W. Bio-nanomaterials for versatile bio-molecules detection technology. *Adv. Mater. Lett.* **2010**, 1, 83–84.

Singh, R.P.; Oh, B.K.; Choi, J.W. Application of peptide nucleic acid towards development of nanobiosensor arrays. *Bioelectrochemistry*. **2010**, 79 (2), 153–161.

Singh, R.P.; Kang, D.Y.; Oh, B.K.; Choi, J.W. Polyaniline based catalase biosensor for the detection of hydrogen peroxide and azide. *Biotechnol. Bioprocess Eng.* **2009**, 14, (4), 443–449.

Singh, R.P.; Oh, B.K.; Choi, J.W. Application of peptide nucleic acid towards development of nanobiosensor arrays. *Bioelectrochemistry*. **2010**, 79(2), 153–161.

Singh, R.P.; Oh, B.K.; Koo, K.K.; Jyoung, J.Y.; Jeong, S.; Choi, J.W. Biosensor arrays for environmental pollutants detection. *Biochip J.* **2009**, 2 (4), 223–234.

Singh, R.P. Prospects of nanobiomaterials for biosensing. *Int. J. Electrochem.* **2011**, 2011, 125487.

Singh, R.P. Prospects of organic conducting polymer modified electrodes, Enzymosensors. *Int. J. Electrochem.* 2012, 2012.

Singh, R.P. Nanocomposites, recent trends, developments and applications. In Advances in Nanostructured Composites, Volume 1, Carbon Nanotube and Graphene Composites, 1st Edition, Mahmood Aliofkhazraei, chap 02, **2019**.

Singh, R.P.; Oh, B.K.; Koo, K.K.; Jyoung, J.Y.; Jeong, S.; Choi, J.W. Biosensor arrays for environmental pollutants detection. *Biochem. J.* **2008**, 2, 4, 223–234.

Singh, R.P.; Kim, Y.J.; Oh, B.K.; Choi, J.W. Glutathione-s-transferase based electrochemical biosensor for the detection of captan. *Electrochem. Commun.* **2009**, 11 181–185.

Singh, R.P.; Choi, J.W. Biosensors development based on potential target of conducting polymers. *Sens. Trans. J.* **2009**, 104(5), 1–18.

Singh, R.P.; Kang, D.Y.; Oh, B.K.; Choi, J.W. Polyaniline based catalase biosensor for the detection of hydrogen peroxide and Azide. *Biotechnol. Bioprocess Eng.* **2009**, 14(4), 443–449.

Singh, R.P.; Kang, D.Y.; Choi, J.W. Electrochemical DNA biosensor for the detection of sanguinarine in adulterated mustard oil. *Adv. Mater. Lett.* **2010a**, 1(1), 48–54.

Singh, R.P.; Oh, B.K.; Choi, J.W. Application of peptide nucleic acid towards development of nanobiosensor arrays. *Bioelectrochemistry*. **2010b**, 79(2), 153–161.

Singh, R.P.; Choi, J.W. Bio-nanomaterials for versatile bio-molecules detection technology. Letter to Editors. *Adv.Mater. Lett.* **2010**, 1(1), 83–84.

Singh, R.P. Prospects of nanobiomaterials for biosensing. *Int. J. Electrochem.* **2011**, 2011, 125487, doi,10.4061/2011/125487.

Singh, R.P.; Tiwari, A.; Pandey, A.C. Silver/polyaniline nanocomposite for the electrocatalytic hydrazine oxidation. *J. Inorg. Organomet. Polym. Mat.* **2011a**, 21, 788–792.

Singh, R.P.; Pandey, A.C. Silver nano-sieve using 1, 2-benzenedicarboxylic acid as a sensor for detecting hydrogen peroxide. *Anal. Methods.* **2011**, 3, 586–592.

Singh, R.P.; Shukla, V.K.; Yadav, R.S.; Sharma, P.K.; Singh, P.K.; Pandey, A.C. Biological approach of zinc oxide nanoparticles formation and its characterization. *Adv. Mater. Lett.* **2011b**, 2(4), 313–317.

Singh, R.P.; Kang, D.Y.; Choi, J.W. Nanofabrication of bio-self assembled monolayer and its electrochemical property for toxicant detection. *J. Nanosc. Nanotech.* **2011c**,11, 408–412.

Singh, R.P.; Choi, J.W.; Tiwari, A.; Pandey, A.C. Utility and potential application of nanomaterials in medicine. In Biomedical Materials and Diagnostic Devices; Eds. Tiwari, A.; Ramalingam, M.; Kobayashi, H.; Turner, A.P.F., John Wiley & Sons, Inc., Hoboken, NJ, USA. **2012a**. doi: 10.1002/9781118523025.ch7.

Singh, R.P.; Kumar, K.; Rai, R.; Tiwari, A.; Choi, J.W.; Pandey, A.C. Synthesis, characterization of Metal oxide based nanomaterials and its application in Biosensing. In Synthesis, characterization and application of Smart material. Nova Science Publishers, Inc., USA. 2012. Chapter 11, 2012b, 225–238.

Singh, R.P.; Choi, J.W.; Tiwari, A.; Pandey, A.C. Biomimetic materials toward application of nanobiodevices, pp. 741–782, In Intelligent Nanomaterials, Processes, Properties, and Applications; Eds. Tiwari, A., Mishra, A.K., Kobayashi, H., and Turner, A.P., John Wiley & Sons, Inc., Hoboken, NJ, USA. Chapter 20. **2012**.

Singh, R.P.; Choi, J.W.; Pandey, A.C. Smart nanomaterials for biosensors, biochips and molecular bioelectronics. Bentham Science Publisher (USA). pp. 3–41. In Smart Nanomaterials for Sensor Application; Eds. Li, S., Ge, Y. and Li, H. Chapter 1. **2012d**.

Singh, R.P.; Choi, J.W.; Tiwari, A.; Pandey, A.C. Functional nanomaterials for multifarious nanomedicine. In, A. Tiwari and A.P.F. Turner (eds.). Biosensors Nanotechnology. John Wiley & Sons, Inc., Hoboken, NJ, USA. **2014**.

Singh, R.P. Nanobiosensors, Potentiality towards bioanalysis. *J. Bioanal. Biomed.* **2016**, 8, e143. doi,10.4172/1948–593X.1000e143.

Simm, A.O.; Banks, C.E.; Ward-Jones, S.; et al., Boron-doped diamond microdisc arrays, electrochemical characterisation and their use as a substrate for the production of microelectrode arrays of diverse metals (Ag, Au, Cu) via electrodeposition. *Analyst* **2005**, 130(9), 1303–1311.

Sassolas, A.; Leca-Bouvier, B.; Beatrice, D.; Blum, L.J. DNA biosensors and microarrays. *Chem. Rev.* **2008**, 108(1), 109–139.

Seeman, N.C. DNA engineering and its application to nanotechnology. *Trends Biotechnol.* **1999**, 17 (11), 437–443.

Song, W.; Li, D.-W.; Li, Y.-T.; Li, Y.; Long, Y.-T. Disposable biosensor based on graphene oxide conjugated with tyrosinase assembled gold nanoparticles. *Biosens. Bioelectron.* **2011**, 26, 7, 3181–3186.

Sawamura, M.; Kuninobu, Y.; Toganoh, M.; Matsuo, Y.; Yamanaka, M.; Nakamura, E. Hybrid of ferrocene and fullerene. *J. Am. Chem. Soc.* **2002**, 124, 32, 9354–9355.

So, H. M.; Won, K.; Kim, Y. H.; et al., Single-walled carbon nanotube biosensors using aptamers as molecular recognition elements. *J. Am. Chem. Soc.* **2005**, 127, 34, 11906–11907.

Song, S.; Wang, L.; Li, J.; Fan, C.; Zhao, J. Aptamer-based biosensors. *Trends Anal. Chem.* **2008**, 27(2), 108–117.

Song, Y.; Wei, C.; He, J.; Li, X.; Lu, X.; Wang, L. Porous Co nanobeads/rGO nanocomposites derived from rGO/Co-metal organic frameworks for glucose sensing. *Sens. Actuators B Chem.* **2015b**, 220, 1056–1063.

Song, H.S.; Kwon, O.S.; Lee, S.H.; Park, S.J.; Kim, U.K.; Jang, J.; Park, T.H. Human taste receptor-functionalized field effect transistor as a human-like nanobioelectronic tongue. *Nano Lett.* **2012**, 13, 172–178.

Saito, R.; Fujita, M.; Dresselhaus, G.; Dresselhaus, M.S. Electronic structure of graphene tubules based on C60. *Phys. Rev. B* **1992**, 46, 3, 1804–1811.

Schon, J. H.; Kloc, C.; Batlogg, B. High-temperature superconductivity in lattice-expanded C60. *Science.* **2001**, 293, 5539, 2432–2434.

Sailor, M.J.; Link, J.R. Smart dust, nanostructured devices in a grain of sand. *Chem. Commun.* **2005**, 11, 1375–1383.

Sperling, R. A.; Parak, W. J. Surface modification, functionalization and bioconjugation of colloidal Inorganic nanoparticles. *Philos. Trans. R. Soc. A* **2010**, 368, 1915, 1333–1383.

Sun, C.; Wang, D.; Geng, Z.; Gao, L.; Zhang, M.; Bian, H.; Liu, F.; Zhu, Y.; Wu, H.; Xu, W. One-step green synthesis of a polypyrrole-Au nanocomposite and its application in myoglobin aptasensor. *Anal. Methods* **2015**, 7, 5262–5268.

Snook, G.A.; Kao, P.; Best, A.S. Conducting-polymer-based supercapacitor devices and electrodes. *J. Power Sources.* **2011**, 196, 1–12.

Shen, J.; Yan, Y.; Yang, Z.; Hong, Y.; Zhiguo, Z.; Shiping, Y. Functionalized Au-Fe3 O4 nanocomposites as a magnetic and colorimetric bimodal sensor for melamine. *Sens. Actuators B Chem.* 2016, 226, 512–517.

Shukla, V.K.; Singh, R.P.; Pandey, A.C. Black pepper assisted biomimetic synthesis of silver nanoparticles. *J. Alloy Comp.* **2010**, 507(1), L13–L16.

Svobodova, L.; Snejdarkova, M.; Hianik, T. Properties of glucose biosensors based on dendrimer layers, Effect of enzyme immobilization. *Anal. Bioanal. Chem.* **2002**, 373 (8) 735–741.

Snejdarkova, M.; Svobodova, L.; Nikolelis, D.P.; Wang, J.; Hianik, T. Acetylcholine biosensor based on dendrimer layers for pesticides detection. *Electroanalysis* **2003**, 15 (14), 1185–1191.

Tabrizi, M.A.A.; Shamsipur, M.; Mostafaie, A. A high sensitive label-free immunosensor for the determination of human serum IgG using overoxidized polypyrrole decorated with gold nanoparticle modified electrode. *Mater. Sci. Eng. C Mater. Biol. Appl.* **2016**, 59, 965–969.

Tao, Y.; Ju, E.; Ren, J.; Qu, X. Polypyrrole nanoparticles as promising enzyme mimics for sensitive hydrogen peroxide detection. *Chem. Commun.* **2014**, 50, 3030–3032.

Tang, J.; Su, B.; Tang, D. Chen, G. Conductive carbon nanoparticles-based electrochemical immunosensor with enhanced sensitivity for α-fetoprotein using irregular-shaped gold nanoparticles-labeled enzyme-linked antibodies as signal improvement. *Biosens. Bioelectron.* **2010**, 25, 12, 2657–2662.

Tang, C.S.; Schmutz, P.; Petronis, S.; Textor, M.; Keller, B.; Voros, J. Locally Addressable Electrochemical Patterning Technique (LAEPT) applied to poly (L-lysine)-graft-poly (ethylene glycol) adlayers on titanium and silicon oxide surfaces. *Biotechnol. Bioeng.* **2005**, 91(3), 285–295.

Tang, H.; Lin, P.; Chan, H.L.; Yan, F. Highly sensitive dopamine biosensors based on organic electrochemical transistors. *Biosens. Bioelectron.* **2011**, 26, 4559–4563.

Tang, H.; Chen, J.; Nie, L.; Kuang, Y.; Yao, S. A label-free electrochemical immunoassay for carcinoembryonic antigen (CEA) based on gold nanoparticles (AuNPs) and nonconductive polymer film. *Biosens. Bioelectron.* **2007**, 22(6), 1061–1067.

Tang. L.; Wang, Y.; Liu, Y.; Li, J. DNA-directed self-assembly of graphene oxide with applications to ultrasensitive oligonucleotide assay. *ACS Nano* **2011**, 5, 3817–3822.

Tiwari, A.; Terada, D.; Kobayashi, H.; Singh, R.P.; Rai, R. Bionanomaterials for emerging biosensors technology. pp. 137–154. In Synthesis, Characterization and Application of Smart Materials; Ed. Rai, Radheshyam. Nova Publishers, Hauppauge, New York, USA. Chapter 7. **2012a**.

Tiwari, A.; Singh, R.P.; Rai, R. Vinyls modified guar gum biodegradable plastics. pp. 125–136. In Synthesis, Characterization and Application of Smart Materials; Ed. Rai, Radheshyam. Nova Publishers, Hauppauge, New York, USA. Chapter 6. **2012b**.

Tiwari, A.; Tiwari, A.; Singh, R.P. Bionanocomposite matrices in electrochemical biosensors. In Biomedical Materials and Diagnostic Devices; Eds. Tiwari, A., Ramalingam, M., Kobayashi, H., and Turner, A.P.F. John Wiley & Sons, Inc., Hoboken, NJ, USA. **2012c**. doi 10.1002/9781118523025.ch10

Tully, D.C.; Frechet, J. M.J. Dendrimers at surfaces and interfaces, chemistry and applications. *Chem. Commun.* **2001**, 14, 1229–1239.

Tombler, T.W.; Zhou, C.; Alexseyev, L.; et al., Reversible electromechanical characteristics of carbon nanotubes under local-probe manipulation. *Nature*. **2000**, 405, 6788, 769–772.

Tombelli, S.; Minunni, M.; Mascini, M. Analytical applications of aptamers. *Biosens. Bioelectron.* **2005**, 20 (12), 2424–2434.

Tombelli, S.; Minunni, M.; Mascini, M. Aptamers-based assays for diagnostics, environmental and food analysis. *Biomol. Eng.* **2007**, 24(2), 191–200.

Tsukruk, V.V.; Rinderspacher, F. Bliznyuk, V.N. Self-assembled multilayer films from dendrimers. *Langmuir* **1997**, 13, 8, 2171–2176.

Tsukruk, V.V. Dendritic macromolecules at interfaces. *Adv. Mater.* **1998**, 10, 3, 253–257.

Umasankar, Y.; Chen, S.M. Recent trends in the application of carbon nanotubes-polymer composite modified electrodes for biosensors, a review. *Anal. Lett.* **2008**. 41(2), 210–243.

Vamvakaki, V.; Hatzimarinaki, M.; Chaniotakis, N. Biomimetically synthesized silica-carbon nanofiber architectures for the development of highly stable electrochemical biosensor systems. *Anal. Chem.* **2008**, 80(15), 5970–5975.

Vasantha, V.; Chen, S.M. Electrocatalysis and simultaneous detection of dopamine and ascorbic acid using poly(3,4-ethylenedioxy) thiophene film modified electrodes. *J. Electroanal. Chem.* **2006**, 592, 77–87.

Wan Y.; Wang, Y.; Wu, J.; Zhang., D. Graphene oxide sheet-mediated silver enhancement for application to electrochemical biosensors. *Anal. Chem.* **2011**, 83, 3, 648–653.

Wang, J. Glucose biosensors, 40 years of advances and challenges. *Electroanalysis* **2001**, 13, 983–988.

Wang, D.; Hu, W.; Xiong, Y.; Xu, Y.; Li, C.M. Multifunctionalized reduced graphene oxide-doped polypyrrole/pyrrolepropylic acid nanocomposite impedimetric immunosensor to ultra-sensitively detect small molecular aflatoxin B-1. *Biosens. Bioelectron.* **2015**, 63, 185–189.

Wang, J.Y.; Chen, L.C.; Ho, K.C. Synthesis of redox polymer nanobeads and nanocomposites for glucose biosensors. *ACS Appl. Mater. Interfaces* **2013**, 5, 7852–7861.

Wang, G.; Ma, Y.; Dong, X.; Tong, Y.; Zhang, L.; Mu, J.; et al., Facile synthesis and magneto-rheological properties of superparamagnetic $CoFe_2O_4$/GO nanocomposites. *Appl. Surf. Sci.* 2015e, 357, 2131–2135.

Wang, Y.H.; Li, F.L.; Wang, Y.Q.; Wu, S.; He, X.X.; Wang, K.M. A TiO2 /CNTs nanocomposites enhanced luminal electrochemiluminescence assay for glucose detection. *Chin. J. Anal. Chem.* **2015f**, 43, 1682–1687.

Wang, L.; Zhang, X.; Xiong, H.; Wang. S. A novel nitromethane biosensor based on biocompatible conductive redox graphene-chitosan/hemoglobin/graphene/room temperature ionic liquid matrix. *Biosens. Bioelectron.* **2010**, 26, 3, 991–995.

Wang, J. Nanomaterial-based amplified transduction of biomolecular interactions. Small. **2005**, 1(11), 1036–1043.

Wang, J. Electrochemical biosensors, towards point-of-care cancer diagnostics. *Biosens. Bioelectron.* **2006**, 21(10), 1887–1892.

Wang, J.; Li, J.; Baca, A. J. et al., Amplified voltammetric detection of DNA hybridization via oxidation of ferrocene caps on gold nanoparticle/streptavidin conjugates. *Anal. Chem.* **2003**, 75(15), 3941–3945.

Wang, J.; Liu, G.; Rivas, G. Encoded beads for electrochemical identification. *Anal. Chem.* **2003**, 75(17), 4667–4671.

Wang, J. Nanomaterial-based electrochemical biosensors. *Analyst*. **2005**. 130, 4, 421–426.

Wang, S. G.; Wang, R.; Sellin, P. J.; Zhang. Q. DNA biosensors based on self-assembled carbon nanotubes. *Biochem. Biophys. Res. Comm.* **2004**, 325, 4, 1433–1437.

Wang, J.; Kawde, A. N.; Jan. M. R. Carbon-nanotube-modified electrodes for amplified enzyme-based electrical detection of DNA hybridization. *Biosens. Bioelectron.* **2004**, 20, 5, 995–1000.

Wang, K.; Liu, Q.; Guan, Q.M.; Wu, J.; Li, H.N.; Yan. J. J. Enhanced direct electrochemistry of glucose oxidase and biosensing for glucose via synergy effect of graphene and CdS nanocrystals. *Biosens. Bioelectron.* **2011**, 26, 5, 2252–2257.

Wang, J.; Xu, D.; Polsky, R. Magnetically-induced solid-state electrochemical detection of DNA hybridization. *J. Am. Chem. Soc.* **2002**, 124(16), 4208–4209.

Wang, Y.; Xu, H.; Zhang, J.; Li, G. Electrochemical sensors for clinic analysis. *Sensors*. **2008**, 8(4), 2043–2081.

Weng, B.; Morrin, A.; Shepherd, R.; Crowley, K.; Killard, A.J.; Innis, P.C.; Wallace, G.G. Wholly printed polypyrrole nanoparticle-based biosensors on flexible substrate. *J. Mater. Chem.* B. **2014**, 2, 793–799.

Wei, Q.; Mukaida, M.; Ding, W.; Ishida, T. Humidity control in a closed system utilizing conducting polymers. *RSC Adv*. **2018**, 8, 12540–12546.

Wilbur, D. S.; Pathare, P. M.; Hamlin, D. K.; Buhler, K. R.; Vessella. R. L. Biotin reagents for antibody pretargeting. 3. Synthesis, radioiodination, and evaluation of biotinylated starburst dendrimers. Bioconjugate Chem. **1998**, 9, 6, 813–825.

Wilson, M. S. Electrochemical immunosensors for the simultaneous detection of two tumor markers. *Anal. Chem.* **2005**, 77(5), 1496–1502.

Wilson, G. S.; Gifford, R. Biosensors for real-time in vivo measurements. *Biosens. Bioelectron.* **2005**, 20 (12), 2388–2403.

Willner, I.; Zayats, M. Electronic aptamer-based sensors. *Angew. Chem.* **2007**, 46(34), 6408–6418.

Wu, H.; Wang, J.; Kang, X.; et al., Glucose biosensor based on immobilization of glucose oxidase in platinum nanoparticles/graphene/chitosan nanocomposite film. *Talanta*. **2009**, 80, 1, 403–406.

Wu, L.; Chu, H. S.; Koh, W. S.; Li, E. P. Highly sensitive graphene biosensors based on surface plasmon resonance. *Opt. Exp.* **2010**, 18, 14, 14395–14400.

Wu, X.; Hu, Y.; Jin, J.; et al., Electrochemical approach for detection of extracellular oxygen released from erythrocytes based on graphene film integrated with laccase and 2, 2-azino-bis(3-ethylbenzothiazoline-6-sulfonic acid). *Anal. Chem.* **2010**, 82, 9, 3588–3596.

Xiao, Y.; Chang, M. L. Nanocomposites, from fabrications to electrochemical bioapplications. *Electroanalysis*. **2008**, 20 (6), 648–662.

Xu, S.; Liu, Y.; Wang, T.; Li, J. Positive potential operation of a cathodic electrogenerated chemiluminescence immunosensor based on luminol and graphene for cancer biomarker detection. *Anal. Chem.* **2011**, 83, 10, 3817–3823.

Yadav, R.S.; Singh, R.P.; Verma, P.; Tiwari, A.; Pandey, A.C. Smart nanomaterials for space and energy applications. pp. 213–250. In Intelligent Nanomaterials, Processes, Properties, and Applications; Eds. Tiwari, A., Mishra, A.K., Kobayashi, H., and Turner, A.P. John Wiley & Sons, Inc., Hoboken, NJ, USA. Chapter 6, **2012**.

Yang, M.; Javadi, A.; Li, H.; Gong, S. Ultrasensitive immunosensor for the detection of cancer biomarker based on graphene sheet. *Biosens. Bioelectron.* **2010**, 26, 2, 560–565.

Yang, G.; Kampstra, K.L.; Abidian, M.R. High performance conducting polymer nanofiber biosensors for detection of biomolecules. *Adv. Mater.* **2014**, 26, 4954–4960.

Yang, L.; Li, Y.B.; Erf, G.F. Interdigitated array microelectrode-based electrochemical impedance immunosensor for detection of Escherichia coli O157, H7. *Anal. Chem.* **2004**, 76(4), 1107–1113.

Yang, L.; Bashir, R. Electrical/electrochemical impedance for rapid detection of foodborne pathogenic bacteria. *Biotechnol. Adv.* **2008**, 26(2), 135–150.

Yogeswaran, U.; Chen, S.M. A review on the electrochemical sensors and biosensors composed of nanowires as sensing material. *Sensors*. **2008**, 8(1), 290–313.

Yoon, H.; Jang, J. Conducting-polymer nanomaterials for high-performance sensor applications, Issues and challenges. *Adv. Funct. Mat.* **2009**, *19*, 1567–1576.

Yoon, H.; Lee, S.H.; Kwon, O.S.; Song, H.S.; Oh, E.H.; Park, T.H.; Jang, J. Polypyrrole nanotubes conjugated with human olfactory receptors, High-performance transducers for FET-type bioelectronic noses. *Angew. Chem. Int. Ed.* **2009**, 48, 2755–2758.

Zheng, G.; Patolsky, F.; Cui, Y.; Wang, W. U.; Lieber, C. M. Multiplexed electrical detection of cancer markers with nanowire sensor arrays. *Nat. Biotechnol.* **2005**, 23(10), 1294–1301.

Zhang, M.; Yin, B.C.; Tan, W.; Ye, B.C. A versatile graphene-based fluorescence "on/off" switch for multiplex detection of various targets. *Biosens. Bioelectron.* **2011**, 26, 7, 3260–3265.

Zhang, Q.; Wu, S.; Zhang, L.; et al., Fabrication of polymeric ionic liquid/graphene nanocomposite for glucose oxidase immobilization and direct electrochemistry. *Biosens. Bioelectron.* **2011**, 26, 5, 2632–2637.

Zhao, X.H.; Kong, R.M.; Zhang, X.B.; et al., Graphene-DNAzyme based biosensor for amplified fluorescence turn-On detection of Pb^{2+} with a high selectivity. *Anal. Chem.* **2011**, 83, 13, 5062–5066.

Zhang, Q.; Qiao, Y.; Hao, F.; et al., Fabrication of a biocompatible and conductive platform based on a singlestranded DNA/graphene nanocomposite for direct electrochemistry and electrocatalysis. *Chemistry*. **2010**, 16, 27, 8133–8139.

Zhai, D.; Liu, B.; Shi, Y.; Pan, L.; Wang, Y.; Li, W.; Zhang, R.; Yu, G. Highly sensitive glucose sensor based on Pt nanoparticle/polyaniline hydrogel heterostructures. *ACS Nano.* **2013**, 7, 3540–3546.

Zhang, X.Z.; Zhou, Y.; Zhang, W.; Zhang, Y.; Gu, N. Polystyrene@Au@prussian blue nanocomposites with enzymelike activity and their application in glucose detection. *Coll. Surf. A Physicochem. Eng. Asp.* **2016b**, 490, 291–299.

Zheng, L. A.; Lairson, B. M.; Barrera, E. V.; Shull, R. D. Formation of nanomagnetic thin films by dispersed fullerenes. *Appl. Phys. Lett.* **2000**, 77, 20, 3242–3244.

Zheng, L. A.; Barrera, E. V.; Shull, R. D. Formation and stabilization of nanosize grains in ferromagnetic thin films by dispersed C60. *J. Appl. Phys.* **2002**, 92, 1, 523.

Zhou, M.; Shang, L.; Li, B.; Huang, L.; Dong, S. Highly ordered mesoporous carbons as electrode material for the construction of electrochemical dehydrogenase and oxidase-based biosensors. *Biosens. Bioelectron.* **2008**, 24(3), 442–447.

Zhuo, Y.; Yi, W. J.; Lian, W. B.; et al., Ultrasensitive electrochemical strategy for NT-proBNP detection with gold nanochains and horseradish peroxidase complex amplification. *Biosens. Bioelectron.* **2011**, 26, 5, 2188–2193.

Zhuo, Y.; Yuan, R.; Chai, Y.Q.; Hong, C. L. Functionalized SiO_2 labeled CA19–9 antibodies, a new strategy for signal amplification of antigen-antibody sensing processes. *Analyst.* **2010**, 135, 8, 2036–2042.

Zhou, M.; Zhai, Y.; Dong, S. Electrochemical sensing and biosensing platform based on chemically reduced graphene oxide. *Anal. Chem.* **2009**, 81, 14, 5603–5613.

Zhou, W. H.; Zhu, C. L.; Lu, C. H.; et al., Amplified detection of protein cancer biomarkers using DNAzyme functionalized nanoprobes. *Chem. Commun.* **2009**, 44, 6845–6847.

Zhao, M.; Sun, L.; Crooks, R. M. Preparation of Cu Nanoclusters within dendrimer templates. *J. Am. Chem. Soc.* **1998**, 120, 4877–4878.

Zhu, N.; Gu, Y.; Chang, Z.; He, P.; Fang, Y. PAMAM dendrimers-based DNA biosensors for electrochemical detection of DNA hybridization. *Electroanalysis.* **2006**, 18, 21, 2107–2114.

Zhu, J.; Morgan, A.B.; Lamelas, F.J.; Wilkie, C.A. Fire properties of polystyrene-clay nanocomposites. *Chem. Mater.* **2001**, 13, 3774–3780.

Zhu, J.; Huo, X.; Liu, X.; Ju, H. Gold nanoparticles deposited polyaniline-TiO_2 nanotube for surface plasmon resonance enhanced photoelectrochemical biosensing. *ACS Appl. Mater. Interfaces.* **2016**, 8, 1, 341–349.

Zhu, Z.T.; Mabeck, J.T.; Zhu, C.; Cady, N.C.; Batt, C.A.; Malliaras, G.G. A simple poly(3,4-ethylene dioxythiophene)/poly(styrene sulfonic acid) transistor for glucose sensing at neutral pH. *Chem. Commun.* **2004**, 13, 1556–1557.

CHAPTER 20

Benefits of Nanomaterials-Based Biosensors

SOURAV MISHRA[1], ROHIT KUMAR SINGH[1], UDAY SURYAKANTA[1], BIJAYANANDA PANIGRAHI[1], and DINDYAL MANDAL[1,2*]

[1]*School of Biotechnology, Kalinga Institute of Industrial Technology Deemed to be University, Campus 11, Patia, Bhubaneswar 751024 Odisha, India*

[2]*School of Pharmacy, Chapman University, Irvine, CA, USA*

[*]*Corresponding author. E-mail: ddmandal@gmail.com*

ABSTRACT

A biosensor is a device which is used to detect the biological analyte that is frequently of natural source including DNA of a virus, bacteria, or proteins produced from the immune systems of unhygienic living organisms. Such analytes also include simple molecules like glucose or pollutants containing a biological receptor. One of the major challenges in biosensor development is transduction, that is, efficiently capturing the signal of the biological recognition event. Nanomaterials can enhance the quality and efficiency of the biosensor by introducing innovative signal transduction technologies. Several nano-biosensor constructions have been developed such as optical resonators, mechanical devices using functionalized nanoparticles, nanowires, nanotubes, and nanofibers. In particular, nanomaterials including carbon nanotubes, gold nanoparticles, magnetic nanoparticles, and quantum dots have been investigated extensively for their potential applications in biosensors, which has unveiled a new interdisciplinary field involving biological detection and material science. Various kinds of nanomaterials have been reported as biosensors in the fields of environment, biomedical, microbial detection, and food industries, respectively to sense and eliminate certain contaminants. Nanoparticles based amperometric, electrochemical, mechanical, optical, enzymatic, bacterial sensors, and DNA sensors are the most common sensors being active today due to their very small size and large surface to volume ratio. This chapter describes the promising applications of nanomaterials and their advantage in the field of biosensing technology to develop a highly sensitive, quick, and cost-effective method of analysis in medical diagnostics, food and drink industry, environmental protection, etc.

20.1 INTRODUCTION

Identification and evaluation of the biomolecules in the atmosphere, water system, and foodstuff are crucial toward avoiding rigorous diseases in the human body (Clark et al., 1962; Zhao et al., 2016). Sensor, a sort of analytical device, produces great response against the changes in physical properties or chemical concentration may be used for this purpose. Biosensors are different from conventional chemical sensors in two different ways: (1) the detecting elements of biological structures, like cells, enzymes, or nucleic acids, (2) the sensors are accustomed to measure biological processes or physical changes. Detection of both chemical and biological moieties bears great importance for the diagnosis of disease, environmental protection, and electrochemical devices. Nanotechnology provides huge advantages in disease diagnosis which is called as "nanodiagnostics" (Lin et al., 2013). However, a number of chemical, physical, and biological procedures have been used for this detection purpose, which shows the use of highly complicated equipment including ICP-MS, AAS (Choudhary et al., 2016). Using the above instruments for this purpose is intense, time taking, and costly. Hence, a quick and cost-efficient procedure for the recognition of biomolecules is needed. Newly emerging colorimetric detection of hazardous biomolecules has fascinated the researchers due to its good biocompatibility along with significant signal amplification, simplicity, and user-friendly nature. Nanoparticles have gained remarkable recognition in the world due to their controlled size, shape, structures, and physicochemical properties which show great potential in optical, electronic, magnetic, and biomedical applications (Nie et al., 2010). Due to their distinctive properties, they reveal extensive application in the detection of biomolecules by a spectral change to the local surroundings of the nanoparticle's surface. Moreover, they are attractive due to their simplicity of monitoring light signals because of their strong scattering or absorption properties (Ray et al., 2010).

Among all the nanomaterials, mostly gold nanoparticles (AuNPs) with a particular size are associated with optical, magnetic, and electrochemical properties which make them a perfect platform to formulate biosensing materials (Zhao et al., 2008). Because of the sensitive spectral change associated with the color change to the local surroundings of the nanoparticle surface, plasmonic metal nanoparticles including gold and silver have tremendous potential as a chemical and biological sensor.

In this chapter, the development and the benefits of biosensors with a long objective in the field of nanotechnology have been reported. This chapter emphasizes on noble metal nanoparticles (gold nanoparticles and silver nanoparticles) which have been widely explored as biosensors.

20.2 GOLD NANOPARTICLES AS BIOSENSOR

Among all the nanomaterials, gold nanoparticles have gained significant importance among researchers due to their wide range of applicability and versatility starting from catalysis to nanomedicine. They have been widely used for biosensor applications (Li et al., 2010), owing to their biocompatibility, optical, electronic properties, their relatively simple production, and modification (Biju et al., 2014).

20.2.1 DETECTION OF PROTEINS

The power of gold nanoparticles as a sensor is based on the binding ability of the gold nanoparticle's surface to the sensing material, which leads to the shifting of large surface plasmon resonance band (LSPR) and change of color of the nanoparticle solution (Rex et al., 2006). Gold nanoparticles or surface-modified gold nanoparticles have been demonstrated to be capable of a potential protein detector. A certain concentration of biomarker proteins or irregular protein existence is an indication of various diseases including cancer (Daniels et al., 2004; Ross et al., 1998; Kim et al., 2016). Kim et al. detected the amyloid protein through citrate reduced gold nanoparticles at acidic conditions (Schofield et al., 2006). They have investigated two types of amyloid β (Aβ) peptides with different sequences, Aβ (1–40) and Aβ (1–42), which are abundant in patients with AD. When the gold nanoparticles were mixed with Aβ (1–40) and Aβ (1–42) in acidic condition, they found a spectral change in each case. With an increase in the concentration of Aβ (1–40), the intensity of the absorption band of AuNPs was found to be decreased followed by a spectral shifting from 520 to 650 nm. This spectral change is associated with a change in color from red to blue, indicating the assembly of AuNPs with Aβ (1–40), which leads to aggregation of AuNPs. This dramatic change in spectra was because of the interaction of positively charged peptides in Aβ (1–40) and negatively charge AuNPs surface. Interestingly, in basic conditions any changes in spectra were not observed, indicating the use of AuNPs as a potential sensor. The spectral change in the case of Aβ (1–42) was found to be concentration dependent; however, the time-dependent spectral change was slow as compared to the Aβ (1–40). Furthermore, they have shown that the amyloid protein could be detected with different sizes of AuNPs and faster detection was found with smaller size nanoparticles.

Russell's group made carbohydrate modified gold nanoparticles to detect lectin concanavalin A (Con A) through the shifting of spectral and color change (Russel et al., 2006). Con A exists monomeric lectin form at pH >7, which binds strongly with the mannose ligands present on the gold nanoparticle surface and leads to aggregation. Red shifting and widening of the LSPR band from 523 nm to 620 nm was found, which leads to the change of color from red to purple with the aggregation of nanoparticles. The detection limit of mannose stabilized gold nanoparticles for lectin Con A was found to be 0.04 μM.

In another study, Narain and coworkers prepared biotinylated polymers functionalized AuNPs to detect streptavidin (Narain et al., 2007). They have prepared the biotinylated chain transfer agent followed by the RAFT polymerization with galactose derivatives. As biotin has an affinity toward streptavidin, they investigated the accessibility of the biotin ligands on the glycol-nanoparticles for binding to a streptavidin-coated sensor chip. They found that the aggregation of gold nanoparticles occurred with a slide blue shift in the SPR band. Further, they checked the sensitivity of the detection by taking nonbiotinylated polymer-stabilized AuNPs and found no aggregation. Previously Ishii and coworkers prepared biotinylated–PEGylated gold nanoparticles and established the interaction of biotin with streptavidin by checking its detection ability with streptavidin (Ishii et al., 2004; Lu et al., 2019). Hereafter, the addition of streptavidin to the polymer-coated AuNPs, a red shift or bathochromic shift of LSPR band with time followed by a color change from pink to blue was observed, indicating the aggregation of AuNPs. Similarly, by taking nonbiotinylated AuNPs as control, they have demonstrated that the aggregation of AuNPs occurs because of the affinity between biotin and streptavidin.

Citrate stabilized AuNPs were coated with poly(styrene-*b*-4-vinylpyridine) (PS-*b*-P4VP) block copolymer brush layer to detect protein human IgG (Doumas et al., 1997). PS-*b*-P4VP block copolymer brush layer was immobilized on AuNPs of different sizes by a simple self-assemble method to build the template for the detection. To compare the detection limit, standard polymers like 3-aminopropyltrimethoxysilane (APTMS), poly(diallyl dimethyl ammonium chloride) (PDDA), poly(allylamine) (PAH), poly(4-styrenesulfonic acid) (PSS) were used as protein biosensors for the sensing of human IgG having anti-human IgG immobilized on the modified AuNPs surface. Among PS-*b*-P4VP, APTMS, and PDDA/PSS/PAH templates on AuNPs surface, PS-b-P4VP displayed much better results than other polymers. When the concentration of human IgG was increased, the absorption band of AuNPs was shifted to 667 nm. PS-*b*-P4VP brush layer-based sensor was found to be about five times more efficient than that of APTMS modified AuNPs and two times more efficient than that of PDDA/PSS/PAH modified AuNPs. Limit of detection (LOD) of the PS-*b*-P4VP-templated AuNPs toward human IgG was found to be 1.2 nM. In contrast, the LODs of APTMS- and PDDA/PSS/PAH-modified AuNPs were found to be about 37 nM and 3.4 nM, respectively.

One of the most lavish proteins present in plasma is human serum albumin (HSA) which maintains the oncotic pressure of blood and acts as a vehicle for multiple substances like vitamins, fatty acids, drugs. When the concentration of HSA in human serine increases more than 20 mg/L, it indicates kidney damage associated with diabetes and cardiovascular diseases (Arques et al., 2011; Fanali et al., 2012; Huang et al., 2015). Therefore, the production of a cost-effective, selective, and sensitive nanosensor system for the detection of HSA is needed. Citrate-capped AuNPs with melamine cross linker can effectively detect HSA protein (Yan et al., 2010). These citrate-capped AuNPs exhibit red color with an SPR band at 518 nm, however, after attaching the cross-linker melamine, the SPR band at 518 nm shifted to 690 nm along with an intense color change related to the agglomeration of AuNPs. Interestingly, when HSA was added to the AuNPs before melamine, there was a light shifting of the SPR band from 518 to 529 nm with no change in color, suggesting the suppression of aggregations. These results were also supported by the respective TEM images with different conditions. It suggests that the adsorption of HSA on the AuNPs can protect the surface of the nanoparticle and stabilize the nanoparticle. However, when melamine is added to AuNPs first before adding HSA into it, it does not help in suppressing the aggregation. This may be because of the weak binding ability of melamine to HSA through H-bonding (Wang et al., 2006). Hence, HSA cannot replace melamine through this weak interaction, suggesting the contribution of HSA on nanoparticle surface in the antiaggregation effect. They have also investigated the selectivity of AuNPs by checking in the presence of other amino acids, small peptides which are often found in human urine and blood as metabolites. It was found that the limit of detection was 1.2 nM.

It is previously established that upon the treatment of biotinylated protein kinase with avidin modified gold nanoparticle, a good colorimetric assay can be developed for the detection of kinase and kinase inhibitor. Wang et al. have prepared peptide-capped gold nanoparticles as specific kinase-substrate which can replace natural substrate for kinase (Wang et al., 2005). When these gold nanoparticles were biotinylated with λ-biotin-ATP as co-substrate followed by the addition of avidin-coated gold nanoparticles, it leads to the agglomeration of nanoparticles due to the binding between biotin and avidin. This was further accompanied by the color change from red to blue, due to the shifting and broadening of the SPR band of gold nanoparticles. Therefore, it can be concluded that the color change may be attributed to the kinase catalyzed biotinylation and no change in color could be due to the inhibition of kinase reaction. To demonstrate this, they have used functionalized gold nanoparticles that were modified by penta-peptides and oligo-peptide sequences which are the common substrate molecules of the cAMP-dependent protein kinase A and the calmodulin-dependent kinase II, respectively. Further, they have repeated the experiments with well-known kinase inhibitors such as H89, KN62, and calmodulin-dependent kinase II, respectively, and the solution was then visually inspected and examined through UV–Vis spectroscopy.

Based on the nanoparticles aggregation mechanism, Stevens and coworkers have developed a simple detection assay for the activity of blood coagulation Factor XIII (Chandrawati et al., 2014). There has been an isopeptide bond formation between peptide chains, which leads to particle aggregation. AuNPs have been shrouded with a Factor XIII reactive peptide domain which contains glutamine or lysine residues. Through the formation of e-(g-glutamyl)-lysine isopeptide bonds, the activity of the blood coagulation factors was studied. They have designed hepta-peptide having lysine and glutamine residue to functionalize the nanoparticle surface which acts as substrate for Factor XIII. As a consequence, it results in aggregation of AuNP associated with a color change and a longitudinal band shifting from 522 nm to 559 nm.

Matrix metalloproteinase matrilysin (MMP-7) is an indicator of salivary gland cancer which helps in invasive growth and metastasis of colon carcinoma and numerous other human cancers (Luukkaa et al., 2010; Aili et al., 2008). Herein, polypeptide (JR2EC) has been capped upon AuNPs which contain 40 amino acid residues having cysteine residue to increase the immobilization. MMP-7 hydrolyses the peptide present on the exterior of AuNPs while incubation with AuNPs. Consequently, the size and net charge of the peptide were found to be decreased, which leads to instability in stable colloidal solution and finally, a bathochromic (red) shift in the SPR band was noticed along with a change in color of the solution (Aili et al., 2009).

It is already known that MMP-7 utilizes both Zn^{2+} and Ca^{2+} as cofactors. However, while incubating AuNPs with MMP-7 in the presence of 10 mM EDTA or 10 mM inhibitor II, no aggregation of AuNPs was observed, which indicates that the aggregation is specific and induced by the cleavage of the immobilized peptide (Chen et al., 2013). The minimum concentration for the hydrolysis of the peptide by MMP-7 was found to be 5 nM (0.1 mg mL^{-1}). Based on this concept, Guarise et al. reported the detection of other proteases (thrombin, lethal; Guarise et al., 2006).

An exclusive, sensitive, and highly specific immunoassay system for anti-protein A antibodies using protein antigen-coated gold nanoparticles has been well established. The sensitivity of the assay was carried out by exploring the effect of the parameters like the pH, the temperature, and the concentration of protein A-coated gold nanoparticles (Thanh et al., 2002). Chen et al. (2010) detected protein using fluorescent dye-coated gold nanoparticles monitored by fluorescence recovery techniques. Initially, they have titrated cationic gold nanoparticles with anionic fluorescent poly(p-phenylene ethynylene) (PPE) derivatives (PPE-CO_2). It was found that the fluorescence of dye was found to be quenched along with a decrease in absorbance and a slight blue shift was observed upon addition of gold nanoparticles. When coating and binding characteristics of PPE-CO_2 and AuNPs were done, these polymer-nanoparticle conjugates have been used for the detection of proteins. For this study, they have chosen different charges and sizes of proteins ranging from 12.3 to 540 kDa, respectively. Initially, fluorescence intensities after mixing PPE-CO_2 (100 nM) and AuNPs were recorded at 465 nm and it was found that >80% PPE-CO_2 was bound to gold nanoparticles. However, upon the addition of 5 μM protein, a good fluorescence response was noticed. On the contrary, individual addition of 5 μM protein into PPE-CO_2 (100 nM) induces marginal fluorescence intensity changes, confirming the interruption of nanoparticle–PPE–CO_2 interactions by proteins (You et al., 2007).

In order to detect proteins, Lo et al. used citrate stabilized AuNPs modified with thiolated single-stranded DNA (ssDNA) through gold-thiol covalent bonding (Lo et al., 2013). Their results indicate that the UV–Vis spectra of gold nanoparticles before and after the addition of ssDNA were found to be similar, which indicates that after the immobilization, the gold colloid was still stable and well dispersed. Therefore, upon addition of Human Topoisomerase I (TOPO) molecule to ssDNA-AuNPs conjugates, the backbone of the negatively charged ssDNA were shielded in spite of the neutral charge of TOPO I, which minimizes the electrostatic repulsion among the nanoparticles. The interaction of ssDNA-AuNPs conjugates with TOPO I was examined through UV-Vis spectroscopy. It was found that the absorbance of AuNPs conjugates was significantly decreased with an increase in TOPO I concentration with a slight red shift of λ_{max} from 570.3 to 575 nm. It was observed that the higher the TOPO I concentration, the longer was the λ_{max} shifting. However, interestingly, with other proteins like BSA as control, change in UV–Vis spectra was found to be negligible. Therefore, these ssDNA-AuNPs conjugates give specificity or selectivity toward TOPO I detection and determination of TOPO-ssDNA interaction.

20.2.2 OLIGONUCLEOTIDE DETECTION

Specific oligonucleotide sequences detection is of great importance in early medical diagnosis, disease detection, drug delivery, as compared to other usual methods. First of all, thiolated oligonucleotides functionalized AuNPs show a new era in DNA sensing. Considering the advantages of different electrostatic properties of double- and single-stranded oligonucleotides (dsDNA and ssDNA), and the electronic properties of gold nanoparticles, Rothberg's group synthesized citrate-capped AuNPs to detect oligonucleotides (Li et al., 2004). In this hybridization assay, they have used fluorescent rhodamine dye tagged ssDNA, dsDNA and incubated with citrate-capped AuNPs. The photoluminescence spectrum of ssDNA was found to be quenched while interacting with AuNPs, however, not in the case of dsDNA. Because of the electronic interactions with the gold nanoparticles, the photoluminescence of ssDNA is quenched with a change in absorption spectra. While checking the colloid stability in the presence of ssDNA, dsDNA and salt/buffer, it is found that when salt is introduced, the adsorption of ssDNA prevents the aggregation of gold nanoparticles and the color of the pink colloid solution remains unchanged. However, in the case of dsDNA, it does not have any effect on preventing aggregation, rather, ends up with aggregation associated with a color change from red to blue.

Considering similar concepts, Yang and co-workers have reported AuNPs-based chemiluminescence (CL) quenching sequence-specific DNA sensors (Xu et al., 2010). They have differentiated ssDNA and dsDNA by using a CL acridinium ester (AE) labeled with ssDNA and dsDNA, respectively. AuNPs were found to be quenched by efficient nonradiative energy transfer to the gold particle during the interaction with DNAs (Figure 20.1). In absence of target DNA or in the presence of noncomplementary DNA, the AE labeled ssDNA probe was adsorbed on the AuNPs surface and the CL quenching was found to be dependent upon the proximity between AE and AuNPs. However, in the presence of target DNA, the AE labeled ssDNA probe was not adsorbed on the AuNPs surface and so no CL quenching was found. Liu et al. reported a colorimetric adenosine biosensor which depends on the aptazyme-directed assembly of gold nanoparticles (Liu et al., 2004).

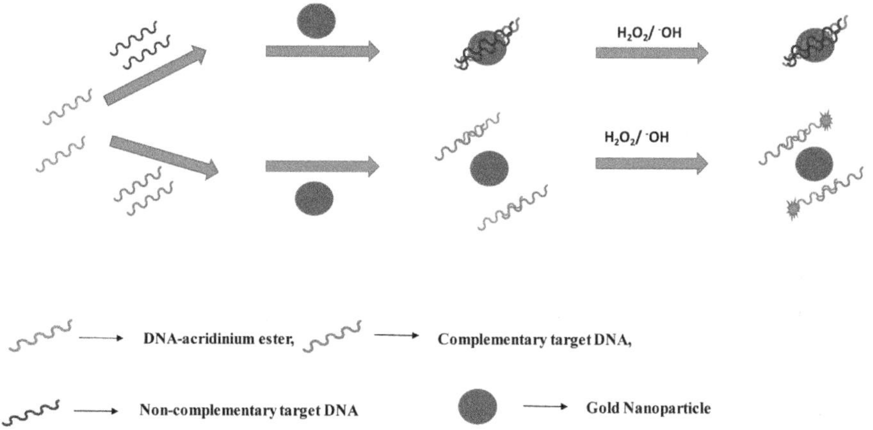

FIGURE 20.1 Schematic illustration for chemiluminescent DNA detection depends on the discrimination ability of AuNPs toward ssDNA/dsDNA, and the quenching ability of AuNPs toward AE CL.

20.2.3 PATHOGEN DETECTION

Peng et al. demonstrated that M13 phages modified citrate reduced AuNPs can be used for the detection of pathogenic bacterial cells (Peng et al., 2018). They have engineered phage M13 to display the receptor-binding protein from a phage which generally targets the desired bacteria. M13 phage was known to express a foreign receptor-binding protein fused to protein p

III, and the chimeric phage is thiolated through EDC chemistry. Then the thiolated chimeric phage is added to the media containing bacteria and they could be attached to the cells. Then the separated phage–cell complexes were incubated with AuNPs solutions. Therefore, the thiolation of phages causes aggregation of AuNPs which leads to color change from red to blue and shifting of SPR band, as we know thiol has a specific affinity toward gold nanoparticles. Hence, by monitoring the color change and shifting of band they have detected strains of *E. coli*, *Pseudomonas aeruginosa* and, *Vibrio cholera*, and *Xanthomonas campestris,* respectively.

20.2.4 GLUCOSE DETECTION

Colorimetric sensors attracted great attention due to their low price, effortlessness, and easy detection through bare eyes (Wei et al., 2008). This technique has been explored for the detection of glucose also. Diabetic patients are unable to self-regulate the blood glucose level which can cause serious health issues. Therefore, the development of a fast, less expensive sensor that can detect glucose levels easily is of great need indeed. Previous reports suggest that the interaction between glucose and amino acid residue like lysine, arginine forms glucose pane, a glycation end product. Many researchers have utilized this glucose pane formation to develop a colorimetric sensor for glucose detection (Oliver et al., 2009; Vlassaea et al., 1986).

Based on the aggregation and dissociation mechanism, Geddes and co-workers have reported a colorimetric sensor for glucose with the help of a large plasmonic band of high molecular weight dextran-coated gold nanoparticles. In the presence of Con A, gold nanoparticles are found to be aggregated due to the interparticle coupling of gold nanoparticles present in close proximity. Therefore, the addition of different concentrations of Con A results in red shift of the LSPR band to 650 nm with a color change from red to blue to purple. This result is further supported by the TEM image which shows that initially, particles are well separated, however, after the addition of Con A, aggregation of particles occurs. However, during incubation with glucose, there is a competition with dextran-coated colloids for Con A binding sites, which ends up with the dissociation of the Con A aggregated nanogold. This results in a decrease in the absorbance of the nanogold while monitored at an arbitrary near-red 650-nm wavelength (Aslan et al., 2004).

Further, citrate-capped gold nanoparticle surface was modified with rat tail collagen fibrils to detect glucose and heparin. As citrate-capped AuNPs are negatively charged and collagen fibrils are positively charged due to the abundance of lysine through collagen, they form a good complex at a 3:5 ratio (Unser et al., 2017). When these complex conjugates were exposed to glucose, it was found that aggregation of gnp-fibrils complexes occur with a large red shift from 517 nm to 635 nm due to some plasmonic coupling. This plasmonic coupling arises due to the formation of glucose pane linkage which closes the proximity of gold nanoparticles and collagen fibril. The change in color and the large plasmonic shift was further confirmed by TEM images. They also checked the activity of particle–collagen complex in mouse serum to see the effect toward cross-linking properties.

Other citrate-capped AuNPs and AgNPs were coupled to detect glucose through naked eyes (Gao et al., 2017). While both the nanoparticles were incubated with glucose in the presence of glucose oxidases (Go_x), the color of mixed nanoparticles disappears. As gold nanoparticles have oxidase mimic activity, it oxidizes glucose and produces H_2O_2. The newly developed H_2O_2 oxidizes AgNPs to Ag (+1) and the LSPR band of AgNPs decreased with color change. The decreased intensity of the LSPR band of AgNPs corresponding to the H_2O_2 concentration can be calculated through spectrophotometry. Further they have coated the gold nanoparticles with L-cysteine to increase the selectivity in the presence of other metal ions or biomolecules.

It was also found that glutathione-capped AuNPs in the presence of glucose and glucose oxidase can detect glucose through the change in fluorescence spectra. In the presence of metal coordination ligand tris(2-carboxyethyl) phosphine, the fluorescence of glutathione-capped AuNPs was found to be quenched, however, in the presence of glucose and Go_x, the fluorescence intensity increased. Glucose was observed to be oxidized in the presence of Go_x and produced glucose acids and H_2O_2. When the glucose concentration was maintained at 1mM, it reached the highest luminescence intensity of glutathione capped-AuNPs without tris(2-carboxyethyl)phosphine. The LOD was found to be 1.1 μM with good selectivity (Lai et al., 2017).

Inspired by the same concept, Radhakumari et al. synthesized thiol-capped gold nanoparticles and glucose oxidase enzyme was immobilized on it to detect glucose by naked eyes. A 100 μg/mL of glucose exhibits a visible color change of glucose oxidase-capped AuNPs from red to blue with a shift of LSPR peak from 535 nm to 569 nm. When they spiked glucose 50 μg/mL into urine sample, the absorption was shifted by 20 nm, demonstrating the feasibility to detect glucose in a real sample (Radhakumary et al., 2011).

20.2.5 HEPARIN DETECTION

Heparin is negatively charged, sulfated polysaccharide macromolecules, generally used in medicine as anticoagulant to prevent blood clots in arteries and veins. It is mainly used in surgery to reduce the risks of blood clots formation and in clinical procedures like cardiac surgery, kidney dialysis (Bromfielsd et al., 2013).

Li et al. and Cao et al. have established a heparin detection assay by using the cysteamine-capped gold nanoparticle with the mechanism of electrostatic interactions (Li et al., 2010; Cao et al., 2011; Gemene et al., 2010). Cysteamine-capped gold nanoparticles are positively charged as the amine of cycteamine is protonated at pH 3.6 and that off heparin is negatively charged. Therefore, this positively charged gold nanoparticle exhibits electrostatic interaction during incubation with negatively charged heparin. Thus, in the presence of different concentrations of heparin, the absorption of AuNPs at 520 nm gradually decreases, whereas, absorption at 670 nm gradually increases with a color change from red to blue. For the mechanistic aspects, negative charge citrate-capped gold nanoparticles have been incubated with heparin, which results in no color change or spectral change. The limit of detection of heparin by these cysteamine-capped gold nanoparticles was found to be 0.03 mg mL^{-1}.

On the basis of self-assembly of AuNPs on graphene oxide (GO), a new colorimetric method for ultrasensitive detection of heparin has been reported (Fu et al., 2012). Citrate-capped AuNPs were synthesized, which displayed a strong absorption band at 520 nm. In the presence of GO or GO and heparin, AuNPs exhibit no change in the SPR band. However, it is known that polycationic protamine induces the self-assembly of AuNPs on GO. Therefore, after the addition of protamine-GO, AuNPs exhibit a color change from red to blue and two SPR bands, at 520 nm and at 655 nm, respectively. However, heparin restricted AuNPs got absorbed onto the surface of GO due to strong interaction between protamine and heparin, which resulted in a color change from blue to red with an intense SPR peak at 520 nm. This phenomenon was further evidenced by TEM images and the limit of detection was found to be 3.0 ng mL^{-1}.

Based on the same electrostatic interaction concept, another positively charged biopolymer chitosan-capped gold nanoparticles have been reported as sensitive nano probe for the detection of negatively charged heparin as depicted in Figure 20.2 (Chen et al., 2013). The chitosan-capped gold nanoparticles exhibit a weak resonance light scattering intensity with a wine-red color in the presence of heparin, the resonance light scattering intensity increases with a color-shifting from red to purple to blue due to the electrostatic interactions between chitosan-stabilized AuNPs and polyanionic heparin. They also found promising results while treated with real human serum samples with a detection limit of 0.8 μM.

Similarly, Wen et al. reported a green synthesized polyethyleneimine-stabilized AuNPs for highly selective and sensitive colorimetric detection of heparin (Wen et al., 2013).

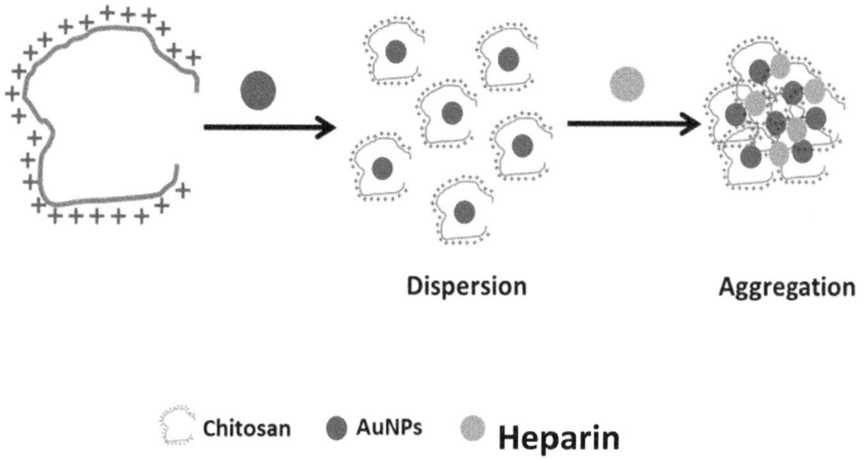

FIGURE 20.2 Schematic presentation of aggregation of chitosan-stabilized AuNPs induced by heparin in colorimetric responses.

Considering the aggregation followed by the dissociation mechanism, a novel colorimetric sensor using polymer nanoparticles (PNPs) and citrate-capped AuNPs has been developed. Citrate-capped gold nanoparticle was incubated with PNP, which results in the color change of the AuNPs from red to blue with an absorption peak shift from 520 nm to 675 nm due to the aggregation of AuNPs. However, after the interaction of positively charged PNP induced AuNPs with negatively charged heparin, it binds to the PNP and dissociates, which results in the drop of 675 nm peak with a color change from blue to pinkish red. They have shown a great selectivity of AuNPs in the presence of amino acids, BSA, HA, etc., with a detection limit of 2.5 nM (Qu et al., 2017).

Early studies already established the presence of heparin-binding sites on collagen using heparin-modified nanoparticles and collagen monomers (San Antonio et al., 1994). Therefore, using these interactions between heparin and collagen, they have detected heparin by using collagen-nanoparticles followed by the heparin functionalized gold nanoparticles. In the absence of heparin in the sample, all the heparin-binding site of heparin-coated

AuNPs are available for binding which results in plasmonic coupling with large LSPR peak shift. When heparin is present in the sample, it binds to the collagen-nanoparticle conjugates. Hence, the addition of heparin gold nanoparticles leads to no shift in SPR peak (Unser et al., 2017). TEM images also confirmed the lack of interaction when the collagen nanoparticles are saturated with heparin and indeed no binding was observed between the heparin-coated gold nanoparticles with collagen.

20.3 AGNPS AS BIOSENSORS

20.3.1 DETECTION OF BIOMOLECULES

AgNPs have also been reported as potential biosensors for the detection of biomolecules. AgNPs based sensor has been reported by Borase et al. using leaf extract of *Calotropisprocera* (Borase et al., 2015) for detection of cysteine. In this work, the synthesized silver nanoparticles a showed surface plasmon resonance peak at 421 nm. The color change from yellow to pink has been observed upon the accumulation of cysteine and thus the intensity of AgNPs has been decreased with time. And the color change occurs because of the aggregation of AgNPs caused by the strong interaction of the surface of AgNPs with the –SH group of cysteine and the color change can be seen by the naked eye. Further, selectivity and sensitivity of AgNPs were evaluated toward cysteine, in the presence of 10 times the high concentration of another amino acid. The results indicate that only cysteine exhibits changes in absorbance peak, while other amino acid does not, demonstrating the distinct character of AgNPs for the recognition of cysteine.

Melamine is a white solid powder which contains trimer of cyanamide mainly nitrogen (66%), which represents the higher protein amount in food/milk product. As a result of this, it is difficult to estimate the melamine. Therefore, estimation of melamine is essential to continue the limit in food products. Thus, the visual presence of melamine in milk was reported using silver nanoparticles synthesized by using leaves extract of *Jatrophagossypifolia* (Borase et al., 2015). The absorbance spectra of AgNPs was found to be at 419 nm and which further decreased gradually by accumulation of melamine and red shifting of the peak was observed with the appearance of new peak at 500 nm, indicating the aggregation of NPs. The limit of detection of melamine was found to be 252 ppb.

Dopamin is a biologically effective molecule as well as known to be the most vital neurotransmitters and several disorders like the neurological and mental result if there are changes in DA levels in a normal human. Therefore, a selective and sensitive technique is required to regulate dopamin concentration in living beings, which will be helpful for pathological research and disease diagnosis. To address this issue, Rostami et al. reported pyridinium-based task-specific ionic liquid (TSIL) functionalized hexagonal silver nanoparticles (TSIL-AgNPs) which can be used in the determination and detection of dopamin. Sensing mechanism here is based on the alteration of particle design and etching process which lead to a change of shape and size of silver nanoparticles.

Herein, a blue shift was observed in LSPR peak of TSIL-AgNPs, as a result of the mentioned mechanism in Figure 20.3. The technique is defined as a colorimetric detection of dopamin at room temperature. It has been confirmed that interaction between dopamin and TSIL-AgNPs is very well occurred in the acidic medium than in alkaline conditions. It is also reported that under the alkaline condition there is no variation in morphology and color changes in alkaline pH. Limit of detection of this technique was found to be 0.031 μM (Rostami et al., 2018).

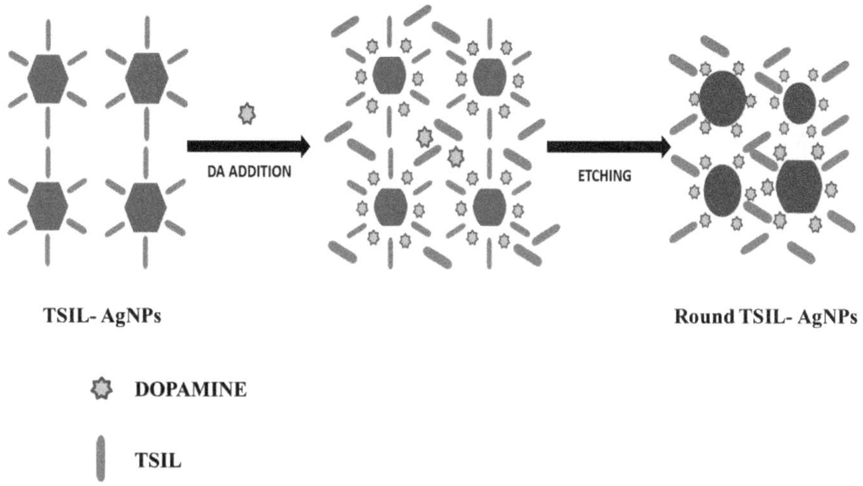

FIGURE 20.3 Schematic presentation of mechanism of TSIL functionalized hexagonal silver nanoparticles with dopamin.

Histidine is a necessary amino acid essential for human growth and tissue regeneration and it also plays an important role in the protection of nerve cells that help in the myelin sheaths maintenance and its metabolization leads to the formation of neurotransmitter histamine. Histidine is also important in the manufacturing of blood cells and also in the protection of tissues from being damaged by radiation and heavy metals (Cholula-Diaz et al., 2018). Li et al. reported a new technique that is efficient in the determination of histidine in neutral aqueous solution using cysteine modified silver nanoparticles (Cys-AgNPs). It is reported that interactions between amino acids and Cys-AgNPs are comparatively weak. However, Cys-AgNPs can attach to metal ions. Metal ions get attached to Cys-AgNPs and amino acid molecules by metal-ligand interactions. The existence of all the three components together leads to the selective determination by the binding affinity between the metal center and the amino acid. Here, the size of the analyte also matters. Thus, Hg^{2+} ions are responsible for strong interaction with Cys-AgNPs. These interactions led to change in color and absorption properties compared to other tested amino acids (Li et al., 2009).

The results suggest that other amino acids tested will not lead to the aggregation of Cys-AgNPs. It is reported that heavy metals like cadmium, lead, and mercury can successfully form the aggregation of the noble nanoparticle. The binding between metal ions and amino acids may be due to cooperative metal-ligand interaction (Shao et al., 2006). Addition of heavy metal ions to the above setup shows no color change when compared to the color of other amino acid solutions that changes from yellow to pink. This specifies that Cys-AgNPs have specificity toward the histidine in the presence of Hg^{2+} ions.

The mechanism shows that the interaction between the free Cys-AgNPs and amino acids is moderately weak; that is why, there is no color change after the addition of Cys-AgNPs into different amino acid solutions. Due to the inducing function of silver nanoparticles, they undergo aggregation in the presence of Hg^{2+} ions. In this reported work both the Cys-AgNPs and histidine molecules are probably bound to Hg^{2+} ions through cooperative metal-ligand interactions which further form a stable Nano complex. The resulting silver nanoparticle solution still remains monodisperse and there is no color change (Li et al., 2009).

Another biomolecule, ascorbic acid is antioxidant and plays a vital function in the human body. Cholula diaz et al. reported green synthesized AgNPs by using potato starch (Cholula-Diaz et al., 2018) for detection of ascorbic acid through the SERS technique. The detection depends upon the interaction of AgNPs with specific wavelengths of light, shape, size. The conductance electrons present on the metal surface experience coherent oscillation, resulting in absorption and scattering. The synthesized silver colloids were yellowish in color and which further when characterized by UV-Vis spectroscopy, showed a typical band at 410 nm. Upon incubation with ascorbic acid, the SPR band shifted to 500 nm due to the development of the cluster in the silver nanoparticle when starch is used in relatively higher concentrations during the synthesis (Zhang et al., 2015; McFarland et al., 2005). An intrinsic Raman peak at 1386 cm^{-1} is seen in starch-capped AgNPs which is due to twisting modes of the starch structure. An enhanced intensity is seen in SERS peak at 1386 cm^{-1} when the ascorbic acid concentration increases and leads to the reduction of the gap between dimers and trimers in silver nanoparticle clusters which are produced because of ascorbic acid in the colloid due to agglomeration of AgNPs. The limit of detection of ascorbic acid is 0.02 mM, which makes it an appropriate method for the detection of ascorbic acid in a biological specimen (Wei et al., 2007).

Recently, polyphenolic mediated nanoparticles (Schwartz et al., 2001) have been used for sensitive and selective detection of kanamycin. Kanamycin is an antibiotic, which inhibits the growth of bacteria and it is mainly used in the food industry and animal health care, and which further stops microbial infections (Singh et al., 2018). In addition, its excess amount causes allergic reactions and kidney toxicity. Singh et al. reported EGCG-mediated AgNPs for the detection of kanamycin in milk, which is based on SPR techniques. The colorimetric change from yellow (AgNPs solution) to reddish pink in the presence of kanamycin (Singh et al., 2018; Mandal et al., 2018) shows the presence of kanamycin in solutions. The LOD for kanamycin was reported to be 0.96 μM.

ACKNOWLEDGMENT

SM thank Indian Council of Medical Research (ICMR), Govt. of India, for providing Senior Research Fellowship (SRF) award, RKS thank Council of Scientific and Industrial Research (CSIR), Govt. of India, for providing Senior Research Fellowship (SRF) award (09/1035(0018)/2019-EMR-1).

KEYWORDS

- **nanotechnology**
- **biosensor**
- **bioimaging**
- **diagnostic**

REFERENCES

Aili, D.; Enander, K.; Baltzer, L; Liedberg, B. Assembly of polypeptide-functionalized gold nanoparticles through a heteroassociation-and folding-dependent bridging. *Nano Lett.* **2008**, 8(8), 2473–2478.

Aili, D.; Selegård, R.; Baltzer, L.; Enander, K.; Liedberg, B. Colorimetric protein sensing by controlled assembly of gold nanoparticles functionalized with synthetic receptors. *Small* **2009**, 5(21), 2445–2452.

Arques, S.; Ambrosi, P. Human serum albumin in the clinical syndrome of heart failure. *J. Card. Fail.* **2011**, 17(6), 451–458.

Aslan, K.; Lakowicz, J.R.; Geddes, C.D. Nanogold-plasmon-resonance-based glucose sensing. *Anal. Biochem.* **2004**, 330(1), 145–155.

Biju, V. Chemical modifications and bioconjugate reactions of nanomaterials for sensing, imaging, drug delivery and therapy. *Chem. Soc. Rev.* **2014**, 43(3), 744–764.

Borase, H.P.; Patil, C.D.; Salunkhe, R.B.; Suryawanshi, R.K.; Kim, B.S.; Bapat, V.A.; Patil, S.V. Bio-functionalized silver nanoparticles: a novel colorimetric probe for cysteine detection. *Appl. Biochem. Biotechnol.* **2015**, 175(7), 3479–3493.

Borase, H.P.; Patil, C.D.; Salunkhe, R.B.; Suryawanshi, R.K.; Salunke, B.K.; Patil, S.V. Biofunctionalized silver nanoparticles as a novel colorimetric probe for melamine detection in raw milk. *Biotechnol. Appl. Biochem.* **2015**, 62(5), 652–662.

Bromfield, S.M.; Wilde, E.; Smith, D.K. Heparin sensing and binding–taking supramolecular chemistry towards clinical applications. *Chem. Soc. Rev.* **2013**, 42(23), 9184–9195.

Cao, R.; Li, B. A simple and sensitive method for visual detection of heparin using positively-charged gold nanoparticles as colorimetric probes. *Chem. Commun.* **2011**, 47(10), 2865–2867.

Chen, C.K.; Huang, C.C.; Chang, H.T. Label-free colorimetric detection of picomolar thrombin in blood plasma using a gold nanoparticle-based assay. *Biosens. Bioelectron.* **2010**, 25(8), 1922–1927.

Chen, P.; Selegård, R.; Aili, D.; Liedberg, B. Peptide functionalized gold nanoparticles for colorimetric detection of matrilysin (MMP-7) activity. *Nanoscale* **2013**, *5*(19), 8973–8976.

Chen, Z.; Wang, Z.; Chen, X.; Xu, H.; Liu, J. Chitosan-capped gold nanoparticles for selective and colorimetric sensing of heparin. *J. Nanopart. Res.* **2013**, 15(9), 1930.

Cholula-Díaz, J.L.; Lomelí-Marroquín, D.; Pramanick, B.; Nieto-Argüello, A.; Cantú-Castillo, L.A.; Hwang, H. Synthesis of colloidal silver nanoparticle clusters and their application in ascorbic acid detection by SERS. *Colloids Surf. B.* **2018**, 163, 329–335.

Choudhary, R.; Patra, S.; Madhuri, R.; Sharma, P.K. Equipment-free, single-step, rapid,"on-site" kit for visual detection of lead ions in soil, water, bacteria, live cells, and solid fruits using fluorescent cube-shaped nitrogen-doped carbon dots. *ACS Sustain. Chem. Eng.* **2016**, 4(10), 5606–5617.

Clark Jr, L.C.; Lyons, C. Electrode systems for continuous monitoring in cardiovascular surgery. *Ann. N.Y. Acad. Sci.* **1962**, 102(1), 29–45.

Daniels, M.J.; Wang, Y.; Lee, M.; Venkitaraman, A.R. Abnormal cytokinesis in cells deficient in the breast cancer susceptibility protein BRCA2. *Science* **2004**, 306(5697), 876–879.

Doumas, B.T.; Peters Jr, T. Serum and urine albumin, a progress report on their measurement and clinical significance. *ClinicaChimica Acta* **1997**, 258(1), 3–20.

Fanali, G.; Di Masi, A.; Trezza, V.; Marino, M.; Fasano, M.; Ascenzi, P. Human serum albumin, from bench to bedside. *Mol. Aspects Med.* **2012**, 33(3), 209–290.

Fu, X.; Chen, L.; Li, J. Ultrasensitive colorimetric detection of heparin based on self-assembly of gold nanoparticles on graphene oxide. *Analyst.* **2012**, 137(16), 3653–3658.
Gao, Y.; Wu, Y.; Di, J. Colorimetric detection of glucose based on gold nanoparticles coupled with silver nanoparticles. *Spectrochim. Acta A Mol. Biomol. Spectrosc.* **2017**, 173, 207–212.
Gemene, K.L.; Meyerhoff, M.E. Reversible detection of heparin and other polyanions by pulsed chronopotentiometric polymer membrane electrode. *Anal. Chem.* **2010**, 82(5), 1612–1615.
Guarise, C.; Pasquato, L.; De Filippis, V.; Scrimin, P. Gold nanoparticles-based protease assay. *Proc. Natl. Acad. Sci.* **2006**, 103(11), 3978–3982.
Huang, Z.; Wang, H.; Yang, W. Gold nanoparticle-based facile detection of human serum albumin and its application as an INHIBIT logic gate. *ACS Appl. Mater. Interfaces* **2015**, 7(17), 8990–8998.
Ishii, T.; Otsuka, H.; Kataoka, K.; Nagasaki, Y. Preparation of functionally PEGylated gold nanoparticles with narrow distribution through autoreduction of auric cation by α-biotinyl-PEG-block-[poly (2-(N, N-dimethylamino) ethyl methacrylate)]. *Langmuir* **2004**, 20(3), 561–564.
Kim, H.Y.; Choi, I. Ultrafast colorimetric determination of predominant protein structure evolution with gold nanoplasmonic particles. *Nanoscale.* **2016**, 8(4), 1952–1959.
Kioi, M.; Yamamoto, K.; Higashi, S.; Koshikawa, N.; Fujita, K.; Miyazaki, K. Matrilysin (MMP-7) induces homotypic adhesion of human colon cancer cells and enhances their metastatic potential in nude mouse model. *Oncogene.* **2003**, 22(54), 8662.
Lai, X.; Luo, F.; Wang, Y.; Su, X.; Liu, J. Coordination-induced decomposition of luminescent gold nanoparticles, sensitive detection of H_2O_2 and glucose. *Anal. Bioanal. Chem.* **2017**, 409(6), 1635–1641.
Li, H.; Bian, Y. Selective colorimetric sensing of histidine in aqueous solutions using cysteine modified silver nanoparticles in the presence of Hg^{2+}. *Nanotechnology* **2009**, *20*(14), 145502.
Li, H.; Rothberg, L. Colorimetric detection of DNA sequences based on electrostatic interactions with unmodified gold nanoparticles. *Proc. Nat. Acad. Sci.* **2004**, 101(39), 14036–14039.
Li, Y.; Schluesener, H.J.; Xu, S. 2010. Gold nanoparticle-based biosensors. *Gold Bull.* **2004**, 43(1), 29–41.
Lin, M.; Pei, H.; Yang, F.; Fan, C.; Zuo, X. Applications of gold nanoparticles in the detection and identification of infectious diseases and biothreats. *Adv. Mater.* **2013**, 25(25), 3490–3496.
Liu, J.; Lu, Y. Adenosine-dependent assembly of aptazyme-functionalized gold nanoparticles and its application as a colorimetric biosensor. *Anal. Chem.* **2004**, 76(6), 1627–1632.
Lo, K.M.; Lai, C.Y.; Chan, H.M.; Ma, D.L.; Li, H.W. Monitoring of DNA–protein interaction with single gold nanoparticles by localized scattering plasmon resonance spectroscopy. *Methods.* **2013**, 64(3), 331–337.
Lu, M.; Zhu, H.; Bazuin, C.G.; Peng, W.; Masson, J.F. Polymer-templated gold nanoparticles on optical fibers for enhanced-sensitivity localized surface plasmon resonance biosensors. *ACS Sens.* **2019**.
Luukkaa, H.; Klemi, P.; Hirsimäki, P.; Vahlberg, T.; Kivisaari, A.; Kähäri, V.M.; Grénman, R. Matrix metalloproteinase (MMP)-7 in salivary gland cancer. *Acta Oncol.* **2010**, 49(1), 85–90.
Mandal, D.; Mishra, S.; Singh, R.K. Green synthesized nanoparticles as potential nanosensors. In *Environmental, Chemical and Medical Sensors.* 2018, pp. 137–164 Springer, Singapore.
McFarland, A.D.; Young, M.A.; Dieringer, J.A.; Van Duyne, R.P. Wavelength-scanned surface-enhanced Raman excitation spectroscopy. *J. Phys. Chem. B.* **2005**, 109(22), 11279–11285.

Narain, R.; Housni, A.; Gody, G.; Boullanger, P.; Charreyre, M.T.; Delair, T. Preparation of biotinylated glyconanoparticles via a photochemical process and study of their bioconjugation to streptavidin. *Langmuir.* **2007**, 23(26), 12835–12841.

Nie, Z.; Petukhova, A.; Kumacheva, E. Properties and emerging applications of self-assembled structures made from inorganic nanoparticles. *Nat. Nanotech.* **2010**, 5(1), p.15.

Oliver, N.S.; Toumazou, C.; Cass, A.E.G.; Johnston, D.G. Glucose sensors, a review of current and emerging technology. *Diabetic Med.* **2009**, 26(3), 197–210.

Peng, H.; Chen, I.A. Rapid colorimetric detection of bacterial species through the capture of gold nanoparticles by chimeric phages. *ACS Nano.* **2018**, 13(2), 1244–1252.

Qu, F.; Liu, Y.; Lao, H.; Wang, Y.; You, J. Colorimetric detection of heparin with high sensitivity based on the aggregation of gold nanoparticles induced by polymer nanoparticles. *New J. Chem.* 2017, 41(19), 10592–10597.

Radhakumary, C.; Sreenivasan, K. Naked eye detection of glucose in urine using glucose oxidase immobilized gold nanoparticles. *Anal. Chem.* **2011**, 83(7), 2829–2833.

Ray, P.C. Size and shape dependent second order nonlinear optical properties of nanomaterials and their application in biological and chemical sensing. *Chem. Rev.* **2010**, 110(9), 5332–5365.

Rex, M.; Hernandez, F.E.; Campiglia, A.D. Pushing the limits of mercury sensors with gold nanorods. *Anal. Chem.* **2006**, 78(2), 445–451.

Ross, J.S.; Fletcher, J.A. The HER2/neu oncogene in breast cancer, prognostic factor, predictive factor, and target for therapy. *Stem Cells.* **1998**, 16(6), 413–428.

Rostami, S.; Mehdinia, A.; Jabbari, A.; Kowsari, E.; Niroumand, R.; Booth, T.J. Colorimetric sensing of dopamin using hexagonal silver nanoparticles decorated by task-specific pyridinum based ionic liquid. *Sens. Actuators B.* **2018**, 271, 64–72.

San Antonio, J.D.; Lander, A.D.; Karnovsky, M.J.; Slayter, H.S. Mapping the heparin-binding sites on type I collagen monomers and fibrils. *J. Cell Bio.* **1994**, 125(5), 1179–1188.

Schofield, C.L.; Haines, A.H.; Field, R.A.; Russell, D.A. Silver and gold glyconanoparticles for colorimetric bioassays. *Langmuir.* **2006**, 22(15), 6707–6711.

Schwarz, S.; Chaslus-Dancla, E. Use of antimicrobials in veterinary medicine and mechanisms of resistance. *Vet. Res.* **2001**, 32(3–4), 201–225.

Shao, N.; Jin, J.Y.; Cheung, S.M.; Yang, R.H.; Chan, W.H.; Mo, T. A spiropyran based ensemble for visual recognition and quantification of cysteine and homocysteine at physiological levels. *Angew. Chem. Int. Ed.* **2006**, 45(30), 4944–4948.

Singh, R.K.; Mishra, S.; Jena, S.; Panigrahi, B.; Das, B.; Jayabalan, R.; Parhi, P.K.; Mandal, D. Rapid colorimetric sensing of gadolinium by EGCG-derived AgNPs, the development of a nanohybrid bioimaging probe. *Chem. Comm.* **2018**, 54(32), 3981–3984.

Singh, R.K.; Panigrahi, B.; Mishra, S.; Das, B.; Jayabalan, R.; Parhi, P.K.; Mandal, D. pH triggered green synthesized silver nanoparticles toward selective colorimetric detection of kanamycin and hazardous sulfide ions. *J. Mol. Liq.* **2018**, 269, 269–277.

Thanh, N.T.K.; Rosenzweig, Z. Development of an aggregation-based immunoassay for anti-protein A using gold nanoparticles. *Anal. Chem.* **2002**, 74(7), 1624–1628.

Unser, S.; Holcomb, S.; Cary, R.; Sagle, L. Collagen-gold nanoparticle conjugates for versatile biosensing. *Sensors* **2017**, 17(2), 378.

Vlassara, H.; Brownlee, M.; Cerami, A. Nonenzymatic glycosylation, role in the pathogenesis of diabetic complications. *Clin. Chem.* **1986**, 32(10), B37–41.

Wang, Z.; Lévy, R.; Fernig, D.G.; Brust, M. Kinase-catalyzed modification of gold nanoparticles, a new approach to colorimetric kinase activity screening. *J. Am. Chem. Soc.* **2006**, 128(7), 2214–2215.

Wang, Z.; Lévy, R.; Fernig, D.G.; Brust, M. The peptide route to multifunctional gold nanoparticles. *Biocon. Chem.* **2005**, 16(3), 497–500.

Wei, H.; Li, B.; Li, J.; Wang, E.; Dong, S. Simple and sensitive aptamer-based colorimetric sensing of protein using unmodified gold nanoparticle probes. *Chem. Comm.* **2007**, (36), 3735–3737.

Wei, H.; Wang, E. Fe_3O_4 magnetic nanoparticles as peroxidase mimetics and their applications in H_2O_2 and glucose detection. *Anal. Chem.* **2008**, 80(6), 2250–2254.

Wen, S.; Zheng, F.; Shen, M.; Shi, X. Synthesis of polyethyleneimine-stabilized gold nanoparticles for colorimetric sensing of heparin. *Coll. Surf. A Physicochem. Eng. Aspects.* **2013**, 419, 80–86.

Xu, Q.; Liu, J.; He, Z.; Yang, S. Superquenching acridinium ester chemiluminescence by gold nanoparticles for DNA detection. *Chem. Comm.* **2010**, 46(46), 8800–8802.

Yan, H.; Wu, J.; Dai, G.; Zhong, A.; Yang, J.; Liang, H.; Pan, F. Interaction between melamine and bovine serum albumin, Spectroscopic approach and density functional theory. *J. Mol. Struct.* **2010**, 967(1–3), 61–64.

You, C.C.; Miranda, O.R.; Gider, B.; Ghosh, P.S.; Kim, I.B.; Erdogan, B.; Krovi, S.A.; Bunz, U.H.; Rotello, V.M. Detection and identification of proteins using nanoparticle–fluorescent polymer 'chemical nose'sensors. *Nat. Nanotech.* **2007**, 2(5), 318.

Zhang, Y.; Walkenfort, B.; Yoon, J.H.; Schlücker, S.; Xie, W. Gold and silver nanoparticle monomers are non-SERS-active, a negative experimental study with silica-encapsulated Raman-reporter-coated metal colloids. *Phys. Chem.* **2015**, 17(33), 21120–21126.

Zhao, W.; Brook, M.A.; Li, Y. Design of gold nanoparticle based colorimetric biosensing assays. *ChemBioChem.* **2008**, 9(15), 2363–2371.

Zhao, D.; Wang, T.; Guo, X.; Kuhlmann, J.; Doepke, A.; Dong, Z.; Shanov, V.N.; Heineman, W.R. Monitoring biodegradation of magnesium implants with sensors. *J. Min., Metals. Mater. Soc.* **2016**, 68(4), 1204–1208.

CHAPTER 21

Role of Nanotechnology in Tissue Engineering and Regenerative Medicine

BIJAYANANDA PANIGRAHI[1], UDAY SURYAKANTA[1], SOURAV MISHRA[1], ROHIT KUMAR SINGH[1], and DINDYAL MANDAL[1,2*]

[1]*School of Biotechnology, Kalinga Institute of Industrial Technology Deemed to be University, Campus 11, Patia, Bhubaneswar, Odisha 751024, India*

[2]*School of Pharmacy, Chapman University, Irvine, CA, USA*

[*]*Corresponding author. E-mail: ddmandal@gmail.com*

ABSTRACT

Tissue engineering is a hybrid strategy consisting of engineering principles, methods, and biological sciences to create implantable tissues for replacement or to re-establish injured tissues and organs. To attain this goal, numerous porous scaffold biomaterials that can serve as support for cell adhesion, growth, and differentiation have been reported till date. However, current biomaterials have certain limitations including inefficient for cell growth, inappropriate structural integrity, and incapable to produce sufficient growth factors. Additionally, current biomaterials are also unable to control the cellular functions and various cellular properties. In order to mimic the proper tissue functionality, scaffolds should establish tissue-specific microenvironment to retain the cell behavior and functions. Nanoparticles show promising results as alternatives to overcome the current issues in tissue engineering (TE) and regenerative medicine due to their size-dependent properties. Nanotechnology-based biomaterials can imitate tissue-specific bio environments and control cellular properties such as biological, mechanical, and electrochemical properties. Despite the tremendous development in the arena of nanoparticles-based TE, it requires an in-depth sympathy of the cellular interaction with nano-based 3D scaffolds materials to implement in the clinic. This chapter describes an overview of potential applications of nanomaterials in TE and their challenges that require to be addressed for their implementation in the medical field.

21.1 INTRODUCTION

Tissue engineering follows a variety of methods from materials engineering, cell biology, biomaterials science, and biotechnology to create artificial constructs of various biomaterial scaffolds for regeneration of new tissue in place of damaged tissue (Williams, 2004). The basic importance of tissue engineering (TE) is to regenerate, sustain or develop tissue and organs function with good biocompatibility and functionality along with very less immune rejection. There are three important constituents of TE including cell, scaffold as well as a growth factor (Ikada, 2006). Scaffold design and fabrication are the key factors in the arena of biomaterial research for TE and regenerative medicine (Langer, 1993). For the importance of TE and repair, scaffolds are considered as three-dimensional solid biomaterials to perform the following functions: (1) cell–nanobiomaterial interaction, cell adhesion, and ECM deposition, (2) providing enough nutrients and growth regulatory factors which will help in cell growth, survival, and differentiation, (3) controlled biodegradable rate which will be helpful in tissue regeneration under the experimental condition, and (4) low inflammation or toxicity levels in in vivo system. Optimal characteristics of biopolymers such as strength, the rate of degradation, porose nature, microstructure, its structure, its form and its dimensions, are more willingly and repeatedly maintained in polymeric scaffolds (Langer, 2004).

21.2 BIODEGRADABLE POLYMERIC NANOMATERIALS FOR TISSUE ENGINEERING

Biomaterials in this arena of TE were very important as synthetic frameworks designated as scaffolds, matrices, or constructs. The purpose of the biomaterials synthesis for biomedical approaches is that they could be implemented to heal and repair the function of damaged part in the human body and also enhance the rate of function of selected tissues. Biomaterials have been utilized as implants in orthopedics, dentistry, cardiology, and also used as medical devices like pacemakers, biosensors (Ramakrishna, 2001; Vert, 2005).

21.2.1 NATURAL AND SYNTHETIC POLYMERS AS SCAFFOLDS IN TISSUE ENGINEERING

Polymers either natural or synthetic have been broadly introduced as biomaterials to invent the medical device and scaffolds (Piskin, 1995; Ji, 2006). Criteria of selecting a suitable material for biomedical implementations are based on molecular weight, material chemistry, solubility, shape and structure, hydrophilicity/hydrophobicity, lubricity, surface energy, water absorption, degradation, and erosion mechanism. Due to their special properties like high surface-to-volume ratio, extremely porous with very less pore size, biodegradability, and mechanical strength polymeric scaffolds are the center of attraction in this field. They show various advantages biocompatibility, the flexibility of biochemical properties, which are important to implement in TE as well as organ transplantation. Researchers have tried to develop skin and cartilage (Eaglstein and Falanga, 1997), bone and cartilage (Dhandayuthapani et al., 2011), liver (Mayer, 2000), heart valves and arteries (Dhandayuthapani et al., 2011), bladder (Oberpenning et al., 1999), pancreas (Tziampazis and Sambanis, 1995), nerves (Mohammad et al., 2000), corneas (Germain et al., 2016), and many other soft tissues (Dhandayuthapani et al., 2011) using polymeric scaffolds.

Polymeric materials can be applied as scaffolds. It could be natural or synthetic, degradable or nondegradable (Ramakrishna et al., 2001). Characteristics of the polymer are dependent on composition, structure, and biological property. Natural polymer, biodegradable/nonbiodegradable synthetic polymers are commonly implemented as biomaterials for TE applications.

Natural polymers were considered to be used as the first biodegradable polymers for clinical polymers (Nair et al., 2007). As natural polymers have good biological properties, they help in better communication with the cells allowing them to increase cell activity in the physiological system. Natural polymers could be differentiated into following: proteins like silk, fibrinogen, keratin, collagen, elastin, gelatin, actin, and myosin, and polysaccharides like cellulose, dextran, amylose, chitin, and glycosaminoglycan and polynucleotides such as DNA and RNA (Ratner et al., 2004)

Synthetic polymers are other biomaterials that help in restoring of structural and functional integration of injured tissues. Some properties like porosity, degradation time periods, and mechanical properties give uniqueness to synthetic polymers in the medical field for exact applications. Synthetic scaffolds are also very cheap than biological scaffolds; they are also created in uniform quantities and have longer storage time. Synthetic polymers are the biggest group in biodegradable polymers and also can be formed in a controlled manner. Synthetic polymers exhibit mechano physical properties like tensile strength, biodegradation and elastic modulus (Gunatillake et al., 2006). Synthetic copolymers including polylactic acid (PLA), poly(glycolic acid) (PGA), and poly(D,L-lactide-*co*-glycolide) (PLGA) are mostly utilized for TE applications (Ma, 2004).

Bioactive ceramics including HAP, TCP, and some silicate compound and phosphate glasses (bioactive glasses) and glass–ceramics materials like apatite–wollastonite are compatible with body fluids and cellular activity (Hench, 1991). However, biocompatibility and biodegradability are the factors that restrict them in clinical implementation. These issues could be resolved by hybridizing synthetic and natural polymers to make composite materials that will enhance scaffold properties. The new material will be degraded in a controlled way and also, exhibit better biocompatibility in TE (Cascone et al., 2001). Furthermore, better mechanical and biological performance in hard tissue can be accomplished by combining degradable polymers and inorganic bioactive particles together (Roether et al., 2002).

21.2.2 APPLICATIONS

The biodegradable polymeric scaffolds are generally utilized in the following applications including Nerve, bone, muscle, tendon, and ligament regeneration.

21.2.2.1 CHITIN NANOFIBRIL APPLICABLE AS MEDICAL DRESSING

The stratum corneum, the outmost layer of keratinized cells, called corneocytes, also a precise skin defensive barrier which is the combination of intracellular lipid and crystalline gel structure, plays the role of mortar between the corneocytes. It is extremely dynamic in lipid enzymatic synthesis and has the capability to acclimate to the environment (Elias and Menon, 1991). Defects are formed by harshly injured skin, which is caused by huge burns or constant wounds. These defects alter the synthesis of the extracellular matrix (ECM) and therefore obstruct the skin's ability to respire, preserve or drive out water, and defend it against the harmful pathogen, oxidants, and toxins. Therefore, the main function of stratum corneum is to protect from external damage. Skin represents itself as the first layer of defense and also as a boundary wall of the body's organs. When the skin is injured beyond a certain limit of the total body area, death may result from such damage. Therefore, it is very necessary to immediately cover with dressing which could save from losing the integrity of tissue, homeostasis level, and prevents from toxic materials along with pathogens. The four main objectives that should be cared of a burnt wound are the following: (1) avoid infection, (2) controlling the moist environment around the wound, (3) protect wounds from external aggressions, and (4) reduction of scar formation (Elias and Menon, 1991).

Hence, here medical dressing plays a vital role in creating a barrier to environmental irritants, impede microbial growth, also maintaining a constantly moist environment, and allows in exchange of gaseous and nutrient ingredients. It should be kept in mind that the dressing material should not adhere to the wound, rather it should allow the growth of cells and must be easily removed. Simple dressing material is not capable to address the above-mentioned difficulties. For overcoming such obstacles, biocompatible, nonallergenic, and nontoxic biomaterials are specifically used for nonwoven tissue development. Alongside they also stimulate wound healing as they can modify ECM synthesis and control bacterial growth. In addition, the application of appropriate fibers is capable to produce nonwoven tissues that are free from binder and chemicals, and also free from active components for tissue repair and tissue regeneration, which are important in the medical sector. For better outcomes in tissue regeneration, natural polysaccharides like chitin and chitosan may be utilized for the synthesis and deposition of ECM in a tissue-specific manner (Morganti et al., 2016).

Chitin is known among the popular biopolymers as they are mainly found in shellfish and insects exoskeletons, and fungus. Chitin nanomaterials have great potential in making scaffolds, drug delivery, dressings, and TE. Morganti et al. created a method for the construction of chitin nanofibrils (CN; Scheme 21.1). When chitin undergoes electrospun, it becomes organized as a porous ECM-like structure that allows cells to be seeded and form biocompatible materials for tissue regeneration (Yang and Young, 2008; Di Martino et al., 2005). Based on the above-mentioned conditions, a combination of CN with hyaluronic acid (HA) was fabricated by the gelation method and electrospinning technology (Figure 21.1; Dvir et al., 2011). The natural chitin nanomaterials based nonwoven tissues have great advantage over other materials in terms of size, composition, porosity of synthesized meshes. Factors that matter the most to fulfill the criteria for biomedical application are the construction of the ECM pore size and density (Morganti et al., 2016).

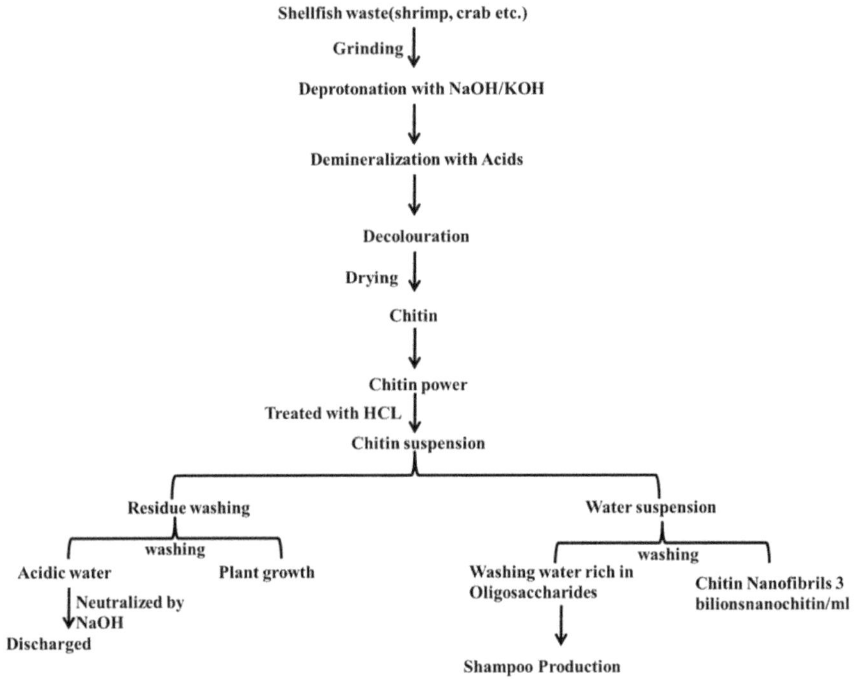

SCHEME 21.1 Chitin nanofibril production cycle.

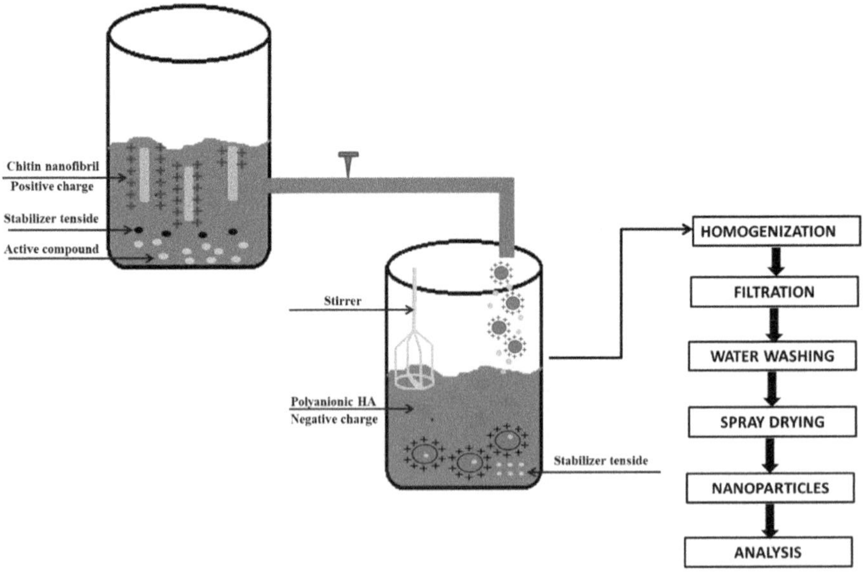

FIGURE 21.1 Nonwoven tissues synthesis by gelation method and electrospinning technology.

21.2.2.2 CHITIN NANOFIBRIL-HYALURONIC ACID NANOPARTICLE (NP)

Usage of CN-HA and their analogues are popular as they are natural polymer derived from fisheries, CN, HA, as both the polymers are completely biodegradable, biocompatible, environment friendly, and compostable. HA present in our body is an important constituent of the ECM. Chitin has the same strength of character and also an important element in human cartilage essential for bone articulation. When the formation of chitin is reduced along with aging, it leads to osteoporosis. CN is nontoxic and safe for both human and the environment when recovered as composites from micro-dimensions or as single component of nanomedicine. As the environment is rich with chitinase enzymes, they are efficient for metabolizing chitin-derived compounds; human consists of 18 different classes of the same enzyme, named chitotriose. They are required to catabolize reaction for chitin/chitosan into components such as glucosamine and acetylglucosamine.

21.2.2.3 FABRICATION OF IN VITRO SCAFFOLDS USING INTEGRATED CIRCUIT COMPATIBLE MICROFABRICATION

The scientists for past many years have resulted in an improved requirement for nanomaterial-based scaffolds for numerous uses. Techniques such as reactive-ion etching, photolithography, chemically based surface modification, and metal NP-assisted etching have been already combined into microfabrication to design scaffolds. These methods allow appropriate planning and produce artificial scaffolds having small-scale characteristics to probe cell-substrate communications (Morganti et al., 2016).

Cell-to-cell communications are very important in the physiological condition for the improvement of organs and tissues in vivo. Knowing the interaction between materials and mammalian cells is meaningful and interesting topic among researchers in different research fields like TE, regenerative medicine, and biosensors. Enough information has been reported in biology-related research field comprising of cell viability, mechanotaxis, and stem cell differentiation by the changes in mammalian cell–material surface interactions (Vermesh et al., 2011; Sato et al., 2004). Cell–material surface interactions are mainly decided by the interaction between adhesive molecules and the cell–material interface. Cell functions such as division, proliferation, differentiation, protein structure distribution, and migration of cells are affected by adhesive molecules (Stevens and

George et al, 2005; Bellin et al., 2009; Jiang et al., 2005). Different molecular mechanisms affecting the way cells differentiate and respond to their surrounding environment have been discovered, along with ligand–integrin interactions (Chen et al., 1997; Pesen et al., 2009), surface hydrophobicity or hydrophilicity (Romanova et al., 2006; Arima et al., 2007), and topography (Webb et al., 1998; Low et al., 2009). As biomedical engineering has a greater significance on cell attachment, the developments of biomimetic biomaterials to enable cell attachment and tolerate desirable architecture are currently great areas of interest for researchers (Choi et al., 2007; Barrera et al., 1993). Substrate-related changes can lead to enhance actin polymerization and also affect focal adhesion complexes (Singhvi et al., 1994).

Several methods are used to produce nanomaterial-based scaffolds for in vitro experiments. Study discovering cell–material surface interactions at the nanoscale can be categorized into two groups, the scrutiny of cellular responses to adhesion molecules on several substrates (chemically based method; Firkowska et al., 2006, Di Mundo et al., 2011), and surface stiffness (Kim et al., 2006; Tanaka et al., 2018) and topographic features (physically based method; Chou et al., 2009; Prodanov et al., 2010). In chemical methods, several chemicals or protein micro patterns are capable of regulating cellular function efficiently which contains cell adhesion and neurite outgrowth (Keselowsky et al., 2007; Poudel, et al., 2013). Physical methods are used to design micro-/nano-topographically reformed surfaces which can impact cytoskeletal transformation, cellular morphology, and cellular spreading (Vogt et al., 2003; Limongi et al., 2015).

In the emerging area of biomedical engineering, it is challenging to define how to arrange cells with a preferred pattern upon an artificial scaffold. These preparations of cells are important for understanding the communication between cell and biomaterials; however, equally it is necessary as a vector for imitating in vivo conditions. For example, natural neural cell cultures are performed for a regular arrangement of neural cells. This is a hard condition for researchers to imitate the regular arrangement of neural cells aggregate in nonuniform groups. Therefore, the use of various microfabrication techniques can help scientists to control neural cell adhesion and growth (Jungbauer et al., 2004). Current successful results in the arena of microfabrication have permitted cellular patterning in chosen regions. The traditional microfabrication procedures for producing micro patterns involve nanoimprint lithography (Merz et al., 2005), micro contact printing (Hoff et al., 2004), and microfluidic-based processes (Kim et al., 1995).

21.3 MAGNETOACTIVE ELECTROSPUN NANOFIBERS IN TISSUE ENGINEERING

Electrospinning was first demonstrated by L. Rayleigh and later it was patented by J.F. Cooley and W.J. Morton (Folch et al., 1998). A formal patent was made on the synthesis technique that permitted the spinning of synthetic polymers by the support of electric charges (Zeleny, 1917; Anton, 1934) for the manufacture of small-sized fibers (Figure 21.2). In 1969, D.G. Taylor reported on the jet formation method by inspecting the activities of the polymer solution droplet that is present at the edge of a capillary in the existence of an electric field. During his investigation, he found the characteristic pattern that represents an elongated conical fluid structure which is called as Taylor cone (Hong et al., 2011). During 90s, Reneker and coworkers' works focused on the usage of electrospinning toward the production of one-dimensional (1D) polymer nanostructures that have been a great area of interest for many researchers (Taylor, 1969; Doshi and Reneker, 1995; Reneker and Chun, 1996).

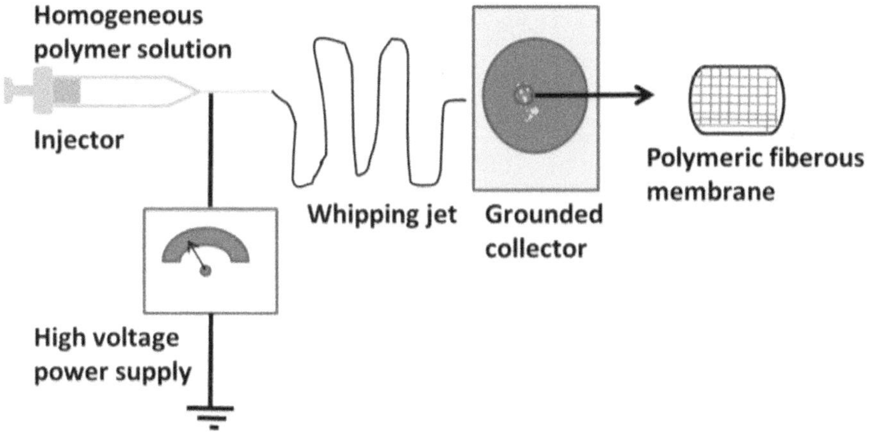

FIGURE 21.2 Schematic presentation of a basic electrospinning setup.

Many new techniques are also reported for assembly of 1D-nanomaterials, namely, drawing (Subbiah et al., 2005), self-assembly (Ondarcuhu et al., 1998; Panigrahi et al., 2018), melt-blow (Smith and Ma, 2004), phase-separation (Ellison et al., 2007), and template synthesis (Vesa et al., 1995). The method developed by Teo et al. may be utilized for generating nonstop production of natural as well as synthetic polymer nanofibers (Teo et al.,

2006; Huang et al., 2003; Deshmukh et al., 2019) with the range of diameters starting from micrometers to nanometers. Electrospinning technique was further used to develop nanocomposites containing polymer and ceramics (Wu et al., 2012; Li et al., 2003).

21.4 ELECTROSPINNING TECHNOLOGY IN TISSUE ENGINEERING

Fibrous nanomaterials are being used in numerous biological applications including biosensing (Ren et al., 2006, Li et al., 2006), drug delivery (Brazel et al., 2009; Jeyaraj et al., 2016), bioseparation (Munaweera et al., 2013), and TE (Walmsley et al., 2015; Ha et al., 2013; Supaphol et al., 2011). It is reported that electrospinning is the most extensive and multiuse nanofiber fabrication technique executed for the production of nanofiber materials which could be synthetically or naturally derived nanofibers (Zhang et al., 2015). Materials produced by this method show a wide-ranging of distinctive features and properties, like very long length, very small diameters resulting in high surface-to-volume ratios, highly porous structures, lightweight properties, and low cost (Deshmukh et al., 2019; Bouasla et al., 2010; Yu et al., 2009; Ahmed et al., 2015). These properties make electrospun polymer-based fibrous materials a preferable choice for various biomedical applications.

Electrospun nanofibers were derived from biopolymers including collagen (Matthew et al., 2010; Rho et al., 2006), alginate (Ma et al., 2012; Chang et al., 2012), HA (Vasita et al., 2006), chitosan (Venugopal et al., 2008), and starch (Kong et al., 2014), and synthetic polymers including polyurethanes, polymethacrylates, and aliphatic polyesters like poly(lactic acid) (PLLA) (Mo et al., 2004; Yoshimoto et al., 2003). Polycaprolactone (PCL) (Xu et al., 2004; Porter et al., 2009) and poly(vinylpyrrolidone), PGA (Fernandes et al., 2014], poly(acrylonitrile), poly(ethylene oxide) (Schiffman et al., 2008), poly(vinyl alcohol) (PVA), poly(ethylene terephthalate) (Abdal-hay et al., 2013) and their combinations with other materials are reported for having a great potential for scaffold formation for musculoskeletal and connective tissue (Jiang et al., 2015), vascular tissue, skin tissue, and neural tissue (Vasita et al., 2006; Nair et al., 2005).

21.5 MAGNETIC NPs IN CELL-BASED THERAPIES

The practice of magnetic nanoparticles (MNP) for different nano medicine applications, which can be subdivided into four themes. First, usage of NPs for magnetic separation and guidance of small molecules and drugs by using NP-bound antibodies second, in imaging techniques by using local magnetic field formed by NPs; third, NMP-directed transfer of therapeutic cells to several organs in the body by applying magnetic fields; and fourth, magnetic thermotherapy where magnetic fluids are used to generate heat for the environment by heat dissipation triggered by an alternating current magnetic field (Shapiro et al., 2014; Šafařík et al., 2011).

In cell therapy MNPs are implemented for the purpose of labeling cells, tracking their path throughout the body, and also improving the targeting of the cell to a specific site (Vaněček et al., 2012). Cell labeling is done to particles earlier to loading into an organism and then bringing the organism in contact with magnetic fields. The human body already holds relatively little magnetic material (due to the presence of iron). So, magnetic fields have less effect on the human body. Some of the recently reported works show that cell magnetic labeling relatively is safe for use inside of the human body (Riegler et al., 2011).

21.5.1 TYPES OF NPs

MNPs are classified into the following: paramagnetic, superparamagnetic, and ferromagnetic. Their magnetic properties are dependent on the magnetic core, amount of manganese, gadolinium, or common iron oxide (Gregory-Evans et al., 2013). Superparamagnetic iron oxide nanoparticles (SPION) are mostly used as MNPs among other iron oxide NPs. The core of SPIONs consists of magnetite (Fe_3O_4) or maghemite (γ-Fe_2O_3). For preventing the formation of agglomeration in colloidal suspension and to improve biocompatibility SPIONs surface are modified with compatible coating. Till date, many polymers including synthetic and natural polymer coatings are reported, such as carboxymethylated dextran, polyethylene glycol, alginate, starch, and poly(D,L-lactide-*co*-glycolide) (PLGA). The surface of SPIONs is altered by using small molecules like amino acids, citrate, hydroxamate, and dimercaptosuccinic acid as coatings (Sharifi et al., 2015).

12.5.2 THERAPEUTIC CELL DELIVERY

Major challenges in therapeutic cell delivery are emerging new ways of delivering cells to target tissue. Parameters that affect the application of cellular transplantation strategies in regenerative medicine are the number of transplanted cells, the therapeutic site of transplantation, and the route of administration. Significant therapeutic efficiency can be obtained by fewer invasive cell transplantation, adequate cell retention, and engraftment in the tissue of interest. MNPs producing magnetic force help in delivering various therapeutic cells to specific organs. Some of the reported studies are mentioned below, where MNPs help in transporting therapeutic cells to diseased tissue: eye disorders (Boughman et al., 1980), spinal cord injury (Sasaki et al., 2011), cancer (Gerbin et al., 2015), heart diseases (Ishii et al., 2011), and respiratory disease (Gonzalez-Molina et al., 2012).

21.6 M13 BACTERIOPHAGE NANOMATERIALS FOR REGENERATIVE MEDICINE BACTERIOPHAGES AS VIRAL NANOMATERIALS

M13 is a bacteriophage, which is a filamentous rod-shaped nanofiber having a diameter of 6.5 nm. This bacteriophage consists of the phage capsid which is a protein molecule surrounded by the phage genetic materials and can be investigated as a nanocarrier which has the ability to deliver the phage genetic materials into the cells. These rod like naturally occurring viruses have unlimited impending to develop various functional nanocarrier and nanostructured materials. These virions also have a great capability of undergoing self-assembly into the nanostructure. By self-assembly, the major structural information remains in phage capsid proteins itself and does not need the participation of other proteins (Yang et al., 2013). The advantage of using bacteriophage for nanotechnology purposes includes the great genetic flexibility that allows variability of modifications upon the membrane surface of their cell. These modifications benefit for building nanomaterials with the potential variety in various grounds of biotechnology and nanotechnology.

Genetic manipulation of phage protein NPs is done by phage display and shows the possibility of having control on the phage particles in biomedical research. George P. Smith introduced a phage display and further used this molecular method for functional expression of external molecules on the phage membrane (Smith et al., 1985). Herein, the foreign oligo nucleotide sequences are cloned in a particular area inside the phage modified protein genes. The hybrid fusion protein, consisting of a foreign peptide, interacts with external molecules.

21.6.1 NANOFILAMENTOUS VIRUSES AS TISSUE ENGINEERING SCAFFOLDS

New nanobiomimetic scaffolds for TE are developed through the application of nanofilamentous virus NPs in neural regeneration. Major aim of neural tissue improvement is to fabricate bioactive constructs with the skill to offer an ideal material for the healing and regeneration of wounded tissues of the network of the nervous system of the body. This process is to restore the complex functions of natural neural tissues (Zhu et al., 2014).

Current scientists showed that phage NPs, mostly M13 are focus points due to their definite synthesis and design of nanomaterial scaffolds for neural TE applications. M13 nanofibers are considered to be the best candidates due to their specific features which will be useful in making building elements for TE applications. M13 virions are capable of undergoing genetic engineering and also able to express the peptide-based functional information on their membrane, this results in the high density of signaling peptides along the exterior of the nanofibrous phage. M13 has the capacity to self-assemble into directionally well-arranged nano filamentous structures which are meant for modification of physiochemical cues. This supports in accomplishing the development of the nerve cells. Long rod-shaped and monodispersed M13 nanofibers here help in synthesis of different self-assembled 2D and 3D structures at the nanoscale (Farr et al., 2014).

21.6.2 NANOMATERIALS FOR NEURAL TISSUE REGENERATION

Bacteriophage nanomaterial is considered as one of the greatest substantial predictions for the neural generation application offered by phage peptide libraries. These libraries give contact to a wide-ranging of peptide ligands that may subsidize the improvement of new therapeutic approaches for the field of neural regeneration. Proper selection of cell-related phage peptide libraries would be useful toward the better choice of unique peptides that can specifically bind to the preferred cells. For specific delivery of therapeutic materials to damaged cells of the nervous system, nerve cell-binding peptides are considered to be the best candidates. An efficient peptide was reported for binding to the motor neurons and dorsal root ganglion cells. This peptide is reported to be an efficient candidate for the handling of motor neuron diseases acting as cellular targeting of neurotherapeutic proteins and gene/drug delivery vectors (Liu et al., 2005).

One of the furthermost notable applications of bacteriophage and phage display technology in neural regeneration is the screening of phage display libraries on neural stem cells in order to know peptide ligands for the precise binding capacity to the stem cells of the central nervous system. Neural stem cells are self-renewing, undifferentiated, and multi-potent progenitor cells that are situated in the subventricular zone and the hippocampal sub granular zone in the adult brain of mammals (Conti et al., 2006).

Recent research works revealed that bacteriophages have shown more prospects in the area of neural regeneration. In the past several years, the usage of bacteriophages in specific phage technology of neural regeneration has increased exponentially. Phage display has produced promising applications in growing necessary scaffolds for neural TE. The brilliant biological safety, nontoxic, and nonimmunogenicity of phage is a sensible attribute for the transformation of phage-based nanomaterial products into the clinic (Bakhshinejad et al., 2014).

21.7 POLYMER AND NP-DERIVED ARTIFICIAL VASCULAR GRAFTS

Vascular disorder, or atherosclerosis, is a prevalent disease worldwide and the patient number of this disease is regularly increasing. Its progress is a lengthy procedure that includes solidification of the artery wall and reduction of its lumen causing the formation of atherosclerotic plaques. Reduction leads to inadequate blood movement into the tissues supplied by a vessel. The main reason for atherosclerosis is a high level of cholesterol and triglycerides in the blood. The vascular graft has to fit well into the arterial system and should completely take over the role of the original arteries. The vascular graft must be biocompatible and must have permanent mechanical support for the formation of a new blood vessel. However, biosecurity is the central concern before the introduction of grafts. Guidelines of biosecurity are strictly defined, to evaluate the protection major of the materials before implementation into the patient's tissues and blood. Till date, most of the experiments are performed in vitro, including cell culture, and in vivo including animal models.

Vascular grafts are divided into three types:

1. Biological vascular grafts
2. Artificial vascular grafts
3. Special types

21.7.1 BIOLOGICAL VASCULAR GRAFTS

Biological grafts are suitable for low risk of infection, immune response and it is biocompatible. However, the biological graft is imitated on the basis of the availability and the dimensions (length, lumen). Biological grafts are divided into the following categories:

- Vascular autograft–patients graft is acquired from one healthy original and fixed into the injured site.
- Vascular allograft–graft is removed from a healthy organism and rooted in another injured organism of the identical species.
- Vascular xenograft–graft is removed from a healthy organism and rooted in another injured organism of an unlike species.

21.7.2 ARTIFICIAL VASCULAR GRAFTS

Currently, artificial vascular grafts can be divided into three different types.

1. **Knitted grafts:** Knitting synthetic fibers are used for the preparation of knitted grafts. Basically, these are polyester fibers and the walls of these grafts are porous in nature. Therefore, blood loss can be minimized.
2. **Woven grafts:** These grafts have low porose walls.
3. **Cast artificial grafts:** Poly(tetrafluoroethylene) are called casted artificial grafts. They are nonporous and hydrophobic walls are appropriate for the lesser flow rates.

21.7.3 SURFACE MODIFICATION OF VASCULAR GRAFTS

In the arena of modern medicine, vascular grafts are a vital element and find their relevance in clinical practice. They are designed in such a nanostructure network from fabric which resembles the ECM in natural blood vessels.

21.7.3.1 MODIFICATION BY NONPOLYMERIC OR POLYMERIC SUBSTANCES

Metal NPs within the range of 1–200 nm have their own properties whereas modification of the surface of NPs may change the property which will be beneficial for TE. Therefore, scientists modified the surface of different NPs with different compounds, such as biopolymers, peptides, chitosan. Surface modified NPs showed altered physical and chemical nature compared with naked metal-based NPs. For orthopedic implants, selenium NPs were found to be a potential material (Wang et al., 2012). Similarly, an important polymeric material such as HA is utilized in the design of complexes for tissue implantation application. It is made up of linear polysaccharide and disaccharide units that are D-glucuronic acid and N-acetylglucosamine (Hrabárová et al., 2012). It exists in ECM, muscle, and nerve tissue (Kenne et al., 2013). They are biocompatible and biodegradable in nature (Bonferoni et al., 2007). HA is used in eye surgery, TE, and cosmetics because of its special organization (Goa et al., 1994).

Traditionally, surgeon favors artificial vascular grafts knitted or woven, which are designed with a compact layer of either collagen or complex of collagen with treated substances. This can enhance the antibacterial property.

21.8 TISSUE ENGINEERING RECREATE BONE FEATURES FROM THE MACROSCALE TO NANOSTRUCTURES

Skeleton plays various vital activities in the human body including formation of structure, protection from injury, providing an internal framework. Besides these roles, the human skeleton is also crucial for various metabolic functions, which include storage of minerals, balance of calcium homeostasis, regulation of endocrine, acid-base balance, and hematopoiesis processes (blood cell formation) which takes place in bone marrow region (Oldknow, 2015).

21.8.1 BONE TISSUE ENGINEERING

Bone injury is a major problem in day-to-day life, bone repairing is a complex process which includes a number of physiological events such as healing by endochondral and intramembranous ossification, inflammation, and remodeling. Bone healing needs to involve a multifarious microenvironment, composed of a network and communication of cells, growth/differentiating factors, mineral, elements, hormone, chemokines, signaling molecules, cytokines, and ECM macro-, micro-, and nano-components. Tissue engineering is intended for developing tissue substitutes which are capable to maintain, restore, and increase tissue functions in vivo. To influence this objective, two main key issues should be measured: (1) the scaffold, which will serve as physical support, and as the provider of molecular factors which will be able to induce cell proliferation. (2) The cells need to penetrate into the scaffold and produce the neo-tissue. A bioreactor which will be able to support neo-tissue formation and growth for the in vitro progress of engineered tissue. The aim of designing the scaffolds in TE for regenerative medicine is to make them sterilizable and biocompatible in order to avoid any initiation of inflammatory and/or immune response.

21.8.1.1 SCAFFOLD COMPOSITION

The key constituents of resources used as scaffolds in bone engineering are different metals, synthetic polymeric materials or natural polymeric materials, and ceramics.

21.8.1.1.1 Metals

Metals introduction leads the advantage to provide instant mechanical support. However, their use is limited due to their poor assimilation with the host tissue, because of the discharge of ions, and by the possible bone collapse. Although, stainless steel was extremely used in the past, titanium and, mostly, titanium alloys are the most extensively planned metal components for bone engineering methods. In order to raise the risk of cell interactions with the metallic supports, metal surfaces are modified by changing their local topography and/or chemistry.

21.8.1.1.2 Natural and Synthetic Polymers

Natural biopolymers can be composed of various molecules including fibrinogen, collagen, alginate, starch, chitosan, silk, fibrin, and poly(hydroxybutyrate), HA. They generally support cell attachment, cell migration, and differentiation; however, they are often too weak to face the biomechanical forces imposed by daily events. To overcome such a limit, the natural biomaterial can be cross-linked, with synthetic polymers or ceramics/bioactive glasses, in order to create bio composite materials. Synthetic polymers indicate the most widely applied biomaterials in TE (Chung et al., 2007). Some of the synthetic biomaterials include poly(α-hydroxy acids), PCL, poly(propylene-fumarates), polycarbonates, polyphosphazenes, and polyanhydrides (Salgdo et al., 2013).

The poly(α-hydroxy acids) group comprises PLA, PGA, and their co-polymer PLGA.

21.8.1.1.3 Ceramics

HA, calcium phosphate, calcium sulfate, aluminum, zirconia, and coralline are also extensively used as scaffolds in bone engineering (Du et al., 2009; Song et al., 2008). They can be elaborated from natural or synthetic materials. Ceramics can be subdivided in inert, bioactive, and resorbable (β-TCP) products, depending on their main properties.

21.9 NANOMATERIALS IN MECHANO-SENSITIVE TISSUES

To repair and replacement to ligament and tendon injuries, the injured tissues are planned and combined with biomaterial support. However, despite the advancement in surgery and biomaterial properties, full restoration of damaged tissue is still challenging. Due to the limitation of surgical management, there is a great demand to create novel material that can show better mechanical properties and mimic the biological environment to support tissue growth.

21.9.1 NANOMATERIALS IN MECHANO-SENSITIVE TISSUES

Nanobiotechnology has played an important role in advancing the field of musculoskeletal TE which includes the regeneration of cartilage, tendon, bone, and ligament. Several materials were investigated since few years for musculoskeletal regeneration application; however, the main factor which prevents the development of successful materials is the failure to mimic the normal hierarchy of the original tissue structure.

21.9.2 BONE TISSUE REGENERATION

There are many uses of bone graft, they are used surgically to repair osseous defects including synthetic, autografts, and allografts (Hosseinkhani et al., 2014). Autografts are being considered as the gold standard in the surgical repair of bone faults due to their low immunogenicity and have all the properties to carry a good bone regeneration effect. However, conventional autograft bone grafts are occupied from the iliac crest, which leads to donor site morbidity. Other limitations include the formation of scar at the site and numerous surgical risks including blood loss and infectivity. Therefore, to overcome such impediment synthetic alternative is being examined and this has started the field of bone TE.

21.9.2.1 EVOLUTION OF NANOMATERIALS FOR BONE TISSUE ENGINEERING

A variety of nonmaterials have been investigated for bone regeneration over the decade including ceramics, polymeric materials, NPs, metals, nanofibers, nanocomposites, a perfect bone scaffold material must have excellent mechanical properties, adequate biocompatibility, low patient morbidity. Moreover, they will be easily accessible to surgeons, economical, and will support bone regeneration.

21.9.3 CARTILAGE TISSUE REGENERATION

Cartilage is an avascular tissue that is difficult to repair (O'Driscoll et al., 1998). Cartilage tissue comprises chondrocytes entrapped in an ECM rich in collagen and proteoglycan macromolecules. Likewise, to bone tissue, the structural organization of cartilage tissue is accountable for its mechanical properties (Temenoff et al., 2000). Making biomaterials for cartilage tissue replacement is determined from the thousands of joint procedures such as knees and hips. Articular cartilages are very less efficient in repairing themselves because of the poor vascular supply. For larger defects, artificial and synthetic knee and hip joints are formed; however, these are not best for young patients due to the restricted time of lifespan of the joint. Although, autografts are available to repair defects, however, they are restricted by contributor site morbidity and other complications. This limitation can be overcome by redeveloping articular tissue by using nanotechnology-based tissue-engineering approaches.

21.9.3.1 NANOCOMPOSITES IMPLEMENTED AS CARTILAGE TISSUE ENGINEERING

ECM is composed of collagen fibrils, noncollagenous proteins, and proteoglycan macromolecules, to create a hierarchical network (Iwasaki et al.,

2004). Polymer-HAp composite was also explored for cartilage repair. For example, PLGA, PVA, and PLLA were electrospun with nHA and establish to provide molecular signaling mechanisms best for cartilage formation (Zhu et al., 2014). nHA/collagen scaffolds confirmed the capability to endure chondrocyte cell attachment, abundant glycosaminoglycan synthesis, and maintain natural morphology.

21.9.3.2 NANOFIBERS FOR CARTILAGE TISSUE ENGINEERING APPLICATION

The size of the nanofibers has been shown to affect the attachment of cells inexistence of serum proteins (Cobum et al., 2012). Coburn et al. explored PVA nanofibers scaffolds and confirmed enhanced chondrogenic differentiation of mesenchymal stem cells as specified by increasing of ECM production and cartilage-specific gene expression while permitting cell proliferation on the fibrous scaffolds. Collagen and PLA nanofibers have also been shown to synergistically stimulate the osteochondral regeneration of MSCs. Osteochondral faults were formed in rabbits and fixed by inserting the collagen-nanofiber scaffold. There is evidence of rapid subchondral bone development and cartilage formation using these scaffolds as demonstrated by histological, biomechanical testing, and microCT testing (Zhang et al., 2013). He et al. established a biodegradable nanofibrous membrane a fusion material of collagen and poly(L-lactic acid-*co*-ε-caprolactone) (75: 25) for cartilage engineering. The scaffold has been demonstrated as cartilage-like tissue after 12 weeks of grafting into mice (Fu et al., 2014). Shafiee et al. demonstrated the effects of PLLA/PCL nanofibers on the cartilaginous capacity of nasal septum resulting in progenitors in vitro (Shafiee et al., 2011). Aligned nanofibers have been exemplified better chondrogenic differentiation than randomly orientated PLLA/PCL scaffolds as shown by the upregulation of collagen II and aggrecan (Shafiee et al., 2011).

21.10 NANOMEDICAL APPLICATIONS OF GRAPHENE AND GRAPHENE OXIDE

Graphene is ordered in a honeycomb lattice is comprised of a monoatomic linear layer of carbon atoms (Novoselov et al., 2005) and it is reported to have a tensile modulus of 1Tpa[2] and tensile strength bigger than 100 GPa

(Lee et al., 2008). Researchers have been showing an anxious interest in these materials because of their unique chemical structure, biomedical properties, and material (Wang et al., 2011). Graphene and graphene oxide (GO) sheets are simply produced via Hummers technique or variants thereof (William et al., 1958). GO sheets are hydrophilic and their surface are modified with various polymers like chitosan (Fan et al., 2010), polyethylene glycol (Ma et al., 2012), poly(ε-caproplactone) (Wojtoniszak et al., 2012), poly-l-lysine (PLL; Yang et al., 2010), and polyvinyl alcohol (Wang and Qiu, 2011). GO consists of a huge quantity of hydrophilic groups on their planes; thus sheets having a small size and lower concentrations show great biocompatibility. These properties make the GO sheets a perfect contender for TE (Wen et al., 2012; Wang et al., 2013) applications. To use these sheets in a clinical facility, toxicity and biocompatibility should be checked in vitro and in vivo studies using definite cell lines, theoretical, and animal models (Dallavalle et al., 2015). These functionalized GO sheets and NPs are commonly used as fillers, complex meshes, and tissue scaffolds in the case of regenerative medicine. The graphene end product is favored more in case of cardio and neuroregeneration. GO sheets mixture with other nanomaterials lead to numerous new areas including drug delivery, bioimaging, cancer therapeutics, TE, and diagnostics.

21.10.1 GO SHEETS IN TISSUE ENGINEERING

GO are used for specific delivery of proteins like bone morphogenetic proteins BMPs and substance P [SP] factors (Gautschi, 2007; Termaat et al., 2005). BMP-2 is a standard growth factor which used for bone regeneration; its overdose leads to side effects including excess bone growth, inflammation at the site, and uncontrolled bone development. La et al. reported of Ti-GO implant where the surface is modified by numerous BMPs and SP. BMP-2 when delivered using GO-Ti and bare Ti, GO-Ti exhibited higher alkaline phosphatase action in cells which is mostly responsible for bone formation. The delivery of BMP-2 and SP showed a higher development of bone growth in mice when compared to other groups.

3D scaffolds with highly ordered and functional properties are of countless value in TE (TE) since natural tissues and organs display highly organized and multifunctional architectures composed of the ECM, different cell types, and chemical and physical signaling clues (Madeira et al., 2015). Myocardium imitating the vascularized structure of different types of cells is still the most challenging in TE. For this purpose, synthesis methods mostly used are bottom-up (Campbell et al., 2015) or layer by layer LBL approach. Shin et al. (2014) reported good conductivity and cellular adhesion in multiple layers of cell build using efficient PLL-GO NPs and the LBL method. Here, the use of 3L constructs made with PLL-GO increases the thickness of tissue growth (65 μm); however, when compared to the build without PLL-GO (control) there is a decrease in thickness of tissue growth (23 μm). The wideness size of the PLL-GO layers is ranged from a few micrometers to 10 μm, and it is thicker than the tissue grown using fibronectin, gelatin (G), and nanofilms (6.2 nm).

Silk fibrin (F) proteins are characteristically active in tissue generation as alternates for bone and skin tissues and blood vessels. Wang et al. reported the benefits of Fibrin and GO by synthesizing a nanocomposite film by simply casting the two components together. Hydroxyapatite can be grown using GO or fibrin-modified GO which can act as a nucleation site (Wang et al., 2014). Deepachitra et al. confirmed that fibrin-graphene hydroxyapatite was a brilliant stage for osteoblast cell proliferation and maturation (Deepachitra et al., 2015).

GO is identified to play a bigger role both in endothelial and hepatocyte cell growth and differentiation of morphology. Zhou et al. also proved that an LBL 3D composite layer made of PSS (polyanion) and polyacrylamide attached to GO show brilliant anticoagulant bioactivities representing heparin-mimicking action (Zhou et al., 2014). Another LBL study composed of GO nanocomposite films was created aiming to develop the mechanical character of polyelectrolyte multilayer (PEM) films comprising PSS and poly(allylamine hydrochloride). Elasticity of PEM film is increased by 181% in the existence of a single layer of GO. For example, fibroblast cells developed faster and also on a larger area, on the PEM/GO composite films (Qi et al., 2014).

KEYWORDS

- **nanotechnology**
- **tissue engineering**
- **scaffolds**
- **regenerative medicine**

REFERENCES

Abdal-hay, A.; Pant, H.R.; Lim, J.K. Super-hydrophilic electrospun nylon-6/hydroxyapatite membrane for bone tissue engineering. *Euro. Polym. J.* **2013**, 49(6), 1314–1321.

Ahmed, F.E.; Lalia, B.S.; Hashaikeh, R. A review on electrospinning for membrane fabrication, challenges and applications. *Desalination.* **2015**, 356, 15–30.

Anton, F.; Gastell, R.S. Process and apparatus for preparing artificial threads. US Patent 1,975,504. 1934.

Arima, Y.; Iwata, H. Effect of wettability and surface functional groups on protein adsorption and cell adhesion using well-defined mixed self-assembled monolayers. *Biomaterials* **2007**, 28(20), 3074–3082.

Keselowsky, B.G.; Bridges, A.W.; Burns, K.L.; Tate, C.C.; Babensee, J.E.; LaPlaca, M.C.; García, A.J. *Biomaterials* **2007**, 25, 3626–3631.

Bakhshinejad, B.; Sadeghizadeh, M. Bacteriophages as vehicles for gene delivery into mammalian cells, prospects and problems. *Exp. Opin. Drug Del.* **2014**, 11(10), 1561–1574.

Barrera, D.A.; Zylstra, E.; Lansbury Jr, P.T.; Langer, R., Synthesis and RGD peptide modification of a new biodegradable copolymer, poly(lactic acid-co-lysine). *J. Am. Chem. Soc.* **1993**, 115(23), 11010–11011.

Bellin, R.M.; Kubicek, J.D.; Frigault, M.J.; Kamien, A.J.; Steward, R.L.; Barnes, H.M.; DiGiacomo, M.B.; Duncan, L.J.; Edgerly, C.K.; Morse, E.M.; Park, C.Y. Defining the role of syndecan-4 in mechanotransduction using surface-modification approaches. *Proc. Nat. Acad. Sci.* **2009**, 106(52), pp.22102–22107.

Bonferoni, M. C.; Sandri, G.; Gavini, E.; Rossi, S.; Ferrari, F.; Caramella, C. Microparticle systems based on polymer–drug interaction for ocular delivery of ciprofloxacin I. In vitro characterization. *J. Drug Del. Sci. Tech.* **2007**, 17(1), 57–62.

Bouasla, C.; Samar, M.E.H.; Ismail, F. Degradation of methyl violet 6B dye by the Fenton process. *Desalination* **2010**, 254(1–3), 35–41.

Boughman, J.A.; Conneally, P.M.; Nance, W.E. Population genetic studies of retinitis pigmentosa. *Am. J. Hum. Gen.* **1980**, 32(2), 223.

Brazel, C.S. Magnetothermally-responsive nanomaterials, combining magnetic nanostructures and thermally-sensitive polymers for triggered drug release. *Pharma. Res.* **2009**, 26(3), 644–656.

Campbell, S.; Maitland, D.; Hoare, T. Enhanced pulsatile drug release from injectable magnetic hydrogels with embedded thermosensitive microgels. *ACS Macro Lett.* **2015**, 4(3), 312–316.

Cascone, M.G.; Barbani, N.; Giusti, P.; Ciardelli, C.C.; Lazzeri, L. Bioartificial polymeric materials based on polysaccharides. *J. Biomater. Sci.* **2001**, 12(3), 267–281.

Chang, J.J.; Lee, Y.H.; Wu, M.H.; Yang, M.C.; Chien, C.T. Preparation of electrospun alginate fibers with chitosan sheath. *Carbohydr. Polym.* **2012**, 87(3), 2357–2361.

Chen, C.S.; Mrksich, M.; Huang, S.; Whitesides, G.M.; Ingber, D.E. Geometric control of cell life and death. *Science***1997**, 276(5317), 1425–1428.

Choi, C.H.; Hagvall, S.H.; Wu, B.M.; Dunn, J.C.; Beygui, R.E. Cell interaction with three-dimensional sharp-tip nanotopography. *Biomaterials* **2007**, 28(9), 1672–1679.

Chou, S.Y.; Cheng, C.M.; LeDuc, P.R. Composite polymer systems with control of local substrate elasticity and their effect on cytoskeletal and morphological characteristics of adherent cells. *Biomaterials* **2009**, 30(18), 3136–3142.

Chung, H. J.; Park, T. G. Surface engineered and drug releasing pre-fabricated scaffolds for tissue engineering. *Adv. Drug Del. Rev.* **2007**, 59(4–5), 249–262.

Coburn, J.M.; Gibson, M. Monagle, S.; Patterson, Z.; Elisseeff, J. H. Bioinspired nanofibers support chondrogenesis for articular cartilage repair. *Proc. Nat. Acad. Sci.* **2012**, 109(25), 10012–10017.

Conti, L.; Reitano, E.; Cattaneo, E. Neural stem cell systems, diversities and properties after transplantation in animal models of diseases. *Brain Pathol.* **2006**, 16(2), 143–154.

Dallavalle, M.; Calvaresi, M.; Bottoni, A.; Melle-Franco, M.; Zerbetto, F. Graphene can wreak havoc with cell membranes. *ACS Appl. Mater. Interface.* **2015**, 7(7), 4406–4414.

Deepachitra, R.; Nigam, R.; Purohit, S.D.; Kumar, B.S.; Hemalatha, T.; Sastry, T.P. In vitro study of hydroxyapatite coatings on fibrin functionalized/pristine graphene oxide for bone grafting. *Mater. Manuf. Proc.* **2015**, 30(6), 804–811.

Deshmukh, K.; Sankaran, S.; Ahamed, M.B.; Pasha, S.K. Biomedical applications of electrospun polymer composite nanofibres. *Poly. Nanocomp. Biomed. Eng.* **2019**, 111–165.

Dhandayuthapani, B.; Yoshida, Y.; Maekawa, T.; Kumar, D.S. Polymeric scaffolds in tissue engineering application, a review. *Int. J. Polym. Sci.* **2011**.

Di Martino, A.; Sittinger, M.; Risbud, M.V. Chitosan, a versatile biopolymer for orthopaedic tissue-engineering. *Biomaterials* **2005**, 26(30), 5983–5990.

Di Mundo, R.; Nardulli, M.; Milella, A.; Favia, P.; d'Agostino, R.; Gristina, R. Cell adhesion on nanotextured slippery superhydrophobic substrates. *Langmuir.* **2011**, 27(8), 4914–4921.

Doshi, J.; Reneker, D.H. Electrospinning process and applications of electrospun fibers. *J. Electrostat.* **1995**, 35(2–3), 151–160.

Du, D.; Furukawa, K. S.; Ushida, T. 3D culture of osteoblast-like cells by unidirectional or oscillatory flow for bone tissue engineering. *Biotech. Bioeng.* **2009**, 102(6), 1670–1678.

Dvir, T.; Timko, B.P.; Kohane, D.S.; Langer, R. Nanotechnological strategies for engineering complex tissues. *Nat. Nanotech.* **2011**, 6(1), 13.

Eaglstein, W.H.; Falanga, V. Tissue engineering and the development of Apligraf®, a human skin equivalent. *Clin. Therapeut.* **1997**, 19(5), 894–905.

Elisa, P.M.; Menon, G.K. Structural and lipid biochemical correlates of the epidermal permeability barrier. *Adv. Lipid Res.* **1991**, 24, 1–26.

Ellison, C.J.; Phatak, A.; Giles, D.W.; Macosko, C.W.; Bates, F.S. Melt blown nanofibers: fiber diameter distributions and onset of fiber breakup. *Polymer.* **2007**, 48(11), 3306–3316.

Fan, H.; Wang, L.; Zhao, K.; Li, N.; Shi, Z.; Ge, Z.; Jin, Z. Fabrication, mechanical properties, and biocompatibility of graphene-reinforced chitosan composites. *Biomacromolecules* **2010**, 11(9), 2345–2351.

Farr, R.; Choi, D.S.; Lee, S.W. Phage-based nanomaterials for biomedical applications. *Acta Biomater.* **2014**, 10(4), 1741–1750.

Fernandes, J.G.; Correia, D.M.; Botelho, G.; Padrão, J.; Dourado, F.; Ribeiro, C.; Lanceros-Méndez, S.; Sencadas, V. PHB-PEO electrospun fiber membranes containing chlorhexidine for drug delivery applications. *Polym. Test.* **2014**, 34, 64–71.

Firkowska, I.; Olek, M.; Pazos-Peréz, N.; Rojas-Chapana, J.; Giersig, M. Highly ordered MWNT-based matrixes, topography at the nanoscale conceived for tissue engineering. *Langmuir.* **2006**, 22(12), 5427–5434.

Folch, A.; Toner, M. Cellular micropatterns on biocompatible materials. *Biotech. Prog.* **1998**, 14(3), 388–392.

Fu, W.; Liu, Z.; Feng, B.; Hu, R.; He, X.; Wang, H.; Wang, W. Electrospun gelatin/PCL and collagen/PLCL scaffolds for vascular tissue engineering. *Int. J. Nanomed.* **2014**, 9, 2335.

Gautschi, O.P.; Frey, S.P.; Zellweger, R. Bone morphogenetic proteins in clinical applications. *ANZ J. Surgery.* **2007**, 77(8), 626–631.

Gerbin, K.A.; Murry, C.E. The winding road to regenerating the human heart. *Cardiovasc. Pathol.* **2015**, 24(3), 133–140.

Germain, L.; Auger, F.A.; Grandbois, E.; Guignard, R.; Giasson, M.; Boisjoly, H.; Guérin, S.L. Reconstructed human cornea produced in vitro by tissue engineering. *Pathobiology* **1999**, 67(3), 140–147.

Goa, K. L.; Benfield, P. Hyaluronic acid. *Drugs*. **1994**, 47(3), 536–566.

Gonzalez-Molina, J.; Riegler, J.; Southern, P.; Ortega, D.; Frangos, C.C.; Angelopoulos, Y.; Husain, S.; Lythgoe, M.F.; Pankhurst, Q.A.; Day, R.M. Rapid magnetic cell delivery for large tubular bioengineered constructs. *J. R. Soc. Interface*. **2012**, 9(76), 3008–3016.

Gregory-Evans, K.; Bashar, A.E.; Laver, C. Use of magnetism to enhance cell transplantation success in regenerative medicine. *Regen. Med.* **2013**, 8(1), 1–3.

Gunatillake, P.; Mayadunne, R.; Adhikari, R. Recent developments in biodegradable synthetic polymers. *Biotech. Annual Rev.* **2006**, 12, 301–347.

Ha, Y.M.; Amna, T.; Kim, M.H.; Kim, H.C.; Hassan, M.S.; Khil, M.S. Novel silicificated PVAc/POSS composite nanofibrous mat via facile electrospinning technique, potential scaffold for hard tissue engineering. *Coll. Surf. B Biointer.* **2013**, 102, 795–802.

Hench, L.L. Bioceramics, from concept to clinic. *J. Am. Ceram. Soc.* **1991**, 74(7), 1487–1510.

Hoff, J.D.; Cheng, L.J.; Meyhöfer, E.; Guo, L.J.; Hunt, A.J. Nanoscale protein patterning by imprint lithography. *Nano Lett.* **2004**, 4(5), 853–857.

Hoff, J.D.; Cheng, L.J.; Meyhöfer, E.; Guo, L.J.; Hunt, A.J. Nanoscale protein patterning by imprint lithography. *Nano Lett.* **2004**, 4(5), 853–857.

Hong, J.K.; Madihally, S.V. Next generation of electrosprayed fibers for tissue regeneration. *Tissu. Eng. Part B Rev.* **2011**, 17(2), 125–142.

Hosseinkhani, M.; Mehrabani, D.; Karimfar, M. H.; Bakhtiyari, S.; Manafi, A.; Shirazi, R. Tissue engineered scaffolds in regenerative medicine. *World J. Plast. Surg.* **2014**, 3(1), 3.

Hrabárová, E.; Rychlý, J.; Sasinková, V.; Valachová, K.; Janigová, I.; Csomorová, K.; Juránek, I.; Šoltés, L. Structural characterisation of thiol-modified hyaluronans. *Cellulose*. **2012**, 19(6), 2093–2104.

Huang, Z.M.; Zhang, Y.Z.; Kotaki, M.; Ramakrishna, A review on polymer nanofibers by electrospinning and their application in nanocoposites. *Compos. Sci. Tech.* **2003**, 63, 2223, 00178–7.

Ikada, Y. Challenges in tissue engineering. *J. R. Soc. Interface*. **2006**, 3 (10), 589–601.

Ishii, M.; Shibata, R.; Numaguchi, Y.; Kito, T.; Suzuki, H.; Shimizu, K.; Ito, A.; Honda, H.; Murohara, T. Enhanced angiogenesis by transplantation of mesenchymal stem cell sheet created by a novel magnetic tissue engineering method. *Arteriosclerosis Thrombosis. Vasc. Bio.* **2011**, 31(10), 2210–2215.

Iwasaki, N.; Yamane, S. T.; Majima, T.; Kasahara, Y.; Minami, A.; Harada, K.; et al. Feasibility of polysaccharide hybrid materials for scaffolds in cartilage tissue engineering, evaluation of chondrocyte adhesion to polyion complex fibers prepared from alginate and chitosan. *Biomacromolecules* **2004**, 5(3), 828–833.

Jeyaraj, M.; Praphakar, R.A.; Rajendran, C.; Ponnamma, D.; Sadasivuni, K.K.; Munusamy, M.A.; Rajan, M. Surface functionalization of natural lignin isolated from Aloe barbadensis Miller biomass by atom transfer radical polymerization for enhanced anticancer efficacy. *RSC Adv.* **2016**, 6(56), 51310–51319.

Ji, Y.; Ghosh, K.; Shu, X.Z.; Li, B.; Sokolov, J.C.; Prestwich, G.D.; Clark, R.A.; Rafailovich, M.H. Electrospun three-dimensional hyaluronic acid nanofibrous scaffolds. *Biomaterials* **2006**, 27(20), 3782–3792.

Jiang, T.; Carbone, E.J.; Lo, K.W.H.; Laurencin, C.T. Electrospinning of polymer nanofibers for tissue regeneration. *Prog. Polym. Sci.* **2015**, 46, 1–24.
Jiang, X.; Bruzewicz, D.A.; Wong, A.P.; Piel, M.; Whitesides, G.M. Directing cell migration with asymmetric micropatterns. *Proc. Nat. Acad. Sci.* **2005**, 102(4), 975–978.
Jungbauer, S.; Kemkemer, R.; Gruler, H.; Kaufmann, D.; Spatz, J.P. Cell shape normalization, dendrite orientation, and melanin production of normal and genetically altered (Haplo insufficient NF1)-melanocytes by microstructured substrate interactions. *Chem. Phys. Chem.* **2004**, 5(1), 85–92.
Kenne, L.; Gohil, S.; Nilsson, E.M.; Karlsson, A.; Ericsson, D.; Kenne, A.H.; Nord, L.I. Modification and cross-linking parameters in hyaluronic acid hydrogels—definitions and analytical methods. *Carbohyd. Polys.* **2013**, 91(1), 410–418.
Kim, D.H.; Kim, P.; Song, I.; Cha, J.M.; Lee, S.H.; Kim, B.; Suh, K.Y. Guided three-dimensional growth of functional cardiomyocytes on polyethylene glycol nanostructures. *Langmuir.* **2006**, 22(12), 5419–5426.
Kim, E.; Xia, Y.; Whitesides, G.M. Making polymeric microstructures, capillary micromolding. *Nature* **1995**, 376, 581–584.
Kong, L.; Ziegler, G.R. Fabrication of pure starch fibers by electrospinning. *Food Hydrocoll.* **2014**, 36, 20–25.
Langer, R.; Tirrell, D.A. Designing materials for biology and medicine. *Nature* **2004**, 428(6982), 487.
Langer, R.; Vacanti, J.P. Tissue engineering, *Science*. 1993, 260 (5110), 920–926.
Lee, C.; Wei, X.; Kysar, J.W.; Hone, J. Measurement of the elastic properties and intrinsic strength of monolayer graphene. *Science* **2008**, 321(5887), 385–388.
Li, D.; Frey, M.W.; Baeumner, A.J. Electrospun polylactic acid nanofiber membranes as substrates for biosensor assemblies. *J. Membr. Sci.* **2006**, 279(1–2), 354–363.
Li, D.; Wang, Y.; Xia, Y. Electrospinning of polymeric and ceramic nanofibers as uniaxially aligned arrays. *Nano Lett.* **2003**, 3(8), 1167–1171.
Limongi, T.; Schipani, R.; Di Vito, A.; Giugni, A.; Francardi, M.; Torre, B.; Allione, M.; Miele, E.; Malara, N.; Alrasheed, S.; Raimondo, R. Photolithography and micromolding techniques for the realization of 3D polycaprolactone scaffolds for tissue engineering applications. *Microelectro. Eng.* **2015**, 141, 135–139.
Liu, J.K.; Teng, Q.; Garrity-Moses, M.; Federici, T.; Tanase, D.; Imperiale, M.J.; Boulis, N.M. A novel peptide defined through phage display for therapeutic protein and vector neuronal targeting. *Neurobio. Dis.* **2005**, 19(3), 407–418.
Low, S.P.; Voelcker, N.H.; Canham, L.T.; Williams, K.A. The biocompatibility of porous silicon in tissues of the eye. *Biomaterials* **2009**, 30(15), 2873–2880.
Ma, G.; Fang, D.; Liu, Y.; Zhu, X.; Nie, J. Electrospun sodium alginate/poly (ethylene oxide) core–shell nanofibers scaffolds potential for tissue engineering applications. *Carbohyd. Polys.* **2012**, 87(1), 737–743.
Ma, J.; Liu, C.; Li, R.; Wang, J. Properties and structural characterization of oxide starch/chitosan/graphene oxide biodegradable nanocomposites. *J. Appl. Polym. Sci.* **2012**, 123(5), 2933–2944.
Ma, P.X. Scaffolds for tissue fabrication. *Mater. Today.* **2004**, 7(5), 30–40.
Madeira, C.; Santhagunam, A.; Salgueiro, J.B.; Cabral, J.M. Advanced cell therapies for articular cartilage regeneration. *Trends Biotech.* **2015**, 33(1), 35–42.
Matthews, J.A.; Wnek, G.E.; Simpson, D.G.; Bowlin, G.L. Electrospinning of collagen nanofibers. *Biomacromolecules* **2002**, 3(2), 232–238.

Mayer, J.; Karamuk, E.; Akaike, T.; Wintermantel, E. Matrices for tissue engineering-scaffold structure for a bioartificial liver support system. *J. Control Rel.* **2000**, 64(1–3), 81–90.

Merz, M.; Fromherz, P. Silicon chip interfaced with a geometrically defined net of snail neurons. *Adv. Funct. Mater.* **2005**, 15(5), 739–744.

Mo, X.M.; Xu, C.Y.; Kotaki, M.E.A.; Ramakrishna, S. Electrospun P (LLA-CL) nanofiber, a biomimetic extracellular matrix for smooth muscle cell and endothelial cell proliferation. *Biomaterials* **2004**, 25(10), 1883–1890.

Mohammad, J.; Shenaq, J.; Rabinovsky, E.; Shenaq, S. Modulation of peripheral nerve regeneration, a tissue-engineering approach. The role of amnion tube nerve conduit across a 1-centimeter nerve gap. *Plastic. Reconstr. Surg.* **2000**, 105(2), 660–666.

Morganti, P.; Del Ciotto, P.; Carezzi, F.; Nunziata, M.L.; Morganti, G. A Chitin Nanofibril-based non-woven tissue as medical dressing. The role of bionanotechnology. *Nanomat. Regene. Med.* **2016**, 123–142.

Munaweera, I.; Aliev, A.; Balkus Jr, K.J. Electrospun cellulose acetate-garnet nanocomposite magnetic fibers for bioseparations. *ACS Appl. Mater. Interfaces*. **2013**, 6(1), 244–251.

Nair, L.S.; Laurencin, C.T. Polymers as biomaterials for tissue engineering and controlled drug delivery. *Tissue Eng.* **2005**, 47–90.

Nair, L.S.; Laurencin, C.T. Biodegradable polymers as biomaterials. *Prog. Polym. Sci.* **2007**, 32(8–9), 762–798.

Novoselov, K.S.; Geim, A.K.; Morozov, S.; Jiang, D.; Katsnelson, M.I.; Grigorieva, I.; Dubonos, S.; Firsov, A.A. Two-dimensional gas of massless Dirac fermions in graphene. *Nature* **2005**, 438(7065), 197.

Oberpenning, F.; Meng, J.; Yoo, J.J.; Atala, A. De novo reconstitution of a functional mammalian urinary bladder by tissue engineering. *Nat. Biotech.* **1999**, 17(2), 149.

O'Driscoll, S. W. Current concepts review—the healing and regeneration of articular cartilage. *JBJS*. **1998**, 80(12), 1795–1812.

Ondarcuhu, T.; Joachim, C. Drawing a single nanofibre over hundreds of microns. *EPL. Lett.* **1998**, 42(2), 215.

Panigrahi, B.; Singh, R.K.; Mishra, S.; Mandal, D. Cyclic peptide-based nanostructures as efficient siRNA carriers. *Artif. Cells, Nanomed. Biotech.* **2018**, 46(3), S763-S773.

Pesen, D.; Haviland, D.B. Modulation of cell adhesion complexes by surface protein patterns. *ACS. Appl. Mater. Interfaces*. **2009**, 1(3), 543–548.

Piskin, E. Biodegradable polymers as biomaterials. *J. Biomater. Sci.* **1995**, 6(9), 775–795.

Porter, J.R.; Ruckh, T.T.; Popat, K.C. Bone tissue engineering, a review in bone biomimetics and drug delivery strategies. *Biotech. Prog.* **2009**, 25(6), 1539–1560.

Poudel, I.; Lee, J.S.; Tan, L.; Lim, J.Y. Micropatterning–retinoic acid co-control of neuronal cell morphology and neurite outgrowth. *Acta Biomater.* **2013**, 9(1), 4592–4598.

Prodanov, L.; Riet, J.; Lamers, E.; Domanski, M.; Luttge, R.; Van Loon, J.J.; Jansen, J.A.; Walboomers, X.F. The interaction between nanoscale surface features and mechanical loading and its effect on osteoblast-like cells behavior. *Biomaterials* **2010**, 31(30), 7758–7765.

Qi, W.; Xuc, Z.; Yuan, W.; Wang, H. Layer-by-layer assembled graphene oxide composite films for enhanced mechanical properties and fibroblast cell affinity. *J. Mater. Chem. B.* **2014**, 2(3), 325–331.

Ramakrishna, S.; Mayer, J.; Wintermantel, E.; Leong, K.W. Biomedical applications of polymer-composite materials, a review. *Compos. Sci. Tech.* **2001**, 61(9), 1189–1224.

Ramakrishna, S.; Mayer, J.; Wintermantel, E.; Leong, K.W. Biomedical applications of polymer-composite materials, a review. *Compos. Sci. Tech.* **2001**, 61(9), 1189–1224.

Ratner, B.D.; Hoffman, A.S.; Schoen, F.J.; Lemons, J.E. Biomaterials Science, An Introduction To Materials in Medicine, Academic Press, 2004.

Ren, G.; Xu, X.; Liu, Q.; Cheng, J.; Yuan, X.; Wu, L.; Wan, Y. Electrospun poly (vinyl alcohol)/ glucose oxidase biocomposite membranes for biosensor applications. *React. Funct. Polym.* **2006**, 66(12), 1559–1564.

Reneker, D.H.; Chun, I. Nanometre diameter fibres of polymer, produced by electrospinning. *Nanotechnology* **1996**, 7(3), 216.

Rho, K.S.; Jeong, L.; Lee, G.; Seo, B.M.; Park, Y.J.; Hong, S.D.; Roh, S.; Cho, J.J.; Park, W.H.; Min, B.M. Electrospinning of collagen nanofibers, effects on the behavior of normal human keratinocytes and early-stage wound healing. *Biomaterials* **2006**, 27(8), 1452–1461.

Riegler, J.; Lau, K.D.; Garcia-Prieto, A.; Price, A.N.; Richards, T.; Pankhurst, Q.A.; Lythgoe, M.F. Magnetic cell delivery for peripheral arterial disease, A theoretical framework. *Med. Phys.* **2011**, 38(7), 3932–3943.

Roether, J.A.; Boccaccini, A.R.; Hench, L.L.; Maquet, V.; Gautier, S.; Jérôme, R. Development and in vitro characterisation of novel bioresorbable and bioactive composite materials based on polylactide foams and Bioglass® for tissue engineering applications. *Biomaterials* **2002**, 23(18), 3871–3878.

Romanova, E.V.; Oxley, S.P.; Rubakhin, S.S.; Bohn, P.W.; Sweedler, J.V. Self-assembled monolayers of alkanethiols on gold modulate electrophysiological parameters and cellular morphology of cultured neurons. *Biomaterials* **2006**, 27(8), 1665–1669.

Šafařík, I.; Horská, K.; Šafaříková, M. Magnetic nanoparticles for biomedicine. *Intracell. Del.* **2011**, 363–372.

Salgado, A. J.; Oliveira, J. M.; Martins, A.; Teixeira, F. G.; Silva, N. A.; Neves, N. M.; et al. Tissue engineering and regenerative medicine, past, present, and future. *Int. Rev. Neurobiol.* **2013**, 108, 1–33.

Sasaki, H.; Tanaka, N.; Nakanishi, K.; Nishida, K.; Hamasaki, T.; Yamada, K.; Ochi, M. Therapeutic effects with magnetic targeting of bone marrow stromal cells in a rat spinal cord injury model. *Spine*. **2011**, 36(12), 933–938.

Sato, M.; Webster, T.J. Nanobiotechnology, implications for the future of nanotechnology in orthopedic applications. *Exp. Rev. Med. Devices.* **2004**, 1(1), 105–114.

Schiffman, J.D.; Schauer, C.L. A review, electrospinning of biopolymer nanofibers and their applications. *Polys. Rev.* **2008**, 48, 317–352.

Shafiee, A.; Soleimani, M.; Chamheidari, G. A.; Seyedjafari, E.; Dodel, M.; Atashi, A.; Gheisari, Y. Electrospun nanofiber-based regeneration of cartilage enhanced by mesenchymal stem cells. *J. Biomed. Mater. Res. A.* **2011**, 99(3), 467–478.

Shapiro, B.; Kulkarni, S.; Nacev, A.; Sarwar, A.; Preciado, D.; Depireux, D.A. Shaping magnetic fields to direct therapy to ears and eyes. *Annu. Rev. Biomed. Eng.* **2014**, 16, 455–481.

Sharifi, S.; Seyednejad, H.; Laurent, S.; Atyabi, F.; Saei, A.A.; Mahmoudi, M. Superparamagnetic iron oxide nanoparticles for in vivo molecular and cellular imaging. *Contrast Media. Mol. Imaging.* **2015**, 10(5), 329–355.

Shin, S.R.; Ghareh-Bolagh, A.B.; Gao, X.; Nikkhah, M.; Jung, S.M.; Dolatshahi-Pirouz, A.; Kim, S.B.; Kim, S.M.; Dokmeci, M.R.; Tang, X.; Khademhosseini, A. Layer by layer assembly of 3D tissue constructs with functionalized graphene. *Adv. Funct. Mater.* **2014**, 24(39), 6136–6144.

Singhvi, R.; Kumar, A.; Lopez, G.P.; Stephanopoulos, G.N.; Wang, D.I.; Whitesides, G.M.; Ingber, D.E. Engineering cell shape and function. *Science* **1994**, 264(5159), 696–698.

Smith, G.P. Filamentous fusion phage, novel expression vectors that display cloned antigens on the virion surface. *Science* **1985**, 228(4705), 1315–1317.
Smith, L.A.; Ma, P.X. Nano-fibrous scaffolds for tissue engineering. *Coll. Surf. B Biointerfaces*. **2004**, 39(3), 125–131.
Song, K.; Liu, T.; Cui, Z.; Li, X.; Ma, X. Three dimensional fabrication of engineered bone with human bio-derived bone scaffolds in a rotating wall vessel bioreactor. *J. Biomed. Mater Res. A*. 2008, 86(2), 323–332.
Stevens, M.M.; George, J.H. Exploring and engineering the cell surface interface. *Science* **2005**, 310(5751), 11135–1138.
Subbiah, T.; Bhat, G.S.; Tock, R.W.; Parameswaran, S.; Ramkumar, S.S. Electrospinning of nanofibers. *J. Appl. Polym. Sci*. **2005**, 96(2), 557–569.
Supaphol, P.; Suwantong, O.; Sangsanoh, P.; Srinivasan, S.; Jayakumar, R.; Nair, S.V. Electrospinning of biocompatible polymers and their potentials in biomedical applications. *Biomed. Appl. Polym. Nanofibers*. **2011**, 213–239.
Tanaka, A.; Fujii, Y.; Kasai, N.; Okajima, T.; Nakashima, H. Regulation of neuritogenesis in hippocampal neurons using stiffness of extracellular microenvironment. *PLoS One*. **2018**, 13(2), p.e0191928.
Taylor, G.I. Electrically driven jets. *Proc. Royal Soc. London. A. Math. Phys. Sci.* **1969**, 313 (1515), 4153–1475.
Temenoff, J. S.; Mikos, A. G. Tissue engineering for regeneration of articular cartilage. *Biomaterials* **2000**, 21(5), 431–440.
Teo, W.E.; Ramakrishna, S. A review on electrospinning design and nanofibre assemblies. *Nanotechnology* **2006**, 17, R89–R106.
Termaat, M.F.; Den Boer, F.C.; Bakker, F.C.; Patka, P.; Haarman, H.T.M. Bone morphogenetic proteins, development and clinical efficacy in the treatment of fractures and bone defects. *JBJS*. **2005**, 87(6), 11367–1378.
Tziampazis, E.; Sambanis, A. Tissue engineering of a bioartificial pancreas, modeling the cell environment and device function. *Biotech. Prog*. **1995**, 11(2), 115–126.
Vaněček, V.; Zablotskii, V.; Forostyak, S.; Růřička, J.; Herynek, V.; Babič, M.; Jendelová, P.; Kubinová, Š.; Dejneka, A.; Syková, E. Highly efficient magnetic targeting of mesenchymal stem cells in spinal cord injury. *Int. J. Nanomed*. **2012**, 7, 3719.
Vasita, R.; Katti, D.S. Nanofibers and their applications in tissue engineering. *Int. J. Nanomed.* **2006**, 1(1), 15.
Venugopal, J.; Low, S.; Choon, A.T.; Ramakrishna, S. Interaction of cells and nanofiber scaffolds in tissue engineering. *J. Biomed. Mat. Res. B, Appl. Biomat*. **2008**, 84(1), 34–48.
Vermesh, U.; Vermesh, O.; Wang, J.; Kwong, G.A.; Ma, C.; Hwang, K.; Heath, J.R. High density, multiplexed patterning of cells at single-cell resolution for tissue engineering and other applications. *Angewandte Chemie Int. Eds*. **2011**, 50(32), 7378–7380.
Vert, M. Aliphatic polyesters, great degradable polymers that cannot do everything. *Biomacromolecules* **2005**, 6(2), 538–546.
Vesa, J.; Hellsten, E.; Verkruyse, L.A.; Camp, L.A.; Rapola, J.; Santavuori, P.; Hofmann, S.L.; Peltonen, L. Mutations in the palmitoyl protein thioesterase gene causing infantile neuronal ceroid lipofuscinosis. *Nature* **1995**, 376(6541), 584.
Vogt, A.K.; Lauer, L.; Knoll, W.; Offenhäusser, A. Micropatterned substrates for the growth of functional neuronal networks of defined geometry. *Biotech. Prog*. **2003**, 19(5), 1562–1568.
Wang, H.; Qiu, Z. Crystallization behaviors of biodegradable poly (l-lactic acid)/graphene oxide nanocomposites from the amorphous state. *Thermochimica Acta*. **2011**, 526(1–2), 229–236.

Wang, L.; Lu, C.; Zhang, B.; Zhao, B.; Wu, F.; Guan, S. Fabrication and characterization of flexible silk fibroin films reinforced with graphene oxide for biomedical applications. *RSC Adv.* **2014**, 4(76), 40312–40320.

Wang, Q.; Webster, T.J. Nanostructured selenium for preventing biofilm formation on polycarbonate medical devices. *J. Biomed. Mater. Res. A.* **2012**, 100(12), 3205–3210.

Wang, Y.; Li, Z.; Wang, J.; Li, J.; Lin, Y. Graphene and graphene oxide, biofunctionalization and applications in biotechnology. *Trends Biotech.* **2011**, 29(5), 205–212.

Wang, Y.; Wang, H.; Liu, D.; Song, S.; Wang, X.; Zhang, H. Graphene oxide covalently grafted upconversion nanoparticles for combined NIR mediated imaging and photothermal/photodynamic cancer therapy. *Biomaterials* **2013**, 34(31), 7715–7724.

Webb, K.; Hlady, V.; Tresco, P.A. Relative importance of surface wettability and charged functional groups on NIH 3T3 fibroblast attachment, spreading, and cytoskeletal organization. *J. Biomed. Mater. Res.* **1998**, 41(3), 422–430.

Wen, H.; Dong, C.; Dong, H.; Shen, A.; Xia, W.; Cai, X.; Song, Y.; Li, X.; Li, Y.; Shi, D. Engineered redox-responsive PEG detachment mechanism in PEGylated nanographene oxide for intracellular drug delivery. *Small.* **2012**, 8(5), 760–769.

William, S.; Hummers, J.R.; Offeman, R.E. Preparation of graphitic oxide. *J. Am. Chem. Soc.* **1958**, 80(6), 1339–1339.

Williams, D. Benefit and risk in tissue engineering. *Mat. Today.* **2004**, 7(5), 24–29.

Wojtoniszak, M.; Chen, X.; Kalenczuk, R.J.; Wajda, A.; Łapczuk, J.; Kurzewski, M.; Drozdzik, M.; Chu, P.K.; Borowiak-Palen, E. Synthesis, dispersion, and cytocompatibility of graphene oxide and reduced graphene oxide. *Coll. Surf. B Biointerfaces.* **2012**, 89, 79–85.

Wu, H.; Pan, W.; Lin, D.; Li, H. Electrospinning of ceramic nanofibers, fabrication, assembly and applications. *J. Adv. Ceram.* **2012**, 1(1), 2–23.

Wu, P.; Dai, Y.; Ye, Y.; Yin, Y.; Dai, L. Fast-speed and high-gain photodetectors of individual single crystalline Zn 3 P 2 nanowires. *J. Mater. Chem.* **2011**, 21(8), 2563–2567.

Xu, C.Y.; Inai, R.; Kotaki, M.; Ramakrishna, S. Aligned biodegradable nanofibrous structure, a potential scaffold for blood vessel engineering. *Biomaterials* **2004**, 25(5), 877–886.

Yang, K.; Wan, J.; Zhang, S.; Zhang, Y.; Lee, S.T.; Liu, Z. In vivo pharmacokinetics, long-term biodistribution, and toxicology of PEGylated graphene in mice. *ACS Nano.* **2010**, 5(1), 516–522.

Yang, S.H.; Chung, W.J.; McFarland, S.; Lee, S.W. Assembly of bacteriophage into functional materials. *Chem. Rec.* **2013**, 13(1), 43–59.

Yang, T.L.; Young, T.H. The enhancement of submandibular gland branch formation on chitosan membranes. *Biomaterials* **2008**, 29(16), 2501–2508.

Yoshimoto, H.; Shin, Y.M.; Terai, H.; Vacanti, J.P. A biodegradable nanofiber scaffold by electrospinning and its potential for bone tissue engineering. *Biomaterials* **2003**, 24(12), 2077–2082.

Yu, D.G.; Zhu, L.M.; White, K.; Branford-White, C. Electrospun nanofiber-based drug delivery systems. *Health.* **2009**, 1(02), 67.

Zeleny, J. Instability of electrified liquid surfaces. *Phys. Rev.* **1917**, 10(1), 1.

Zhang, C.; Salick, M.R.; Cordie, T.M.; Ellingham, T.; Dan, Y.; Turng, L.S. Incorporation of poly (ethylene glycol) grafted cellulose nanocrystals in poly (lactic acid) electrospun nanocomposite fibers as potential scaffolds for bone tissue engineering. *Mater. Sci. Eng. C.* **2015**, 49, 463–471.

Zhang, S.; Chen, L.; Jiang, Y.; Cai, Y.; Xu, G.; Tong, T.; et al. Bi-layer collagen/microporous electrospun nanofiber scaffold improves the osteochondral regeneration. *Acta Biomater.* **2013**, 9(7), 7236–7247.

Zhou, H.; Cheng, C.; Qin, H.; Ma, L.; He, C.; Nie, S.; Zhang, X.; Fu, Q.; Zhao, C. Self-assembled 3D biocompatible and bioactive layer at the macro-interface via graphene-based supermolecules. *Polym. Chem.* **2014**, 5(11), 3563–3575.

Zhu, W.; Chen, K.; Lu, W.; et al. In vitro study of nano-HA/PLLA composite scaffold for rabbit BMSC differentiation under TGF-β1 induction. *In Vitro Cell. Develop. Bio. Anim.* **2014**, 50(3), 214–220.

Zhu, W.; O'Brien, C.; O'Brien, J.R.; Zhang, L.G. 3D nano/microfabrication techniques and nanobiomaterials for neural tissue regeneration. *Nanomedicine* **2014**, 9(6), 859–875.

CHAPTER 22

Protein-Based Nanosystems as Emerging Bioavailability Enhancers for Nutraceuticals

ROHINI SAMADARSI and DEBJANI DUTTA*

Department of Biotechnology, National Institute of Technology Durgapur, Mahatma Gandhi Avenue, Durgapur 713209, West Bengal, India

**Corresponding author. E-mail: debjani.dutta@bt.nitdgp.ac.in*

ABSTRACT

Various nutraceuticals and bioactive components have therapeutic properties to prevent health disorders. The oral route of delivery is the preferred routes for the administration of nutraceuticals, but their poor solubility and bioavailability reduces their efficacy, which could be enhanced by nanoencapsulation strategy. Hence, delivery through the nanoparticle-mediated mode of action is considered ideal. To obtain efficiency in delivering nutraceuticals, it is essential to understand the interactions of nanoparticles with the biological environment at the specific site. Biocompatible polymers used for delivery purposes, such as polysaccharides, lipids, and proteins, could enhance the therapeutic benefit and reduce unwanted side effects. Protein-based nanoparticles have achieved importance in the field of medicine. Proteins such as albumin, gliadin, elastin, zein, milk protein that have several unique advantages over the other macromolecules such as ease of extraction from natural sources, abundantly present in plants and animals, amphiphilic in nature are biocompatible and biodegradable, and can be easily surface modifiable. Protein nanoparticles encapsulating nutraceuticals can be effectively prepared by methods such as electrospray, desolvation technique, emulsification, and complex coacervation. These nanoparticles are characterized by techniques such as dynamic light scattering, SEM, AFM, to evaluate the properties like particle size, structure of nanoparticles, surface charge, nutraceutical loading, entrapment efficiency, and mechanism of in vitro release. Several anticancer drugs, as well as nutraceuticals, have been successfully formulated using nanomaterials. These nanoparticles can be incorporated within food systems for the efficient delivery of nutraceuticals. This chapter discusses the preparation, characterization, release mechanism of nanoparticles for the efficient delivery of nutraceuticals.

22.1 INTRODUCTION

Due to the rapidly changing food habits and lifestyle, human health is continuously compromised. The phytochemicals, as well as nutraceuticals extracted from plants and other sources, have numerous therapeutic properties to fight against health disorders. Hence, several innovative functional foods that have been derived from plants are consumed around the world. Vimang, a commercial mangiferin product is one such compound. Nutraceuticals are favored over conventional pharmaceutical drugs to prevent diseases owing to their health-promoting properties and nontoxicity (Ashwini et al., 2013). Various nutraceutical compounds such as curcumin, resveratrol, blueberry, rutin, mangiferin, and carotenoids have attracted researchers as well as consumers not only due to their ability to prevent and treat many diseases like cancers, cardiovascular disorders, and neurodegenerative disorders but also because of their well-built antioxidant property and immune-boosting ability, resulting in increased consumption of nutraceuticals (Liang et al., 2017).

The intake of sufficient numbers of bioactive compounds or nutraceuticals through the consumption of fruits, nuts, and vegetables might not be practical. Fortification of food products with extracted nutraceuticals might be a plausible alternative. However, some of the limitations occurring during fortification are the reaction of these components with other food compounds during processing conditions and also the low solubility of nutraceuticals. This might affect the release of bioactive components in the digestive tract leading to their incomplete absorption and in turn, would also affect bioavailability and bioaccessibility (Sari et al., 2015; Teng et al., 2016). Significant improvements in the delivery patterns were made for enhancing the bioavailability and bioaccessibility of the nutraceuticals. However, delivery systems that utilize metal or chemical ingredients might not be preferable in the food systems as they might produce unwanted side effects, hence the materials for nutraceutical formulation requires to have GRAS status. Delivery systems are designed in a way to enhance the physic-chemical properties of bioactive such the solubility, sustained release, and protection from the external environment (pH, temperature, gastrointestinal tract enzymes) (Jones et al., 2011; Luo et al., 2012). Using the nanotechnological strategy, the bioavailability of hydrophilic bioactive components can be considerably improved (Chen et al., 2006; Teng et al., 2016).

22.2 NANOTECHNOLOGY

Nanotechnology has developed as an area for the incorporation of potent nutraceutical carriers, which would behave like a complete unit concerning its properties and transport mechanism. Remarkably these systems can enhance cellular uptake and body distribution. Nanoparticles have several advantages in nutraceutical delivery to target sites owing to their significant surface area to weight ratio, making nanoparticles ideal as a drug/nutraceutical carrier. Numerous biocompatible polymers are used in the synthesis of nanoparticles including, polysaccharides, lipids, and proteins (Alfadul and Elneshwy, 2010; Joye et al., 2015). Several nanosystems have been successfully produced for the efficient delivery of nutraceuticals (Gonza, 2007; Zuidam and Shimoni, 2010). Curcumin, a bioactive component extracted from turmeric, solubility, and bioavailability, was enhanced by different approaches of nanoformulations such as micelles, liposomes, and protein-based nanoparticles (Malsuno and Adachi, 1993; Jahanshahi and Babaei, 2008).

22.2.1 NANOPARTICLE SELECTION CRITERIA

To design a nanoparticle, the following criteria need to be studied:

1. The average size of the nanoparticle.
2. The release profile of nutraceutical.
3. The solubility of the nutraceutical/drug.
4. Stability of the nanoparticle under processing conditions (temperature, pH, and ionic strength).
5. The biodegradability and toxicity of the nanoparticles.

Protein-based nanoparticles have gained much attention in food industries owing to their low toxicity and biodegradability (Luo et al., 2011; Larsson et al., 2012).

22.3 PROTEINS-BASED NANOPARTICLES

The gel-forming properties of proteins are commonly exploited in developing nanoparticles for the localization of therapeutic agents to their target sites. The gel structure act as the matrix for encapsulating the bioactive compounds (Ko and Gunasekaran, 2006). Protein-based nanoparticles have not been developed for nutraceutical deliveries; however, few exceptions include vitamins, quercetin, curcumin, mangiferin, and resveratrol. The challenge in the development of protein-based nanosystems is the reduction of gel matrix size to a nanoscale. Several methods such as electrospray, desolvation, coacervation methods are employed for nanoparticle preparation. In a study, nanoparticles were developed with whey proteins with a particle size of 40 nm by heat denaturing the proteins at 55 °C and maintaining a low ionic strength and concentration of protein, and in vitro studies were conducted to examine their ability to deliver nutraceuticals (Chen et al., 2006). Bovine serum albumin (BSA) based nanoparticles were also prepared by denaturing the proteins in heat, encapsulating with a photosensitizer, which was intended for photodynamic site-directed cancer therapy (Gibbs et al., 1999; Gonza, 2007; Zuidam and Shimoni, 2010). The limitation in preparing nanoparticles developed by heat treatments is that they cannot be applied for thermolabile nutraceutical encapsulation.

This complication could be addressed by cold-induced gelation of whey and soy proteins, making it a more suitable technique for encapsulating heat-sensitive nutraceuticals. However, obtaining a reduced size nanosystem, in this case, is a major concern because of the self-assembly of proteins into random elongated structures (fibrils), mechanism of which varies and is specific to each protein. β-lactoglobulin, an abundant protein in milk can form semiflexible aggregates (fibrils) when undergone heat denaturation at lower pH and ionic strength, leading to the conversion of dimmers into monomers and finally to fibrils. When α-lactalbumin is partially hydrolyzed by proteases, it produces peptides that self-assemble into tubular structures. Both fibrils of β-lactoglobulin and tubular structures of α-lactalbumin can be exploited as a delivery vehicle for nutraceuticals (Haug et al., 2009). Casein micelle was used to entrap hydrophobic nutraceutical compounds like vitamin D2 and curcumin, which could protect against UV-light-induced denaturation of vitamin D2. Self-assembly properties of proteins should be further studied for the synthesis of novel nutraceutical delivery systems (Saadati and Razzaghi, 2012).

22.4 TYPES OF PROTEINS USED FOR NANOPARTICLES

22.4.1 ALBUMIN

Albumin can be efficiently used for nanoparticle preparation due to its water solubility (up to 40% w/v at pH 7.4), heat stability (up to 60 °C for 10 h), and nutraceutical binding capacity. Sources of albumin include ovalbumin, human serum albumin (HSA), and BSA. Albumin nanoparticles have various advantages which include nontoxicity, biodegradability, and anti-immunogenicity. Reactive groups present on albumins such as thiol, amines, and carboxyl can act as a ligand-binding site, which can be covalent linked. A limitation is that albumin nanoparticles are prone to proteolytic enzyme digestion. Various albumin-based nanoparticles are synthesized and developed for the localization of nutraceuticals for cancer therapy (Larsen et al., 2016; Verma et al., 2018).

Studies showed that the albumin nanoparticles enhanced the bioavailability of nutraceuticals, improve pharmacokinetics properties, and also controlled the release of therapeutic components. A study by Joung et al. (2016) showed that curcumin entrapped albumin nanoparticles could be efficiently applied for breast cancer therapy.

22.4.2 GELATIN

Gelatin nanoparticles are one of the earliest protein-based nanoparticles, derived from hydrolysis of collagen and insoluble protein. Gelatin can be isolated from different sources including bones, skin, and connective tissue. Advantages of gelatine, include biodegradability, nontoxicity, sterilizable, low antigenicity, which can easily be crosslinked with nutraceuticals (Foox and Zilberman, 2015). The ionizable group present on gelatine makes it an ideal protein for the colloidal nutraceutical delivery system. The limitation of gelatine-based nanoparticles is that the outer surface of gelatin causes intermolecular and intramolecular crosslinking with an increase in time, temperature, and humidity. To overcome this issue, crosslinkers like glutaraldehyde (GA) are used in the formulation to give stability. Two types of gelatine (gelatin type A with pH 7–9 and gelatin type B with pH 4–5) can be synthesized by hydrolysis of either acid or base, leading to the change in properties such as viscosity, molecular weight, amino acid composition, pH, and isoelectric points. Gelatin nanoparticles are prepared by precipitation method with either BSA or HSA or in combination (BSA/HSA). The encapsulation efficiency while using BSA was 90% with a controlled release profile and for HSA encapsulation efficiency with a linear release profile (Salata, 2003).

22.4.3 GLIADIN AND LEGUMIN

Gluten-based protein like gliadin is extracted from wheat, which is a dynamic biopolymer mostly suited for oral delivery of bioactive components and nutraceuticals. In addition to the advantages of protein nanoparticles, gliadin has much adhesivity, which is found in polysaccharides like chitosan and greater tropism (indicative growth toward uppermost regions of the gastrointestinal tract and has shown lesser existence toward intestinal region). The hydrophobicity and solubility of gliadin-based nanoparticles can protect the entrapped nutraceutical for its sustained release. For a protein-based nanoparticle, neutral amino acids present in them nurture hydrogen bonding with mucous membranes while lipophilic amino acids present on nanoparticles bind with biological tissue via hydrophobic interactions. Gliadin is rich in both lipophilic and neutral amino acids suggesting that it would have both hydrophobic and hydrogen bonding interactions (Joye et al., 2015; Peng et al., 2017). Peng et al. (2017) developed gliadin-based nanoparticles encapsulated with an anticancer drug for targetting breast cancer cell lines. Gliadin-gelatin-based nanoparticles were also developed with the electrospray technique for the sustained releasing of drugs for cancer therapy.

22.4.4 ELASTIN

Elastin an essential component in connective tissue enables maintaining the elasticity of the skin and tensile strength of the underlying tissues. Formation of desmosine and isodesmosine results from oxidative deamination of lysine that can crosslink elastin. Two types of polypeptides derived from elastin are used in nutraceutical delivery system, namely, (1) α-o elastin, which undergoes aggregation under certain selective conditions, and (2) elastin-like polypeptides, derived from tropoelastin, a precursor synthesized during the development of desmosine and isodesmosine which aggregates at elevated temperature. The elastin-like polypeptides are highly soluble. Wu et al. (2009) synthesized elastin nanoparticles for the sustained release nutraceutical by electrospraying method.

22.4.5 ZEIN

Zein protein is rich in amino acids, such as proline and glutamine. Zein nanoparticles are ideal for the preparation of films and coatings. Zein nanoparticles have become popular in the delivery of nutraceuticals like curcumin and are widely used for the prevention of health disorders. Zein nanoparticles are efficiently used in the food system for the detection of azorubine (Luo et al. 2012). Zein-pectin nanoparticle consisting of zein core (hydrophobic) and a pectin shell (hydrophilic) was developed for the smooth delivery of curcumin (Labib, 2018).

22.4.6 SOY PROTEIN

Soybean is a legume and an abundant source of proteins called soy protein isolates that have gained popularity due to their high nutritional value. The important component for soy protein isolate is glycinin and β conglycinin. Soy protein isolates based nanoparticles are commonly synthesized by desolvation technique by adding crosslinkers such as tripolyphosphate (TPP). When subjected to heat denaturation, these protein aggregates, and links to the crosslinkers forming nanoparticles. Soy protein nanoparticles were developed by Verma et al. (2018) for the targeted delivery of bioactive components like curcumin. Stabilizers for oil in water emulsion were developed with soy protein nanoparticles by Singh et al. (2017). Two types of soy protein nanoparticles were prepared with heat treatment and without heat treatment and from the results, it was indicated that the heat-denatured protein of soy isolate showed better results. A gel-like network was formed by the protein which could encapsulate oil droplets and therefore these protein emulsions could stabilize Pickering.

22.4.7 MILK PROTEIN

Milk contains numerous different types of proteins with varied functions and properties. The use of milk protein as a nutraceutical delivery system is an emerging trend that has received much consideration. Two types of protein in milk widely used for the delivery of nutraceuticals are casein and *β*-lactoglobulin (βLG). *β*-lactoglobulin is mainly used in the delivery of hydrophobic molecules such as curcumin and polyphenols. βLG is a smaller protein, with 162 amino acid residues and a molecular weight of 18.4 kDa (Sakurai and Oobatake, 2001). Conditions like high temperature (above 60 °C), alkaline pH, and ionic strength denature βLG, triggering the monomer formation and these monomers aggregates resulting in transparent "fine-stranded" gels. These protein molecules further assemble to form extended stiff fibers under continuous heating at low pH as well as low ionic strength conditions (Haug et al., 2009). Samadarsi and Dutta (2019) successfully synthesized βLG-based nanoparticles for the target-specific delivery of nutraceutical, mangiferin by desolvation technique, which has high stability in the stomach. Due to the rigid β-sheet structures, βLG protein has significant resistance toward the digestion of proteolytic enzymes like pepsin, but βLG protein could digest in the small intestine (Saadati and Razzaghi, 2012). Therefore, βLG protein-based nanoparticles can be efficient for the controlled delivery of nutraceuticals by oral administration. βLG nanoparticles are used for the successful delivery of nutraceuticals such as curcumin (Sari et al., 2015; Teng et al., 2016) and epigallocatechin-3-gallate (EGCG) (Liang et al., 2017). Another milk protein used as a delivery system is casein existing as micelles. Casein is efficient in the delivery of calcium and amino acid. Casein nanoparticles are stable at elevated temperatures and mechanical stress (Luo et al. 2011, 2012).

22.5 MAIN CONSIDERATIONS FOR DESIGNING ENCAPSULATING AGENTS AND DELIVERY SYSTEMS

Various parameters determine the stability and efficiency of a nutraceutical delivery system (Foox and Zilberman, 2015).

1. The interaction of protein matrix with nutraceuticals and the external environment.
2. The physical–chemical characteristics of the protein-based encapsulant could influence the encapsulation capacity of the nutraceuticals. In different physiological processes, the surface properties of the protein encapsulant have a significant influence in interacting with the external environment and nutraceuticals and are summarized in Table 22.1.

TABLE 22.1 Properties for the Encapsulation of Nutraceuticals

Properties	Description	Contributing Factors
Loading capacity	Weight (or molar) ratio between the encapsulated compound and the encapsulant	Indicates the efficiency of encapsulation; compound-matrix interaction (electrostatic, hydrophobic, hydrogen bonding, | Van der Waals, etc.)
Dispersion stability	Stability against precipitatio	Contributes to the solubility and absorption of encapsulated compounds; electric charge, hydrophilic groups, and steric hindrance on the surface
Controlled release	Release at desired time or locales, or upon exposure to certain stimuli	Improves the efficacy of delivery and minimizes the possible side effect; suitable polymers or functional groups responsive to certain environmental changes (e.g., pH or enzymes)
Mucoadhesion	Adhesion to the mucosa in the gastrointestinal-tract.	Contributes to the absorption of entrapped compounds; positive charges on the surface; abundance of hydrogen bond forming groups (e.g., hydroxyl groups)
Prolonged circulation	Extended dwelling time in the circulative system	Reduces the loss of bioactive compounds due to opsonisation; steric hindrance or biomimetic polymers on the surface
Cellular uptake	Delivery at the cellular level	Reduced size of the delivery system; positive surface charge, high surface hydrophobicity, existence of target-specific ligands

22.6 METHODS OF PREPARATION OF BIOACTIVE COMPOUND-PROTEIN NANOPARTICLES

22.6.1 COMPLEX COACERVATION

Complex coacervation is a widely used method in the synthesis of nanoparticles, which involves a solvent to be associated, a nutraceutical/bioactive component to be entrapped, and protein-based coating material. The process consists of four steps: (1) formation of an aqueous solution with two or more protein polymers, prepared at higher gelling temperature, and isoelectric point; (2) blending the hydrophobic phase of the protein–polymer to the aqueous solution of another polymer and mixing the obtained mixture, for a stable emulsion formation; (3) adjusting the temperature and pH to activate coacervation and phase separation; and (4) the surface of protein–polymer is hardened by either using high temperature, desolvating agents or cross-linkers, which is depicted in Figure 22.1. The coacervation is induced by altering the pH and temperature or by adding salt. Fine coacervate particles are formed by the process which is called microcoacervation. Continuation of the coacervation process leads to the aggregation of nanoparticles resulting in larger particles, which in turn possess higher density resulting in the sedimentation of particles. The definite separation of two systems is produced after a certain period (Figure 22.1). This stage is called macrocoacervation (Timilsena et al., 2019).

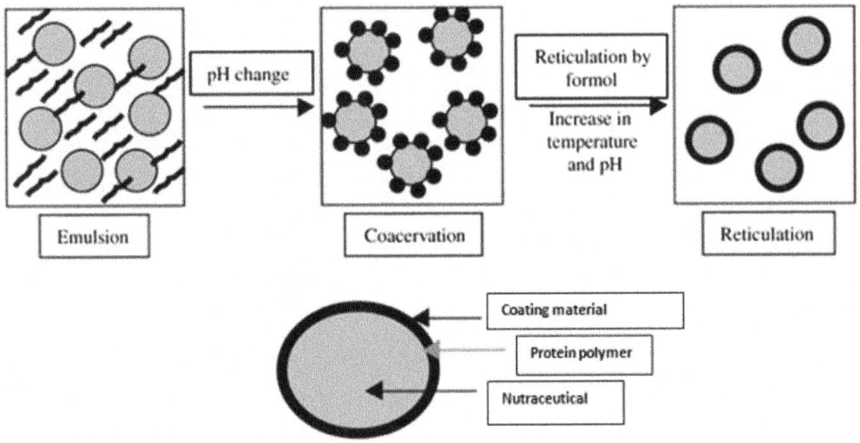

FIGURE 22.1 Process of coacervation.

22.6.2 DESOLVATION METHOD

This comes under the coacervation method (Ko and Gunasekaran, 2006). In this method, to the aqueous state of protein solution, a desolvating agent, such as salt or acetone, or alcohol is added under continuous stirring, which would result in protein is structural change. To these crosslinkers such as TPP or GA can be added to mediate the nanoparticle formation (Samadarsi and Dutta, 2019). The turbidity of the system can be increased to separate the nanoparticles. A flowchart presentation of protein nanoparticles synthesis with desolvation/coacervation technique is discussed in Figure 22.2.

22.6.3 METHOD OF NANOPARTICLE PREPARATION BY EMULSIFICATION

The selection of nanoparticle preparation method depends on the nature and function of the protein encapsulant and the nutraceutical to be encapsulated.

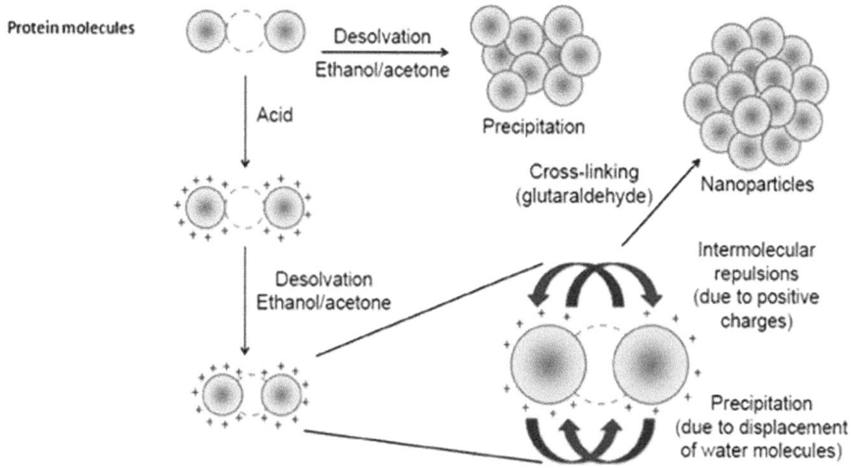

FIGURE 22.2 Preparation of protein nanoparticles by desolvation method.

22.6.3.1 EMULSION-SOLVENT EVAPORATION METHOD

The emulsion-solvent evaporation method is the most commonly used techniques for the synthesis of protein nanoparticles. This technique involves two-steps, the first step is the emulsification of the protein into an aqueous state. The next step is the evaporation of polymer-solvent, activating polymer precipitation as nanoparticles (Figure 22.3.). These nanoparticles are purified by the centrifugation process, and free unencapsulated nutraceuticals and stabilizers are removed. Further, the nanoparticles are lyophilized for storage (Luo et al., 2011; Larsson et al., 2012).

22.6.3.2 DOUBLE EMULSION AND EVAPORATION METHOD

Most of the techniques discussed above are used for the entrapment of hydrophobic nutraceutical molecules, for hydrophilic molecules, the limitation of these methods is the poor encapsulation efficiency. Hence for the entrapment of hydrophilic nutraceuticals, the double emulsion technique is preferred. The steps involving this process are the preparation of nutraceutical solution and adding it to the polymer solution under constant magnetic stirring forming emulsions (w/o). The first polymer solution is added to the aqueous phase also with continuous magnetic stirring resulting in w/o/w emulsion (Figure 22.3). The solvent from the emulsion is removed by evaporation, and purified nanoparticles are obtained by several centrifugation cycles and washed before lyophilization (Teng et al., 2016). The parameters that affect the nanoparticle size are the concentration of nutraceutical, polymer, and stabilizer used, and the volume of aqueous solution (Liang et al., 2017).

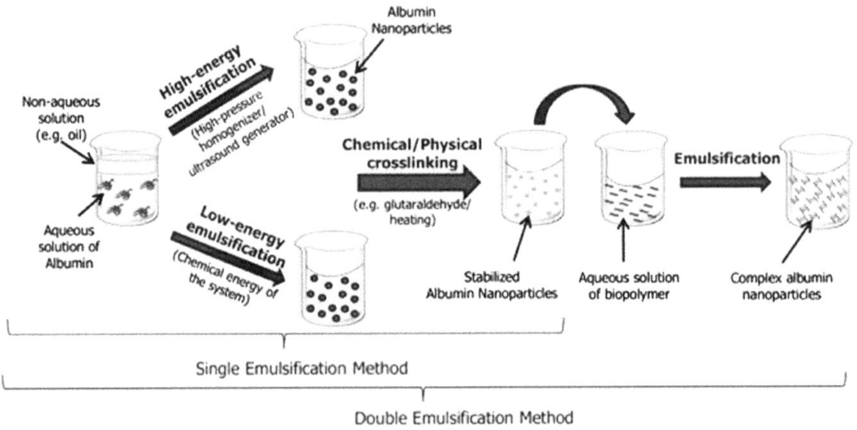

FIGURE 22.3 Preparation o\f nanoparticles by emulsification method.

22.6.4 EMULSIONS–DIFFUSION METHOD

This is another most commonly used method for nanoparticle synthesis. Into a partially water-miscible solvent like propylene carbonate or benzyl alcohol, the protein solution used for entrapment is dissolved in the water-saturated solution to assure the thermodynamic equilibrium of both solvents initially. Finally, the solution saturated with protein–water solvent emulsion is prepared in an aqueous solution of stabilizer, resulting in the diffusion of solvent to the external phase and thus forming the nanoparticles, conferring to the oil-to-polymer ratio. Subsequently, the solvent is removed by evaporation. Merits of this method include high encapsulation efficiencies (above 70%), nonrequirement of homogenization, high reproducibility and can be easily scaled up, simplicity, and uniform size distribution. Disadvantages include the removal of a high quantity of water from the suspension resulting in the loss of water-soluble nutraceuticals (Mirshahi et al., 1996). Several nutraceuticals have been entrapped by this technique, including curcumin/β-lactoglobulin nanoparticles, EGCG-loaded protein nanoparticles (Sari et al., 2015).

22.6.5 PRECIPITATION TECHNIQUE OR SOLVENT DISPLACEMENT TECHNIQUE

In this method, protein in the organic solution is precipitated and the organic solution is diffused in an aqueous medium, with surfactants. In a water-miscible solvent such as acetone or alcohol, protein solution, bioactive components to be encapsulated, and or lipophilic surfactant are dissolved. Under continuous magnetic stirring, the abovementioned solution was added to the aqueous solution composed of a stabilizer. Nanoparticles are formed spontaneously, by rapid solvent diffusion, further removal of solvent from the suspension subjected toa lowered pressure is carried out. The nanoparticle size is affected by the rate of organic phase addition into the aqueous phase. A reduction in both size and nutraceutical encapsulation was observed at an increase in the rate of homogenizing of the two phases (Labib, 2018). Nano precipitation method is efficient for poorly soluble nutraceuticals. Nanoparticle size, the release of the nutraceutical from the encapsulant, and the yield can be controlled by optimizing preparation parameters (Larsen et al., 2016).

22.6.6 ELECTROSPRAY TECHNIQUE

Electrospray technique (Figure 22.4) is another widely preferred method of nanoparticle preparation. The process involves a syringe pump having a protein–polymer solution that is connected to a high voltage power supply, composed of the functional electrode. A metal foil collector placed opposite functions as the ground electrode. Depending on the flow rate of the solution, the voltage was optimized for electrospraying. Due to the surface tension, the liquid coming from the nozzle into the electric field forms a tailor cone. An increased electric field, results in the breaking of Taylor cone into highly charged droplets, optimizing the parameters, droplets can be made into nanoparticles. In the electrospraying technique, a very high voltage is given to a protein solution so that the protein–polymer is induced to come out from the syringe pumps nanoparticles. Electro spraying uses integrated technology for the development of nanostructures. EGCG protein nanoparticles based on the electron spray method were found to stable and do not lead to the loss in the bioactivity of bioactive components (Sari et al., 2015). For the intention of inhaling medicines, for respiratory disorders, electron spray nebulizers, producing micro- and nano-particles were developed. These "breathable size range particles" are designed to deliver the medicine into the lower airways without loss of activity of the encapsulated medicine. Electro sprayed nanoparticles can encapsulate nutraceuticals and can be specific nutraceutical carriers because of their active surface absorption, binding, or complexation. The size and the shape of the nanoparticle are the two important parameters in deciding the action mode and therapeutic potential of the nanostructure as it selects the specific binding to the target molecule and reactivity toward the environment (Sridhar and Ramakrishna, 2013). Therefore, nanoparticles prepared by the electron spray method have an enormous scope in food nanotechnology and nutraceutical delivery for therapeutic purposes. The advantages of electrospraying techniques include the production of the lowest and uniform particle size, easy to control the operation parameters, fast preparation, and ability to be synthesized in bulk amount. One of the major disadvantages of this method is the degradation of bioactive components owing to the stress in operational parameters (Sridhar and Ramakrishna, 2013).

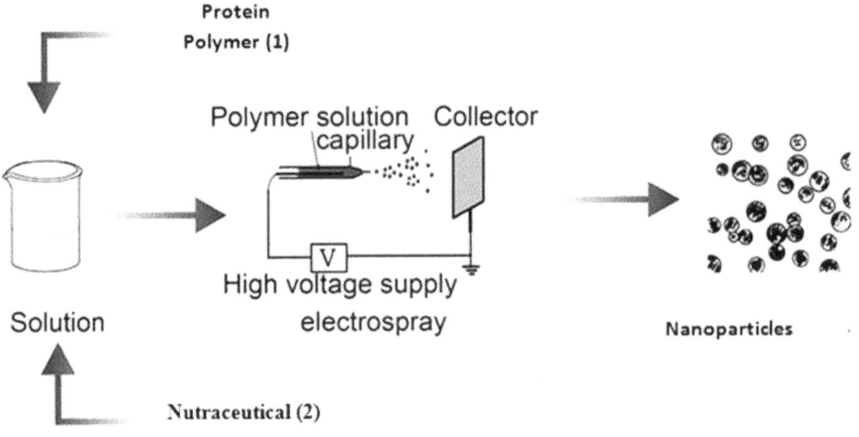

FIGURE 22.4 Preparation of nanoparti\cles by electro spray technique.

22.6.7 COMPARISON OF NANOPARTICLES PREPARATION METHODS

Merits, demerits, and application of various nanoparticle preparation techniques are given in Table 22.2.

TABLE 22.2 Comparison of Nanoparticle Synthesis Methods

Type	Methods	Mechanisms	Merits	Demerits	Application
Bottom-up synthesis	Emulsion	Precursor vaporization, nucleation, growth	Low cost, continuous operation, high yield	More chances of polymer, nutraceutical, or biomaterial degradation	Calcium phosphate micro-spheres and Au nanoparticles for drug delivery
	Desolvation	Precursor solution	Low cost, continuous operation	Needs removal of catalyst components, may involve excess solvent usage for scale ups	Highly explored technology for nutraceutical delivery and biomedical applications
Top-down synthesis	Lithography, etching, milling, or machining	Breaking down of large piece of material	Offers reliability and device complexity	Higher energy spent and more waste produced	Biomedical industry
	Electrospraying	Driven by difference in electric potential and surface viscosity	Increased drug encapsulation efficiency, simplicity, low cost, continuous operation, high yield	Chances of shear or thermal stress in some biomaterial nanoparticle	Has good prospects in drug/nutraceutical delivery and pharma-based

22.7 CHARACTERIZATION OF NANOPARTICLES

Characterization of the nanostructure is based on their average particle size, size distribution, surface charge, physical stability, morphology, the in vivo distribution, and toxicity of the nanoparticles. Various microscopic techniques are used for the morphological analysis of the nanoparticles such as atomic force microscopy (AFM), scanning electron microscopy (SEM), and transmission electron microscopy (TEM), which can be used in understanding the structure of nanoparticles, size, and surface charge can be evaluated by dynamic light scattering (DLS) method (Samadarsi and Dutta, 2019).

22.7.1 PARTICLE SIZE

The major application of nanoparticles is in nutraceutical release and targeting. According to previous investigations, particle size affects the release. Smaller particles offer a larger surface area. Most of the nutraceuticals entrapped in smaller size nanoparticles are on the surface, resulting in the faster release of the nutraceuticals, or bioactive components once the protein encapsulant starts degrading. The particle size of the nanostructure is also determined by the polydispersity (PDI) value of the nanoparticle and zeta potential. Mostly for nanoparticles, a lower PDI value of 0.2–0.3 is preferred. Zeta potential evaluates the surface charge of particles in a colloidal solution. Surface charge either above +25 or below −25 mV is preferred, as it would prevent the nanoparticles from aggregating and thus the overall increase in particle size (Gonza, 2007). Particle size also influences the rate of polymer degradation. For example, as the particle size increased, the rate of degradation of polylactic-co-glycolic acid increased in vitro studies (Gibbs et al., 1999). A study conducted by Samadarsi and Dutta (2019) on β-lactoglobulin nanoparticle encapsulated with mangiferin showed that during in-vitro digestion, the zeta potential of the nanoparticle was reduced due to the change in pH in the intestinal tract. There are various techniques for measure nanoparticle size, as discussed below.

22.7.1.1 DYNAMIC LIGHT SCATTERING

Currently, the most rapid and widely used technique of evaluating the particle size of a nanoparticle is by photon-correlation spectroscopy (PCS), also termed as DLS. DLS determines the particle size of the Brownian particles in colloidal suspension. Nanoparticles undergo a Doppler shift when subjected to a laser beam (monochromatic light) on the surface of particles in Brownian motion, which in turn changes the wavelength of the light. This method helps to evaluate the average size distribution of the particles, motion of the nanoparticles in the colloidal suspension, and also the coefficient of diffusion of the nanoparticle by the autocorrelation function. The PCS represents the most commonly used technique for accurate estimation of the particle size and size distribution based on DLS (Gibbs et al., 1999; Gonza, 2007; Zuidam and Shimoni, 2010).

22.7.1.2 SEM FOR SIZE DETERMINATION AND MORPHOLOGICAL ANALYSIS OF THE NANOPARTICLES

SEM is used to determine the particle size and morphology of nanoparticles; however, the technique gives minimum information about the size distribution and true population average. Limitation of SEM characterization includes the need for nanoparticles solution to be presented in a dry state and nonconductive nanomaterial need to be sputter coated with conductive metals such as platinum or gold, following which sample is scanned with a focused fine beam of electrons (Gonza, 2007; Zuidam and Shimoni, 2010). The surface characteristics of the sample are obtained from the secondary electrons emitted from the sample surface, which could cause damages to the surface of the protein encapsulants. The mean size obtained by SEM is comparable with results obtained by DLS; however, the SEM analysis is mostly done in the dry state causing the nanoparticles to shrink. Therefore, usually in SEM results in the particle size of the nanoparticles could be lower than that obtained in DLS (Samadarsi and Dutta, 2019). Moreover, these techniques are time-consuming, costly, and frequently need complementary information about sizing distribution (Sari et al., 2015; Teng et al., 2016).

22.7.1.3 TRANSMISSION ELECTRON MICROSCOPE

Similar results are produced by SEM and TEM though their principles are different. The sample preparation for TEM is more time-consuming than SEM and is a complicated procedure, as an ultra-thin electron bean needs to be transmitted through an ultra-thin sample, which would interact with the nanoparticles. The dispersion of nanoparticles is placed on either a grid or films and fixed using negative staining material (phosphotungstic acid) or by exposure to liquid nitrogen posttreatment in vitreous ice (Sakurai and Oobatake, 2001). Most of the nanoparticle preparation and characterization studies have used anyone of the morphological characterization techniques either SEM, TEM, or AFM.

22.7.1.4 ATOMIC FORCE MICROSCOPY

AFM can produce ultra-high resolution in particle size measurement. AFM is based on the scanning of samples physically at a submicron level with the help of a probe tip of the atomic scale (Luo et al., 2011). This technique has several advantages over SEM and TEM. Using AFM, the nonconducting samples, like biocompatible polymers, can be imaged without the requirement of conducting coating material, thus allowing imaging of biocompatible nanoparticles. A three-dimensional view of the nanoparticle samples can be obtained with the help of AFM along with the topographical image. Nanoparticle samples that need to be imaged can be done in contact or noncontact mode. In the contact mode, the topographical map is developed by patting the probe onto the surface across the sample, and probe drifts over the conducting surface in noncontact mode. AFM imaging also finds out the interaction between nanoparticles and their entrapped bioactive components as well as their interaction with the biological environment (Larsson et al., 2012).

22.7.2 SURFACE CHARGE

The nanoparticle communicates with the biological environment, through various food systems in association with the bioactive components encapsulated within it and can be evaluated by its surface charge. The stability is measured through zeta potential which measures the indirect surface charge of nanoparticles. It correlates to the potential difference between the outer Helmholtz plane and the surface of shear. By evaluating the zeta potential of the nanoparticles, the storage stability and PDI of the nanoparticles can be assimilated. The zeta potential value above +25 mv or below −25 mv prevents particle aggregation. In β-latctoglobulin nanoparticles used for the efficient delivery of nutraceutical mangiferin, it was seen that as zeta potential increased particle aggregation decreased, and stability was enhanced (Samadarsi and Dutta, 2019). The extent of surface hydrophobicity can then be predicted from the values of zeta potential. The zeta potential also gives an idea about the physicochemical properties of the encapsulant and the molecule entrapped (Gibbs et al., 1999; Gonza, 2007; Zuidam and Shimoni, 2010).

22.7.3 SURFACE HYDROPHOBICITY

The surface hydrophobicity of a nanoparticle can be measured by diverse techniques like biphasic partitioning, adsorption of probes, hydrophobic interaction chromatography, contact angle measurements, etc. Lately, numerous complicated analytical techniques are reported in the research studies for surface characterization of nanoparticles. X-ray PCS allows the documentation of specific chemical groups on the surface of nanoparticles (Saadati and Razzaghi, 2012).

22.7.4 NUTRACEUTICAL RELEASE

The main reason for following nanoencapsulation is to deliver drugs or nutraceuticals; hence, understanding the aspect and extent of release of the drug or nutraceutical is pivotal (Teng et al., 2016). To determine the release of molecules loading capacity and encapsulation efficiency of the nanoparticles needs to be analyzed. The nutraceutical loading capacity of the nanoparticles can be defined as the amount of nutraceutical bound per mass of nanomaterial. Analysis can be carried out by UV spectroscopy or high-performance liquid chromatography after centrifugal ultrafiltration, ultracentrifugation, ultrafiltration, or gel filtration (Sari et al., 2015).

22.8 ROUTES

There is different relevance for protein nanosystems as carriers for the delivery of nutraceuticals, bioactive components, and drugs via various routes of administration which have been discussed further.

22.8.1 ORAL ROUTE

Oral administration is among the most preferred routes for drug administrations, as this route has several advantages such as avoiding contaminations, patient convenience, and compliance, and infections (Jones et al., 2011; Luo et al., 2012). Protein-based nanoparticles gave a poor performance in the gastrointestinal system, owing to their denaturation ability due to various proteases and pH changes in the tract resulting in aggregation, thus affecting oral bioavailability. Hindrance to oral administration of protein and peptide can be categorized as physical, chemical, and enzymatic barriers (Chen et al., 2006; Teng et al., 2016). The advantage of exploring protein nanoparticles is that they could be modified to facilitate their transport to the targeted site (Haug et al., 2009). Milk protein-based nanoparticles like β-lactoglobulin and α-lactoglobulin have shown excellent bioavailability and accessibility as they are resistant to proteolytic enzyme degradation and can be ideal for oral delivery of nutraceuticals (Samadarsi and Dutta, 2019).

22.8.2 NASAL ROUTE

Another preferred route for noninvasive nutraceutical delivery is the nasal route. This is an effective way to treat various respiratory-related disorders. Recent up gradation in inhalation devices and targeting patterns has remarkably increased research interest in drug and nutraceutical delivery via this route (Atta et al., 2017). Advantages of nutraceutical delivery through nasal route include a highly vascularized mucosal layer, porous endothelial membrane, large surface area, lower enzymatic activity as compared to oral delivery, and evasion of the first-pass metabolism. However, the arrangement of dismissal and size characterization through the delivery system and nasal clearance mechanisms might be a challenge to nutraceutical delivery. Post-intranasal administration, nutraceuticals can be directly absorbed into the systemic circulation, into the central nervous system, or the gastrointestinal tract. DDAVP®, Miacalcin®, Fortical®, Synarel®, and Syntocinon® are some of the marketed nutraceuticals and peptides for nasal administration (Verma et al., 2018).

22.8.3 PULMONARY ROUTE

Among the most routinely examined noninvasive routes, to enhance the absorption of nutraceuticals and peptides, are the pulmonary route. These routes provide several advantages such as large absorptive surface area ($100\,m^2$), huge vascularization, thin epithelial membrane ($0.1–0.2\,\mu m$), and reduced enzymatic activity. Even with these merits, numerous factors might regulate pulmonary nutraceutical absorption. PGLA nanoparticles encapsulated with Alpha 1-antitrypsin-loaded PLGA have been intercepted as a material for treating respiratory disorders shortly (Joung et al., 2016).

22.8.4 BLOOD–BRAIN BARRIER ROUTE

Protein nanoparticles can cross the blood–brain barrier that could not be crossed by drugs through parenteral administration. Loperamide, doxorubicin, and tubocurarine are some of the drugs encapsulated by protein nanosystems (Joye et al., 2015).

22.8.5 OCULAR THERAPY

Protein nanoparticles-based eye formulations have a remarkably longer half-life when compared to that of commercially available eye drops. Gelatin nanoparticles encapsulated with pilocarpine considerably reduced the intraocular pressure in rabbits with glaucoma in comparison to commercially available pilocarpine eye-drop solution in-vitro experiments (Salata, 2003; Suri et al., 2007).

22.9 CONCLUSION

The preparation of nanoparticles for nutraceutical and drug delivery system is accepted to play a major role in the therapy of various life-threatening diseases as it assures secure, effective, and stable treatment effects, while protein nanoparticles retain promising results in various administration routes like oral, nasal, ocular, pulmonary, and blood–brain barrier route. Also, various protein nanosystems like albumin and gelatin show elevation at a commercial level. New proteins obtained from natural sources are used in nanotechnology for the synthesis of new drugs and nutraceuticals, which results in some effective results. Growth in various fields of nanosystems can lead to the improvement in the application of protein-based nanoparticles for the therapy of diseases, though the involvement of protein nanoparticles for several health disorders have shown some astonishing results and also clench greater promise in the medical field and food \ technologies in the future.

KEYWORDS

- **emulsification**
- **desolvation**
- **complex coacervation**
- **electrospray particle size**
- **in vitro release**

REFERENCES

Alfadul, S.; Elneshwy, A. Use of nanotechnology in food processing, packaging, and safety review. *African J. Food, Agric. Nutr. Dev.* **2010**, 10(6), 2719–2739.

Ashwini, C.; Vaishali, K.; Digambar, N. Role of nutraceuticals in various diseases, a comprehensive review. *Int. J. Res. Pharma. Chem.* **2013**, 3(2), 290–299.

Atta, E. M.; Mohamed, N. H.; Abdelgawad, A. A. M. Antioxidants, an overview on the natural and synthetic types. *Eur. Chem. Bull.* **2017**, 6(8), 365. doi, 10.17628/ecb.2017.6.365-375.

Chen, L.; Remondetto, E.; Subirade, M. Food protein-based materials as nutraceutical delivery systems. *Trends Food Sci. Technol.* **2006**, 17, 272–283. doi, 10.1016/j.tifs.2005.12.011.

Foox, M.; Zilberman, M. Drug delivery from gelatin-based systems. *Expert Opin. Drug Delivery* **2015**, 12(9), 1547–1563. doi, 10.1517/17425247.2015.1037272.

Gibbs, B. F.; Kermasha, S.; Alli, I.; Mulligan, C. N. Encapsulation in the food industry: a review. *Int. J. Food Sci. Nutr.* **1999**, 50, 213–224.

Gonza, A. F. Effect of β-lactoglobulin A and B whey protein variants on the rennet-induced gelation of skim milk gels in a model reconstituted skim milk system. *J. Dairy Sci.* **2007**, 90, 582–593. doi, 10.3168/jds.S0022-0302(07)71541-2.

Haug, I. J.; Skar, H. M.; Vegarud, G. E.; Langsrud, T.; Draget, K. I. Electrostatic effects on β-lactoglobulin transitions during heat denaturation as studied by differential scanning calorimetry. *Food Hydrocolloids* **2009**, 23(8), 2287–2293. doi, 10.1016/j.foodhyd.2009.06.006.

Jahanshahi, M.; Babaei, Z. Protein nanoparticle: a unique system as drug delivery vehicles. *Afr. J. Biotechnol.* **2008**, 7(25), 4926–4934.

Jones, O. G.; Handschin, S.; Adamcik, J.; Harnau, L.; Bolisetty, S.; Mezzenga, R. Complexation of β-lactoglobulin fibrils and sulfated polysaccharides. *Biomacromolecules* **2011**, 12, 3056–3065. doi, 10.1021/bm200686r.

Joung, H. J.; Choi, M.-J.; Kim, J. T.; Park, S. H.; Park, H. J.; Shin, G. H. Development of food-grade curcumin nanoemulsion and its potential application to food beverage system, antioxidant property and in vitro digestion. *J. Food Sci.* **2016**, 81(3), N745–N753.

Joye, I. J.; Nelis, V. A.; McClements, D. J. Gliadin-based nanoparticles, fabrication and stability of food-grade colloidal delivery systems. *Food Hydrocolloids* **2015**, 44, 86–93.

Ko, S.; Gunasekaran, S. Preparation of sub-100-nm β-lacto globulin (BLG) nanoparticles. *J. Microencapsulation* **2006**, 23, 887–898. doi, 10.1080/02652040601035143.

Labib, G. Overview on zein protein, a promising pharmaceutical excipient in drug delivery systems and tissue engineering. *Expert Opin. Drug Delivery* **2018**, 15(1), 65–75. doi, 10.1080/17425247.2017.1349752.

Larsen, M. T.; Kuhlmann, M.; Hvam, M. L.; Howard, K. A. Albumin-based drug delivery, harnessing nature to cure disease. *Mol. Cell. Ther.* **2016**, 4(1), 1–12. doi, 10.1186/s40591-016-0048-8.

Larsson, M.; Hill, A.; Duffy, J. Suspension stability: why particle size, zeta potential, and rheology are important. *Annu. Trans. Nord. Rheol. Soc.* **2012**, 20, 209–214.

Liang, J.; Yan, H.; Wang, X.; Zhou, Y.; Gao, X.; Puligundla, P.; Wan, X. Encapsulation of epigallocatechin gallate in zein/chitosan nanoparticles for controlled applications in food systems. *Food Chem.* **2017**, 231, 19–24. doi, 10.1016/j.foodchem.2017.02.106.

Luo, Y.; Zhang, B.; Whent, M.; Yu, L.; Wang, Q. Preparation and characterization of zein/chitosan complex for encapsulation of α-tocopherol, and its in vitro controlled release study. *Colloids Surf. B, Biointerfaces* **2011**, 85(2), 145–152. doi, 10.1016/j.colsurfb.2011.02.020.

Luo, Y.; Teng, Z.; Wang, Q. Development of zein nanoparticles coated with carboxymethyl chitosan for encapsulation and controlled release of vitamin D3. *J. Agric. Food Chem.* **2012**, 60, 836–843. doi, 10.1021/jf204194z.

Malsuno, R.; Adachi, S. Lipid encapsulation technology—techniques and applications to food. *Trends Food Sci. Technol.* **1993**, 4, 781–785.

Mirshahi, T.; Irache, J. M.; Gueguen, J.; Orecchioni, A. M. Development of drug delivery systems from vegetal proteins, legumin nanoparticles. *Drug Dev. Ind. Pharm.* **1996**, 22(8), 841–846. doi, 10.3109/03639049609065914.

Peng, D.; Jin, W.; Li, J.; Xiong, W.; Pei, Y.; Wang, Y.; Li, Y.; Li, B. Adsorption and distribution of edible gliadin nanoparticles at the air/water interface. *J. Agric. Food Chem.* **2017**, 65(11), 2454–2460. doi, 10.1021/acs.jafc.6b05757.

Saadati, Z.; Razzaghi, A. Comparative analysis of chemical and thermal denatured β-lactoglobulin AB in the presence of casein. *Adv. Stud. Bio.* **2012**, 4(6), 255–264.

Sakurai, K.; Oobatake, M. Salt-dependent monomer-dimer equilibrium of bovine-lactoglobulin at pH 3. *Protein Sci.* **2001**, 10, 2325–2335. doi, 10.1101/ps.17001.milk.

Salata, O. Applications of nanoparticles in biology and medicine. *J. Nanobiotechnol.* **2003**, IV(5), 1–6.

Samadarsi, R.; Dutta, D. Design and characterization of mangiferin nanoparticles for oral delivery. *J. Food Eng.* **2019**, 247, 80–94. doi, 10.1016/J.JFOODENG.2018.11.020.

Sari, T. P.; Mann, B.; Kumar, R.; Singh, R. R. B.; Sharma, R.; Bhardwaj, M.; Athira, S. Preparation and characterization of nanoemulsion encapsulating curcumin. *Food Hydrocolloids* **2015**, 43, 540–546. doi, 10.1016/j.foodhyd.2014.07.011.

Singh, T.; Shukla, S.; Kumar, P.; Wahla, V.; Bajpai, V. K. Rather, I. A. Application of nanotechnology in food science, perception and overview. *Front. Microbiol.* **2017**, 8, 1–7. doi, 10.3389/fmicb.2017.01501.

Sridhar, R.; Ramakrishna, S. Electrosprayed nanoparticles for drug delivery and pharmaceutical applications. *Biomatter* **2013**, 3(3), 37–41. doi, 10.4161/biom.24281.

Suri, S. S.; Fenniri, H.; Singh, B. Nanotechnology-based drug delivery systems. *J. Occup. Med. Toxicol.* **2007**, 2(16). doi, 10.1007/s12030-009-9028-2.

Teng, Z.; Luo, Y.; Li, Y.; Wang, Q. Cationic b-lactoglobulin nanoparticles as a bioavailability enhancer, effect of surface properties and size on the transport and delivery in vitro. *Food Chem.* **2016**, 204, 391–399. doi, 10.1016/j.foodchem.2016.02.139.

Timilsena, Y. P.; Akanbi, T. O.; Khalid, N.; Adhikari, B.; Barrow, C. J. Complex coacervation, principles, mechanisms and applications in microencapsulation. *Int. J. Biol. Macromol.* **2018**, 121, 1276–1286. doi, 10.1016/j.ijbiomac.2018.10.144.

Verma, D.; Gulati, N.; Kaul, S.; Mukherjee, S.; Nagaich, U. Protein based nanostructures for drug delivery. *Int. J. Pharm.* **2018**, 2018, 1–18. doi, 10.1155/2018/9285854.

Wu, Y.; MacKay, J. A.; McDaniel, J. R.; Chilkoti, A.; Clark, R. L. Fabrication of elastin-like polypeptide nanoparticles for drug delivery by electrospraying. *Biomacromolecules* **2009**, 10(1), 19–24. doi, 10.1021/bm801033f.

Zuidam, N. J.; Shimoni, E. Overview of microencapsulates for use in food products or processes and methods to make them. In: *Encapsulation Technologies for Active Food Ingredients and Food Processing*, Springer: New York, NY, **2010**, 3–29.

CHAPTER 23

Application of Nanomaterials in Environmental Pollution Abatement and Their Impact on Ecological Sustainability: Recent Status and Future Perspective

SYED NIKHAT AHMED[1], SUBHASHREE SUBHADARSINI MISHRA[1], JAYANTA KUMAR SAHU[2], SABITA SHROFF[3], PRAJNA PARAMITA NAIK[4], ISWAR BAITHARU[1*], and SANJAT KUMAR SAHU[1]

[1]P.G. Department of Environmental Sciences, Sambalpur University, Jyoti Vihar, Burla, Odisha, India

[2]School of Life Sciences, Sambalpur University, Jyoti Vihar, Burla, Odisha, India

[3]School of Chemistry, Sambalpur University, Jyoti Vihar, Burla, Odisha, India

[4]P.G. Department of Zoology, Vikram Deb Autonomous College, Jeypore, Odisha, India

[*]Corresponding author. E-mail: iswarbaitharu@suniv.ac.in

ABSTRACT

The use of nanomaterials is increasing at a very rapid rate in almost every field. Accumulation of nanomaterials in air, water, and soil to toxic levels following their end use can impair ecological functioning and human health since most of them are nondegradable and persistent in nature. While nanotechnology is established as an emerging technology of the 21st century, toxicological studies of the nanomaterials to human health and other ecological components are still at a nascent stage. The lack of standardized up-to-date methods for the detection and monitoring of nanomaterials in the environmental compartments and nonavailability of adequate databases related to toxicity of nanomaterials hinders the progress of environmental risk assessment. On the other hand, implications of nanomaterials in environmental pollution control through source reduction, degradation, and sensing of pollutants are gaining wide acceptance. Nanoscale zero-valent iron, carbon nanotubes, and nanofibers are now used for remediation of hydrocarbons, organic compounds, heavy metals, polyaromatic hydrocarbons, dioxins, and polychlorinated biphenyls from water. However, despite their promising efficiency in environmental cleanup is a necessity for the synthesis of smarter and readily degradable nanomaterials, bionanomaterials derived from plants and microbial sources to make them ecologically sustainable.

23.1 INTRODUCTION

Ever since its inception, nanotechnology has found its applications in almost all major sectors like engineering, medical, pharmaceutical, agriculture, and environment (Bhattacharyya et al., 2009). Nanomaterials like titanium dioxide and zinc oxide are being used widely in cosmetic products, sunscreens, and surface coating. Silver nanoparticles are used in clothing, disinfectants, household appliances, food packing, bandages, and water purification systems. Recently, the use of carbon nanotube (CNT) in the manufacturing of solar panels has transformed the global commercial scenario of nanotechnology. The incorporation of nanotechnology as key functional components in several products has increased many folds in the recent past. Extensive studies are being carried out to explore the potential implications of nanotechnology in electronics, automotive, healthcare, energy, and consumer products sector. On the contrary, most nanomaterials are nonbiodegradable and tend to accumulate in environmental compartments after their intended use. Production of nanomaterials has increased manifolds in the recent past and increasing rapidly each year all over the world because of its widespread use in various products. It has been estimated that the C_{60} fullerene is produced at a rate of 2000 tons per annum, whereas CNT production has gone up to several hundreds of tons per year. However, the production and commercial use of TiO_2 nanoparticles and silica nanoparticles have developed manifold more compared to C_{60} or CNT. Studies based on industrial survey and life cycle analysis of the various nanomaterials suggest that 17% of these products may enter to soils, 21% to water, 2.5% to air, and the rest to landfills (Keller and Lazareva, 2013). Exposure to nanomaterials during fabrication, their processing to produce derivative products and after their disposal as waste products have been reported to be associated with the risk of the number of diseases.

The protection of the environment from pollution has become a crucial worldwide issue. Numerous point and nonpoint sources of pollutants such as manufacturing and mineral industries, oil fields, landfill sites are increasing globally at an alarming rate. Contamination of aquifers with pesticide residues, oil and greases, heavy metals, and other toxic chemicals have polluted groundwater in various regions. Analysis of leachate from municipal solid waste landfill sites indicates increased contamination of aquifers with heavy metals like Al, Cd, Cr, Fe, Zn, Ni, and Pb in water (Kale et al., 2010). As nanomaterials have several properties suitable for remediating polluted air, water, and soil, the use of nanomaterials to abate environmental pollution has given birth to a new modality of remediation termed nanoremediation. Nanoremediation is now transforming the environmental cleanup modalities by exhibiting a crucial part in the prevention of pollution, detection of pollutants and their monitoring, and finally incurring effective and promising results in remediation (Rajan, 2011). Nanoremediation offers a cost-effective solution for some of the challenging on-site remediation (Ball, 2013). Nanoremediation is now rapidly replacing the existing expensive, tedious, and less effective soil and groundwater remediation technologies such as thermal treatment, chemical oxidation, and surfactant cosolvent flushing (Ball, 2013; Löffler and Edwards, 2006). However, there is the paucity of available data on implications of the effective environmental nanoremediative method in the environmental sector indicating the necessity of a major boost in innovative research in environmental nanotechnology. Though nanotechnology has been projected as a next-generation technique for environmental cleanup (Gavaskar et al., 2005), the potential environmental and health risks associated with nanomaterials are largely unknown (Keller and Lazareva, 2013).

23.2 IMPLICATIONS OF NANOMATERIALS IN ENVIRONMENTAL PROTECTION

The world today is facing the harmful impact of environmental contamination that is getting exaggerated each passing year. Devising appropriate, convenient, and affordable pollution abatement technologies is the need of the day. The nanomaterials have emerged as a new generation of technologies for the prevention and control of environmental pollution and the protection of public health. The nanomaterials possess several unique and novel size-dependent properties that make them potent candidates for pollution abatement compared to larger molecules in current use. The properties such as large surface area, higher reactivity, solubility, and high sorption capacity can further be modulated using the functionalization process through the incorporation of suitable selective materials. The use of nanomaterials further prevents environmental pollution through the minimization of the use of raw materials and the subsequent elimination of waste generation (Ibrahim et al., 2016).

Nanocatalyst, nanoabsorbent, and nanomaterials based membrane techniques are the advanced methods that are preferably used for remediation of groundwater pollution and control of pollution of air and water. Apart from the direct implications of nanomaterials in pollution abatement, nanomaterials are also used in the energy generation and transmission sector that are more environment friendly compared to the available highly polluting conventional technologies. New nanomaterials are being developed and used in fuel cells, photovoltaics, and electrical transmission providing an inspiring solution with higher efficiency. Reactive nanomaterials like bimetallic nanoparticles, metal oxides, CNT, and fibers, nanoscale zeolites are the best options for detoxification and transformation of both organic as well as inorganic pollutants in water. Among these reactive nanomaterials, due to high reactivity and larger surface area, nanoscale zero-valent iron (nZVI) has found the most wider application for transforming or degrading contaminants in soil and water (Garner and Keller, 2014).

23.2.1 IMPLICATIONS OF NANOMATERIALS IN THE ABATEMENT OF WATER POLLUTION

Nanomaterials have promising prospects as treatment modalities of the water and wastewater. Various forms of nanomaterials such as dendrimers, bioactive nanoparticles, nanocomposite membrane, zeolites are effective in the treatment of polluted water (Baruah et al., 2016; Ghasemzadeh et al., 2014; Palit, 2017; Santhosh et al., 2016; Zhang et al., 2016). Unlike the treatment of surface water, remediation of groundwater is a tedious and costly process. At present, environmental engineers generally use pump and in situ treatment methods to treat and remove contaminants from groundwater followed by their reinjection making the cost prohibitively high. It does not satisfy the cleanup goal. Application of nZVI as catalysts that enhance redox reactions responsible for the degradation of contaminants as an in situ treatment of groundwater has been demonstrated to display promising and cost-effective results. Apart from nZVI, a number of engineered nanoparticles have been developed that promote such reactions at high rates; however, more innovative researches are required to control the mobility of nanoparticles, their reactivity, and specificity in degrading contaminants in groundwater (Table 23.1).

TABLE 23.1 List of the Nanomaterials Used for Environmental Pollution Abatement

	Nanomaterials Used	Phase	Target Pollutant
Metal-Based Nanoparticles Used for Water Treatment			
1	Ag-nanoparticles (NP)	Water	*Disinfection of Escherichia coli*
2	TiO_2-NP	Water	*E. coli*, hepatitis B virus, aromatic hydrocarbons
3	Metal-doped TiO_2	Water	2-chlorophenol, endoxin, rhodamine
4	Binary mixed oxides	Water	Methylene blue dye
5	Iron -based nanomaterials	Water	Heavy metals, chlorinated organic solvents
6	Bimetallic NPs	Water	Chlorinated and brominated contaminates
Silica-Based Nanomaterials Used for Water Treatment			
7	Carboxylic acid-functionalized mesoporous silica	Wastewater	Cationic dyes, heavy metals
8	Amino-functionalized mesoporous silica	Wastewater	Heavy metals
9	Thiol-functionalized mesoporous silica	Wastewater	Heavy metals
Graphene-Based Nanomaterials			
10	Pristine graphene	Water	Fluoride
11	Graphene oxide	Water	Pesticides
12	ZnO-graphene/CDS-graphene	Water	Heavy metals
Dendrimers and Nanocomposites			
13	PANAM dendrimer	Water	Heavy metals
14	Polymeric nanocomposites	Water	Metal ions, dyes, microorganisms

23.2.1.1 NANOADSORBENT

Wastewater treatment system based on nanomaterials offers a cost-effective and sustainable solution that does not require the establishment of large infrastructures. The larger available reactive surface area of nanomaterials make them a multifunctional and highly efficient candidate to be used as adsorbents (Tang et al., 2014). Polymeric nanomaterials are used to develop adsorption sites having the capacity to bind a specific contaminant through molecular templating methods. Advanced water treatment systems with higher competency can be developed implicating different forms of nanomaterials like zeolites, CNTs (Kyzas and Matis, 2015; Zhang et al., 2016). Nanomaterials like magnesium oxide (MgO) nanoparticles adsorb halogen compounds up to 20% by weight. CNT adsorbs numerous organic contaminants with higher affinity compared to activated carbon. Metal ions have higher competency to bind with CNTs and have a higher capacity to adsorb heavy metals and radionuclides.

Iron oxide magnetic nanoparticles can easily be produced in larger quantities using physicochemical methods as they have higher adsorption capacity and affinity for pollutants because of the availability of larger surface area. The external magnetic field can be used to separate metal-loaded nanomaterials from treated wastewater and hence can be regenerated and reused (Shirsath and Shirivastava, 2015). Further modification of the surface of the iron oxide nanoparticle by coating with organic compounds can elevate its sorption efficiency and also improve specificity to bind selective contaminants from water (Gutierrez et al., 2017).

23.2.1.2 PHOTOCATALYTIC DEGRADATION OF ORGANIC POLLUTANTS IN WATER USING TIO_2 NANOPARTICLES

Photocatalytic oxidation of the contaminants in the water is mediated by nanomaterials like TiO_2 nanoparticles (6–20 nm) shows promising efficiency as reported by many investigators (Babu et al., 2015; Karthik et al., 2015; Kumar et al., 2017). Nanoparticle TiO_2 adsorb and initiate photocatalytic degradation of organic pollutants in the water when irradiated with UV light (Prevot et al., 2001). Further doping of transitional metal belonging to d and f block elements in the periodic table with TiO_2 nanomaterials bring in the separation of charge leading to a decrease in bandgap and results in enhanced oxidative degradation of organic pollutants in water.

23.2.1.3 NANOMEMBRANE

Membrane technologies play a key role in the prevention of pollution, recovery of resources from water, and monitoring of environmental pollution.

Nanomembranes are widely used for the removal of particles, organic pollutants, and also in removing salt from water (Han et al., 2008). Fuel cell technologies used for the generation of electricity from wastewater also implicate nanomembrane to enhance its performance. Modification of membrane architecture using nanomaterials improvises membrane selectivity for contaminants in water and reduces the cost of water treatment (Wang et al., 2018).

Membrane with nanofibers can be used to remove microsized particles efficiently and rapidly from the aqueous phase without fouling, which very often occurs in conventional methods. Nanocomposite membranes developed by the addition of nanomaterials into polymeric membranes are multifunctional in nature. The hardness of water due to the presence of a higher level of multivalent cations can be removed by nanofilters that eliminate organic solutes of size larger than 1000 Da (Zeman and Zydney, 2017). Further, the addition of metal oxides like zeolite and silica on polymeric ultrafiltration membrane makes them more hydrophilic and resistant to fouling. Doping of antimicrobial nanomaterials like nanosilver can efficiently prevent attachment of bacteria and formation of biofilm on the surface of polymeric membrane.

23.2.1.4 NANOCATALYSTS

Nanocatalytic substances such as zero-valent metals, semiconductor materials, and bimetallic nanoparticles accelerate the reactivity and degradation of various environmental contaminants such as organochlorine pesticides, halogenated herbicides, polychlorinated biphenyls (PCBs), and nitroaromatics (Zhao et al., 2011). The nanocatalysts such as silver nanoparticles, N-doped TiO_2, and ZrO_2 nanoparticles have been reported to accelerate the degradation of contaminants in water (Chaturvedi et al., 2012).

23.2.1.5 NANOSTRUCTURE CATALYTIC MEMBRANE

Nanostructured catalytic membranes accelerate multiple ordered reactions by providing homogeneous catalytic sites, multifunctionality with less contact time. These catalytic membranes can easily be commercialized for water treatment. Nanostructured TiO_2 membranes and films are widely used for the degradation of organic pollutants and inactivation of microorganisms (Choi et al., 2009).

23.2.1.6 DENDRIMERS

Dendrimers are symmetrical and spherical nanoparticles, constituted of a relatively dense shell. Dendritic polymers have the capacity to remove dissolved organic and inorganic solutes in water with molecular mass less 3 kDa that is not possible by nanofilters or reverse osmosis (Abbasi et al., 2014). Dendrimers efficiently eliminate toxic metal ions, radionuclide, organic and inorganic dissolved solutes, organic pollutants including bacteria and viruses. Dendrimers based on silver complexes and nanocomposites are effective antimicrobial agents in vitro (Balogh et al., 2001).

23.2.2 ROLE OF NANOPARTICLES IN AIR POLLUTION ABATEMENT

Increased pollution of air due to huge fossil fuel burning and rapid industrialization has deteriorated the air quality in the atmosphere all over the world. Pollutants like CO, chlorofluorocarbons, volatile organic compounds (VOCs), hydrocarbons, sulfur oxides, and nitrogen oxides are on the rise in the atmosphere. Numerous investigations have been carried out exploring the potential applications of nanomaterials for monitoring and cleaning up the environmental pollution.

23.2.2.1 NANOTECHNOLOGY FOR THE ADSORPTION OF TOXIC GASES

Toxic gases like dioxin and related compounds such as polychlorinated dibenzofuran and PCBs are mainly generated from the combustion of organic compounds in waste incineration and released to the ambient air from different sources. The stability, persistence, and toxicity of these compounds depend on the number of chlorine atom attached to its structure. Dioxin compound such as 2,3,7,8-Tetrachlorodibenzo-*p*-dioxin is a known carcinogen that affects the immune and endocrine systems and fetal development. The surface of CNTs interacts strongly with the benzene rings structure of dioxin and is used for removal of dioxins from the polluted air (Kulkarni et al., 2008). Both the single-walled nanotubes and multiwalled nanotubes have one-dimensional structure, high thermal stability, and unique chemical properties for suitable adsorption and removal of various types of organic and inorganic pollutants including dioxins and related compounds (Long and Yang, 2001).

23.2.2.2 NO_x ADSORPTION

Extensive research works have been conducted in the recent past to devise technologies to eliminate NO_x released from fossil fuel burning. Conventional methods involving the use of ion exchange zeolites activated carbon and FeOOH dispersed on active carbon fiber are not effective in removing NO_x from polluted air. CNTs pose unique structural features and functional groups suitable for adsorption of NO_x species. NO undergoes oxidation to form NO_2 when passes together with O_2 through the CNTs and these nitrate species get adsorbed to its surface (Dai et al., 2009).

23.2.2.3 REMOVAL OF VOLATILE ORGANIC COMPOUNDS FROM THE AIR

In addition to nitrogen oxides and sulfur oxides, many chemicals are formed by atmospheric reactions, such as nitrous acid, polyaromatic hydrocarbons from vinyl radical, and acetylene. Air pollution prevention and control act in India have become very stringent since these pollutants are extremely dangerous for public health even in the case of exposure at very low concentrations. Photocatalysis, ozonolysis, and adsorption using activated carbon play a crucial role in the conventional air purification system. However, conventional systems are not very good at getting rid of organic pollutants at room temperature. Recently, a new technology involving manganese oxide and gold nanoparticles has been developed by Japanese scientists that is found to be very effective in removing VOCs, sulfur oxide, and nitrogen oxides from the air at room temperature. The nanomaterials can effectively remove indoor air pollutants like hexane and toluene (Sinha et al., 2007).

23.2.3 NANOMATERIALS IN SOIL POLLUTION

In the last few decades, the quality and health of soil all over the world have undergone rapid degradation due to the increased pressure of higher productivity using synthetic fertilizers, pesticides, accumulation of toxic metals and metalloids, and other persistent organic pollutants. However, technologies available to remediate soil pollution are mostly ineffective in maintaining and reversing deteriorated soil quality (Chen et al., 2015). Modern agricultural practices and other anthropogenic activities that cause contamination of soil with pesticides, metals, explosives, solvents, and many other pollutants can be remediated using nanoremediation technology. The remediation of toxicants in soil occurs through the direct reaction or the transformation of the toxicants into less toxic forms (Adeley et al., 2013). nZVI is a highly reactive material that enhances oxidation–reduction reaction and causes immobilization of contaminants and transformation products. nZVI catalyzes transformation reactions through the process of adsorption, speciation, and precipitation. The dual property of nZVIs is useful to both separate and transform many different contaminants like trinitrotoluene, chlorinated solvents, and pesticides (Sun et al., 2006). Nanoremediation techniques can further be combined with phytoremediation processes that lead to effective remediation of endosulfan pesticide in soil. Nanophytoremediation can also remove trinitrotoluene from the soil within 60 days. Plants like *Alpinia calcarata* hyperaccumulate endosulfan residues with the potential of removing 81% of endosulfan from the soil. Lemongrass (*Cymbopogon citratus*) and Tulsi are also known to remove endosulfan residues by accumulating it.

23.3 NANOMATERIALS AS ENVIRONMENTAL POLLUTANTS

The last decade has seen a dramatic rise in the use of engineered nanomaterials in almost all the major sector including medicine, textile, paint, electronic devices, consumer products, and cosmetics. Their increased applications have inevitably given birth to a unique class of environmental pollutants that necessitate extensive research to predict their safe limit in environmental compartments like air, water, and soil. Recently progress has been made to understand factors affecting the fate and transport of nanomaterials in the environment. Studies show that unlike many other chemicals nanomaterials behave differently when dissolved in water. However, accurate exposure assessment of nanomaterials in the environments is essential to develop predictive capabilities and nanomaterials related risk management.

The conventional analytical methods used for environmental sample analysis are not sensitive enough for detecting various physicochemical forms of engineered nanomaterials. Because of the inadequacy and lower sensitivity of existing analytical instruments, the progress of studies targeting the detection and quantification of nanomaterials at environmentally relevant concentrations in complex media is getting hindered. Most of the nanomaterials released to the environmental compartments exist in colloidal form and their properties get influenced by surrounding environmental conditions leading to the development of a number of artifacts. Therefore there is an urgent necessity for the development of techniques suitable for extraction, cleanup, separation, and sample storage that will have minimal artifacts with increased sensitivity and specificity. Innovative research works are also essential to develop techniques that can differentiate between naturally occurring particles and manufactured nanoparticles (Guerra et al., 2018).

23.3.1 POTENTIAL HEALTH RISKS ASSOCIATED WITH NANOMATERIALS ON HUMAN AND OTHER ORGANISMS

One of the most adverse environmental impacts of nanoparticles that have become a global concern is the potentiality of nanoparticles to infiltrate environmental compartment such as water bodies and air sources causing serious health hazards both in animals and humans (Maynard et al., 2006). Nanoparticles pose health risks as it can disperse to a larger area, cause ecotoxicity, have higher persistence, and can bioaccumulate. Nanotechnology has emerged as a breakthrough discovery expanding its utility in various fields as a beneficial and advantageous practice but little intervention has been done regarding its health and environmental effects, which is the need of the hour (Borm et al., 2006).

Among various modes of exposures, inhalation is one of the greater concerns when it comes to the fabrication of nanomaterials. Numerous simulation studies involving inhalation of CNT, silica, and asbestos have been carried out to demonstrate its possible roles in damaging lungs (Muller et al., 2005). As CNTs form large aggregates, it appears as if there is a lower possibility of its transport into the lungs and subsequent incurrence of adverse effects in the lungs. However, it has been seen that smaller nanomaterial concentration in the air does not permit them to interact much among themselves to cluster together allowing their easy transport to the lungs. In several experiments, the lungs of laboratory animals are being introduced with nanotubes by washing them in CNT solutions. In this mode of exposure, pulmonary accumulation of nanomaterials and consequent suffocation proffer the major risk factor (Taquahashi et al., 2013). Furthermore, dermal exposure presents another significant risk associated with the fabrication and use of nanomaterial.

One of the highly used nanoparticles TiO_2-NPs is known to affect human health through all three routes of entry, that is, oral (via consumption of contaminated food), dermal (through the application of cosmetics and sunscreen), and inhalation (under occupational and manufacturing conditions). TiO_2-NPs cause cytotoxicity, genotoxicity, and oxidative stress leading to inflammation and subsequently to apoptosis as suggested by a number of in vitro studies (Kim et al., 2010; Kisin et al., 2007). One of the major concerns in the nanotoxicity is that NPs can cause DNA damage and hence are mutagenic and have the potential to induce teratogenesis (Ahamed et al., 2008). TiO_2-NPs have been reported to induce lung cancer in rats and are classified as a possible carcinogen to humans under group 2B by the International Agency for Research on Cancer. National Institute for Occupational Safety and Health also designated TiO_2-NPs as potential occupational carcinogens (Scarino et al., 2012).

Nanomaterials-associated human health risks are primarily through three modes: dermal exposure through the application of cosmetics and personal care products; exposure through oral mode during ingestion of water or food; and inhalation of air contaminated with NPs in the workplace. NP exposure in human can trigger many pathophysiological responses such as genotoxicity, oxidative stress, lipid peroxidation, inflammation, pulmonary pathological changes; and lung diseases (Manke et al., 2013; Jaeger et al., 2012). Surprisingly, NMs can be easily taken up by mitochondria and nucleus in the cell. Therefore exposure to NPs is associated with DNA mutations and alterations in mitochondria structure leading to apoptosis (Bacchetta et al., 2016). Further research and concern over the NM-associated health hazards may constrain its uncontrolled usage and pave the way toward environmental remediation. Monitoring and intervention measures must be taken sooner than later to render nanotechnology more advantageous than unsafe.

23.3.2 MECHANISMS OF CELLULAR TOXICITY OF NANOMATERIALS

23.3.2.1 INTERACTION OF NANOMATERIAL WITH CELLULAR COMPONENTS

The influence of nanomaterial on living cells can be analyzed by studying their interactions with individual cell components. Different NMs may invoke similar overall observed effect; however, the underlying mechanism inflicting such effect can be determined by elucidating how they interact with a particular cell component. Nanomaterials interact with proteins, nucleic acids, and membranes inside the cell differently to exert their unique toxic effects. To monitor and visualize the details in cells, some of these interactions are engineered by labeling the proteins or DNA (Dinesh et al., 2012).

23.3.2.2 MEMBRANE DISRUPTION

Nanomaterials that incur physical injury to a cell when it comes in contact can be bactericidal in nature (Neal, 2008). Survivability of a bacterium under such conditions depends mainly on the extent of damage to the cell membrane. Severe irreparable damage to cell membrane leads to cell death. One of the examples reflecting such a condition is the exposure of Gram-positive bacteria to carboxyfullerene. Carboxyfullerene punctures the bacterial cell membrane resulting in leakage of cellular contents leading to cell death (Mashino et al., 2003). Some nanomaterials alter lipid components of the cell membrane through the production of reactive oxygen species (ROS) causing membrane damage by lipid peroxidation (Pal et al., 2015).

23.3.2.3 GENOTOXICITY

Nanomaterials like fullerene that cleaves the double-stranded DNA molecule are mostly genotoxic in nature. The cleaving of DNA initiates the repair process inside a cell that itself is a risky venture. The improper repair itself incorporates a number of alterations in the nucleic acid sequence causing mutation (Magdolenova et al., 2014).

23.3.2.4 REACTIVE OXYGEN SPECIES (ROS) PRODUCTION

The most common mechanism of nanomaterials toxicity is the generation of free radicals in the cell by inducing an imbalance in its generation and scavenging system. Damage to mitochondrial functioning is responsible for the generation of huge free radicals in the form of singlet oxygen, superoxide ions, and hydroxyl radicals. Once these radicals are formed, they can activate a series of cellular events leading to apoptosis of the cell. Nanomaterials like TiO_2, SiO_2, and ZnO are some of the examples that are known to induce oxidative stress on the cells (Dalai et al., 2012).

23.4 ECOTOXICOLOGICAL IMPACT OF NANOMATERIALS

The standard methodologies followed to detect and quantify nanoparticle in air, water, and soil are not up to date. The major drawback of available instruments for the measurement of nanoparticles in air, water, and the soil is that it cannot differentiate the background nanoparticles from the nanomaterials added to it. Further, most of the devices used in nanoscience research differentiate nanomaterials based on their size but not density (Roduner, 2006). Due to the complex nature of these nanomaterials and the lack of proper instrumentation, the rate of deposition and subsequent aggregation of engineered nanomaterials in the environment are largely unknown. Therefore accurate assessment of the toxicity of nanomaterials once incorporated into the air, water, and soil becomes very difficult. The development of sensitive and more specialized instruments through innovative research and innovation in this field is the utmost necessity (Shen et al., 2015; Moore, 2006).

Ecotoxicological study of nanomaterials is at its infancy. Most of the studies conducted on the human cell model to investigate the toxicity of various forms of nanomaterials reveal inconsistent and contradictory results. The inconsistency in the result of nanomaterials toxicity studies could be because of the addition of organic solvents as a procedural requirement. For example, some researchers reported buckyball as highly toxic to the brain cells in the fish model and inferred that it is toxic to human cells too, whereas some other investigators did not observe significant toxicity in their studies. Out of the various forms of nanomaterials understudy, much of the focus was laid initially on the toxicological assessment of fullerene-based nanomaterials because of their widespread application (Sumi and Chitra, 2016).

23.4.1 ECOLOGICAL FATE, TRANSPORT, AND TRANSFORMATION OF NANOPARTICLES

The nanomaterials eventually end up in the environmental compartment in different forms after their manufacture, use, and disposal. Most nanomaterials get mixed either in the water or in the soil. The majority of the nanomaterials interact with their new surrounding but few may remain inert (Selck et al., 2016). Nanomaterials can bind to soil, be transported or diluted in water, or react with water or other chemicals or may undergo weathering. The phase in which the nanomaterial resides to determine how it interacts with organisms in the environment plays a major role in their ecotoxic effect (Van Koetsem et al., 2015). In soil–water systems, the main property that determines bioavailability is the desorption of the chemical from soil particles. Bioavailability of the nanomaterials depends mainly on the propensity of these particles to attach to the mineral surface or to form aggregate. Nanomaterials that tend to attach to the surface of the minerals generally are less mobile even in the porous media such as groundwater aquifers, and sand filters. However, smaller particles could be less mobile due to their relatively large diffusivity that produces more frequent contacts with the surfaces of aquifer porous media (Wang et al., 2018; Solovitch et al., 2010).

Alteration of physicochemical properties of engineered nanoparticles (ENPs) such as aggregation, stability, transport, and sedimentation in the presence of humic substances have been studied in natural water under ambient conditions. The fate and behavior of ENPs are attributed to their physicochemical changes in the environment. Besides, uptake by aquatic organisms and the mechanisms of entry may be influenced by adsorption of synthetic or biogenic organic matter from the environment onto the surface of ENPs. Release of different metal ions from metal and metal oxide nanoparticles to aquatic/soil systems could be toxic to the living being in that medium. It has been found that the toxicity of fullerenes occurs not because of the fullerene but due to chemical residues that remain on their surface during fabrication. Further, microbes play a crucial role in biotransformation reactions that affect the mobility of nanoparticles in soil and toxicity of these materials to the environment. However, there is a paucity of information related to interactions between fullerenes, CNTs, metal oxides, and microbial populations. This lack of information and understanding can be detrimental in maintaining sustainable soil health because of the potential impacts of nanomaterials on microorganisms as well as the mineralization of complex organic matter essential for nutrient cycling.

23.4.2 ECOLOGICAL IMPLICATIONS

Nanomaterials are the class of contaminants that may be intentionally produced or maybe a result of unintentional contamination from human activities. The risk associated with exposure to nanomaterials in the environment cannot be assessed without understanding the dose-response effect and route of entry of nanomaterials to the body system. Evaluation of toxicity of nanomaterials includes wildlife, human, and also the lower group of organisms that participate in energy flow through the food chain (Colman et al., 2014). Initially, research on nanotoxicity was based on its evaluation in pure culture of bacteria like *Escherichia coli*, *Pseudomonas fluorescens*, and *Bacillus subtilis* var. *niger* and production of ROS on exposure to nanoparticles (Pietroiusti et al., 2016; Bondarenko et al., 2013). Iron in its nanoform nZVI has been reported to enter the cell and induce oxidative stress leading to oxidative damage of cell membrane and consequent leakage of intracellular materials and cell death (Chen et al., 2015; Kadar et al.,

2011). The similar harmful effect of nZVI has also been observed in the plant system. nZVI significantly decreased the transpiration and growth of hybrid poplars. The decline in the population of soil invertebrates such as earthworm, aquatic macroflora like zoo-planktons, and phytoplankton have been observed when exposed to nanomaterial nZVI in higher concentration (Dev et al., 2018; Tolaymat et al., 2017). Soil microbial communities are highly susceptible to the accumulation of nanoparticles like CuO and Fe_3O_4 in the soil that directly impact the mineralization process and nutrient cycling (Simonin and Richaume, 2015). However, some nanomaterials such as C_{60} have been reported to show no toxicity on the microbial system even at a higher concentration of 100 μg/g and cause no effect on the biological activity of the soil microbial population. Although production and uses of nanomaterials in the recent past have increased 100 fold, only three nanoparticles namely silver, nZVI, and ZnO are of major concern (Garner and Keller, 2014). ZnO among the three nanomaterials showed toxicity to almost all species (Ma et al., 2013). Therefore more stringent regulations and monitored release of nanomaterials to the environment have to be devised to minimize the negative effect of nanomaterials used in soil.

The use of nanotechnology for in situ remediations of contaminated soil and groundwater has gained more attention in the recent past. The implementation of nanoremediation depends on the risk-benefit analysis. The distance traveled by the nanoparticle and their ecotoxicological effect determines the environmental risks associated with it. Particles that can migrate long distances and contaminate unaffected groundwater, surface water, or ecosystem are of major concern. Nanoremediation restricts the movement of the nanomaterials inhibiting spreading and contamination of a larger area of water, soil, and air (Corsi et al., 2018; Bardos et al., 2015).

Nanomaterials are nonbiodegradable in nature and hence are conservatory and tend to accumulate in the air, water, and soil (Lead et al., 2018). In soil, these nanomaterials can be transformed into some less toxic intermediate products mostly due to microbial activities. Microbes are the key mediators of the mineralization process where the complex organic molecules are converted into their simple bioavailable forms and mixed with soil to participate in the cyclization of nutrients such as nitrogen, phosphorus, carbon, and other minerals. A large fraction of biomass in the aquatic medium is constituted by microbial organisms. Survivability of plants in terrestrial systems is dependent on the activity carried out by the microbes by breaking down dead organic matter and release of nutrients to the soil. The presence of nanomaterials in higher quantity can directly influence the metabolic activities of microbial species in the soil ecosystem leading to microbial toxicity (He et al., 2016; Mohanty et al., 2014). As microorganisms play a crucial role in food webs and are the primary mediator of biogeochemical cycles, the negative effect of nanomaterials in the soil can directly hamper the ecological processes.

23.5 NANOMATERIALS AND SUSTAINABLE DEVELOPMENT: PRESENT STATUS AND FUTURE SCOPE

Proper use of nanotechnology has the potential to make the world environment friendly by replacing widely used toxic matters with less toxic and safer nanomaterials. Some of the products that have a higher impact on creating lesser pollution include biodegradable plastics, a nanocrystalline composite that replaces the lithium-granite electrodes in rechargeable batteries, glass with self-cleaning ability. Nanomaterials are now being used to produce plastic that is biodegradable and undergo rapid decomposition. Silver nanomaterials are used as antifouling surface coatings that protect medical equipment and instruments from being contaminated with microbial organisms (Chen et al., 2015).

However, the major concern is the synthesis of nanomaterials that require the use of reactants and volatile organic and inorganic chemicals that are toxic in nature (Palit, 2017). Though water treatment using nanotechnology is far better and efficient than the conventional water treatment system, the environmental fate of used nanomaterials, their transport in environmental compartments and interaction with living being are poorly understood. Numbers of scientific and innovative works are going on to substitute these toxic chemicals used in the synthesis of nanomaterials. Regulatory bodies are framing stringent regulation to curtail the use of the toxic chemical in nanomaterials synthesis so that their concentrations in the environment remain within a safe limit. Fabrication of nanomaterials must follow the greener route so that the use of nanotechnology in the various fields will be ecologically sustainable and will not cause the addition of any toxic and hazardous materials into the environment during their course of manufacturing and after the end of their intended use.

An alternative method of nanomaterials synthesis that does not involve the use of toxic chemicals is the biogenic method (Figure 23.1). The biogenic method involves natural substances derived from plants, bacteria, algae, fungi, yeast, actinomycetes that produce reducing, capping, and stabilizing agents required for the synthesis of the nanomaterials. The biogenic method is an ecologically sustainable as well as an economically viable option.

Manufacturing metallic nanomaterials using naturally occurring vitamins, polyphenols, carbohydrates, amino acids, and natural surfactants are gradually gaining wider acceptability (Dhillon et al., 2012). The biological method of nanomaterial synthesis is rapid, eco-friendly, and suitable for large scale production compared to available conventional synthesis methods. Numbers of different species of bacteria have been reported to catalyze biogenic production of various inorganic nanomaterials that are otherwise very difficult to synthesize using chemical methods. Certain magnetotactic bacteria have the ability to produce magnetic nanoparticles, or magnetosomes. Nanomaterials such as nano- and microZnO rods are synthesized by *Magnetospirillum magnetotacticum*, Incubation of *E. coli* bacterial species with cadmium chloride and sodium sulfide can lead to the generation of cadmium sulfide nanomaterials. *Pseudomonas stutzeri* is known to synthesize silver-based nanocrystals. Sulfate-reducing bacteria are used to produce sphalerite (ZnS) nanoparticles (Table 23.2).

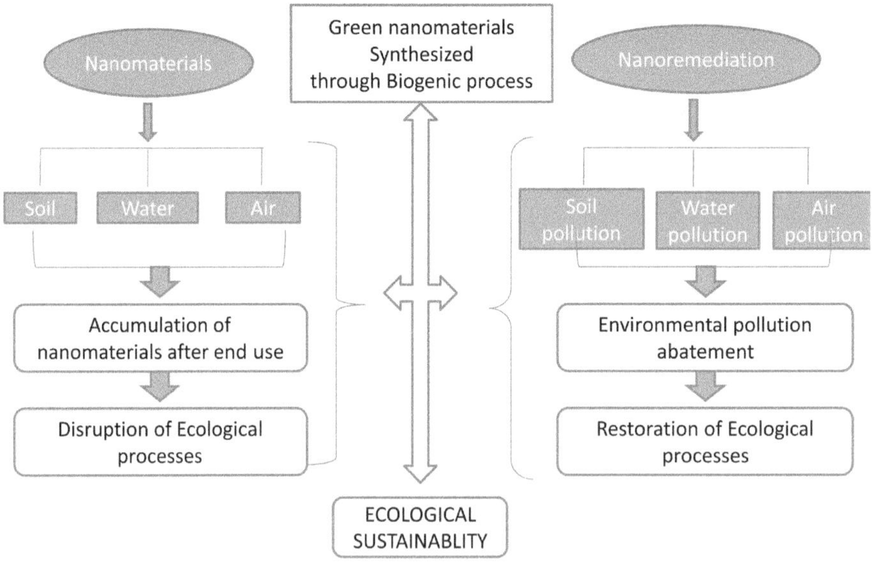

FIGURE 23.1 Showing sustainable use of nanomaterials.

23.5.1 EXISTING SUSTAINABLE SOLUTIONS

Synthesis of nanoparticles like nZVI from plants or plant parts based natural products such as green tea leaves, oak, pomegranate are the greener approach, which have gained wider acceptability in the last few years (Palit, 2017).

The green chemistry principle used in nanomaterials synthesis has numerous benefits that include decreased toxicity, high process efficacy, cost-effective, and biodegradability of the nanomaterials (Hoag et al., 2009). Reports show that nZVIs synthesized using plant parts degraded 54%–66% ibuprofen in aqueous solutions that attained degradation capacity of approximately 95% compared to nZVI generated in Fenton like reaction (Machado et al., 2013).

TABLE 23.2 List of Biogenic Nanomaterials

Name of Biogenic Nanomaterials	Plant/Microbial Organisms Used	Botanical Name	Applications
Gold nanoparticles (NPs)	Alfalfa (Biomass)	*Medicago sativa*	Phytomining gold from soils, synthesis of quantum dots
Silver NP	Rice (leaf)	*Oryza sativa*	Antibacterial activity against *Escherichia coli*
CuO NP	Aloe vera	*Aloe barbadensis*	Antibacterial activity against all the fish pathogens
Pt NP	Persimmon	*Diopyros kaki*	Biomedical application
Silver NP	pomegranate	*Punica granatum*	Green catalyst
CdS NP	Fungus	*Fusarium oxysporum*	Semiconductor
Au and Ag NPs	Magnetotactic bacteria	*Breynia rhamnoides*	Green catalyst
FeS NP	Magnetotactic bacteria	*Magnetospirillum magnetotacticum*	Adsorbents for heavy metals

Many metal nanoparticles like iron, gold, silver, oxides, and salts are also synthesized from plant parts of banana, coffee, sugar as redundant and capping agents using green synthesis processes (Palit, 2017). Complete decolorization of aqueous solutions with methyl orange can be obtained using iron nanoparticles (GT–Fe) synthesized from green tea leaves (Shahwan et al., 2011). GT–Fe nanoparticles showed higher decolorizing ability compared to the nanoparticles produced by borohydride reduction.

23.5.1.1 BIONANOPARTICLES

Bacteria, fungi, actinomycetes, yeasts, and algae can produce nanoparticle that has the potential to degrade organic matter. Clostridia bacteria have been reported to reduce palladium, Pd (II) ions to its metallic form (bio-Pd) in the microbial cell wall (Johnson et al., 2013). The *Clostridium pasteurinum* culture reduced the azo dyes, Evans blue, and can be potentially used to degrade large quantities of spent dyes discharged from textile industries as wastewater. Among the eukaryotes, fungi have a very high potential for the synthesis of a metal nanoparticle and hence are designated as nanofactories (Dhillon et al., 2012). Fungal species like *Fusarium semitectum, Fusarium oxysporum*, and *Aspergillus fumigates* are widely used for the synthesis of nanoparticle with defined dimensions (Shedbalkar et al., 2014).

Biogenic synthesis of nanomaterials is generally made through bioreduction and bioprecipitation by polyphenols, peptides, and other bioactive compounds. Commonly plants are used for the synthesis of bioactive nanomaterials (Park et al., 2011). Bacterial species *Bacillus cereus* has been reported to synthesize silver nanoparticles having very high antibacterial activity. Bacteria, fungi, and actinomycetes secrete a number of different chemicals that strongly reduce or oxidize metal ions to produce zero-valent or magnetic nanoparticles. The plant extract has been documented to be associated with the reduction of the metallic salts for the synthesis of Fe, Zn, Au, Ag, and Cu NPs of different shapes and sizes (Kuppusamy et al., 2016).

23.5.1.2 EMULSIFIED ZERO-VALENT IRON

Emulsified zero-valent iron (E-ZVI) is one of the eco-friendly nanomaterials used widely for the degradation of insoluble toxicants into harmless compounds. E-ZVI is a low-cost method for in situ remediations of halogenated hydrocarbons. Groundwater decontamination capacity of the E-ZVI increase many folds because iron remains encapsulated and does not come in direct contact with contaminants. However, products demonstrating efficient decontamination as greener nanoparticles for remediation are coming up gradually. Large scale productions of greener nanoparticle with potential for nanoremediation of environmental pollution need extensive research and innovation (Patil et al., 2016).

23.6 CONCLUSION

The increase in the use of nanomaterials in almost all major industrial sectors has resulted in an elevated concentration of disposed nanomaterials in all the three environmental compartments. As the existing equipment used for the analysis of the environmental sample cannot differentiate between the background/natural nanomaterials from the nanomaterials added as a result of anthropogenic activities, the exact rate, route, and level of these nanomaterials entering to environmental compartment are not known with certainty. The existing data related to the ecotoxicological impacts of nanomaterials are either inconclusive or contradictory in nature. However, preliminary reports based on studies on few laboratory animal models and small ecosystem-based studies indicate that most of the nanomaterials are toxic to human health and environment and can cause disturbances in ecological functioning. On the contrary, nanoremediation provides a cost-effective, eco-friendly and efficient system for in situ cleanup of large scale contaminated sites. Ironically, the beneficial and most desirable properties of nanomaterials that make them unique and capable of wider use in the medical field are smaller size, a reactive surface that facilitates easy accessibility into the cell and are often also the major cause of concern when it comes to their environmental impact and toxicity. The conventional nanomaterials are environmentally not sustainable and there is a necessity for synthesis and development of safer and greener smart nanomaterials such as biogenic nanomaterials and E-ZVI. Extensive research and innovations are the necessity of the day to attain ecologically sustainable and human health-friendly nanotechnology without polluting the environment.

KEYWORDS

- **nanomaterials**
- **sustainability**
- **pollution abatement**
- **toxicity**

REFERENCES

Abbasi, E.; Aval, S.F. Akbarzadeh, A.; Milani, M.; Nasrabadi, H.T.; Joo, S.W.; Hanifehpour, Y.; Nejati-Koshki, K.; Pashaei-Asl, R. Dendrimers, synthesis, applications, and properties. *Nanoscale Res. Lett.* **2014**, 9, 247.

Adeleye, A.S.; Keller, A.A.; Miller, R.J.; Lenihan, H.S. Persistence of commercial nano-scaled zero-valent iron (nZVI) and by-products. *J. Nanopart. Res.* **2013**, 15, 1418.

Ahamed, M.; Karns, M.; Goodson, M.; Rowe, J.; Hussain, S.M.; Schlager, J.J.; Hong, Y. DNA damage response to different surface chemistry of silver nanoparticles in mammalian cells. *Toxicol. App. Pharmacol.* **2008**, 233, 404.

Arakha, M.; Pal, S.; Samantarrai, D.; Panigrahi, T.K.; Mallick, B.C.; Pramanik, K.; Mallick, B.; Jha, S. Antimicrobial activity of iron oxide nanoparticle upon modulation of nanoparticle-bacteria interface. *Sci. Rep.* **2015**, 5.14813.

Babu, S.G.; Neppolian, B.; Ashokkumar, M. Ultrasound-assisted synthesis of nanoparticles for energy and environmental applications. *Handbook of Ultrasonics and Sonochemistry* **2015**, 1–34.

Bacchetta, R.; Maran, B.; Marelli, M.; Santo, N.; Tremolada, P. Role of soluble zinc in ZnO nanoparticle cytotoxicity in *Daphnia magna*, a morphological approach. *Environ. Res.* **2016**, 148, 376.

Ball, A. The use of microorganisms for bioremediation. Proceedings of Organization for Economic Co-operation and Development (OECD) on Environmental Use of Microorganisms. **2013**.

Balogh, L.; Swanson, D.R.; Tomalia, D.A.; Hagnauer, G.L.; McManus, A.T. Dendrimer–silver complexes and nanocomposites as antimicrobial agents. *Nano Lett.* **2001**, 1,18.

Bardos, P.; Bone, B.; Černík, M.; Elliott, D.W.; Jones, S.; Merly, C. Nanoremediation and international environmental restoration markets. *Remed. J.* **2015**, 25, 83.

Baruah, S.; Khan, M.N.; Dutta, J. Perspectives and applications of nanotechnology in water treatment. *Environ. Chem. Lett.* **2016**, 14,1.

Bhattacharyya, D.; Singh, S.; Satnalika, N.; Khandelwal, A.; Jeon, S.H. Nanotechnology, big things from a tiny world, a review. *Int. J. Sci. Technol.* **2009**, 2, 29.

Bondarenko, O.; Ivask, A.; Käkinen, A.; Kurvet, I.; Kahru, A. Particle-cell contact enhances antibacterial activity of silver nanoparticles. *PLoS One*. **2013**, 8,e64060.

Borm, P.J.; Robbins, D.,; Haubold, S.; Kuhlbusch, T.; Fissan, H.; Donaldson, K.; Schins, R.; Stone, V.; Kreyling, W.; Lademann, J. The potential risks of nanomaterials, a review carried out for ECETOC. *Part. Fibre Toxicol.* **2006**, 3,11.

Chaturvedi, S.; Dave, P.N.; Shah, N. Applications of nano-catalyst in new era. *J. Saudi Chem. Soc.* **2012**, 16, 307.

Chen, M.; Xu, P.; Zeng, G.; Yang, C.; Huang, D.; Zhang, J. Bioremediation of soils contaminated with polycyclic aromatic hydrocarbons, petroleum, pesticides, chlorophenols and heavy metals by composting, applications, microbes and future research needs. *Biotechnol. Adv.* **2015a**, 33, 745.

Chen, M.; Xu, P.; Zeng, G.; Yang, C.; Huang, D.; Zhang, J. Bioremediation of soils contaminated with polycyclic aromatic hydrocarbons, petroleum, pesticides, chlorophenols and heavy metals by composting, Applications, microbes and future research needs. *Biotechnol. Adv.* **2015b**, 33, 745.

Choi, H.; Al-Abed, S.R.; Dionysiou, D.D. Nanostructured titanium oxide film-and membrane-based photocatalysis for water treatment. *Nanotechnol. Appl. Clean Water*, **2009**.

Colman, B.P.; Espinasse, B.; Richardson, C.J.; Matson, C.W.; Lowry, G.V.; Hunt, D.E.; Wiesner, M.R.; Bernhardt, E.S. Emerging contaminant or an old toxin in disguise? Silver nanoparticle impacts on ecosystems. *Environ. Sci. Technol.* **2014**, 48,5229.

Corsi, I.; Winther-Nielsen, M.; Sethi, R.; Punta, C.; Della Torre, C.; Libralato, G.; Lofrano, G.; Sabatini, L.; Aiello, M.; Fiordi, L. Ecofriendly nanotechnologies and nanomaterials for environmental applications, Key issue and consensus recommendations for sustainable and ecosafe nanoremediation. *Ecotoxicol. Environ. Saf.* **2018**, 154, 237.

Dai, J.; Giannozzi, P.; Yuan, J. Adsorption of pairs of NOx molecules on single-walled carbon nanotubes and formation of $NO^+ NO_3$ from NO_2. *Surf. Sci.* **2009**, 603, 3234.

Dalai, S.; Pakrashi, S.; Kumar, R.S.; Chandrasekaran, N.; Mukherjee, A. A comparative cytotoxicity study of TiO_2 nanoparticles under light and dark conditions at low exposure concentrations. *Toxicol. Res.* **2012**, 1,116.

Dev, A.; Srivastava, A.K.; Karmakar, S. Nanomaterial toxicity for plants. *Env. Chem. Lett.* **2018**, 16,85.

Dhillon, G.S.; Brar, S.K.; Kaur, S.; Verma, M. Green approach for nanoparticle biosynthesis by fungi, current trends and applications. *Critic. Rev. Biotechnol.* **2012**, 32, 49.

Dinesh, R.; Anandaraj, M.; Srinivasan, V.; Hamza, S. Engineered nanoparticles in the soil and their potential implications to microbial activity. *Geoderma.* **2012**, 173, 19.

Garner, K.L.; Keller, A.A. Emerging patterns for engineered nanomaterials in the environment, a review of fate and toxicity studies. *J. Nanopart. Res.* **2014**, 16(8), 2503.

Gavaskar, A.; Tatar, L.; Condit, W. Cost and performance report nanoscale zero-valent iron technologies for source remediation. *Naval Facilities Engineering Command*, California, **2005**, pp.1–54.

Ghasemzadeh, G.; Momenpour, M.; Omidi, F.; Hosseini, M.R.; Ahani, M.; Barzegari, A. Applications of nanomaterials in water treatment and environmental remediation. *Front. Environ. Sci. Eng.* **2014**, 8,471.

Guerra, F.; Attia, M.; Whitehead, D.; Alexis, F. Nanotechnology for environmental remediation, materials and applications. *Molecules* **2018**, 23, 1760.

Gutierrez, A.M.; Dziubla, T.D.; Hilt, J.Z. Recent advances on iron oxide magnetic nanoparticles as sorbents of organic pollutants in water and wastewater treatment. *Rev. Environ. Health.* **2017**, 32, 111.

Han, J.; Fu, J.; Schoch, R.B. Molecular sieving using nanofilters, past, present and future. *Lab Chip.* **2008**, 8,23.

He, S.; Feng, Y.; Ni, J.; Sun, Y.; Xue, L.; Feng, Y.; Yu, Y.; Lin, X.; Yang, L. Different responses of soil microbial metabolic activity to silver and iron oxide nanoparticles. *Chemosphere.* **2016**, 147,195.

Hoag, G.E.; Collins, J.B.; Holcomb, J.L.; Hoag, J.R.; Nadagouda, M.N.; Varma, R.S. Degradation of bromothymol blue by 'greener' nano-scale zero-valent iron synthesized using tea polyphenols. *J. Mater. Chem.* **2009**, 19, 8671.

Ibrahim, R.K.; Hayyan, M.; AlSaadi, M.A.; Hayyan, A.; Ibrahim, S. Environmental application of nanotechnology, air, soil, and water. *Environ. Sci. Pollut. Res.* **2016**, 23, 13754.

Jaeger, A.; Weiss, D.G.; Jonas, L.; Kriehuber, R. Oxidative stress-induced cytotoxic and genotoxic effects of nano-sized titanium dioxide particles in human HaCaT keratinocytes. *Toxicology.* **2012**, 296, 27–36.

Johnson, A.; Merilis, G.; Hastings, J.; Palmer, M.E.; Fitts, J.P.; Chidambaram, D. Reductive degradation of organic compounds using microbial nanotechnology. *J. Electrochem. Soc.* **2013**, 160, G27.

Kadar, E.; Tarran, G.A.; Jha, A.N.; Al-Subiai, S.N. Stabilization of engineered zero-valent nanoiron with Na-acrylic copolymer enhances spermiotoxicity. *Environ. Sci. Technol.* **2011**, 45, 3245.

Kale, S.S.; Kadam, A.K.; Kumar, S.; Pawar, N. Evaluating pollution potential of leachate from landfill site, from the Pune metropolitan city and its impact on shallow basaltic aquifers. *Environ. Monitor. Assess.* **2010**, 162,327.

Karthik, P.; Vinoth, R.; Babu, S.G.; Wen, M.; Kamegawa, T.; Yamashita, H.; Neppolian, B. Synthesis of highly visible light active TiO_2-2-naphthol surface complex and its application in photocatalytic chromium(vi) reduction. *RSC Adv.* **2015**, 5, 39752.

Keller, A.A.; Lazareva, A. Predicted releases of engineered nanomaterials, from global to regional to local. *Environ. Sci. Technol. Lett.* **2013**, 1, 65.

Kim, I.S.; Baek, M.; Choi, S.J. Comparative cytotoxicity of Al_2O_3, CeO_2, TiO_2 and ZnO nanoparticles to human lung cells. *J. Nanosci. Nanotechnol.* **2010**, 10, 3453.

Kisin, E.R.; Murray, A.R.; Keane, M.J.; Shi, X.C.; Schwegler-Berry, D.; Gorelik, O.; Arepalli, S.; Castranova, V.; Wallace, W.E.; Kagan, V.E. Single-walled carbon nanotubes, geno-and cytotoxic effects in lung fibroblast V79 cells. *J. Toxicol. Environ. Health.* **2007**, Part A, 70, 2071.

Kulkarni, P.S.; Crespo, J.G.; Afonso, C.A. Dioxins sources and current remediation technologies—a review. *Environ. Int.* **2008**, 34, 139.

Kumar, P.S.; Selvakumar, M.; Babu, S.G.; Induja, S.; Karuthapandian, S. CuO/ZnO nanorods, an affordable efficient pn heterojunction and morphology dependent photocatalytic activity against organic contaminants. *J. Alloys Comp.* **2017**, 701, 562.

Kuppusamy, P.; Yusoff, M.M.; Maniam, G.P.; Govindan, N. Biosynthesis of metallic nanoparticles using plant derivatives and their new avenues in pharmacological applications—an updated report. *Saudi Pharm. J.* **2016**, 24, 473.

Kyzas, G.Z.; Matis, K.A. Nanoadsorbents for pollutants removal, a review. *J. Mol. Liq.* **2015**, 203,159.

Lead, J.R.; Batley, G.E.; Alvarez, P.J.; Croteau, M.N.; Handy, R.D.; McLaughlin, M.J.; Judy, J.D.; Schirmer, K. Nanomaterials in the environment, behavior, fate, bioavailability, and effects—an updated review. *Environ. Toxicol. Chem.* **2018**, 37, 2029.

Löffler, F.E.; Edwards, E.A. Harnessing microbial activities for environmental cleanup. *Curr. Opin. Biotechnol.* **2006**, 17, 274.

Long, R.Q.; Yang, R.T. Carbon nanotubes as superior sorbent for dioxin removal. *J. Am. Chem. Soc.* **2001**, 123, 2058.

Ma, H.; Williams, P.L.; Diamond, S.A. Ecotoxicity of manufactured ZnO nanoparticles—a review. *Environ. Pollut.* **2013**, 172, 76.

Machado, S.; Stawiński, W.; Slonina, P.; Pinto, A.; Grosso, J.; Nouws, H.; Albergaria, J.T.; Delerue-Matos, C. Application of green zero-valent iron nanoparticles to the remediation of soils contaminated with ibuprofen. *Sci. Total Environ.* **2013**, 461, 323–329.

Magdolenova, Z.; Collins, A.; Kumar, A.; Dhawan, A.; Stone, V.; Dusinska, M. Mechanisms of genotoxicity. A review of in vitro and in vivo studies with engineered nanoparticles. *Nanotoxicology.* **2014**, 8, 233.

Manke, A.; Wang, L.; Rojanasakul, Y. Mechanisms of nanoparticle-induced oxidative stress and toxicity. *BioMed. Res. Int.* **2013**, 2013, 942916

Mashino, T.; Nishikawa, D.; Takahashi, K.; Usui, N.; Yamori, T.; Seki, M.; Endo, T.; Mochizuki, M. Antibacterial and antiproliferative activity of cationic fullerene derivatives. *Bioorg. Med. Chem. Lett.* **2003**, 13, 4395.

Maynard, A.D.; Aitken, R.J.; Butz, T.; Colvin, V.; Donaldson, K.; Oberdörster, G.; Philbert, M.A.; Ryan, J.; Seaton, A.; Stone, V. Safe handling of nanotechnology. *Nature.* **2006**, 444, 267.

Mohanty, A.; Wu, Y.; Cao, B. Impacts of engineered nanomaterials on microbial community structure and function in natural and engineered ecosystems. *Appl. Microbiol. Biotechnol.* **2014**, 98, 8457.

Moore, M. Do nanoparticles present ecotoxicological risks for the health of the aquatic environment. *Environ. Int.* **2006**, 32, 967.

Muller, J.; Huaux, F.; Moreau, N.; Misson, P.; Heilier, J.F.; Delos, M.; Arras, M.; Fonseca, A.; Nagy, J.B.; Lison, D. Respiratory toxicity of multi-wall carbon nanotubes. *Toxicol. Appl. Pharm.* **2005**, 207, 221–231.

Neal, A.L. What can be inferred from bacterium–nanoparticle interactions about the potential consequences of environmental exposure to nanoparticles? *Ecotoxicology.* **2008**, 17, 362.

Palit, S. Nanomaterials for industrial wastewater treatment and water purification. *Handbook of Ecomaterials*. **2017**, pp. 1–41.

Park, Y.; Hong, Y.; Weyers, A.; Kim, Y.; Linhardt, R. Polysaccharides and phytochemicals, a natural reservoir for the green synthesis of gold and silver nanoparticles. *IET Nanobiotechnol.* **2011**, 5, 69.

Patil, S.S.; Shedbalkar, U.U.; Truskewycz, A.; Chopade, B.A.; Ball, A.S. Nanoparticles for environmental clean-up, a review of potential risks and emerging solutions. *Environ. Technol. Innov.* **2016**, 5, 10.

Pietroiusti, A.; Magrini, A.; Campagnolo, L. New frontiers in nanotoxicology, gut microbiota/ microbiome-mediated effects of engineered nanomaterials. *Toxicol. Appl. Pharm.* **2016**, 299, 90.

Prevot, B. A.; Baiocchi, C.; Brussino, M.C.; Pramauro, E.; Savarino, P.; Augugliaro, V.; Marci, G.; Palmisano, L. Photocatalytic degradation of acid blue 80 in aqueous solutions containing TiO_2 suspensions. *Environ. Sci. Technol.* **2001**, 35, 971.

Rajan, C. Nanotechnology in groundwater remediation. *Int. J. Environ. Sci. Dev.* **2011**, 2, 182.

Roduner, E. Size matters, why nanomaterials are different. *Chem. Soc. Rev.* **2006**, 35, 583.

Santhosh, C.; Velmurugan, V.; Jacob, G.; Jeong, S.K.; Grace, A.N.; Bhatnagar, A. Role of nanomaterials in water treatment applications, a review. *Chem. Eng. J.* **2016**, 306, 1116.

Scarino, A.; Noel, A.; Renzi, P.; Cloutier, Y.; Vincent, R.; Truchon, G.; Tardif, R.; Charbonneau, M. Impact of emerging pollutants on pulmonary inflammation in asthmatic rats, ethanol vapors and agglomerated TiO_2 nanoparticles. *Inhal. Toxicol.* **2012**, 24, 528.

Selck, H.; Handy, R.D.; Fernandes, T.F.; Klaine, S.J.; Petersen, E.J. Nanomaterials in the aquatic environment, A European Union–United States perspective on the status of ecotoxicity testing, research priorities, and challenges ahead. *Environ. Toxicol. Chem.* **2016**, 35, 1055.

Shahwan, T.; Sirriah, S.A.; Nairat, M.; Boyacı, E.; Eroğlu, A.E.; Scott, T.B.; Hallam, K.R. Green synthesis of iron nanoparticles and their application as a Fenton-like catalyst for the degradation of aqueous cationic and anionic dyes. *Chem. Eng. J.* **2011**, 172, 258.

Shedbalkar, U.; Singh, R.; Wadhwani, S.; Gaidhani, S.; Chopade, B. Microbial synthesis of gold nanoparticles, current status and future prospects. *Adv. Coll. Interface Sci.* **2014**, 209, 40.

Shen, Z.; Chen, Z.; Hou, Z.; Li, T.; Lu, X. Ecotoxicological effect of zinc oxide nanoparticles on soil microorganisms. *Front. Environ. Sci. Eng.* **2015**, 9, 912.

Shirsath, D.; Shirivastava, V. Adsorptive removal of heavy metals by magnetic nanoadsorbent, an equilibrium and thermodynamic study. *Appl. Nanosci.* **2015**, 5, 927.

Simonin, M.; Richaume, A. Impact of engineered nanoparticles on the activity, abundance, and diversity of soil microbial communities, a review. *Environ. Sci. Pollut. Res.* **2015**, 22,13710.

Sinha, A.K.; Suzuki, K.; Takahara, M.; Azuma, H.; Nonaka, T.; Fukumoto, K. Mesostructured manganese oxide/gold nanoparticle composites for extensive air purification. *Angew. Chem.* **2007**, 46,2891.

Solovitch, N.; Labille, J.; Rose, J.; Chaurand, P.; Borschneck, D.; Wiesner, M.R.; Bottero, J.Y. Concurrent aggregation and deposition of TiO_2 nanoparticles in a sandy porous media. *Environ. Sci. Technol.* **2010**, 44, 4897.

Sumi, N.; Chitra, K. Effects of fullerene (C60) on antioxidant enzyme activities and lipid peroxidation in gill of the cichlid fish, *Etroplus maculatus*. *J. Zool. Stud.* **2016**, 3, 31.

Sun, Y.P.; Li, X.Q.; Cao, J.; Zhang, W.X.; Wang, H.P. Characterization of zero-valent iron nanoparticles. *Adv. Coll. Interface Sci.* **2006**, 120, 47.

Tang, X.; Zhang, Q.; Liu, Z.; Pan, K.; Dong, Y.; Li, Y. Removal of Cu (II) by loofah fibers as a natural and low-cost adsorbent from aqueous solutions. *J. Mol. Liq.* **2014**, 199, 401.

Taquahashi, Y.; Ogawa, Y.; Takagi, A.; Tsuji, M.; Morita, K.; Kanno, J. An improved dispersion method of multi-wall carbon nanotube for inhalation toxicity studies of experimenta l animals. *J. Toxicol. Sci.* **2013**, 38, 619.

Tolaymat, T.; Genaidy, A.; Abdelraheem, W.; Dionysiou, D.; Andersen, C. The effects of metallic engineered nanoparticles upon plant systems, an analytic examination of scientific evidence. *Sci. Total Environ.* **2017**, 579, 93.

Van Koetsem, F.; Geremew, T.T.; Wallaert, E.; Verbeken, K.; Van der Meeren, P.; Du Laing, G. Fate of engineered nanomaterials in surface water, factors affecting interactions of Ag and CeO_2 nanoparticles with (re) suspended sediments. *Ecol. Eng.* **2015**, 80, 140.

Wang, Z.; Wu, A.; Colombi Ciacchi, L.; Wei, G. Recent advances in nanoporous membranes for water purification. *Nanomaterials.* **2018**, 8, 65.

Zeman, L.J.; Zydney, A.L. *Microfiltration and Ultrafiltration, Principles and Applications.* CRC Press. **2017**.

Zhang, Y.; Wu, B.; Xu, H.; Liu, H.; Wang, M.; He, Y.; Pan, B. Nanomaterials-enabled water and wastewater treatment. *NanoImpact.* **2016**, 3, 22.

Zhao, X.; Lv, L.; Pan, B.; Zhang, W.; Zhang, S.; Zhang, Q. Polymer-supported nanocomposites for environmental application, a review. *Chem. Eng. J.* **2011**, 170, 381.

Index

β

β-carotene, 87
β-cell dysfunctions, 193
β-lactoglobulin (βLG), 608, 612, 616, 620, 624

A

Abiotic stress, 4, 10, 16, 121, 122
  tolerance, 12
Acetogenesis, 124, 125
Acetoin, 12
Acquired immunodeficiency syndrome (AIDS), 85, 384, 388, 446, 463, 485
Acridine orange (AO), 412, 413
Acridinium ester (AE), 559
Actinobacteria, 4, 81, 91
Actinomycetes, 99, 121, 123, 645, 647, 648
Active
  biomolecules, 104
  targeted drug delivery, 432
    aptamers, 433
    monoclonal antibodies, 432
    peptides, 434
    small molecules, 433
Acute pancreatitis, 92, 93
Adeno-associated virus (AAV), 66, 267
Adenovirus, 72, 263, 302
Adhesions, 97
Adult stem cells (ASCs), 284, 301
Advanced
  glycosylation end (AGE), 193, 194, 206
  oxidation protein products (AOPPs), 196
Aerobic respiration, 228
Aerosol generation, 154
Affinity column chromatography, 67
Age-related macular degeneration (AMD), 199, 380
Agrobacterium, 84, 122, 269
Airborne particle transmission, 166
Airway epithelial cell barrier functions, 192
Aldolase enzymes, 411, 417, 418
Algae, 26, 39, 52–56, 58–60, 121, 645, 647
  bioremediation, 50, 53
    experimental setup, 55
    microscopic study, 54
    modeling and computer simulation, 56
    physicochemical characteristics, 56
    study area, 54
    water quality prediction, 58
  treatment, 56, 58
Alzheimer's disease (AD), 40, 185, 187, 201, 206, 287, 385, 466, 467, 533, 553
Amino acid, 26, 68, 101, 104, 125, 185, 200, 276, 362, 556, 563, 566, 583, 610, 611, 646
  sequences, 33
Aminocyclopropane-1-carboxylic acid (ACC), 9
  deaminase, 7, 9, 12
Aminoglycosides, 123
Ammonia, 52, 53, 124, 129, 136, 395
Amplicons, 28
Amyloid β (Aβ), 201, 204, 466, 553, 554
Amyotrophic lateral sclerosis (ALS), 187, 201, 202, 287, 288, 291, 387
Anemia, 42, 77, 78, 102, 193, 289, 371, 384
Anionic exchange, 68
Anthropogenic
  factors, 44
  spots, 40
Antibiotic, 11, 78, 79, 81, 86, 90, 92, 96, 98, 99, 101, 103, 106, 107, 121–123, 271, 272, 430, 435, 462, 495, 519, 534
  agent, 104
Anticancer
  action, 72
  drugs, 80, 389, 457, 487, 521, 606
  therapy, 100
Anticancerous drug, 72
Anti-CRISPR mechanism, 252
Antifungal, 97
  activity, 96, 97
  compounds, 97

Antigen, 69, 71, 97, 100, 123, 132, 196, 330, 366, 384, 387, 389, 416, 432, 451, 479, 480, 483, 493, 521, 522, 525, 527, 557
binding antibody fragments, 69
specific antibodies, 123
Antigenotoxicity, 130
Antihelminthic activity, 80
Antiherbivory compounds, 3
Antihypocholesterolemic drugs, 98, 99
Anti-inflammatory drugs, 91
Antimicrobial, 81, 123, 271, 272, 529
adjuvants, 97
compounds, 320
peptides (AMPs), 101
resistance, 97, 271, 272
Antimutagenicity, 130
Antioxidant, 40, 78, 81, 83, 87, 102, 179–181, 189, 201, 203, 205, 227, 229, 230, 232, 237–239, 467, 497, 529, 566, 606
activity (AOA), 81, 229, 238, 239, 497
Antitumor drugs, 99, 123, 124, 495
Antivascular epithelium protein, 531
Antiviral activity, 77, 78
Antivirulence factors, 104
Apoplast, 6
Apoptosis, 100, 183, 184, 189, 190, 194, 196, 197, 200, 201, 227, 228, 335, 341, 388, 457, 458, 466, 490, 640, 641
Aquatic
macrophytes, 43, 44
plants bioremediation, 44
Arachidonic acid release, 191
Archaea, 4, 26, 246, 247, 357
Aromatic
bonds, 133
compounds, 129
Arsenic efflux membrane protein, 15
Arsenicosis, 42
Artificial vascular grafts, 586–588
Ascorbic acid, 229
Atherosclerosis, 189, 190, 192, 194, 206, 230, 394, 532, 586
Atmospheric nitrogen, 10
Atomic
absorption spectrometer, 46
force microscopy (AFM), 605, 620, 622
Atopic dermatitis, 78, 198
Atorvastatin, 98
Autoimmune diseases, 84, 196, 381, 387
Autointoxication, 129
Autophagy, 183, 190, 191, 198
Avian polyomavirus (APV), 67

B

Bacterial
bacteriocins, 101
endophytes, 16
Bacteriocins, 96, 101, 130
Bacteriocolicins, 124
Bacteriophage, 71, 72, 246, 272, 273, 584, 586
nanomaterials, 584
Bacteroidetes, 4, 107
Baculovirus, 66
expression vector, 67
Basidiomycetes, 87
B-cell epitope, 66
Begomoviruses, 269
Beneficial microorganisms, 5, 17
Benzothio carboxypurine (BCP), 412, 413
Bifidobacteria, 83, 100, 102, 103, 124, 129
Bioaccumulation, 39, 40, 43, 44, 50, 51, 60, 134
factor (BAF), 43, 46, 51
Bioactive
compounds, 34, 80, 81, 85, 87, 93, 96, 107, 529, 606, 608, 613, 648
molecules, 87, 490, 500, 519
Bioactivity, 77, 78, 238, 617
Biochemical pathways, 35
Biocompatibility, 426–428, 433, 440, 445, 456, 469, 517, 520, 523, 529, 531, 552, 553, 574–576, 583, 591, 593
Bioconductor statistical analysis, 29
Biocontrol agent, 11
Biodegradable polymeric nanomaterials, 574
Biodiesel, 119, 125, 128, 129
Bioenhancers, 16
Bioenzymes, 227
Bioethanol, 119, 125–127, 137
Bioethics, 171
Biofertilizer, 9, 15, 121, 122
Biofilms, 135, 320
Biofilter, 44
Biofunctionalization, 425–427, 434, 435, 438–441

Biogas, 119, 124, 125
Biogeochemical cycles, 120, 645
Biohazard, 148, 154, 159, 173
Bioimaging, 431, 499, 532, 567, 593
Bioinformatics, 351–353, 369, 371, 383
Bioinoculant formulations, 16
Bioleaching, 133, 135
Biological
  databases significance, 356
  functions, 30, 34, 434
  oxygen demand, 45, 53
  safety cabinet (BSC), 154–159, 162–167
  vascular grafts, 587
Biomarker, 180, 185, 186, 206, 238, 341, 351, 352, 375, 377, 382, 384, 387, 394, 396, 436, 458, 494, 525, 528
  discovery, 34, 353, 396
Biomass, 40, 44, 50, 52, 53, 56, 120, 125, 126, 128, 644
  metabolism production, 40
Biomethane, 124
Biomineralization, 133–135
Biomolecules, 73, 78, 124, 181, 184, 198, 311, 312, 353–355, 367, 382, 430, 431, 435, 437, 440, 493, 517–519, 521, 522, 526, 531, 534, 552, 561, 564
Bionanoparticles, 647
Biopesticides, 121, 122
Biopolymers, 78, 79, 446, 492, 574, 577, 582, 588
Bioremediation, 34, 39, 43, 44, 46, 50, 52, 54, 56, 105, 119, 132–134
Biosafety, 66, 147–155, 157, 158, 162–173, 490
  cabinet handling, 158
  committee, 169, 170, 172
  laboratory techniques and practices, 151
  level (BSL), 66, 153, 158, 159, 162–168, 172, 173
  materials, 154
  microbiological risk assessment, 152
  officer, 169, 170
  team, 152, 170
Biosecurity, 148, 171, 528, 586
Biosensing, 426, 499, 517–523, 525–531, 533–535, 552, 553, 582
Biosensor, 120, 130–132, 137, 276, 343, 435, 517–523, 525–535, 551, 553, 555, 559, 564, 567, 574, 579
Biosorbents, 53
Biosorption, 50, 133–135
Biosynthesis, 12–14, 83, 187
Biosynthetic pathways, 8
Biotechnology, 44, 148, 170, 253, 426, 517, 574, 584
Biotic stress, 4, 5, 269
Bleomycin, 72
Blood
  brain barrier (BBB), 451, 453, 456, 464–468, 477, 499
  diseases, 289
Bone tissue
  engineering, 588
  regeneration, 590
Bovine serum albumin (BSA), 466, 558, 563, 608–610
Breast cancer gene1/2 (BRCA1/2), 368
Bronchial inflammation, 191
Bulk dopamine, 468
Burkholderia, 9–12, 133
Burkholderiales, 4
Butyric acid, 125

C

Cancer
  cell reprogramming, 289
  research and therapy, 254
Cannibalism, 104
Capsid, 65–73, 262, 325, 584
Capsomere, 66–68
Carbohydrates, 433, 435, 646
Carbon
  dioxide, 124, 125, 127, 128, 137, 334
  nanomaterials, 523
  nanotube (CNT), 460, 490, 494, 517, 520, 523–525, 531, 534, 551, 630, 632, 634, 639
Carcinogen, 100, 130, 517, 534, 636, 640
Carcinogenesis, 195, 232, 241, 289
Carcinogenic compound cyclohexylamine, 91
Carcinogenicity, 81, 307
Cardiac diseases, 264, 265, 300
Cardiomyocytes (CM), 189, 194, 264, 284, 288, 289, 335, 336
Cardiomyopathy, 192, 234, 264, 288
Cardiovascular (CV), 67, 105, 184, 188–191, 194, 205, 230, 235, 264, 265,

283, 288, 320, 339, 341, 375, 394, 396, 436, 446, 484, 526, 555, 606
diseases (CVD), 105, 188, 189, 193, 194, 203, 230, 234, 235, 238, 264, 288, 375, 394, 436, 446, 484, 555
Carotenoids, 81, 104, 606
Cartilage tissue
engineering, 591, 592
regeneration, 591
Cell
biology, 297, 320
culture media, 66
death, 40, 41, 181, 192, 195, 202, 205, 235, 260, 287, 290, 321, 465, 494, 641, 643
differentiation, 8
elongation, 8
entry, 73
genome, 254, 277, 278, 286, 289
migration, 195, 288, 314, 326, 335, 340, 341, 589
penetrating peptides (CPPs), 71
signaling, 228, 437
wall
degrading enzymes, 16
lysis enzymes, 121
Cellobiase, 124
Cellular
components, 179, 314, 320, 328, 330, 498, 640
defense mechanisms, 8
imaging, 313
membranes, 230
metabolism, 34, 179
toxicity, 640
Cellulose, 68, 79, 126, 127, 414, 575
Central nervous system (CNS), 204, 232, 266, 267, 339, 446, 464, 483, 490, 586, 624
Cephalosporin, 90, 96, 123
Charge-coupled device (CCD), 318, 319, 328, 339
Chelation, 43
Chemical
fertilizer, 16, 122
application, 11
oxygen demand (COD), 45, 52, 53
Chemiluminescence (CL), 526, 559
Chemoattractants, 5
Chemotherapeutic
agents, 99
drug, 73
Chemotherapy, 72, 93, 100, 123, 260, 374, 394, 456–458, 462, 487
Chitin nanofibril (CN), 576, 577, 579
Chitosan, 445, 446, 465, 467, 469, 480, 483, 484, 523, 527, 533, 562, 563, 577, 579, 582, 588, 589, 593, 610
Chloramination, 136
Chlorella vulgaris, 51, 52
Cholesterol oxidation, 190
Chronic
arsenic toxicity, 42
brucellosis, 85
hyperglycemia, 230
infectious diseases, 101
inflammation, 193
kidney disease (CKD), 193, 194, 230
lung disease, 42
obstructive pulmonary disease (COPD), 191, 192, 230
oxidative stress, 191
Chymostatin, 100
Chymotrypsin, 99
Ciclosporin, 86
Circulating tumor cells (CTCs), 372, 384, 436
Classical microbial techniques, 27
Clavicipitaceae, 4
Clindamycin, 90
Clinical diagnosis, 381, 384, 385
Clostridium difficile infection (CDI), 88, 90
Clustered regularly interspaced short palindromic repeats (CRISPRs), 29, 245–248, 251–254, 259–278, 291, 501
action mechanism, 248
advantage and limitations, 276
application and effectivity, 253
editing, 270
historical aspects, 247
implications, 261, 265
mediated agricultural advances, 268
Cobalamin, 12, 83
biosynthesis, 14
Cobaltochelatase, 14
Colloidal stability, 438

Colonic fermentation, 102
Colonization, 5, 6, 103, 121
Compartmentalization, 43
Complex
coacervation, 605, 613, 626
natural ecosystem, 33
Compound annual growth rate (CAGR), 122–124, 127, 129, 130, 132
Computational tools, 30, 36
Computed tomography (CT), 92, 338, 339, 343, 418
Concanavalin A (Con A), 554, 560
Conducting polymer (CP), 71, 529–533
Confocal microscopy, 71, 314, 317, 319–322, 331, 332, 344
advantages, 321
application, 319
instrumentation and technical features, 317
limitations, 322
overview, 317
principle, 317
types, 319
disk-based systems, 319
spectral scanning, 319
Conjunctival congestion, 42
Consortia, 5, 15, 16
formulation development, 15
Conventional
breeding, 11
chemical fertilizer, 10
method, 6, 131
treatment systems, 60
Copper
homeostasis, 15
sulfide (CuS), 497, 498
tolerant plants, 9
Copy number variants (CNV), 369
Coronary artery disease (CAD), 189, 192, 234, 264, 379
Cosmeceuticals, 79, 107
Crop
improvement, 3–6, 12, 15–17, 276
productivity, 3, 16, 121
yield, 4
Crucial metabolic processes, 30
Cryoelectron microscopy (Cryo-EM), 312, 344
Cryo-EM
advantage, 325
limitations, 325
Cyclosporin, 78, 80, 84–86, 104
Cysteine modified silver nanoparticles (Cys-AgNPs), 566
Cystic fibrosis, 368, 371, 376
Cytochrome p-450 3A4, 479
Cytokinin, 122
Cytoplasm, 270, 379, 501
Cytoplasmic incapability (CI), 88
Cytosolic protein cyclophilin, 84
Cytotoxicity, 181, 329, 438, 440, 469, 639

D

Database, 30, 354, 356, 357, 369, 391
Decision-making analysis, 58
Dendrimer, 429, 430, 454, 459, 464, 476, 490, 491, 517, 528, 529, 632, 636
Dendritic cell activation, 192
Desolvation, 605, 608, 611, 612, 614, 615, 626
Detoxification, 127, 229, 632
Dexamethasone (Dex), 481
Diabetes, 42, 77, 78, 85, 86, 102, 105, 106, 184, 185, 189, 192, 193, 206, 230, 234, 283, 288, 300, 351, 352, 369, 371, 372, 387, 391, 393, 394, 434, 522, 525, 555
mellitus (DM), 42, 105, 189, 193, 200, 230, 234
treatment, 105
Diarrhea, 41, 78, 85, 90, 105, 106, 130
Dichlorodiphenyltrichloroethane, 133
Differential interference contrast (DIC), 331, 332
Differentiation, 8, 180, 183, 184, 196, 198, 206, 284, 288–290, 297, 300, 301, 305, 308, 309, 335, 531, 573, 574, 579, 589, 592, 594
Disease diagnosis, 351, 352, 367, 368, 375, 376, 378, 381, 387, 393, 395, 396, 552, 565
Disinfection, 173, 633
Division, 8, 147, 151, 245, 266, 268, 297, 298, 309, 311, 312, 376, 457, 458, 579
Double
emulsion, 615
strand break, 246, 247, 278

stranded DNA (ds DNA), 68, 263, 641
Doxorubicin (DOX), 72, 460, 479, 480, 487, 490, 491, 502, 625
Draft genomes, 30
Dried blood spots (DBSs), 411, 414–416, 421
Drosophila, 265, 337
Drought, 10, 12
Drug
delivery system (DDS), 314, 453, 458, 460, 478, 481, 482, 484, 485, 489, 491, 495, 497, 500, 502, 625
applications and advantages, 451
resistant solid tumor, 73
Dry mass, 45
Duchenne muscular dystrophy (DMD), 253, 254, 265, 306, 371, 393
Dynamic light scattering (DLS), 71, 452, 605, 620, 621
Dyslipidemia, 190, 192, 379
Dystrophin (DYS), 253, 306, 371

E

Echinocandins, 97, 98
Electron, 40, 71, 125, 134, 181, 182, 185, 193, 228, 238, 313, 316, 319, 322, 323, 326, 344, 522, 528, 617, 618, 620, 621
microscope, 313, 316, 322, 323
applications, 324
cryo-EM, 323
instrumentation and technical features, 322
overview, 322
principle, 322
SEM, 323
transmission electron microscope (TEM), 6, 316, 323, 555, 560–562, 564, 620–622
types, 323
multiplying charge-coupled device, 319
Electrophoresis, 33, 356, 366, 382, 383
Electrophysiological polarity, 264
Electroporation, 267, 268, 286, 476, 480, 481
Electrospinning, 489, 577, 578, 581, 582
Electrospray
ionization (ESI), 384, 385, 390
particle size, 626
technique, 610, 617
Electrospun, 489, 577, 582, 592
Embryogenesis, 88, 277
Embryonic stem cells (ESCs), 283–285, 287, 289, 292, 298–300, 302, 304, 436
Emulsification, 605, 614–616, 626
Emulsified zero-valent iron (E-ZVI), 648, 649
Emulsions
diffusion method, 616
solvent evaporation method, 615
Endemic pemphigus foliaceus (EPF), 198
Endometabolomics, 34
Endophytes, 3–9, 11, 12, 15–17
action mechanism, 7
phytohormone production and regulation, 7
biocontrol properties, 11
isolation and screening, 6
nitrogen fixation and mineral solubilization, 9
Endophytic
application, 15
association, 4
bacteria, 9–13
colonization, 5, 6
fungi, 4, 5, 10–12, 15, 16
isolation, 6
microbial distribution, 3
microorganisms, 4, 5, 7
nature, 4
Endoplasmic reticulum (ER), 181, 182, 318, 389
Endothelial cells, 101, 190, 199, 200, 203, 284, 285, 288, 464
Endotoxins, 69, 194
Enhanced permeability and retention (EPR), 432, 436, 438, 451, 455, 460
Enterobacteriaceae, 15, 101
Enterobacterial infections, 103
Environmental
damage, 9
pollutants, 132, 638
stress, 5, 34, 367
Enzyme
inhibitors, 78, 99
linked immunosorbent assay (ELISA), 365, 366, 384
Enzymosensor, 517, 535
Epidemiological evolution, 371

Epithelial cells, 91, 103, 191, 199, 302
Erythrocytes, 193, 199, 417, 527
Erythromycin, 103
Escherichia coli (*E. coli*), 84, 86, 101, 104, 123, 124, 172, 247, 248, 251, 435, 560, 633, 643, 646, 647
Estrogen receptor (ER), 389
Ethionamide (ETH), 380, 461, 462
Ethylene, 9, 122, 467, 478, 481, 491, 582
Eukaryotes, 66, 69, 648
Eukaryotic
  cells, 65, 67, 69
  hosts, 69
  peroxisome cells, 229
Eutrophication, 43, 135
Evaporation method, 615
Exometabolomics, 34
Exopolysaccharides (EPS), 77, 79
Extracellular matrix (ECM), 195, 196, 198, 305, 457, 465, 489, 574, 576, 577, 579, 587, 588, 591–593
Extremophiles, 29

F

Fatty acids, 86, 87, 101, 125, 128, 129, 555
Fecal
  microbiota transplant (FMT), 79, 90, 91, 93
  transplantation, 79, 88
Fermentation, 78, 83, 84, 126, 127, 131
Fibroblasts, 195, 285–287, 302, 306, 336
Fidaxomicin, 90
Field-effect transistor (FET), 523, 526, 531, 532
Firmicutes, 4, 90, 107, 135
Fluorescence, 6, 71, 260, 263, 313–315, 317, 319, 320, 326, 328, 329, 331, 338, 339, 412, 413, 418, 429, 430, 434, 435, 437, 492, 493, 499, 528, 557, 561
  lifetime imaging (FLIM), 320, 332
  microscopy diagnosis, 412
  recovery after photobleaching (FRAP), 320, 332
  resonance energy transfer (FRET), 320, 332, 340, 341
Fluorophore, 331, 492, 493
Foliar endophytes, 6
Folic acid, 102, 460
Food
  additives, 129
  adulterants, 131
  analysis, 518, 521, 535
  industry, 106, 131, 320, 567
Free radical, 40, 79, 81, 179–181, 184, 185, 192, 203, 207, 228–230, 233, 236–238, 641
Fullerenes, 495, 503, 523, 643
Functional
  dynamics, 26, 35
  proteomics, 32, 353, 382
Fungal
  spores, 88
  stress, 269
Fusarium wilt, 16

G

Galactose, 92, 126, 554
Gastroscopy, 90
Gelling adhesion, 79
Gene
  databases, 30
  expression, 31–33, 72, 181, 195, 228, 254, 265–267, 285, 286, 311, 312, 356, 362, 374, 379, 381, 592
  molecular analysis, 12
  polymorphisms, 192
  sequence, 66, 362
  therapy, 73, 245, 263, 277, 305, 306, 491, 529
Genetic
  disorder, 247, 253, 277, 371
  engineering, 11, 70, 245, 275, 585
  mutations, 264, 308
Genetically modified organisms (GMOs), 150, 151
Genome
  dynamics, 31
  editing, 246–248, 253, 254, 261, 264, 265, 267, 268, 270, 271, 275–278, 289, 376
  engineering, 245–247, 253, 265, 277
  sequencing, 28, 372, 382
  wide association studies (GWAS), 369
Genomic, 25, 27, 34, 36, 351–355, 357, 367–369, 373, 374, 376, 378, 382, 389, 391, 393, 395, 396

data, 27, 28, 361, 362, 377
fragments, 28
medicine challenges and opportunities, 375
Genosensor, 517, 518, 521, 525, 535
Genospecies, 28
Genotoxicity, 42, 639–641
Geriatric diseases, 200
Gibberellins, 122
Glassy carbon electrode (GCE), 527, 529, 532
Glioblastoma multiforme (GM), 328, 329, 341
Gliotoxin, 86
Global
biogas production, 125
biosensor market, 132
population, 3
Glomerular filtration barrier, 194
Glomerulosclerosis, 85
Glucose oxidases (Gox), 561
Glutamine synthetase gene, 13
Glutaraldehyde (GA), 55, 609, 614
Glutathione, 67, 81, 102, 181, 200, 229, 230, 233, 393, 561
reductase (GR), 81
S-transferases, 67
Gluten, 93
Glycosidases, 92
Goeppert-Mayer (GM), 328
Gold nanoparticles (AuNPs), 267, 520–523, 525, 527, 528, 532, 551–564, 637, 647
glucose detection, 560
heparin detection, 562
oligonucleotide detection, 558
pathogen detection, 559
proteins detection, 553
Gram-positive bacteria, 81, 129, 641
Graphene oxide (GO), 526, 527, 562, 593, 594, 633
Green
bioconversions, 120
fluorescent protein (GFP), 71, 260, 263, 272, 318, 329, 333, 336, 337, 339, 429
revolution, 9
technology, 121
Greenhouse gas, 125, 128, 137
Guide RNA (g RNA), 247, 260–263, 265, 267, 269, 275–278
Gut
bacteria, 105, 106
microflora, 93

H

Haemophilus influenzae, 27
Hantavirus, 168
Haplotype map (HapMap), 369
Heavy metal, 12, 16, 39–44, 46, 47, 51, 53, 59, 93, 129, 133, 566, 630, 631, 633, 634, 647
toxicity, 12, 40, 41, 44, 60
Hematopoietic stem cells (HSCs), 285, 301
Hemicellulose, 126, 127
Hemoglobin, 289, 330, 416, 435, 527
Hepatitis B virus (HBV), 66, 163, 172, 633
Herbaceous crops, 125
Heterocyclic compounds, 132
Heterotrophs, 136
High-efficiency particulate air (HEPA), 155, 156, 158
High-throughput sequencing (HTS), 27, 31
Homogenization, 501, 616
Horseradish peroxidase (HRP), 418, 419, 520, 527, 532
Host
cell, 33, 66–68, 70, 72, 73, 263
receptors, 90, 463
genome sequences, 30
immunity, 66, 330
parasite interactions, 85
Human
artificial chromosome (HAC), 306
cell lines, 67, 163
embryonic stem cells (hESCs), 285, 289, 298–300
genome project (HGP), 122, 355, 368, 38`1
immune deficiency virus (HIV), 78, 163, 206, 261–263, 291, 325, 373–375, 380, 384, 388, 446, 463, 464, 485, 487, 532
serum albumin (HSA), 490, 555, 556, 609, 610
somatic cell reprogramming, 289
Humanized Cas9 (hCas9), 263
Huntington disease, 266, 371
Hyacinth, 50

Hyaluronic acid (HA), 79, 196, 502, 563, 577, 579, 582, 588–590
  nanoparticle, 579
Hybridization, 6, 31, 320, 361, 362, 372, 377, 522, 530, 531, 558
Hydrocarbons, 39, 132, 533, 630, 633, 636, 637, 648
Hydrogen
  bond, 67, 613
  peroxide, 6, 81, 135, 180, 188, 228, 229, 527
    molecule, 229
Hydrolysis, 85, 91, 92, 104, 124–126, 495, 557, 609
Hydrolytic enzyme, 92
Hydrosphere, 93
Hydroxyl radicals, 81, 181, 228, 641
Hyperglycemia, 105, 192, 193
Hyperplasia, 191, 198
Hypertension, 42, 189, 194, 204, 230, 232, 394
Hypertrophic cardiomyopathy (HCM), 264
Hypocholesterolemic agents, 78
Hypomyces, 99

I

Immobilization, 131, 382, 519, 520, 526, 557, 558, 638
Immune
  chromatographic capture, 409
  dysfunction, 42
Immunoassays, 531
Immunochromatographic test, 411, 417–419, 421
Immunogenicity, 65, 66, 73, 433, 434, 499, 590, 609
Immunoglobulin, 69
Immunomodulation, 130
Immunomodulators, 78, 107
Immunosensor, 525–527, 531, 532, 535
Immunosuppressants, 78, 84, 85
Immunosuppression, 84, 85
  microbes potential, 83
Immunosuppressive drugs, 84
In vitro
  condition, 66
  neural cells, 266
  release, 606, 626
  scaffolds, 579
In vivo
  applications, 73, 261
  methods, 6
  studies, 72, 149, 469, 593
Indole
  3-acetic acid (IAA), 7–9, 12, 13, 15, 16
  acetic acid, 122, 393
Induced pluripotent stem cells (iPSCs), 264, 267, 284–292, 299, 301–309
  applications, 304
    disease modeling, 305
    regenerative medicine, 305
  generation, 302
  research challenges, 307
Inflammatory
  bowel disease (IBD), 88, 90, 93, 130, 196, 197, 369
  genes, 193
Injectable nanoparticle generator (iNPG), 502
Innate defense mechanisms, 4
Inorganic
  filters, 104
  nitrogen, 53
Insect cell lines, 67
Insecticides, 87, 410
Integrated water quality index (IWQI), 56, 60
Intercellular
  regions, 5
  spaces, 6
Internal standards, 35
Intravital
  microscopy (IVM), 311, 313, 314, 326, 329
  multiphoton
    imaging, 342, 344
    microscopy, 312, 314
Ischemia/reperfusion (IR), 189, 205, 493
Isobaric tags for relative and absolute quantification (iTRAQ), 385
Isoniazid (INH), 375, 461–463

J

John Cunningham virus (JCV), 67, 71

K

Keratinocytes, 198, 302
Keratosis, 42
Kidney diseases, 193, 230

L

Laboratory
- acquired infections (LAI), 150
- biosafety, 169

Lactobacillus, 83, 84, 101–103, 125, 129
Lactoferrins, 468
Lantibiotics, 101
Lapatinib (LAPA), 375, 389, 458
Large surface plasmon resonance (LSPR), 553, 554, 560, 561, 564, 565
Lateral root hairs, 5
Lead efflux transporter phosphatase, 14
Leaf endophyte, 6
Left ventricular hypertrophy (LVH), 194, 204
Lesions, 40–42, 123, 185–187, 195, 198, 205, 467, 468
Leukemia, 72, 132, 134, 260, 388, 394, 487
Ligands, 71, 135, 426, 427, 431–434, 437–439, 453, 455, 495, 528, 554, 585, 586, 613
Lignocellulosic
- biomass, 125
- waste, 126, 129

Limit of detection (LOD), 555, 556, 561, 562, 564, 567
Lipid, 40, 41, 81, 128, 136, 181, 189, 190, 194, 199, 206, 367, 462, 487, 605, 607
- membranes, 65
- oxidation, 81
- oxidative damage, 183
- peroxidation, 183, 189, 191, 192, 196, 198, 199, 203, 205, 206, 233, 640, 641

Lipopolysaccharides (LPS), 97, 106, 135
Liposomes, 70, 425, 426, 428, 429, 433, 446, 451, 459, 464, 476, 486–488, 493, 500, 529, 607
Liquid chromatography, 33, 34, 365, 382–385, 392, 623
Live
- animal imaging, 338, 340–344
  - advantages, 343
  - application, 340
  - instrumentation and technical features, 339
  - limitations, 343
  - overview, 338
  - principle, 338
- cell imaging, 331–335, 338, 344
  - applications, 335
  - chambers, 332
  - overview, 331
  - principle, 331

Lovastatin, 98
Low-density lipoprotein (LDL), 190, 192–194, 196, 206, 466
Lung cell apoptosis, 191
Lymphocytes, 84, 191, 196, 260, 289, 463
Lymphokine, 84
Lymphoma, 72

M

Maceration, 6
Macroalgae, 87
Macromolecules, 40, 70, 232, 299, 430, 431, 459, 482, 484, 488, 489, 498, 522, 562, 591, 605
Macrophages, 190, 191, 228, 388, 461–463
Magnesium, 48, 53, 333, 336, 634
Magnetic
- barcoding, 435
- nanoparticle (MNP), 435, 478, 498, 517, 525, 551, 582, 634, 646, 648
- resonance
  - cholangiopancreatography, 92
  - imaging (MRI), 338, 339, 341, 343, 436, 437, 440, 491, 493, 498, 500, 529

Magnetoactive electrospun nanofibers, 581
Malaria, 409, 410
- diagnosis, 409–411, 414, 416, 417, 419, 421

Mammalian cells, 69, 440, 579
Mannoproteins, 97
Mannose, 126, 554
Marine microbial bioactive compounds, 93
Mass spectrometry (MS), 32, 33, 354–357, 365, 367, 368, 382–386, 388–393, 395, 396, 552
Matrix
- material proteinases (MMP), 192, 195, 197, 203, 465, 466, 556, 557
- metalloproteinase matrilysin (MMP-7), 191, 556

Mean intensity projection, 318
Mechano-sensitive tissues, 590

Medical dressing, 576, 577
Membrane
  disruption, 203, 641
  function, 10
Mendelian disorder, 370, 371
Mercuric reductase, 93
Metabolic
  activities, 83, 644
  engineering, 86
  enzymes, 35
  networks, 26, 30
  pathways, 27, 30, 33, 35, 44, 125, 228, 247, 393, 395
  profiles, 28
Metabolites, 3–5, 7, 11, 12, 34, 35, 77, 78, 80, 81, 93, 96, 97, 102, 103, 130, 135, 204, 352–354, 367, 368, 391–396, 556
Metabolome, 34, 367, 391
Metabolomics, 26, 33–36, 351–357, 367, 391–393, 395, 396, 478
  challenges and limitations, 395
  metabolic signature, 391
  therapeutics, 394
Metagene, 29
Metagenomic, 25, 27–30, 36, 137
  approaches, 28, 30
  technologies, 30
Metalloids, 53, 494, 637
Metaproteomics, 25, 28, 32, 33, 36
Metatranscriptomics, 6, 25, 31, 32
Methanogenesis, 124, 125
Microalgae, 51, 52, 78, 87, 128, 129
Microbes
  allergies treatment, 103
  application, 120
    agricultural sector, 120
    biodegrader, 132
    biodiesel production, 128
    bioethanol production, 125
    biogas production, 124
    biosensors, 130
    healthcare and medicine, 122
    probiotics, 129
    sewage and wastewater treatment, 135
Microbial
  action, 93
  agents and toxins, 168
    food borne diseases, 169
  analysis, 30
  antibiotics, 96
  antioxidants, 81
  biochemicals, 78
  candidates, 5
  cells, 27, 34, 272, 313, 647
  communities, 25, 26, 28–33, 35, 644
  consortia, 15, 32
  cooperation, 33
  cultivation technologies, 27
  derivatization, 79
  diversity, 6, 25–31, 34–36
  drug resistance, 103
  ecology, 28
  endophytic community, 4
  enzymes, 93
  flora, 106
  genomes, 28
  growth, 26, 80, 153, 577
  interactions, 35
  laboratory agents, 149
  metabolites, 80
  population, 25, 26, 30, 103, 124, 272, 643, 644
  products, 78–81, 84, 107, 119, 122, 137
  species, 34, 99, 104, 644
  systems, 34
Microbialnutraceuticals, 107
Microbiological operations, 162, 164–166
Microbiology, 27, 30, 149, 162, 164, 170, 351–353, 384
Microbiome, 4, 15, 29, 30, 85, 91, 106, 107, 120, 122, 378, 395
Microcins, 101
Microfabrication, 579, 580
Microflora, 88, 90, 91
Micronutrients, 51, 83
Microscopic slide examination, 411
Microscopy, 69, 71, 311–317, 319–328, 330–332, 343, 344, 410, 411, 416, 418, 421, 620
Mildew disease, 270
Molecular
  compositions, 27
  identification, 12, 26, 27
  level, 32, 338, 368
  networks, 35, 305
  probes, 33

weight, 11, 34, 71, 93, 96, 482, 484, 560, 575, 609, 612
Monocultures, 15
Monocytes, 190, 191, 196, 230
Morphoanatomical
features, 26
identification, 26
Mosquito-borne diseases, 87–89
Mucous membrane exposure, 172
Multicriteria decision-making approach, 56
Multidimensional images, 343
Multidrug-resistant bacterial strains, 271
Multimechanistic features, 3
Multimeric antigens, 73
Multiomics, 351, 352, 391, 396
Multiphoton intravital microscopy (IVMPM), 311, 312, 314, 316, 326–330, 338, 341, 343, 344
advantages, 330
application, 329
instrumentation and technical features, 328
limitations, 330
overview, 326
principle, 327
Municipal
solid waste, 124, 631
waste composts, 40
Murine polyomavirus, 67
Muscular dystrophy, 40, 254, 287, 300, 371
Mutagenesis, 186, 252, 253, 266, 268, 277
Mutations, 195, 196, 202, 233, 261, 264–269, 276, 357, 371, 435, 640
Mutualistic interaction, 5
Mycelial elongation, 8
Mycobacterium tuberculosis, 85
Mycophenolic acid, 86
Mycotoxins, 96, 270
Myelin oxidation, 193
Myeloperoxidase (MPO), 194, 232, 341, 342
Myocardial infarction (MI), 188, 189, 234, 389

N

Nanobiocontrol delivery system, 16
Nanobiosensor, 517, 519, 520, 522, 529, 534, 535
Nanocatalysts, 635
Nanoclew, 486, 501, 503
Nanocomposites, 518, 521, 525, 528, 532–534, 582, 591, 633, 636
Nanodroplets, 502
Nanofilamentous viruses, 585
Nanoghost, 486, 501, 503
Nanomedical applications, 592
Nanomembrane, 634, 635
Nanoparticle (NP), 427
based drug delivery, 456
carbon nanotubes, 460
dendrimers, 459
hydrogels, 457
liposomes and micelles, 459
bioconjugation, 427
dendrimer, 429
liposomes, 428
nanoshell, 430
nucleic acid-based nanoparticles, 431
polymer-drug conjugates, 428
polymeric nanoparticle, 430
quantum dots, 431
biological nanoparticles, 499
exosome, 499
nanobubble, 499
nanoclew, 501
nanocluster, 499
nanoghost, 501
nanoneedle, 500
ceramic nanoparticle, 494
characterization, 620
dendrimer, 490
fullerenes, 495
hybrid nanoparticles, 502
injectable NP generator, 502
nano-terminator, 502
inorganic nanoparticles, 497
copper, 497
gold, 497
magnetic iron oxide nanoparticles, 498
silver, 497
liposomes, 486
nanofiber, 489
nanopore, 489
nanoshells, 494
nanotube, 490
niosomes, 487
polymeric nanoparticle, 488

preparation methods, 618
protein nanoparticles, 495
quantum dots (QDs), 431, 437, 492–494
types, 486, 583
Nanoscale zero-valent iron (nZVI), 630, 632, 638, 643, 644, 646, 647
Nanosensors, 435
Nanostructure catalytic membrane, 635
Nanotechnology, 426, 436, 446, 451, 464, 478, 518, 535, 552, 553, 567, 573, 584, 591, 594, 607, 618, 625, 629–631, 636, 639, 640, 644, 645, 649
nanoparticle selection criteria, 607
Nano-terminators, 502
Nanowires (NWs), 520, 522, 525, 551
National institutes of health (NIH), 162, 170, 355, 376
Natural
antioxidants, 81
fermentation, 79
Naturopathy, 107
Necrosis, 92, 183, 189, 191, 196, 389
Neovascularization, 288, 340
Nephropathy, 85, 192, 194, 205
Nerve growth factors (NGF), 467
Nervous system diseases, 288
Neural tissue regeneration, 585
Neurodegenerative diseases, 201, 232, 266, 267, 287
Neuronal
cell membrane, 201
degeneration, 467
stem cells, 335
Neuropathy, 192, 200, 461
Neuropsychiatric disorder, 267
Neurotransmitter, 202, 205, 468, 533, 565
Neutrophils, 191, 198, 330, 341, 342
Next generation sequencing (NGS), 29, 355, 367, 369, 370, 372, 396
Niche adaptation, 32, 33
Nickel-nitrilotriacetic acid (Ni-NTA), 70
Niosomes, 487, 488, 503
Nitric oxide (NO), 180, 185, 189, 194, 198, 232, 233, 463, 523, 637
Nitrogen fixation, 4, 7, 12, 13, 15, 122
Nitrogenase metallocluster biosynthesis protein, 13
N-methyl-D-aspartate (NMDA), 201, 205, 267
Noncoding, 27, 28, 248, 259–261, 263, 277, 288, 361, 372
transcriptome (ncRNA), 31, 361, 377
Nonconventional methods, 6
Nonmetallic toxicities, 53
Nonpathogenic microorganisms, 26, 83
Non-pitting oedema, 42
Nonprotein-coding RNA, 31
Nonrenewable resources, 119
Novel drug, 78
Nuclear magnetic resonance (NMR), 34, 253, 324, 325, 355, 356, 367, 382, 391, 395
Nucleic acid, 70, 181, 206, 429, 430, 432, 434, 451, 483, 489, 492, 500, 552, 640
acids oxidative damage, 185
DNA, 185
RNA, 187
sequences, 409, 411
Nucleocapsid, 66
Nutraceuticals, 78, 85–87, 605–612, 615–618, 620, 623–625

O

Off-target mutation, 268–270, 278
Oleaginous microorganisms, 128
Oligonucleotides, 71, 259, 380, 431, 525, 558
Omics, 25, 351, 353, 354, 368
approaches, 26, 35, 367
overview, 353
technology, 26
types, 357
genomics, 357
metabolomics, 367
proteomics, 362
Optical imaging, 339–341, 343
Organic
acids, 10, 78, 102, 124, 125, 127, 128, 135
agricultural practices, 4
carbon, 45, 128
compounds, 132, 134, 135, 630, 634, 636
electrochemical transistor, 532
nitrogen, 52
pollutants, 45, 60, 132, 135, 634–637
waste, 124, 128, 129
Osmotic shock, 68
Osteoblast cells, 71

Overpopulation, 120
Oxidative
homeostasis, 192
imbalance, 181, 182, 184, 190, 195, 200–202
status, 102
stress, 14, 40–42, 81, 102, 179–207, 227–238, 288, 340, 465, 639–641, 643
acute pathologies, 188
chronic pathologies, 189
diseases, 187
induced complications, 187, 193, 204

P

Paclitaxel (PTX), 72, 458, 459, 489, 490
Parasite lactate dehydrogenase (pLDH), 416, 417, 419
Parasitemia, 411, 412, 414, 416–419, 421
Parkinson's disease (PD), 40, 254, 267, 287, 288, 391, 467
Passive targeted drug delivery, 432
Pathogenesis, 101, 187, 190–192, 197, 201, 227, 234, 305, 340, 378, 393, 463
Penicillin, 78, 80, 96, 103, 122
Penile cancer, 72
Pentamers, 68
Peptide, 70
Peripheral
cutaneous anaphylaxis, 192
vascular disease, 42
Personalized medicines, 373, 396
Pest resistance, 5
Pesticides, 39, 41, 59, 133, 519, 534, 635, 637, 638
pH, 12, 31, 46–49, 53, 56, 57, 77, 79, 102, 131, 134, 314, 321, 334, 411, 415, 416, 432, 438, 441, 454, 460, 476, 478, 479, 482, 485, 487, 498, 502, 518, 554, 557, 562, 565, 607–609, 612, 613, 620, 624
Pharmacogenomics (PGx), 91, 370–373, 375
Pharmacokinetics, 91, 314, 373, 437, 464, 469, 490, 609
Phenazine, 11, 12
Phenolic compounds, 86, 127
Phenols, 129, 238
Phosphate, 10, 184
buffer saline (PBS), 341, 342, 415, 416
solubilization, 10, 12
Phospholamban, 265
Phosphorus (P), 10
Photoacoustic
degradation, 634
imaging (PAI), 13, 340
Photoemission, 130
Photomultiplier tubes (PMTs), 317, 328
Photon-correlation spectroscopy (PCS), 452, 620, 621, 623
Photo-protective compounds, 104
Photosynthesis, 51
Phylogenomic studies, 28
Phytochemicals, 86, 227, 238, 606
Phytoextraction, 43
Phytohormone, 4, 7, 10, 121, 378
Phytopathogen, 7, 11
Phytophtora capsici, 5
Phytoremediation, 12, 43, 44, 46, 47, 638
Phytostabilization, 43
Phytotransformation, 43
Pigmentation, 42, 198
Piper nigrum, 9
Plague, 169
Plant
associated microorganisms (PAMs), 3, 4
growth promotion (PGP), 4, 7, 8, 12, 13, 15–17
bacteria (PGPB), 121, 122
rhizobacteria (PGPR), 15, 16, 121, 122
pathogens, 3, 168
Plasma cells, 97
Plasmid DNA transfer, 302
Plasmodium
falciparum (Pf), 100, 410, 412, 417–419
vivax (Pv), 410, 412, 418, 419
Plethora, 3, 78, 83, 107, 199
Pluripotent stem cell production, 289
Pneumocandins, 97, 98
Pollutants, 39, 43, 44, 51, 131, 132, 134, 136, 192, 199, 498, 517, 533, 534, 551, 630–634, 636–638
Pollution abatement, 629, 631–633, 649
Poly(4-styrenesulfonic acid) (PSS), 531–533, 555, 594
Poly(allylamine) (PAH), 133, 555
Poly(amidoamine) (PAMAM), 529
Poly(D,L-lactide-*co*-glycolide) (PLGA), 430, 462, 467, 468, 480, 483, 490, 576, 583, 590, 592, 625

Poly(diallyl dimethyl ammonium chloride) (PDDA), 555
Poly(lactic acid) (PLLA), 467, 582, 592
Poly(*p*-phenylene ethynylene) (PPE), 557
Poly(vinyl alcohol) (PVA), 582, 592
Polyaromatic hydrocarbon, 132
Polycaprolactone (PCL), 492, 582, 589, 592
Polychlorinated biphenyls (PCB), 133, 630, 635, 636
Polycyclic aromatic hydrocarbons (PAH), 133
Polydispersity (PDI), 182, 620, 622
Polyelectrolyte multilayer (PEM), 594
Polyethene glycol (PEG), 427, 428, 430, 453, 468, 487, 490, 492
Polymerase chain reaction (PCR), 6, 71, 149, 300, 355, 361, 362, 372, 413, 414, 416, 418, 419, 421
Polymeric substances, 587
Polymers, 68, 428, 437, 451, 453, 476, 478, 484, 488, 491, 493, 497–499, 528, 530, 531, 554, 555, 575, 576, 579, 581–583, 589, 593, 605, 607, 613, 622, 636
Polyneuropathy, 42
Polyol pathway flux, 193
Polyomavirus, 68, 69, 72
Polyphosphate kinase, 13
Polyposis, 371
Polysaccharides, 77, 87, 101, 127, 136, 427, 575, 577, 605, 607, 610
Polystyrene sulfonate (PSS), 531
Polythiophene (PT), 530–532, 534
Polyunsaturated fatty acids (PUFAs), 87, 183
Positron emission tomography (PET), 339–341, 527
Postmitotic neurons, 266, 267
Post-translational modifications (PTMs), 32, 33, 69, 101, 382, 390
Potassium, 10
Potential
  health risks, 639
  protein detector, 553
Prebiotics, 86, 103
Pregnancy and developmental abnormality, 202
Primary molecular imaging techniques, 343
Principal component analysis, 34
Probiotic, 81, 87, 93, 102, 103, 106, 107, 120, 129, 130, 137, 272
  bacteria, 81, 83
Profiling, 25, 29, 32, 206, 291, 368, 376, 377, 384, 389, 390
Prokaryotic cells, 65
Propidium iodide, 69
Propionibacterium, 84, 102, 103
Propranolol hydrochloride (PNL), 483
Protein
  based nanoparticles, 608
  disulfide isomerase (PDI), 182
  expression, 32, 33, 67, 354, 381, 384
  kinase C (PKC), 192–195
  oxidative damage, 184
  protein interaction, 32, 311, 312, 355, 384
  quantification, 33, 365
  synthesis, 10, 181
  types, 609
    albumin, 609
    elastin, 610
    gelatin, 609
    gliadin and legumin, 610
    milk protein, 611
    soy protein, 611
    zein, 611
Proteobacteria, 4, 91
Proteome, 25, 32, 353, 356, 362, 381, 384, 387
  challenges and opportunities, 390
Proteomics, 25, 27, 31–33, 36, 351–357, 362, 368, 381, 382, 384, 385, 387–391, 393, 395, 396
  diagnostic application, 388
  disease diagnosis, 387
  general classification, 382
  targeted therapeutics, 389
  techniques, 384, 390
Pseudomonas, 4, 6, 10, 11, 15, 16, 84, 102, 122, 124, 131–133, 136, 252, 275, 560, 643, 646
Pseudotetrasaccharide, 105
Pseudovitamins, 83
Putrescine, 14

Q

Quality
  guideline index, 47
  index, 56, 58
Quantum dot, 313, 329, 331, 426, 431, 454, 492, 493, 503, 517, 526, 551, 647

R

Radioactive contaminants, 135
Radioisotopes, 135
Radionuclides, 134, 135, 157, 634
Ramsar convention, 43
Rapid diagnostic tests (RDTs), 411, 416, 419, 421
Raw data, 35
Reactive
 nitrogen species (RNS), 180, 188, 192, 196, 198, 202, 206, 232
 oxygen species (ROS), 40, 41, 180–185, 188–206, 227–229, 232–238, 461, 532, 641, 643
Recombinant
 baculovirus system, 67
 proteins, 287, 302
Red blood cell disorder, 289
Redox
 homeostasis, 179–181, 207
 reactions, 44, 632
Reduced graphene oxide (rGO), 526, 527, 532–534
Regenerative medicine, 284, 291, 298, 305, 308, 573, 574, 579, 584, 589, 593, 594
Repulsive electrostatic forces, 438
Retinal
 ganglion cells (RGCs), 199
 pigment epithelium (RPE), 199, 200
Retinopathy, 192, 200, 204
Rhizobiales, 4
Rhizofiltration, 43
Rhizome, 11, 12
Rhizosphere, 5, 6, 43, 44, 121
Rhodanese, 12
Ribonucleoprotein, 267, 269
Rice endophytes, 15
Root
 elongation, 9, 12
 endophytes, 6
Rot diseases, 270
Rotaviruses, 103
Routes, 5, 91, 437, 463, 475, 482, 484, 486–488, 492, 605, 623–625, 639
 blood-brain barrier route, 625
 nasal route, 624
 ocular therapy, 625
 oral route, 623
 pulmonary route, 624

S

Saccharification, 126
Saccharolytic bacteria, 85
Saccharomyces cerevisiae, 79, 84, 98, 127
Salmonella typhimurium, 101, 124
Sanguinivorous, 87
Saprophytic association, 4
Scaffold, 79, 92, 573–577, 579, 580, 585, 586, 589, 590, 592–594
 composition, 589
Scytonemins, 104
Sediment, 44–48, 50, 51, 133, 134, 136
Sedimentation, 43, 136, 614, 643
Selenium accumulators, 87
Self-renewal, 284, 297, 300, 305, 309
Sendaivirus, 302
Sequestration, 122, 135
Serial analysis of gene expression (SAGE), 356, 361, 362, 378
Serotonin, 192
Severe acute respiratory syndrome (SARS), 149
Sewage
 treatment, 51
 wastewater, 52
Shoot meristematic tissues, 6
Short-chain fatty acids, 79, 125
Shotgun sequencing, 29
Siderophore, 4, 7, 11, 16, 121
Single
 nucleotide polymorphism (SNP), 361, 371, 526
 photon emission computed tomography (SPECT), 339, 341, 343
 stranded DNA (ssDNA), 68, 264, 501, 517, 526, 557–559
Sodium
 hypochlorite, 6
 thiosulphate, 6
Soil
 fertility, 121
 niche, 5
 pollution, 637
Solid lipid nanoparticles, 462
Solvent displacement technique, 617
Sphingomonadales, 4
Spore germination, 8
Sporulation suspending protein, 104

Stable isotope labeling with amino acid in cell culture (SILAC), 385
Standard
 microscopic methodology, 55
 operating procedure (SOP), 150, 421
Staphylococcus aureus, 69, 81, 147, 373
Stem cells, 262, 264, 267, 278, 283–287, 289–292, 297–301, 306, 309, 320, 335, 341, 436, 586, 592
Sterilization, 6, 173
Streptococcus
 pyrogens, 248, 262, 270, 271, 276
 thermophilus, 81
Streptomyces koyangensis, 86
Streptomycin, 96
Stress
 conditions, 3–5, 9, 12, 96
 tolerance, 4, 5, 12, 270
Substantia nigra (SN), 202, 305, 467, 468
Superoxide
 anion, 229, 232, 236
  radicals, 81
 dismutase (SOD), 86, 181, 229, 236
Superparamagnetic iron oxide nanoparticles (SPION), 583
Suppressed autophagy, 190
Surface
 hydrophobicity, 580, 613, 622, 623
 modification, 433, 441, 497, 531, 579
 sterilization, 6
Sustainability, 60, 119, 120, 629, 649
Sustainable development goals (SDG), 119, 120
Syndromic diagnosis, 411, 413

T

Tacrolimus, 85, 86, 480
Tafazzin (TAZ), 264, 265
Taxonomic
 classification, 28, 30
 information, 29
Terpenoids, 86
Testicular cancer, 72
Tetracycline, 96, 123
T-helper cell activation, 192
Therapeutic, 86, 301, 375, 376, 378–381, 389, 394–396, 427, 428, 431, 433, 435, 437, 440, 441, 445, 451, 466–469, 477, 481, 486, 490, 491, 498, 527, 593
 approaches, 311, 312, 395, 431, 464, 585
 nanoparticles characteristics, 452
  drug loading and release, 454
  surface properties, 453
  targeted drug delivery, 455
 proteins, 492
 site, 73, 584
 techniques, 105, 381
Thermal power plants, 44
Thrombosis, 190, 234
Tight junctions (TJs), 451, 456, 464, 465
Tilapia nilotica, 42
Tissue engineering, 79, 531, 532, 573–575, 581, 582, 585, 588, 591–594
Tolypocladium nivenum, 84
Total hardness (TH), 46, 48, 56, 49
Toxin degradation, 26
Transcription factors, 181, 183, 184, 194, 197, 206, 227, 283, 284, 286, 292, 302, 303, 365
Transcriptome, 31, 32, 361, 362, 377–381
Transcriptomics, 25, 27, 31, 32, 36, 351–356, 361, 376–379, 381, 382, 389, 391, 396
 current scenario, 379
 disease marker discovery, 378
 limitation and challenges, 381
Transesterification, 128
Transferrin receptor (TfR), 465
Transforming growth factor (TGF), 85, 466
Transgenes, 302
Transmission electron microscope, 316, 323, 621
Triacylglycerol, 129
Tricalcium phosphate (TCP), 10, 576, 590
Trichoderma, 5, 12, 86, 99, 126
Tripolyphosphate (TPP), 483, 611, 614
Tropism, 8, 65, 66, 73, 374, 499, 610
Trypsin, 92, 99, 365, 385
Tryptophan, 8, 393
 synthase subunit alpha, 13
Tuberculosis (TB), 85, 86, 123, 164, 172, 375, 378–380, 388, 430, 446, 461, 462
Tumor necrosis factor (TNF), 192, 197, 198, 201, 389, 466, 497
Tween 20, 6

U

Ulcerative colitis (UC), 90, 485
Ultra-low penetration air (ULPA), 158
Ultrasonography/ultrasound (US), 98, 122, 129, 132, 248, 307, 338–340, 343, 368, 376, 379, 380, 389, 394
Ultrasound, 92, 476, 481, 499
Ultraviolet (UV), 79, 104, 136, 159, 199, 366, 412, 413, 493, 556, 558, 566, 608, 623, 634
  absorbing elements, 104
Uniform suspension, 54
Universal precautions, 173

V

Vaccine, 65, 73, 97, 103, 123, 166, 382, 480
Valosin-containing protein (VCP), 388
Value-added nutraceuticals, 85
Van der Waals forces, 67, 438
Vancomycin, 90, 103, 435, 488
Vascular
  grafts surface modification, 587
  inflammation, 190, 193
Vigna mungo, 16
Viral proteins, 65, 67, 325
Virus-like particle (VLP), 65–73
  delivery systems, 70
  modifications, 70
  production, 66, 69
    loading, 68
    VLPs targeting, 69
Volatile
  compounds, 12
  organic compounds (VOCs), 636, 637

W

Waste minimization, 120
Wastewater treatment, 39, 43, 51, 53, 60, 119, 120, 136, 137, 633
Water
  bodies, 39, 42, 131, 135, 136, 639
  pollution, 47, 60, 135, 632
  quality indices, 56, 58
Weissleder group, 435
Wetland plants, 39, 44, 60
White-rot fungi, 126
Whole-exome sequencing (WES), 369, 371, 372, 396
Whole-genome
  analysis, 12, 13
  sequencing (WGS), 369–372, 396

X

Xanthomonadales, 4
Xenobiotics, 91, 92
  chemical transformation, 91
Xylanase, 124
Xylem vessels, 6
Xylobiose, 127
Xyloglucans, 122

Y

Yarrowia lipolytica, 102, 129
Yeast, 67, 69, 78, 84, 102, 120, 127, 128, 129, 130, 131, 132, 136, 147, 253, 367, 645, 647,

Z

Zoogloea, 136
Zoospore, 88
Zygosaccharomyces bailii, 127
Zymomonas mobilis, 126